Design of Rockets and Space Launch Vehicles

Design of Rockets and Space Launch Vehicles

Don Edberg
Consultant & Professor of Aerospace Engineering
California State Polytechnic University, Pomona

Willie Costa
Engineering & Management Consultant

AIAA EDUCATION SERIES
Joseph A. Schetz, Editor-in-Chief
Virginia Polytechnic Institute and State University
Blacksburg, Virginia

American Institute of Aeronautics and Astronautics, Inc.

American Institute of Aeronautics and Astronautics, Inc., Reston, Virginia

1 2 3 4 5

Library of Congress Cataloging-in-Publication Data

Cataloging in Publication data on file.

ISBN: 978-1-62410-593-7

Cover image credit: NASA. Original scan by Kipp Teague.

CONTENTS

Chapter 9   Surface and Launch Environments, Launch and Flight Loads Analysis   493

Chapter 10   Launch Vehicle Stress Analysis   569

Chapter 11   Launch Vehicle and Payload Environments: Vibration, Shock, Acoustic, and Thermal Issues   611

PREFACE

This book is based on a course on space launch vehicle design that has been presented as an undergraduate class at California State Polytechnic University, Pomona (Cal Poly Pomona or CPP) in the Aerospace Engineering department, and as a graduate class for the Astronautics Department at the University of Southern California. The material has also been presented as a short course.

The authors created this textbook, **Design of Rockets and Space Launch Vehicles** ("DRSLV") as a new text to fill the gap between aircraft and spacecraft design textbooks. Until its introduction, there was no comprehensive work covering the breadth of launch vehicle design. When the authors sought a means to teach the subject of launch vehicle (LV) design, they quickly found that there was nothing comprehensive enough – other books covered part of the subject, but none of them was satisfactory for an introductory course.

Some of the possibilities had chapters on rocket engines, others on vehicle performance, others had chapters on trajectories & aerodynamics, a few had stability and control. There was little history included, and most were silent about important but "non-engineering" topics such as manufacturing, launch pad and range facilities, testing and reliability, and cost estimation.

This text, which we shall refer to as DRSLV, is our attempt to "put it all together" in one comprehensive volume. DRSLV is intended for senior-level capstone engineering design classes at colleges and universities, but will work equally well for graduate classes. In particular, it may be used by readers that do not have a degree in aerospace engineering. However, based on the content – which not only includes the theoretical bases for many design principles but also includes many illustrated examples of these bases – it is also recommended for practicing engineers as well as engineering management who are interested in how things work in either the "big picture" sense, or in areas other than one's specialty. This book will serve to answer many questions people have as far as "why things are done this way" or "what do I need to know about the subject of ___?" A lot of background information is presented in the book's extensive technical history chapter (Ch. 2), as well as the chapters on testing and reliability (Ch. 15) and on failures and lessons learned (Ch. 16).

There are many worked numerical examples of all the aspects of the process of engineering and producing LVs, which clearly serve to introduce principles of these aspects and provide a "feel" for what the answer should look like. The authors believe that part of engineering and design is to have a feel for "approximately what" answer should be expected for a given problem.

Materials in this book are a combination of original materials generated by the authors, along with material extracted from publicly-available sources: books, papers, symposium presentations, corporate sources, and the Internet. All non-original material is credited to its owner and is printed with permission. No classified or controlled materials have been incorporated.

The copy of DRSLV that you hold in your hands has a wide scope, all the way from requirements to performance to analysis, production, testing, and cost estimation. That being said, to save space, this book does not include many comprehensive derivations of theoretical formulae and related material. Such material is available in dedicated texts — and, as Oliver Heaviside (the developer of the Heaviside Step Function) said, "Am I to refuse to eat because I do not fully understand the mechanism of digestion?" Instead of derivations, we provide the pertinent results along with external references at chapter ends providing more in-depth information. This book concentrates on the application of well-known engineering knowledge to the design and simulation of these fascinating vehicles.

Therefore, we provide the caveat that this is an introductory text and as the experts will tell you, "the devil is in the details." This text will introduce the reader to the basic concepts associated with almost all aspects of the design and engineering processes of a launch vehicle, but it will only provide a snapshot of the details or knowledge that would be available to an experienced practitioner in some aspect of the art. There is no substitute for experience; however, this book will guide the reader in the necessary direction to find further information if it is needed.

Launch vehicle and rocket design is a continuously-evolving field, and topics that seemed far-fetched at the time we began collecting information for the text have become common-place, such as the recovery and reuse of first steps of rockets to reduce costs and production rates, a now common-place event by SpaceX. So, this book should be considered a "snapshot" of the current state of LV affairs at the time of its publishing, and will hopefully remain pertinent as time goes on. If it gets horribly out of date – unlikely because the physics of rocketry will remain constant – well, there's always the possibility of a Second Edition!

The materials inside this book not attributed to others are the authors' own work. The authors would appreciate readers' inputs including potential topics or subjects; corrections or typographical errors, or other pertinent information.

So now: "Launch is served!" Catherine Edberg, September 8, 2013.

About the book: DRSLV is laid out in a sequence that mimics the particular sequence of events that might be used to develop the initial or conceptual design of a LV from scratch; exceptions are pointed in the summary of contents of each chapter immediately below. The majority of chapters have exercises or problems that are intended to highlight the major parts of the chapters' contents.

Chapter 1 provides a broad overview of the "world" of launch vehicles. It provides basic information on what they are used for and the space industry in general, basic launch vehicle layouts, and typical launch trajectories. We then describe the design process with a flowchart which indicates the sequence of events beginning with a set of mission requirements, then carrying out initial vehicle sizing, aerodynamics, performance analysis, ground and launch environment, structure, stability, and control analyses, design iterations, and ending with a cost analysis. Finally, we present a list of the world's launch sites.

Chapter 2 is a background chapter* providing the technical history of launch vehicles. Knowing this history provides background for current design practices and also gives knowledge of the vast design space possible for LVs. The chapter spans early history to the pre-WWII period, where the Nazis developed the first ballistic missile, to the development of the ICBM, its evolution into launch vehicles, the moon race, and the development of more modern LVs. It ends with a survey of current LVs and some future developments, including small LVs.

Chapter 3 begins the LV design process by considering how the LV's payload mass and orbital requirements will flow down to specify the performance of the LV. We include a brief explanation of common orbits and their required orbital mechanics. With a specified orbit, we show how to calculate the required LV shutdown performance in terms of shutdown speed and direction. Then, the gravity, drag, and steering losses associated with the trajectory must be vectorially added to the shutdown conditions to determine the LV's design speed. Finally, we discuss launch azimuth directions, orbit injection strategies, and launch windows for both direct and indirect orbits.

Chapter 4 discusses rocket propulsion: how thrust is generated, how the rocket's performance is calculated, and different types and propellants used. We provide details, advantages, and disadvantages of solid and liquid propulsive methods, and some insights on the liquid propulsion 'cycles'. We also discuss hybrid rockets and aerospikes. This information is useful

---

* By "background chapter," we mean a chapter that is provided for background and historical purposes. Typically these background chapters do not provide assignments at the end and are strictly for educational purposes.

for understanding available propulsion options, and for selecting the propulsion system for a particular mission or requirement.

Chapter 5 is an important chapter on LV performance. We show that there are just three parts of a rocket for performance calculations: payload, propellant, and inert mass. The rocket equation is introduced, and we discuss the need for multi-stage vehicles to achieve needed speeds. We discuss theoretical optimization processes (where optimization here means minimum mass to fly a specified payload) and "brute force" methods to optimize. We consider the effect of different propellants to accomplish a specific mission, and consider what an all-hydrolox Saturn V would have looked like (compared to the existing version which used kerolox in its first stage). We discuss and show how to analyze both serial and parallel staging methods, and look at design sensitivities, which indicate the best places to add propellant, reduce mass, and improve rocket performance. Finally, we provide a set of useful formulas that will be used to calculate useful rocket parameters such as propellant mass, structure mass, etc. for sizing purposes.

Once we have some idea of performance and staging requirements, we look at trajectory analysis in **Chapter 6**. This chapter looks at vertical and inclined flight near bodies with and without an atmosphere. We introduce LV aerodynamics and their effect on ascent. We show a simple way to do trajectory simulations using a simple spreadsheet or numerical integration approach. The elegant mathematics of trajectory optimization are introduced, and various methods and techniques of optimization are demonstrated by providing a number of examples of actual launch vehicle trajectories.

Chapter 7 is a background chapter intended to provide a comprehensive introduction to the structures and possible layouts that might be used on various LVs with different propulsion systems. We present a number of ICBMs that evolved into space launchers, as well as dedicated launchers such as the Saturn family, the Space Shuttle and SLS, Atlas V, and Delta IV. We discuss the geometry of tanks, domes, feedlines, and layouts and their ramifications on layout. Finally, we present a survey of structure types and commonly-used structural materials.

With structure and layout knowledge available, **Chapter 8** begins the physical layout process of the inboard profile of a LV and finishes with an estimate of the LV's mass, center of mass location, and mass moments of inertia. First, the required speeds along with the rocket equation are used to determine the ideal amounts of solid or liquid propellants needed to accomplish the mission. Then, "real life" additions are made for factors such as engine startup, unusable propellant, tank shrinkage, and so on. The tanks or casings are sized to incorporate these factors, and guidelines are given for interconnecting structure such as intertanks, interstages, payload fairings, and thrust structures. Then, we illustrate the process of mass estimation, and show how to calculate the LV's center of mass (CM).

This is used to find the LV's mass moments of inertia, which are critical to the maneuvering and stability of a LV.

Chapter 9 introduces the calculation of aerodynamic loads on the LV's structure caused by ground winds on the pad, as well as crosswind "wind shears" that strike the LV during ascent. We show how to estimate the magnitude of the wind loads, their point of application, and how to "march" from top to bottom of the launch vehicle to assess the shear and bending moment distribution along the LV's length, and the axial load distribution. As an example, these processes are used to calculate the wind loads on a Saturn V in flight during the max-q condition. We finish by discussing some "load relief" techniques.

Chapter 10 continues with the calculation of stresses that are induced by the aerodynamic loads from Ch. 9. After providing the terminology and definitions, we show how to calculate axial, shear, and bending stresses using the simple cylinder analysis. We then consider structural stability or "buckling" and show seven ways to improve stability. Pressurization has a strong stabilizing and unloading quality that makes it ideal for LVs in flight. We then show how to size the walls of the propellant containers and their interconnecting structures.

Chapter 11 deals with the transient environments on board a LV which include vibration and shock, acoustic, and thermal loading. We discuss each of these environments and show where and when they are most critical. Finally, we discuss the process of "coupled-loads analysis" and the environment that the payload, the LV's paying customer, will be exposed to.

Chapter 12 is fairly involved exposition of LV stability and control, and also a discussion of many interesting dynamical and stability problems that can occur. We show how the LV knows at any given time where it is, what direction it's pointing, and how fast its direction is changing. With this knowledge and the LV's aerodynamic loading, we can predict the motion of the LV. By gimbaling the engine(s) or using other control effectors, we can "close the loop" and write the equations of motion to assess the LV's stability and controllability. After this controls assessment, we consider several instabilities that can occur, such as structure flexibility, "tail wags dog" instability, propellant sloshing, and propulsion instabilities such as pogo and resonant burn phenomena.

Chapter 13 is a background chapter providing information on the process of LV manufacturing. We show examples of basic machining needed to assemble LVs, and provide a detailed description of the fabrication of the Saturn I's S-IV second step, whose structure is primarily aluminum alloys. We then consider composite materials such as fiberglass-epoxy and graphite-epoxy structures and the techniques used to manufacture them. Next, we consider advanced techniques such as filament winding, fiber placement, and additive manufacturing (AKA "3D printing"). Once the parts are manufactured, we discuss how the LV is assembled, also known as vehicle "stacking".

Launch vehicle systems and launch pad facilities are discussed in **Chapter 14**, another background chapter. Here the Saturn V first step is used to demonstrate many of the systems needed for a successful LV; as the expression goes, "the devil is in the details." Propellant management, power, environmental controls, data management, telemetry, and many other aspects of LVs are presented. We also provide ground accommodation information, including pad services (holddowns, technician access, umbilicals, propellant storage, acoustic suppression, lightning protection, engine startup sequence, and more).

Chapter 15 is a collection of information related to LV testing. We show many of the types of testing that is done to best try to verify that the vehicle will operate successfully, including mechanical, shock, vibration, and acoustics; thermal; RF and electrical; wind tunnel testing on the ground and in flight; and software testing procedures. We look at scale model testing for launch pad acoustics and exhaust noise assessment. Then we move on to discuss design procedures to improve reliability. *Redundancy* is one method used; we show how to calculate reliability with series and parallel connections, and we show reliability for 'k-out-of-n' systems, applicable to multiple-engine LVs. Finally, we show the reliability of a LV range-safety system with multiple failure probabilities.

Chapter 16 is background chapter which provides a number of examples of LV failures in many disciplines, such as structures, separation systems, propulsion systems, GN&C, and software. The text provides some lessons learned and a number of suggestions on questions evaluators can ask to try to prevent failures in future vehicles. We show how range safety is enhanced by flight termination systems, and how those systems operate. We close with a list of best practices to avoid failure.

Chapter 17 is the least "engineering" of the chapters in the book, and also the last, but it's certainly NOT the least! Economics of engineering and systems is critically important in today's world, and this chapter introduces the principal aspects of launch vehicle economics. We begin by describing the design cycle and how decision-making early in the cycle can be a "make-or-break" contribution to the system's overall costs. We discuss cost engineering and considerations that can affect it, and discuss cost estimation methods through statistical fits and modelling. Finally, we provide some actual cost modeling examples. Several exercises challenge the reader to estimate costs on developing their own launch systems, and we provide access to an older graphical but still useful cost estimating model.

A comprehensive **Glossary** provides a list of abbreviations and definitions, followed by an **Index** which completes the book.

Corrections and supplemental materials such as syllabi and curricula will be available from AIAA's electronic library, Aerospace Research Central

(ARC), at arc.aiaa.org: use the menu bar at the top to navigate to Books. AIAA Education Series; then, navigate alphanumerically to this book title's "landing page."

Suggested curricula for courses of different student level and duration:
The material in this book has been presented for two different types of classes: a one-year long undergraduate senior capstone design class, and a one-quarter-long or one-semester-long survey class. It has worked well for both, but the scope of the class may be limited for the survey classes.

For the year-long course, we recommend going through the chapters in numerical order, however the presenter may wish to cover Chapter 17 (cost analysis) before presenting Chapters 13–16 (which are primarily "background" chapters) so as to allow time for any Ch. 17 homework assignment to be completed before the course ends. The assignments in Chapters 6, 9, and 10 are significantly time-consuming, so allotting extra completion time is recommended.

For a survey course, the presenter should follow a similar schedule, except the following chapters or sections may be omitted without Too much loss of content: 3.8, 4.8, 4.9, 5.8, 5.12, 8.9, Ch. 10 (or its homework), 12.4, 12.5, Ch. 13, Ch. 14, 15.2, 16.5. Much of the omitted material is interesting in its own right, and it should be suggested that the reader examine it on their own time.

Suggested curricula for one- and two-semester versions of this course, as well as class syllabi, may be found at the AIAA's electronic library, listed above.

Disclaimer:
The material presented in this book has been carefully selected to describe ideas and principles of rocket design and operation. However, please note that none of the materials provided — engineering formulae, data, material, results, or conclusions — can be absolutely guaranteed, even though every effort has been made to insure accuracy and reliability. This book may include technical inaccuracies as well as typographical errors.

The authors assume no responsibility for errors or omissions in this book, or other documents which are referenced by or linked to this book. In no event shall the authors be liable for any special, incidental, indirect, or consequential damages whatsoever, including but not limited to, those resulting from loss of use, data, or profits, and on any theory of liability, arising out of or in connection with the use or reliance on this information.

THE FINAL RESPONSIBILITY OF VERIFYING ACCURACY AND CORRECTNESS OF ANY AND ALL RESULTS REMAINS WITH THE USER OF THE BOOK, and not with the authors.

Changes made to the information contained herein will be incorporated in new editions of this book. The authors reserve the right to make improvements and/or changes in the book at any time without notice.

ABOUT THE AUTHORS

AIAA Associate Fellow **Don Edberg** has been teaching aerospace vehicle design since 2001 at California State Polytechnic University, Pomona (CPP); University of Southern California (USC); and UCLA. CPP student teams he has advised have placed over 20 times in AIAA student design competitions in Missile Systems, Launch Vehicle design, Space Transportation Systems, Spacecraft Design, and Aircraft Design. At CPP, he has received the *Provost's Excellence in Teaching* award, College of Engineering's *Outstanding Teaching Award*, and the Northrop Grumman *Faculty Teaching* award. He has presented short courses in launch vehicle and spacecraft design to several NASA centers, National Transportation Safety Board, Northrop Grumman, Boeing Satellite Systems, and several small aerospace businesses. He was a *Technical Fellow* at Boeing and McDonnell Douglas, where he authored or co-authored ten US patents, and received the *Silver Eagle* award from McDonnell Douglas. He has also worked at Convair, AeroVironment, JPL, NASA MSFC, and US Air Force Research Lab. He received B.A. in Applied Mechanics from the University of California, San Diego, and M.S. and Ph.D. engineering degrees in Aeronautical and Astronautical Sciences from Stanford University. He is an Eminent Engineer in Tau Beta Pi and was Engineer of the Year of the AIAA Orange County, CA section.

Guillermo Costa is an aerospace engineer and Lean Six Sigma Black Belt who has worked predominantly with small aircraft, high-power rockets, and missiles. He is an alumnus of California Polytechnic University, Pomona and Penn State. He has served as a NASA Student Ambassador, Penn State STEM Fellow, and Bunton-Waller Fellow, and has been awarded the NASA Associate Administrator for Aeronautics Award for his aircraft design work. He has worked at Boeing, NAVAIR, NASA Ames Research Center, NASA Armstrong Flight Research Center, NASA Langley Research Center, and at several startups and small enterprises. He is also a pilot, logging over 300 hours in seven aircraft. He has received two graduate degrees from Penn State – a Master of Science in Aerospace Engineering and a Master of Management and Organizational Leadership, with a concentration in finance – and currently serves as President of Vuong Enterprises, a multinational management consulting and private equity firm.

Testimonials

"The course was overall very interesting. The physical model of the Saturn V and the lecture videos were very helpful." P.M., Northrop-Grumman Corp.

"This was a good overall review of SLV design and systems engineering. This is a tough subject because of the many facets of SLV design." A.H., Scitor Corp.

"Thanks again for the seminar. It was very informative and hopefully useful in our future efforts with Launch Vehicle Programs." M.E., Staff Design Engineer, Moog Corp.

"Thanks so much for sending us a copy of your LV Design Course notes ... I'm quite impressed by the depth of the notes; a novice could perform a Level 1 design of a rocket just by using your notes." Jeff Kwong, Senior Systems Analyst, Virgin Orbit & Vox Space.

"The design for the vehicle ... follows the design processes prescribed by Dr. Don Edberg in his [USC ASTE574] Space Launch Vehicle design course in conjunction with his book, **Design of Rockets and Space Launch Vehicles**. A great degree of credit is owed to Dr. Edberg for the insight and resources that allowed for the launch vehicle design." Chris Eldridge, Graduate Student '2020, USC.

At the end of the 2018–2019 academic year and the first year the course was taught at Cal Poly Pomona using this text, *undergraduate* student design teams placed both 1st and 3rd in AIAA's Graduate Missile Systems design competition, besting a number of teams of *graduate* students.

ACKNOWLEDGEMENTS

The authors have many folks to thank!

Author Edberg would like to thank:
* *For stimulating and supporting DE's early interest in model rocketry:*
  Hugh Debberthine, adviser for the *Covina Skylighter Model Rocket Club*,
  Dane Boles, advisor for *West Covina Model Rocket Society*, and his
  former neighbor and colleague **Doug Malewicki**, who wrote technical
  engineering articles in *Model Rocketry* magazine that captured his
  attention in high school, providing a solid reason to excel in mathematics
  and physics. Edberg only met Malewicki in person some 30 years later in
  Irvine, CA!
* *For providing DE an excellent academic background before college:*
  DE's Covina High School math, physics, and chemistry teachers: Mr.
  Haldi, Mr. **Stuart Bates**, Mr. **Smith**, as well as Mt. San Antonio College,
  the local community college, which allowed DE to take courses in
  computer programming while still in high school.
* *For providing an engineering stimulus for DE's interest in space and rockets:*
  DE's undergrad education at UC San Diego was a mix of engineering
  classes and daily trips to Torrey Pines to fly model airplanes and learn
  first-hand about design, structures, stability, and control. I was privileged
  to study under **Profs. Hugh Bradner, Paul Libby, David Miller** and
  many others in a Caltech-like atmosphere (only 8,000 students back
  then), and received a wonderful background in applied mechanics. At
  Stanford, I was lucky enough to be able to study under world-renowned
  professors **Arthur Bryson, Donald Baganoff, Brian Cantwell, Dan
  DeBra, George Springer, Steele**. For his never-ending support as *advisor*
  for DE's Ph.D. studies, **Prof. Holt Ashley**, who magically arranged for DE
  to carry out his Ph.D. research in a vacuum facility at NASA Ames and
  the U.S. Army Aeromechanics Lab, kindly tolerated DE's distractions of
  field trips to many aviation sites and to model airplane competitions
  (including two world championships) while eventually completing his
  research, and also providing a model of how to be an outstanding
  professor, a true gentleman, and a textbook author. Dr. **Gary A. Allen**,
  DE's grad student office-mate at Stanford who inspired DE to join him on

a trip to see the first Space Shuttle launch at Kennedy Space Center and its landing at Edwards Air Force Base in 1981, and appreciate Saturn Vs ever since.

- *For professional development:*
DE's first supervisor at McDonnell Douglas in Huntington Beach, **Dr. John Tracy**, who encouraged DE to "walk the production lines on the floor and understand how rockets work" while working at MacDac. **Dr. Andy Bicos**, colleague, who encouraged pushing the state of the art in research and supported work on LV vibration isolation. **Bruce Wilson**, who monitored, supported, and assisted with MacDac R&D work that led to DE's active isolation system that flew on the Space Shuttle in 1995 and development of microgravity isolation system umbilicals for ISS payload racks.

- *Academic support of teaching this subject:*
California State Polytechnic University Pomona (CPP) Aerospace Engineering Department Chair **Ali Ahmadi**, who supported the development of the course by DE that led to a year-long senior-level capstone design course in the Aerospace Engineering department catalog as ARO4711L-4721L and ultimately to this book. Dr. **Bill Goodin**, who was willing to offer DE's short course on the subject of this book in the curriculum of UCLA Extension. Dr. **Mike Gruntman** at USC for accepting DE as a part-time lecturer for a class on this subject, now listed in the catalog of Astronautical Engineering as ASTE574.

- *For the writing of this book:*
McDonnell Douglas colleagues **Tony Straw** and **Dave Bonnar** who graciously shared their design materials and writings in support of DE's teaching, leading directly to parts of this book. Dr. **Mark Whorton**, formerly at NASA MSFC, who worked with DE on joint MacDac/MSFC STABLE flight experiment that flew on the Space Shuttle, and later encouraged and supported the writing of this book. **Mike Huggins** of the Air Force Research Lab, who arranged for DE's summer visit to AFRL in summer 2014, where he was fortunate to be able to study trajectory optimization, leading to part of Chapter 6 of this book. And, **NASA Marshall Space Flight Center**, who took on DE as a Faculty Fellow in Summer 2016, allowing DE to learn much more about the LV business in-person. All the folks who supported my summer faculty visit to MSFC during the summer of 2016 and answered all of my numerous questions: **Mark Phillips** and **John Rakoczy** and many of the staff in the MSFC Guidance, Navigation, & Mission Analysis Branch who provided input materials and had many fruitful GN&C conversations with DE. The research librarians at the **University of Alabama Huntsville's *M. Louis Salmon Library*** Special Collection and the librarians at **Redstone Arsenal's *Redstone Scientific Information Center*** (RSIC) collection (which was closed by the Army in 2019 ☹), both of whom provided access to unique, rare books and materials as well as current works.

* *Student support:*
Many students received drafts and provided excellent feedback for the
material which now forms this book. (Former) CPP students and USC
students **Miguel Maya, Andrew Fung, Daniel Roseguo, Ryan Robinson**,
and many others for providing wonderful feedback and assistance in
making the text clearer, and correcting typographic errors.
* *Technical information, illustrations, and artwork:*
Mike Jetzer and his website **heroicrelics.com**, provided super-quality
scans of some of the University of Alabama Huntsville library's collection.
Richard Kruse and his great site **historicspacecraft.com** graciously
provided almost all of the excellent "rocket family" drawings found mainly
in Chapter 2 (History) and scattered around elsewhere in the book. **Ed
LeBouthillier** provided numerous photos, scans, papers, and
presentations containing valuable documentation for many launch
vehicles. Norbert Brügge and his info-packed website **www.b14643.de**
* *Family:*
My parents **Joe and Sophie Edberg**, who are no longer living, nurtured my
interest in science and engineering and encouraged all my activities (that
they knew about) and supported my scholarly pursuits. This interest must
be in my blood, because my father Joseph worked at Aerojet and Lockheed
and was a member of the American Rocket Society before I was born. My
mother **Sophie** was a technical editor for several engineering firms, and my
grandfather **Jacob Edberg**, a cabinetmaker, built and carved propellers for
NACA to test in wind tunnels in the 1930s.
* *And last, but most certainly not least:*
My wonderful wife **Cathy**, who in addition to providing significant and
invaluable assistance with the book's organization, illustrations, and layout,
provided invaluable advice and has cheerfully put up with all the time and
activity I spent creating the course and writing this book, when I should
have been spending more time with her.

Author Costa would like to thank:
* *For stimulating and supporting my early interest in model rocketry:*
Dr. Don Edberg, first and foremost, for reminding me that the sky is not
the limit, but merely the beginning; the folks at **Sport Rocketry** magazine
for proving that the road to high-power rocketry fun is not only simple, but
also educational.
* *For providing me an excellent academic background before college:*
Mrs. Rockwell, my high school physics teacher, for actually taking the
time to do hands-on experiments in the classroom instead of simply
making us bored with endless equations – without you I'd have likely not
gotten anywhere near as far, and many thanks for showing us how
energetic certain metals can be when dropped in water – I'll never forget to
treat chemistry with respect thanks to you, and I'm sure the ceiling tiles in
chem lab will never forget, either ☺; my calculus teacher, **Mrs. Smith**

("Friends. . .!"), for making a difficult subject even harder so that I learnt the value of hard work and dedication; and **Mrs. Coley**, my high school English teacher, for trying (and occasionally succeeding) to make me understand that there's always a human cost to everything. I don't know what you're all doing these days, and I don't know if you'll ever get a chance to read this, but thank you.

• *For professional development:*
This section would never be complete without an acknowledgement to **Prof. Ray Hudson** of Cal Poly Pomona – your "tender advisory sessions" were tear-jerking, but damned if you weren't always proven to be right. To **Dean Jamie Campbell** and **Pauline McCarl** at Penn State, for helping me find ways to temper my fiery exuberance, even if sometimes lessons needed to be learnt the hard way. To **Dr. William Warmbrodt** of NASA Ames Research Center, for always being a reliable source of encouragement no matter how dark things might appear. To **Drs. Elizabeth Ward** and **Walt Silva** of NASA Langley Research Center, who always encouraged me to follow my dreams. . . no matter how unorthodox and unconventional that path might be.

• *Technical information, illustrations, and artwork:*
Ed LeBouthillier, as always, has remained an invaluable and ready source of technical information and "boots on the ground" intel for how technical challenges in the custom-built rocket community are approached and solved in practical and effective manners.

• *Family:*
My parents have nurtured my love of science and engineering since an early age, despite several chemical, ballistic, and pyrotechnic incidents during my youth which my neighbours surely did not appreciate; however, I still have all of my fingers, so I will celebrate these youthful hijinks as educational moments. My father, who has unfortunately passed on, was an electrical engineer of some capability, and worked in the telephony industry in multiple countries for several decades; my mother, a mathematician, specialized in mathematical concepts that I honestly can't even fathom and sound like sorcery. I clearly took a hard left turn and eschewed all of those heady things to instead pursue interests in high-speed tubes that ride on large fireballs of propulsive thrust. **Vinh Vuong**, you have always supported me when no one else would, and I love you like a brother.

• *And last, but most certainly not least:*
My wonderful wife **Dana**, who in addition to demonstrating a seemingly endless supply of patience with my temperament during this book's creation, put on a happy face as I spent years helping create this book when I should have been spending time with her.

Chapter 1

Launch Vehicles: Introduction, Operation, and the Design Process

1.1 Introduction to Launch Vehicles

Since October 1957, mankind has demonstrated the capability to place artificial satellites into orbit around the Earth, and has since sent them to all planets in the solar system. The flight of such satellites depends on the development of space launch vehicles (SLVs or LVs). At first, launch vehicles were derived from weapons of war, primarily modified versions of ballistic missiles. Later, newer launch vehicles were designed from the start with the specific intent of taking objects into orbit or farther; today's SLVs are a mix of the two types. In this book, the development of a number of LVs is described in Chapter 2.

A space launch vehicle has a simple objective: to *accelerate* and *position* a mass (the payload, usually a satellite or spacecraft) up to the *velocity* and *location* required for the payload's trajectory. The term *velocity*, in this instance, refers to the definition of velocity found in any physics textbook—namely, that velocity is a vector that has both speed *and* direction. To fulfill the "position" part of its objective, the LV must raise the payload from the launch site location (either the surface of the planet from which the LV is launching or from where the LV is dropped from a carrier aircraft) to a height that is above the planet's sensible atmosphere. This means that the LV must provide the kinetic energy of the necessary orbital speed, as well as the potential energy for the raising of the payload to the desired orbital altitude. The LV must also do the additional work needed to overcome the aerodynamic drag produced by passing through the atmosphere (if any), and other losses such as thrust variations due to ambient pressure and steering losses due to the thrust vector not being aligned with the velocity vector. The calculation of the energy values required to accomplish these goals will be covered in detail in Chapter 3.

There are many ways to accomplish this objective, including traditional chemically propelled rockets, air-breathing propulsion, cannons, magnetic accelerators, nuclear propulsion, laser propulsion, and others. So far, only chemical propulsion has been demonstrated on a common and practical basis, and so this book will focus exclusively on launch vehicles that rely on chemical propulsion. An introduction to the analysis and operation of liquid-propellant rocket engines and solid-propellant rocket motors is presented in Chapter 4. Note that we follow the industry's traditional nomenclature to distinguish between liquids and hybrids.

Throughout this book, we will refer to "steps" as individual propulsive "chunks" of a rocket, and "stages" as assemblies of steps and payload. The *first stage* consists of the entire stack at liftoff, including the first step, second step, third step, additional steps (if present), and payload. The *second stage* consists of the entire rocket after the 1st step has separated and includes the second step, third step, additional steps (if present), and payload. The *third stage* (if present) is the entire rocket after the 2nd step separates: third step, any additional steps, and payload. And so on. The definitions of *stage* and *step* are visually illustrated in Figure 5.7 of Chapter 5.

An LV may be *expendable*—that is, a given LV is utilized only one time and then discarded—or it may be *reusable*—where parts are recovered and refurbished in order to be used multiple times. With very few exceptions, launch vehicles have been expendable, so this volume will focus on expendable launch vehicles (ELVs). One example of a reusable launch vehicle was the NASA Space Shuttle, which operated from 1981 until the system was decommissioned in 2011. At the time of this writing, Space Exploration Technologies (SpaceX) has demonstrated a recoverable first step for its Falcon 9 LV and has re-used several Falcon boosters multiple times. With this and other proposed ventures in mind, the authors might consider writing a second volume on reusable launch vehicles in the future.

A single propulsion system for an LV would be very desirable on an economic basis (i.e., reduced complexity, increased reliability, etc.); however, because of the performance limitations of chemical propellants, combined with the current state of the art in materials engineering and structural design, it is not practical to send an LV into orbit using a single propulsion system. Therefore, it is currently necessary to use a process called *staging*. In a staged LV design, the LV is broken into several separate propulsive units (sometimes called *steps*) that are assembled together for a mission. Each propulsive unit is typically separated and discarded after it has depleted its fuel, which improves the performance of the vehicle's remaining steps because the inert mass of the discarded step does not need to be accelerated or carried to higher altitudes. To get a perspective on the sensitivity of a launch vehicle to inert mass, there have been cases where a failure to separate a depleted solid rocket booster or even a payload fairing led to a deficiency in speed and ultimately to a "subterranean orbit." The sensitivities of launch vehicles to changes in structure, payload, propellant mass, and propulsive efficiency are discussed in later Chapters 3 and 5.

Steps can sometimes fire one at a time in sequence, referred to as *series staging or firing in series*, or they may fire at the same time, known as *parallel staging*. Strap-on boosters typically fire in parallel with lower steps, whereas upper steps are fired sequentially in series. A launch vehicle such as a Delta II has both: parallel staging (which uses three to nine strap-on boosters at liftoff) and series staging, where the second (and third, if present) steps fire sequentially after the first step is expended. In subsequent chapters, all of the different types of staging will be examined, and the advantages and disadvantages of each type will be presented. Knowing the propulsive requirements imposed on each step of the LV allows the calculation of the amount of required propellants and initial sizing. The complete, fully assembled launch vehicle—including strap-ons—is often referred to as a *stack*.

1.2   Anatomy of a Launch Vehicle

Launch vehicles are very complex and have many specialized design features. Additionally, a unique vernacular is used to describe the different parts, operations, and functions of a launch vehicle. Like most aerospace disciplines, launch vehicles use many abbreviations and acronyms to increase conciseness (and often reduce comprehensibility!). This section is intended to introduce the reader to some of the more important components of a launch vehicle, as well as the vocabulary of space launch vehicles. A complete list of names and abbreviations commonly used in the business is presented in the Glossary.

1.2.1   Surface-Launched Vehicle

A cutaway of the United Launch Alliance Atlas V 551 is shown in Fig. 1.1. This vehicle is shown because it contains elements of three different types of rocket propellants: solid, liquid cryogenic (liquid oxygen and liquid hydrogen propellants, also known as LOx/LH_2 or hydrolox), and room-temperature RP-1 (essentially kerosene) along with LOx (known also as kerolox). The AV551 vehicle will be used to demonstrate many examples of components commonly found on launch vehicles.

The most important part of the AV551 vehicle is the payload (item 20 in Fig. 1.1). The payload is usually a spacecraft (or multiple spacecraft) whose mission requires the launch vehicle to place it into a specific condition, usually defined by speed, heading, and altitude. The payload represents the paying customer (quite literally, the "load that pays"), and thus is the launch vehicle's entire reason for being. The payload can be commercial, military, exploratory, scientific, experimental, or some combination of

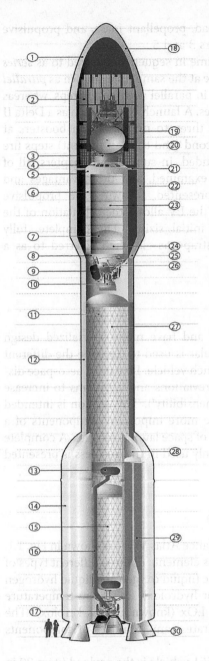

these. Of course, it is vitally important that the payload survive the launch, a consideration discussed in later chapters.

The payload is physically attached to the top of the launch vehicle with a separation system. The separation system is commanded by the LV's computer to fire and release the payload when the vehicle has achieved desired launch performance and required velocity (speed and direction) at required altitude. The system often uses pyrotechnic devices to either release a clamp band (sometimes known as a Marman clamp) or fire explosive fasteners (bolts or nuts) to "de-attach" the spacecraft. Once free, small compression springs called separation springs are used to gently push the payload away from the launch vehicle. It's not well-known, but the Marman clamp was invented by Zeppo Marx, one of the famous 20th-century Marx Brothers Hollywood movie actors (see [1]).

The separation system is attached to a structure either directly on top of the upper step of the launch vehicle or forward of the upper steps forward propellant tank. This structure, item 3 in the figure, is known as a payload attach fitting (PAF) or a launch vehicle adapter (LVA). Details and drawings of the PAFs offered by different LV suppliers are usually available in the vehicle's payload planner's guide or payload user's manual. A number of these guides are given as references in later chapters of the book.

Fig. 1.1 Features of the ULA Atlas V 551.

The payload fairing (PLF), item 1, serves to protect the payload from aerodynamic and heating loads as the launch vehicle climbs out of the atmosphere. The duties of the PLF, also known as a shroud, include protecting against foreign object contamination of the payload and serving as a container for a specially conditioned atmosphere to meet the payload's thermal requirements. The PLF forward section (item 18) may be conical, ogive-shaped, or have a shape specified as a mathematical curve. It connects to a uniform-diameter cylindrical section (item 19), both of which are commonly treated with acoustic panels (item 2) made from a material to attenuate sound created by air rushing over the PLF and the sounds coming from the engines during liftoff.

In order to maximize launch performance, the PLF is generally jettisoned from the launch vehicle as soon as possible, because its mass detrimentally impacts the performance of the vehicle. Separation occurs when the thermal heating from the atmosphere rushing by the spacecraft is below an acceptable value, and is usually near the end of the first-step burn or the beginning of the second-step burn for a multistage vehicle. Of course, the PLF must separate from the LV before the payload can be released. A special pyrotechnic method is used to forcefully eject the payload upon command. These PLF separation systems are described in more detail in Chapter 11.

Note that the Atlas V vehicle shown in Fig. 1.1 has a PLF that encloses its entire Centaur second step (item 6) and is mounted atop its first step, which is not a common configuration. Because of this arrangement, the payload sits on top of a long, relatively flexible second step (the Centaur) and would undergo potentially large lateral motions, were it not restricted by the Centaur Forward Load Reactor Deck (item 21).

The majority of launch vehicles mount their PLF at the top of the second step. In these vehicles, the bottom of the PLF and the bottom of the PAF are often in close proximity, so that the shock load associated with PLF jettison can be significant. It will be shown later in this book (Fig. 11.2, Chapter 11) that the shocks associated with PLF jettison can be among the most significant transient design environments that the spacecraft must endure. The trailing portion of the PLF that tapers down to the smaller diameter shown (item 25) is known as a *boattail* or *skirt*.

Item 5 is a cylindrical adapter, situated underneath the PAF. Attached to the bottom of the adaptor is a dome or bulkhead (not numbered) that forms the forward boundary of the Centaur's liquid hydrogen (LH_2) fuel tank (item 23). The liquid hydrogen is at a temperature of $-253°C$ or 20 K, requiring thermal insulation to keep it from boiling off due to higher temperatures outside the tank. The lower boundary of the LH_2 fuel tank is a common bulkhead (item 7), so called because it is the common or shared boundary between the hydrogen tank above and the liquid oxygen (LOx) tank below (item 24). Because LOx is stored at a temperature of $-183°C$ or 90 K, the resulting 70 K of temperature difference between the LOx and the LH_2 requires that the common bulkhead be well-insulated to prevent as much heat flow as possible.

The bottom or aft end of the second-step hydrogen tank is known as the aft bulkhead (item 8). Bulkheads may also be referred to as domes. The structure attached to the perimeter of the aft bulkhead is known as the second step's thrust structure, which also mounts the Centaur's RL-10 rocket engine (item 10) and a high-pressure container of gaseous helium (item 9), often called a *bottle*. A tank of hydrazine (item 26) is also mounted to the thrust structure and is used for attitude control.

The Atlas V's second step is attached, via another pyrotechnic separation system, to the top of the rocket's first step. Below the exhaust nozzle of the RL-10 is the forward bulkhead (not numbered) of the first step's LOx tank (item 27). A long cylindrical section, made from isogrid (a plate of aluminum machined with triangular reinforcements and rolled into a curved shape), provides the volume needed for the LOx with an aft bulkhead at its bottom. A helium bottle (item 13) is located inside the LOx tank just in front of its aft bulkhead.

The cylindrical section between the first step's upper LOx tank and its lower RP-1 fuel tank (Item 15) is known as an intertank adapter or center body. (RP-1 is a specially refined type of kerosene similar to jet engine fuel.) This center-body structure, along with the LOx tank's aft dome and the fuel tank's forward dome, will likely have considerably more mass than a single common bulkhead, such as the one used in the second step (item 7); however, as will be demonstrated later, the penalties associated with mass added to the first step are quite a bit smaller than those associated with additional mass loaded onto the second step. A LOx feed line (item 16) connects the bottom of the LOx tank with a turbopump (not shown) that forces the oxidizer into the step's motor.

The feed line shown in item 16 runs outside of the fuel tank; however, many vehicles use the more direct route that passes through the lower tank. The decision of whether to pass through the tank or to pass around it as shown hinges on the extra mass of the "sideways" runs of the bypassed line vs the difficulties associated with penetrating the pressurized tank. The latter include the masses and difficulties associated with feed-through openings, which require reinforcement, the issues with sealing the penetrations, and the long, unsupported length of feed line that can experience undesired vibration. There are also cooling issues to consider (i.e., running one cryogenic propellant in close proximity to another, which could be at a higher temperature) and friction losses through the length of the ducting (one of the reasons why the Saturn V's S-IC first step had five lines go right through the lower tank, pumping 1 ton each per second).

Beneath the aft dome of the fuel tank is the thrust structure, which mounts the first step's RD-180 engine (item 17). A vertical fuel feed line provides fuel to a turbopump that forces the fuel into the motor. A thrust vector control (TVC) system is also deployed there, to be used for steering the vehicle during launch. (Thrust vectoring is described in the Chapter 12 on guidance, navigation, and control.) The thrust structure is enclosed inside an aft transition skirt, which contains thermal protection blankets and materials to reduce the thermal exposure of the rocket's aft end to the high thermal radiation levels associated with the motor's exhaust plume.

Interestingly, although the Atlas V is critical to carrying out missions for the United States, its first-step RD-180 engine is purchased from Russia. The irony here is that the original Atlas missiles were designed to fly to Russia as a target; the original designers of the Atlas would hardly believe that the latest version sports rocket engines manufactured by the country's former adversary!

The final portion of the Atlas V 551 vehicle is the solid rocket booster motors (item 14). A number of these external solid rocket motors (SRMs) are needed because the RD-180 motor does not have sufficient thrust to get the vehicle off the pad without assistance. Known as strap-ons, these motors are fired at liftoff so that their added thrust gets the rocket off the pad. The number of strap-ons is determined by the payload's mass and the required launch performance; Atlas Vs can take from one to five SRMs. Item 28 shows the nose cone or forward fairing of the SRMs, and item 29 indicates the internal arrangement of the solid propellant, often described as its grain. Note that the long, internal cavity bored into the grain has a specially designed shape to provide a tailored thrust profile that varies with time in order to maximize the vehicle's performance.

Once the SRMs have consumed all of their propellant, they are jettisoned. In many cases, they are allowed to impact downrange and are not recovered. Other times (as with the Space Shuttle), the SRMs will have a parachute recovery system, allowing the motors to be recovered and then transported to a refurbishment facility where they will be made ready for later flights. Note that strap-ons may use either solid or liquid propellant. The Delta IV Heavy launch vehicle utilizes two strap-ons that use the same liquid propellants (LOx/LH_2) as the core vehicle to which they are strapped. Just like their solid counterparts, they are jettisoned when their propellants are consumed.

1.2.2 Air-Launched Vehicle

So far, we have taken a look at ground-launched vehicles, which make up the majority of LVs; however, there are a few vehicles that are air-launched, which is to say, they are dropped from a flying carrier vehicle. This method of launching provides several benefits over surface launching, including:

* Increased payload capability due to launching kilometers above the ground. In other words, they have a head start on gravitational potential energy that does not need to be supplied by the vehicle itself.
* Increased payload capability due to launching with the carrier vehicle's speed rather than zero speed on the ground. In other words, they start with positive kinetic energy that does not need to be supplied by the vehicle itself.
* The ability to launch wherever the carrier aircraft can carry the launch vehicle (meaning less dependence on traditional rocket launch sites, where

scheduling and paperwork can be burdensome) and whenever the carrier vehicle is available (allowing more flexibility of launch times and the ability to launch rapidly if desired).

However, air launching has its own set of disadvantages. These include:

- The need for a dedicated, specialized carrier aircraft, pilots and crew, and hangar facilities, and their associated costs
- The need to design the launch vehicle to be able to withstand the side loads provided by a suspension system underneath the carrier vehicle, and the mass of the suspension system's fittings on the launch vehicle
- The added mass and complexity of a means to allow the air-launched vehicle, dropped in a horizontal attitude, to transition to a relatively vertical climb to ascend out of the atmosphere

These advantages and disadvantages must be weighed against each other as a design study progresses, and will be discussed in Chapter 3. So far, only the Orbital-ATK Pegasus (Fig. 1.2) is made for air-launching; however, at the time of this writing, there are a number of other programs underway to examine and create new air-launched vehicles such as the Virgin Orbit LauncherOne LV.

The first observation one would make about the Pegasus is the presence of a triangular-shaped delta wing mounted above the "traditional" cylindrical structure of the launch vehicle. It should be fairly obvious that the added wing structure requires additional mass that would not be present (or needed) on a surface-launched vehicle. Additionally, the Pegasus has three fins, each containing an articulated control surface. These surfaces are used to control the vehicle and, in conjunction with the delta wing, generate a pitch-up maneuver using the wing's lift to rotate the vehicle into a near-vertical climb.

Because the first step fires within the atmosphere, the articulated fins may be use to provide all required pitch, yaw, and roll commands during boost.

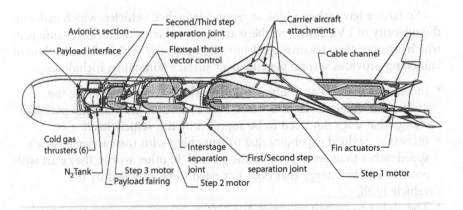

Fig. 1.2 The Orbital-ATK Pegasus air-launched vehicle. Source: [2].

Like an aircraft, pitch and yaw are commanded by "mirror-image" fin deflections on either side of the body, and roll is provided by deflections by all three control surfaces moving in the same direction. This aerodynamic control system eliminates the additional mass and complexity of a thrust vectoring system for the first step.

Between the first and second steps (and the second and third steps as well) is a location where the steps separate, known as a separation joint or separation plane. The structure of the separation plane and its vicinity needs to be strong and stiff in order to withstand flight loads, yet it has to also be detachable upon command. Therefore many, but not all, consist of pyrotechnic (explosive) devices used to cut or sever the two structures. The Space Exploration Technologies (SpaceX) Falcon vehicles use a nonpyrotechnic pneumatic separation system, which has the advantage that it can be tested before flight, as opposed to a pyrotechnic separation system, which can only be used once before requiring replacement.

Upstream of the first/second-step separation joint are the step 2 and step 3 motors, and then an avionics section along with the payload and its interface. (*Avionics* is a portmanteau of *aviation electronics*; rarely, the term *astrionics* may be used.) Like the solid rocket motors on the Atlas V discussed earlier, the Pegasus vehicle's motors utilize solid propellant; however, they must also provide steering via thrust vector control (TVC) for the Pegasus upper steps. With a single motor and exhaust nozzle, pitch and yaw control is provided by gimbaling the nozzle; roll control is provided by a roll torque produced by the cold-gas thrusters shown near the nose. This is true for both of the vehicle's upper steps. The payload fairing serves the same purpose as for a surface-launched vehicle and is jettisoned as quickly as possible.

1.3   The Phases of Launch and Ascent

As previously mentioned, all launch vehicles are basically *accelerators* and *lifters*: their main purpose is to accelerate a payload to a certain speed and direction at a certain height above the ground. Assuming that the LV has been built, delivered, and assembled at its launch site, the ascent phase of flight has a series of events or milestones. As described earlier, the two typical ways of launching vehicles are either to launch the LV from the Earth's surface (ground or sea launch) or to drop the LV from a carrier aircraft.

The typical phases of flight for a ground-launched LV are shown in Fig. 1.3. As seen in Fig. 1.3, the generic launch vehicle consists of two (or more) strap-on boosters connected to a central core; for this example, the central core consists of two steps. Thus, this example LV is a parallel-and series-burn vehicle: the parallel portion, sometimes referred to as the "zeroth" stage, occurs when the strap-on boosters are firing (in addition to the first-step engine firing), before the strap-ons are jettisoned. The series portion occurs after the strap-ons have been jettisoned, where the first step continues its burn and is then jettisoned before the second step ignites.

The LV step numbering terminology is an artifact of the development of many historical vehicles, which began life as LVs without strap-ons; the initial step of these vehicles was known as the first step. Once strap-on boosters began to be added to the vehicle core in order to increase payload mass capabilities, the traditional numbering remained, with the "first" step becoming the "core" stage, and the strap-on boosters being termed the "zeroth" stage. For a ground-launched LV, a signal is sent to ignite the motors, and an automatic determination is made as to whether the LV is in a proper condition for flight; the restraint mechanisms are deactivated (or the release mechanisms are activated), and the LV is released to begin its ascent from the launch pad. For ground launches, the launch pad may include additional services, such as the ability to supply a water "blanket" to protect the launch pad from the LV's acoustic energy, and often include mechanisms by which to detach and retract umbilicals and service arms from the LV, as well as restraint mechanisms that prevent the launch vehicle from lifting off until system checks are completed. These ground-based subsystems must also include methods to protect themselves from the launch plume of the LV during ascent; these additional services are vital elements of the ground-based LV system and are discussed in greater depth in the Chapter 14 on launch facilities.

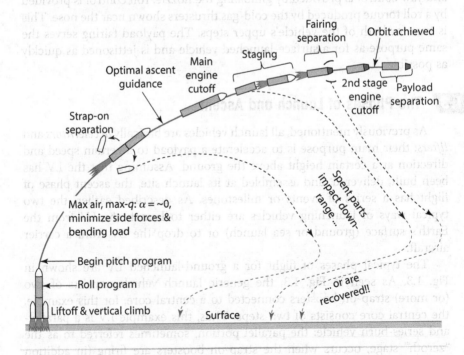

Fig. 1.3 Example lift and ascent profile for a ground-launched LV.

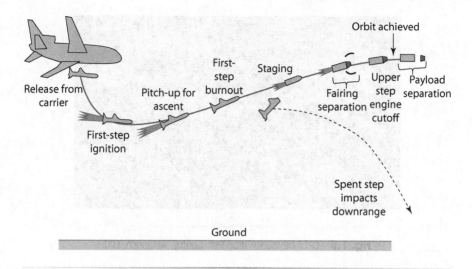

Fig. 1.4 Example lift and ascent profile for an air-launched LV.

The ascent profile for an air-launched LV is shown in Fig. 1.4. Once the air-launched vehicle has been released and has accomplished its pitch-up maneuver, the events during its ascent phase are very similar to those of a ground-launched vehicle.

The release sequence for an air-launched LV is similar to that of a ground-launched LV. After the determination has been made that the LV is in a proper condition for flight, a signal is sent to deactivate the restraint mechanisms (or to activate the release mechanism) and ignite the LV's engine(s). The LV is then free to begin its ascent.

1.3.1 Vertical Climb

A ground-launched LV must climb a specified vertical distance in order to clear any service towers, masts, or other obstacles that may be in the vicinity of the launch pad. The actual flight path will depend on constraints and boundary conditions imposed during flight, and will also be dependent on ground winds at the launch pad as well as the necessary heading or azimuth angle for the desired orbit.

An air-launched LV will, ideally, be dropped from its carrier aircraft with a heading that is very close to the desired launch azimuth. Unlike a ground-launched system, an air-launched vehicle must transition to a climbing attitude; in the case of the Northrop Grumman Pegasus, the first step uses a wing and tail to execute a pitch-up maneuver in a manner similar to a fixed-wing aircraft. Another possibility is the nonaerodynamic pitch-up provided by proper drop conditions, which was demonstrated by the Airlaunch drop test from a U.S. Air Force C-17 in 2005, as shown in Fig. 1.5. Such a

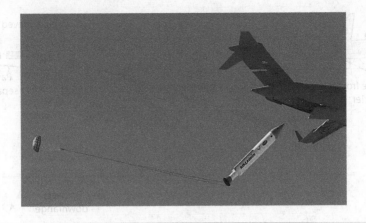

Fig. 1.5   C-17 air launch drop test. Source: Airlaunch, LLC.

system does not require a separate wing and tail and has the potential to save considerable mass.

1.3.2 Roll Program

As the ground-launched LV is climbing, it must "roll" or rotate about its long axis, so that its reference axes are pointed in the direction required by the mission. The final direction is related to a heading angle called the *launch azimuth angle*. This direction is determined by the launch site location and the details of the payload's orbital requirements. The calculation of the azimuth angle is discussed in Chapter 3.

1.3.3 Pitch Program and Vehicle Turning

Most ground-launched LVs lift off vertically, although some (such as Japan's Mu-3-S-2) will launch with an off-vertical tilt for range safety reasons. The final LV stage and its attached payload, however, must attain burnout or enter orbit in an approximately horizontal attitude. This means that at some point during ascent, the vehicle needs to make a turn or rotate from a vertical or nearly vertical attitude to an approximately horizontal attitude. A *pitch program* from the vehicle's flight computer commands the vehicle's attitude control system to carry out the turning process. Note that, as used here, the term *pitching* refers to tilting the vehicle's long axis upwards and downwards so as to climb or dive, respectively.

The pitch program begins with a pitch maneuver where the LV is commanded to tilt so it is no longer oriented vertically. This maneuver is done relatively early in the climb (typically within the first 10–20 s) when speed is low, so as to minimize any aerodynamic loads that result from the vehicle's long axis no longer being pointed in the direction of flight. This difference in pointing angles is referred to as an *angle of attack* and produces side forces that, in turn, create internal loads within the vehicle. Aerodynamic loads due to angle of attack are covered extensively in a later Chapter 9.

Once the pitch program is initiated, the slight change in pitch is the beginning of a very gradual turn that ends as the vehicle reaches its final trajectory conditions at burnout. There are several methods that are used to achieve this gradual turn; these methods will be discussed in some detail.

One key goal of any turning program is to minimize aerodynamic loads while in the atmosphere. This usually means flying with an angle of attack as close to zero as possible. Once the dynamic pressure drops below a preset value, on the order of $3{,}500\,\mathrm{N/m^2}$ or 0.035 atm, the guidance system assumes aerodynamic loads may be neglected and flies at whatever angle of attack is needed to execute the optimal trajectory to orbit or escape, usually with the minimum amount of propellant required to do so. Optimal launch trajectories will be discussed in greater detail in Chapter 6. One common pitch program is called a *gravity turn*. This maneuver is based on the same phenomenon that occurs when a tall object tips over: gravity causes it to tip. It will be shown Chapter 6, section 6.4.7 that the turn rate slows down as vehicle speed increases, and a properly executed gravity turn requires no control system inputs once it is initiated.

Normally, in the initial design period, the determination of the necessary propellant is the focus, and as such it is assumed that the launch trajectory is flown at zero lift. A maximum angle of attack value that would be encountered by the LV when flying at *max-q* (maximum dynamic pressure) is also assumed, typically on the order of 2–5 degrees. This assumption, in turn, is used to define the vehicle's maximum bending loads. These assumptions would be set as the ground rules of the design and would become part of the system's design requirements.

As mentioned earlier, air-launched LVs must initiate a pitch program to transition into climbing flight, in order to accelerate while climbing out of the atmosphere. An air-launched vehicle will need to factor in the same aerodynamic and bending forces as a ground-launched LV; however, an air-launched LV will need to minimize its period of horizontal acceleration in order to minimize the drag losses encountered during ascent.

1.3.4 Strap-on Separation

Much of the world's current fleet of LVs utilize some type of propulsive stage augmentation to increase the amount of payload mass that a vehicle can deliver to a given orbit. Augmentation devices are often self-contained rocket propulsion systems that are attached to the vehicle and are commonly called solid rocket boosters (SRBs), solid rocket motors (SRMs), or liquid rocket boosters (LRBs). These stage-augmentation units are often referred to as *strap-ons*, because they are "strapped onto" the core of the vehicle and are "unstrapped" (or jettisoned) when they are no longer producing effective thrust, while the core vehicle continues to fire during and after jettison. LVs in the past have had anywhere from one to nine strap-ons.

The separation of the strap-ons may be a significant shock event and is often of major importance to vehicle guidance and loads. For safety, the dropped mass(es) must separate cleanly, not recontact the LV, and not land on a populated area.

1.3.5 Max-Air/Max-q/Buffet

It is simply impossible to *completely* predict the entirety of weather conditions through which the launch vehicle will pass during its ascent. In particular, there can be gusts of wind (or wind shears) that act to provide a (momentary) relative angle of attack that is different from the value commanded by the LV's pitch program. This momentary change in angle of attack can produce substantial loads for the vehicle and payload. These loads are particularly important during the period of *max-q*, or maximum dynamic pressure, where the aerodynamic loads on the vehicle are at their highest. Max-*q* usually occurs just after the vehicle has passed Mach 1, typically around 70–80 s into the flight. *Buffeting* loads (the irregular oscillations of the vehicle, caused by turbulence) are also highest near max-*q*. Once the vehicle has survived max-*q*, the aerodynamic loads decrease as the vehicle ascends out of the atmosphere.

If the LV continues to accelerate, why doesn't the dynamic pressure continue to rise? The reason is that the atmospheric density falls faster than the speed rises. This concept is shown in Fig. 1.6. Later, in Chapters 9 and 10, it will be shown that during the entire ascent and turning maneuver, minimizing the bending loads on the LV is of vital importance; these loads will determine the dimensions and material that comprise the vehicle's structural elements, which in turn will dictate the launch vehicle's mass and thereby constrain its performance. Reasonable limits to bending loads are used to set preliminary design specifications for the vehicle.

Before, during, or after max-*q*, it's common for the motors to be throttled back or cut off to reduce the vehicle's accelerations or loads. The Space Shuttle, for instance, throttled its main motors during max-*q* to reduce aerodynamic loads. This is shown in Fig. 1.7; typically, a man-rated launch vehicle will need to limit its axial acceleration to $3\,g$ due to the frailty of its human cargo.

1.3.6 First-Step Motor Shutdown

The first-step shutdown is defined as the moment when the first main propulsive step shuts down. For simple multistep LVs, this is the burnout of the first step and occurs after the strap-ons have separated. If the LV

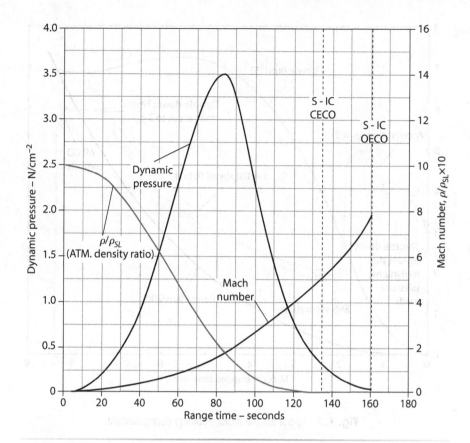

Fig. 1.6   Mach number (*M*), dynamic pressure (*q*), and density ratio (*ρ/ρ0*) for Apollo XI launch; density ratio curve added by author from mission data at scale height = 8.5 km. Source: [7].

has a single main motor, this event may be referred to as *main engine cutoff (MECO)*. The required speed at burnout, less gravity and drag losses, is used to size the amount of useful propellant for the first step, including proper allowances for any additional strap-ons and their contributions to the vehicle's velocity.

The sudden cutoff of the first step's thrust impulsively unloads the compressed vehicle stack, which can then oscillate (or ring) longitudinally and produce transient axial loading. If the vehicle utilizes multiple engines, they must all shut down simultaneously, or else a moment may be imposed on the base of the vehicle, which in turn can cause lateral loads on the payload.

1.3.7   Staging and Separation

It has previously been shown that to increase the performance of a launch vehicle—specifically, to increase the amount of payload that can be accelerated and lifted to a given orbit—strap-ons are utilized, and spent steps are

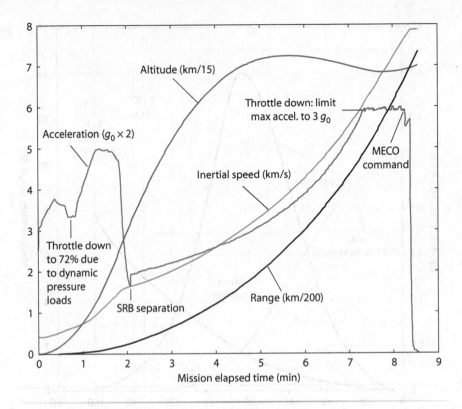

Fig. 1.7 Space Shuttle motor throttling during descent.

jettisoned during flight. This concept of jettisoning applies to any piece of the vehicle that has served its purpose. Jettisoning mass allows the vehicle designer to utilize different motors and propellants for different steps, which can improve the overall performance of the vehicle. Again, for safety, all dropped steps must not land on a populated area. Spent steps are usually separated pyrotechnically. The pyrotechnic shock that occurs can impose additional loads on the LV and its payload.

1.3.8 Upper-Step Ignition

After the first step has been jettisoned, the vehicle's next step is ignited. The time delay after separation is chosen to ensure that the jettisoned step is physically clear of the remaining vehicle, so there is no possibility of contact between the jettisoned step and the rest of the vehicle. In some instances, the upper step may be fired while still attached to the lower step. This procedure is known as *fire-in-the-hole*, and requires either venting or an open truss interstage structure to prevent pressure from destroying the structures.

1.3.9 Payload Fairing Jettison

Once the value of aerodynamic loading due to passage through the atmosphere has dropped below a certain threshold, the payload fairing (PLF) that serves to protect the payload from effects of ram air and aerodynamic heating is no longer needed for the mission and is jettisoned. PLF jettison often occurs soon after MECO and staging.

1.3.10 Upper-Step Shutdown

This is the moment when the propulsion system for an upper step shuts down. If there is a single, "sustainer" motor, it may be referred to as *sustainer engine cutoff (SECO)*. The difference in the vehicle's speed at startup and at burnout (the Δv required for the mission) is used to size the useful propellant mass for the upper steps.

Figures 1.3 and 1.4 illustrate typical ascent profiles with generic vehicles that have only two main steps. Note that for required mission performance, some vehicles have more than two steps. The descriptions about separation, ignition, and burnout apply to these vehicles as well.

1.4 Typical Launch Vehicle Mission and Mission Elements

The makeup of a typical LV mission is shown symbolically in Fig. 1.8, including the components required to perform the mission elements. Not only is the launch vehicle hardware required to perform the mission, but so are other necessary elements including the actual launch trajectory, the launch operations personnel and launch controllers, tracking, data relay,

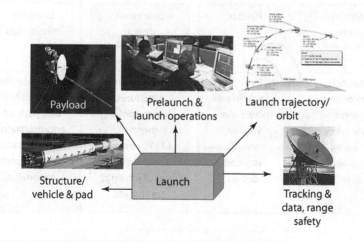

Fig. 1.8 Typical launch vehicle mission elements.

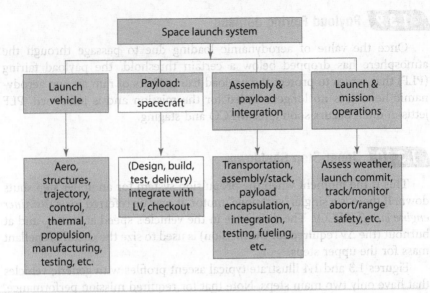

Fig. 1.9 The major elements of a launch vehicle design process.

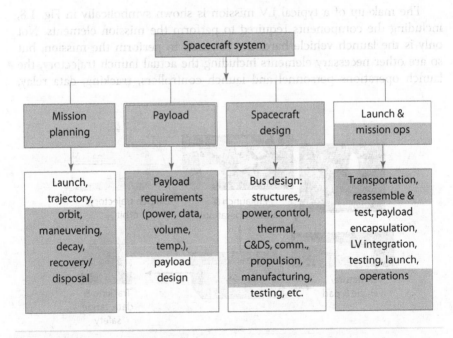

Fig. 1.10 The design and operation of many spacecraft systems, shown in blue shading, depend heavily on the launch vehicle.

and range safety operations. The launch *system* will not be successful unless all of these components individually operate as designed.

A different breakdown of a space mission may be seen in Fig. 1.9. This figure indicates the major actions in a vertical format, reading approximately left to right. The payload is shown in a double-line borders to indicate its independence from the launch vehicle; however, the spacecraft's design and operation are heavily dependent on the capabilities of the launch vehicle. The dependence is symbolically indicated in Fig. 1.10.

1.5 The Typical Launch Vehicle Design Process

The typical LV design process is best illustrated as a flowchart similar to the one shown in Fig. 1.11. The entire process begins with the development of mission requirements and design constraints (top left of figure). As with any other engineering design effort, it is necessary to specify the requirements the vehicle must meet, as well as any constraints to which the design must adhere. The requirements could be to inject a specified mass into a specified orbit or trajectory, and constraints could be on propellants, available motors, and so forth.

Several requirements (called *system requirements* or *originating requirements*) will be mandated by the end user of the launch vehicle, because they

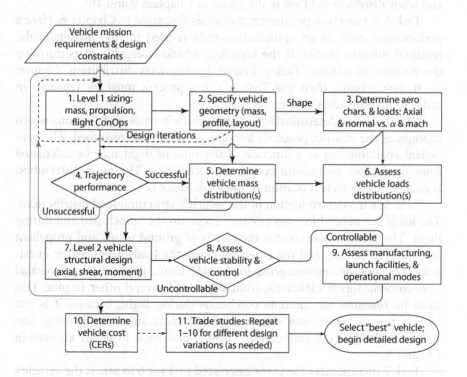

Fig. 1.11 Launch vehicle design process.

will likely be the ones paying for the LV's development and deployment. These originating requirements typically evolve from a need to remedy a perceived shortfall or deficiency within the end user's capabilities or to improve upon the end user's current capabilities. Other requirements, called *derived requirements*, are so called because the design constraints are derived from them. Stated simply, originating requirements specify *what* needs to be done, whereas the derived requirements specify *how* things should be done.

Task 1 is the Level 1 initial sizing. By knowing the payload, propellants, and required velocity, and estimating gravity and drag losses, it is possible to come up with a "first cut" of the vehicle's gross characteristics: namely, the masses of the steps. This first-cut concept is discussed in Chapters 3 and 5, and will be refined later as the design progresses.

Task 2 is to utilize the masses of the steps to design tanks that will contain the requisite amounts of propellant, and to come up with a layout—specifically, step lengths and diameters. Extra volume is allotted to margins, unusable propellant, and other "real" engineering needs. These topics, and the physical layout of the vehicle, are discussed in Chapter 8.

Task 3 is to determine the vehicle's aerodynamic characteristics. These include axial and normal force coefficients as a function of angle of attack α and Mach number M and are discussed in Chapter 9. The results are used in trajectory performance evaluation (Task 4, discussed in Chapter 6) and loads distribution (Task 6, discussed in Chapters 9 and 10).

Task 4 is trajectory performance analysis discussed in Chapter 6. Here a performance code or an optimization code is used to "fly" (simulate) the required mission profile. If the assumed vehicle is capable of performing the mission simulation, Task 5 follows (assess mass distribution, Chapter 9). If unsuccessful, then the Task 1 sizing process must be repeated in order to provide additional margin.

Task 5 is the determination of the vehicle's mass distribution, which changes as the mission proceeds because propellant is consumed. The propellant consumption as a function of the time of flight may be calculated from the known propulsion system characteristics. The mass distribution is needed to calculate a portion of the loads, Task 6.

Task 6 is the determination of the vehicle's structural load distributions. The loads are needed for two circumstances: on the launch pad and during flight. The former incorporates the effects of ground winds and propellant loads on the vehicle, and the latter combines the loads produced by quasi-static acceleration, maneuvering loads, side loads, pressurization, thermal environment, thrust variations, wind shear, and several other factors. This must be repeated for multiple conditions during flight, because it is not known, per se, which one will produce the most strenuous loading, also called the design's extreme case(s). Ground and flight loads are covered in Chapter 9.

Task 7 incorporates the loads calculated in Task 6 to assess the vehicle's stresses, which (along with allowable levels) determine the vehicle's structural design. This leads to details such as tank materials and thicknesses, pressurization schedules, and the "placard," which dictates if a vehicle can launch due to environmental considerations. This material may be found in Chapter 10.

Task 8 is the vehicle's flight controllability assessment. With the details of the structural design from the previous task, the mission can now be simulated with the appropriate hardware characteristics. If the assessment of the stability of the vehicle is satisfactory for all flight conditions, a baseline design may now be specified. This process is described in Chapter 12.

Task 9 is an assessment of the vehicle's operations. With the details of the structural design from previous steps, the mission's timeline of assembly, stacking, prelaunch testing, fueling, and so forth can be planned. These topics are covered in Chapters 13 and 14.

Task 10 is a first cut at estimating the vehicle's cost. Design, fabrication, testing, and operations are all estimated using statistical cost estimating relationships (CERs). Vehicle costing is discussed in Chapter 17.

Task 11 is the vehicle's optimization. During tasks 1–10, much information has been learned about the new vehicle. The added knowledge allows the designers to tweak and redesign aspects of the vehicle in order to improve its performance. The best resulting design moves forward as the baseline for more detailed analysis and design activities.

There are many characteristics of LVs that need to be discussed for proper design and manufacturing practices. Some of these topics are discussed in Chapters 6 (trajectory optimization), 7 (structures), 8 (sizing and inboard profile), 12 (stability and control, higher-order effects), 14 (vehicle systems and launch pad facilities), 15 (testing, reliability, and redundancy), and 16 (launch vehicle failures, range safety, flight termination systems, and best practices to avoid failure). Acquiring the knowledge contained in these chapters is recommended, even if the reader does not formally study these topics or cover them in a design class.

1.6 Launch Sites

Perhaps the most important mission element is the selection of the launch site itself. As will be discussed in later chapters, the selection of a launch site for a given mission cannot be an arbitrary or ill-informed decision. The selection of a launch site will directly constrain the orbits that a launch vehicle can attain during its ascent trajectory, and will thus have a direct and measurable impact on the amount of propellant required for a given mission. Some launch sites will naturally lend themselves to attaining certain orbits efficiently; for instance, the Kodiak Launch Complex in Alaska and the Woomera range in Australia provide efficient access to polar orbits. It should also be noted that candidate launch sites are constrained by legal, political, and a wide variety of other, purely non-technical considerations; for example, in the United States the Federal

Table 1.1   List of Global Launch Sites

Site Name	Country	Lat.	Long.	Orbit Types
Alcântara Launch Center	Brazil	2.3 deg S	44.4 deg W	Suborbital
Baikonour	Kazakhstan	45.6 deg N	63.4 deg W	LEO, GTO, PEO, SSO
Barking Sands, HI	USA	22.0 deg N	159.8 deg W	Suborbital
Centre Spatial Guyanais	French Guiana	5.2 deg N	52.8 deg W	LEO, GTO, SSO, PEO
Dombarovsky ICBM base	Russia	50.8 deg	59.52 deg E	64.5-deg orbits
Eastern Test Range	USA	28.5 deg	81 deg W	LEO, GTO, GEO, ISS
Iran Satellite Launch Center	Iran	35.2 deg N	52.9 deg E	
Jiuquan Space Launch Center	China	40.6 deg N	99.9 deg E	57–70-deg orbits
Kagoshima	Japan	31.2 deg N	131.1 deg E	
Kapustin Yar	Russia	48.4 deg N	45.8 deg E	Suborbital
Kodiak Launch Complex	USA	67.5 deg N	146 deg W	PEO, SSO
Kwajelein Missile Range	Marshall Islands	9.40 deg N	167.48 deg E	
Naro Space Center	S. Korea	34.03 deg N	127.54 deg E	
SeaLaunch Odyssey Platform	Equator	0 deg N	154 deg W	GTO
Palmachim AFB	Israel	31.5 deg N	34.5 deg E	Retrograde
Plesetsk Cosmodrome	Russia	62.8 deg N	40.1 deg E	
San Marco	Kenya	2.9 deg S	40.3 deg E	
Sohae Station	N. Korea	39 deg N	124 deg W	
Sriharikota Island	India	13.9 deg N	80.4 deg E	PEO, GTO
Svobodny	Russia	51.7 deg N	128.0 deg E	
Taiyuan Space Launch Center	China	37.5 deg N	112.6 deg E	PEO
Tanegashima Island	Japan	30.4 deg N	131.0 deg E	
Vandenberg AFB	USA	34.4 deg N	120.35 deg W	SSO, PEO
Vostochny Cosmodrome	Russia	51.82 deg N	128.25 deg E	
Wallops Flight Facility	USA	37.8 deg N	75.5 deg W	ISS, suborbital
White Sands Missile Range	USA	32.3 deg N	106.8 deg W	Suborbital
Woomera	Australia	31.1 deg S	136.8 deg E	PEO
Xichang Space Launch Center	China	28.25 deg	102.0 deg E	

Lat. = Latitude, Long. = Longitude, GTO = Geosynchronous Transfer Orbit, PEO = Polar Earth Orbit, SSO = Sun-Synchronous Orbit, ISS = International Space Station orbit, NSEW = North South East West.

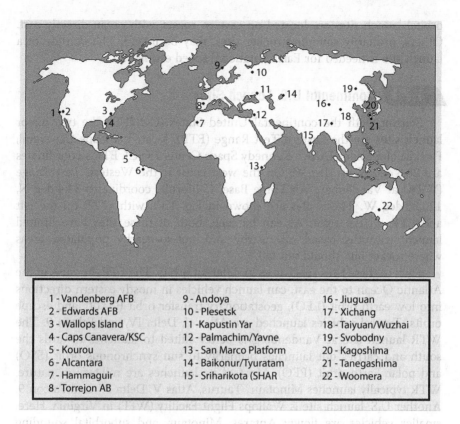

1 - Vandenberg AFB	9 - Andoya	16 - Jiuguan
2 - Edwards AFB	10 - Plesetsk	17 - Xichang
3 - Wallops Island	11 -Kapustin Yar	18 - Taiyuan/Wuzhai
4 - Caps Canavera/KSC	12 - Palmachim/Yavne	19 - Svobodny
5 - Kourou	13 - San Marco Platform	20 - Kagoshima
6 - Alcantara	14 - Baikonur/Tyuratam	21 - Tanegashima
7 - Hammaguir	15 - Sriharikota (SHAR	22 - Woomera
8 - Torrejon AB		

Fig. 1.12   Global launch sites. Source: [8]

Aviation Administration requires a number of policy, safety, and environmental reviews as part of the process it uses to issue launch licenses and
experimental permits. Other countries have similar licensing and permitting
processes managed by their own civil aviation authorities. Still other launch
sites face political constraints that prevent certain types of launches
altogether; for instance, Palmachim Air Force Base in Israel must rely
almost exclusively on retrograde launches due to the political climate
immediately to the country's east. A short list of launch sites from
around the world is given in Table 1.1, including the types of orbits attain-
able from a given site. A map of launch sites from around the world is given
in Fig. 1.12.

1.7   Launch Site Selection Criteria

As previously mentioned, there are numerous launch facilities worldwide.
These sites have a direct impact on the performance requirements of
a launch vehicle, so an examination of them is appropriate. Many of the

orbital launch sites are located on eastern coasts, with ocean directly east. Others, primarily sounding rocket sites, may be inland. The latitude of a launch site is needed for Earth rotational speed estimation.

1.7.1 Continental U.S. Launch Sites

Starting with the continental United States (CONUS), the two major launch sites are the Eastern Test Range (ETR), located at Cape Canaveral, Florida, alongside NASA's Kennedy Space Center (KSC). ETR's coordinates are 28.5 deg N, 81 deg W. On the west coast is the Western Test Range (WTR) at Vandenberg Air Force Base, California, coordinates 34.4 deg N, 120.35 deg W. These sites are shown in Fig. 1.13, with WTR on the left and ETR on the right. As can be seen, both of these sites have limited launch azimuths based on nearby and not-so-nearby populated areas where rocket bits should not fall.

Each of these sites specializes in different orbits. The ETR, because of the Atlantic Ocean to the east, can launch vehicles in mostly eastern directions into low earth orbit (LEO), geostationary transfer orbit (GTO), and escape orbits. Typical vehicles launched are Atlas V, Delta IV, and Falcon 9. The WTR launch site at Vandenberg is better suited to launches towards the south and carries out launches primarily to sun synchronous orbit (SSO) and polar Earth orbit (PEO). Most of its launches are military in nature. WTR typically launches Minotaur, Taurus, Atlas V, Delta IV, and Falcon 9. Another U.S. launch site is Wallops Flight Facility (WFF) in Virginia. Here, smaller vehicles are flown: Antares, Minotaur, and suborbital sounding rockets. Rocket Lab is building an Electron launch pad here. White Sands

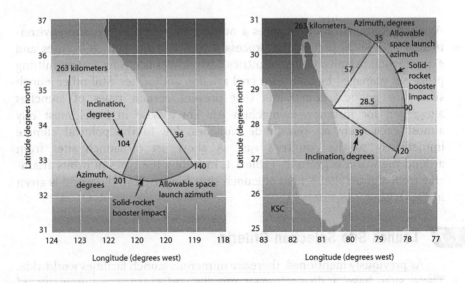

Fig. 1.13 Left: Western Test Range; right: Eastern Test Range. Source: [3].

Fig. 1.14 Kodiak Island launch site. Source: [4].

Missile Range (WSMR) in New Mexico is used primarily for suborbital sounding rockets and testing.

1.7.2 Other U.S. Launch Sites

Kodiak Island, Alaska, is the host to the Kodiak Launch Complex (KLC), approximately 30 miles off the coast and south of Anchorage. It is also known as Pacific Spaceport Complex Alaska (PSCA). It has a wide-open, unobstructed launch corridor towards its south. The location is shown in Fig. 1.14.

Some other U.S. launch sites are located in the Pacific Ocean. Kwajalein Island was the launch site of the SpaceX Falcon 1 and also launches interceptors towards test vehicles originating at the WTR. The Pacific Missile Range Facility at Barking Sands, Hawaii, launches interceptors and sounding rockets. Its only orbital attempt, a small vehicle called Super Strypi, ended in failure. A number of other launch sites are currently or will be licensed by the U.S. Federal Aviation Administration (FAA) Office of Commercial Space Transportation for commercial space activities in the near future. A list of these licenses and their details is provided in Table 1.2.

Should the reader be curious, the FAA concerns itself with the following in the United States [5]:

> The FAA issues a commercial space transportation license or experimental permit when we determine that your launch or reentry proposal, your proposal to operate a launch or reentry site, or your proposal to test equipment, design or operating techniques will not jeopardize public health and safety, property, U.S. national security or foreign policy interests, or international obligations of the United States.

Table 1.2 FAA Active Launch Site Operator Licenses

Site	Operator	Loc.
California Spaceport	Harris Corporation	CA
Mid-Atlantic Regional Spaceport	Virginia Commercial Space Flight Authority	VA
Pacific Spaceport Complex	Alaska Aerospace Corp.	AK
Oklahoma Spaceport	Oklahoma Space Industry Development Authority	OK
Florida Spaceport	Space Florida	FL
Ellington Airport	Houston Airport System	TX
Cecil Field Spaceport	Jacksonville Aviation Authority	FL
Midland International Airport	Midland International Airport	TX
Mojave Air & Space Port	Mojave Air & Space Port	CA
Spaceport America	New Mexico Spaceflight Authority	NM

Source: [5].

Launch or Reentry Vehicles
* Specific
* Expendable
* Reusable
* Operator

Launch site

* Pre-application consultation
* Policy review and approval
* Safety review and approval
* Environmental review
* Compliance monitoring (post-issuance of license)

Experimental permits for reusable suborbital rockets

* Experimental permits

Safety approvals

* FAA safety approvals for commercial launch operators

Now, let's get away from all of this paperwork and move onto non-U.S. launch sites!

1.7.3 European Launch Sites

Kourou, French Guiana, is the location of the European Space Agency's launch site. It's called Centre Spatial Guyanais (CSG), and is located on the northeast coast of South America with location latitude 5.2 deg N, longitude 52.8 deg W, as shown in Fig. 1.15. Vehicles launched include Ariane 5, Russia's Soyuz, and Vega. This near-equatorial site favors launches to LEO, GTO, and SSO.

Fig. 1.15 The European Space Agency launches from this site near Kourou, French Guiana.

The Russian launch vehicles at Kourou are relatively recent. The majority of Russia's launches take place from a site called Baikonur (formerly known as Tyuratam), in Kazakhstan, located at latitude 45.6 deg N, longitude 63.4 deg E. Among the vehicles launched are Soyuz, Proton, Zenit, Tsyklon, and others. This site can launch into LEO, GTO, PEO, and SSO, although its relatively high latitude favors the latter two. Due to political issues, and the fact that Kazakhstan is not a part of the Russian Federation, Russia has built a new launch site on its own territory in eastern Russia, near Vostochny. This site is controlled by Russia, and operations are expected to move there in the future. Launch sites used by the Russian Federation are shown in Fig. 1.16.

China has several launch sites. Xichang is located at 28.25 deg N, 102.0 deg E. Taiyuan Space Launch Center near Beijing is located at 37.5 deg N, 112.6 deg E. The vehicles launched are primarily Long March of various sizes and payloads. Launch sites within China are shown in Fig. 1.17.

India has a launch site on its eastern coast at Sriharikota, latitude 13.9 deg N, longitude 80.4 deg E. Because of its location on the Indian Ocean, it can easily launch to both equatorial and polar inclinations.

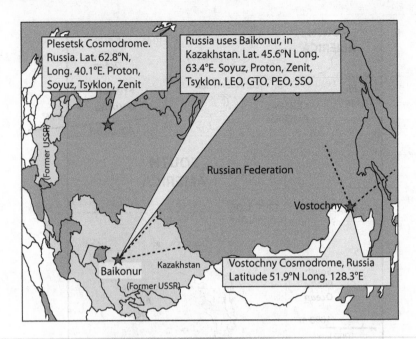

Fig. 1.16 Russian Federation launch sites.

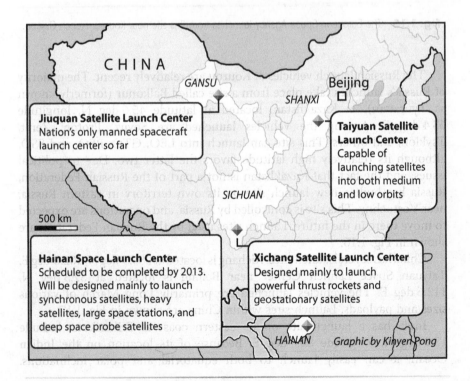

Fig. 1.17 Chinese launch sites.

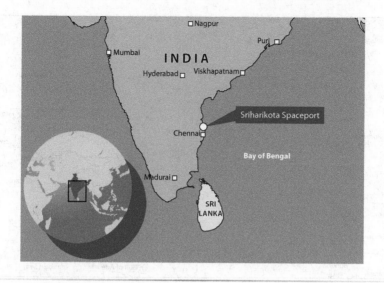

Fig. 1.18 Indian launch site at Sriharikota on the Bay of Bengal.

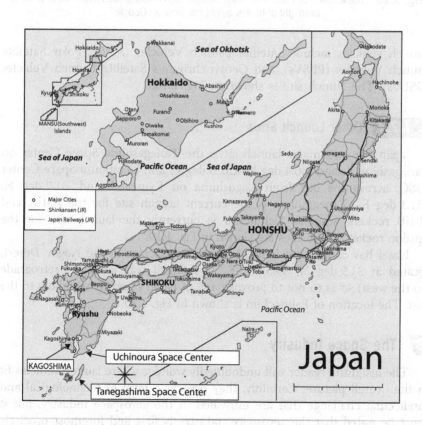

Fig. 1.19 Japan's launch sites are in the southern part of the main islands. Source: NASA.

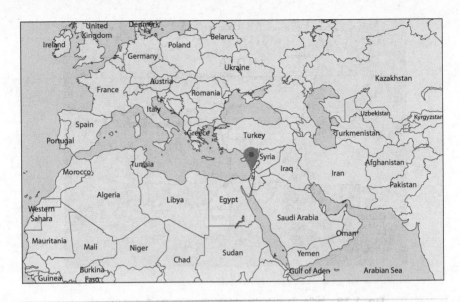

Fig. 1.20 The location of Palachim, Israel, necessitates a retrograde launch so as to avoid overflight of hostile neighbors. Source: Google.

Launch vehicles include Satellite Launch Vehicles (SLVs), Polar Satellite Launch Vehicles (PLSVs), and Geosynchronous Satellite Launch Vehicles (GSLVs). The launch site is shown in Fig. 1.18.

1.7.4 Other Launch Sites

Japan has two major launch sites: the Tanegashima Space Center on Tanegashima Island (30.4 deg N, 131.0 deg E) and Uchinoura Space Center (USC, across the bay from Kagoshima on Kyushu Island, 31.2 deg N, 131.1 deg E). Tanegashima is the current launch site for the H-IIA and H-IIB rockets, whereas Uchinoura is currently the launch site for the Epsilon rocket. These sites are shown in Fig. 1.19.

Israel has carried out launches from Palmachim in the Negev Desert, located at 31.9 deg N, 34.5 deg E. The Shavit was launched retrograde (to the west) so as to not to provoke any of Israel's hostile neighbors to the east. The location of Palmachim is shown in Fig. 1.20.

1.8 The Space Industry

The insightful reader will undoubtedly wonder where launch vehicles fit in the overall picture. Certainly, they present a level of technological and intellectual challenge that are unrivaled in the aerospace industry, but it must be noted that the aerospace industry is first and foremost precisely that—an *industry*—and commercial applications present an enormous opportunity to enterprising aerospace companies and the future-minded people who work there. Space transportation itself represents only a small fraction ($5.5 billion, or about 1.6%) of the $345 billion space economy [6], of which approximately $261 billion was revenue generated by corporations providing services such as television, mobile communications, broadband communications, remote sensing, satellite and ground system manufacturing and sales, and launch services. Of the $5.5 billion generated by launch services, only about 20% was generated by U.S. providers, demonstrating that the design and fabrication of launch vehicles is a truly international endeavor. The remaining $84 billion—about 34% of the total space economy—consists of government expenditure, including defense spending, and a further $2 billion or so consisted of human spaceflight. In the words of the Federal Aviation Administration [6], "Space transportation is an enabling capability, one that makes it possible to send national security and commercial satellites into orbit, probes into the solar system, and humans on exploration missions."

At the time of this writing, large commercial spacecraft are typically launched by the Ariane 5, Proton, and SpaceX's Falcon 9. A number of new systems are under development and are mentioned in Chapter 2.

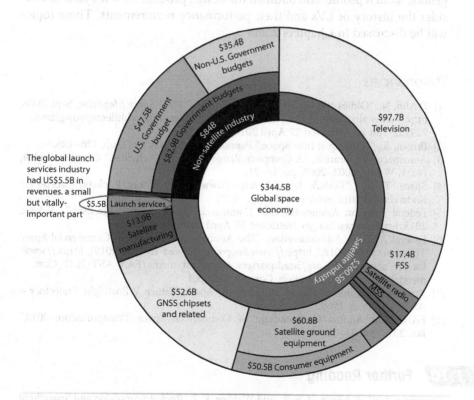

Fig. 1.21 A breakdown of the space economy. Source: [6].

Another emerging market is that of launch services for smaller payloads, who were almost always forced to fly as secondary payloads on a larger launch vehicle. These smaller launchers are intended to provide primary launch services to the smaller payloads.

It is thus arguable that, despite the enormous military and research potential of launch vehicles and the satellites they carry, "spacelifters" are primarily economic vehicles. Launch vehicles are the only currently available method of supporting the global space economy. Graphically, the enormity of the space industry is shown in Fig. 1.21, from [6], which presents compiled industry statistics circa 2018. When one considers the awesome power of some launch vehicles—such as the ULA Vulcan (over 19,000 lb to GTO)—as well as the awesome amount of resources necessary (up to $260 million per launch in some cases), the need for a multitude of highly trained, highly motivated launch vehicle experts becomes clearly evident. It is the intent of this book to assist the reader in taking the next step toward that future.

1.9 Summary

We have briefly discussed the need for space launch vehicles, provided a generic launch profile, and outlined the design process. Now it's time to consider the history of LVs and their performance requirements. These topics will be discussed in Chapters 2 and 3.

References

[1] D'Alto, N., "Oldies and Oddities: Zeppo's Gizmo," *Air & Space Magazine*, Sept. 2008, http://www.airspacemag.com/history-of-flight/oldies-amp-odditieszeppos-gizmo-729132/?no-ist [retrieved 22 April 2016].
[2] Brown, S., "Winging It Into Space," *Popular Science*, May 1989, pp. 126–128.
[3] Aerospace Corporation, "A Complete Range of Launch Activities," *Crosslink*, Vol. 4, No. 1, Winter 2002–2003, pp. 16–21.
[4] Space Today, "Kodiak Island," http://www.spacetoday.org/Rockets/Spaceports/KodiakIsland.html [retrieved 10 Aug. 2018].
[5] Federal Aviation Administration, "Commercial Space Launch Licenses," 17 Sept. 2013, https://www.faa.gov [retrieved 21 April 2018].
[6] Federal Aviation Administration, "The Annual Compendium of Commercial Space Transportation: 2018," https://www.faa.gov [retrieved 15 June 2019], https://www.faa.gov/about/office_org/headquarters_offices/ast/media/FAA_AST_2017_Commercial_Space_Transportation_Compendium.pdf
[7] Krausse, S. C., Boeing Report D5-15560-6 "Apollo/Saturn V Postflight Trajectory – AS-506," Oct. 6, 1969.
[8] FAA, "The Annual Compendium of Commercial Space Transportation: 2018," Jan. 2018, pp. 32–33.

1.10 Further Reading

Cornelisse, J. W., Schöyer, H. F. R., and Wakker, K. F., Rocket Propulsion and Spaceflight Dynamics, ISBN 0-273-01141-3, 1979.

Lange, O. H., and Stein, R. J., Space Carrier Vehicles, Academic Press, New York, NY, 1963.

Leondes, C. T., and Vance, R. W. (Eds.), Lunar Missions and Explorations, John Wiley and Sons, Inc., New York, NY, 1964.

Linshu, H. (Ed.), *Ballistic Missiles and Launch Vehicles Design*, ISBN: 781077-187-6, 2002.

Seifert, H. S. (Ed.), Space Technology, John Wiley and Sons, Inc., New York, NY, 1959.

Sforza, P. M., *Manned Spacecraft Design Principles*, ISBN: 978-0-12-8044254, 2016.

Suresh, B. N., and Sivan, K., *Integrated Design for Space Transportation System*, ISBN 978-81-322-2531-7, 2015.

U.S. Federal Aviation Administration (FAA), Office of Commercial Space Transportation Reports and Studies Library, https://ipv6.faa.gov/about/office_org/headquarters_of fices/ast/reports_studies/

1.11 Assignment: Launch Vehicle System Report

Prepare a PowerPoint® briefing on some rocket-powered vehicle that has gone into (or attempted to go into) Earth orbit or beyond (no suborbital vehicles other than the X-15 or other NASA X-vehicles). Any vehicle that has flown is acceptable; do not report on any "paper designs." The purpose is to extend your knowledge of space launch vehicles and their technical attributes. Your briefing must provide the following information:

1. Title slide with a good photo or drawing of vehicle, title, your name, date, course, and so on.

2. Historical and political context: Give the need that drove the concept. Was it politics, profit, or something else entirely?

3. Responsible parties: Give the designer and manufacturer. Tell: (a) how long it took from idea to proposal to contract to delivery to flight, (b) the number produced, and (c) whether it is still in service.

4. Provide a photo, and include a three-view (front, side, and top) with significant dimensions (length, diameter(s), wing span, etc.). Show/ discuss any configuration(s)/options/models, etc. What launch site(s)?

5. Design heritage, design philosophy: Important dates: start of design, first flight, and so on. Give any special features, innovations, or technologies (e.g., first use of LH_2, densified LOx, electric-powered turbopumps, common bulkhead, etc.). Compare the selected design with contemporary designs carrying out a similar task or performance, if any. Anecdotal is okay, but quantitative is preferred.

6. Performance: Provide a table or list of the vehicle's/vehicles' performance (payload mass and orbit parameters), dimensions, masses (for each step: empty, propellants, etc.), payload fraction and propellant fraction, propellant descriptions, engine(s) designation(s) and description(s) including I_{sp} (specific impulse), and other useful or unique technological information.

7. Describe the vehicle's operational experience. How well did it work? Describe any interesting issues or problems that were (or weren't) overcome.

8. What happened to the program? Was it considered a success or failure? What was the cost? (Give the year of the cost figure.) For ongoing programs, what's the current status? If I were a government entity or a private company needing a launch, why should (or shouldn't) I buy this vehicle?

9. List a few technical things that you learned about launch vehicles from your investigation.

10. Provide a list of your sources in proper format; bibme.org is recommended. As for the text of your slides, *do not* cut and paste from any sources; everything must be in your own words.

Format: Write your report in a clear black font (≥ 18 pt. Times, Helvetica, etc.) on a white background. Enlarge illustrations enough to fill the majority of the slide. There is no expected minimum or maximum slide count—use enough slides to make your report concise; don't omit interesting information for the sake of space. A length of 8–20 slides is fine, including title page.

Chapter 2   A Technical History of Space Launch Vehicles

Most of this book is technical in nature; however, it's important to be exposed to some of the history of launch vehicles in order to understand current designs and practices. Accordingly, this chapter is dedicated to providing the reader with pertinent details of historical rockets, as well as current and near-term launch vehicles. Information on some of the improvements in technical aspects of historical rockets and missiles will make the information provided in subsequent chapters much more useful to the reader.

This chapter is *not* intended to be a comprehensive history of rocketry and launch vehicles; rather, it is intended to highlight some of the engineering and innovation that has appeared during their development. The reader interested in a more detailed history of launch vehicles is referred to any of the many historical texts on the subject, such as [1] and [2], among others. An online review of rocket history is available at http://www.spaceline.org, and another can be found courtesy of NASA Marshall Space Flight Center (http://history.msfc.nasa.gov). An extremely comprehensive online collection of rocket knowledge is available at http://astronautix.com, and a useful collection of rocket history is available at Historic Spacecraft (http://historicspacecraft.com).

2.1   Rockets in the Early 20th Century

The design of rockets, in the modern sense of the word, began in earnest with the work of Russian theoretician and schoolteacher Konstantin Tsiolkovsky (1857–1935). Tsiolkovsky carried out a great deal of theoretical work, and in 1903 a magazine called *Nauchnoe Obozrenie* (*Scientific Review*) published Tsiolkovsky's report entitled "Exploration of the World Space with Reaction Machines" (see Fig. 2.1). This report was a landmark in the history of rocket development, because it demonstrated that rocket performance was determined in part by the exhaust velocity of the gases escaping the rocket motor. The governing equation from this report has come to be known as the *Tsiolkovsky rocket equation.* This equation predicts the performance of rockets in an ideal situation—when they are firing in a

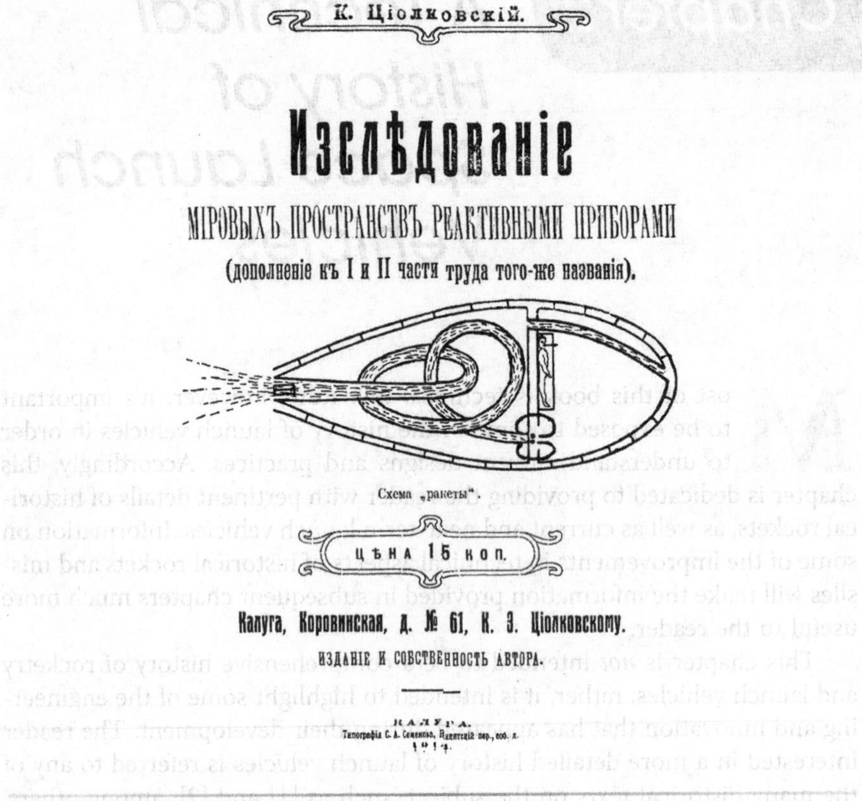

К. Ціолковскій.

Изслѣдованіе

МІРОВЫХЪ ПРОСТРАНСТВЪ РЕАКТИВНЫМИ ПРИБОРАМИ

(дополненіе къ I и II части труда того-же названія).

Схема „ракеты".

ЦѢНА 15 КОП.

Калуга, Коровинская, д. № 61, К. Э. Ціолковскому.

ИЗДАНІЕ И СОБСТВЕННОСТЬ АВТОРА.

КАЛУГА.
Типографія С. А. Семенова, Никитскій пер., соб. д.
1914.

Fig. 2.1 The 1914 edition of Tsiolkovsky's revolutionary work, "Exploration of the World Space with Reaction Machines."

vacuum and are not under the influence of outside forces such as gravity and atmospheric drag, and is discussed in Chapter 3. Tsiolkovsky also drew sketches of rockets and anticipated many engineering features (e.g., streamlined body, narrow throat, etc.) that nowadays are taken for granted.

Another prominent figure in the development of rocketry is Robert Hutchings Goddard, an American professor of physics at Clark University. Goddard also pursued the development of rocketry in the early 20th century, although his work was much more experimental than Tsiolkovsky's. Goddard is widely regarded as the "father of modern rocketry," due to his many innovations and inventions (over 200 patents) that supported the development of rockets [3]. Some of Goddard's work, specifically his research on recoilless projectiles, supported the military. A self-propelled projectile (as opposed to a gun with a barrel) has no recoil, and therefore provides some critical advantages as a weapon. Goddard's work in this discipline led to a device that came to be known as the *bazooka*, shown in Fig. 2.2. Goddard developed the basic idea behind the bazooka and demonstrated a prototype at the Aberdeen Proving Ground in 1918, just two days before the World

War I armistice. For this demonstration, Goddard's launch platform was a music rack that supported the launch tube of the bazooka.

In 1920, Goddard wrote about the possibility of a rocket attaining a speed fast enough to leave the Earth's atmosphere and reach the moon. This necessitated, among other things, the establishment of an *escape velocity* (the speed required to break away from a planet's gravitational pull) and the need for rockets to provide thrust in a vacuum. The *New York Times* promptly ridiculed Goddard's idea of space flight, and the government paid little attention to his work. Nevertheless, Goddard continued his research. In 1926, Goddard launched the world's first liquid propellant rocket in a field near Worcester, Massachusetts. Powered by gasoline and liquid oxygen, the rocket flew for only 2.5 s, during which time it climbed to a height of 41 ft and landed 184 ft away, in a cabbage patch. This flight of the humble "cabbage patch rocket" set the stage for modern rocketry. Up to this point, all rockets used solid propellant; however, the use of liquid propellant provides a number of advantages (including higher performance), which will be detailed in Chapter 4.

Goddard's revolutionary liquid propellant rocket is shown in Fig. 2.3. Figure 2.3a is a photo of the rocket and its launch stand; the rocket itself is in the center of the trapezoid-shaped stand. Figure 2.3b provides some engineering details of Goddard's design, much of which will appear unusual to the modern eye. For instance, the rocket engine in Goddard's design is located at the top of the vehicle, rather than at the bottom as in today's designs. Goddard believed this arrangement would provide stability by suspending the majority of the rocket's mass beneath the engine, similar to the mass

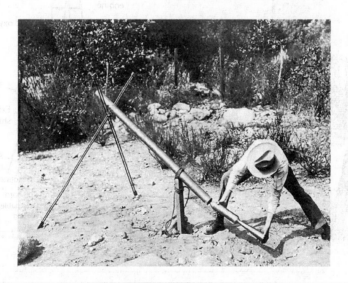

Fig. 2.2   Goddard's work in recoilless projectiles led to the bazooka. Source: NASA.

arrangement of a pendulum. This arrangement necessitated long propellant lines from the rocket motor to a set of tanks, and the tanks had to be placed beneath a shield to protect them from the exhaust plume of the rocket nozzle. Thus, the propellant had to be pumped "uphill" to the motor; again, this is the opposite of modern designs, where the propellant flows "downhill" to the motor. The engine shown in Fig. 2.3 used pressure feeding, where high-pressure gas is used to force the propellants into the combustion chamber.

Goddard's later work became more complex and versatile. He was not only the first to build and launch a liquid propellant rocket, but also was the first to equip a rocket with scientific instruments. Goddard began to incorporate gyroscopic measurement systems to stabilize his rockets, steering through the use of movable vanes in the rocket's exhaust, high-pressure propellant pumps, film cooling (using a rocket's liquid propellant to cool the motor, thus preventing the motor from melting), and many other aspects of rocketry that are still commonly used. One of Goddard's more complex rockets is shown in Fig. 2.4.

During World War II, Goddard offered his services to the U.S. government and was tasked by the U.S. Navy with developing practical "jet"-assisted takeoff (JATO) rocket motors, as well as liquid propellant engine capable of variable thrust. JATO rocket motors were uniquely advantageous in wartime,

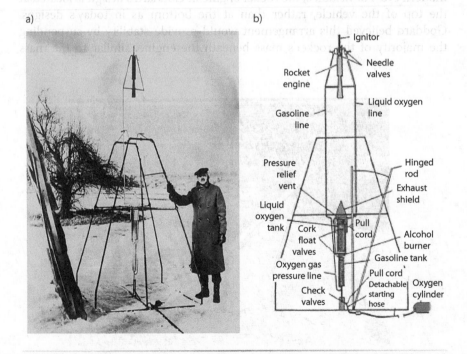

Fig. 2.3   a) Goddard and the world's first liquid propellant rocket, 1926; b) schematic of the rocket's components. Source: NASA.

Fig. 2.4 A view of the complex mechanisms in Goddard's 1936 rocket. Note the unique use of a stack of books as a shop support at the lower right corner of the photo. Source: NASA.

because they helped heavily loaded aircraft become airborne by providing additional takeoff thrust (More properly, these should have been called RATO, for rocket-assisted takeoff, but there was an aversion to using the word "rocket" in connection with "serious" endeavors at the time. Indeed, the name of the world-famous Jet Propulsion Laboratory in Pasadena, Calif. was deliberately chosen to avoid the word "rocket"). The use of JATO-like devices actually began in 1920s Germany, and such devices were used extensively by both Germany and the Allied powers during World War II. The $1,000 grant awarded to the Rocket Research Group by the U.S. National Academy of Sciences was the first instance of rocket research to receive financial assistance from the U.S. government since World War I. The JATO work also culminated in the first instance of an American flying an aircraft propelled solely by rocket power: Army Captain Homer Boushey flew an Ercoupe light aircraft with six JATO motors attached to the wings and with the airplane's propeller removed [4].

Unfortunately, Goddard was not well-recognized for his pioneering work in rocketry until after his death in 1945. He was posthumously awarded a Congressional Gold Medal in 1959—the highest civilian award in the United States and equal in prestige to the Presidential Medal of Freedom—and was only then honored as the "father of space flight." That same year, NASA named the Goddard Space Flight Center (formerly the Beltsville Space Center) in his honor. In 1960, the government awarded Goddard's estate $1 million for the use of his many rocket patents without formal permission. Finally, the *New York Times*—having famously ridiculed Goddard in 1920—issued a formal apology after Apollo 11 lifted off on its way to the moon in 1969.

2.2 World War II and the Development of the V-2

After World War I, the Germans continued to pursue the development of rocket technology. A group led by Dr. Wernher von Braun developed the famous V-2 (Vergeltungswaffe 2, or Retribution Weapon 2), which was the world's first guided ballistic missile. The V-2, also known as the A-4 (Aggregat 4), is shown in Fig. 2.5.

A young von Braun was first inspired to pursue rocketry in the 1920s, when he acquired a copy of Hermann Oberth's book, *Die Rakete zu den Planetenräumen* (*The Rocket into Interplanetary Spaces*). Wernher von Braun assisted Oberth in liquid propellant rocket motor tests beginning in 1930; von Braun's thesis, "Construction, Theoretical, and Experimental Solution to the Problem of the Liquid Propellant Rocket" (16 April 1934), was actually kept classified and not published until 1960. It was an artillery captain, Walter Dornberger, who arranged an Ordnance Department research grant for von Braun, who at the time was working adjacent to Dornberger's solid propellant rocket test site [5]. Goddard's research also did not go unnoticed by the Germans. Prior to 1939, German engineers and scientists—including von Braun—would occasionally contact Goddard directly with technical questions, and Goddard's various publications were used by von Braun as the basis for building the Aggregat series of rockets.

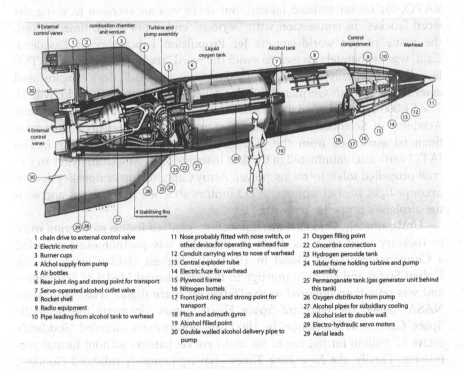

1 chain drive to external control valve
2 Electric motor
3 Burner cups
4 Alchol supply from pump
5 Air bottles
6 Rear joint ring and strong point for transport
7 Servo-operated alcohol outlet valve
8 Rocket shell
9 Radio equipment
10 Pipe leading from alcohol tank to warhead
11 Nose probably fitted with nose switch, or other device for operating warhead fuze
12 Conduit carrying wires to nose of warhead
13 Central exploder tube
14 Electric fuze for warhead
15 Plywood frame
16 Nitrogen bottels
17 Front joint ring and strong point for transport
18 Pitch and azimuth gyros
19 Alcohol filled point
20 Double walled alcohol delivery pipe to pump
21 Oxygen filling point
22 Concertina connections
23 Hydrogen peroxide tank
24 Tublar frame holding turbine and pump assembly
25 Permanganate tank (gas generator unit behind this tank)
26 Oxygen distributor from pump
27 Alcohol pipes for subsidiary cooling
28 Alcohol inlet to double wall
29 Electro-hydraulic servo motors
29 Aerial leads

Fig. 2.5 The German V-2, the world's first guided ballistic missile, was used as a terror weapon in the later years of World War II. Source: National Museum of the U.S. Air Force.

The development of the A-4 was originally postponed due to the unfavorable aerodynamic stability characteristics of its predecessor; however, by late 1941, the German Army Research Center at Peenemünde possessed the technologies necessary for the success of the A-4; these four key technologies were large, liquid propellant rocket motors; refined supersonic aerodynamic theory; gyroscopic guidance; and engine exhaust vane control surfaces. It should be noted that at the time, Hitler was not impressed by the A-4, calling it "an artillery shell with a longer range and much higher cost" [6]. Even by mid-1944, the complete parts list for the A-4 was unavailable, and it was not until Hitler became impressed by the enthusiasm of the A-4's developers—and the need for a "wonder weapon" to maintain German morale—that large-scale deployment of the V-2 was finally authorized.

The V-2 used liquid oxygen and alcohol for propellants. At launch, the V-2 would be self-propelled for up to 65 s, at which point the motor would shut down and the rocket would continue along a ballistic trajectory, with an apex of up to 50 miles [7]. The propellant pumps were driven by a steam turbine, with concentrated hydrogen peroxide producing the steam in the presence of a sodium permanganate catalyst. Both propellant tanks were constructed of an aluminum–magnesium alloy [8].

The rocket motor was *regeneratively cooled*, meaning that the cool propellants were pumped through the double wall of the combustion chamber to extract heat, and were then consumed to produce thrust. The necessity of this is apparent when one considers that the combustion chamber regularly reached temperatures of 2,500–2,700°C (4,530–4,890°F). This arrangement not only cooled the combustion chamber and nozzle, but also served to preheat the propellants prior to combustion. The injector head of the motor featured 18 prechambers (or pots), which ensured the correct mixture of fuel to oxidizer at all times. The oxidizer was supplied directly to the center of each prechamber and distributed through a shower-style, 120-hole injector (or vaporizer) into the bell-shaped mixing chamber. Swirl nozzles would impinge the fuel flow onto the flow of the oxidizer, and the mixed propellants would travel downstream for combustion. This arrangement is shown in Fig. 2.6, modified from a still frame of a video on *V-2* injectors produced by Astronomy and Nature TV (https://goo.gl/K4sCRW).

The V-2's warhead was an amatol 60/40 (60% TNT, 40% amatol) explosive detonated by an electric contact fuse. Although the warhead was protected by a thick layer of fiberglass, it could still detonate during the descent phase of the rocket's flight. The warhead's percentage by weight of explosive was on the order of 93%, which is a very high number compared to other types of munition, which are usually only on the order of 50% explosive by weight.

The V-2 had a guidance system utilizing a platform of two gyroscopes (a horizontal and a vertical) for lateral stabilization, and a pendulous integrating gyroscopic accelerometer (PIGA) to command engine cutoff at a specified velocity; needless to say, this arrangement was very advanced for the time.

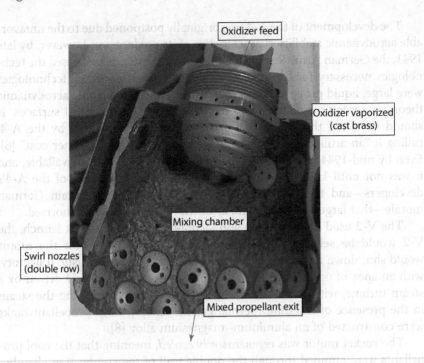

Fig. 2.6   Sample V-2 mixing prechamber.

The gyros fed data to a control system that moved both aerodynamic surfaces (located at the tips of the four stabilizing fins) and graphite thrust vanes (located just aft of the nozzle) so that the rocket could be steered along the proper trajectory. Because the V-2 was launched from presurveyed locations, the distance and azimuth to the target were known. Some later V-2s used radio signals to correct the missile's trajectory midflight, but earlier models used a simple analog computer to correct for azimuthal drift; approximately 20% of all operational V-2s were beam-guided [9].

The V-2's propellant tanks were independent of its outer airframe. This combination made the vehicle heavier than it needed to be, and later sections of this chapter will examine vehicles that had combined tanks and airframes. Each V-2 was carried on a towed trailer and could be set up to launch from almost any site relatively quickly. Each missile could deliver a 738-kg high explosive warhead approximately 320–360 km. The rocket had terrible accuracy, but served as an effective terror weapon. The firing of about 3,172 V-2s is estimated to have caused the deaths of some 3,000 people in England and many others in Belgium and other targets. The V-2 production was eventually transitioned to the Mittelwerk site, which was staffed by prisoners from the Mittelbau-Dora concentration camp, and it has been said that the V-2 is perhaps the only weapon system to have caused more deaths by its production than its deployment [10]. After the war's end, its engineering continued to influence the development of both Soviet and U.S. missiles. During postwar tests conducted by the United States, it became the first manmade object to cross the boundary of space.

2.3   The Cold War, ICBMs, and the First Space Launch Vehicles

Before the end of World War II, both the Allies and the Soviet Union realized the strategic usefulness of rocketry, and sought to capture German rocket experts. Both countries were successful in "harvesting" these individuals to jump-start indigenous rocket programs. Under Operation Paperclip, the U.S. military brought a number of German V-2 rocket scientists, led by Dr. Wernher von Braun, to the United States; these scientists were eventually settled near the Army's Redstone Arsenal in Huntsville, Alabama. The Soviets brought a number of German experts to their country as well. Both countries then initiated programs to develop ballistic missiles that could be used to deliver nuclear warheads to their opponents. During the development of these vehicles, both countries realized that such a missile could also serve as a satellite launcher.

2.3.1   Soviet Developments

The Soviet Union developed the R-7 Semyorka as its first intercontinental ballistic missile (ICBM) to carry nuclear warheads. The R-7 used liquid oxygen and kerosene as propellants and was designed for *parallel-burn* operation. This means that the center, core vehicle and its four external strap-ons were all burning simultaneously (i.e., in parallel) during liftoff and the initial ascent phase. When the strap-ons (shown above and below the core vehicle in Fig. 2.7) ran out of propellant, they were jettisoned, and the core stage continued burning all the way to orbit. The parallel burn at launch was done because at the time it was felt that attempting to ignite a rocket motor at altitude (an *air start*) was unreliable. This same reasoning led to the "stage-and-a-half" design of the U.S. Atlas intercontinental ballistic missile (ICBM), discussed later in this chapter.

All ballistic missiles may be turned into space launchers if suitably modified. These modifications may include a reduced payload mass, added upper stage(s), additional strap-on boosters, larger/longer propellant tanks, higher-performance motors, and so forth. As will be shown later in this chapter, virtually all of the ballistic missiles developed in the 1950s and 1960s went on to become space launch vehicles (SLVs), including ICBMs as well as intermediate-range ballistic missiles (IRBMs).

The Soviet Union modified one of its R-7 ICBMs to carry the first manmade object ever placed into orbit, called Sputnik, in October 1957. This technical achievement surprised most of the world and triggered the so-called "space race" between the United States and the Soviet Union. Realizing the propaganda value of space led the Soviets to pursue and claim

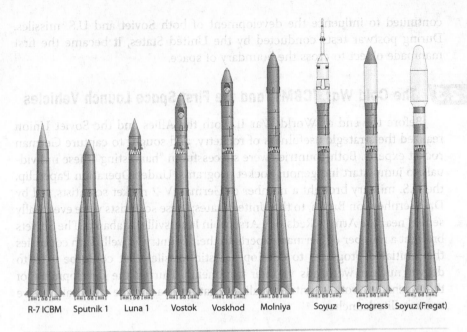

R-7 ICBM   Sputnik 1   Luna 1   Vostok   Voskhod   Molniya   Soyuz   Progress   Soyuz (Fregat)

Fig. 2.7   Russian R-7 launch vehicle evolution, 1957–present. Courtesy: Richard Kruse.

several firsts in space, and the United States spent many years trying to catch up to the Soviets.

Another important milestone first achieved by the Soviet Union was to put a human into orbit, which was accomplished by adding an upper stage to the R-7. The vehicle fourth from the left of Fig. 2.7, known as the Vostok (8K72K), was used by the Soviets to orbit the world's first cosmonaut, Yuri Gagarin, in 1961. This additional rocket stage provided enough energy to significantly increase the payload capacity. To reduce the overall length of this vehicle, the upper stage has toroidal ("donut-shaped") tanks. Launch vehicle propellant tanks and internal arrangements are discussed in Chapter 8.

A common expression in the United States is, "If it ain't broke, don't fix it." The rocket designers in the Soviet Union apparently have a similar expression, and it is seen in their rocket designs to this day: the Russians continue to use a launch vehicle derived from the R-7 (now called Soyuz, not to be confused with the Soyuz space capsule used to launch three crew into space). The evolution of the R-7 is shown in Fig. 2.7; more than 2,200 of these vehicles have been launched, making the R-7 the most versatile and commonly used SLV in history. The European Space Agency (ESA) has even built a Soyuz launch pad in French Guiana so the Russian-built rocket may launch from there as well.

The current version of the Soyuz launches from either the Baikonur Cosmodrome in Kazakhstan or from French Guiana. They function both as crew and cargo carriers to the International Space Station and as launch platforms for commercial and military missions. A photo of the current version is shown in Fig. 2.8. Note that these vehicles are carried on their sides on railroad tracks, which allows rapid transport to the launch pad and quick erection for launch.

2.3.2 U.S. Developments

The United States began a number of developments in parallel, including the civilian Vanguard and Scout launchers, as well as the military Redstone, Jupiter, Thor, Minuteman, and Polaris IRBMs and the Atlas and Titan ICBMs. Atlas, Titan, and Minuteman were silo-based, whereas Polaris was submarine-based. These basing schemes were selected to provide protection to the missiles in case of nuclear attack by the enemy, which at the time was the Soviet Union and the People's Republic of China.

2.3.2.1 The Vanguard

At the time of the Vanguard's development, the United States already had a longstanding interest in space. Long before Sputnik, the U.S. plan was to orbit a small satellite in the 1956–1957 timeframe to support the International Geophysical Year activities. This satellite and launch vehicle were known as Vanguard. The U.S. government did not want to be known for militarizing space, and so the Vanguard launch vehicle was ostensibly to be built by civilians. Ironically, the program was managed by the (military) Naval

Fig. 2.8 The current version of the R-7, now called Soyuz, some 50 years after first flight.
Source: NASA.

Fig. 2.9 a) the Vanguard on the pad; b) failed first launch attempt. Vanguard eventually went on to successfully orbit payloads three times. Source: NASA.

Research Laboratory. The Vanguard SLV and its infamous unsuccessful first attempt at an orbital launch in Dec. 1957 are shown in Fig. 2.9.

During the first (rushed) launch attempt (rushed because of the pressure from the recent Sputnik launch), the Vanguard reached a height of approximately one meter before falling back onto the pad and exploding. This caused national embarrassment for the Vanguard team and paved the way for Wernher von Braun's U.S. Army team that had earlier proposed an orbital attempt (and was turned down) in 1956. A total of 11 Vanguard launches were attempted between 1957 and 1959; of these 11 launch attempts, the third, eighth, and eleventh launches successfully delivered a satellite into orbit (payload weights of 3 lbs., 3 lbs., and 22 lbs., respectively). The remainder of the launches failed.

The Vanguard launch vehicle consisted of several steps with quite different propellants. The first step was a modified Viking sounding rocket powered by liquid oxygen and kerosene, with thrust vectoring and roll thrusters; the second step was powered by an Aerojet engine that used nitric acid and hydrazine propellants, with pitch, yaw, and roll thrusters; the Altair third step used solid propellant and was spin-stabilized. The Vanguard internal layout is shown in Fig. 2.10. More information may be found in [11] and [12].

2.3.2.2 The U.S. Ballistic Missiles

Although the development of the Vanguard was in public view, a number of long-range ballistic missiles were being developed in parallel in secret "crash" programs by different branches of the U.S. military. These programs were the Redstone and Thor IRBMs, and the Atlas, Titan, Minuteman, and Polaris ICBMs. Most of these later evolved into space launch vehicles after suitable modification; this trend is shown in Table 2.1, which shows the military roots of several successful launch vehicles.

2.3.3   The Redstone IRBM and Jupiter Launch Vehicle

After World War II, a team of U.S. Army engineers and scientists at the Army Ballistic Missile Agency (ABMA) at Redstone Arsenal set out to develop an IRBM. Because this IRBM was developed under the leadership of Dr. Wernher von Braun, it could trace its roots directly to the development of the German V-2. The vehicle that emerged from this development

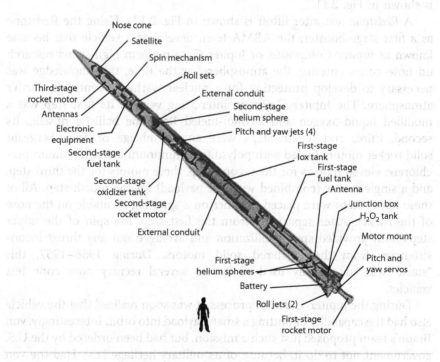

Fig. 2.10 Vanguard engineering details.

Table 2.1   The Evolution of Ballistic Missiles into Launch Vehicles

R-7 → Sputnik → Soyuz
Redstone → Juno I → Mercury-Redstone
Thor → Thor-Delta → Delta → Delta II → Delta III
Atlas → Atlas-Mercury → Atlas II → Atlas III
Titan → Titan II → Gemini-Titan II → Titan 23 G SLV → Titan III → Titan IV
Minuteman → Minotaur
Zenit → SeaLaunch and Antares

effort was the PGM-11 Redstone. This rocket showed a number of improvements over the German V-2: the Redstone had integral propellant tanks, meaning that the tank walls also served as part of the outer aerodynamic and structural airframe (thus saving weight); the payload, an explosive warhead, contained a radar altimeter fuse and was designed to separate from the spent booster, improving accuracy and decreasing the descent time needed to reach the target; and the inertial guidance system was also greatly improved over the V-2. All told, the Redstone could deliver a 2,860-kg (6,305-lb) warhead up to 200 miles. It was to become a reliable lower stage for space launch vehicles. An exploded view of the Redstone is shown in Fig. 2.11.

A *Redstone* just after liftoff is shown in Fig. 2.12a. Using the Redstone as a first-stage booster, the ABMA team developed a vehicle that became known as Jupiter-Composite, or Jupiter-C, to perform high-speed research on nose cones entering the atmosphere; at the time, this knowledge was necessary to develop protection for a nuclear warhead reentering Earth's atmosphere. The Jupiter-C was an interesting vehicle. Its first step was a modified liquid-oxygen and alcohol-fueled Redstone ballistic missile. Its second, third, and fourth steps were an assemblage of Baby Sergeant solid rocket motors fueled with polysulfide-aluminum and ammonium perchlorate: eleven motors for the second step, three motors for the third step, and a single motor (combined with the payload) for the fourth step. All of these upper steps were placed together on a spinning turntable on the nose of the rocket. After separation from the first step, the spin of the upper steps both provided spin stabilization and averaged out any thrust inconsistencies from the clustered solid motors. During 1956–1957, this "stack" was successfully used to launch several reentry nose cone test vehicles.

During the Jupiter-C design process, it was soon realized that the vehicle also had the capability of putting a small payload into orbit. Interestingly, von Braun's team proposed just such a mission, but had been ordered by the U.S. government not to do it, because of its military heritage [13]. Had the von Braun team been given permission in 1956, there is a pretty good chance

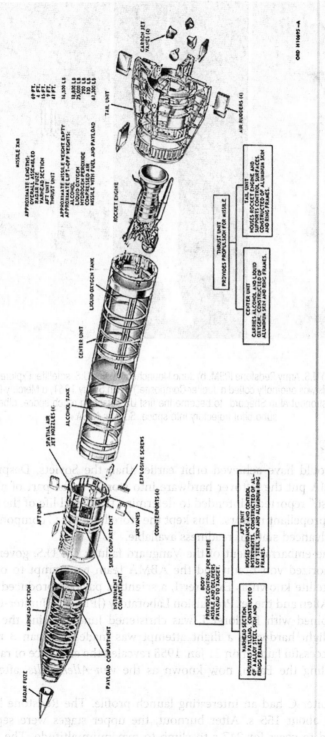

Fig. 2.11 Exploded view of Redstone missile. Source: U.S. Army TM 9-1410340-14/2.

Fig. 2.12 a) U.S. Army Redstone IRBM. b) Juno I launched the first U.S. satellite, Explorer I, on 31 Jan. 1958. (This was originally called a Jupiter-Composite.) c) On 5 May 1961, a Mercury-Redstone lofted U.S. astronaut Alan Shepard to become the first US person to reach space, albeit on a suborbital trajectory into space. Source: NASA.

that they could have achieved orbit earlier than the Soviets. Despite this order, ABMA put the leftover hardware into storage to be part of a "long-term life test," reportedly intended to determine the useful life of the upper-stage solid propellant motors. This kept the stored Jupiter-C components in the most advanced state of readiness available.

After the embarrassment of the Vanguard failure, the U.S. government finally authorized von Braun and the ABMA team to attempt to orbit an artificial satellite known as Explorer I, a scientific payload produced by Dr. James van Allen and the Jet Propulsion Laboratory (JPL). The Jupiter-C hardware, combined with Explorer I, was christened Juno I. Using the stored Jupiter-C flight hardware, a flight attempt was made less than 3 months later; its successful launch on 21 Jan. 1958 revealed the existence of radiation "belts" circling the Earth, now known as the *van Allen belts*, after their discoverer.

The Jupiter-C had an interesting launch profile. The Redstone booster burned for about 155 s. After burnout, the upper stages were separated and allowed to coast for 247 s to climb to maximum altitude. The second step's 11 motors were fired for 6 s, followed by a 2-s pause for separation; the third step's 3 motors then fired for 6 s, followed by another 2-s pause. The fourth step's single motor fired and placed the 13-kg (30-lb) payload into orbit. This is quite different from the common strategy of powered flight all the way up to orbit.

The letters *UE* painted on the booster of Explorer I were another unique feature. The numbering scheme was chosen in order to hide the actual number of Redstone rockets produced. This code was taken from the numerical values given to the letters in the name *Huntsville* (where the ABMA was based), and formed a simple substitution cypher (see Table 2.2). Thus, the *UE* painted onto the Juno I booster indicates that the booster was the twenty-ninth Redstone rocket.

The launching of the Explorer I satellite did not mark the end of the Redstone vehicles. The Redstone was called up to serve once again as the initial launcher for the Mercury capsule in NASA's new manned space program.

This vehicle is shown in Fig. 2.12c. Because of the mass of the Mercury capsule, it could not achieve orbit, but this allowed the United States to fly the capsule on a suborbital trajectory, with astronaut Alan Shepard becoming the first U.S. astronaut to reach space on 5 May 1961. Once the Atlas rocket became available, its larger payload capabilities were used to attain orbital Mercury flights.

2.3.4   The Jupiter A and Juno SLV

The Jupiter A, shown in Fig. 2.13a, was initially designated SM-78 and represented a logical follow-on to the Redstone. It had a range of 3,180 km (1,976 miles) and was designed to carry a 1-megaton warhead. Before

Table 2.2 Redstone Numbering Cypher

Letter	Value
H	1
U	2
N	3
T	4
S	5
V	6
I	7
L	8
E	9
X	0

Fig. 2.13 a) The U.S. Army's Jupiter A ballistic missile. b) The Jupiter A evolved into the Juno II, which was used to launch payloads, such as the Explorer VII in this photo, into space.
Source: NASA MSFC.

launch, the missile was shrouded by "petal" segments at its base to provide environmental protection; these petals were opened prior to launch. Unlike the Redstone, which had movable vanes inside the exhaust for control, the Jupiter first-step engine was gimbaled in two directions for pitch and yaw control, so called "thrust vector control" or TVC. Roll motion was controlled by gimbaling the vernier engine thrust. The early versions of the Juno 1 suffered from a propellant sloshing instability, which was solved with the inclusion of numerous metal spheres that floated on the top of the liquid propellants. Later vehicles used *slosh baffles* built inside the tanks to achieve the same purpose. Slosh baffles are discussed in Chapter 7.

With the addition of the Juno I's upper stages, the Jupiter missile evolved into the Juno II launch vehicle, shown in Fig. 2.13b. Unlike the Juno I, this vehicle's spinning upper stages were carried inside of an outer fairing for improved aerodynamics. This vehicle could carry a 45-kg (100-lb) payload to low Earth orbit.

2.3.5 Thor IRBM

The U.S. Air Force's Thor missile (Fig. 2.14a) was developed in the 1950s as an IRBM. The Thor was originally built by Douglas Aircraft and had machined *isogrid* propellant tanks; the isogrid pattern consisted of equilateral triangles machined out of aluminum sheeting. Thor used RP-1 (kerosene) and liquid oxygen as propellants, and could deliver its payload around 2,400 km (1,500 miles) downrange. The missile was propelled and controlled in pitch and yaw by a gimbaling Rocketdyne LR79 engine and two steerable LR-101 vernier rockets for roll control, mounted at the missile's base on either side of the main engine. The first successful launch took place on 20 Sept. 1957. Like other ballistic missiles, the Thor became the first stage booster of a number of space launch vehicles, eventually culminating in the Delta launch vehicle family.

As has been seen with other missiles, upper stages were added to the Thor for improved performance. The Thor-Able, shown in Fig. 2.14b, combined a Thor first step with an Able upper step derived from the Vanguard's second step. Early Thor-Able launches were suborbital and flew reentry test vehicles; later Thor-Able rockets launched a number of early orbital satellites,

Fig. 2.14 a) Thor IRBM; b) Thor-Able space launch vehicle. Source: U.S. Air Force.

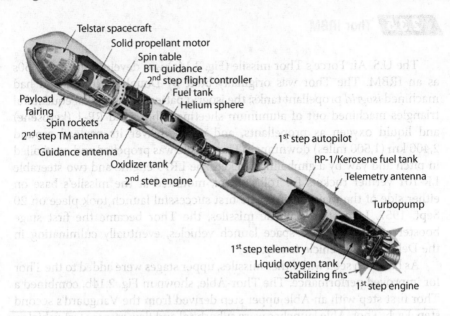

Telstar spacecraft
Solid propellant motor
Spin table
BTL guidance
2nd step flight controller
Fuel tank
Helium sphere
Payload fairing
Spin rockets
2nd step TM antenna
Guidance antenna
Oxidizer tank
2nd step engine
1st step autopilot
RP-1/Kerosene fuel tank
Telemetry antenna
Turbopump
1st step telemetry
Liquid oxygen tank
Stabilizing fins
1st step engine

Fig. 2.15 Delta-Telstar vehicle. (Courtesy NASA.)

including Explorer 6, Tiros 1, and early Pioneer missions. The vehicle could carry approximately 135 kg (300 lb) to low Earth orbit.

The inboard profile of a three-stage Thor-Delta B is shown in Fig. 2.15. The Thor continued to be the basis for better-performing space launch vehicles. It received higher energy upper stages; then, when the core vehicle had insufficient thrust for adequate liftoff acceleration, auxiliary strap-on motors were added to the first stage of the vehicle. Continued improvements included increased tank size for larger propellant capacity. Eventually, the Thor-based series of space launch vehicles was renamed Delta. One explanation of this name was that it was based on the Greek letter Δ, which in engineering equations represents a change in a value: the vehicle kept changing, as can be seen from Fig. 2.16. Delta models are available with varying numbers of strap-on boosters, upper stages, and different sizes of payload fairings, all meant to accommodate payloads of widely varying size and mass. Delta vehicles were built by McDonnell Douglas, which was eventually bought out by Boeing.

The Delta II-7925 launch vehicle, seen second from right in Fig. 2.16, represents the penultimate model in the Delta II series and can carry as much as 5,100 kg (11,300 lb) to low Earth orbit, or 1,300 kg (2,900 lb) to an interplanetary (Earth-escape) trajectory. The Delta II vehicle shown in Fig. 2.17 launched the Mars Exploration Rover Opportunity in July 2003.

Wondering about the Delta numbering scheme? Each digit in the vehicle name identifies different aspects of the launch vehicle, including the type of engine on each stage, augmentation motors (strap-ons), and the fairing size.

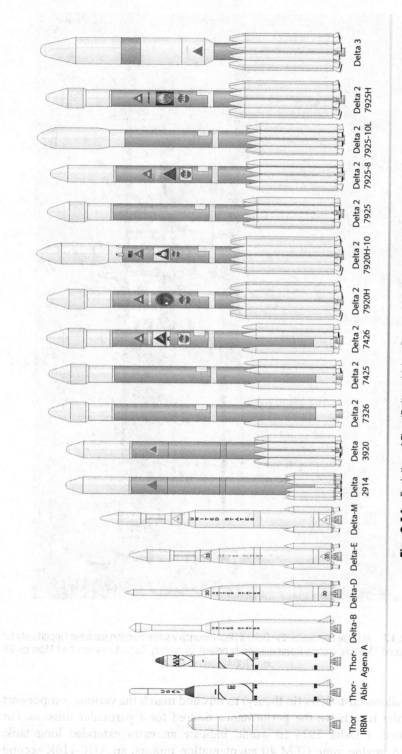

Fig. 2.16   Evolution of Thor/Delta vehicles. Source: historicspacecraft.com

Fig. 2.17   At Cape Canaveral, the Delta II Heavy launch vehicle carrying the rover Opportunity for the second Mars Exploration Rover mission is poised for launch. Opportunity reached Mars on 25 Jan. 2004. Source: NASA.

This allows customers (in theory) to mix and match the various components in order to achieve the performance desired for a particular mission. For instance, a Delta 7925-10 would indicate an extra extended long tank, RS27A engine, nine GEM 40 augmentation motors, an AJ10-118K second step, a Star 48B third step, and a 10-ft-diameter fairing. Detailed summaries of the history of the Delta launch vehicle may be found in [14] and [15].

The astute reader will undoubtedly be aware of higher number Delta models, in particular the Delta III and Delta IV. The Delta III is shown on the far right of Fig. 2.16 and was designed with a new, liquid hydrogen (LH_2) and liquid oxygen (LOx) upper step. This upper step eventually went into service as an upper step on the Delta IV. Additionally, the Delta III had higher performance strap-ons with thrust vectoring (gimbaled exhaust nozzles). Both the high-energy upper step and the larger strap-ons increased its payload performance remarkably; unfortunately, due to a failure during its first launch (and some performance shortfalls in later launches), the Delta III was discontinued after only three launches. Meanwhile, Delta IV shares its name with its predecessors, but is related to its previous brethren by name only. It is an entirely new launch vehicle designed with a larger body diameter, different first-step propellants, and larger and newer engines; it is a completely different vehicle from the Delta I–III line. The numbering scheme for the Delta rockets is shown in Table 2.3.

2.3.6 Atlas ICBM

The General Dynamics (now Lockheed Martin) Atlas ICBM was the first operational U.S. ICBM. It was a revolutionary design with a number of unique features, including its structural design and staging system. From [16], the Atlas D had a mass of 116,100 kg (255,960 lb) without payload and an empty weight of only 5,395 kg (11,894 lb); the other 95.35% was propellant. Dropping the 3,048-kg (6,720-lb) booster engine and fairing reduced the dry weight of the vehicle to 2,347 kg (5,174 lb), which was a mere 2.02% of the gross weight of the vehicle (again, excluding payload). This very low dry weight allowed the Atlas to send a thermonuclear warhead (a hydrogen bomb, not to be confused with the more conventional nuclear or atomic bomb) up to 14,500 km (9,000 miles) downrange, or payloads to orbit without an upper step.

Atlas achieved this tremendously lightweight structure by using an airframe that was "pressure-stabilized," meaning that the propellant tanks had to be continuously pressurized (to at least 0.34 atm, or 5 psi) in order to avoid collapsing under their own weight. This type of structure is popularly called a *balloon structure* because its behavior is similar to that of a balloon; however, it is not inaccurate to refer to the original Atlas as a stainless steel balloon. The reason for this balloon structure was that the structural mass would be very low; it will be shown in Chapter 5 that the lower the dry mass of a rocket, the better its performance. Except for the Centaur upper step (also manufactured by General Dynamics), no other space launch vehicle used or uses a pressure-stabilized structure: all use internal reinforcements that allow the vehicle to support its own weight

Table 2.3   Delta Numbering Scheme

First Digit: First Step and Augmentation	
0	Long tank, MB-3 engine, Castor II motors (1968 baseline)
1	Extended long tank, MB-3 engine, Castor II motors (1972)
2	Extended long tank, RS-27 engine, Castor II motors (1974)
3	Extended long tank, RS-27 engine, Castor IV motors (1975)
4	Extended long tank, MB-3 engine, Castor IVA motors (1989–1990)
5	Extended long tank, RS-27 engine, Castor IVA motors (1989)
6	Extra extended long tank, RS-27 engine, Castor IVA motors (1989)
7	Extra extended long tank, RS-27A engine, GEM 40 motors (1990)
8	Delta III tank, RS-27A engine, GEM 46 motors (1998)
Second Digit: Number of Augmentation Motors	
3, 4, 6, or 9	Number of strap-on rocket motors
Third Digit: Type of Second Step	
0	Aerojet AJ10-118F (DSV-3N-4; Titan 3C transtage derivative (1972)
1	TR-201 (1972)
2	Aerojet AJ10-118K (1982)
3	Pratt & Whitney RL10B-2 (1998)
Fourth Digit: Type of Third Step	
0	No third step
2	FW-4 (1968)
3	Star 37D (TE-364-3) (1968)
4	Star 37E (TE-364-4) (1972)
5	Star 48B (TE-M-799; PAM-D derivative) (1989)
6	Star 37FM (TE-M-783) (1998)
Dash Number: Fairing Size	
-8	8-ft cylindrical (standard on 1000- through 3000-series, discontinued)
-9.5	9.5-ft (standard on Delta II)
-10	10-ft metal skin-and-stringer (discontinued)
-10C	10-ft composite
-10L	Lengthened 10-ft composite
-H	Heavy

without internal pressurization. The inboard profile of the Atlas D ICBM is shown in Fig. 2.18.

To understand the lengths to which the Atlas engineers went to minimize the dry weight of the vehicle, the entire thickness of the vehicle's tank skins was about that of a U.S. dime, or about 0.8 mm (0.032 in.)! The whole vehicle was fabricated from sheets of stainless steel that were rolled into cylinders and then joined to each other. It's said that the skin weighed less than 2%

of the fuel it carried, and if the skin's thickness varied by as little as 0.025 mm (0.001 in.), the missile's weight could increase by 45 kg (100 lb), and its range could decrease by 161 km (100 miles). To get a feel for these numbers, a sheet of paper is about 0.2 mm (0.008 in) thick.

Unfortunately, the Atlas's lightweight structure meant that either it needed constant pressurization or an external frame had to be connected to keep the structure stretched so it wouldn't collapse. These needs made all Atlas logistics very complex and were the likely motivation to switch to a more conventional stiffened structure for the subsequent Atlas V series.

To understand the second notable design feature of the Atlas, it is necessary to have an understanding of what "staging" means. The term *staging* refers to the procedure where a rocket jettisons a *step*, (an assembly of tanks and rocket engines) in order to reduce its mass once the propellants in a given step are consumed. Normally, this would mean that an upper step would be ignited after a lower step had been jettisoned; however, at the time the Atlas was being designed, these so-called "air starts" of rocket engines were unreliable, so the designers came up with a staging scheme that would ensure all engines were operating before liftoff (similar to the Soviet R-7 discussed earlier) while still ensuring a way to reduce mass during the flight.

What the Atlas engineers came up with was a unique staging system known as a "stage-and-a-half" propulsion system. This is a variation on the conventional staging scheme, where instead of dropping both empty propellant tanks and engines, only the engines (boosters) are dropped after a few minutes of flight. This was done after enough propellant mass had been consumed that a single engine (the *sustainer*) could fly the remaining rocket all the way to orbit. Because *something* was jettisoned, it was considered staging; but, because the tanks were not jettisoned, this wasn't

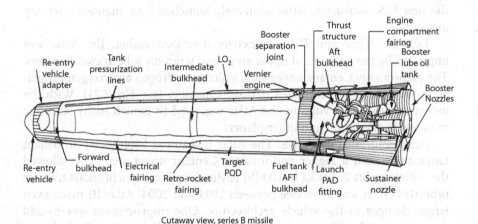

Fig. 2.18   Cutaway of Atlas ICBM.

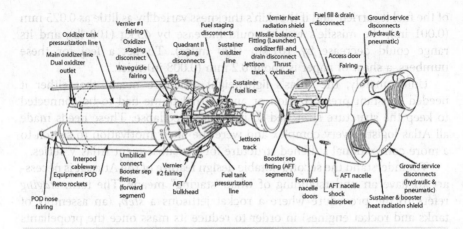

Interpod cableway
Equipment POD
Retro rockets
POD nose fairing

Oxidizer tank pressurization line
Main oxidizer line
Dual oxidizer outlet

Vernier #1 fairing
Oxidizer staging disconnect
Waveguide fairing

Fuel staging disconnects
Quadrant II staging disconnects

Vernier heat radiation shield
Sustainer oxidizer line
Sustainer fuel line

Fuel fill & drain disconnect
Missile balance Fitting (Launcher)
oxidizer fill and drain disconnect
Jettison Thrust track cyclinder

Ground service disconnects (hydraulic & pneumatic)
Access door
Fairing

Umbilical connect
Booster sep fitting (forward segment)

Vernier #2 fairing
AFT bulkhead

Jettison track

Fuel tank pressurization line

Booster sep fitting (AFT segments)
Forward nacelle
AFT nacelle doors

AFT nacelle
AFT nacelle shock absorber

Ground service disconnects (hydraulic & pneumatic)
Sustainer & booster heat radiation shield

Fig. 2.19 The Atlas ICBM utilized stage-and-a-half construction, allowing the outer two booster engines to be jettisoned during flight, while the center engine continued to provide propulsive thrust.

considered a "full" stage. It was therefore christened a *half stage*; thus, the expression *stage-and-a-half* referred to the sustainer as the full stage and the jettisoned booster engines as the half stage. This allowed all engines to be started on the ground, and the required performance was met by jettisoning mass. This design architecture is shown in Fig. 2.19.

An Atlas ICBM is shown launching in Fig. 2.20a. Although none was ever fired in anger, many Atlas vehicles were used for research. NASA used the Atlas to launch its Mercury capsules, which placed the first U.S. astronauts in space. A Mercury launch is shown in Fig. 2.20b. One of the most famous payloads for Atlas was the 1962 Friendship 7 Mercury spacecraft, which carried astronaut John Glenn, the third man to orbit the Earth and the first U.S. astronaut. Atlas ultimately launched four manned Mercury missions.

Like the Thor and Delta launchers described earlier, the Atlas was improved by the addition of upper steps and strap-on solid rocket boosters. The high-energy Centaur upper step, utilizing hydrogen and oxygen propellants, was added for improved payload capability (see Fig. 2.21). With the increased propellant volume, the Atlas was used to launch many scientific missions to the moon and other planets.

Atlas continued to evolve. The Atlas II featured lengthened propellant tanks and added a higher performance Centaur upper step. This allowed the vehicle to lift 2,770-kg (6,100-lb) payloads into geosynchronous transfer orbit (GTO). It was in service between 1991 and 2004. Atlas III made even larger changes to the vehicle architecture. One improvement was to add droppable solid rocket motor strap-ons like the Delta. A radical change was to replace the stage-and-a-half staging scheme with a single, more powerful rocket engine that eliminated the need to jettison booster engines, turning the vehicle back into a "conventional" launcher. Even more radical was the fact that the new single, more powerful engine was the RD-180 and was purchased from Russia! So, the missile originally designed to lob nuclear warheads at the Soviet Union had evolved to the point where it used purchased Russian engines to perform its mission. The Atlas III flew between 2000 and 2005 and could launch up to 4,500 kg (9,900 lb.) into GTO. This was the final pressure-stabilized version of the Atlas that was flown.

The Atlas—much like the Delta—was the basis for a long series of successful space launch vehicles. It received the high-energy Centaur upper steps, and auxiliary strap-on boosters provided additional thrust. Continued improvements included increased tank size for a larger propellant capacity. The Atlas kept changing, as can be seen in Fig. 2.22. Atlas vehicles were built by General Dynamics, which was eventually taken over by Martin Marietta, which was itself later absorbed to become Lockheed Martin. A very nice summary of the development of the Atlas and Centaur vehicles, complete with many personal stories, may be found in [17].

a)   b)

Fig. 2.20 a) Atlas ICBM; b) Mercury-Atlas launch vehicle. The Atlas was also used to carry NASA's manned Mercury capsule, placing the U.S.'s first astronauts into orbit during 1960–1962. Source: U.S. Air Force.

Fig. 2.21 Adding the high-energy LH$_2$/LOx Centaur upper step to the Atlas created the Atlas-Centaur space launch vehicle. Source: NASA MSFC.

The astute reader is likely aware of higher number models of the Atlas, in particular the Atlas V; however, the Atlas V is related to its predecessors solely in name. It is an entirely new launch vehicle design that is no longer pressure-stabilized, has a much larger body diameter, has Russian-made engines, and is basically a completely different vehicle compared to the previous Atlas models. And what happened to the Atlas IV? No one seems to know why this number was skipped in the naming sequence. Anecdotes suggest that Lockheed wanted to have a "higher model number" compared to the Delta IV, so they took IV + I, and Atlas V was the result.

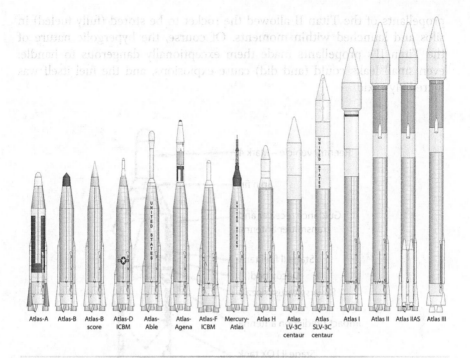

Atlas-A Atlas-B Atlas-B score Atlas-D ICBM Atlas-Able Atlas-Agena Atlas-F ICBM Mercury-Atlas Atlas H Atlas LV-3C centaur Atlas SLV-3C centaur Atlas I Atlas II Atlas IIAS Atlas III

Fig. 2.22 Evolution of the Atlas launch vehicle. Courtesy: Richard Kruse.

2.3.7 Titan ICBM

The Titan family of launch vehicles began with the Martin Marietta SM68A/HGM-25A Titan I (1959–1965), which was the United States' first multistage ICBM. The Titan I provided an additional nuclear deterrent and was designed to complement the Atlas missile. The Titan I was unique in that its LR-87 engine was the only rocket in the Titan family to use liquid oxygen and RP-1 as propellants; later versions all used storable fuels. A total of 368 Titan missiles were launched, including all manned flights of Project Gemini. The basic layout of the Titan I ICBM is shown in Fig. 2.23.

Most of the Titan rockets produced were of the Titan II variety (Fig. 2.24), as well as their civilian derivatives for NASA. The Titan II used the LR-875 engine, which was a modified version of the LR-87 found on the Titan I and which used hypergolic propellants (i.e., propellants that combust automatically when in contact with each other, without the need for an ignition source); specifically, the Titan II used Aerozine 50 [a 1:1 mixture of hydrazine and unsymmetrical dimethylhydrazine (UDMH)] as a fuel and dinitrogen tetroxide [more commonly known as red fuming nitric acid (RFNA)] as an oxidizer. Unlike the Titan I, which relied on liquid oxygen that had to be loaded into the rocket immediately prior to launch, the hypergolic propellants of the Titan II allowed the rocket to be stored (fully fueled) in silos and launched within moments. Of course, the hypergolic nature of the Titan II's propellants made them exceptionally dangerous to handle: even small leaks could (and did) cause explosions, and the fuel itself was incredibly toxic.

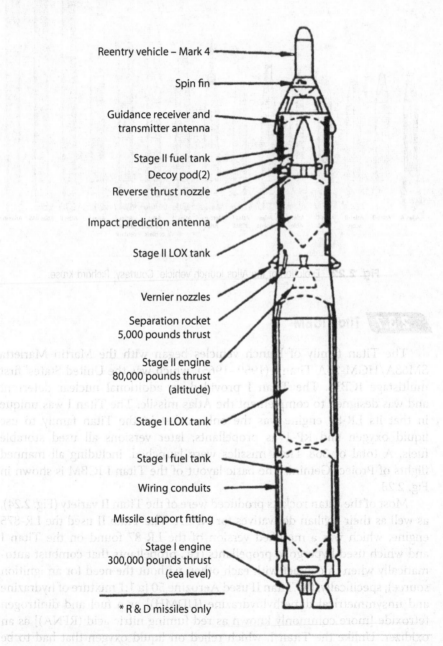

Reentry vehicle – Mark 4

Spin fin

Guidance receiver and transmitter antenna

Stage II fuel tank
Decoy pod(2)
Reverse thrust nozzle

Impact prediction antenna

Stage II LOX tank

Vernier nozzles

Separation rocket 5,000 pounds thrust

Stage II engine 80,000 pounds thrust (altitude)

Stage I LOX tank

Stage I fuel tank

Wiring conduits

Missile support fitting

Stage I engine 300,000 pounds thrust (sea level)

* R & D missiles only

Fig. 2.23   Titan I missile. Source: U.S. National Parks Service.

Fig. 2.24 Titan II, a two-stage, hypergolic, silo-based ICBM. Source: U.S. Air Force.

The ground guidance system for the Titan was the Unisys ATHENA computer, which was located in a hardened underground bunker and used radar data to calculate necessary course corrections during the burn phase of the rocket. As a point of historical note, the ATHENA computer was designed by Seymour Cray, the "father of supercomputing" and designer of the Cray mainframe, who for decades designed and built the world's fastest computers.

The first onboard guidance system to fly on a Titan was an inertial measurement unit derived from designs originally conceived at MIT Draper Labs and subsequently built by AC Spark Plug. The IBMASC-15 served as the missile guidance computer. When spare parts for this computer became difficult to obtain, it was replaced with the Delco Universal Space Guidance System (a Carousel IV inertial measurement unit and a Magic 352 computer), which was subsequently carried over into the Titan III.

The most famous civilian use of the Titan was during the NASA Gemini manned space program in the mid-1960s. Twelve Titan II rockets were used to launch unmanned Gemini test articles and 10 manned capsules, each with a two-man crew. All of these launches were successful although NASA and the Air Force had to solve a longitudinal vibration problem called "pogo," before crewed flight could begin. Pogo and its mitigation is discussed in Chapter 12. During the 1980s, some of the decommissioned Titan II rockets were converted into launch vehicles for government payloads; the final launch for these converted vehicles was the Defense Meteorological Satellite Program (DMSP), which was weather satellites launched from Vandenberg Air Force Base on 18 Oct. 2003.

The Titan III (Fig. 2.25) originally began as a modified Titan II with optional solid propellant motors and was originally developed to launch military and civilian intelligence payloads (including the Vela Hotel satellite, used to monitor compliance with the 1963 Partial Test Ban Treaty, which sought to limit nuclear testing), intelligence-gathering satellites, and various defense communications satellites. The Titan III consisted of a standard Titan II with a Transtage for its upper step. The Titan IIIB and its variants maintained the Titan III core, but used an Agena D for its upper step. The maximum payload of the Titan III was approximately 3,000 kg (6,600 lbs.).

The more powerful Titan IIIC maintained the Titan III core, but used two large strap-on boosters to increase its maximum launch thrust (and therefore its payload capacity). The boosters used on the Titan IIIC represented a significantly more complex engineering challenge than previous boosters, due to not only their large size and thrust, but also their ability to vector thrust after liftoff. The removal of the upper Transtage from the Titan IIIC produced the Titan IIID, which was used to launch the Key Hole series of reconnaissance satellites. The more advanced Titan IIIE added the high-energy Centaur upper step and was used to launch several scientific payloads, including both of NASA's Voyager space probes to Jupiter, Saturn, and beyond, and both of the Viking missions that placed orbiters around Mars and instrumented landers on its surface.

The final version of the Titan rocket family was the Titan IV, which was developed to launch Space Shuttle–class payloads for the U.S. Air Force. The Titan IV had several staging options, including no upper step, the Inertial Upper Stage (IUS), or the Centaur upper step. The Titan IV consisted of two solid propellant boosters and a liquid propellant core similar to the Aerozine/RFNA-propellant core of the Titan II. The Titan IV could be launched from either Cape Canaveral in Florida or Vandenberg Air Force Base in California. An example of a Titan IV on the launch pad is shown in Fig. 2.26.

Although the Titan family of launch vehicles was successful, the operational history of the Titan IV was not without its failures. In 1993, a Titan IV was destroyed during ascent due to a solid propellant booster malfunction, and a U.S. Navy signals intelligence (SIGINT) payload was lost in

Fig. 2.25 First launch of the Titan III. Source: U.S. Air Force.

the process. Ironically, this accident was found to have been caused by pro-cedures implemented to improve the reliability of the Titan's boosters after two Titan 34Ds (similar to the Titan III) were lost in the 1980s due to booster failures; an investigation revealed that repairs made to the third segment of the solid boosters were not carried out correctly, allowing burn-through during ascent. Another Titan IV accident occurred in 1998, which caused the loss of a U.S. Navy electronic intelligence (ELINT) satellite. An electrical short circuit caused the flight computer to re-boot and impart an undesired pitching moment to the launch vehicle, and the resulting aerody-namic loads caused a booster to separate. This triggered an automatic reaction from the onboard destruct system, terminating the mission approximately 40 s after liftoff. This mission was actually intended to be the final launch of the Titan IVA, even before its failure.

Perhaps the most famous mission carried out by the Titan IV was the 1997 launch of Cassini-Huygens, a pair of scientific payloads sent to Saturn. This was also the only civilian mission of the Titan IV. The Huygens probe landed on Saturn's moon Titan on 14 Jan. 2005; the Cassini spacecraft that orbited Saturn was "retired" in 2017 by being commanded to enter Saturn's atmosphere and destroyed.

The advent of the Atlas V and Delta IV rockets heralded the end of the Titan IVB, the last variant of the Titan family. A Titan IV has been

Fig. 2.26 The penultimate Titan: Titan IV was used to launch Cassini to Saturn in 1997.
Source: Lockheed Martin.

restored and is on display at the National Museum of the U.S. Air Force in Dayton, Ohio. The evolution of the Titan family of launch vehicles is shown in Fig. 2.27.

2.3.8 Minuteman ICBM Family

The Minuteman (LGM-30) is a land-based ICBM that forms one third of the U.S. "nuclear triad," the other two parts of which are the Trident submarine-launched ballistic missile (discussed later in this chapter) and the nuclear weapons carried by strategic bombers. Each Minuteman missile can carry up to three independently targetable nuclear warheads [called multiple independently targetable reentry vehicles (MIRVs)], with a yield in the 300 to 500-kiloton range. The current U.S. stockpile consists of approximately 400 armed Minuteman III missiles and 50 unarmed missiles to be held in reserve. Current plans are to leave the Minuteman missiles in service until at least 2030. The three generations of Minuteman are shown in Fig. 2.28.

The Minuteman owes its name to the Minutemen of the Revolutionary War, implying a state of constant readiness and fast reaction time. When a valid launch order is received, the Minuteman can be launched within minutes [18]. Beginning in 1957, a series of intelligence reports suggested that missile developments underway in the Soviet Union would soon allow them to overwhelm the United States by the 1960s. Although it was later demonstrated that this "missile gap" was fictitious, it spawned serious concerns in the U.S. Air Force, which then began to push the Minuteman program for an accelerated "crash" development beginning in Sept. 1958 [19]. The Minuteman family of missiles owes its existence to Edward N. Hall, who in 1956 was an Air Force colonel commanding the solid propellant programs of the Western Development Division (originally formed to lead the development of the Atlas and Titan missiles). Although Hall's superiors were interested in using solid propellants for short- and medium-range missiles, Hall remained convinced that a true ICBM—with a range in excess of 5,000 nautical miles—could be developed using solid propellants [20]. To achieve the energy required for such a feat, Hall funded research programs at Boeing and Thiokol to investigate the use of an ammonium perchlorate propellant that had been adapted from a British design: by casting the fuel into large cylinders with a star-shaped plug along the center, the final fuel grain retained the central star shape. This increased the effective burn area of the fuel, which increased the burn rate and increased thrust. This also prevented the heat given off during combustion from reaching the walls of the missile until the fuel had been completely burnt [21].

Despite the advantages of this ingenious development, the Air Force saw no pressing need for a solid propellant ICBM, especially not with the progress of Atlas and Titan. "Storable" (noncryogenic) liquid propellants were being developed that would permit missiles to be stored for prolonged periods in

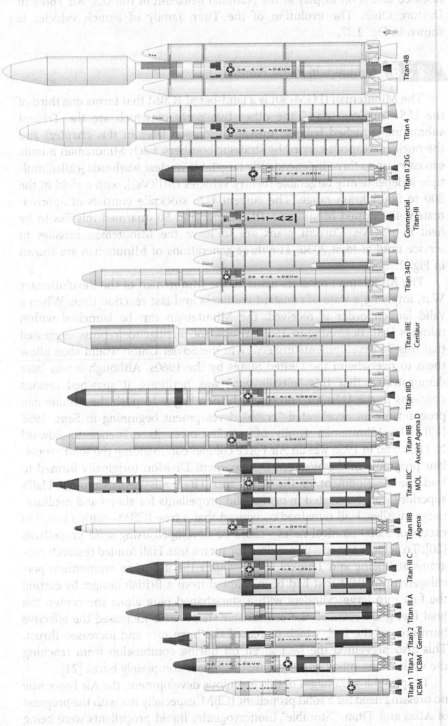

Fig. 2.27 Evolution of the Titan launch vehicles. Courtesy: Richard Kruse.

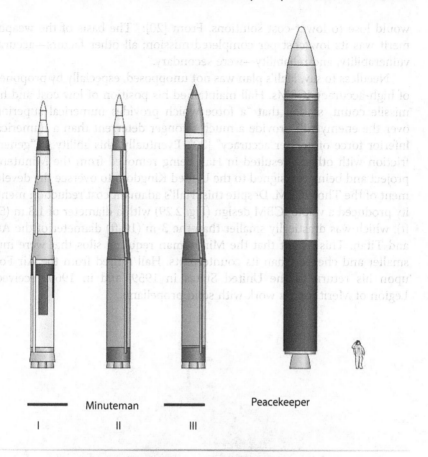

Minuteman

I   II   III   Peacekeeper

Fig. 2.28   Minuteman and Peacekeeper missile comparison, showing the Minuteman I, II, and III variants. Source: historicspacecraft.com.

a ready to shoot state; Hall, however, saw solid propellants not only as a method of potentially improving the response times and safety of missiles, but also as part of a plan to reduce the cost of ICBMs. This would not only greatly increase the number of missiles that could be built, but also reduce the manpower necessary to oversee operations on computerized assembly lines for dozens or hundreds of missiles. Solid propellant missiles would therefore not only be easier to build, but also easier to service and maintain than their liquid propellant counterparts [20].

Hall's ultimate vision was a number of "missile farms," which would include factories, silos, logistics, maintenance, and recycling all in one integrated facility. Each farm would continuously manufacture missiles in a low-rate cycle, detecting failures during the manufacturing process so defective components could be removed and recycled while a newly built missile was loaded into the silos. Cost was king in the missile farm concept, and even mission elements that might have seemed important in other missile designs would lose to lower-cost solutions. From [20]: "The basis of the weapon's merit was its low cost per completed mission; all other factors—accuracy, vulnerability, and reliability—were secondary."

Needless to say, Hall's plan was not unopposed, especially by proponents of high-accuracy ICBMs. Hall maintained his position of low cost and high missile count, stating that "a force which provides numerical superiority over the enemy will provide a much stronger deterrent than a numerically inferior force of greater accuracy" [20]. Eventually, this ability to "generate friction with others" resulted in Hall being removed from the Minuteman project and being reassigned to the United Kingdom to oversee the development of the Thor ICBM. Despite this, Hall's adamant cost reduction mentality produced a viable ICBM design (Fig. 2.29) with a diameter of 1.8 m (5.92 ft), which was drastically smaller than the 3-m (10-ft) diameter of the Atlas and Titan. This meant that the Minuteman required silos that were much smaller and cheaper than its counterparts. Hall retired from the Air Force upon his return to the United States in 1959, and in 1960 received a Legion of Merit for his work with solid propellants.

Fig. 2.29 Minuteman ICBM launch: three warheads (150 kilotons each), 5,300+-mile range. Source: U.S. Air Force.

The Minuteman II (LGM-30F) was an improvement over the original Minuteman I. Development of the Minuteman II began in 1962, and the missile was deployed in 1965. Relative to its predecessor, the Minuteman II had greater range, increased payload capacity, and an improved guidance system, all of which combined to provide military strategists with greater accuracy and a much wider range of targets. The Minuteman II's payload consisted of a single reentry vehicle, which contained a 1.2-megaton W56 nuclear warhead. This larger warhead increased the vehicle's kill probability.

An improved first-step motor increased the missile's reliability; a liquid-injected, fixed nozzle with thrust vector control on the second step increased the missile's range; miniaturized, discrete electronic components (a first in the history of U.S. missile design) allowed for an improved guidance system, with greater accuracy as well as reduced size and weight and increased survivability. The Minuteman II was, in fact, the first program to use computers constructed entirely from integrated circuits (the Autonetics D-37C); the only other major customer of these early integrated circuits was the Apollo Guidance Computer. The primary difference between the two integrated circuits was in their architecture: the Minuteman II's circuits were diode-transistor logic (the direct ancestor to modern transistor-transistor logic), whereas the Apollo Guidance Computer's circuits were designed with the slightly older resistor-transistor logic. The Minuteman II also featured "penetration aids" that were designed to camouflage the warhead during reentry into hostile airspace, including stealth features to reduce its radar signature (making the warhead more difficult to distinguish from decoys); this low-signature requirement, in fact, was one of the reasons why the reentry vehicle was specifically *not* made of titanium [22].

The modifications that led to the Minuteman III (LGM-30G) primarily focused on the upper step and reentry system. The third step was improved with a fluid-injected motor, which gave finer control relative to the four-nozzle system on the Minuteman II. Improvements to the reentry system focused on improved survivability after a nuclear attack. The Minuteman III's W62 warhead had a yield of "only" 170 kilotons, compared to the 1.2-megaton yield of the Minuteman II's W56 warhead. The Minuteman III was the first missile to feature a MIRV payload; this meant that a single missile could be fired and independently target multiple locations (three, in the case of the Minuteman III). The reentry vehicles were propelled by a *postboost* step (or *bus*), which was a liquid propellant engine that was used to adjust the trajectory of the reentry vehicle; for the Minuteman III, this bus was a Rocketdyne RS14 engine. Thus, although the nominal yield of the Minuteman III warheads was three orders of magnitude less than the single warhead of the Minuteman II, the Minuteman III could provide better operational flexibility with regard to destroying targets of strategic importance. The reentry vehicle was also capable of deploying penetration aids such as chaff and decoys, in addition to the warheads.

2.3.9 Peacekeeper

For all of the advances of the Minuteman family of missiles, certain considerations borne from their design limited their potential usefulness during a wartime scenario. First and foremost, the limited accuracy of the Minuteman missiles—with a circular error probable of 0.6–0.8 nautical miles [20]—and kiloton-class warheads on the Minuteman III meant that the missiles would be ineffective against hardened targets, such as missile silos. The circular error probable (CEP) is an important measurement of a weapon's precision; it is defined as the radius of a circle whose boundary includes the landing points of 50% of the rounds fired. A lower CEP is indicative of higher accuracy. For the Minuteman, a CEP of 0.6–0.8 nautical miles means that 50% of all of the missiles fired would land within 0.6–0.8 nautical miles of the intended target. This meant that the Minuteman missiles would only be effective against targets such as cities, and the missile had virtually zero capability for counterstrike scenarios. At the time, the U.S. Air Force relied heavily upon its bomber fleet for attacks against hardened targets and relegated ICBMs to the role of survivable deterrents that would prevent attacks on its bombers.

When the Kennedy administration took power, Robert McNamara, the new Secretary of Defense, was tasked with what seemed to be an impossible assignment: make the U.S. military the most powerful in the world, while simultaneously reducing its expenditures. This was ultimately accomplished by severely undercutting the reliance on bombers and increasing the reliance on the Minuteman—a missile that, as stated previously, sought to reduce the all-up cost of missions by any means necessary. In fact, by 1964, the United States had more ICBMs on active alert than strategic bombers [23]. However, the intelligence reports from the late 1950s that suggested the Soviet advances in missile technology would one day quickly outpace those of the United States (the so-called "missile gap," which was later found to have been as fictitious as the "bomber gap" that preceded it) led to the worrying scenario, whether real or imagined, that a Soviet first strike with a limited number of warheads could cripple the U.S. ICBM fleet. At the time, a "limited number of warheads" would have been all that the Soviets could have used to attack the United States: its missiles had the same accuracy problems, and so only a very small number of missiles carrying very large warheads could be used to attack U.S. silos. Such an attack would have been damaging, but not critical.

In the event of a Soviet first strike, the United States faced a difficult decision: immediately retaliate with its own missiles, or wait and determine the targets of the Soviet launch. Because the U.S. ICBM fleet was primarily targeted against cities, firing early could have meant striking civilian targets when the Soviets had only struck military installations, which was politically untenable; by contrast, waiting to retaliate might have meant the loss of a significant portion of the land-based ICBM fleet. The submarine-launched Polaris, by contrast, was essentially invulnerable to a Soviet first strike. This was of grave concern to the Air Force, because the near-invulnerability of the Polaris meant that the U.S. Navy could be handed the counterstrike mission outright. Thus, the Air Force shifted its focus away from deterrence and more toward counterforce. The improved inertial navigation system on the Minuteman II improved its CEP to 0.34 nautical miles [20], which—when combined with the Minuteman II's 1.2-megaton warhead—allowed the missile to attack hardened targets. The Minuteman II's guidance computer, more importantly, allowed for the inclusion of up to eight preprogrammed targets; this would allow the United States to absorb the damage from a Soviet first strike and then counterattack against the appropriate targets (military or civilian). If the Soviets launched only a limited attack, this doctrine would give the United States a strategic advantage; of course, the Soviets could (and did) improve their own missiles' CEP, effectively turning their ICBMs into a counterforce fleet as well.

The logical conclusion of this philosophy was obvious. In 1971, the Air Force began the requirements development process for a new program called *Missile, Experimental (MX)*. This new missile would be so accurate, carry so many warheads, and have such a high survivability that even a small handful of them would be able to obliterate any surviving Soviet forces. At the time, the MX missile was to be based in Minuteman silos, in keeping with the original concept of the MX as being essentially a larger Minuteman. The improvements to tracking accuracy alone were substantial: for the MX program, the Draper Laboratory developed a guidance system with a drift rate of 0.000015 degrees per hour. This guidance system [called the Advanced Inertial Reference Sphere (AIRS)] consisted of a beryllium sphere floating in a fluid; jet nozzles stabilized the inertial platform as commanded by the onboard sensors [a set of floating gas-bearing gyroscopes and specific force integrating receiver (SFIR) accelerometers]. This design also eliminated gimbal lock (the loss of one degree of freedom from a gimbal-stabilized platform, caused by two of the three gimbals being driven into a parallel configuration). Over the period of the missile's entire flight, the accuracy of the guidance system would be within 1% of the warhead's final accuracy; any other inaccuracies would be due to timing of the rocket engines, variabilities in the construction of the warhead, and atmospheric turbulence [20]. The AIRS system reliance on SFIR accelerometers meant that its pinpoint accuracy came at a heavy cost: approximately $300,000 for each SFIR accelerometer [24], of which each AIRS unit needed three. This solved the accuracy problem, and the AIRS was transferred to Northrop for further development; the political problems, however, continued to plague the MX program.

The concept of land-based silo vulnerability continued to haunt MX, and in 1976 Congress halted the project. Several alternate basing concepts were explored, including mobile basing via rail cars and even submerged bunkers that could dig themselves out of the ground after an attack. The program was eventually reinstated by President Carter in 1979, and 200 missiles were to be deployed throughout Nevada and Utah in several "shelters," which were inter-connected via aboveground and underground roads. Political outcry among the residents of Utah eventually forestalled this deployment, until President Reagan canceled Carter's shelter system in 1981. Reagan proposed an initial deployment of MX missiles in the 60 or so Titan II silos in the area, removing the outdated Titan missiles from service. The Titan II silos were modified for increased strength, and a similar modification of several Minuteman III silos brought the total force to 100 missiles.

The Reagan administration announced on 22 Nov. 1982 that the MX missile was to be known as the Peacekeeper (although it was originally to be called the Peace*maker*), and was first fired on 17 June 1983 from Vandenberg Test Pad 01. The missile traveled 7,800 km (4,800 miles) to strike the Kwajalein Test Range in the Pacific Ocean. A total of 50 flight tests were performed. The first operational missile was deployed in Dec. 1986 to a refurbished Minuteman silo in Cheyenne, Wyoming; however, the highly accurate AIRS guidance system—each of which contained approximately 19,000 parts, some of which required 11,000 test and cali-bration steps—was not yet ready, and so these first missiles were deployed with inoperative guidance units. The excruciating government procurement process was too much for some managers, who began bypassing official channels and buying replacement parts however they could (including sour-cing parts via possibly apocryphal trips to Radio Shack) or creating shell com-panies to order the requisite test equipment [25], [26]. When these allegations were made public, Northrop was immediately fined $130 million for late delivery, and was then countersued in whistleblower suits when it retaliated against employees. The Air Force admitted that 11 of the 29 missiles that had been deployed were nonoperational [26].

The first AIRS was finally delivered in May 1986 (a mere 203 days late), although it was not until July 1987 that the first production units were finally ready to ship; the complete supply of AIRS [now known as the more generic inertial measurement unit (IMU)] for the first 50 Peacekeeper mis-siles was delivered in Dec. 1988 [26]. These delays, combined with the increased performance of the Trident II (UGM-133), caused Congress to cancel the 100-missile Peacekeeper option in July 1985; the 50 Peacekeeper ICBMs that had already deployed were to be the last of the Peacekeepers deployed until a more "survivable" basing option could be developed. The final flyaway cost of each Peacekeeper was approximately $20–70 million [25], [26].

2.3.10 Minotaur and Taurus

The Minotaur rockets are the direct descendants of the Minuteman and Peacekeeper ICBMs, built by Orbital Sciences Corporation (now Northrop

Grumman Space Systems) under contract to the U.S. Air Force Space and Missile System Center's Space Development and Test Directorate. The Minotaur rockets were converted from retired ICBMs as part of the Air Force's Rocket Systems Launch Program. Minotaur variants are used for a variety of missions, including launching small satellites to low Earth orbit (LEO), launching to GTO and translunar trajectories, and suborbital flights as a target for tracking and anti–ballistic missile tests. The Minotaur I and II were derived from the Minuteman missile, whereas the Minotaur III, IV, and V are derived from the Peacekeeper. The Minotaur family of rockets is shown in Fig. 2.30.

The Minotaur I is a four-step rocket that consists of the M55A1 first step and SR19 second step of the Minuteman missile; the third and fourth steps consist, respectively, of the Orion 50XL and Orion 38 motors from the Orbital Sciences Pegasus. The Minotaur I can place up to 580 kg (1,280 lb) of payload into LEO at up to 28.5 degrees of inclination [27]. An optional hydrazine auxiliary propulsion system (HAPS) upper step can be flown for missions requiring greater precision or if the mission

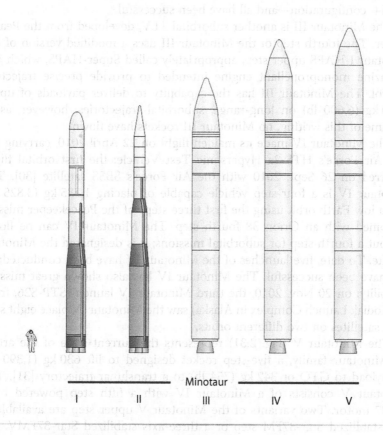

Fig. 2.30   Minotaur launch vehicles. Credit: historicspacecraft.com

requires the rocket to maneuver to release multiple payloads [28]. To date, there have been 10 launches of the Minotaur I, all of which have been successful [28].

The Minotaur II, derived from the Minuteman II missile, is primarily used for long-range, suborbital launches. The Minotaur II is also known as the Chimera, the Minuteman II TLV, or simply the Target Launch Vehicle. The rocket consists of the same M55A1 first step and SR19AJ1 second step of the Minuteman II. The Minotaur II comes in several "flavors," wherein the third step will vary depending on the payload. For instance, the baseline Minotaur II third step uses the same M57A1 step from the Minuteman II, whereas a Minotaur II+ third step uses an Aerojet SR-73-AJ1 motor from the Minuteman III third step. The Minotaur II Lite, by contrast, doesn't use a third step at all. A Minotaur II Heavy is also available, with the Orion 50XL third step used on the Minotaur I. The baseline Minotaur II has a range of 4,000 km (2,500 miles) with a 400-kg (880-lb) payload, whereas the Minotaur II Heavy has a range of 8,000 km (5,000 miles) with a 1,400-kg (3,100-lb) payload capacity [29]. Eight launches of the Minotaur II have taken place as of 2009—six with the baseline configuration and two with the II+ configuration—and all have been successful.

The Minotaur III is another suborbital TLV, developed from the Peacekeeper. The fourth step of the Minotaur III uses a modified version of the Minotaur I HAPS upper step, appropriately called Super-HAPS, which is a hydrazine monopropellant engine intended to provide precise trajectory control. The Minotaur III has the capability to deliver payloads of up to 3,000 kg (6,600 lb) on long-range, suborbital trajectories; however, as of the time of this writing, no Minotaur III rockets have flown.

The Minotaur IV made its maiden flight on 22 April 2010, carrying the U.S. Air Force's HTS-2a Hypersonic Test Vehicle; the first orbital flight occurred on 26 Sept. 2010 with the Air Force's SBSS satellite [30]. The Minotaur IV is a four-step vehicle capable of placing 1,735 kg (3,825 lb) into a low Earth orbit using the first three steps of the Peacekeeper missile, combined with an Orion 38 fourth step. The Minotaur IV can be flown without a fourth step for suborbital missions; it is designated the Minotaur IV Lite. To date, five launches of the Minotaur IV have been conducted; all five have been successful. The Minotaur IV has also shown great mission flexibility: on 20 Nov. 2010, the third Minotaur IV launch (STP-S26, from the Kodiak Launch Complex in Alaska) saw the Minotaur IV place eight separate satellites on two different orbits.

The Minotaur V (Fig. 2.31) represents the current state of the art in the Minotaur family, a five-step rocket designed to lift 630 kg (1,390 lb) of payload to GTO or 342 kg (754 lb) to a translunar trajectory [31]. The Minotaur V consists of a Minotaur IV with a fifth step powered by a Star37 motor. Two variants of the Minotaur V upper step are available: a spin-stabilized Star-37FM step or a three-axis stabilized Star-37FMV, the latter of which is heavier but more maneuverable.

Fig. 2.31   Minotaur V on launch pad. Source: Northrop-Grumman.

The initial launch of the Minotaur V occurred on 7 Sept. 2013, carrying the NASA Lunar Atmosphere and Dust Environment Explorer (LADEE) spacecraft. The LADEE mission was successful, although the upper steps of the launch vehicle met with slightly unconventional ends: both the fourth and fifth steps of the Minotaur V are currently derelict satellites in orbit around the Earth [32]. Presently, all Minotaur V launches are scheduled from the Mid-Atlantic Regional Spaceport (MARS), located at the southern tip of the NASA Wallops Flight Facility.

Fig. 2.32   Minotaur-C, formerly Taurus XL. Source: Northrop Grumman.

The Taurus rockets, now known as Minotaur-C, originally began life as four-step, solid propellant vehicles developed by Orbital Sciences Corporation (now Northrop Grumman) from its Pegasus air-launched booster (Fig. 2.32). The Castor 120 first step is based on the Peacekeeper, whereas the second and third steps are powered by the Orion-50 motor (similar to the Pegasus 1). The fourth step is powered by an Orion-38 derived from the Pegasus 3. First launched in 1994, Taurus was capable of launching 1,350 kg (2,970 lb) to low Earth orbit.

Six of the nine Taurus missions were successful [33]. Three of the four launches between 2001 and 2011 were unsuccessful: Orbview-4, Orbiting Carbon Observatory, and Glory. The failures resulted in $700 million worth of losses for NASA (not including the cost of the rockets themselves), of which $424 million were due solely to the loss of Glory [34]. The reason for the OCO and Glory failures was identical: the payload fairing refused to separate due to faulty materials provided by an aluminum supplier, and the added weight prevented the rocket from reaching orbit [34]. The rocket was upgraded with new avionics from the Minotaur family, and the Taurus was rebranded Minotaur-C (for Minotaur-commercial) [33] and is currently used for commercial launches.

2.3.11 Polaris, Poseidon, and Trident

The UGM-27 Polaris (Fig. 2.33) was a nuclear-tipped, two-step, submarine-launched ballistic missile (SLBM) first flown from Cape Canaveral in Jan. 1960. The Polaris formed the second branch of the U.S. nuclear triangle, complementing the ICBM and nuclear bomber fleets. Several new project management techniques were also created for the Polaris program,

Fig. 2.33 Polaris A-3 on a launch pad at Cape Canaveral, ca. 1964.

including the Program Evaluation and Review Technique (PERT) still used, in one fashion or another, by many programs today. Like the early Minuteman missiles, the CEP of the Polaris was insufficient for use as a first-strike weapon.

The Polaris missile replaced an earlier plan to develop a naval version of the Army's Jupiter IRBM; this argument claims that the Jupiter missile's large diameter was a byproduct of the need to keep the overall length short, so that the missile could fit within a reasonably sized submarine [35]. There is some contention over this "Naval Jupiter" assertion, claiming that the Navy's program was completely unrelated to the Army's and that the large submarine missiles said to have been in development were likely the SSM-N2 Triton, none of which was ever built. The official history of the Army's Jupiter program (specifically, [35]) states that the Navy was involved during the early stages of the program, but later withdrew.

The key drawback of the original naval missiles, such as the Regulus cruise missiles deployed on the Grayback-class submarines, was the need for the submarine to surface—and to remain surfaced for some time—in order to launch. Needless to say, the vulnerability of a submarine is very high when surfaced, and having fueled (or partially fueled) nuclear missiles on deck during this time was an extreme hazard. The advantages of solid-fueled missiles was immediately apparent: unlike the Jupiter missile (and unlike cruise missiles of the day), a solid propellant ballistic missile could be launched while the submarine was still submerged, greatly improving survivability.

The Polaris began development in 1956, and the first missile was successfully launched from the submerged USS *George Washington* (the first U.S. missile submarine) on 20 July 1960. The upgraded A-2 version entered service in 1961, serving on 13 submarines until June 1974. Problems with the arming and safety equipment of the W-47 warhead led to numerous recalls. During the recall, the Navy began seeking a replacement for the warhead that would have either a larger yield or equivalent destructive power. This led to the "clustering" of three warheads on the A-3, which was the final model of the Polaris produced.

A study began in 1963, seeking to develop a longer-range version of the Polaris by enlarging it as much as possible, within the constraint of the existing launch tubes. The possibility of this was primarily due to the fact that tests had previously shown that the Polaris missiles could be launched without incident if the launch tubes had their fiberglass liners and locating rings removed [36]. The modified missiles were originally given the name Polaris B3, but this was changed to Poseidon in order to emphasize that the new missile had numerous technical advances over its predecessor. The Poseidon (Fig. 2.34) was given the designation UGM-73A.

The Poseidon was slightly longer than the Polaris, although the Poseidon was significantly wider. Both missiles had the same 4,600-km (2,500-mile)

Fig. 2.34   A Poseidon missile fired from the submerged USS *Ulysses S. Grant* in May 1979.

range, although the Poseidon had a greater payload capacity, improved accuracy, and MIRV capability: a single Poseidon could deliver up to 14 W-68 thermonuclear warheads, each with a nominal yield of 40–50 kilotons. As with Polaris, the Poseidon missile was ejected from the submarine via high-pressure steam, and the rocket would automatically ignite when the missile had climbed 10 m (33 ft) above the submarine. The Poseidon officially entered service on 31 March 1972, and equipped 31 Lafayette-, James Madison–, and Benjamin Franklin–class submarines. The Poseidon remained in service until 1992, when the collapse of the Soviet Union and the terms of the first Strategic Arms Reduction Treaty (START I) saw the Poseidon-equipped submarines disarmed.

The U.S. Navy had its eyes firmly fixed on the future even before the Poseidon officially deployed. Beginning in 1971, almost a full year before the Poseidon missile entered service, the Navy began a study of an advanced Undersea Long-range Missile System (ULMS). This study sought to develop a longer-range missile, dubbed ULMS II, which would possess twice the range of the Poseidon (termed ULMS I in the study). In addition to a more capable missile, a more capable class of "boomers" (ballistic missile submarines, labeled SSBN for ship, submersible, ballistic, nuclear), called the Ohio-class, would replace the aging James Madison– and Benjamin Franklin–class fleet in 1978. ULMS II would be designed to be retrofitted into existing ballistic missile submarines, while also fitting the proposed Ohio-class fleet. The ULMS II term was replaced in 1972 with Trident, which would be a larger, higher performance weapon with a range in excess of 9,600 km (6,000 miles). The Trident I was deployed from 1979 until 2005 [37]; its mission was similar to the Poseidon, but with an extended range. The Trident II was first deployed in 1990 and will remain in service until 2027; the Trident II improved upon the earlier Trident and Poseidon missiles with a lower CEP. All Trident missiles are MIRV-equipped and tipped with thermonuclear warheads. Fourteen U.S. Ohio-class submarines would carry the weapon, and four Royal Navy Vanguard-class submarines would be armed with a variant of the Trident that carried British warheads. As was the case with its predecessors, the Trident is launched while the submarine is still submerged; the missile is ejected from its tube by high-pressure steam, which carries it clear of the water's surface. The sudden acceleration from the steam activates inertial sensors aboard the missile. Once the water surface has been breached, the inertial sensors detect the missile's downward acceleration due to gravity, and the first-step motor ignites. This is shown in Fig. 2.35. A telescoping aerospike (not to be confused with the aerospike propulsion system discussed in Chapter 4) then deploys from the nose of the missile, halving the amount of aerodynamic drag that would otherwise be caused by the blunt shape of the missile's nose. The third step of the missile fires approximately 2 min after the missile was launched, at which point the missile is already traveling in excess of 20,000 ft/ s, which is the equivalent of approximately Mach 18. Although the Trident is not an orbital weapon, it is able to attain a temporary orbital altitude during flight.

As with the AIRS system on the Peacekeeper, the inertial guidance system for the Trident was originally developed by the Draper Laboratory and is maintained by a Draper/General Dynamics facility. In addition to the inertial guidance system, the Trident carries a star-sighting system, forming a combination known as astro-inertial guidance. This scheme corrects small measurement errors in the missile's position and velocity (thus improving CEP performance), which are the result of position and velocity errors at the time of launch, and further errors that can accumulate within the guidance system during flight due to imperfect

Fig. 2.35   A Trident missile launched from a submerged Royal Navy Vanguard-class submarine.

instrument calibration. The missile has used the Global Positioning System (GPS) on some test flights to automatically ascertain its position [36], but the availability and reliability of GPS during a nuclear exchange is uncertain; hence, the astro-inertial guidance scheme remains. Once the star-sighting has been completed, the missile maneuvers itself to the necessary orientation to send its MIRVs toward their intended targets; the exact attitude and altitude of the missile during this star-sighting, as well as the downrange and crossrange capabilities of the MIRVs, remains classified.

The two Trident variants—C4 (UGM-96A) and D5 (UGM-133A)—have almost nothing in common, and as shown in Fig. 2.36, are not even similar classes of vehicles. Whereas the C4 is merely an improved version of the Poseidon, the D5 is a clean-sheet design, with limited technologies carried over from the C4. Both Trident variants are three-step missiles with astro-inertial guidance and solid propellant motors. The U.S. Navy announced in 2002 that the life of the D5 Tridents (and the submarines that carry them) would be extended to 2040 [38].

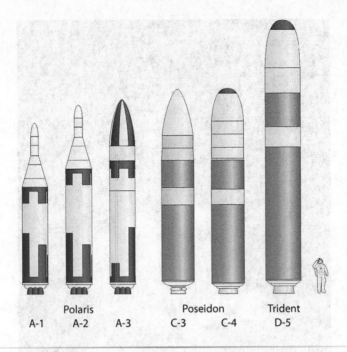

Fig. 2.36 Evolution of the U.S. Navy fleet ballistic missiles, from Polaris to Trident. Credit: historicspacecraft.com.

The main focus of the D5 Life Extension Program currently underway is to replace obsolete components with commercially available hardware while maintaining the performance of the missiles. In theory, this will minimize the cost of operating and maintaining the missile fleet without sacrificing performance. A contract for $848 million was awarded to Lockheed Martin in 2007 to perform this work, including upgrades to the missiles' reentry systems; the Draper Laboratory was also awarded $318 million to upgrade the guidance system [39]. The first flight of a D5LEP Trident took place on 22 Feb. 2012, almost exactly 22 years after the first flight of the Trident D5 missile. A video of the D5LEP test is available online at https://www.youtube.com/watch?v=V1eFhUMSJ9s. This video also shows the deployment of the Trident aerospike, near the 10-s mark. The total cost of the Trident program was $39.56 billion in 2011 dollars, amortizing to a cost of approximately $70 million per missile [40].

2.3.12 Scout

The Scout rocket was a four-step, solid propellant launch vehicle operated between 1960 and 1994. It was capable of lifting 175 kg (385 lb) to an orbital height of 800 km (500 miles). Some have claimed that the name *Scout* was an abbreviation for Solid Controlled Orbital Utility Test, although this is a "backronym": the launch vehicle was named in the spirit of the Explorer series of satellites with which the rocket would usually be paired [41]. Contrary to many of the launch vehicles discussed so far, the Scout was almost comically small: barely 1.01 m (3.32 ft) in diameter and 25 m (82 ft) long [41]; a sense of how small this actually is can be gleaned from Fig. 2.37, which shows two technicians to the lower left of the vehicle for scale. The Scout, like the Minuteman, represented a development program that kept costs and reliability in mind; from [42]:

> The vehicle was built with off-the-shelf hardware. Chance Vought Aircraft (later LTV Missiles and Electronics Group and subsequently Loral Vought Systems) of Dallas was the prime contractor for the development of Scout systems. Designers selected from an inventory of solid propellant rocket motors produced for military programs: the first stage motor was a combination of the Navy Polaris and the Jupiter Senior; the second stage came from the Army Sergeant; and the third and fourth stage motors were designed by Langley engineers who adapted a version of the Navy Vanguard.

The four parts of off-the-shelf hardware making up the Scout rocket are shown in Fig. 2.38. There were a total of 118 Scout launches during its 34-year lifetime, with an overall success rate of 96%. During the initial phase of launch, steering was provided by movable jet vanes and fin tips at the bottom of the first step; the second and third steps were

Fig. 2.37 a) NASA's diminutive Scout launch vehicle being erected for launch; note size of technicians in lower left relative to size of rocket (Source: FAS.org). b) The Scout just after liftoff (Source: NASA).

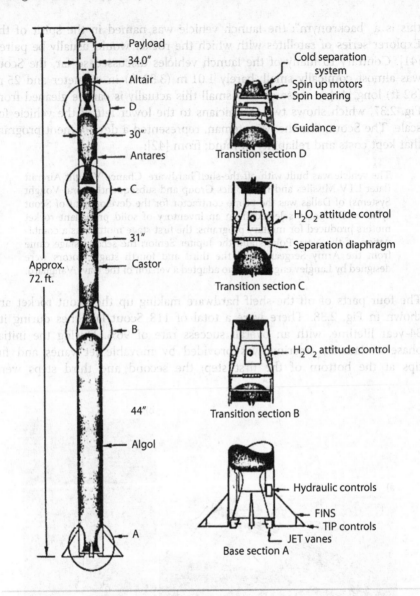

Payload
34.0"
Altair
D
30"
Antares
C
31"
Castor
B
44"
Algol
A

Approx. 72. ft.

Cold separation system
Spin up motors
Spin bearing
Guidance
Transition section D

H_2O_2 attitude control
Separation diaphragm
Transition section C

H_2O_2 attitude control

Transition section B

Hydraulic controls
FINS
TIP controls
JET vanes
Base section A

Fig. 2.38 Scout engineering drawing. Source: NASA.

controlled by hydrogen peroxide jets, and the fourth step was passively spin-stabilized. The Scout was the smallest orbital launcher in NASA's inventory, but its capability grew dramatically over the years. Originally able to place 59 kg (131 lb) in a 552-km (345-mile) circular orbit, the Scout's performance was continually improved upon. The heaviest satellite ever placed in orbit by the Scout was an Italian payload that weighed more than 270 kg (600 lb) and was launched out of Africa. The Scout increased its payload capability 350% over that of the original vehicle, with little increase in the size of its steps [43]. The Scout was replaced by the Orbital Sciences (now Northrop Grumman) Pegasus when NASA was told it could no longer maintain its own rockets and instead had to use commercially available launchers. A documentary entitled *The Scout Rocket, Unsung Hero of Space* is available at https://www.youtube.com/watch?v=VQ7fM5FvRVM.

2.3.13 Pegasus

At its heart, a rocket exists merely to provide the payload with enough energy to maintain the desired orbit, a concept that will be discussed in much greater detail in Chapter 3. One method of providing the payload with a "head start" to this energy is to give it a head start by mounting the rocket onto a carrier aircraft and launching it high above much of the thick, drag-inducing atmosphere found at lower altitudes. Essentially, an aircraft becomes the first "stage" of the rocket. The Pegasus is one such vehicle, an air-launched rocket developed by Orbital Sciences Corporation (now Northrop Grumman). Pegasus first flew in 1990 and is capable of launching up to 443 kg (977 lb) into low Earth orbit. It is capable of launching up to 443 kg (977 lb) into low Earth orbit. Pegasus is a three-step vehicle with each step powered by solid propellant; an optional monopropellant fourth step is available.

When released from its carrier aircraft at an altitude of approximately 12,000 m (40,000 ft), a small wing and empennage provide lift and attitude control (Fig. 2.39). The three Orion solid propellant motors were custom-designed for the Pegasus by Hercules Aerospace. Most of the vehicle itself was designed by Dr. Antonio Elias, although the wing was designed by Burt Rutan [44]. Several of Orbital's internal projects—including the Orbcomm satellite constellation, the OrbView satellites, and an Orbcomm-derived constellation called Microstar—served as guaranteed customers for the Pegasus during its initial development, although the retirement of the Scout rocket saw several government and military payloads in need of a launch vehicle.

First flight of the Pegasus occurred on 5 April 1990. During initial flights, a B-52 Stratofortress (NB-0008) served as the carrier aircraft, although Orbital Sciences later transitioned to a converted Lockheed L-1011 TriStar airliner. The carrier aircraft is able to provide only a small amount of the lift and velocity needed for a low Earth orbit. At an altitude of 40,000 ft, the carrier aircraft provides only 4% of the orbital altitude and 3% of the orbital velocity, but launching from this altitude circumvents enough launch losses that an expensive first step booster is not needed. Air-launching also reduces weather concerns. Bad weather can still scrub a launch during takeoff, ascent, and travel to the launch point; however, the carrier aircraft is capable of climbing above much of what is conventionally thought of as "weather," which is limited to the lower part of the troposphere. Thus, the

Fig. 2.39 Pegasus launch vehicle prior to being loaded on its carrier aircraft. Source: Northrop-Grumman.

Pegasus is less vulnerable to the weather-related phenomena that could delay a ground-based launch. Additionally, much of the costs associated with a launch site can be eliminated, because air-launching obviates the need for a launch pad, blockhouse, acoustic damping, and other support infrastructure; this allows Pegasus to be launched from virtually anywhere, although launching a 20-ton rocket filled with explosive propellant over an ocean still carries a much lower insurance premium than launching the same vehicle over a populated city, which can represent a significant reduction in costs.

The Pegasus XL, first flown in 1994, has longer first and second stages relative to the original Pegasus, and as such has a higher payload capacity. Flight operations between the two vehicles are virtually identical. The Pegasus rockets have flown 44 missions to date, in both the original and XL configurations; of these missions, 39 were considered successful—a success rate of 88.64%. The initial launch price was $6 million, but that did not include the optional HAPS maneuvering stage. Prices can quickly escalate once the XL configuration and support services are considered: NASA's Ionospheric Connection Explorer had a total launch cost of approximately $56 million, which included "firm-fixed launch service costs, spacecraft processing, payload integration, tracking, data and telemetry and other launch support requirements" [45]. Dual payloads can also be launched, although for many satellites it is advantageous to be the primary (upper) payload, because this guarantees the desired orbit if the mission is successful, whereas the secondary (lower) payload may be placed in a compromised orbit.

2.4 The Moon Race

The so-called "moon race" officially began on 25 May 1961, when U.S. President John F. Kennedy spoke before a joint session of Congress, stating that "this nation should commit itself to achieving the goal, before this decade is out, of landing a man on the moon and returning him safely to the Earth." Thus, the United States issued a challenge to the Soviet Union—a challenge that the United States believed it could win.

Going to the moon called for a launch vehicle with a high payload capacity and a very high thrust capacity—a vehicle that, at the beginning of the moon race, did not exist anywhere in the world. Designers and engineers in the ABMA were transferred to the newly created National Aeronautics and Space Administration (NASA) Marshall Space Flight Center (MSFC) in Huntsville, Alabama. No very-high-thrust rocket engines yet existed, so the designers looked toward grouping several smaller engines together, called *clustering*, and attaching them to the back of the rockets then being designed. Clustering has several advantages and disadvantages; the chief advantage is that a mission can still potentially succeed even if an engine is lost. Of course, the main disadvantage is that the plumbing is much more complicated: eight engines require eight sets of plumbing, and for multiple engines, the chances of a failure are multiplied by the number of engines present.

Nevertheless, the initial large rocket designs used clusters, such as the Saturn IB shown in Fig. 2.40. This vehicle had eight Rocketdyne H-1 engines clustered on its rear, producing a total thrust of 7,580 kN (1.7 million lb); when the clustered arrangement was first introduced, critics jokingly referred to it as "Cluster's Last Stand"; however, the design proved sound and very flexible. Although the Saturn I was originally planned as a "universal" booster for military payloads, only 10 Saturn I rockets were flown before being replaced by the more capable Saturn IB.

The Saturn IB was originally commissioned by NASA for the Apollo program. The key difference is that the S-IV step of the Saturn I was replaced by the more powerful S-IVB step, which was capable of launching the Apollo Command/Service Module (CSM) or lunar module (LM) into low Earth orbit for flight testing. The S-IVB step was shared between the Saturn IB and the larger Saturn V needed for lunar flight; this provided a common interface for the Apollo spacecraft. The key difference between the S-IVB step used for the Saturn IB and Saturn V flights was that, when used in the Saturn V, the S-IVB step burned only a portion of its propellant to reach low Earth orbit, so that the remainder of the propellant could be used for translunar injection. By contrast, the S-IVB step on the Saturn IB needed

Fig. 2.40 The Saturn IB rocket with eight Rocketdyne H-1 engines clustered to deliver 1.7 million lb of thrust.

every last bit of propellant to attain low Earth orbit. The Saturn IB launched the first manned CSM orbital mission, first planned as Apollo 1 but later flown as Apollo 7, after several unmanned CSM and LM flights. The Saturn IB also launched AS-203, an orbital mission that did not contain a payload; in this fashion, the S-IVB step would have residual propellant once on orbit. This allowed engineers to observe the behavior of the cryogenic propellants in microgravity, supporting the design of the restartable version of the S-IVB step that was eventually used on the Saturn V.

The Saturn IB was a two-step vehicle consisting of a LOx/RP-1 S-IB first step and a LH$_2$/LOx S-IVB second step. Nine propellant tanks fed the eight engines, the outer four of which were gimbaled to provide steering. The first step burned for approximately 2.5 min, separating at an altitude of 68 km (42 miles). The S-IVB upper step was a single-engine version of the S-IV step and served as the second step on the Saturn IB and as the third step on the Saturn V. Two versions of the S-IVB were built—the 200 series and the 500 series—the primary difference between the two being that the 200 series did not have a flared interstage and had less helium pressurization aboard. As such, the 200 series could not be restarted in flight. The 200 series also had three solid rocket motors for separating from the S-IB lower step, whereas the 500 series used only two. During later Apollo missions, the S-IVB step was intentionally crashed into the moon to perform seismic measurements, which were used to characterize the moon's interior. After the moon landing, a surplus S-IVB step was converted into Skylab, the United States' first space station (see Fig. 2.43).

As time went on, it became apparent that the development of higher thrust engines was a necessity if the moon race was to be won; the alternative, as explored by the Soviets, could involve up to 30 engines clustered together, as was the case with the unsuccessful N-1 moon rocket (Fig. 2.41). (The N-1 was unsuccessful in four launch attempts.) The answer was to be found in the

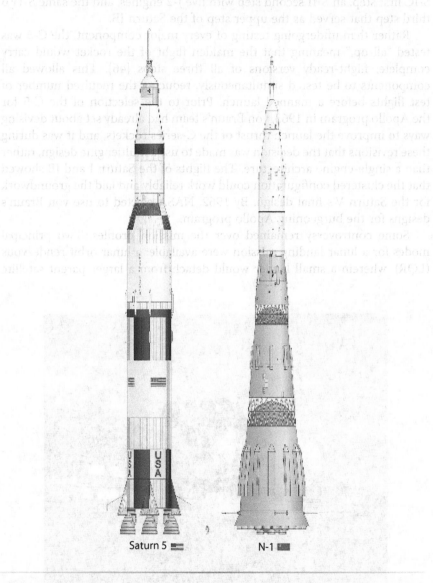

Saturn 5   N-1

Fig. 2.41   The Soviet N-1 moon rocket, 1968, along with the U.S. Saturn V, for scale. Note the spherical tanks, which make "stretches" (i.e., changes in tank volume) much more difficult than cylindrical tanks, which can be simply lengthened. Courtesy: Richard Kruse.

Saturn V (Fig. 2.41, 2.42, 2.43 and 2.44). The Saturn V rocket emerged from the Jupiter rocket family and began life as the C series of rockets; the configuration that would come to be named Saturn V was originally developed as the C-5. The Rocketdyne F-1 engine, originally designed to meet a 1955 U.S. Air Force specification, was selected to power the first step of the Saturn V. Specifically, five F-1 engines would power the Saturn V first step, as shown in Fig. 2.42. The three-step rocket would consist of the five-engined S-IC first step, an S-II second step with five J-2 engines, and the same S-IVB third step that served as the upper step of the Saturn IB.

Rather than undergoing testing of every major component, the C-5 was tested "all-up," meaning that the maiden flight of the rocket would carry complete, flight-ready versions of all three steps [46]. This allowed all components to be tested simultaneously, reducing the required number of test flights before a manned launch. Prior to the selection of the C-5 for the Apollo program in 1963, von Braun's team had already set about devising ways to improve the launch thrust of the C-series rockets, and it was during these revisions that the decision was made to use a multiengine design, rather than a single-engine architecture. The flights of the Saturn I and IB showed that the clustered configuration could work reliably and laid the groundwork for the Saturn V's final design. By 1962, NASA elected to use von Braun's designs for the burgeoning Apollo program.

Some controversy remained over the mission profiles. Two principal modes for a lunar landing mission were available: a lunar orbit rendezvous (LOR), wherein a small lander would detach from a larger parent satellite

Fig. 2.42 The cluster of Rocketdyne F-1 engines powering the Saturn V first step. Credit: NASA.

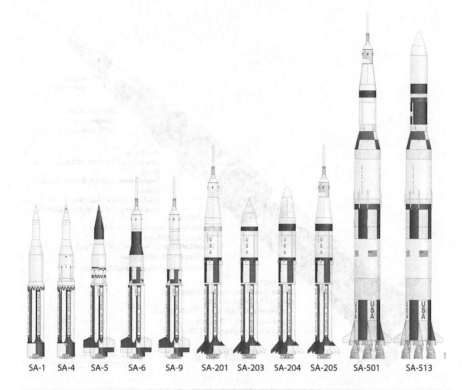

SA-1   SA-4   SA-5   SA-6   SA-9   SA-201 SA-203 SA-204 SA-205   SA-501   SA-513

Fig. 2.43 The Saturn rockets, 1960–1975. Left to right: Saturn I SA-103, SA-104, SA-105, SA-106; Saturn I-B AS-203, Apollo 5, Apollo 7; Saturn V moon rocket; and Saturn V Skylab. Courtesy: Richard Kruse.

that remained in lunar orbit and then rejoin the parent satellite for the trip back to Earth, or an Earth orbit rendezvous (EOR), whereby a larger vehicle would be landed directly on the lunar surface, eliminating the need for separate ascent and descent stages. The EOR mode had the added complication of using multiple launch vehicles and assembling the "satellite" lander/ascender while in orbit. Although the LOR mode was eventually selected, the various bureaucratic skirmishes within NASA demonstrated the managerial roadblocks the agency faced when one center had oversight of the launch vehicle and another managed the payload [47]. Needless to say, the Apollo program was successful, landing 24 astronauts on the moon from 1968 to 1972. The Saturn V was launched 13 times from Kennedy Space Center in Florida, with zero loss of crew or payload; it remains the tallest, most massive, and most powerful launch vehicle to ever reach operational status. The evolution of the Saturn rocket family is shown in Fig. 2.43.

As should have been understood before this point, rockets tend to be *massive* vehicles, and as demonstrated in later chapters, a good portion of a launch vehicle's total weight at liftoff is propellant. So much of a launch

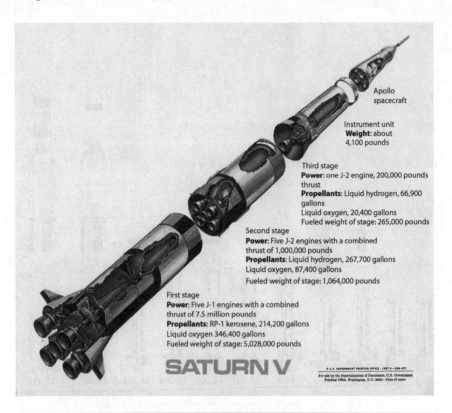

Apollo
spacecraft

Instrument unit
Weight: about
4,100 pounds

Third stage
Power: one J-2 engine, 200,000 pounds
thrust
Propellants: Liquid hydrogen, 66,900
gallons
Liquid oxygen, 20,400 gallons
Fueled weight of stage: 265,000 pounds

Second stage
Power: Five J-2 engines with a combined
thrust of 1,000,000 pounds
Propellants: Liquid hydrogen, 267,700 gallons
Liquid oxygen, 87,400 gallons
Fueled weight of stage: 1,064,000 pounds

First stage
Power: Five J-1 engines with a combined
thrust of 7.5 million pounds
Propellants: RP-1 kerosene, 214,200 gallons
Liquid oxygen 346,400 gallons
Fueled weight of stage: 5,028,000 pounds

SATURN V

Fig. 2.44   Saturn V components. Source: NASA.

vehicle is propellant, in fact, that the actual payload mass can be a ridiculously small fraction of the launch vehicle's total mass. For example, the Saturn V had a maximum liftoff weight of approximately 6 million lb; that much launch vehicle would only put about 300,000 lb into low Earth orbit (less than one third of that amount to a translunar orbit). This equates to a payload fraction on the order of 3–5%.

2.5   The Space Shuttle

The Space Shuttle was a partially reusable launch system operated by NASA from 1981 to 2011, during which time 135 missions were launched from the Kennedy Space Center in Florida. The Shuttle fleet spent a total of 1322 days, 19 hours, 21 minutes, and 23 seconds on missions [48]. The official program name of the Shuttle was the Space Transportation System, which was taken from a Space Task Group proposal that sought to give NASA direction in the post-Apollo days of the agency [49]. Of the recommendations listed in this report, the Shuttle was the only item to receive funding for further development.

The first orbital test flight was performed in 1981, and the Shuttle entered service in 1982. The first Shuttle, *Enterprise*, was designed purely for approach and landing tests; it had no orbital capability. The rest of the fleet consisted of four orbiters: *Columbia, Challenger, Discovery*, and *Atlantis*. Of these, *Challenger* and *Columbia* suffered catastrophic accidents and were destroyed while on mission in 1986 and 2003, respectively. A fifth orbiter, *Endeavour*, was built in 1991 to replace *Challenger*. The fleet was retired from service at the conclusion of *Atlantis*'s final flight on 21 July 2011. During its three decades of service, the Shuttle fleet launched numerous scientific missions, including interplanetary probes, on-orbit experiments, and the Hubble Space Telescope; the Shuttle was also instrumental in the construction and servicing of the International Space Station.

The Shuttle was a highly complex vehicle consisting of the orbiter vehicle (OV) carrying the Space Shuttle Main Engines (SSMEs), two solid rocket boosters (SRBs), and an external tank (ET) that contained the liquid oxygen and liquid hydrogen for the Shuttle's main engines. The two SRBs were recoverable; however, the ET was not. The Shuttle launched vertically, like a typical rocket, with the parallel-staged SRBs igniting just after the main engines ignition and checkout before liftoff; the SRBs jettisoned during ascent, whereas the ET was jettisoned prior to orbital insertion. Somewhat unique to the Shuttle was its flyback capability: at the conclusion of a mission, the OV would glide to an unpowered landing on a runway at the Shuttle Landing Facility at Kennedy Space Center. Rogers Dry Lake at Edwards Air Force Base in California served as a backup landing site, after which the Shuttle would be transported back to Kennedy aboard a specially modified Boeing 747 Shuttle Carrier Aircraft. Note that the Shuttle had no propulsive options during the landing phase: the main engines were unusable (primarily due to the fact that the ET, which carried the main engines' fuel, had already been jettisoned), and the orbital maneuvering system (OMS) that the Shuttle used for positioning and attitude control while in space were useless within the atmosphere. Thus, every single one of the Shuttle landings had to be performed "dead-stick," making the Space Shuttle not only one of history's most complex launch vehicles, but also the world's largest and heaviest glider!

The Soviet Union, of course, was not entirely inactive during this time. The chief components of the Space Shuttle system are shown in Fig. 2.45. Its own reusable orbiter, called *Buran* ("Snowstorm," although the name was used to refer to both the program and the OK-1K1 orbiter) was also undergoing development during this time. This program would become the largest and most expensive in the history of Soviet space exploration. The program officially began in 1974 and was formally discontinued in 1993; the single orbiter produced during this program was destroyed in 2002 when the roof of its hangar at Baikonur Cosmodrome collapsed due to poor maintenance. To date, the Buran remains the sole reusable Soviet spacecraft ever launched.

Fig. 2.45   Liftoff of *Discovery* on STS-120.

The sole orbital flight of the Buran—an uncrewed or uninhabited mission—launched on 15 Nov. 1988; the vehicle completed two orbits and landed exactly 206 minutes after launch [50]. As was the case with the N-1 moon rocket, the Soviets seemed to favor the clustering of a large number of engines to achieve the required launch thrust, as shown in Fig. 2.46. Note that on the Buran, the central structure is not a fuel tank, but a complete Energia launch vehicle.

The similarities between the Buran and the Space Shuttle are striking (Fig. 2.47), and the fact that the Buran debuted after the Shuttle—combined with the similarities of previous Soviet projects to their Western counterparts (e.g., the British/French Concorde and the Tupolev Tu-144 Charger, mockingly referred to as *Concordeski*)—created a great deal of speculation that the design of the Buran was influenced as much by Cold War espionage as by engineering. On the surface, this was justifiable, because the external similarities between the two vehicles are remarkable; however, several key differences existed between the Buran and the Shuttle, both structurally and operationally, and it is likely that if espionage had played a part in the Buran's design, it would likely have been limited to photography of the external airframe or using early orbiter designs as a basis for subsequent work. One CIA study on the topic corroborates this conclusion, stating that the Buran was based on an early Shuttle design that was rejected by NASA [51].

One of the key differences between the Shuttle and the Buran was in its configuration. The Shuttle was an integral part of the launch system, and the stack itself was configured to always operate together; by contrast, the Soviet Energia launch vehicle could be configured to lift payloads other than the Buran. Despite the ungainly appearance of 23 rocket nozzles at the aft end of the Buran, the Soviet vehicle did contain what could be considered design improvements relative to the Shuttle.

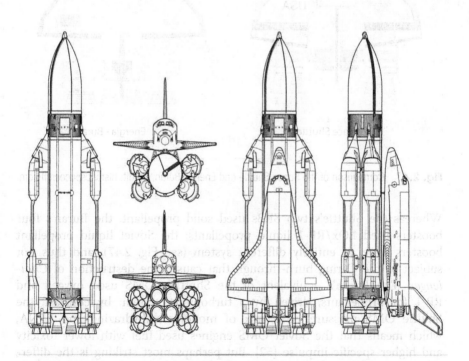

Fig. 2.46 Fully assembled Buran launch system. Source: Voertiguen Blueprints, martworkshop.com.

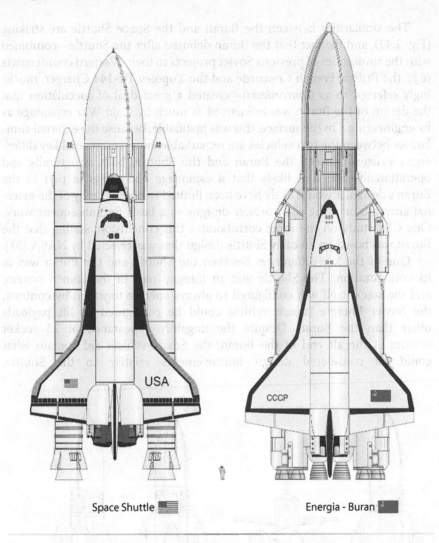

Fig. 2.47   Comparison of the Space Shuttle and Energia Buran. Credit: historicspacecraft.com.

Whereas the Shuttle's two SRBs used solid propellant, the Buran's four boosters used LOx/RP-1 liquid propellants; the Soviet liquid propellant boosters were an entirely different system (see Fig. 2.47), and thus not subject to the O-ring burn-through that caused the destruction of *Challenger*. The Buran's equivalent of the Shuttle's OMS used oxygen and RP-1 for propellants, driven by a turbopump system; by contrast, the Shuttle OMS pressure-fed a mix of monomethylhydrazine and RFNA, which means that the Soviet OMS engines used fuel with lower toxicity and higher specific impulse [52]. But perhaps most striking is the difference in gliding performance: the Buran's subsonic L/D is cited as 6.5, compared to the Shuttle's subsonic L/D of 4.5—an improvement of over 44% [53]. There is some skepticism as to the accuracy of the Buran's L/D ratio, however.

2.6 Launch Vehicle Oddities and Dead-Ends

Up to this point, the focus has been almost exclusively on rockets and missiles that may be considered "conventional": staged, typically launched vertically, and carrying a scientific or nuclear payload that makes up a small fraction of its gross liftoff weight. However, there have been a number of noteworthy applications of rocket propulsion that do not fit into this broad generalization. Some of these "odds and ends" of the rocketry world are discussed in the following sections.

2.6.1 German A-9/A-10 Amerika

It has already been well-established that the Germans invented the field of guided ballistic missiles during World War II; however, the Germans also anticipated combat with the United States, and they began to consider long-range rockets to address this perceived strategic inevitability. One of these designs was the A-9/A-10 Amerika, a ballistic missile intended to have transatlantic range and a high-explosive warhead. This two-step, boosted glide bomb is arguably the world's first practical design for an ICBM, aside from the minor inconvenience that the upper step was to be piloted, and thus the pilot would have been part of the expendable warhead. Work on the Amerika was prohibited after 1943, after which all efforts were to be spent on the perfection and production of the V-2. Despite this prohibition, Wernher von Braun managed to continue the development and flight testing of the A-9 under the pseudonym of A-4b; because this was believed to have been a modification of a production design, von Braun's work was allowed to continue.

The vehicle that would eventually bear the nickname Projekt Amerika was in reality two separate vehicles: the A-9 piloted warhead and the A-10 boost step. The A-10 step was originally designed to use six A-4 combustion chambers feeding into a single expansion nozzle; however, this changed to a massive single-chamber/single-nozzle arrangement as the vehicle evolved. Engine stands at Peenemünde would handle the testing of the final engine design, with thrust levels on the order of 2,300 kN (518,600 lbf). The A-9 step began development as a refined A-4 with swept wings; this design evolved into the final configuration, featuring fuselage strakes. The change from swept wings to strakes was found to improve the aerodynamic performance of the A-9 in the supersonic regime and solved the issue of the center of lift shifting in the transonic regime (the "Mach tuck"). Like its V-2 predecessor, the A-9/A-10 was designed to burn liquid oxygen and alcohol; total burnout time was on the order of 1 min for the boost step and 115 s for

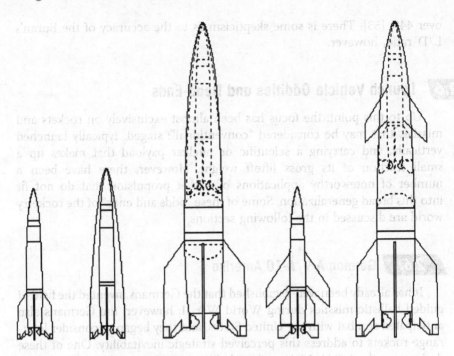

Fig. 2.48 From left to right: A-4, early concept of an A-9, early concept of an A-9/A-10, flight test configuration of an A-4b, and final concept of the A-9/A-10. Source: astronautix.com.

the upper step. Some concepts of the German A-9 and A-10 rockets are given in Fig. 2.48.

Despite the fact that the V-2's guidance system was adequate for its mission, the guidance technology of the time would have been almost comically inaccurate over the course of the Amerika's 5,000-km (3,100-mile) flight. Thus, it was decided that the A-9 would require a pilot. After engine cutoff at a speed of 3,400 m/s (7,600 mph) and a peak altitude of 390 km (240 miles), which is roughly four times as high as the officially demarcated edge of space, the A-9 would reenter the Earth's atmosphere and begin an extended glide toward its target. German submarines in the Atlantic Ocean would have surfaced to act as radio beacons and thereby provide navigation cues to the pilot. Once the target was within visual range, the pilot would lock the vehicle's controls and bail out of the craft. It was expected that the pilot would then either die or be captured as a prisoner of war.

Several other designs based on the A-9/A-10 arrangement were also considered. Adding a third ("A-11") step would have resulted in a satellite launcher, whereas a fourth ("A-12") step would have allowed the A-9 to become the world's first manned orbiting shuttle. The German design efforts were not simply abandoned at the end of World War II, however: research found that an A-9-like second step equipped with a ramjet could extend the warhead's range to the 10,000-km (6,200-mile) minimum necessary for the United States and Soviet Union to attack one another. This led to the American SM-64 Navaho and Soviet Burya supersonic cruise missiles. Conventional rocket technologies—specifically, improvements to structural and propulsive efficiencies—eventually made possible the design of purely ballistic missiles with burnout speeds more than twice as high as the A-9's and true intercontinental range. The speed of these missiles (on the order of Mach 12 or faster during reentry) made them all but impossible to intercept once launched, thus sounding the death knell for the designs inspired by the A-9.

2.6.2 North American X-15

The X-15 (see Fig. 2.49) was a hypersonic, rocket-powered aircraft designed by North American Aviation (now part of Boeing) as part of the US's X-plane series of experimental and research aircraft. The X-15 set numerous speed, altitude, and time-to-climb records in the 1960s, and through its many flight tests collected data that were invaluable to the evolution of aircraft and spacecraft design. During the X-15 program, eight U.S. Air Force pilots exceeded an altitude of 80 km (50 miles) over the course of 13 flights, thus qualifying them for astronaut status; by contrast, the sole Navy pilot in the program never took the aircraft high enough to qualify for this distinction. The X-15 still holds the official world record for the highest speed attained by a powered, manned aircraft: 7,274 km/h (4,520 mph), or Mach 6.2 [54]. Of the 199 total X-15 flights, two flights (by the same pilot, U.S. Air Force Captain Joseph A. Walker) qualified as spaceflights per the definition of

Fig. 2.49 X-15 research aircraft. Source: NASA.

spaceflight set forth by the Féderation Aéronautique Internationale, which defines spaceflight as any flight exceeding 100 km (62.1 miles) in altitude.

The X-15 began life as a concept study conducted by Walter Dornberger for the National Advisory Committee for Aeronautics (NACA), the predecessor of NASA. Two manufacturers were selected for the project: North American Aviation would build the airframe, and Reaction Motors would build the rocket engines. The aircraft was designed to be air-dropped, and two B-52 bombers—NB-52A (sn 52-0003, also known as *The Mighty One* and *Balls Three*) and NB-52B (sn 52-0008, also known as *The Challenger* and *Balls Eight*)—served as the exclusive carrier aircraft for all X-15 flights (see Fig. 2.50). A thick *wedge tail* consisting of an upper and lower vertical stabilizer (sometimes referred to, respectively, as the dorsal and ventral fins) was necessary for stability at hypersonic speeds; however, this empennage configuration produced an enormous amount of drag at lower speeds—so much drag, in fact, that the fin's blunt trailing edge could produce as much drag as an entire F-104 Starfighter [55]. The X-15 was released at an altitude of 13.7 km (45,000 ft) at a speed of approximately 800 km/h (500 mph). Several parts of the X-15 fuselage were constructed of a nickel superalloy known as Inconel-X 750, designed to withstand the high thermal loads that the aircraft would experience at hypersonic speeds. Landing gear consisted of a nose wheel and two main gear skids; because these skids could not extend past the ventral fin, the pilot would need to jettison the ventral fin (fitted with a parachute) before landing [54].

Fig. 2.50   X-15 being carried by B-52B (52-0008) in captive flight. Source: NASA.

The X-15 primarily flew two types of missions, which investigated the aircraft's ability to meet or exceed either speed or altitude goals. The mission elements, however, remained relatively constant: captive carry to drop altitude, drop from the carrier aircraft, engine start, and acceleration; at this point, the aircraft would either continue accelerating to investigate the hypersonic speed regime or it would begin to climb. If the rocket engine could not successfully start, the aircraft would glide directly to a landing. Although the rocket engine operated for only a short period of time, it was able to accelerate the X-15 to tremendous speeds and altitudes: one popular anecdote at NASA Dryden Flight Research Center (now NASA Armstrong Flight Research Center, colocated with Edwards Air Force Base) stated that if the X-15 were dropped at the northeast corner of Nevada, it would land at the Edwards dry lake bed in southern California 10 minutes later.

The X-15 spent most of its flights in an environment where the surrounding atmosphere was too thin for conventional aerodynamic controls to be of much use; as such, it made extensive use of a reaction control system (RCS) that utilized high-test peroxide (which, when in the presence of a catalyst, decomposes into water and oxygen) for attitude control. The original control system of the X-15 consisted of three separate control sticks: a traditional stick-and-rudder setup for low-speed flight; a second joystick on the left of the pilot, which sent commands to the RCS; and a third joystick, to the right of the pilot, which augmented the center stick during high-g maneuvers [56]. A stability augmentation system was also present, to help the pilot maintain proper attitude control. This arrangement was eventually replaced by the MH96 flight control system, an arguably lower-workload setup that used a single control stick and automatically blended aerodynamic and RCS controls based on how effectively each control system was at controlling the aircraft within a given flight regime [56]. The pilot also manually controlled the throttling of the rocket engine and had a separate control system for jettisoning the ventral fin. Early test flights used two XLR11 engines, which burned ethyl alcohol and liquid oxygen. These were eventually replaced with the more powerful XLR99 engine, which burned liquid oxygen and anhydrous ammonia to generate 250 kN (57,000 lbf) of thrust; a peroxide-driven, high-speed turbopump fed the propellants to the engine, which could consume 6,800 kg (15,000 lb) of propellant in 80 s [57]. The XLR99 could also be throttled, and was the first man-rated, variable-throttle, liquid propellant rocket motor.

2.6.3   McDonnell Douglas DC-X

The DC-X (Delta Clipper Experimental) was an unmanned prototype of a single-stage-to-orbit (SSTO) reusable launch vehicle built by McDonnell Douglas in the early 1990s. Keen to not repeat the same mistakes that doomed the Rockwell (formerly North American Aviation, now part of Boeing) X-30 National Aero-Space Plane (another SSTO concept billed as a passenger spaceliner) the decade before, the development of the DC-X focused on the use of low-cost, commercially available "off the shelf" (COTS) parts that had high technological maturity; the development plan for the DCX was the gradual exploration of the limits of such COTS technology through a systematic "fly a little, break a little" approach [65]. As experience with fully reusable SSTO vehicles increased, a larger prototype (the "DC-Y") could be built for suborbital and orbital flight testing; a commercial version (the "DC-1") would then follow. The *Delta Clipper* monicker was chosen deliberately, in honor of the Douglas DC-3 aircraft that first made passenger air travel affordable; in a similar fashion, the DC-1 would make passenger space travel affordable.

The DC-X was never intended to achieve orbital flight; rather, it was built to demonstrate the capability of a launch vehicle to take off and land vertically. This vertical takeoff and landing (VTOL) arrangement would require the DC-X to use control thrusters and retro-firing rockets to control its attitude during descent; the vehicle would reenter the atmosphere nose-first, but then arrange itself for a tail-first landing—an advantageous characteristic, because the aircraft could then refuel and immediately be ready for another launch. A tail-first entry would have been more manageable for several reasons: the thermal protection necessary for reentry was already partially present in order for the vehicle to survive the engine exhaust; the vehicle could still be controlled during descent, without the need for large pitch changes in preparation for landing; and most importantly, because the base of the vehicle is much larger than the nose, it would experience lower peak temperatures during reentry, because the thermal load would be spread over a larger area. Several operational requirements prevented this tail-first reentry arrangement, specifically the need for a launch vehicle to possess an "abort once around" failure mode (to return and land after completing a single orbit); this requirement mandates that the vehicle have a large cross-range capability, which the DC-X could only attain with a nose-first reentry attitude.

Construction of the DC-X began in 1991 at McDonnell Douglas's facility in Huntington Beach, California. The outer shell was custom-fabricated by Scaled Composites of Mojave, California; however, the vast majority of the aircraft was built from COTS components. This included the flight control systems and the four Rocketdyne RL-10 engines used on the Centaur. The DC-X first flew at the White Sands Missile Range in New Mexico on 18 Aug. 1993 for 59 s, and again on 11 and 30 Sept. of that year (see Fig. 2.51). The program was then halted due to funding cuts, until NASA provided sufficient funds to restart the program in June 1994, with a 136-s flight. Four more flights were conducted throughout 1994 and 1995; two of these flights suffered malfunctions—a minor in-flight explosion and a hard landing that cracked the aeroshell. Because the funding for the program had already been cut, there was no way to make repairs.

The entirety of the DC-X program was transferred to NASA after the last flight in 1995. Several modifications were made to the vehicle (including a

Fig. 2.51 Photo montage of a DC-X test flight (ca. 1993). Source: New Mexico Museum of Space History.

composite fuel tank and a lightweight, Russian-made aluminum-lithium alloy oxidizer tank that was so defective it didn't meet flight readiness standards), and the vehicle (renamed the DC-XA) resumed flight tests in 1996. The first flight of the DC-XA overheated the aeroshell and caused a fire during landing. Two subsequent flights (on 7 and 8 June 1996) demonstrated that the vehicle could be flown, landed, refueled, and reflown within 26 h. The flight of 7 July 1996 proved to the be vehicle's last: one of the LOx tanks had cracked during previous flights, and when a disconnected hydraulic line prevented one landing strut from extending, the DC-XA toppled and spilled the contents of the cracked LOx tank. This spill caused an explosion that damaged the vehicle beyond the point of practical repair [58], and the program was permanently shuttered. The Brand Commission's postaccident report cited crew fatigue due to the "on-again/off-again" nature of the project's funding, combined with the constant threats of project cancellation, as a key contributor to the accident. Much of the crew was from the original DC-X program, and they found the demands of NASA's paperwork requirements as part of the testing regimen frustrating and ineffectual. The Commission stated that NASA had taken on the project begrudgingly, and only after having been "shamed" by the very public success of the DC-X under purely commercial and military development. A new DC-XA, estimated to cost a mere $50 million, was shelved due to "budget constraints"; however, a more likely reason that the DC-XA was never replaced was due to the infighting within NASA caused by the potential success of the DC-XA threatening the agency's "home-grown" X-33 VentureStar project [58].

All was not lost when the DC-XA program finally ended. Several engineers from the DC-X program relocated to smaller commercial space ventures after the program's cancellation; these included Blue Origin, Armadillo Aerospace, Masten Space Systems, and TGV Rockets. Several design features of these manufacturers' vehicles can be traced directly to the DC-X, including the reusable VTOL launch vehicle architecture.

2.6.4 Lockheed Martin X-33 VentureStar

The X-33 was an unmanned, suborbital, one-third-scale technology demonstrator first developed in the 1990s for the planned VentureStar orbital space-plane (see Fig. 2.52). The X-33 was funded under the Space Launch Initiative program. As a technology demonstrator, the X-33 was intended to serve as a flight test article for the technologies that NASA believed would be critical to the success of a reusable SSTO vehicle, including metallic thermal protection, composite propellant tanks, autonomous flight control, lifting-body aerodynamics, an aerospike engine, and the capability for rapid operational turnaround. Several failures plagued the program, the most notable of which were the structural failures of the multilobed composite tanks, and the project was ultimately cancelled in 2001.

Like the DC-X, the X-33 was designed to be an SSTO vehicle; this would do away with the need to stage a launch vehicle, and because there would be no components to retrieve (or, in the case of expendable components,

Fig. 2.52 Lockheed Martin X-33 VentureStar. Source: NASA.

to rebuild) prior to the next launch, the vehicle could theoretically offer savings in terms of material, labor, and operational costs. The X-33 was designed to launch vertically, like a normal rocket, but land horizontally at the end of a mission. Initial suborbital test flights were planned from Edwards Air Force Base to the Dugway Proving Grounds in Utah; subsequent tests would gather more data on reentry heating and engine performance along a flight track from Edwards to Malmstrom Air Force Base in Montana. Lockheed Martin was selected as the prime contractor for the X-33, and the vehicle was designed and built at the company's Skunk Works facility in Palmdale, California.

The unmanned X-33 was originally scheduled to fly 15 suborbital flights to an altitude of 75.8 km (248,000 ft, 47 miles); unlike a conventional aircraft, which has a flight profile that includes a long cruise-climb segment, the X-33 would launch vertically and then climb diagonally for half of its flight, before gliding diagonally down for the remainder of its mission. The X-33 was never intended to fly higher than 100 km (328,000 ft, 62.1 miles), nor faster than 50% of orbital velocity. Lockheed Martin attempted to build a business case for VentureStar as a privatized, reusable launch vehicle that would allow NASA to purchase launch services from a commercial provider. VentureStar was originally planned to begin ferrying passengers on suborbital, intercontinental routes by 2012; however, this much more ambitious project was never officially begun nor funded. The X-33 program was canceled in 2001. At this point in the project, 100% of the launch facility had been constructed, and the prototype had been 85% assembled from 96% of the necessary parts. Prior to cancellation, NASA had invested $922 million in the project, and Lockheed Martin had invested an additional $357 million. The program was continually hampered by technical difficulties, including instability and excess weight. The most damning of these failures, however, was the composite tank designed to hold the cryogenic liquid hydrogen fuel. In order for the X-33 to successfully be a true SSTO vehicle, the operating empty weight of the vehicle (its weight without propellant) had to be on the order of 10% of the vehicle's maximum takeoff weight. This required every part of the vehicle's structure to be as lightweight as possible, and the propellant tanks were an obvious choice for structural optimization.

The oxidizer tank for the X-33 was built out of the same aluminum-lithium alloy that had been used on the DC-X; however, the X-33's fuel tank was designed as a composite "sandwich panel" consisting of a honeycomb core and fiber-reinforced skin (a variant of IM7 carbon fiber). Complicating matters further, the liquid hydrogen tank was not designed as a conventional (i.e., cylindrical) pressure vessel, but rather as a multilobed structure. The first composite LH_2 tank fabricated by ATK (now part of Northrop Grumman) was rife with debonds and delaminations, and the very personnel who were trying to find solutions to the problems with the composite tanks had no experience with composite pressure vessels themselves [59]. A second composite tank was shipped to NASA Marshall Space Flight Center, where it failed during fueling and pressurization tests; subsequent investigation revealed that microcracking had caused the tank's failure. The proposed solution was to fill the sandwich core with closed-cell foam; however, this would add another 500 kg (1,100 lb) to the aft end of a vehicle whose configuration was already suffering from center of gravity and stability issues. Faced with impending failure, NASA and Lockheed approved the use of a metallic fuel tank made of the same aluminum-lithium alloy as the oxidizer tank; ironically, this tank was found to be lighter than the composite tank, which obviated the primary reason for using composites in the first place. On 11 April 2001, former NASA Director Ivan Bekey testified before the House Subcommittee on Space and Aeronautics, stressing the need for the X-33 to continue development with composite fuel tanks. Bekey's comments [59] doomed the X-33 program:

> The principal purpose of the X-33 program is to fly all the new technologies that interact with each other together on one vehicle, so that they can be fully tested in an interactive flight environment. If that is not done, the principal reason for the flight program disappears. Since the biggest set of unknowns in this vehicle configuration have to do with the structure-tankage-aeroshell-TPS-airflow interactions, it is my belief that to fly the vehicle with an aluminum tank makes little sense from a technical point of view. Worse yet, flight of an X-33 with an aluminum tank will increase the difficulty of raising private capital for a commercially developed VentureStar from the merely very difficult to the essentially impossible. What I would recommend is that NASA and Lockheed Martin face up to the risks inherent in an experimental flight program and renegotiate the X-33 cooperative agreement so as to delay the flight milestone until a replacement composite tank can be confidently flown. Both NASA and Lockheed Martin should make the investments required to build another composite tank and to absorb the program costs of the delay, because only then will the X-33 program be able to meet its objectives. To do anything less is flying for flying's sake, wastes the funds already expended, and makes little sense.

History is not without a sense of irony, so it is worth mentioning that even after the X-33 program was canceled, a composite tank for cryogenic fuel was finally developed. On 7 Sept. 2004, Northrop Grumman and NASA demonstrated a composite LH_2 tank that withstood repeated cycles of fueling and launch loads; these tests paved the way for low-energy, out-of-autoclave curing of thermoset resins that can even be used on conformal tanks [60].

Not all efforts on the X-33 were for naught. For instance, the XRS-2200 linear aerospike engine developed by Rocketdyne (now Aerojet Rocketdyne) for the project (Fig. 2.53) was successfully tested. These engines are unique in that their altitude-compensating nozzles maintain high efficiency at a greater range of altitudes, as opposed to conventional nozzles, which are "point-design" components and suffer off-design performance penalties. Aerospike engines will be discussed in greater detail in Chapter 4.

Fig. 2.53   Rocketdyne XRS-2200 linear aerospike engine. Source: NASA.

2.7   Other Launch Vehicles from Around the World

Although the United States and the former Soviet Union typically receive the lion's share of attention among launch vehicle historians and aficionados, they were far from the only countries making strides toward space access. In the annals of history, several other regions developed launch vehicles of note. They are briefly summarized here.

2.7.1   Chinese Launch Vehicles

In 2003, China became the third country in the world to achieve human spaceflight. Major General Yang Liwei became China's first *taikonaut*, flying into space aboard a Shenzhou capsule lofted aboard a Long March CZ-2F. Prior to Liwei's flight, the Long March 2F rocket flew four unmanned Shenzhou capsules, beginning in 1999. As has been periodically demonstrated in this chapter, rockets and politics are often intertwined, and the Shenzhou 5 launch that carried Liwei into space was no different: the Chinese state media heralded the successful flight as a triumph of Chinese nationalism, although it has been noted that Liwei flew the flags of both the People's Republic of China and the United Nations, and that crop seeds from Taiwan were brought aboard the spacecraft [61]. The Chinese continue to fly manned spacecraft and intend to build an orbiting space station, or cooperate with the International Space Station, in the future.

One of the most important and influential figures in Chinese rocketry actually got his start at MIT and Caltech, and it was only political machinations that saw him leave the United States for China. Qian Xuesen left China in 1935 to study mechanical engineering at MIT, obtaining his Master of Science degree in one year before joining Theodore von Kármán's research group (nicknamed the "suicide squad" for the sometimes explosive nature of their work) at Caltech, where he received his Ph.D. in 1939. His work with the Caltech team led to the development of the Private A rocket (first flight in 1944), Corporal, WAC Corporal, and other designs [62]. Allegations that Xuesen was a communist led to his security clearance being stripped in 1950, and he returned to China after five years of house arrest [63]. Xuesen's return to China resulted in his helping to lead the country's nuclear and thermonuclear weapons programs, making China the fifth nuclear weapons state and achieving the fastest fission-to-fusion development cycle in history (32 months, compared to 86 months for the United States and 75 months for the USSR [63]).

Xuesen's success in China's nuclear research programs gave him the opportunity to lead the development programs of the Dongfeng missile and the Long March series of rockets. The Long March launch vehicles (usually abbreviated as LM-) are named after the Chengzheng ("Long March") year-long military retreat of 1934–1935, during which the Red Army (forerunner of the modern People's Liberation Army) fled the pursuit of the Kuomintang (KMT, or Chinese Nationalist Party). The Red Army, under the eventual leadership of Mao Zedong, reportedly walked in excess of 9,000 km (6,000 miles) over 370 days to eventually reach safety in Shaanxi.

The LM rockets, much like their western counterparts, were derived from missile designs: the LM1 from the Dongfeng 4 ("East Wind") IRBM, and the LM-2, -3, and -4 rockets from the DF-5 ICBM. Both the DF-4 and DF-5 were tipped with megaton-class nuclear warheads, and the missiles had a range, respectively, of up to 7,000 km (4,300 miles) and 13,000 km (8,000 miles). All of the Long March rockets to date (see Fig. 2.54) have liquid propellant lower steps, although the type of fuel varies by series. For instance, the first two steps of the LM-1 use UDMH/RFNA fuel (much like the Titan), but it is the only Long March rocket to date that uses solid propellant for the spin-stabilized upper step. The LM-2, -3, and -4 all use the same UDMH/RFNA propellant combination for the lower steps, but an LH_2/LOx upper step. The new generation of Long March rockets (LM-5, -6, and -7), by contrast, will use LOx/RP-1 lower steps and maintain an LH_2/LOx upper step. Today, more than 20 variants of the Long March series of rockets have been developed (see Table 2.4 and Fig. 2.54) as China continues to develop its indigenous space-faring capabilities.

The destruction of the Space Shuttle *Challenger* gave China's burgeoning space program an unexpected boost, and the growing commercial backlog eventually convinced U.S. President Ronald Reagan to allow U.S. satellites

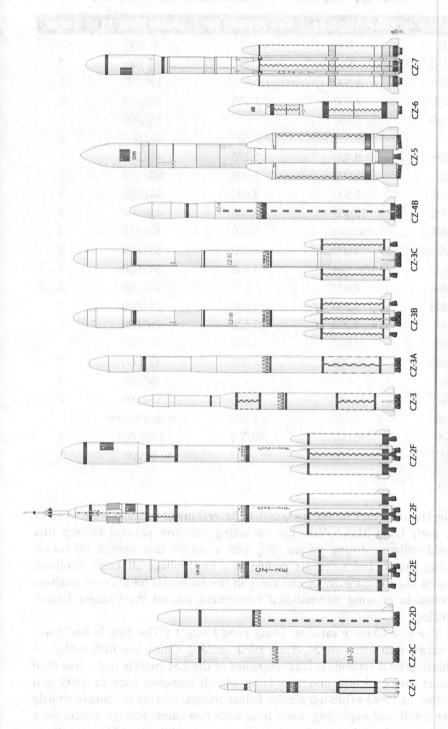

Fig. 2.54 Chinese Long March series. Courtesy: Richard Kruse.

Table 2.4 Long March Rocket Variants; Retired Variants Are Shaded

Vehicle	LEO Payload, kg	GTO Payload, kg	Launch Mass, kg	Stages
LM1	300	–	81,600	3
LM1D	930	–	81,100	3
LM2A	1,800	–	190,000	2
LM2C	2,400	1,250	192,000	2
LM2D	3,100	–	232,000	2
LM2E	9,500	3,500	462,000	2
LM2F	8,400	3,370	480,000	2
LM3	5,000	1,500	202,000	3
LM3A	8,500	2,600	241,000	3
LM3B	12,000	5,100	426,000	3
LM3B/E	Unknown	5,500	458,970	3
LM3B(A)	13,000	6,000	580,000	3
LM3C	Unknown	3,800	345,000	3
LM4A	4,000	–	249,000	3
LM4B	4,200	1,500	254,000	3
LM4C	4,200	1,500	250,000	3
LM5	25,000	14,000	867,000	
LM5B	25,000	14,000	837,500	2
LM6	–	–	103,000	3
LM7	13,500	7,000	594,000	2
LM8	7,600	2,500	In development	
LM9	140,000	50,000	3,000,000	3
LM11	700	–	58,000	3

to be launched on Chinese rockets [64]. Several major setbacks were faced by the early Long March launches, including defective payload fairings that would collapse during ascent [65] and a rocket that veered off-course during launch and killed six people in a local village although unofficial reports put the quantity of casualties in the hundreds [66]. These mishaps resulted in growing international resentment toward the Chinese launch vehicles.

The first Chinese satellite, Dong Fang Hong 1 ("The East Is Red") was launched into orbit on 24 April 1970, making China the fifth nation to achieve orbital capability. Early launches of the LM rockets had a less than perfect safety record, and two Long March launches (one in 1995 and another in 1996) exhibited similar failure modes, veering off-course shortly after liftoff and exploding, each time with casualties. Foreign media were sequestered for 5 hours after the second crash while the Chinese military allegedly attempted to reduce the visible extent of the damage [61]. U.S. satellite manufacturer Loral Space and Communication (maker of the Intelsat 708 that was lost in the 1996 LM-2 crash) shared information that permitted the Chinese to determine that the launch failures had been caused by improper welds, a gesture that earned Loral a $14 million fine from the U.S. government [67]. Five satellites were irreparably damaged or destroyed during Long March launches between 1990 and 1996, and the resulting U.S. embargo on Chinese launches has forced China to seek technological development via other avenues.

The reliability of the Long March rockets began to improve. Between August 1996 and August 2009, 75 consecutive successful launches were conducted, culminating with the launch of the Indosat Palapa D1 geostationary communications satellite on 31 Aug. 2009. (This launch was considered a partial failure due to a malfunction in the launch vehicle's third step.) On 24 Oct. 2007, the LM-3A launched another Chinese first: China's first lunar orbiter, the Chang'e 1, named after the moon goddess of Chinese mythology. On 7 Dec. 2014, China launched the Sino-Brazilian remote sensing satellite CBERS-4 (China-Brazilian Earth Resources Satellite) aboard an LM-4B, which marked the 200th launch of the Long March family.

2.7.2 European Efforts

The success of the United States in landing men on the moon was an elation not necessarily shared by the European space community. The British had begun developing the Black Arrow (nicknamed the "Lipstick rocket" for its bright-red colored nose fairing, see Fig. 2.55) in 1964. Originally intended to launch a 317-lb payload in low Earth orbit, its fourth and final flight—and only successful orbital launch—occurred in 1971. The rocket was then retired in favor of using the American Scout rockets, which were deemed cheaper than continuing to fund the program. Thus Great Britain became the first country to develop orbital launch capability before abandoning it due to budget concerns.

But the Black Arrow was not the first time that Europe intended to take its place on the launch vehicle stage. The United Kingdom first put forward proposals to France and Germany beginning in 1961 to create a launcher based on the de Havilland Blue Streak medium-range ballistic missile. The missile's first step was readily available and would form the first step of the Europa 1 project, intended to launch 1 metric ton (1,000 kg, or 2,205 lbs.) to low orbit. This created a bit of a logistical issue: the British would supply a first step that was already in production, and hence had a design and operating characteristics that were frozen; France and Germany would develop the second and third steps, respectively, but the two countries developed their respective steps independently of the other and with little communication overall. As such, there was little (if any) effort at what might resemble systems engineering or an overall optimization effort, with

Fig. 2.55 Black Arrow rocket at Woomera rocket park. Source: Magnus Manske.

predictable results. The resulting technical and financial problems stalled the program, and the British threatened to withdraw from the program. Two years later, in 1968, they did precisely that in favor of developing the Black Arrow [68].

The exit of the British heralded a restructuring of the program into Europa 2, which changed the payload specification to 270 kg to geostationary orbit and added a solid propellant fourth step. The Australian launch range at Woomera was also abandoned in favor of a new launch complex to be built in French Guiana. The first launch from this new complex failed, exploding 150 s after liftoff on 5 Nov. 1971. Germany subsequently withdrew from the Europa 2 program in Dec. 1972, citing the belief that European space transportation efforts could be limited to technological development in the post-

Apollo world. The European effort was again restructured into Europa 3, with the intent of placing payloads of up to 1,500 kg (3,307 lbs.) into geostationary transfer orbit. Few lessons seem to have been learned from the Europa 1 failures, because cost overruns and technical issues continued to stall these efforts: although the first step was developed from France's experiences with the Diamant program, the second step required the development of a liquid propellant engine far too ambitious for Europe's technical capabilities at the time.

Politically, Europe's space launch efforts fared little better. The Europa failures, combined with Germany's stance of technical development support, soured negotiations between European representatives and NASA: although first invited to be involved in major technological development efforts for the Space Shuttle, European contributions narrowed over time and were eventually limited to a small science module that would fit within the Space Shuttle cargo bay [68]. France's ambitions to acquire absolute control of space launch operations and dominate geostationary launches were hampered by two Diamant failures in 1971 and 1973 despite six previous launch successes, and the adversarial negotiations undertaken with NASA for the launch of the two Symphonie telecommunications satellites strengthened France's resolve to attain launch autonomy [69]. It was decided that the European effort would focus on three programs:

1. The development of the L3S heavy launch vehicle
2. The Spacelab program, in cooperation with NASA (backed by Germany)
3. The MARECS maritime telecommunications program (backed by the United Kingdom)

The principles of the European Space Agency (ESA) were also defined, and the ESA was officially established in 1975 [69]. France's commitment to this agreement was considerable; the country agreed to fund more than 60% of the ESA effort. In return, ESA would delegate program management responsibilities to the French Space Agency, *Centre national d'études spatiales (CNES)*, and the countries participating in ESA programs would be ensured a workload return in proportion to their contributions. The L3S designation was a French abbreviation signifying a third-generation launcher, but the rocket is perhaps better known by its commercial name: Ariane.

To date, six variants of the increasingly powerful Ariane line of rockets have been developed, and five have flown [70]; the sixth is scheduled to fly in 2020, and features the ability to bring the upper step out of orbit after satellite insertion (in an attempt to mitigate the ever-growing problem of "space junk" [71]) and an estimated 40% reduction in costs [72]. The family of rockets eschewed its modified-missile roots with the clean-sheet design of the Ariane 5, which replaced the two hypergolic lower steps of previous models with a single LOx/LH_2 step; strap-on boosters can be recovered for examination as a sort of launch *postmortem*, but are not reused [70].

Table 2.5 General Data of Ariane Rockets; Maximum Masses Shown

Vehicle	LEO Payload (kg)	GTO Payload (kg)	Launch Mass (kg)	Steps
Ariane 1	—	1,850	211,500	3
Ariane 2	—	2,180	220,950	3
Ariane 3	—	2,700	234,000	3
Ariane 4	7,600	4,300	470,000	3
Ariane 5	21,000	10,500	780,000	2
Ariane 6 (projected)	20,000	10,500	900,000	2

Summary-level data of the Ariane series of rockets are given in Table 2.5. A lineup of European rocketry is shown in Fig. 2.56.

2.7.3 India

The Indian Space Research Organization (ISRO) began development of indigenous launch vehicles in the 1960s, first with sounding rockets and then attaining complete launch autonomy (including complete operational support infrastructure) by the 1980s. India's first launch vehicle capable of inserting a payload into orbit was the Satellite Launch Vehicle, or SLV-3, a solid propellant launcher capable of carrying an 88-lb payload to a height of 310 miles; unfortunately, only two of the four SLV launches were successful [73]. The addition of two solid rocket boosters (SRBs) (which had similar performance to the SLV first step) created the Augmented Satellite Launch Vehicle (ASLV), which was a five-step solid propellant rocket (the SRBs would act as the first step, with the core motor igniting after the SRBs had burned out) designed to put payloads of up to 150 kg (330 lb) into low Earth orbit. The ASLV made four launch attempts, of which only one was successful. SLV and ASLV flights were launched from Sriharikota Range on India's eastern coast, a site that is still used today. The failures of the SLV and ASLV led to the creation of the Polar Satellite Launch Vehicle (PSLV), designed to launch payloads into heliosynchronous orbits and small payloads to GTO. The PSLV exhibited significantly better performance than the SLV and ASLV, and of the 50 PSLV launches to date, 47 have been successful. This paved the way for the Geosynchronous Satellite Launch Vehicle (GSLV) series, the latest of which is the GSLV Mark III. The GSLV series first used Russian liquid propellant motors (the UDMH-dinitrogen nitrogen tetroxide Vikas engine). The energy requirements for geosynchronous orbit required the development of a liquid oxygen–liquid hydrogen upper step, which at the time India did not have the technological capability to manufacture. India originally intended to procure the cryogenic engines from the Russian company Glavcosmos, but the

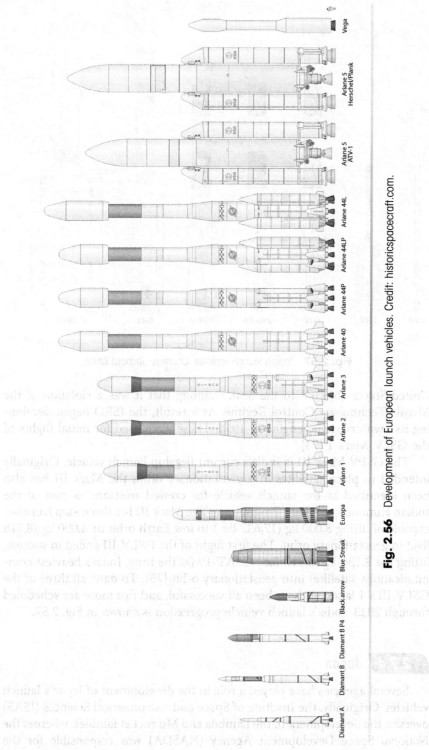

Fig. 2.56 Development of European launch vehicles. Credit: historicspacecraft.com.

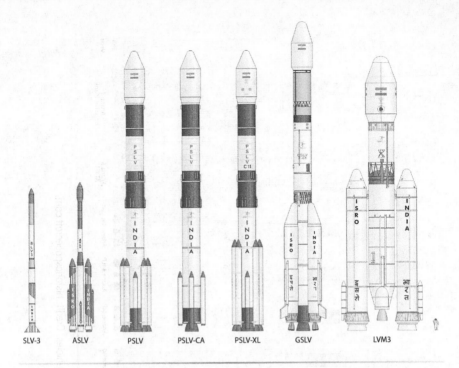

Fig. 2.57 Indian launch vehicles. Courtesy: Richard Kruse.

United States objected to the deal, claiming that it was a violation of the Missile Technology Control Regime. As a result, the ISRO began developing its own cryogenic engine in 1994. These were used for initial flights of the GSLV Mark I [74].

The GSLV Mark III is India's current flagship launch vehicle. Originally intended to place satellites into geostationary orbit, the Mark III has also been identified as the launch vehicle for crewed missions as part of the Indian Human Spaceflight Programme. The Mark III is a three-step launcher capable of lifting 8,000 kg (17,637 lbs.) to low Earth orbit or 4,000 kg (8,818 lbs.) to geostationary orbit. The first flight of the GSLV-III ended in success, lifting the 3,136-kg (6,917-lbs.) GSAT-19 (at the time, India's heaviest communications satellite) into geostationary orbit [75]. To date, all three of the GSLV-III's 4 launches have been all successful, and five more are scheduled through 2023. India's launch vehicle progression is shown in Fig. 2.57.

2.7.4   Japan

Several agencies have played a role in the development of Japan's launch vehicles. Originally, the Institute of Space and Astronautical Sciences (ISAS) oversaw the development of the Lambda and Mu rocket families, whereas the National Space Development Agency (NASDA) was responsible for the development of the N-1, N-2, and H series of launch vehicles. As of 2003, ISAS, NASDA, and the National Aerospace Laboratory of Japan combined to form the Japan Aerospace Exploration Agency (JAXA). As the "Japanese NASA," JAXA now oversees all Japanese launch operations. A lineup of Japan's launch vehicle families is shown in Fig. 2.58.

The Lambda 4S (or L-4S) was a four-step rocket using only solid propellant (with the first step augmented by two SRBs), which was designed to place 26 kg (57 lb) of payload into low orbit; however, the first four launches of the Lambda ended in failure; only the fifth launch successfully placed the Ohsumi-5—Japan's first satellite—into orbit [76]. The L-4S was eventually retired in 1970, although a sounding rocket derived from it (the Lambda 4-SC) successfully completed three flights in order to test technologies for the Mu series of rockets to follow, which eventually replaced the Lambda series for orbital launches.

The Mu series of rockets first began flying in 1966 from Uchinoura and flew 30 missions in several variants (see Fig. 2.59) until 2006, of which only 4 ended in failure. Although the Mu-1 was suborbital, the Mu-3 and Mu-4 saw decades of service before the introduction of the M-V (or Mu-5) in 1997. Typically, the Mu-5 flew in a three-step configuration; however, a four-step configuration (the M-V KM) was used three times between 1997 and 2003. Although the three-step Mu configurations had modest payload capacities [1,800 kg (3,968 lbs.) to a 30-degree inclination, or 1,300 kg (2,866 lbs.) to polar orbit], the MV KM could lift the three-step's 1,800-kg (3,968-lbs.) payloads to more than twice the orbital altitude [402 km (250 miles) for the KM, 193 km (120 miles) for other Mu rockets].

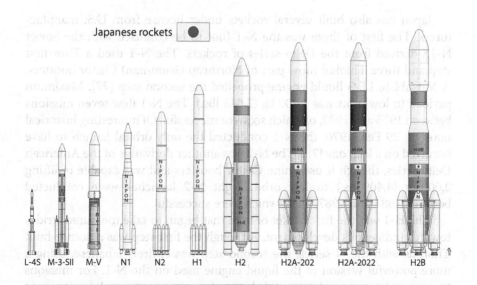

Fig. 2.58 Japanese launch vehicles. Courtesy: Richard Kruse.

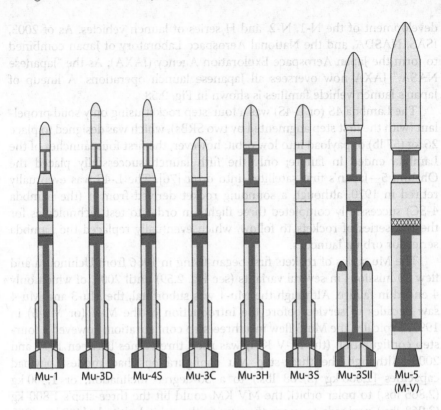

Fig. 2.59 Mu family of rockets. Source: GW Simulations.

Japan has also built several rockets under license from U.S. manufacturers. The first of these was the N-1 (not to be confused with the Soviet N-1), derived from the Delta series of rockets. The N-1 used a Thor first step and three Thiokol (now part of Northrop Grumman) Castor boosters. A Mitsubishi LE-3 liquid engine propelled the second step [77]. Maximum payload to low orbit was 1,200 kg (2,646 lbs.). The N-1 flew seven missions between 1975 and 1982, of which six were successful. Of interesting historical note: on 29 Feb. 1976, the N-1 conducted the only orbital launch to have occurred on a leap day [78]. The N-2 was another derivative of the American Delta series, though it used nine Castor boosters and was capable of lifting 2,000 kg (4,409 lbs.) to low orbit. Eight N-2 launches were conducted between 1981 and 1987, all of which were successful.

The H-1 was the first rocket of note that began to take measured strides towards indigenous development. Although the first step was a license-built Thor, as with the N series, the second step was entirely Japanese, using a more powerful version of the liquid engine used on the N-1. For missions to geosynchronous orbits, a third step (consisting of a solid propellant motor built by Nissan) was used. Depending on the payload, up to nine

Castor boosters could be used. The H-1 was capable of launching 3,200 kg (7,055 lbs.) to low orbit, or up to 1,100 kg (2,425 lbs.) to geosynchronous transfer orbit. All nine H-1 launches between 1986 and 1992 were successful.

The H-2 (or H-II) was the first truly Japanese rocket. The use of the LE7 liquid engine meant that this rocket was the first one developed in Japan without the use of manufacturing licenses from U.S. companies. Although development of the LE-7 engine was fraught with challenges and setbacks (including the death of one worker in an accidental explosion), the engine was eventually completed in 1990. The H-2 was capable of lifting over 10,000 kg (22,046 lb) to low orbit or 3,930 kg (8,664 lbs.) to geosynchronous orbit. Between 1994 and 1999, seven launches were conducted with the H-2, of which five were successful; however, each launch cost approximately $190 million in 1997 dollars, which was far in excess of the cost of launching an Ariane. The successor to the H-2 was developed specifically with reducing launch costs in mind. That successor was the H-2A (or H-IIA), which was not only significantly more powerful [15,000 kg (33,069) to low Earth orbit, 6,000 kg (13,228 lb) to geosynchronous orbit), but also significantly cheaper to operate: each launch of the H-2A cost "only" $90 million—less than half the cost of the H-2 [79]. The pinnacle of the H-series rockets was the H-2B (or H-IIB), capable of carrying 8,000 kg (17,637 lbs.) to geosynchronous orbit and having the honor of being the first Japanese launch vehicle to send supplies to the International Space Station [80].

Japan has also developed a series of rockets that, although smaller than the H2B, are no less impressive in their ingenuity. Among these are the J-1 sounding rocket, created as a way to study ways to reduce development costs (sadly, it only flew once, and not in an orbital configuration); the S-series of rockets, which in 2018 became the smallest and lightest launch vehicle to ever send a payload to orbit (the TRICOM-1R cubesat) [81]; and the Epsilon family of rockets, developed to launch up to 590 kg (1,301 lbs.) of scientific payload into heliosynchronous orbit for only $38 million per launch, or roughly half the cost of its predecessor [82].

2.8 Commercial Launch Vehicles: The Future?

The development of launch vehicles did not simply cease when the Cold War ended; if anything, their development has only accelerated in a world where threats have become decentralized, technological advancement has miniaturized ever-more-powerful sensing and computing capabilities, and economies of scale have increased the popularity of once-rare developments (e.g., composite structures, adaptive control systems, etc.) while simultaneously driving down their cost. Many nations, in fact, simply began their exploration of commercial space by launching commercial payloads on military boosters with varying levels of modification; other entities, such as Space Exploration Technologies (SpaceX), began building new launch vehicles from what were effectively clean-sheet designs. Even companies such as Blue Origin, XCOR, Masten Space Systems, Armadillo Aerospace, Virgin Galactic, and others (most of which can be considered "micro-entities" compared to the heavily funded Cold War programs) have benefitted from past advances and found market niches for their services. A (very) brief summary of commercial launch vehicles' capabilities is tabulated in Table 2.6, using information and specifications compiled from a variety of sources. Note that this list is by no means exhaustive, because properly comparing commercial launch vehicle capabilities is an endeavor that would likely require an entire book. Note that Table 2.6 excludes launch vehicles that have been retired or decommissioned.

It should become immediately apparent that the payload capabilities of commercial launch vehicles vary widely, both among themselves and relative to the military rockets that came before. On the surface, this shouldn't come as a surprise, because most of the IRBMs and ICBMs were developed to perform missions that were very specific and somewhat restricted in scope: namely, to deliver nuclear warheads to enemy targets in a reliable and expedient fashion. The various rocket parameters (e.g., basing, propellant type, staging, etc.) may have varied, but the tasks that the vehicles needed to perform (i.e., the operational requirements) were relatively constant. Because the Outer Space Treaty of 1967 forbade the use of space (or celestial bodies, such as the moon) as a staging area for weapons of mass destruction—including nuclear weapons—many of the performance parameters (e.g., translunar injection and planetary escape trajectories) that might be part of a modern mission would not have been possible using unmodified IRBM/ICBM launch vehicles.

2.9 Small Launch Vehicles

One of the most interesting developments of the past few decades has been the growing capability of "smallsats," also known as microsats, nanosats, picosats, and the like. The continuing evolution of digital electronics and increasing amount of computing power available from smaller and smaller form factors have created an entirely new application for launch vehicles unlike anything that has come before. The true offspring of rocket science and the digital age, smallsats have created demand for a whole new class of launch vehicles. Although many smallsat developers are content to piggyback onto existing launches (for a fraction of the cost of having a launch all to themselves), several launch providers have sprung up in recent years under the belief that when it comes to launches, "now" is better than "later." As can be imagined, large problems become no less challenging at a smaller scale, and would-be small-launch developers are facing many of the same problems as those suffered by the rocketry pioneers. Still, at the time of this writing, Rocket Lab's Electron vehicle has demonstrated orbital capability, while several companies, including Astra Space,

Table 2.6   Comparison of Commercial Launch Vehicle Families

Family	Country	Manufacturer	LEO Payload, kg
Angara	Russia	Khrunichev	14,600–35,000
Antares	U.S.A.	Northrop Grumman	6,000
Ariane 5	E.U.	Airbus	21,000
Athena	U.S.A.	Lockheed Martin	2,065
Atlas V	U.S.A.	ULA	18,850
Delta IV	U.S.A.	ULA	23,040
Dnepr	Russia/Ukraine	Yuzhmash	3,600
Electron	New Zealand	Rocket Lab	225
Epsilon	Japan	IHI Corp.	1,200
Falcon 1	U.S.A.	SpaceX	420
Falcon 9	U.S.A.	SpaceX	13,150
GSLV	India	ISRO	5,000–8,000
H-II	Japan	Mitsubishi	19,000
Haas	Romania	ARCA	400
Long March	China	CALT	1,500–20,000
Minotaur I	U.S.A.	Northrop Grumman	580
Minotaur IV & V	U.S.A.	Northrop Grumman	1,735
Naro	South Korea	Khrunichev KARI	100
Pegasus	U.S.A.	Northrop Grumman	450
UR-500 Proton	Russia	Khrunichev	23,000
PSLV	India	ISRO	3,800
UR-100N Rokot Strela	Russia	Eurokot/Khrunichev	2,100
Safir	Iran	ISA	50
Shavit	Israel	IAI	225
R-29 Shtil Volna	Russia	Makeyev	430
R-7 Semyorka Soyuz	Russia	RSC Energia/ TsSKB-Progress	5,500
Simorgh	Iran	ISA	100
Start-1	Russia	MITT	532
Taurus	U.S.A.	Northrop Grumman	1,450
R-36 Tsyklon	Russia/Ukraine	Yuzhmash	4,100
Unified Launch Vehicle	India	ISRO	15,000
Unha	North Korea	KCST	100
Vega	E.U.	ESA ASI Avio	2,300
VLS-1	Brazil	CTA	380
Zenit	Russia/Ukraine	Yuzhnoye	13,740

Virgin Orbit, and Relativity (Vector bankrupt Dec. 2019) [83], remain tantalizingly close to achieving orbital capability. Some companies are even turning to additive manufacturing [84] to reduce the cost and development cycle of flight-critical components. Overseas, several smallsat launchers are also being developed, including the Miura 1 and Miura 5 [85] from PLD Space, to service the sub-100-kg (220-lb) market. To date, only the Rocket Lab small launch company has have attained the widespread success and notoriety of the larger rocket developers, but many seem determined to combine decades of historical best practices with modern, venture-backed entrepreneurial gusto to succeed regardless. An on-line compendium of small launch vehicles may be found at http://www.b14643.de/Spacerock ets_3/index.htm.

References

[1] Baker, D., *The Rocket: The History and Development of Rocket and Missile Technology*, Crown, New York, 1978.

[2] Gruntman, M., *Blazing The Trail: The Early History Of Spacecraft And Rocketry*, AIAA, Reston, VA, 2004.

[3] National Museum of the U.S. Air Force, "Dr. Robert H. Goddard," http://www.natio nalmuseum.af.mil [retrieved 15 Sept. 2015].

[4] Milton, L., *Robert H. Goddard*, Da Capo Press, New York, 1988.

[5] King, B., and Kutta, T. J., *Impact: The History of Germany's V-Weapons in World War II*, Sarpedon, New York, 1998.

[6] Neufeld, M. J., *The Rocket and the Reich: Peenemünde and the Coming of the Ballistic Missile Era*, The Free Press, New York, 1995.

[7] The History Channel, *V-2 Factory*, Nordhausen 070723.

[8] Kennedy, G. P., *Vengeance Weapon 2: The V-2 Guided Missile*, Smithsonian Institution Press, Washington, DC, 1983.

[9] Irving, D., *The Mare's Nest*, William Kimber and Co., London, 1964.

[10] Béon, Y., *Planet Dora: A Memoir of the Holocaust and the Birth of the Space Age*, Harper Collins, New York, 1997.

[11] Escher, W. J. D., and Foster, R. W., "A Sequence Diagram Analysis of the Vanguard Satellite Launching Vehicle," NASA TN D-782, 1961.

[12] Green, C. M., and Lomask, M., "Vanguard—A History," NASA SP-4202, 1970.

[13] White, J. T., "Prelude to Orbit," 29 Sept. 2011, http://www.whiteeagleaerospace. com/prelude-to-orbit/

[14] Forsyth, K. S., "History of the Delta Launch Vehicle," http://kevinforsyth.net/delta/ backgrnd.htm [retrieved 18 Sept. 2015].

[15] Space Launch Report, "Thunder God: U.S. Space Launch Workhorse," http://www. spacelaunchreport.com/thorh.html, [retrieved 18 Sept. 2015].

[16] Lethbridge, C., "Atlas D Fact Sheet," http://www.spaceline.org/rocketsum/atlas-d. html [retrieved 20 Sept. 2015].

[17] Walker, C., *Atlas: The Ultimate Weapon by Those Who Built It*, Collector's Guide Publishing, Ontario, Canada, 2005.

[18] Bott, M., "Unique and Complementary Characteristics of the U.S. ICBM and SLBM Weapon Systems," Center for Strategic and International Studies, 9 Sept. 2009.

[19] Yengst, W., *Lightning Bolts: First Maneuvering Reentry Vehicles*, Tate, Mustang, OK, 2010.

[20] MacKenzie, D., *Inventing Accuracy: A Historical Sociology of Missile Guidance*, MIT Press, 1993, Cambridge, MA.

[21] Maugh, T. H., II, "Edward N. Hall, 91; Rocket Pioneer Seen as the Father of Minuteman ICBM," *Los Angeles Times*, 16 Jan. 2006.

[22] Isaacson, W., *The Innovators: How a Group of Inventors, Hackers, Geniuses, and Geeks Created the Digital Revolution*, Simon & Schuster, New York, 2014.

[23] Pomeroy, S., "Echos That Never Were: American Mobile Intercontinental Ballistic Missiles, 1956-1983," U.S. Air Force, Aug. 2006.

[24] McMurran, M. W., *Achieving Accuracy: A Legacy of Computers and Missiles*, Xlibris, Bloomington, IN, 2008.

[25] Hansen, C., *The Swords of Armageddon*, Chukelea, Sunnyvale, CA, 2007.

[26] Hansen, C., *U.S. Nuclear Weapons: The Secret History*, Crown, New York, 1988.

[27] Orbital Sciences Corporation, "Minotaur Fact Sheet," 2012.

[28] Orbital Sciences Corporation, "Minotaur User's Guide," 2002.

[29] Orbital Sciences Corporation, "Minotaur II Fact Sheet," 2006.

[30] Orbital ATK, "Minotaur IV Fact Sheet," 2015.

[31] Orbital ATK, "Minotaur V Fact Sheet," 2015.

[32] Graham, W., "Orbital's *Minotaur V* Launches LADEE Mission to the Moon," https://www.nasaspaceflight.com/2013/09/orbitals-minotaur-v-launch-ladee-mission-moon/ [retrieved 8 Sept. 2013].

[33] Clark, S., "Taurus Rocket on the Market with New Name, Upgrades," Spaceflight Now, 24 Feb. 2014, https://www.spaceflightnow.com/news/n1402/24minotaurc/ [retrieved 22 Sept. 2015].

[34] CBS News, "NASA Launch Mishap: Satellite Crashes into Ocean," 4 March 2011, http://www.cbsnews.com/news/nasa-launch-mishap-satellite-crashes-into-ocean/ [retrieved 22 Sept. 2015].

[35] Grimwood, J. M., and Strowd, F., "History of the Jupiter Missile System," U.S. Army Ordnance Missile Command, 1962, http://heroicrelics.org/info/jupiter/jupiter-hist/History%20of%20the%20Jupiter%20Missile%20System.pdf

[36] Graham, S., *From Polaris to Trident: The Development of U.S. Fleet Ballistic Missile Technology*, Cambridge University Press, Cambridge, UK, 1994.

[37] Popejoy, M., "USS Alabama Offloads Last of C4 Trident Missiles," U.S. Navy, 5 Nov. 2005, https://www.navy.mil/submit/display.asp?story_id=20913 [retrieved 22 Sept. 2015].

[38] Lockheed Martin Space Systems, "Navy Awards Lockheed Martin $248 Million Contract for Trident II D5 Missile Production and D5 Service Life Extension," Lockheed Martin press release, 29 Jan. 2002.

[39] U.S. Navy, "Back to the Future with Trident Life Extension," *Undersea Warfare Magazine*, Spring 2012.

[40] National Priorities Project, "Analysis of the Fiscal Year 2012 Pentagon Spending Request," *Cost of War*, 15 Feb. 2011.

[41] Hansen, J. R., "Spaceflight Revolution: NASA Langley Research Center from Sputnik to Apollo," NASA SP-4308, 1995.

[42] National Aeronautics and Space Administration, "Scout Fact Sheet," NASA Langley Research Center, http://www.nasa.gov/centers/langley/news/factsheets/Scout.html [retrieved 22 Sept. 2015].

[43] Savage, D., "Scout Launch Vehicle to Retire After 34 Years of Service," NASA press release, 6 May 1994, http://www.nasa.gov/home/hqnews/1994/94-072.txt [retrieved 22 Sept. 2015].

[44] Brown, S., "Winging It Into Space," *Popular Science*, May 1989.

[45] National Aeronautics and Space Administration, "NASA Awards Launch Services Contract for Ionospheric Connection Explorer," NASA Contract Release C14-047, 20 Nov. NASA SP-350, 2014.

[46] Cortright, E. M., "Apollo Expeditions to the Moon," NASA Langley Research Center, 1975, https://www.hq.nasa.gov/office/pao/History/SP-350/cover.html

[47] Bilstein, R. E., "Stages to Saturn: A Technological History of the Apollo/Saturn Launch Vehicles," NASA SP-4206, 1980.

[48] Zak, A., "Disaster at Xichang," *Air & Space Magazine*, 2013, https://www.airspacemag.com/history-of-flight/disaster-at-xichang-2873673/

[49] National Aeronautics and Space Administration, *Biographies of Aerospace Officials and Policymakers, T-Z*, 6 Feb. 2013.

[50] Wines, M., "Qian Xuesen, Father of China's Space Program, Dies at 98," *The New York Times*, 3 Nov. 2009.

[51] Stevenson, R. W., "Shaky Start for Rocket Business," *The New York Times*, 16 Sept. 1988.

[52] Zinger, K. J., "An Overreaction that Destroyed an Industry: The Past, Present, and Future of U.S. Satellite Export Controls," *Colorado Law Review*, 13 July 2015.

[53] Lan, C., "Mist Around the CZ-3B Disaster," *The Space Review*, 1 July 2013.

[54] Crock, S., and Brull, S., "Hughes and Loral: Too Eager to Help China?," *Business Week*, 13 Sept. 1999.

[55] Malik, T., "NASA's Space Shuttle by the Numbers: 30 Years of a Spaceflight Icon," *space.com*, 21 July 2011.

[56] National Aeronautics and Space Administration, "Report of the Space Task Group," https://www.hq.nasa.gov/office/pao/History/taskgrp.html [retrieved 29 Sept. 2015].

[57] "Russia Starts Ambitious Super-Heavy Space Rocket Project," *Space Daily*, 19 Nov. 2013, https://www.rt.com/news/russiabooster-rocket-energia-817/

[58] Weiss, G. W., "The Farewell Dossier: Duping the Soviets—A Deception Operation," *Studies in Intelligence*, Vol. 39, No. 5, 1996.

[59] Lukashevich, V., *Joint Propulsion System*, 21 Nov. 2013.

[60] Chaffee, N., "Space Shuttle Technical Conference," NASA CP2342, 1985.

[61] Jenkins, D. R., *Space Shuttle: The History of the National Space Transportation System: The First 100 Missions*, Voyageur Press, Stillwater, MN, 2001.

[62] Stillwell, W. H., "X-15 Research Results: With a Selected Bibliography," NASA SP-60, 1965.

[63] Jarvis, C. R., and Lock, W. P., "Operational Experience with the X-15 Reaction Control and Reaction Augmentation Systems," NASA TN D-2864, 1965.

[64] Raveling, P., "X-15 Pilot Report, Part 1: X-15 General Description & Walkaround," , https://www.sierrafoot.org/x-15/pirep1.html [retrieved 30 Sept. 2015].

[65] Butrica, A. J., "La Force Mortice of Reusable Launcher Development: The Rise and Fall of the SDIO's SSTO Program, from the X-Rocket to the Delta Clipper," https://www.hq.nasa.gov/office/pao/History/x-33/nasm.htm [retrieved 30 Sept. 2015].

[66] Bergen, C., "X-33/VentureStar: What Really Happened," http://www.nasaspaceflight.com/2006/01/x-33venturestar-what-really-happened/ [retrieved 30 Sept. 2015].

[67] Black, S., "An Update on Composite Tanks for Cryogens," http://www.compositesworld.com/articles/an-update-on-composite-tanks-for-cryogens [retrieved 30 Sept. 2015].

[68] Hill, C. N., *A Vertical Empire: The History of the UK Rocket and Space Programme, 1950–1971*, World Scientific, Godalming, UK. 2001.

[69] What-When-How, "Ariane Rocket Program," http://what-whenhow.com/space-science-and-technology/ariane-rocket-program/

[70] BBC News, "European Rocket Powers to Record," 4 May 2007.

[71] Guteri, F., "Space Junk: Earth Is Being Engulfed in a Dense Cloud of Hazardous Debris That Won't Stop Growing," *Newsweek*, 1 Aug. 2009.

[72] Amos, J., "Full Thrust on Europe's New Ariane 6 Rocket," *BBC News*, 22 June 2017.

[73] Indian Space Research Organization, *Milestones*, 2007.

[74] Raj, N. G., "The Long Road to Cryogenic Technology," *The Hindu*, 12 Nov. 2016.

[75] Singh, S., "GSLV Mk. III Breaks ISRO's Jinx of Failure in Debut Rocket Launches," *Times of India*, 5 June 2017.

[76] McDowell, J., "Lambda," *Orbital and Suborbital Launch Database*, 8 May 2009.

[77] Japan Aerospace Exploration Agency, *JAXA Digital Archives*, 9 Sept. 2009.

[78] Pearlman, R., "Space Station Command Change is One Giant Leap (Day) for Space History," https://news.yahoo.com/space-station-command-change-one-giant-leap-day-213819429.html [retrieved 2 Nov. 2017].

[79] U.S. Government Accountability Office, "Surplus Missile Motors: Sale Price Drives Potential Effects on DOD and Commercial Launch Providers," GAO-17-609, 16 Aug. 2017.

[80] Amos, J., "Japan's Space Freighter in Orbit," *BBC News*, 10 Aug. 2009.

[81] Graham, W., "Japanese Sounding Rocket Claims Record-Breaking Orbital Launch," *NASA Spaceflight*, 2 Feb. 2018, https://www.nasaspaceflight.com/2018/02/japanese-rocket-record-borbital-launch/

[82] Clark, S., "Japan's 'Affordable' Epsilon Rocket Triumphs on First Flight," Spaceflight Now, 16 Sept. 2013, https://spaceflightnow.com/epsilon/sprinta/130914launch/

[83] Foust, J., "For the Small Launch Industry, Just Wait Until Next Year," *The Space Review*, 10 Dec. 2018, https://thespacereview.com/article/3620/1

[84] Osborne, T., "Skyrora Smallsat Launcher to Use 3D-Printed Engines," *Aerospace Daily*, 30 Oct. 2018.

[85] European Space Agency, "*Winning Ideas for New Space Transport Services*," 18 Dec. 2018.

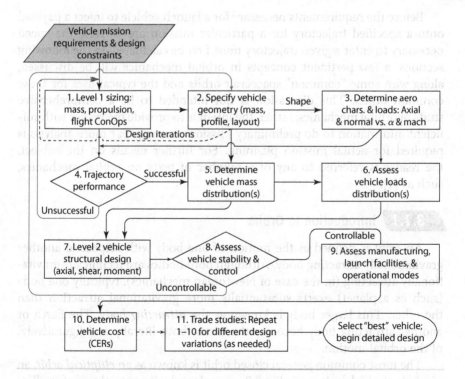

Chapter 3   Missions, Orbits, and Energy Requirements

W e begin the design process for our space launch vehicle by determining the technical requirements it must meet to carry out its mission, as well as any constraints that can affect the mission's execution. This is indicated by the highlighted "signpost" on the upper left of the schematic of the design process shown in Fig. 3.1. In subsequent chapters, we will carry out many of the design tasks as indicated by the numbered shapes connected by lines in the schematic.

The requirements imposed by the payload, the customer of the launch vehicle, directly affect its performance. This chapter will cover the methods used to quantify the launch vehicle's performance based on the requirements

Vehicle mission requirements & design constraints

1. Level 1 sizing: mass, propulsion, flight ConOps

2. Specify vehicle geometry (mass, profile, layout)

Shape

3. Determine aero chars. & loads: Axial & normal vs. α & mach

Design iterations

4. Trajectory performance

Successful

5. Determine vehicle mass distribution(s)

6. Assess vehicle loads distribution(s)

Unsuccessful

7. Level 2 vehicle structural design (axial, shear, moment)

8. Assess vehicle stability & control

Controllable

9. Assess manufacturing, launch facilities, & operational modes

Uncontrollable

10. Determine vehicle cost (CERs)

11. Trade studies: Repeat 1–10 for different design variations (as needed)

Select "best" vehicle; begin detailed design

Fig. 3.1 We begin the design process by defining the requirements that the launch vehicle must meet to accomplish its mission, which is indicated by the shaded box at the top left.

it must meet. Once the requirements are understood, we will examine rocket propulsion in Chapter 4, as propulsion dictates vehicle performance. Initial sizing or Item 1 in the flowchart, the next step the design process, will be covered in Chapter 5.

A launch vehicle's reason for being is the payload, which is a satellite or spacecraft whose mission is to be accomplished after being placed into its required orbit or trajectory. This is specified by

* Speed
* Direction
* Altitude

These are the parameters the spacecraft must attain at burnout or engine shutdown. The amount of energy the launch vehicle must provide is based on these required values. As one might expect, the energy the launch vehicle delivers is strongly dependent on the mission profile and the mass of the spacecraft. This chapter discusses how the basic requirements on a launch vehicle flow down from the mission requirements of the payload.

3.1   Orbits, Orbital Parameters, and Trajectories

Before the requirements necessary for a launch vehicle to inject a payload onto a specified trajectory for a particular mission are derived, the speed necessary to enter a given trajectory must first be calculated. In the following sections, a few pertinent concepts in orbital mechanics will be discussed, along with some "common" spacecraft orbits and the typical uses for these common orbits. This discussion is not intended to be a comprehensive study of orbital mechanics; rather, the intent is to provide the reader with sufficient information to do preliminary mission planning. Far more analysis is required for actual mission planning. For further details on the subject, the reader is referred to any of a number of texts on orbital mechanics, such as [1].

3.1.1   Introduction to Orbits

An orbit is defined as the motion of one body with respect to another gravitationally attracting body. Although both bodies are technically gravitationally attracting (in the case of Newtonian mechanics), typically one body (such as a planet) exerts substantially more gravitational attraction than the other. This larger body is known as the *attracting body*. The Earth or another attracting body becomes the focus (both literally and figuratively) of the orbital motion.

The most common general closed orbit is known as an *elliptical orbit*, an example of which is shown in Fig. 3.2; a circular orbit is a special case of an elliptical orbit and will be discussed later in this chapter. As illustrated in Fig. 3.2,

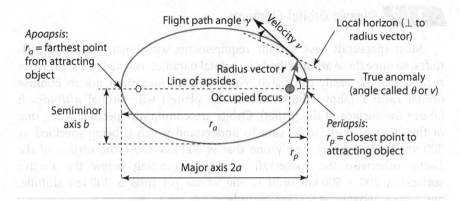

Fig. 3.2 Elements of a typical elliptical orbit.

the attracting object (Earth, sun, etc.) occupies the right-hand focus of the ellipse; the other focus is vacant. For now, the orbit is assumed to occur around a spherical planet with radius R. For Earth, $R = R_e = 6{,}378$ km.

Figure 3.2 shows the major parameters that define the geometry of an orbit, many of which are terms derived from Latin. These parameters are used to define the requirements that a launch vehicle must meet in order to perform a given mission. The orbital parameters are:

- *Apoapsis radius r_a*: The distance from the *center* of the attracting object to the *farthest* point in the orbit. (When orbiting Earth, apoapsis is often referred to as apogee.)
- *Periapsis radius r_p*: The distance from the *center* of the attracting object to the *closest* point in the orbit. (When orbiting Earth, periapsis is often referred to as perigee.)
- *Line of apsides*: An imaginary line connecting the periapsis and apoapsis.
- *Periapsis altitude $h_p = r_p - R$*: The distance from the *surface* of the planet to the periapsis.
- *Apoapsis altitude $h_a = r_a - R$*: The distance from the *surface* of the planet to the apoapsis.
- *Semimajor axis $a = (r_p + r_a)/2$*: A measure of the overall size of an orbit. (For circular orbits, the semimajor axis is equal to the radius.)
- *Eccentricity $e = (r_a - r_p)/(r_a + r_p)$*: An indication of how circular the shape of an orbit happens to be; $e = 0$ for a circular orbit, and $0 < e < 1$ for an elliptical orbit. The higher the eccentricity, the more "squashed" the orbit becomes.

Note that the terms *periapsis* and *apoapsis* refer to an orbit around any attracting body; these terms become *perigee* and *apogee* when an orbit around the Earth is described, and *perihelion* and *apohelion* when an orbit around the sun is described.

3.1.2 Classic Orbital Elements

Most spacecraft have specific requirements with regard to orbital altitudes, so once these are specified, the orbital parameters listed in the previous section may be easily calculated. Note that it's important not to confuse orbital *radii r* (above the center of the planet) with orbital *altitudes h* (above the surface of the planet). Orbits are commonly specified using one or the other or both, so be sure to understand which is being specified. A 300-km Earth orbit is clearly one that is 300 km *above the surface of the Earth*—otherwise the spacecraft would be orbiting below the Earth's surface! A 200 × 500 km orbit is one whose periapsis is 200 km altitude, and whose apoapsis is 500 km altitude.

When a vehicle is in an elliptical orbit, it climbs with a positive flight path angle γ between periapsis and apoapsis, and descends with a negative flight path angle between apoapsis and periapsis. The only locations in the orbit where the flight path angle is zero (no climbing or diving) is at periapsis and apoapsis. For the rest of this chapter, it will be assumed that the vehicle in question is orbiting Earth. Of course, the principles of Earth orbits are applicable to other celestial bodies, so long as the proper parameters are utilized for the body being investigated.

In general, many desired orbits are *tilted* relative to the equatorial plane. The angle of tilt of the orbit relative to the equatorial plane is known as its inclination *i*. Inclination angles can vary from 0 degrees (known as equatorial orbits) to 90 degrees (known as polar orbits) up to 180 degrees. Four orbital categories, along with their corresponding inclinations, are shown in Fig. 3.3. Inclination values between 0 and 90 degrees are known as *prograde*, because they orbit the Earth in the same direction as the Earth rotates. Inclination angles between 90 and 180 degrees are known as *retrograde* because they orbit the Earth in the opposite direction that the Earth rotates. *Polar orbits* with *i* = 90 degrees are the dividing line between the two: neither prograde nor retrograde.

A general orbit must be oriented in space with respect to a coordinate system, as shown in Fig. 3.4. The orbit itself is shown as the light-blue ellipse. Notice that the elliptical orbit is tilted at an angle *i* with respect to the equatorial plane the darker-blue rectangle shape. This causes the angular momentum vector of the orbit **h** to be tilted the same angle *i* from the North Pole *Z*.

Once the orbit is allowed to tilt, two additional parameters are needed to describe the orientation of the orbit relative to the equatorial plane. The first of these is called the *right ascension of the ascending node (RAAN)*, and is the angle Ω measured from the vernal equinox line ♈ to the location where the orbit passes through the equatorial plane heading north, called the *line of nodes*. "Right ascension" therefore has the same meaning as "longitude." A second parameter, ω, is called the *argument of the periapsis* and indicates the clocking angle from the line of nodes to where the closest point in the orbit is to the planet—in other words, the "rotation" of the ellipse in its own orbital plane.

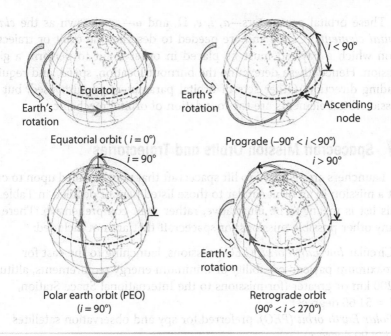

Fig. 3.3 Four categories of orbits based on their inclinations. Prograde orbits rotate in the same direction as the Earth's rotation, while retrograde orbits rotate opposite the Earth. Polar orbits form the dividing line between the two inclinations.

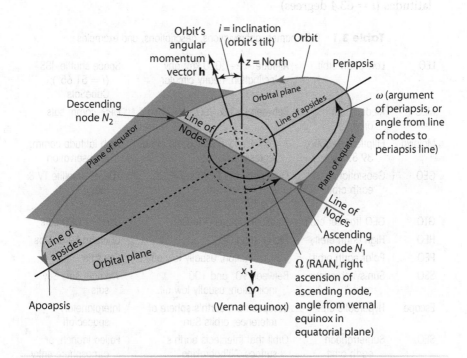

Fig. 3.4 Diagram of an elliptical orbit.

These orbital parameters—*a*, *i*, *e*, Ω, and ω—are known as the *classic orbital elements (COEs)* and are needed to describe the orbit or trajectory upon which a payload must be placed in order for it to perform a given mission. Hence, these determine the burnout location, speed, and required heading direction. Other sets of orbital parameters are also used, but the classic orbital elements are the foundation of orbital mechanics.

3.2 Spacecraft Mission Orbits and Trajectories

Launchers are required to lift spacecraft that may be called upon to carry out a mission in an orbit similar to those listed here or detailed in Table 3.1. This list is meant to be illustrative, rather than comprehensive. There are many other possible missions for spacecraft that are not included:

* Circular *low-Earth orbit (LEO)* missions, launching to the east for maximum payload capability or minimum energy requirements, altitude 200 km or greater (for missions to the International Space Station, $i = 51.65$ degrees)
* *Polar Earth orbit (PEO)*, preferred for spy and observation satellites ($i \cong 90$ degrees)
* *Sun-synchronous orbit (SSO)*, preferred for weather and observation satellites ($i \cong 95$ degrees)
* *Molniya orbit*, used to communicate with and observe areas at high latitudes ($i = 63.4$ degrees)

Table 3.1 Orbit names, abbreviations, descriptions, and examples

LEO	Low earth orbit	Altitude < ~1,000 km; any inclination, nearly circular	Space shuttle, ISS ($i = 51.65°$), CubeSats
MEO	Medium earth orbit	Between LEO & GEO	GPS, comm sats
Molniya	Elliptical, ~500–39,873 km	$i = 63.4°$, 12 h period, no nodal regression	High latitude comm. & observation
GEO	Geostationary earth orbit	Orbit period = Earth's, $i = 0°$, $R = 42,164$ km, $h = 35,786$ km	TDRSS, satellite TV & radio
GTO	GEO transfer orbit	Elliptical orbit: LEO to GEO	Used by GEO sats
HEO	High earth orbit	Higher than GEO	Chandra X-Rax Obs
PEO	Polar earth orbit	90° inclination, usually low alt	Spy sats
SSO	Sun-synchronous orbit	Between 90° and 100° inclination, usually low alt.	Weather sats Spy sats
Escape	Hyperbolic orbit	Escapes from earth's sphere of reference, orbits Sun	Interplanetary spacecraft
SEO	Subterranean earth orbit	Orbit that intersects earth's surface, "lithobraking"	Failed launch, or atmosphere entry

- *Geostationary Earth orbit (GEO)*, preferred location for communication satellites ($i = 0°$)
- *Geostationary transfer orbit (GTO)*, used by communication satellites to transfer from LEO to GEO (typically $i = \lambda$, or the orbit is inclined with the same angle as the launch site latitude)
- *Escape orbit*, in order to leave the gravitational pull of Earth and transfer to another planet

Note that the last orbit given in Table 3.1 is not an "official" designation.

3.2.1 Required Orbital Injection Speed

Once the mission or orbit is specified, it is necessary to calculate the required orbital speed, which may be calculated easily if the orbit parameters are known. This will be demonstrated for circular, elliptical, and escape orbits. Note that these sections are deliberately concise; for more information on orbital mechanics calculations, please refer to the standard texts on spacecraft design, such as [1] or [2].

3.2.1.1 Circular Orbit Speed

For a circular orbit, the required speed may be calculated as follows:

$$v = \sqrt{\frac{g_0 R^2}{r}} \tag{3.1}$$

where g_0 is the gravitational acceleration at the planet's surface, R is the planet radius, and r is the orbital radius. For Earth, sea level $g_0 = 9.80665 \text{ m/s}^2$, $R = 6{,}378$ km, and $g_0 R^2 = 398{,}924 \text{ km}^3/\text{s}^2$. The term $\mu = g_0 R^2$ is sometimes used as an abbreviation. Note that the circular orbit radius r is equal to the sum of the planet radius R and the orbital height h

$$r = R + h \tag{3.2}$$

Example 3.1

Determine v_{ISS}, the circular orbit speed of the International Space Station, at an altitude of 407 km above the surface of the Earth.

$$v_{ISS} = \sqrt{\frac{g_0 R^2}{r}}$$

$$v_{ISS} = \sqrt{\frac{398{,}924 \text{ km}^3/\text{s}^2}{(6{,}378 + 407) \text{ km}}}$$

$$v_{ISS} = 7.668 \text{ km/s}$$

3.2.1.2 Elliptical Orbit Speed

For elliptical orbits whose semimajor axis is a, the required speed at radius r at the desired injection point of the orbit may be calculated by using the *vis viva* relation

$$\frac{v^2}{2} - \frac{g_0 R^2}{r} = -\frac{g_0 R^2}{2a} \tag{3.3}$$

Rearranging and substituting for the velocity,

$$v = \sqrt{g_0 R^2 \left(\frac{2}{r} - \frac{1}{a}\right)} \tag{3.4}$$

It may be noted that for circular orbits, the semimajor axis a is identical to the radius r, and Eq. (3.4) reduces to Eq. (3.1).

Low Earth orbits (LEOs) may be reached by a direct ascent. Medium

---

Example 3.2

Determine $v_{molniya}$, the orbital speed required for a Molniya satellite at periapsis altitude of 500 km, orbiting the Earth. The semimajor axis is 26,562 km.

$$v = \sqrt{g_0 R^2 \left(\frac{2}{r} - \frac{1}{a}\right)}$$

$$v = \sqrt{(398{,}924\ \mathrm{km^3/s^2})\left(\frac{2}{(6{,}378 + 500)\ \mathrm{km}} - \frac{1}{26{,}562\ \mathrm{km}}\right)}$$

$$v = 10.05\ \mathrm{km/s}$$

Notice the larger, higher-energy Molniya orbit requires a significantly larger speed at periapsis than a circular orbit with the same radius.

Earth orbits (MEOs) are most efficiently reached by a powered ascent to approximate LEO altitude, followed by a coast period and a circularization burn. Both of these concepts are shown in Fig. 3.5. Ordway et al. [3] provide a convenient formula for the required burnout speed for a coasting transfer into a general ascending transfer orbit, as shown in Fig. 3.6. For the orbit shown, the required injection speed is

$$v_1 = \sqrt{\frac{g_0 R^2}{r_1}\left[\frac{2(r_2 - r_1)r_2 \sin^2 \theta_2}{r_2^2 \sin^2 \theta_2 - r_1^2 \sin^2 \theta_1}\right]} \tag{3.5}$$

For a Hohmann transfer orbit (both injection and final orbit velocity

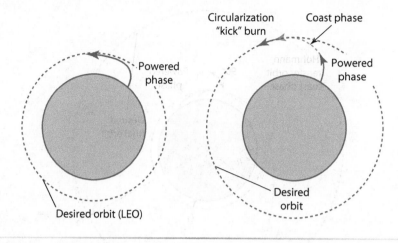

Fig. 3.5 LEOs and MEOs are reached by direct ascent or by a powered phase shutting down at LEO altitude and coasting up to the MEO altitude, where a circularization burn is performed.

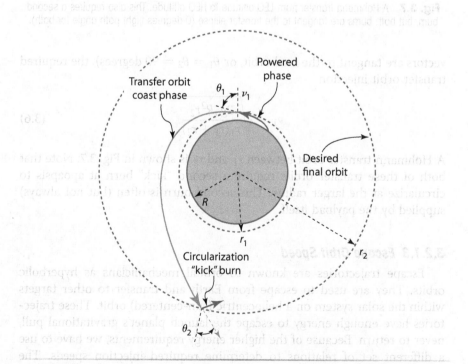

Fig. 3.6 Higher orbits are reached using by a powered phase shutting down at LEO altitude and coasting up to the MEO altitude, where a circularization burn is performed. This figure depicts a general transfer orbit, whose starting and ending angles θ_1 and θ_2 are not 90°.

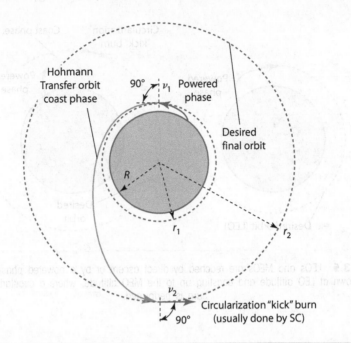

Fig. 3.7 A Hohmann transfer from LEO altitude to HEO altitude. This also requires a second burn, but both burns are tangent to the transfer ellipse (0-degrees flight path angle for both).

vectors are tangent to the final orbit, or $\theta_1 = \theta_2 = 90$ degrees), the required transfer orbit injection

$$v_1 = \sqrt{\frac{2g_0 R^2 r_2}{r_1(r_1 + r_2)}} \qquad (3.6)$$

A Hohmann transfer orbit between r_1 and r_2 is shown in Fig. 3.7. Note that both of these transfer orbits require a second "kick" burn at apoapsis to circularize at the larger radius. This second burn is often (but not always) supplied by the payload itself.

3.2.1.3 Escape Orbit Speed

Escape trajectories are known to orbital mechanicians as hyperbolic orbits. They are used to escape from Earth and transfer to other targets within the solar system on a heliocentric (sun-centered) orbit. These trajectories have enough energy to escape the launch planet's gravitational pull, never to return. Because of the higher energy requirements, we have to use a different set of relations to determine required injection speeds. The speed is determined by the orbital radius (planet radius + orbital altitude) and the desired hyperbolic escape speed (v_{HE}) that needs to be delivered. The escape speed is determined by the transfer orbit around the sun.

CHAPTER 3 Missions, Orbits, and Energy Requirements **141**

The required injection speed at periapse radius r_p may be calculated as

$$v_p = \sqrt{\frac{2g_0 R^2}{r_p} + v_{HE}^2} \tag{3.7}$$

The minimum escape speed is the speed that allows a spacecraft to escape an orbit with zero speed at infinite distance, or $v_{HE} = 0$. In this case, the required injection speed at periapse radius r_p is calculated as

$$v_p = \sqrt{\frac{2g_0 R^2}{r_p}} \tag{3.8}$$

Trajectory planners often use a parameter called C3, which is simply the square of the speed far from the planet, v_∞, or $C3 = v_\infty^2$. C3 is twice the specific kinetic energy of the vehicle at escape. The effects of C3 on the payload a launch vehicle can carry on a specific mission is discussed later in this chapter.

Example 3.3

Determine v_{escape}, the speed required for a spacecraft at periapsis altitude of 200 km to escape the Earth with a v_{HE} of 2.94 km/s. What is the C3?

$$v_p = \sqrt{\frac{2g_0 R^2}{r_p} + v_{HE}}$$

$$v_p = \sqrt{\frac{2(398,924\,\text{km}^3/\text{s}^2)}{(6,378 + 200)\,\text{km}} + (2.94\,\text{km/s})^2}$$

$$v_p = 11.40\,\text{km/s}$$

$$C3 = (2.94\,\text{km/s})^2 = 8.64\,\text{km}^2/\text{s}^2$$

Note that the v_{escape} speed is significantly higher (46.4%) than that required for a 200-km circular orbit. Note also that the change in speed is:

$$v_{escape} - v_{circular} = 11.40\,\text{km/s} - 7.79\,\text{km/s} = 3.61\,\text{km/s} \tag{3.9}$$

Why is the difference not 2.94 km/s?

The reason is that the payload has to have enough additional energy to "coast uphill" out of Earth's gravity well. This is depicted in Fig. 3.8. As shown in this figure, the launch vehicle first enters a parking orbit at an altitude of 200 km, waits until the proper time, and then performs a second burn to achieve the needed hyperbolic escape speed, shown as v_∞ (note $v_p = v_{HE} = v_\infty$). The launch vehicle could just as easily have done the

(Continued)

Example 3.3 *(Continued)*

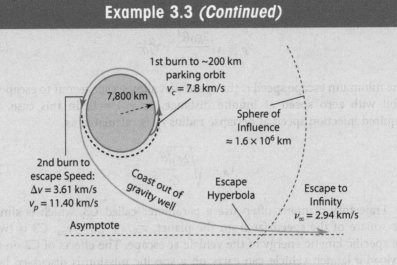

Fig. 3.8 Escape trajectory from Earth.

maneuver in a single burn, but this would have required very precise timing of the liftoff event. In either case, the escape is timed so that the spacecraft's velocity far from Earth is parallel to the Earth's speed around the sun. The interested reader is referred to Example 7.1 of [4] for a more detailed exposition.

3.3 Required Energy to Be Delivered for Orbit

As previously discussed, the launch vehicle must provide the proper speed, direction, and altitude to inject the spacecraft into its desired trajectory. These parameters are specified by the desired orbit. However, because of the losses due to overcoming gravity, aerodynamic drag, and other sources (to be discussed later in this chapter), it is convenient to lump the speed, direction, altitude, and loss values into a single "total energy" value based on a fictitious total speed Δv that the launch vehicle must deliver. This speed is fictitious because the launch vehicle will actually deliver less speed due to overcoming gravity and drag losses. It is easy to see that because of these losses, it is necessary to design considerably more capability into the launch vehicle than what would be required to simply accelerate to orbital velocity. How to estimate these losses will be covered in the following sections, and later chapters will show how to calculate them exactly.

3.3.1 Benefits from the Rotation of the Earth

The Earth rotates once per day about an axis connecting the north and south poles. This means that every point on the Earth is moving towards the east at a speed that is dependent on its distance from the axis of rotation. One can easily use the rotation rate and the equatorial radius to show that the speed of the equator is 465.1 m/s $\cong$ 1,525.9 ft/s toward the east, because the Earth rotates counterclockwise when viewed from the north pole. The speed at any particular latitude λ may be calculated as

$$v_{\text{launch site}} = v_{\text{equator}} \cos \lambda \qquad (3.10)$$

This rotational speed must be accounted for in the calculation of launch speed, because it can reduce or increase the required Δv depending on the launch location and direction. For launches towards the east, it subtracts from required Δv because it provides an initial speed before the launch vehicle begins its flight; however, the Earth's rotation *adds* to the required Δv for launches towards the west (against the Earth's rotation). Furthermore, if a launch requires an inclination near 90 deg, the eastbound speed must be counteracted because for these launches the entire speed is north–south in direction. That being said, the following general guidelines are given:

• *For launches to low orbital inclinations*: Launching at or near the equator to the east is favored. This is why the SeaLaunch and European Space Agency (ESA) located their launch sites near the equator. Launches occurring from high latitudes do not get this benefit.
• *For launches to near-polar inclinations*: It is necessary to eliminate eastward motion, which favors high-latitude launch sites such as Kodiak, Alaska.

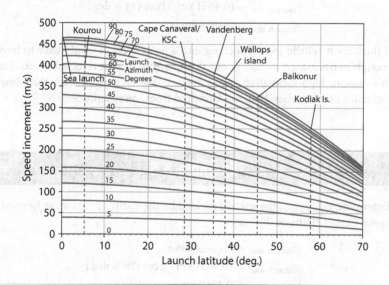

Fig. 3.9 Benefit/penalty provided by Earth rotation.

- To maximize the payload capability of the launch vehicle, the launch should be due east.
- If the launch vehicle launches due east, the inclination of the orbit achieved is equal to the latitude of the burnout location. For example, a launch from Kennedy Space Center (latitude 28.5 deg) due east will result in an orbit with an inclination $i = 28.5$ deg.

For launches, the important speed is the speed achieved at *engine shutdown*. This speed may also be referred to as the *burnout* speed. The former refers to a commanded thrust termination, while the latter is caused by propellant exhaustion. As the end result is the same, we use the terms interchangeably. The use of the word *shutdown* implies that there may be additional, unconsumed propellants available for later burns. These benefits and penalties are shown graphically in Fig. 3.9 for different launch latitudes (horizontal axis) and for varying launch directions or azimuths (the individual curved lines on the graph). A list of common launch sites and their details is provided in Chapter 1.

Example 3.4

Determine the speed that a launch vehicle ascending into a polar orbit would have to cancel out if the vehicle were launching from Vandenberg Air Force Base (California), latitude 34.4 deg.

$$v_{\text{launch site}} = v_{\text{equator}} \cos \lambda$$
$$v_{\text{launch site}} = (0.4651 \,\text{km/s}) \cos (34.4 \,\text{deg})$$
$$v_{\text{launch site}} = 0.3838 \,\text{km/s}$$

If the launch vehicle were launching into a polar orbit, it would have to have enough performance to cancel out this quantity of east-directed speed. For this reason, high-inclination orbits require higher performance than low-inclination orbits when both are launched from lower latitudes.

Example 3.5

Determine $v_{\text{launch site}}$, the speed provided by the Earth's rotation, at Kennedy Space Center (latitude 28.5 deg).

$$v_{\text{launch site}} = v_{\text{equator}} \cos \lambda$$
$$v_{\text{launch site}} = (0.4651 \,\text{km/s}) \cos (28.5 \,\text{deg})$$
$$v_{\text{launch site}} = 0.4087 \,\text{km/s}$$

3.3.2 Estimating Gravity Loss

To accurately estimate the amount of energy needed to overcome gravity during the launch vehicle's ascent, the gravitational force acting throughout the ascent must be integrated. In mathematical terms, $\int dv = \int g \sin \gamma \, dt$ must be calculated over the entire duration of the launch. In this formula, g is the (time-varying) gravitational acceleration, and γ is the (time-varying) flight path angle. Note that when $\gamma = 90$ deg, the vehicle is vertical and the thrust is doing 100% of the work against gravity; however, at this point the trajectory (viz. the flight path angle γ as a function of time) is unknown. Therefore, the losses due to gravity must be estimated. Two common methods used to do this are discussed in the following sections.

3.3.2.1 Estimating Gravity Loss: Conservation of Energy

The first method uses conservation of energy. We wish to find the speed $\Delta v_{grav\ loss}$ that is required to coast upwards from starting radius r_0 to zero speed at a final radius r_f in a vacuum. To find this speed, we equate the total energy (kinetic + potential) at start to the total energy (kinetic = 0 + potential) at finish. The kinetic energy per unit mass is $K = (\frac{1}{2}mv^2/m) = v^2/2$. The gravitational potential energy for Earth (radius R_e) at a distance r from its center is defined as $U = -g_0 R_e^2/r$. Therefore, conservation of total energy ($K_0 + U_0 = K_f + U_f$) states

$$\frac{\Delta v_{grav\ loss}^2}{2} - \frac{g_0 R_e^2}{r_0} = 0 - \frac{g_0 R_e^2}{r_f} \tag{3.11}$$

When Eq. 3.11 is solved for the gravity loss speed $v_{grav\ loss}$, we find

$$\Delta v_{grav\ loss} = \sqrt{2g_0 R_e^2 \left(\frac{1}{r_0} - \frac{1}{r_f}\right)} \tag{3.12}$$

If the vehicle is to launch to an altitude h starting from the Earth's surface at sea level (SL), then $r_0 = R_e$, and $h = r_f - R_e =$ altitude above Earth's surface, and the gravity loss expression (3.12) reduces to

$$\Delta v_{grav\ loss-SL} = \sqrt{\frac{2g_0 R_e h}{R_e + h}} = \sqrt{\frac{2g_0 h}{1 + h/R_e}} \tag{3.13}$$

The first expression, Eq. (3.12) may be used for any launches that occur somewhere other than sea level or planetary surface radius, i.e. an air launch from a flying vehicle, or a launch from an elevated area (for example, Mt. Everest). The second, Eq. (3.13), may only be used for launches originating at ground or sea-level.

This assumption of energy conservation provides a very rough approximation, because it does not take into account how *long* the ascent takes: recall that because the launch vehicle consumes propellants during its ascent, it requires less energy the higher it climbs. In addition, it assumes that the thrust is antiparallel to the gravity vector for the *entire* boost, which is clearly incorrect if the rocket is injecting into a near-circular orbit with a horizontal flight path. As an extreme example, a launch vehicle with minimal thrust would accelerate very slowly and suffer from large gravity losses compared to a "sportier" launch vehicle that accelerates much faster, but this distinction is not reflected in Eq. (3.11). It is recommended that the results of this method be multiplied by 0.80 to obtain a more realistic estimate. The second method, discussed in the next section, provides a much more accurate result.

3.3.2.2 Estimating Gravity Loss: Statistical Estimation

This method comes from several sources, such as [6], [8] and [11], and is based on numerous calculations of the performance of generic vehicles, as discussed in [8] and [9]. This method *does* take into account the duration of the ascent and the pitch-over of the thrust vector, so it would be expected to provide a better approximation of gravity losses.

To use this method, a rough idea of the actual performance of the launch vehicle is needed. This may be difficult to estimate during the initial design phase, when almost zero information about the launch vehicle is available (and it's actually what the initial design phase is trying to find!). What is needed is an estimation of the launch vehicle's thrust/weight ratio and a specification of the performance of the launch vehicle's engines as specified by their overall vacuum specific impulse. With these two parameters, one can utilize the plots in Figs. 3.10 and 3.11 to determine values for the constants

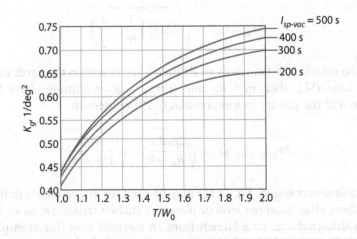

Fig. 3.10 Gravity loss estimation plots for K_g. Source: [6].

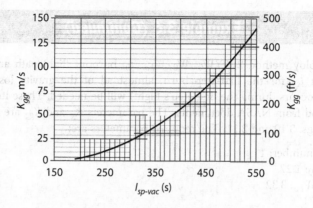

Fig. 3.11   Gravity loss estimation plot for K_{gg}. Source: [6], redrawn by author.

K_g and K_{gg}. Those values are then utilized within the gravity loss equation

$$\Delta v_{\text{gravity loss}} = (g_0 \tau_b - K_{gg}) \left[1 - K_g \left(1 - \frac{W_{bo}}{W_0} \right) \left(\frac{\beta_{bo}}{90 \deg} \right)^2 \right] \quad (3.14)$$

where τ_b is the total burn time, β_{bo} is the burnout flight path angle, K_g and K_{gg} are obtained from the aforementioned plots, and W_0 is the liftoff weight. The subscript 0 denotes a quantity at liftoff, and the subscript bo denotes a quantity at burnout. Of course, consistent units must be used throughout. The flight path burnout angle is zero for horizontal and 90 degrees for vertical. Values for K_g and K_{gg} can be estimated from Figs. 3.10 and 3.11.

Example 3.6

Estimate gravity loss for Saturn V launch. The burnout height given is 105 NM = 195 km. Using method 1,

$$\Delta v_{\text{gravity loss}} = \sqrt{\frac{2 g_0 h}{1 + \dfrac{h}{R}}}$$

$$\Delta v_{\text{gravity loss}} = \sqrt{\frac{2(0.0098 \text{ km/s}^2)(195 \text{ km})}{1 + \dfrac{195 \text{ km}}{6,378 \text{ km}}}}$$

$$\Delta v_{\text{gravity loss}} = 1.93 \text{ km/s}$$

(Continued)

Example 3.6 *(Continued)*

To employ method 2, T/W_0, W_0/W_{bo}, τ_b, burnout flight path angle, and vacuum I_{sp} for the first-step burn (almost all of the gravity loss of the Saturn occurs during the first step's flight) will be needed. These items may be found from NASA documents. The parameters K_g and K_{gg} are obtained from Figs. 3.10 and 3.11. The required parameters are:

- Step number: 1
- T/W_0: 1.27
- W_0/W_{bo}: 3.32
- τ_b: 171
- I_{sp}: 300
- β_{bo}: 23 deg
- K_g: 0.53
- K_{gg}: 23 m/s

Now, from Eq. (3.14):

$$\Delta v_{\text{gravity loss}} = (g_0\tau_b - K_{gg})\left[1 - K_g\left(1 - \frac{W_{bo}}{W_0}\right)\left(\frac{\beta_{bo}}{90\deg}\right)^2\right]$$

$$\Delta v_{\text{gravity loss}} = 1{,}517\,\text{m/s}$$

This is substantially less—about 21.4% less—than the estimate provided by method 1, hence the suggestion to use method 1 and multiply the result by 0.8, as mentioned earlier. Also note that the actual gravity loss of the Saturn V is 1,534 m/s [9], so the methods presented in the previous example seem to yield very good results. The statistical estimation method will usually result in a conservative estimation (overestimation) of gravity loss of approximately 5%. The energy method of [4] is much easier to use but considerably less accurate unless it's reduced by the 80% knockdown factor. More data on gravity losses for different launch vehicles can be found later in this chapter.

3.3.2.3 Estimating Gravity Loss: Exact Methods

In reality, the gravity loss is a function of the flight path angle integrated over time:

$$\Delta v_{\text{gravity loss}} = \int_{t_0}^{t_{\text{final}}} g\sin\gamma\,dt \qquad (3.15)$$

Note that this method is difficult to use unless the trajectory information $\gamma(t)$ is known beforehand. For reference, the $\Delta v_{\text{gravity}}$ loss of the Ariane is approximately 1.08 km/s for LEO, and $\Delta v_{\text{gravity}}$ loss for the Saturn V was approximately 1.68 km/s to escape Earth orbit.

It may be surmised that the easiest way to minimize gravity losses is to have the flight path angle $\gamma \to 0$ as soon as possible, suggesting horizontal flight into or out of the atmosphere. It will be shown later that this conflicts with minimizing the aerodynamic loading on the vehicle, suggesting that there may be an optimum way to ascend out of the atmosphere. The concept of optimizing the ascent to minimize the total energy require-ment—and thus minimizing the amount of propellant and rocket required—is examined as a trajectory optimization problem in Chapter 6.

3.3.3 Estimating Aerodynamic Drag Loss

Aerodynamic drag is the resistance offered by air to a body moving through it. The drag force D acts in the opposite direction of the velocity vector and is given by the equation

$$D = C_D \frac{1}{2} \rho v^2 S_{\text{ref}} \qquad (3.16)$$

where C_D is the drag coefficient, ρ is the air density, v is the speed of the launch vehicle, and S_{ref} is a reference area of the body; for launch vehicles, this reference area is the maximum cross-sectional area A of the body normal to the flow. C_D is usually taken to be some percentage of the maximum drag coefficient $C_{D_{\text{max}}}$, which varies with Mach number. From Fig. 3.12, $C_{D_{\text{max}}}$ is usually about 0.42.

To accurately estimate the amount of energy needed to overcome drag during the launch vehicle's ascent, the deceleration done by the drag force per unit mass must be integrated through the powered portion of the ascent. In other words,

$$\Delta v_{\text{drag loss}} = \int_0^{t_{\text{burn}}} \frac{D(t)}{m(t)} \, dt \qquad (3.17)$$

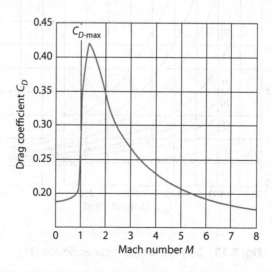

Fig. 3.12   Drag coefficient as a function of Mach number. Source: [6].

where t_{burn} is the burn time, $D(t)$ is the time-varying drag force, and $m(t)$ is the time-varying vehicle mass. Just as with the gravitational calculations, at this time the velocity and altitude information as a function of time are unknown, so an explicit calculation is impossible. Therefore, an approximate method must again be used; such a method is presented in [11]. The appropriate estimation curve is shown in Fig. 3.13.

Using Fig. 3.13 to obtain the parameter K_D, the velocity loss due to drag is then

$$\Delta v_{\text{drag loss}} = \frac{K_D C_D S_{\text{ref}}}{W_0} \qquad (3.18)$$

The other values come from the launch vehicle's physical dimensions, with the exception of the drag coefficient C_D, which must be estimated. A general guideline is to use $C_D = 0.2$, although it is significantly higher in the transonic region where losses are largest. The drag losses themselves are usually small compared to the gravity loss, on the order of 40–200 m/s (0.04–0.20 km/s).

Occasionally, the cross-sectional area is difficult to calculate. Consider, for example, the first step of the Saturn V, shown in Fig. 3.14. In addition to the step core, there are conical fairings covering the outboard engines, as well as fins for stabilization. The step core diameter for the Saturn V was 10.06 m, yielding a cross-sectional area of 79.49 m^2. The radius of the

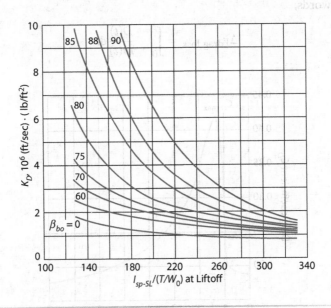

Fig. 3.13 Drag loss estimation curves. Source: [11].

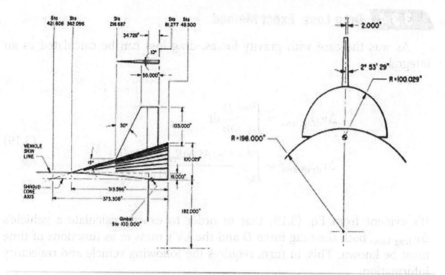

Fig. 3.14   Dimensions of Saturn V S-IC first step fins and engine fairings. Source: [19].

four curved fairings was 100 in. (2.54 m), and the aft tip of the fairings spanned a 180-deg arc (a semicircle), giving the fairings a total area of 40.54 m^2. Each fin was 14.4 in. (0.366 m) thick at the root, 4 in. (0.102 m) thick at the tip, and spanned 103 in. (2.62 m), providing an area of 3.52 m^2. Based on these dimensions, the total step cross-sectional area is calculated to be 123.55 m^2 (1,330 ft^2).

Example 3.7

Let's calculate the drag loss for the Saturn V launch carrying Apollo 11. For the first step, $I_{sp-SL} = 265$ s, liftoff thrust $= 7.891 \times 10^6$ lb, liftoff weight $= 6.54 \times 10^6$ lb, and $T/W_0 = 1.207$. The parameter to be input on the bottom of Fig. 3.13 is $I_{sp-SL}/(T/W_0) = 265$ s/$1.207 \approx 220$ s.

From Fig. 3.13 with a burnout angle $\beta_{bo} = 23$ deg (per NASA report), $K_D \approx 1.25 \times 10^6$ fps · psf.

Thus, the drag loss is estimated using Eq. (3.18) with $C_{D-max} = 0.42$ and $S_{ref} = 1,330$ ft^2:

$$\Delta v_{\text{drag loss}} = K_D C_D S_{ref}/W_0 = (1.25 \times 10^6 \text{ fps} \cdot \text{psf})(0.42)(1,330 \text{ ft}^2)/$$

$$(6.478 \times 10^6 \text{ lb}) = 108 \text{ ft/s} = 33 \text{ m/s} = 0.033 \text{ km/s}$$

This value is fairly close to the drag loss value of 0.040 km/s found in Table 3.4 later in this chapter. Thus, the plots and calculations in [6] seem to yield a reasonable result.

3.3.4 Drag Loss: Exact Method

As was the case with gravity losses, drag loss can be calculated as an integral

$$\Delta v_{\text{drag loss}} = \int_{t_0}^{t_{\text{final}}} \frac{D}{m} dt$$

$$\Delta v_{\text{drag loss}} = \int_{t_0}^{t_{\text{final}}} \frac{C_D(M,\alpha)S_{\text{ref}}\frac{1}{2}\rho(h)v^2(t)}{m(t)} dt \tag{3.19}$$

It's evident from Eq. (3.19) that in order to exactly calculate a vehicle's $\Delta v_{\text{drag loss}}$, both the drag force D and the LV's mass m as functions of time must be known. This, in turn, requires the following vehicle and trajectory information:

- The LV's time-varying dimensionless drag coefficient C_D, which in turn depends on the LV's time-varying angle of attack α and its time-varying Mach number $M(t)$
- The trajectory's time-varying air density ρ, which depends on $h(t)$, the LV's altitude
- The time-varying speed $v(t)$ of the vehicle, and
- The time-varying mass $m(t)$ of the launch vehicle, which depends on the LV's propellant consumption rate, which is determined by the LV's thrust and specific impulse.
- The reference area S_{ref} is, mercifully, constant. Reference area is, in point of fact, the only variable that remains constant with time; the rest of the variables will vary as the launch proceeds.

It stands to reason, then, that aerodynamic losses may be minimized by having $\rho \to 0$ as quickly as possible, suggesting a purely vertical climb. Such a trajectory contradicts a horizontal flight direction, which is desired to minimize gravity losses. Thus, there is a conflict between minimizing drag loss and minimizing gravity loss, which suggests a trajectory optimization problem which is discussed in Chapter 6.

In reality, the drag losses are usually quite a bit smaller than the gravity losses, often a factor of 5 or more. For the Ariane and Saturn V, drag loss values are 220 m/s and 46 m/s, respectively; compare these to gravity losses of 1,080 m/s and 1,676 m/s respectively.

3.3.4.1 Thoughts on Minimizing Drag Losses

Generally speaking, a longer, thinner LV will have lower drag than a shorter, "fatter" LV. Let's take a look at the relationship between drag loss and vehicle slenderness. This will be done by using the definition and a bit of algebra, as follows (t_b represents the burn time):

$$\Delta v_{\text{drag loss}} = \int_0^{t_b} \frac{D}{m}\,dt$$

$$= \int_0^{t_b} \frac{\frac{1}{2}\rho v^2 C_D A}{m_0}\frac{m_0}{m}\,dt$$

$$= \frac{A}{m_0}\int_0^{t_b} \frac{1}{2}\rho v^2 C_D \frac{m_0}{m}\,dt \qquad (3.20)$$

$$= \frac{A}{m_0} I_D$$

The integral on the right, I_D, is independent of size because it contains only dynamic pressure, drag coefficient, and a mass ratio. It may be assumed that the vehicle's initial mass m_0 is going to be proportional to the vehicle's volume, which in turn is proportional to its length L multiplied by its cross-sectional area A, so $\Delta v_{\text{drag loss}} \propto \frac{1}{L}$. This suggests that the vehicle should be made as long as possible to minimize drag loss; however, there are practical limits (with regard to structures, loads, and dynamics) that constrain how long and thin the vehicle can be made. These limits will be discussed in Chapters 11 and 12.

Some other comments apply to geometrically similar, or scaled, launch vehicles. Using the *cube-square law*—an object's area A goes as the square of the principal dimension L ($A \propto L^2$), while the volume goes as the cube of that dimension ($V \propto L^3$)—it may be seen that the aerodynamic reference area is $A \propto m^{2/3}$. Earlier it was demonstrated that the drag loss is approximately $A/m_0 = m_0^{-1/3}$ if the drag coefficient C_D is roughly constant with scale. From this we can conclude that drag losses are relatively larger for vehicles of smaller size, which suggests these vehicles will have tighter payload fairing constraints, and in general, larger vehicles will have relatively smaller drag losses.

3.3.5 Propulsion Losses

Propulsion losses occur due to nonoptimum nozzle expansion that happens whenever the nozzle exit pressure is not equal to ambient pressure, which is almost all the time because most rocket engines can only be optimized at a single exhaust pressure. Rocket engine performance will be discussed in detail in Chapter 4.

The propulsion system performance may be estimated by using the nozzle's exhaust pressure P_e, the exhaust area A_e, and the ambient pressure P_∞ to calculate the total thrust. The total thrust T is the sum of momentum thrust and the pressure thrust during boost

$$T = \dot{m}v_e + (P_e - P_\infty)A_e \qquad (3.21)$$

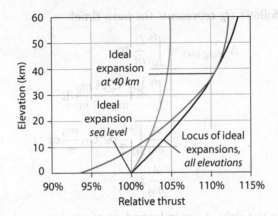

Fig. 3.15   Thrust variation for differing ideal expansion pressure values. This number is fixed by the engine geometry, so a single value must be selected.

The effects of this may be seen in Fig. 3.15, which plots the relative thrust as a function of elevation (altitude) above sea level. As a good design rule, the initial design of a nozzle for a rocket operating from sea level to near-vacuum should undergo ideal expansion at an elevation of approximately 40 km.

Because ambient pressure changes during ascent, so too does thrust. Note that the propulsion loss is typically less than the gravity loss, but larger than the drag loss estimated earlier. The propulsion loss may be estimated using another empirical curve, shown in Fig. 3.16. Figure 3.16 is used to obtain

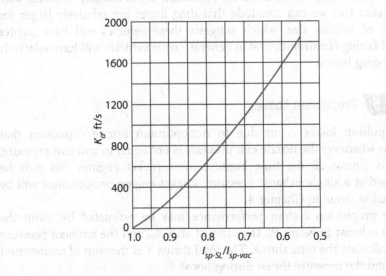

Fig. 3.16   Thrust loss approximation curve. Source: [6].

the thrust loss parameter K_a. Note that the units of K_a are ft/s and need to be converted to m/s. Thus, for "bookkeeping" launch losses,

$$\Delta v_{\text{thrust loss}} = K_a \qquad (3.22)$$

Example 3.8

Let's calculate the thrust loss for a Saturn V launch. For the vehicle, the sea level $I_{sp} = 264$ s, the vacuum $I_{sp\text{-}vac} = 304$ s, and $T/W_0 = 1.183$ from a previous example. The parameter on the bottom of Fig. 3.16 is

$$\frac{I_{sp\text{-}SL}}{I_{sp\text{-}vac}} = \frac{264\,\text{s}}{304\,\text{s}} = 0.8684$$

Extrapolating from the figure, $K_a = 0.15$ km/s.

Note that this 'thrust loss' value need not be calculated if one uses the vehicle's *effective* specific impulse or *effective* exit speed for Δv calculations. It is only included here to illustrate the thrust variation behavior shown in Fig. 3.15.

3.3.6 Application to Multiple Steps

The results presented so far have focused on single-step vehicles; however, the same methodology may be applied to multistep vehicles. Gravity losses remain the most significant factor. As a guideline, the losses for the second step can be calculated using the burnout angle at the time of second-step separation, known as the intermediate velocity angle β_2. The velocity loss after separation, Δv_2, is then

$$\Delta v_2 = g_0 \left(\frac{R_e}{R_e + h_2} \right)^2 t_{b_2} \cos \beta_2 \qquad (3.23)$$

where h_2 is the average altitude for the flight of the second step, and t_{b2} is the burn time for the second step.

For calculating drag losses, if the first-step burnout occurs at an altitude at or above 61 km (approximately 200,000 ft) at a velocity angle of less than 75 deg, it is safe to assume that *all* drag losses have occurred during the first-step burn. K_D may be assumed constant during first-step ascent, but must be recalculated for second-step ascent at a velocity angle roughly 5–15 deg lower than the velocity angle at first-step burnout. For initial design purposes, it may be assumed that all nozzle pressure losses occur during the first-step burn. Typical multistep losses are given in Tables 3.2 and 3.3 for two- and three-step launches.

Table 3.2 Two-Stage Loss Factors

Stage	$\dfrac{\Delta v_{actual}}{\Delta v_{ideal}}$	Δv_{ideal}	Propulsion	Steering	Drag	Gravity
1	0.63	1.00	−0.02	0.00	−0.03	−0.32
2	0.93	1.00	0.00	−0.02	−.00	−0.05
Escape	0.81					

$T/W = 1.25$, staging at 8,000 ft/s (2.44 km/s). Source: [20].

3.3.7 Steering Losses

Steering losses occur when a vehicle's thrust vector is not parallel to its velocity vector. This can be understood by considering that the definition of delivered mechanical power P is the scalar or "dot" product of thrust vector $\mathbf{T}$ and velocity vector $\mathbf{v}$, or $P = \mathbf{T} \cdot \mathbf{v}$. Thus, it can be seen that any angular separation between thrust and velocity directions will lead to a loss of thrust power delivered to the vehicle. The angular separation may be seen in Fig. (3.17) to be the sum of the angle of attack α and any thrust vector control angle δ; symbolically

$$\Delta v_{T \& v \ not \ parallel} = \int_0^{t_b} Tv[1 - \cos(\delta + \alpha)]dt$$

In general, the steering losses are on the order of 30–400 m/s. The actual value usually depends greatly on the design of the upper step, specifically its thrust-to-weight ratio ψ (whether the upper step is 'overpowered' or not). Upper steps possessing lower ψ values need longer burn times to accelerate to orbital speed, and tend to have to spend more time "lofted," or intentionally sent to a higher-than-preferred altitude, after which they gradually descend vertically while accelerating horizontally. During this period, they are flying at an angle of attack with their thrust vectors above the velocity direction to partially offset gravity, leading to higher steering losses.

The calculation of steering losses requires knowledge of the vehicle's angle of attack and thrust vector angles as a function of time; this information is usually not known at early stages of the design and is difficult to estimate, so we often use values from vehicles with similar expected performance.

Table 3.3 Saturn V Three-Stage Translunar Injection Loss Factors [5]

Stage	Δv_{ideal}	Gravity Loss		Drag Loss		Steering Loss		Δv_{actual}	$\dfrac{\Delta v_{actual}}{\Delta v_{ideal}}$
	m/s	m/s	%	m/s	%	m/s	%	m/s	%
1	4,923	1,219	25.0	46	0.9	0	0	3,658	74
2	5,242	335	6.4	0	0	183	3.5	4,724	90
3	4,242	122	2.9	0	0	5	0.1	4,115	97
Total	14,407	1,676	34.3	46	0.9	188	3.6	12,497	86.7

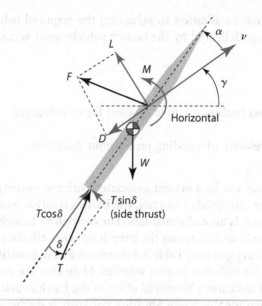

Fig. 3.17 Forces on a launch vehicle in flight.

Steering losses are discussed further in Chapter 6, section 6.4.4; details of trajectories and lofting are discussed in more detail later in Ch. 6.

3.3.8 Summing Up the Losses

It should now be clear that a launch vehicle must supply considerably more energy than that required to accelerate the payload to orbital speed. This idea is illustrated in Fig. 3.18, showing the performance breakdown for the Saturn V. Here one can see that a significant fraction of the impulse the launch vehicle provides goes to overcome the gravity, drag, and steering losses we detailed previously, and does not contribute to the orbital speed. (Note that this figure does not depict propulsion losses.)

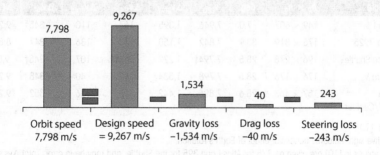

Fig. 3.18 Illustration of the launch performance and losses of the Saturn V launch vehicle launching to Earth orbit. Source: [10].

To summarize: in addition to achieving the required orbital speed, the impulse or energy delivered by the launch vehicle must account for:

* Gravity loss
* Drag loss
* Propulsion loss (unless effective I_{sp} used for calculations)
* Steering loss
* Launch site velocity (depending on burnout direction)

There may or may not be a benefit associated with the velocity of the launch site. For polar or retrograde launches, the Earth's rotation works *against* the launch vehicle and is an additional loss. For eastbound launches, the Earth's rotation is a benefit and *increases* the launch vehicle's effective payload mass. To put things in perspective, Table 3.4 summarizes the quantitative values of all these losses for different launch vehicles. Note that the negative sign for v_{rot} in Table 3.4 indicates a beneficial effect of the Earth's rotation. The negative height shown for the Space Shuttle's periapsis is deliberately chosen to ensure the jettisoned External Tank will re-enter the atmosphere for disposal. An additional Δv of 144 m/s is required to circularize the orbit at a height of 278 km.

3.3.9 Combined Launch Vehicle Performance Estimation

The methods of [6] and [9] provide fairly simple methods of estimating performance, using curves to predict the total required speed based on curve-fitting of approximately 100 launch vehicle configurations, with T/W ranges from 1.2 to 2.0 and I_{sp} values of 250–430 s. An analytical expression for the

Table 3.4 Design Velocities (in m/s) By Launch Vehicle

Vehicle	$h_p \times h_a$	i, deg	v_{LEO}	Δv_{grav}	Δv_{steer}	Δv_{drag}	Δv_{rot}	$\Sigma(\Delta v)$
Atlas I	149 × 607	7.0	7,946	1,395	167	110	−345†	9,243
Delta 7925	175 × 319	33.9	7,842	1,150	33	136	−347	8,814
Space Shuttle	−196 × 278	28.5	7,794‡	1,222	358†	107	−345†	9,086**
Saturn V	176 × 176	28.5	7,798	1,534	243	40	−348	9,267
Titan IV/Centaur	157 × 463	28.6	7,896	1,442	65	156	−352	9,207

Source: [10].
*Negative sign indicates beneficial effect of Earth's rotation.
†Δv_{rot} values in [10] are given as 375 for Atlas I and 395 for the Shuttle, and may be in error. Total Δvs shown include the corrected values for Δv_{rot}.
‡Injection occurs at approximately 111 km.
**Additional $\Delta v = 144$ m/s needed to circularize orbit at apoapsis height $h_a = 278$ km.

prediction of total Δv required is given by [12] as

$$\Delta v_{\text{total}} = \sqrt{v_{\text{circ}}^2 + 2g_0 h \left(\frac{R}{R+h}\right)^2 + \frac{0.0015}{s^2} T_a^2 + \frac{0.0882}{s} T_a + 1{,}036 \,\text{m/s}}$$

(3.24)

where v_{circ} is the circular orbit speed for altitude h, and T_a is the ascent time. A better estimate can be obtained by utilizing a penalty function that incorporates both altitude and ascent time, where a curve fit results in the relation

$$\Delta v_{\text{penalty}} = K_1 + K_2 T_a$$

(3.25)

where K_1 and K_2 are given as

$$K_1 = 662.1 + 1.602 H_p + 0.001224 H_p^2$$

$$K_2 = 1.7871 - 0.0009687 H_p$$

(3.26)

$$H_p = \text{parking orbit altitude in km}$$

Thus, the total Δv required is

$$\Delta v_{\text{total}} = \Delta v_{\text{circ}} + \Delta v_{\text{penalty}} + \Delta v_{\text{rot}}$$

(3.27)

where Δv_{rot} is the contribution or penalty due to the Earth's rotation relative to the burnout azimuth. If the parking orbit altitude is unknown, a value of 185 km or 100 nautical miles is a reasonable estimate.

The time required for the direct-ascent phase of flight must also be known. For the most part, this is a simple burn time calculation, although delays resulting from staging and ignition should be considered. If the vehicle is capable of a coasting period prior to the ignition of the upper step, the most accurate estimate is generally achieved using the minimum possible coast period.

Modern vehicles tend to use lower thrust-to-weight ratios on the upper steps, so the method presented here is not as appropriate, because the assumptions of similar (or identical) mass ratios, T/W, and specific impulses are not applicable. Optimal designs often have a high-thrust first step (often augmented by strap-on boosters) to rapidly ascend through the initial flight regime (where gravity losses are highest), and a relatively low-thrust but high-I_{sp} upper step to add Δv in the most efficient manner once loss terms have become less critical.

A further modification to the performance estimation utilizes a weighted average of boost time and the boost time of a hypothetical three-step vehicle, where each step has identical mass ratios, specific impulses, and T/W. The equivalent ascent time of such a vehicle, T_{3s}, is

$$T_{3s} = \frac{3 m_0 g_0 I_{sp}}{T} \left(1 - e^{\left(\frac{-\Delta v_p}{3 g_0 I_{sp}}\right)}\right)$$

(3.28)

where Δv_p is the Δv required to attain a parking orbit, and T/m_0 is the liftoff acceleration without gravity.

Another source of "penalty" Δv is due to steering or trajectory-shaping losses. Most often, these stem from a requirement to have the initial trajectory loft the vehicle to an altitude higher than the parking orbit altitude, to ensure that the vehicle does not re-enter the atmosphere before attaining orbital velocity. This is strongly correlated with ascent time, because a longer ascent requires greater excess vertical velocity. This will be discussed in greater detail in Chapter 6.

A more accurate method of obtaining $\Delta v_{penalty}$, from a survey of over 1,000 data points representing the performance of 17 modern launch vehicles, is the best-fit relation

$$\Delta v_{penalty} = K_3 + K_4 T_{mix} \tag{3.29}$$

where

$$K_3 = 429.9 + 1.602 h_p + 0.001224 h_p^2$$

$$K_4 = 2.328 - 0.0009687 h_p \tag{3.30}$$

$$T_{mix} = 0.405 T_a + 0.595 T_{3s}$$

This latter method typically produces an RMS error of approximately 260 m/s (less than 3% in total mission Δv for a circular LEO launch), and "usually" a <10% error in payload. Ready-made functions that execute these calculations, including a handy online calculator [13], are readily available at http://www.silverbirdastronautics.com.

3.4 Determining the Launch Vehicle Velocity Vector

So far, the discussion has focused on the *speed* that a payload has to achieve, but no mention has yet been made as to the *direction* in which it needs to be sent. Knowing the direction allows the designer to take the launch site speed into account when calculating the Δv needed for a particular mission with a specified altitude, speed, and launch direction. The total velocity the launch vehicle needs to provide is noted by the variable Δv_{needed}. Note that this variable does *not* include the gravity, drag, propulsion, and steering losses that were estimated in the previous sections, so those losses will be added later.

The magnitude of Δv_{needed} is the root sum square (RSS) value of its vector components. For this calculation, it's convenient to use the south-east-zenith coordinate system, depicted in Fig. 3.19, where the south direction is the principal direction (abbreviated with capital S), the zenith is vertically upwards (abbreviated with capital Z), and east completes the system (abbreviated as E). This method makes the "flat Earth" assumption: the ascent trajectory occurs over a short enough distance that the curvature of the Earth may be neglected.

In the figure, the angle β represents the burnout azimuth angle, measured clockwise from north using the horizontal projection of the velocity. If the launch is directly east, the burnout azimuth angle $\beta = 90$ degrees. The angle γ represents the flight path angle and is measured from local horizontal. For a circular orbit or for burns ending at periapsis or apoapsis, the flight path angle $\gamma = 0$ degrees.

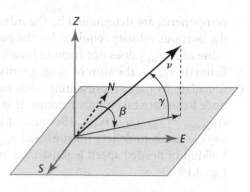

Fig. 3.19 SEZ coordinate system. The azimuth angle β is measured clockwise from North in the horizontal plane, and the flight path angle γ is measured up from the horizontal plane.

In order to fly a certain mission, there is an associated required burnout speed that the launch vehicle must provide. These speeds were calculated earlier and resulted in the required trajectory $v_{burnout}$ *magnitude*. The *direction* that speed has to be (the direction specified by the azimuth angle β, measured from north, and the flight path angle γ, measured from horizontal and indicating any climb) now must be calculated. The following definitions are used for the speed and velocity calculations:

- $v_{burnout}$: Burnout inertial velocity needed to be in desired orbit (mission dictated).
- $v_{launch\ site}$: Velocity of launch site due to Earth's rotation (may or may not assist, as discussed previously).
- v_{needed}: Velocity needed for the mission with the velocity of the launch site taken into account.
- v_{losses}: Speed equivalent (scalar) of energy needed to overcome the sum of gravity, drag, propulsion, and steering losses.
- Δv_{design}: Total speed change (scalar) the launch vehicle must generate to meet the mission requirements, including all losses. *This is the speed change that the launch vehicle must be designed to meet in order to complete the mission successfully!*

The value of v_{design} may be calculated using the following procedure:

1. Calculate the SEZ components of $v_{burnout}$ based on mission requirements.
2. Determine that magnitude of $v_{launch\ site}$ based on latitude λ, as discussed previously.
3. Calculate $\vec{v}_{needed}$ for the mission. Note that this is a vector equation: $\vec{v}_{needed} = \vec{v}_{burnout} - \vec{v}_{launch\ site}$. Only the eastward component of $\vec{v}_{needed}$ is different.
4. Calculate the magnitude $\|\vec{v}_{needed}\| = v_{needed}$. Using vector arithmetic, $v_{needed} = \sqrt{v_{needed,\ South}^2 + v_{needed,\ East}^2 + v_{needed,\ Zenith}^2}$. The three

components are determined by the mission requirements, particularly by the burnout velocity required for the payload. As stated previously, this value of v_{needed} does not include losses.

5. Estimate v_{losses}, the sum of drag, gravity, propulsion, and steering losses.
6. Calculate Δv_{design}, representing what the launch vehicle must provide once losses are taken into account. It is estimated as $\Delta v_{design} = v_{needed} + v_{losses}$. Because of losses, the speed the launch vehicle needs to deliver must exceed v_{needed}, or the mission will fail. This buildup of needed speed is illustrated by the speed bar graph shown in Fig. 3.18.

3.4.1 Determining the Required Launch Vector

The burnout speeds in the south, east, zenith (SEZ) coordinate system must be determined with the SEZ coordinate system fixed at the burnout site because there is an existing speed due to the Earth's rotation, which has to be dealt with in a vector fashion. Therefore, the speed components will be determined in the SEZ coordinate system. Remember that the Earth's rotation produces an eastward velocity, which must be accounted for in the east speed component. Referring to Fig. 3.20 and using simple trigonometry to project the burnout velocity onto the SEZ axes, the three components of the burnout velocity are determined as:

* $v_{burnout, South} = -v_{burnout} \cos \gamma \cos \beta$
* $v_{burnout, East} = v_{burnout} \cos \gamma \sin \beta$
* $v_{burnout, Zenith} = v_{burnout} \sin \gamma$

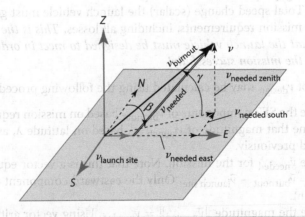

Fig. 3.20 Required launch velocity vector.

where v_{burnout} is the magnitude of the required burnout velocity, γ is the flight path angle (between v and horizontal, positive upwards), and β is the burnout azimuth angle (measured clockwise from north).

Now, v_{needed} will be calculated, which is the speed that the launch vehicle must provide with the launch site's speed taken into account. The needed velocity may be written as $v_{\text{needed}} = v_{\text{burnout}} - v_{\text{launch site}}$. This is a vector calculation, so each component will need a separate equation:

* $v_{\text{needed, South}} = v_{\text{burnout, South}}$
* $v_{\text{needed, East}} = v_{\text{burnout, East}} - v_{\text{launch site}}$
* $v_{\text{needed, Zenith}} = v_{\text{burnout, Zenith}}$

The individual components of v_{needed} are the three components of velocity needed by the launch vehicle to get to orbit in the south, east, and zenith directions. The components of v_{burnout} are components of the burnout velocity in the south, east, and zenith directions. $v_{\text{launch site}}$ is the eastward velocity of the launch site due to the Earth's rotation, and has no S or Z component. For Earth, $v_{\text{launch site}} = 465.1 \cos \lambda$, where λ is the launch site latitude, and $v_{\text{launch site}}$ is given in m/s.

3.4.2 Air-Launch Systems

It has been demonstrated that gravitational losses and drag losses add a substantial amount to the required speed for any launch vehicle launching from the Earth. One might ask: what can be done to reduce these losses?

Example 3.9

Find Δv_{design} for a Saturn V launch from Kennedy Space Center (KSC) ($\lambda = 28.5$ deg) to a 195-km circular orbit with $i = 28.5$ deg. Using the steps outlined in the previous section,

1. $v_{\text{burnout}} = \sqrt{\dfrac{g_0 R^2}{r}} = \sqrt{\dfrac{3.986E5\,\text{km}^3\text{s}^2}{(6{,}378 + 195)\,\text{km}}} = 7.787$ km/s. Because the inclination of the orbit is equal to the latitude of the launch site, the launch vehicle launches directly due east. Therefore, $\beta = 90$ deg . Because the orbit is circular, $\gamma = 0$. Computing the SEZ components of v_{burnout}:

$$v_{\text{burnout, South}} = -v_{\text{burnout}} \cos \gamma \cos \beta = 0$$
$$v_{\text{burnout, East}} = v_{\text{burnout}} \cos \gamma \cos \beta = v_{\text{burnout}} = 7.787\,\text{km/s} \quad (3.31)$$
$$v_{\text{burnout, Zenith}} = v_{\text{burnout}} \sin \gamma = 0$$

2. Compute the launch site velocity: $v_{\text{launch site}} = 0.4087$ km/s due east.

(Continued)

Example 3.9 (Continued)

3. Compute the SEZ components of v_{needed}:

$$v_{needed, South} = v_{burnout, South} = 0$$

$$v_{needed, East} = v_{burnout, East} - v_{launch\ site} = 7.379\ km/s \qquad (3.32)$$

$$v_{needed, Zenith} = v_{burnout, Zenith} = 0$$

4. $\|v_{needed}\| = \sqrt{(0)^2 + (7.379)^2 + (0)^2} = 7.379\ km/s.$

5. $v_{gravity\ loss} = 1.670\ km/s$ from a previous example; $v_{drag\ loss} = 0.035\ km/s$ from a previous example; $v_{propulsion\ loss} = 0.150\ km/s$, as before. Summing these losses together, $v_{loss} = 1.855\ km/s.$

6. $\Delta v_{needed} = v_{needed} + v_{losses} \Rightarrow \Delta v_{needed} = 9.236\ km/s.$

The result, 9.236 km/s, compares well with the value of 9.267 km/s for the Saturn V given in Table 3.4. This estimate would be the starting value for the speed delivery needed to meet the launch requirements given for this example, with the three-stage vehicle.

Note that [4] provides a different method to calculate the value of Δv_{design}. This method incorporates the gravitational potential energy version of $v_{grav\ loss}$ as a vertical component of v_{needed}, rather than adding it as a loss afterwards. This approach tends to underpredict the design velocity, and so it's necessary to add an artificially high drag loss value to come up with a reasonable answer. In Example 9.2 of [4], a drag loss of 1,500 m/s is used; this value is much larger than the actual drag loss of about 35–40 m/s. It makes more sense to add all of the losses after finding the needed speed to get to the desired orbit.

It is important to note that all of the calculations up to this point represent the conditions that have to be met at *burnout* or *shutdown*. Because liftoff occurs minutes before burnout, to do this launch properly, liftoff has to be timed so that the burnout conditions occur when the launch vehicle has reached the desired physical location. This is another way of saying that the timing of the launch is important to "anchor" the time at which burnout occurs. To do this properly requires an estimate of the time needed between liftoff and burnout.

Gravity and drag losses are large, so starting out at a higher altitude and passing through a lower-density atmosphere would help. One simple way to do this is to drop the launch vehicle from an aircraft at altitude and begin the launch vehicle's mission after dropping. As might be expected, starting at a higher altitude and speed does improve things. Some advantages are:

- Less atmosphere to fly through at launch (smaller drag loss).
- Launch at high altitude (smaller gravity loss).

* Nonzero initial velocity saves propellant (less Δv_{design} needed, smaller vehicle for same payload).
* With a movable launchpad, there is much more ability to launch at the desired location and time; for example, it would be (conceivably) much easier to launch near the equator to maximize payload on an eastbound launch with no ground launch site scheduling issues.

Some disadvantages of air launching are:

* The launch vehicle may have to carry wings and an aircraft-style control system in order to transition from its horizontal drop configuration to a climbing ascent, increasing mass (although experiments have demonstrated a wingless air-drop launch system).
* The launch vehicle may be subjected to substantial bending loads produced as a result of its suspension underneath or inside the launch aircraft; these loads increase airframe stress, thereby increasing the mass of the launch vehicle.
* A dedicated aircraft is needed, along with a crew, which means that the launch aircraft spends most of its time on the ground unless a large number of launches is planned.

Air launching has been used for several current suborbital aircraft, most notably the Orbital Sciences Corp. Pegasus launch vehicle (Fig. 3.21). Scaled Composites' SpaceShip 1 and SpaceShip 2 are also air-launched (although both are suborbital); however, Virgin Orbit's air-dropped LauncherOne orbital launch vehicle is expected to carry out its first missions in 2020. The exercises contain problems on the performance of air-launched vehicles.

Fig. 3.21 Pegasus LV attached to the bottom of a modified L-1011 carrier aircraft.
Source: Northrop Grumman.

3.5 Direct Orbit

Now that the amount of energy needed to orbit or escape has been established, the discussion now turns toward entering orbits. What is called a *direct orbit* must pass over the latitude of the launch site. It's called "direct" because the launch vehicle can fly the payload directly to the orbit. There are two types of direct orbits, discussed in the following sections.

3.5.1 Launch Directly East

If a launch vehicle injects into orbit directly to the east, the burnout site will be the highest latitude that the orbit will achieve. By definition, the inclination of this orbit will be the burnout site latitude, or $i = \lambda$. Because the burnout vector is tangent to the orbit, there is exactly one opportunity per orbit to launch into this particular orbit; that is, there is one launch window per orbit. In this situation, shown in Fig. 3.22, the burnout azimuth angle is $\beta = 90$ degrees. The direct east injection is often the selected type of launch because it yields the highest payload into orbit. This is because it takes the maximum possible advantage of the Earth's rotational speed.

The result of launching due east from KSC is shown in Fig. 3.23. The orbital plane is inclined 28.5 degrees, and its trace (the projection of the orbit vertically downward on the Earth) is tangent to the ± 28.5-degrees latitude lines as shown. From this illustration, it is readily apparent that if a spacecraft must fly over a certain portion of the Earth, the inclination of the orbit must be greater than or equal to the latitude of the target area: $i \geq \lambda_{target}$. Certain target areas will result in inclination requirements, which in turn will result in launch vehicle requirements.

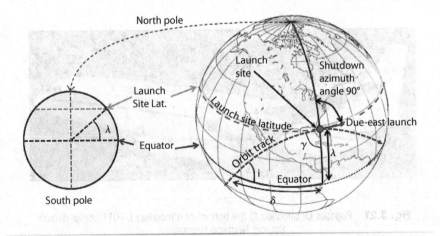

Fig. 3.22 λ and i for direct-east launch.

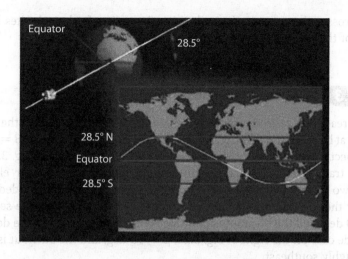

Fig. 3.23   Due east launch from KSC with $i = \lambda = 28.5$ deg. Source: NASA.

3.5.2   Launch in Other Directions

If an orbit is desired having an inclination that is greater than launch latitude, it can be shown that there will be two launch opportunities per orbit. The launch opportunities occur when the orbit passes over the launch site. This situation is illustrated in Fig. 3.24, where the reader can see that there is one opportunity after the orbit passes over the equator heading north (left), and a second opportunity when the orbit passes over the launch site heading south towards the equator (right). The left and right opportunities are known as the ascending node and descending node cases, respectively, where *ascending* or *descending* refers to whether the orbit is heading north or south, and *node* refers to the point where the

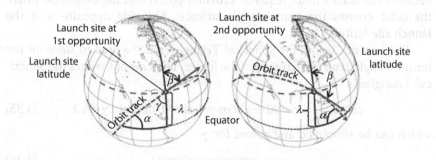

Fig. 3.24   Direct orbit launch with $i > \lambda$. There are two opportunities to orbit, one at the orbit's ascending node, and a second at the descending node.

orbit crosses the equator. A similar situation exists for launch sites located south of the equator.

3.5.3 Calculation of Burnout Azimuth Angles

To reach the desired orbit with $i > \lambda$, the required direction of the launch vehicle at burnout must be determined. For the eastbound launch, $\beta = 90$ deg by inspection; however, for the cases where $i > \lambda$, as shown in Fig. 3.24, the orbital trace goes to a higher latitude than the launch site. For either of those two cases, calculation of the burnout azimuth angle β is needed.

For the ascending node case on the left of Fig. 3.24, it can be seen that $\beta < 90$ deg, and the burnout is pointing roughly northeast. For the descending node case on the right of Fig. 3.24, $\beta > 90$ deg, and the burnout is pointing roughly southeast.

Determination of the burnout azimuth angle β requires reference to a special type of trigonometry, called *spherical trigonometry*. Spherical trigonometry deals with angles and sides of triangles that exist on the surface of a sphere (rather than on a flat plane as in regular trigonometry). The laws of spherical trigonometry are well-known, but derivation of the spherical trigonometry formulae used is beyond the scope of this book; the reader is referred to an appendix in [4]. Also of note is that the Earth is not a perfect sphere for two reasons: first, its one-per-day rotation rate causes it to "flatten out" into an ellipsoid; second, its surface is nonuniform (e.g., ocean, mountains, etc.). The shape is, however, close enough to a sphere to use this geometry for calculations. Additional information on the nonuniformity of the Earth (and its effect on orbiting spacecraft) is included in section 3.3.2 of [2].

Some additional angles are first defined, as shown in Fig. 3.24, so that the laws of spherical trigonometry may be applied. The angle α in this instance is known as the *inclination auxiliary angle* of the orbit. The angle γ is called the *launch window location angle* of the orbit, measured along the equator between the orbit's node (equator crossing point) and the longitude where the orbit crosses the launch site's latitude. The side opposite α is the launch site latitude λ.

The Law of Cosines for Spherical Triangles may be used to solve for the burnout angle β on the left side of the figure. The Law of Cosines for Spherical Triangles is

$$\cos \alpha = -\cos(90 \deg) \cos \gamma + \sin(90 \deg) \sin \gamma \cos \lambda \qquad (3.33)$$

which can be simplified and solved for γ

$$\gamma = \arcsin (\cos \alpha / \cos \lambda) \qquad (3.34)$$

For the ascending node (AN) burnout case, it is immediately apparent that the angle γ is equal to the burnout azimuth angle β, so

$$\beta_{AN} = \arcsin(\cos\alpha/\cos\lambda) \qquad (3.35)$$

For the descending node (DN) case (right side of Fig. 3.24), the triangle is a mirror image of the ascending node case. To determine the burnout azimuth angle, the launch window location angle is subtracted from 180 deg

$$\beta_{DN} = 180 \ \text{deg} - \gamma \qquad (3.36)$$

Example 3.10

Calculate the ascending node azimuth angle for a launch to ISS at an inclination of 51.65 deg from KSC. For the ascending node case, the inclination auxiliary angle α is equal to the inclination i, or $\alpha \cong i = 51.65$ deg. From this, $\gamma = \arcsin(\cos(51.65 \ \text{deg})/\cos(28.5 \ \text{deg})) = \arcsin(0.706) = 44.91$ deg. For an ascending node burnout, this is also the burnout azimuth angle—almost directly northeast.

Spherical trigonometry may also be used to determine the time of burnout, based on the local time at the launch site. Of course, the launch time will depend on the desired orbit as well as the time between liftoff and burnout. These calculations don't affect the launch vehicle's performance, so they are not presented here. The inquisitive reader is referred to [4] for a more detailed discussion of launch latitudes, azimuths, and launch timing.

3.5.4 Polar and Retrograde Orbits

Polar and retrograde orbits are used for special missions such as surveillance, weather, ground observation, and the like. For these types of orbits, the launch vehicle must cancel out the Earth's velocity and, in the case of retrograde orbits, provide speed in the opposite direction. Naturally this process will consume more of the launch vehicle's energy and will reduce the possible payload.

The effects of prograde vs retrograde may be assessed by applying the previously discussed methods of calculating v_{needed} along with the proper determination of the burnout azimuth angle. Careful attention is required because the orbital direction is reversed for a retrograde orbit. Some examples of this may be found in the problems at the end of this chapter.

3.6 Desired Inclination Less than Launch Latitude

It has been shown that it's possible to launch directly into orbits with inclination equal to, or greater than, the burnout latitude; however, there is a large number of possible launches to orbits with an inclination angle that is less than, or smaller than the launch latitude. One large class is the equatorial orbit with $i = 0$, which is used by many commercial spacecraft. For the case of $i < \lambda$, the desired orbital path does not ever pass over the launch site, and there is no direct route to the desired orbit. In these instances, there are two methods to attain the desired orbit: the lateral maneuver and the inclination change.

3.6.1 Launch Vehicle Lateral Maneuver

The first possibility is to have the launch vehicle *yaw*, or fly sideways to perform an orbital plane change to the desired orbital plane. This maneuver is sometimes called a *yaw-torquing* technique or a *dogleg* maneuver, and may be used to either increase or decrease the orbit inclination obtained by the direct injection method for a given launch azimuth. For the situation we are discussing here, the sideways maneuver is used to reduce inclination; however, there are certain launch sites where the desired increased inclination cannot be reached near liftoff due to range safety considerations. In this case, the launch vehicle doglegs to increase inclination.

The yaw option results in less delivered payload mass because of the inefficiency of the dogleg maneuver. This maneuver must use some of the launch vehicle's launch energy for the sideways maneuver, and then may also have to cancel out the sideways velocity; that energy could instead be used for the needed orbit energy itself. As shown in Fig. 3.25, energy is lost pointing

Fig. 3.25 Dogleg trajectory begins with a southern heading towards the equator (dashed line) and ends due east over the equator.

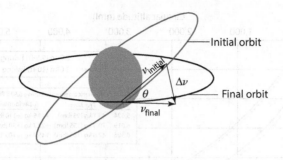

Fig. 3.26   Orbital plane change.

the vehicle towards south and then later pointing it towards north to cancel out the southbound velocity at the equator. From this, it's easy to see why a nonequatorial launch site is not preferred for launches to equatorial orbits. This is why launch systems such as SeaLaunch (which launches off a converted oil drilling platform that has sailed to the equator) or Pegasus (which can be dropped from a host aircraft flying over the equator) can provide a significant payload advantage compared to other launch systems. It is also for this reason that ESA's Ariane launch complexes are located in French Guiana, very close to the equator.

3.6.2 Orbital Inclination Change

The second possibility, and the more commonly used one, is to change the orbital inclination after reaching orbit. There are two cases for this option. For the first case, if the desired final orbit is similar in size to the initial orbit injection, either the launch vehicle (before spacecraft separation) or the spacecraft itself (after separation) fires thrusters to provide a plane change Δv, which inclines the orbital plane or removes the inclination, as desired. The spacecraft (or launch vehicle upper step) fires a thruster to produce the "cross" velocity change shown, to change the inclined $v_{initial}$ to the noninclined v_{final}. This procedure is shown in Fig. 3.26. This method of changing the inclination of the orbit once the payload is on orbit is sometimes known as *cranking* the orbit.

For the second case, the desired orbit is much larger than the initial one, such as would be the case for a geostationary spacecraft. In this situation, the launch vehicle injects the spacecraft into a transfer orbit whose apoapsis is at or near the desired final orbit radius, as shown in Fig. 3.27. At the top of the transfer orbit, the spacecraft fires its own *kick motor* (so-called because it provides the final "kick" into GEO) or onboard thrusters to circularize the orbit. Any maneuver executed by the spacecraft (as opposed to the launch vehicle) requires the spacecraft to carry enough propellant to make the costly (in terms of Δv) plane-change maneuver to the desired inclination. Because

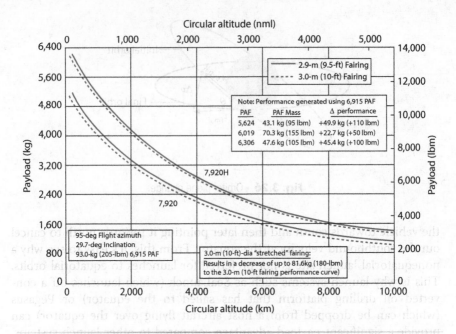

Fig. 3.27 Circular orbit performance for Delta-II. Source: [14].

the main subject of this book is not orbital mechanics, this possibility will not be discussed further; the interested reader is referred to sections 3.4.1.1–3.4.1.4 of [2] for a discussion of the velocity changes required for an orbital plane change. Note that the velocity requirements vary substantially depending on the sequence and timing of the plane-change burns.

3.7 Launch Vehicle Performance Curves

Launch vehicle performance curves are graphs that describe a launch vehicle's performance. They are provided by the launch vehicle manufacturer, usually within the payload planner's guide or user's guide, and typically provide payload delivery performance under a set of specific orbit or escape parameters. Performance curves for the ULA Delta 7,920 and 7,920H circular orbit altitudes are given in Fig. 3.27 for vehicles launching from the Eastern Test Range (Cape Canaveral Air Force Station). As one would expect, the larger the payload mass, the lower the possible circular orbit altitude.

The penalties associated with launching from a latitude λ that was larger than the desired inclination *i* have been discussed previously. It has been shown that the larger the deviation between the two angles, the more performance was needed for the plane change, leaving less performance for orbiting the payload (meaning a smaller-mass payload). Verification of this is shown in the GTO inclination vs payload mass performance curves for the ULA Delta 7,925 and 7,925H vehicles launching from the Eastern Test

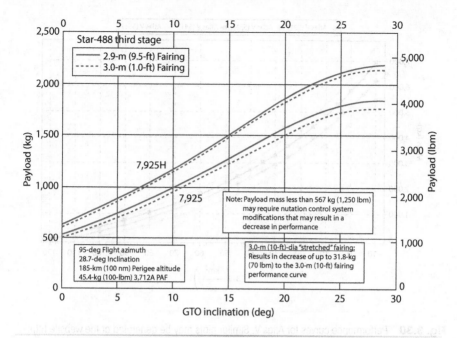

Fig. 3.28   GTO performance of Delta 7,925/7,925H. Source: [14].

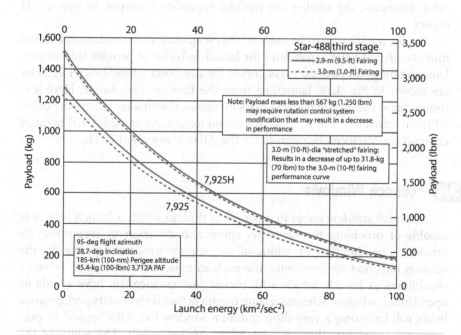

Fig. 3.29   Launch energy (C3) curves for Delta 7925/7925H. Source: [14].

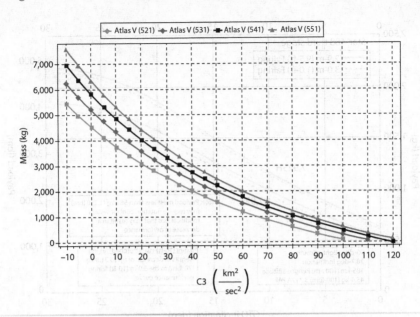

Fig. 3.30 Performance curves for Atlas V. Similar plots may be generated at the website http://elvperf.ksc.nasa.gov/Pages/Query.aspx. Source: NASA.

Range, given in Fig. 3.28, which shows that as the inclination of the transfer orbit decreases, the smaller the payload capability becomes, as one would expect.

Finally, mention must be made of the escape trajectory, and consideration must be given to the capabilities for launch vehicles to provide this service. Launch energy vs payload mass curves for the Delta 7,925/7,925H vehicles are shown in Fig. 3.29, launching from the Eastern Test Range. Here it is shown again that as the payload mass increases, the smaller the energy capability (in terms of $C3 = v_\infty^2$), as one would expect. As an additional point of comparison, performance curves for the Atlas V are given in Fig. 3.30.

3.8 Launch Windows

A *launch window* refers to a period of time in which a launch vehicle is capable of providing the necessary speed and direction to accomplish its mission. The duration or width of the launch window depends on the mission injection requirements, the payload mass, and the launch vehicle's capabilities. A launch vehicle with excess performance will have a wide or open launch window, whereas a launch vehicle that is right at its performance limits will have only a very short duration window (or, if it's beyond its performance capability, there is no launch window at all). The width of the launch window is calculated as follows.

3.8.1 Launch Window Duration I: Orbital Missions

It has previously been shown that the ideal situation for a launch vehicle is to achieve burnout directly into the desired orbit; however, launch delays, off-nominal performance, and other factors make it difficult to time things just right. Therefore, the launch window is a measure of the performance margin of the selected launch vehicle.

As an example, transfer orbit injections with significant inclination changes require either less payload mass or more launch vehicle performance, as shown earlier in Figs. 3.27–3.29. A similar situation applies where the desired orbit is not located exactly at the burnout location; in this case, the launch vehicle has to dogleg to one side or the other. This, too, incurs a penalty. Simply stated, the orbital launch window is a measure of the performance margin of the launch vehicle. If it has a large margin, it will be able to dogleg significantly and thus tolerate significant delays (or launch early) and still make it to the desired orbit. Conversely, a launch vehicle that is near its performance limits will have a very narrow launch window, perhaps only a few seconds on either side of nominal.

So, all one needs to know to find the launch window width is the launch vehicle's plane change capabilities with the chosen payload. This is a curve showing launch vehicle capability vs deviation from nominal launch time, as shown in Fig. 3.31. The figure shows that the launch window results from the widest deviation that the launch vehicle can accommodate.

There may be other, nonlaunch performance issues that affect the launch window. A good example of this was the Space Shuttle, which had a number of constraints pertaining to visibility for ascent photography, having daylight on transatlantic abort landing sites, changes to orbital altitude, added mission

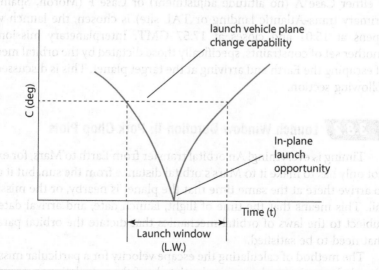

Fig. 3.31   Launch window based on launch vehicle performance. Source: [15].

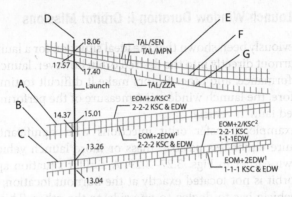

A. Sunrise at KSC & KAFB for EOM, EOM + 1 day, and EOM + 2. "2-2-2" means ther will be daylight
   for 2 potential landings at each site for each of 3 days. valid at the nominal mission altitude.
B. Same as A, except that the information is valid with an orbit altitude adjustment during the mission.
C. Similar to A, except there is only 1 daylight landing opportunity at KSC for EOM+2 (2-2-1), and 1
   opportunity each day at Edwards (1-1-1).
D. Vertical bar represents 9-21-1995, USML-2 launch date. if launch is delayed, bar slides to right to
   new launch date
E. The time the sun sets at TAL site Ben Guerir, Morocco, a potential closing of a launch window.
F. Same as E, but for moron, spain
G. Same as E, but for Zaragoza, spain

Fig. 3.32 Space Shuttle launch window constraints. Source: NASA.

days, an alternate landing site (Edwards Air Force Base, in addition to KSC),
and so forth. This complex group of constraints is shown in Fig. 3.32, which
also contains explanations of each curve.

With all of these constraints, what is the Space Shuttle's launch window?
If either Case A (no altitude adjustment) or Case F (Morón, Spain as the
primary trans-Atlantic landing or TAL site) is chosen, the launch window
opens at 15:01 and closes at 17:57 GMT. Interplanetary missions pose
another set of constraints, specifically those dictated by the orbital mechanics
of escaping the Earth and arriving at the target planet. This is discussed in the
following section.

3.8.2 Launch Window Duration II: Pork Chop Plots

Timing is everything! An orbital transfer from Earth to Mars, for example,
not only has to make it to Mars's orbital distance from the sun, but it also has
to arrive there at the same time that the planet is nearby, or the mission will
fail. This means that the time of flight, launch date, and arrival date are all
subject to the laws of orbital mechanics that dictate the orbital parameters
that need to be satisfied.

The method of calculating the escape velocity for a particular mission has
already been shown; however, the details of the calculations were not provided. It turns out that there are an infinite number of different possible

transfer orbits that go from Earth to Mars; what differentiates them are the time of flight (TOF), the required energy of escape (*C*3), and the arrival speed (which dictates how much energy the spacecraft needs to provide or dissipate in order to be captured by the target planet). TOF and *C*3 are often plotted on a graph that is known as a *pork chop plot* because it resembles the appearance of a cut through a piece of meat. An example pork chop plot showing transfers from Earth to Mars is shown in Fig. 3.33. This is a compact way to illustrate a huge number of orbital possibilities on a single figure.

Looking at Fig. 3.33, we see that the horizontal and vertical axes correspond to the dates of departure from Earth and arrival at Mars. There are contours indicating the required *C*3 to initiate the transfer orbit. There are also lines slanting up from left to right that indicate the time of flight in days. What is seen in orbital mechanics is much the same as is seen in anything else: the faster that things need to happen, the more energy it takes.

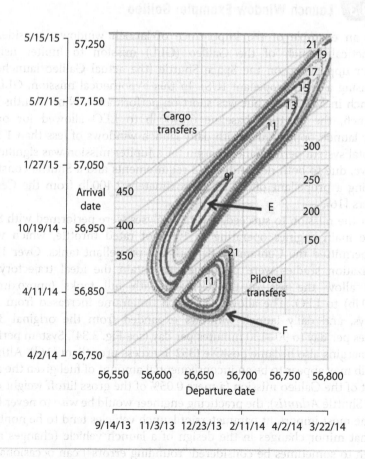

Fig. 3.33   Pork chop plot for Earth–Mars transfer. Source: [17].

The fastest transfer (shown as F on the plot) takes a bit less than 150 days and requires $C3 \cong 19 \text{ km}^2/\text{s}^2$. The most efficient (lowest $C3$) transfer is shown as E on the plot and takes about 310 days and $C3 \cong 9 \text{ km}^2/\text{s}^2$. The choice depends on the $C3$ capability of the launch vehicle. For example, if the selected launch vehicle can provide a $C3 = 10 \text{ km}^2/\text{s}^2$ carrying a payload of the desired size, it may be possible do a "slow" transfer taking about 280 days, but that's it. If the launch vehicle's $C3 = 20 \text{ km}^2/\text{s}^2$, then it's possible to send the spacecraft on a fast transfer.

As mentioned previously, the trajectory planner also needs to consider the *arrival* speed at the target. If the arrival speed is prohibitively high, then there will be a high cost for propellants needed to capture the spacecraft. Clearly the chosen trajectory is a compromise among $C3$, TOF, and arrival speed. For more information on pork chop plots, refer to the NASA pork chop plot page at http://marsprogram.jpl.nasa.gov/spotlight/porkchopAll.html.

3.8.3 Launch Window Example: Galileo

As an example of the importance of launch windows, consider the hypothetical launch of the Galileo (GLL) mission to Jupiter using a Centaur upper step on the Space Shuttle (the actual Galileo launched in 1989 using a solid-propellant IUS). In this hypothetical mission, GLL was to launch in May 1986, but was short on performance. Nine months prior to launch, the Shuttle's baseline 65,400 lb to LEO allowed for only a 14-day launch opportunity, with daily launch windows of less than 1 hour. The total system performance margin for a Jupiter mission was significantly negative, due to both demanding $C3$ requirements and a Shuttle constraint requiring a propellant offoad of approximately 2,400 lb from the Centaur boosters [16].

For the mission to succeed, several analyses were performed with Space Shuttle main engines operating at 109% of rated throttle, which would have permitted the Centaur to fly with full propellant tanks. Over 15,000 optimization studies were performed to obtain the ideal trajectory that would allow the use of the Centaur with full tanks (approximately 67,750 lb) to LEO. The allowable launch timeframe increased from 14 to 22 days, and daily launch windows expanded from the original 30–50 minutes per day to 90–130 minutes per day (see Fig. 3.34). System performance margins also became positive (352 lb excess propellant [16]). Although 2,400 lb may appear to be an inconsequential amount of fuel given the all-up weight of the Galileo mission (a mere 0.05% of the gross liftoff weight of the Space Shuttle *Atlantis*), the practicing engineer would be wise to never forget that the most important parameters of launch vehicles tend to be nonlinear, and that minor changes in the design of a launch vehicle (changes minor enough to sometimes be considered "rounding errors") can occasionally be what determines the success or failure of a mission.

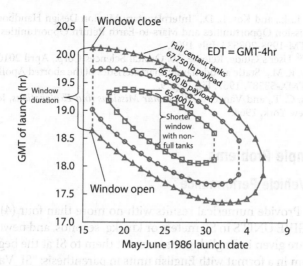

Fig. 3.34   Launch window for Galileo mission. Source: NASA.

References

[1] Curtis, H., *Orbital Mechanics for Engineering Students*, Butterworth-Heinemann, Oxford, UK, 2013.

[2] Brown, C., *Elements of Spacecraft Design*, AIAA, Reston, VA, 2003.

[3] Ordway, F. I., Gardner, J. P., and Sharpe, M. R., *Basic Astronautics: An Introduction to Space Science, Engineering and Medicine*, Englewood Cliffs. NJ, Prentice-Hall, 1962.

[4] Sellers, J., Astore, W., Giffen, R., and Larson, W., *Understanding Space: An Introduction to Astronautics*, Pearson, Upper Saddle River, NJ, 2007.

[5] Logsdon, T., *Orbital Mechanics: Theory and Applications*, Wiley, New York, 1997.

[6] Dergarabedian, P., and Ten Dyke, R. P., *Estimating Performance Capabilities of Boost Rockets*, STL TR-59-00792, Space Technology Labs, Los Angeles, 1959.

[7] Haviland, R. P., and House, C. M., *Handbook of Satellites and Space Vehicles*, van Nostrand, New York, 1965.

[8] White, J. F., *Flight Performance Handbook for Powered Flight Operations*, 1962, John Wiley & Sons, Inc., New York, 1963.

[9] Jensen, J., Townsend, G. E., Jr., Kraft, J. D., and York, J., *Design Guide to Orbital Flight*, McGraw-Hill, New York, 1962.

[10] Griffin, M. D., and French, J. R., *Space Vehicle Design*, AIAA, Reston, VA, 2004.

[11] Northrop Grumman Corp., *Space Data*, Northrop Grumman, El Segundo, CA, 5th edition, 2003. Out of print.

[12] Schilling, J., "Launch Vehicle Performance Estimation," http://www.silverbirdastronautics.com/LaunchMethodology.pdf [retrieved 24 March 2018].

[13] Silverbird Astronautics, http://www.silverbirdastronautics.com [retrieved 10 Aug. 2018].

[14] Boeing Publication 06H0214, "Delta II Payload Planner's Guide," December 2006.

[15] Chobotov, V. A., *Orbital Mechanics*, AIAA, Reston, VA, 2002.

[16] National Aeronautics and Space Administration, "DUKSUP: A Computer Program for High Thrust Launch Vehicle Trajectory Design and Optimization," NASA TM 2015-218753, 2015.

[17] George, L. E., and Kos, L. D., "Interplanetary Mission Design Handbook: Earth-to-Mars Mission Opportunities and Mars-to-Earth Return Opportunities 2009-2024," NASA TM-1998-208533, July 1998.

[18] Pegasus® User's Guide, Release 7.0. Orbital Sciences Corp., April 2010.

[19] Glasgow, R. M., "Static aerodynamic characteristics of the aborted Apollo-Saturn V," NASA TMX-53587, 1967, Fig. 3, p. 9.

[20] Leondes, C. T., and Vance, R. W., *Lunar Missions and Explorations*, John Wiley & Sons, New York, 1964.

3.9 Example Problems

Required Vehicle Performance

NOTE: Provide numerical results with no more than four (4) significant digits and GIVE UNITS in SI: meters or km, kg, seconds, and newtons. If specifications are given in English units, convert them to SI at the beginning and provide them in a format with English units in parenthesis: "SI_Value SI_unit (English_value English unit)". Include a summary PPTX along with XLSX copies of spreadsheets that are used and/or PDFs of any codes. NOTE: you may find it helpful to use MS Excel®'s "goal seek" function to determine payload mass for a given speed.

1. The State of Israel needs to orbit spy satellites to "keep an eye" on her hostile neighbors. Since all neighbors to her east are hostile, it is impossible to launch to the east to take advantage of the earth's rotation speed, and all satellites must be launched from Israel in a retrograde orbit, to the west. The launch location is Palmachim AFB, Israel, latitude: 31.88° N, longitude: 34.68° E. After liftoff, the LV travels down the Mediterranean before a "left turn" at Gibraltar to reach the azimuth angle needed for the desired inclination. (39 pts.)

a) Enter the necessary formulae into the Excel® spreadsheet template shown below to calculate design speed (with losses) necessary to place the satellite into a **250 km** circular orbit with **143°** inclination, Δv_{143}. For losses, use **80%** of the "conservation of energy" gravity loss (eq. 3.11) and **200 m/s** aerodynamic drag loss. Ignore the losses associated with the long path east of Gibraltar, and that

Spreadsheet Template by D. Edberg last mod. 2018 APR 29			Your_name	
Earth Launch Velocity Calculator adding grav loss AFTER v_{needed}			Launch site	Latitude
Constants			KSC	28.50
g_0	0.00980665	km/s²	Kourou	5.53
μ_{earth}	398600	km³/s²	Vandenburg	34.60
R_{earth}	6378.00	km	Baikonur	45.90
R_{GEO}	42164	km	Palmachim	31.90
Orbital calculations (user inputs in RED)			Sriharikota	13.62
(user input) orbit altitude	500	km	with 180 inclination Δ:	
$v_{circular}$	7.613	km/s		
(user input) orbit inclination i	143.00°	retrograde		prograde
Launch azimuth β	is 0 if $i = L_0$			
β given for ascending node (1st) opportunity.				
(user input or select above) latitude L_0	31.90°		same	
Launch dir aux angle γ or τ				
use 180°– this for descending node.				
Launch window loc ang δ				
(user input) flight path angle ϕ or γ		(0 if circular)	same	
$v_{launch\ site}$		km/s East	same	
SEZ components of burnout velocity				
$v_{burnout\ south}$		km/s South		km/s South
$v_{burnout\ east}$		km/s East		km/s East
$v_{burnout\ zenith}$		km/s Zenith		km/s Zenith
SEZ Components of needed velocity change				
$\Delta v_{needed\ south}$		km/s South		km/s South
$\Delta v_{needed\ east}$		km/s East		km/s East
$\Delta v_{needed\ zenith} = \Delta v_{burnout\ zenith}$		km/s Zenith		km/s Zenith
Δv needed				
Δv_{needed}		km/s		km/s
88% of $v_{loss\ gravity}$		km/s Zenith	same	
(user input) $v_{drag\ losses}$	0.200	km/s	same	
LV design speed inc. grav & drag Losses				
Δv_{design}	0.200	km/s		0.200 km/s
$\Delta(\Delta v_{design})$	0	m/s		
Upper Stage Calculations				
m_0	2048	kg		
m_f	170	kg		
I_{sp}	298	s		
m_{PL}	200	kg		
Δv		km/s		
v_e		km/s		
new Δv		km/s		
$MR = m_0/m_f$		=exp($\Delta v/v_e$)		
new mass		kg		
Δm_{PL}	-200.0	kg		
% m_{PL} increase		%		

shutdown/burnout occurs immediately after the left turn.
Provide values for 1) azimuth angle needed, 2) gravity loss and 3) Δv_{143} (in m/s).

b) Determine the inclination of a *prograde* orbit having the same "tilt" as the retrograde one in part a. Use the prograde inclination and part a's altitude to determine the Δv_{pro} necessary to place the satellite into a prograde orbit, using the same gravity and drag losses specified in part a.

c) How much *additional* Δv, call it $\Delta v_{difference} = \Delta v_{143} - \Delta v_{pro}$, does the retrograde orbit require over the prograde orbit with the same inclination?

d) There is a significant Δv disadvantage for the retrograde orbit, but the orbit has at least one advantage for a reconnaissance satellite. Try to think of one advantage of the retrograde orbit as compared to the preferred prograde orbit, and explain your answer using non-mathematical arguments about the satellite's ability to image the ground.

e) The Israeli launch vehicle *Shavit* has a third step with fueled mass $m_0 = 2,048$ kg, an empty mass $m_s = 170$ kg (neither include payload), and $I_{sp} = 298$ s. Its thrust is $T = 58,800$ N, and its burn time is **94 sec**. Assume a payload mass of **200 kg** to calculate the Δv_3 the third step (only) provides.

f) Calculate m_{PL-pro}, the mass of the payload the third step could launch to the **prograde** orbit, assuming that the lower steps produce the same Δv regardless of "small" changes in payload mass. To do this, use the value of $\Delta v_{difference}$ to calculate the (larger) payload mass the third step could accelerate in the *prograde* direction. Supply 1) m_{PL-pro}, the resulting *prograde* payload mass, and 2) the *percentage increase* in mass compared to item e.

2. Engineering data for the Northrop Grumman *Pegasus* are provided below (table from 2010 **Pegasus User's Guide** [18], p. 6):

Assume that a circular orbit is required at **250 km** altitude. Neglect the change in effective specific impulse due to change in ambient pressure. Use the above data to determine the launch performance of NGC's *Pegasus* under two sets of conditions. Use the spreadsheet template below, or modify the spreadsheet you used for the previous problem.

Typical Pegasus XL Motor Characteristics in Metric (English) Units.

Parameter	Units	Stage 1 Motor Orion 50S XL	Stage 2 Motor Orion 50 XL	Stage 3 Motor Orion 38
Overall length	cm (in)	1,027 (404)	311 (122)	134 (53)
Diameter	cm (in)	128 (50)	128 (50)	97 (38)
Inert weight (1)	kg (lb)	1,369 (3,019)	416 (918)	126 (278)
Propellant weight (2)	kg (lb)	15,014 (33,105)	3,925 (8,655)	770 (1,697)
Total vacuum impulse (3)	kN-sec (lbf-sec)	43,586 (9,799,080)	11,218 (2,522,070)	2,185 (491,200)
Average pressure	kPa(psia)	7,515 (1,090)	7,026 (1,019)	4,523 (656)
Burn time (3) (4)	sec	68.6	69.4	68.5
Maximum vacuum thrust (3)	kN (lbf)	726 (163,247)	196 (44,171)	36 (8,062)
Vacuum specific impulse effective (5)	Nsec/kg (lbf-sec/lbm)	2,846 (295)	2,838 (289)	2,817 (287)
TVC deflection	deg	NA	±3	±3

Notes: (1) Including wing saddle, Truss, and associated fasteners
(2) Includes igniter propellants
(3) At 21°C (70° F)
(4) To 207 kPa (30 psi)
(5) Delivered (includes expended inerts)

a) Assume it's dropped at the equator from a carrier vehicle heading due east. The carrier vehicle is flying at $M = 0.82$ at **11.9 km** altitude (use any atmosphere table to look up the speed of sound a at this altitude,

Air Drop Velocity, adding grav loss AFTER v_{needed} (user inputs in RED)		Launch site	Latitude
g_0	0.00980665 km/s²	KSC/CCAFS	28.50
μ_{earth}	398600 km³/s²	Vandenberg	34.60
	AIR LAUNCH	GROUND LAUNCH	
launch alt (km) above SL	11.90 km	0.00 km	
R_{earth}	6378.14 km	6378.14 km	
launch radius r	6390.04 km	6378.14 km	
aircraft horiz. speed	0.82 Mach	0.00 Mach	
sound speed (www.digitaldutch.com/atmoscalc/)	km/s	km/s	
Orbital calculations			
(user input) orbit altitude	250 km		
$v_{circular}$	km/s	same	
(user input) orbit inclination i	0.00° prograde	0.00° prograde	
Launch azimuth β			
(user input) flight path angle ϕ or γ	(0 if circular)		
$v_{launch\ site}$	km/s East	same	
total initial speed = $v_{drop} + v_{launch\ site}$	km/s	km/s	
SEZ components of burnout velocity			
$v_{burnout\ south}$	km/s South	km/s South	
$v_{burnout\ east}$	km/s East	km/s East	
$v_{burnout\ zenith}$	km/s Zenith	km/s Zenith	
SEZ Components of needed velocity change			
$\Delta v_{needed\ south}$	km/s South	km/s South	
$\Delta v_{needed\ east}$	km/s East	km/s East	
$\Delta v_{needed\ zenith} = \Delta v_{burnout\ zenith}$	km/s Zenith	km/s Zenith	
Δv needed			
Δv_{needed}	km/s	km/s	
80% of energy conservation gravity loss	km/s Zenith		
(user input) $v_{drag\ losses}$	0.040 km/s	0.200 km/s	
Launch vehicle design velocity including 88% gravity & assumed drag Losses			
Δv_{design}	40.0 m/s	200.0 m/s	
$\Delta(\Delta v)$		-160.0 m/s	

Pegasus Air Drop performance (a.) Equator Airdrop: 11.9 km alt, 0.82 M, 0° i, 40 m/s drag	Below based on Pegasus User's guide, figure 2-4	Pegasus Ground launch (b.) i=alt=M=0, 200 m/s drag
Step 1 inert mass m_{i1}	1369 kg	
Step 1 propellant mass m_{p1}	15014 kg	
Step 1 loaded mass m_{o1}	kg	
Step 1 exit velocity v_{e1}	2846 m/s	
Step 2 inert mass	416 kg	
Step 2 propellant mass	3925 kg	
Step 2 loaded mass	kg	
Step 2 exit velocity	2838 m/s	
Step 3 inert mass	126 kg	
Step 3 propellant mass	770 kg	
Step 3 loaded mass	kg	
Step 3 exit velocity	2817 m/s	
Payload mass m_{PL}	kg (a)	kg (b.)
Stage 1 Δv	m/s	m/s
Stage 2 Δv	m/s	m/s
Stage 3 Δv	m/s	m/s
total Δv	0 m/s	0 m/s
	Payload difference	0.0% (c.)

then vehicle speed = Ma). After calculating the magnitude of $v_{required}$, add the 80% factor on gravity loss from an air drop (Eq. 3.9), and 40 m/s as the drag loss. Determine m_{PL-air}, the air-dropped payload (kg) that may be delivered to this orbit.

b) Assume that the *same* vehicle is launched vertically from a fixed pad located at sea level at the equator to the same circular orbit as part a. Even though the real version doing this would not use wings, *use the same values* for the first step as shown the table above. Also, use the same **80%** factor on the gravity loss, but for this case, assume the drag loss is **200 m/s** due to the ascent from the bottom of the atmosphere. It's likely the payload mass needs to be reduced to achieve the same orbit. What is the new payload mass a $m_{PL-ground}$ (kg)?

c) What percentage larger $= (m_{PL-air} - m_{PL-ground})/m_{PL-air} \times 100\%$ is the air-launched payload mass compared to the ground-launched payload mass?

d) Write a few sentences commenting about the advantages and disadvantages of these two approaches. Also, describe what changes you might make to provide a more realistic estimate.

3. **Calculate the required LV design speed(s) for the vehicle(s) and required orbit(s) provided by your instructor, or by the assigned mission.** Your instructor will assign a particular mission or set of requirements. Do this/these calculation(s) by following the same procedures requested for executing Problems 1 and 2 above. You will need to calculate the values for the exact orbits stated in the RFP. For each of the orbits specified in your RFP, provide 1) launch azimuth angle(s), 2 & 3) periapse & apoapse altitudes in km, 4 & 5) periapse & apoapse speeds in m/s, 6) required design Δv in m/s, 7) the assumed drag loss in m/s, and 8) the calculated gravity loss value in m/s and knockdown factor (%), before your calculations. Include any other pertinent RFP orbital requirements such as additional speeds (m/s) for added speeds for a transfer orbit, plane change, or kick burns if required. If any information needed for your calculations is missing, use those provided below. Otherwise, provide estimates for it/them so you can carry out the calculations, and provide a few words of justification for your estimate.

If the mission specifies three or less orbits or trajectories, do the calculations for all specified. If more than three are specified, than provide data for the three you think will require the highest performance.

a) If launching from the ground, provide the launch location's latitude(s) in degrees, elevation(s) above SL in m, and local Earth rotational speed(s) in m/s. If launching from air, provide the drop vehicle's air speed (m/s) and altitude (m).

Your trajectories should include one or more of b) through e) following:

b) If your required orbit is circular with altitude ≤ 250 km, assume a direct ascent with horizontal burnout at the required altitude, with the required inclination. State all needed shutdown orbital speeds.

c) If your required orbit is circular with altitude > 250 km, assume an ascent that shuts down at the proper speed at 250 km to coast elliptically up to the required altitude, where a second circularization burn takes place. If an inclination change is needed, carry it out as a combined plane change/apogee kick burn at apoapsis. State all needed shutdown and kick burn speeds.

d) If your required orbit is elliptical or a Hohmann transfer to GEO at $0°$ i, assume your vehicle shuts down with the correct speed at 200 km altitude to inject into an elliptical orbit with the desired apoapsis altitude. If the periapsis is specified to be other than 200 km, provide the proper periapsis speed for its altitude. If an inclination change is needed, combine any plane change burn with circularization burn at apoapsis. State all needed kick burn speeds.

e) If your required orbit is an escape trajectory, assume your vehicle's shutdown occurs with the correct hyperbolic escape speed at 200 km altitude. You may launch directly towards the east, since escape can be done from any orbital inclination and this provides maximum payload. State the trajectory's $v_{periapsis}$, v_∞, and C3 values.

Only consider closed orbits or escape trajectories for this assignment. You may ignore any ballistic or subterranean trajectories.

Chapter 4 Propulsion

F rom the standpoint of atmospheric flight, thrust is one of the four
principal forces with which the engineer is likely to be familiar. (The
other three principal forces are lift, drag, and weight.) Rockets are
unique in this regard, because all nonwinged launch vehicles must use
their engine(s) to produce both lift and thrust. Generally speaking, a *propulsive element* can be defined as any assemblage of components that is designed
to impart an acceleration to the launch vehicle in order to change its position
or velocity with respect to a frame of reference. The rocket engine that accelerates a launch vehicle upwards towards its final orbit is the most obvious
example, although reaction control thrusters are also generally considered
propulsive elements. Note that some devices, such as reaction wheels,
change a vehicle's *orientation* with respect to a reference frame, but not its
position or velocity with respect to that reference frame; as such, they are
not considered propulsive elements and are not discussed here.

As is the case with other aerospace vehicles, the mission requirements of
"how much," "how far," and "how fast" will constrain the design space of the
propulsion system. Launch vehicles are also likely to encounter additional
constraints that are unique among aerospace vehicles; these constraints
usually deal with the large amount of thrust required to deliver a relatively
small amount of payload to orbit (which drives weight fractions and performance requirements), as well as the extreme toxicity of several propellants.
This toxicity not only impacts the performance of the propulsion system
(due to chemistry and thermodynamics), but also drives material selection
(due to corrosion) and storage and handling of propellants. The notion of
propellant toxicity should not be taken lightly: jet fuel, for instance, can be
safely handled with a minimum of protective equipment; should a few
drops spill onto the ground (or onto the unlucky technician who's fueling
the aircraft), the spill can be mitigated fairly easily. By contrast, spilling a
few drops of hydrazine can result in liver and kidney damage, seizures, and
death. Once consideration of propellant toxicity is taken into account,
other real-world factors—specifically, the economy and lifecycle considerations of a launch vehicle design—become immediately relevant, rather
than being a deprioritized, "also-need" design feature.

Several types of propulsive elements will be covered in this chapter. This
chapter is devoted to chemical rockets (i.e., rockets that use combustion to
provide thrust), because at the time of this writing they are still the most
common form of propulsion for launch vehicles. Because chemical rockets
derive their performance chiefly from the combustion of their propellants,
an overview of the combustion process is presented first; this overview is

by no means extensive, and the interested reader is directed to [1, 2, 3] for a more in-depth treatment of the subject.

A note on terminology: the word *engine* is typically used to refer to propulsive elements that "breathe" air (or oxygen) as part of the combustion cycle. For launch vehicles, the term *engine* refers to liquid- or hybrid-architecture propulsive units (e.g., LOx/RP-1), whereas the term *motor* is reserved for propulsive units that utilize solid propellant. Although the *Oxford English Dictionary* defines *motor* as a machine that supplies motive force (and motive force is certainly something that launch vehicles have in ample supply), the terminology used herein differentiates between the two terms in accordance with current standard practice within the industry.

4.1   Combustion

Chemical rockets are unique among aerospace propulsion solutions in that the chemical rocket is typically not an air-breathing engine (because there is no sensible "air" above about 100 km or at orbital altitudes), and thus must carry its own "air" (in the form of an oxidizer) with it throughout its flight. This not only adds to the all-up mass of the launch vehicle, but also opens new design spaces for the propulsion engineer: after all, a jet engine is forced to breathe whatever composition of air is immediately available (regardless of its designer's wishes), but the designer of a chemical rocket can, during the design phase, tailor the performance of the vehicle's oxidizer to a certain degree. These propellants are consumed during flight, and it is the mechanics of combustion that impart to the launch vehicle the necessary momentum to perform its mission.

The calculation of the total temperature of a working fluid experiencing combustion under conditions of constant total pressure is a useful tool to predict the behavior of the combustion products as they enter the nozzle; as such, the majority of this section will be devoted to calculating this parameter. *Combustion* refers to the high-temperature chemical reaction that occurs between two elements, which produces heat (usually in the form of a flame). The entire purpose of combustion is to convert chemical energy (stored within the molecules of the propellants) into thermal energy, which can then be used to generate thrust. Within a launch vehicle's engine, these chemical elements are the propellants (which may be in solid, liquid, or gaseous form), which when combusted produce gaseous products. Once a specific amount of energy (called the *activation energy*) is supplied to the chemicals, a reaction can be sustained. The chemicals that "enter" a reaction are called *reactants*, whereas the results of a chemical reaction are called the *products* of the reaction.

Flames are classified according to three characteristics: the composition of the reactants, the character of the flame, and the steadiness of the flame. The first characteristic, the composition, identifies the composition of the reactants upon entering the reaction zone. If the reactants are uniformly mixed together (or very nearly so), the flame is called *premixed*; otherwise, it is known as a *diffusion* flame, because the mixing of the propellants is accomplished via diffusion. The second characteristic identifies whether the flame is laminar or turbulent; for laminar flames (i.e., those occurring at low Reynolds numbers) the mixing and transport are accomplished through purely molecular processes, whereas turbulent flames (i.e., those occurring at high Reynolds numbers) substantially enhance the mixing and transport phenomena via macroscopic turbulent phenomena (e.g., eddy flow). The third characteristic of a flame identifies whether it is steady or unsteady, determined by whether the flame structure changes relative to the initial phase of the reactants.

To simplify the design cycle of propulsion systems, a number of assumptions must be made with regard to combustion. These assumptions are commonplace and are especially prevalent in classical combustion models:

* Reacting fluid is a continuum.
* Chemical equilibrium (stoichiometry) is always attained.
* Chemical reactions occur infinitely fast.
* Chemical reactions are irreversible.
* Ideal gas assumptions are valid.
* All species have equal mass diffusivity.
* Lewis, Schmidt, and Prandtl numbers are approximately unity.
* Fick's law of diffusion is valid.
* All gas phases have constant specific heats.
* Reacting solid surfaces are energetically homogeneous.
* Low-speed combustion occurs at uniform pressure.
* Dufour effect (energy flux due to mass concentration gradient) and thermophoresis (Soret effect, wherein particle types in a mixture exhibit different responses to thermal gradients) are negligible.
* When flowing through a nozzle, all chemistry is assumed "frozen" (i.e., having an infinitely slow reaction rate).

Despite the often-complicated sequence of chemical reactions that occurs during combustion, it is important to remember that the heat produced by combustion often makes the combustion self-sustaining. Combustion also requires fairly high temperatures, especially when catalysts (substances that increase the rate of chemical reactions without undergoing significant chemical changes themselves) are not used. Combustion is said to be "complete" when it is *stoichiometric*—that is, when there is no remaining fuel, there is also no remaining oxidant. In addition, stoichiometry implies that the products of a chemical reaction are stable. In practice, this chemical equilibrium is virtually impossible to achieve; therefore, complete combustion is similarly difficult to achieve. Real combustion reactions often contain unburned reactants that are carried off by the products of the combustion (e.g., smoke), which means that these products will usually be toxic. Unburned reactants, by definition, have not undergone combustion, and have therefore contributed nothing to the overall energy output of the engine; thus, attaining and maintaining chemical equilibrium while minimizing the amount of unburned reactants is a crucial effort in the design of rocket engines.

Note that for any combination of fuel and oxidizer used, the ratio of fuel to oxidizer (commonly denoted as f) that results in a stoichiometric reaction is unique. It is often beneficial to describe the oxidizer–fuel mixture in terms of this unique stoichiometric ratio, typically through the use of an *equivalence ratio*, commonly referred to as φ

$$\varphi = \frac{f}{f_{\text{stoich}}} \tag{4.1}$$

where f_{stoich} represents the stoichiometric oxidizer–fuel ratio. For $\varphi < 1$, the mixture is said to be fuel-lean (oxidizer-rich), whereas $\varphi > 1$ denotes a fuel-rich (oxidizer-lean) mixture. Combustion behavior is very dependent on equivalence ratio and flame temperature, and these terms have a very real impact on the performance of an engine. A lean mixture, for instance, will typically result in a lower combustion temperature, but a rich mixture of hydrocarbon fuel can produce soot or oxides of nitrogen within certain temperature regimes.

Another useful fuel metric is known as *bulk density*, which is a measure of the density of the individual propellants relative to the total volume they occupy; it can be thought of as the "average" density of the mixed propellant, and is defined as

$$\rho = \frac{f+1}{\dfrac{f}{\rho_o} + \dfrac{1}{\rho_f}} \tag{4.2}$$

Note that the bulk density is entirely and only dependent upon both the individual densities of the oxidizer (ρ_o) and fuel (ρ_f), as well as the oxidizer–fuel ratio. Generally speaking, a higher propellant bulk density will result in smaller propellant tanks.

The maximum possible flame temperature for a combustion reaction is known as the *adiabatic flame temperature*, and is the temperature that a flame reaches if there is no energy lost to an external environment. This flame temperature is usually reached at equivalence ratios between 1.0 and 1.1, although it should be noted that the adiabatic flame temperature (which usually ranges from 2,000 to 5,000°C or 3,600 to 9,000°F) is *theoretical*, and that material properties will usually limit the maximum operating temperature of an engine. Additionally, a limit on flame temperature may be imposed due to the fact that some products of chemical reactions will, at sufficiently high temperature, begin to dissociate back into their constituent reactants. The implications of this will be discussed later in this section.

The gases that comprise the combustion "charge" (also known as the working fluid) can, for the purposes of initial design, be treated as ideal gases. There are various forms of the ideal gas law, which relates the gas pressure P, gas volume V, gas mass m, and gas temperature T through the use of a specific gas constant R or the universal gas constant $\mathcal{R}$ (8.314462 J/mol-K), and the molecular weight of the gas MW

$$PV = mRT = \rho RT = m\frac{\mathcal{R}}{MW}T = n\mathcal{R}T \qquad (4.3)$$

where the universal gas constant is taken to be $\mathcal{R} \cong 8.314\frac{J}{mol-K}$. Equation (4.3) is also known as the equation of state of an ideal gas. Note that the specific gas constant is a function of the universal gas constant and the molecular weight of the gas,

$$R = \frac{\mathcal{R}}{MW} \qquad (4.4)$$

and will therefore be different for different gases.

Recall from basic thermodynamics, the gas specific heat at constant volume c_v (also known as the isochoric specific heat) and the specific heat at constant pressure c_p (also known as the isobaric specific heat) are related by the specific gas constant

$$c_p - c_v = R \qquad (4.5)$$

The behavior of a flowing gas is strongly influenced by its ratio of specific heats γ, which is simply

$$\gamma = \frac{c_p}{c_v} \qquad (4.6)$$

4.2 The Thrust Equation and Rocket Equation

The thrust provided to a rocket can be derived from Newton's second law, $F = ma$. Converting this to a more immediately applicable form and defining momentum as the product of mass and velocity, $p = mv$,

$$F = ma$$

$$F = \frac{dp}{dt}$$

$$F = m\frac{dv}{dt} \qquad (4.7)$$

$$F = v\frac{dm}{dt}$$

In rocketry, this final form of $F = ma$ is usually written as $F = \dot{m}v_e$, indicating that thrust is proportional to the mass flow rate of propellants (fuel and oxidizer) multiplied by the exit velocity of the combusted propellants, which for initial design may be assumed constant. This *jet thrust*, or *momentum thrust*, is produced solely by the expulsion of combusted propellant mass from the rocket nozzle (Fig. 4.1). A more accurate form of thrust must also include contribution from the exhaust pressure acting on the exit plane of the nozzle; this is commonly known as the *thrust equation*,

$$F = \dot{m}v_e + (P_e - P_\infty)A_e \qquad (4.8)$$

where P_e and P_∞ are, respectively, the exit pressure of the nozzle and the freestream pressure of the surrounding atmosphere, and A_e is the cross-sectional exit area of the nozzle. The total thrust is the sum of the momentum thrust less the pressure losses acting upon the nozzle exit plane (Fig. 4.1). The pressure term from the rocket equation represents the pressure loss incurred when the exhaust flow is not properly expanded.

Generally speaking, the momentum term ($\dot{m}v_e$) is much greater than the pressure term (the pressure term is a small percentage of the momentum term or less), and can often be neglected. It should also be apparent that, with conventional nozzles, it is possible to ideally expand the exhaust flow at only one flight condition (that being where the exhaust pressure Pe is equal to the freestream pressure P_∞); at every other flight condition, the exhaust flow will be either over- or underexpanded, as shown in Fig. 4.2. As seen in Fig. 4.3, the difference in thrust from Apollo 11 due to properly expanding the exhaust flow was on the order of almost 100,000 lb$_f$, which is certainly not insignificant!

It should be immediately obvious that there are only so many "knobs to turn" to increase thrust. Increasing the pressure of the combustion chamber is certainly possible, although this will also increase the combustion temperature and therefore affect the oxidizer–fuel ratio. Increasing the mass flow rate of the propellants increases the momentum term and will provide a

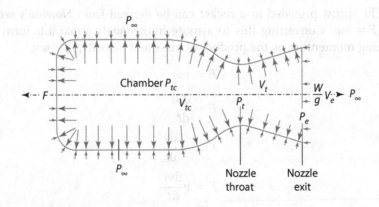

Fig. 4.1   Rocket thrust control volume.

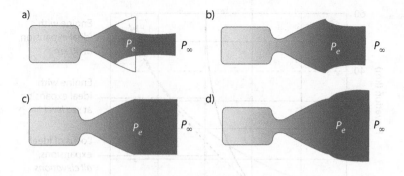

Fig. 4.2 a) Grossly overexpanded flow; b) overexpanded flow; c) ideally expanded flow; d) underexpanded flow.

much higher effect on the thrust produced; however, only so much propellant can be carried for a given mission. Properly expanding the flow such that the pressure term becomes zero will certainly increase thrust, but the change will be both negligible and applicable at only one flight condition, because the freestream pressure will constantly change during the vehicle's ascent. (The exception to this is the aerospike engine, which is discussed later in

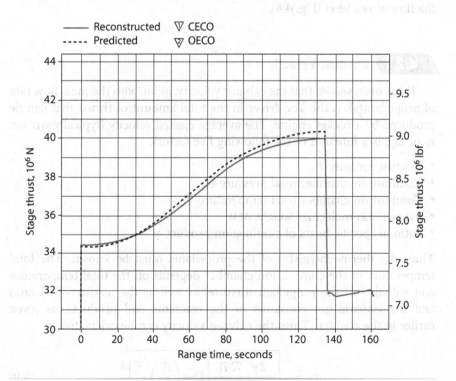

Fig. 4.3 Thrust increase from ambient pressure drop, Apollo 11. Source: NASA.

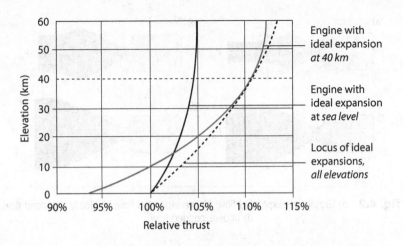

Fig. 4.4 Thrust vs altitude for various nozzle expansions. Source: [4].

the chapter.) Ideally, the nozzle should be designed to yield ideal expansion at approximately two thirds of the total flight altitude desired, which will yield more favorable thrust throughout the flight than a nozzle that ideally expands the flow at sea level (Fig. 4.4).

4.2.1 Exhaust Velocity

It has been shown that the exhaust velocity, along with the mass flow rate of propellant(s), is the key driver in the total amount of thrust that can be produced by a rocket engine. The average ejected velocity (typically written v_e or U_e) is a function of the following five factors:

- Exhaust pressure P_e
- Combustion chamber total pressure P_0
- Combustion chamber total temperature T_0
- Exhaust gas molecular weight MW
- Ratio of specific heats of combustion product γ

Thus, the thermochemistry of the propellants must be known. The total temperature of the combustion chamber depends on the total temperature and enthalpy of the propellant mixture, as well as the oxidizer–fuel ratio and the isobaric gas constants of the reactants and products, as given earlier in the chapter. Thus, the exhaust velocity can be calculated as

$$v_e = \sqrt{\frac{2\gamma}{\gamma - 1}\frac{\mathcal{R}T_0}{MW}\left[1 - \left(\frac{P_e}{P_0}\right)^{\frac{\gamma-1}{\gamma}}\right]} \qquad (4.9)$$

For an ideally expanded nozzle, the exhaust backpressure is approximately zero, and the exhaust velocity can be approximated as

$$v_e \cong \sqrt{\frac{2\gamma}{\gamma-1}\frac{RT_0}{MW}} \tag{4.10}$$

This approximation assumes that all of the thermal energy of the exhaust has been converted to kinetic energy, and thus the exhaust velocity is solely a function of combustion temperature and molecular weight of the combustion products. This also reveals another important point: the importance of molecular weight. As shown previously, a chemical reaction must be balanced; this means that the total mass of reactants prior to combustion must equal the total mass of products after combustion; however, it must also be noted that heavier, more complex fuels require more oxygen to combust (cf., methane vs hexane), and the molar mass of the products will also be heavier and more complex. This impacts the *molar efficiency* (i.e., total thrust produced per unit mass of propellant) of the vehicle. Higher molar mass equates to a lower exhaust velocity, and thus reduced thrust per unit mass of propellant. Therefore, a *lighter* fuel will, on the basis of molar efficiency, be preferable. Liquid hydrogen is favored as a propellant for precisely this purpose, although it is a cryogenic fuel and therefore not necessarily the best solution for every design when operating and cost constraints are considered. Regardless, it is generally preferable to use the lightest fuel possible for a given application, in order to maximize the thrust efficiency of the vehicle.

The exhaust velocity can also be rewritten as a *specific* exhaust velocity c, which is the total thrust divided by the mass flow rate of propellant

$$c = \frac{F}{\dot{m}} = v_e + \frac{A_e(P_e - P_\infty)}{\dot{m}} \tag{4.11}$$

which is, essentially, the total impulse of the rocket engine normalized by the mass flow rate. Rewriting the specific exhaust velocity in terms of Mach number at the exit plane of the nozzle,

$$c = v_e\left[1 + \frac{1}{\gamma M_e^2}\left(1 - \frac{P_\infty}{P_e}\right)\right] \tag{4.12}$$

For a large exhaust area with a high Mach number, the pressure term is a negligible portion of the overall thrust produced.

4.2.2 Specific Impulse

A useful analysis tool is the impulse provided by the propellant, which is equal to the area under the thrust-time curve (if the thrust is constant, the total impulse is the thrust multiplied by the burn time):

$$I = \int_0^{t_b} F(t)\mathrm{d}t = F_{\text{ave}}t_b \qquad (4.13)$$

This is the *total impulse*, which is a measure of the momentum that can be supplied to the vehicle by combustion of the propellants. Solid propellant hobby model rocket motors are categorized by their total impulse on an alphabetic scale, i.e. an "A" motor has a total impulse between 1.26–2.5 Ns, a "B" motor has twice the total impulse, between 2.51–5.0 Ns, and so on, doubling with each successive letter of the alphabet. Under such a designation, the Mercury-Redstone rocket would have a classification of letter Z, between 41.9 and 83.9 × 10^6 (million) Ns!

We will also use a term called *specific impulse* defined as the amount of thrust produced per unit of propellant consumed per unit time. Specific impulse is a measure of the efficiency of a rocket engine, and *higher is better*.

In the English/Imperial system, propellant consumed (w_p) is measured by weight (lb$_f$), and in the SI system, propellant consumed (m_p) is measured by mass (kg). Hence, if g_0 is the (standard) acceleration of gravity at Earth sea level ($g_0 = 9.80665$ m/s^2 = 32.17405 ft/s^2), for the specific impulse in the two sets of units we have the following relations:

$$I_{sp-\text{English}} = \frac{T}{w_p/t} = \frac{T}{\dot{w}_p} = \frac{T}{g_0\dot{m}_p} = \frac{v_e}{g_0} \qquad (4.14)$$

and

$$I_{sp-SI} = \frac{T}{m_p/t} = \frac{T}{\dot{m}_p} = v_e \qquad (4.15)$$

Note that there is a difference in I_{sp} depending on the unit system used: the units of I_{sp} could be in *seconds* or in *meters per second*, depending on whether the Imperial or SI systems of units is used, respectively. The SI version of specific impulse happens to be particularly useful because it is precisely the engine's effective exhaust speed. (If the English/Imperial specific impulse also used its mass units of slugs, then its specific impulse would be the effective exhaust speed in ft/sec.)

In this text, I_{sp} will be given using both sets of units, either seconds (which is consistent with U.S. industry practice), or in meters/second (the rest of the world). The units will immediately indicate which system is being used. Typical I_{sp} values (Imperial units) for various propellants are given in Table 4.1.

The specific impulse of a propellant can also be used to determine the overall size of the launch vehicle through the use of *volumetric specific impulse*, usually written as $I_{sp_{\text{vol}}}$, which is equal to the specific impulse

Table 4.1 Typical Specific Impulse Values

Propellants	Application	Exhaust Products	I_{sp}, s
Black powder	Amateur rocketry	(various)	80
Zinc, sulfur	Amateur rocketry	ZnS	240
Al, NH_4ClO_4	Shuttle SRM	(various)	287
Hydrazine, RFNA	OMS, Titan	H_2O, NH_3	313
Ethyl alcohol, LOx	V-2	H_2O, CO_2	330
RP-1, LOx	Atlas V, Saturn IC	H_2O, CO_2	350
Methane, LOx	(various)	H_2O, CO_2	370
LH_2, LOx	Delta IV, Space Shuttle	H_2O	450

multiplied by the specific gravity of the propellant

$$I_{sp_{vol}} = I_{sp}SG \qquad (4.16)$$

The *higher* the volumetric specific impulse is, the *less* space will be needed to store the propellants, which in turn equates to a smaller (and likely lighter) launch vehicle. For a fixed volume and mass, increasing the volumetric specific impulse increases the velocity increment delivered during flight. A very high volumetric specific impulse is desirable for the first stage of a multistage launch vehicle, in order to minimize both weight and aerodynamic penalties; if this cannot be achieved, strap-on boosters are often an economical way to increase the payload fraction of the launch vehicle. Typical volumetric specific impulse values for the Saturn V and Space Shuttle are given in Table 4.2. Notice that hydrolox volumetric specific impulses are a factor of three or four lower than those of kerolox, meaning the former will need much larger tanks than the latter.

Three other factors govern a rocket's specific impulse: oxidizer–fuel ratio (mixture ratio), chamber pressure, and expansion ratio (ratio of exit area A_e to throat area A_t), as shown in Fig. 4.5b. As can be seen, the specific impulse will usually depend more on the expansion ratio of the nozzle than on the chamber pressure. This would lead one to conclude that larger nozzles are

Table 4.2 Typical Sea-level and Vacuum Volumetric Specific Impulse Values

Propellants	SG	SL I_{sp}, s	Vac I_{sp}, s	SL $I_{sp_{vol}}$, s	Vac $I_{sp_{vol}}$, s
RP-1, LOx	1.3	265	304	345	395
LH_2, LOx (Saturn V)	0.28	360	424	101	119
LH_2, LOx (Shuttle)	0.28	390	455	109	127
Solid (Shuttle SRM)	1.35	242	262	327	354

a)

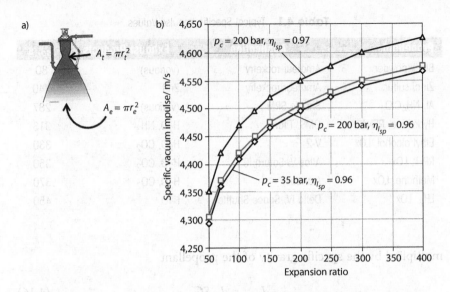

b)

Fig. 4.5 Specific impulse as a function of nozzle expansion ratio and chamber pressure. (a) Nozzle throat (t) and exit (e) dimensions; (b) I_{sp} vs chamber pressure expansion ratio. Source: Ley et al. [5].

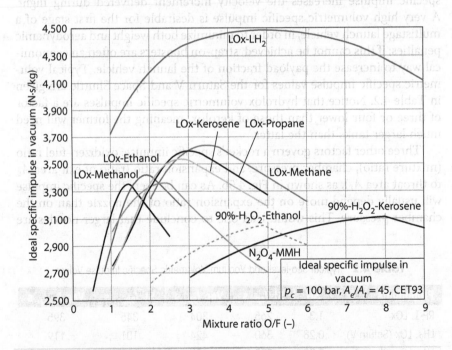

Fig. 4.6 Specific impulse as a function of propellant type and oxidizer–fuel ratio. Source: Ley et al. [5].

more desirable. Generally speaking, this is true; however, incorrectly expanding the nozzle results in detrimental effects on the thrust produced, as discussed previously.

Specific impulse also changes as a function of the oxidizer–fuel ratio of the propellant, as shown in Fig. 4.6. It should be immediately apparent that, for any combination of fuel and oxidizer used as propellants, there exists an optimum oxidizer–fuel ratio for which the I_{sp} will be at a maximum. The engine itself may not be able to *operate* at this oxidizer–fuel ratio (due to other considerations, such as thermal limits, performance of pumps, etc.), but the specific impulse does not increase without bound as the oxidizer–fuel ratio increases.

4.3 The Rocket Equation

The ultimate performance requirement for any launch vehicle or spacecraft is the change in velocity Δv (as discussed in Chapter 3), and is given as the *Tsiolkovsky rocket equation*, or

$$\Delta v = g_0 I_{sp-\text{English}} \ln \frac{m_0}{m_f} = v_e \ln \frac{m_0}{m_f} \qquad (4.17)$$

where m_0/m_f is called the *mass ratio* (usually given as μ or MR) and refers to the initial mass of the launch vehicle (typically its gross liftoff mass, or GLOM) divided by the final mass of the system (typically the mass of everything except the consumed propellants). Generally speaking, a higher mass ratio is more desirable, because it indicates that more of the rocket's all-up mass is "useful" (i.e., less inert mass). The rocket equation can, therefore, be applied to multiple-stage or single-stage launch vehicles with equal ease. The first form of the rocket equation is typically more useful in design, because the specific impulse of various propellant combinations can often be found in tables, whereas the exhaust velocity of a motor is typically not known before the motor has been designed. Note that the gravitational constant is equal to the standard value of 9.80665 m/s² no matter where the Δv is being calculated.

4.3.1 Propellant Mass Fraction and Total Impulse

Occasionally, a *propellant mass fraction* is used, especially for situations requiring multiple burns of varying durations or thrust levels, such as multiple burns of an upper- or lower-step engine. This propellant mass fraction is given as

$$\zeta_p = \frac{m_p}{m_0} \qquad (4.18)$$

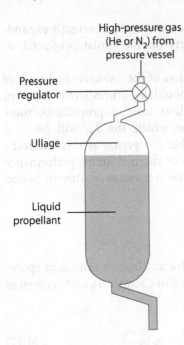

High-pressure gas (He or N$_2$) from pressure vessel

Pressure regulator

Ullage

Liquid propellant

Fig. 4.7 Typical arrangement of liquid propellant with ullage space for pressurizing gas.

where m_p is the mass of the propellant and m_0 is the all-up mass of the vehicle prior to the burn. Note that the propellant mass fraction only accounts for the amount of usable propellant, and does not account for residual propellant, which is never consumed.

Rarely are propellant tanks fully filled; instead, as shown in Fig. 4.7, there is usually "ullage space" for a pressurizing gas to enter the propellant tank. Note that

$$MR = \mu = \frac{1}{1 - \zeta_p} \quad (4.19)$$

and

$$\zeta_p = 1 - \frac{1}{\mu} \quad (4.20)$$

Typical propellant mass fractions are in the range of approximately 90%.

4.3.2 Thrust-to-weight ratio and burn time

The total impulse of a rocket is calculated as the area below the thrust-time curve, by integrating (symbolically or numerically) the time-varying thrust over the total burn time:

$$I_{tot} = \int_0^{t_b} T(t)\mathrm{d}t = T_{ave}t_b \quad (4.21)$$

The average thrust is defined as the total impulse divided by the burn time, or $T_{ave} = I_{tot}/t_b$. Hence the total impulse for a constant-thrust rocket is the (constant) thrust $\overline{T}$ multiplied by the burn time, or $I_{tot} = \overline{T}t_b$.

The thrust to weight ratio of the rocket is equal to the total thrust produced divided by the all-up mass of the rocket:

$$\frac{T}{W} = \frac{T}{m_0 g_0} \quad (4.22)$$

This can be related to the mass flow rate of propellant and the specific impulse of the rocket by

$$\frac{T}{W} = \frac{\dot{m}_p c}{m_0 g_0} = \frac{\dot{m}_p I_{sp}}{m_0} \quad (4.23)$$

Thus, for a given engine or motor, the higher the exit Mach number, the higher the propellant mass flow rate, and the higher the specific impulse, the greater the thrust-to-weight ratio of the rocket. Higher values of T/W are preferred, but a $T/W > 1$ is required for the rocket to be able to lift off at all. The initial thrust-to-weight ratio is usually defined as

$$\frac{T}{W_0} = \psi_0 \qquad (4.24)$$

For the purposes of initial design, the propellant mass flow rate can be averaged by dividing the total mass of the available propellant by the burn time, t_b:

$$\dot{m}_p = \frac{m_p}{t_b} \qquad (4.25)$$

This equation reveals a useful design metric. Solving for the burn time and using the definitions of propellant mass fraction and specific impulse,

$$t_b = \frac{\zeta_p I_{sp}}{T/W_0} = \frac{\zeta_p I_{sp}}{\psi_0} \qquad (4.26)$$

The propellant mass fraction, ζ_p, must always be less than one (rockets are rarely just propellant devoid of *any* structure); conversely, the thrust-to-weight ratio must always be greater than one (in order to lift off). Therefore, the theoretical burn time, in seconds, can never be greater than the engineering (weight normalized) specific impulse!

Example 4.2 Spacecraft with a partial burn cycle

A spacecraft with the following properties combusts a portion of its propellants. Find 1) the exhaust velocity, 2) the specific impulse, 3) the total impulse, and 4) the propellant mass fractions before and after the burn.

Item	Value
Total spacecraft mass	27,320 kg
Propellant mass before burn	1,521 kg
Propellant mass after burn	862 kg
Burn time	2 sec.
Δv	56 m/s

To find the exhaust velocity during the burn,

$$\Delta v = v_e \ln \frac{m_0}{m_f} \Rightarrow v_e = \frac{\Delta v}{\ln \frac{m_0}{m_f}} = \frac{56 \text{ m/s}}{\ln\left(\frac{27{,}320 \text{ kg}}{27{,}320 \text{ kg} - (1{,}521 - 862)\text{kg}}\right)}$$

$$= 2{,}293 \text{ m/s} \tag{4.27}$$

The specific impulse is simply the exhaust velocity divided by the gravitational constant,

$$I_{sp} = \frac{v_e}{g_0} = \frac{2{,}293 \text{ m/s}}{9.80665 \text{ m/s}^2} = 233.9 \text{ s} \tag{4.28}$$

The total impulse is found by using the mass flow rate of the propellant, the burn time, and the exhaust velocity

$$I_t = Ft_b = \dot{m}v_e t_b$$

$$I_t = \frac{\Delta m}{\Delta t} v_e t_b$$

$$I_t = \frac{(1{,}521 - 862) \text{ kg}}{2 \text{ s}} (2{,}294 \text{ m/s})(2 \text{ s}) \tag{4.29}$$

$$I_t = 1{,}511 \text{ kNs}$$

The propellant mass fractions are calculated as follows:

$$\zeta_{p1} = \frac{m_{p1}}{m_0} = \frac{1{,}521 \text{ kg}}{27{,}320 \text{ kg}} \Rightarrow \zeta_{p1} = 0.05567$$

$$\zeta_{p2} = \frac{862 \text{ kg}}{26{,}661 \text{ kg}} \Rightarrow \zeta_{p2} = 0.03233 \tag{4.30}$$

4.3.3 Summary of Rocket Engine Parameters

It is often useful to have an easily accessible reference of various rocket engine performance parameters. These are tabulated in Table 4.3.

Note that although rocket engines may appear to be complex (and they are!), the design of a propulsion system is not insurmountable. As with all design exercises, the selection of a rocket engine architecture is determined by a small number of performance parameters. These are:

- Δv required for mission
- Thrust required for launch
- Propellant types and chemistry (which determine exhaust products, storability, logistics, etc.)
- Propellant density (which constrains tank volume and inert mass)
- Chamber pressure and temperature (which constrains motor material, thermal management, etc.)

Table 4.3   Editable Rocket Engine Parameter Conversions

Name & Symbol	c	c^*	C_T	T	I_{sp}
c, Effective exhaust speed	—	$c^* \cdot C_T = \dfrac{c^* \cdot C_T}{P_c A_T}$	$\dfrac{C_T P_c A_T}{\dot m}$	$\dfrac{T}{\dot m}$	$g_0 I_{sp}$
c^*, characteristic exhaust speed	$\dfrac{c}{C_T}$	—	$\dfrac{T}{\dot m C_T} = \dfrac{g_0 I_{sp}}{C_T}$	$\dfrac{P_c A_T}{\dot m}$	$\dfrac{g_0 I_{sp}}{C_T}$
C_T, Thrust coefficient $= T/(P_c A_T)$	$\dfrac{c}{c^*}$	$\dfrac{g_0 I_{sp}}{c^*}$	—	$\dfrac{T}{P_c A_T}$	$\dfrac{g_0 I_{sp}}{c^*}$
T, Thrust	$\dot m c$	$\dfrac{P_c A_T g_0 I_{sp}}{c^*}$	$C_T P_c A_T$	—	$\dot m g_0 I_{sp}$
I_{sp}, specific impulse	$\dfrac{c}{g_0}$	$c^* \dfrac{C_T}{g_0} = \dfrac{c^*}{g_0}\dfrac{T}{P_c A_T}$	$\dfrac{C_T P_c A_T}{\dot m g_0}$	$\dfrac{T}{\dot m g_0}$	—

A_T = nozzle throat cross-section area, P_c = combustion chamber pressure, T = thrust
g_0 = Earth sea level gravity acceleration = 9.80665 m/s^2 = 32.174 ft/s^2
(*) is not a multiplication sign! Definitions in black.

4.4   Solid-Propellant Motors

Given the complexities of liquid-propellant rocket engines (discussed in greater detail in Section 4.5), it is perhaps unsurprising that a simpler option would exist. Solid-propellant motors [commonly referred to as solid rocket motors (SRMs)] certainly have a longer history than their liquid-propellant counterparts, dating back to China circa 200 BC. It therefore makes logical sense, from the perspectives of both simplicity and history, to examine SRMs first.

The solid-propellant motor utilizes a solid propellant that deflagrates (combusts subsonically) after ignition. It has several distinct advantages over its liquid-propellant counterpart: it is (relatively) inexpensive, storable, and simple in design; has high volumetric specific impulse (i.e., dense fuel); and does not require a propellant delivery system. This makes the SRM ideal for small- or medium-sized launchers, or as a reliable upper stage for orbital injection. The ease of storability of the propellant means that a solid-propellant rocket will likely need a minimum of support infrastructure relative to a liquid-propellant equivalent, simplifying the deployment strategy of an SRM. The SRM is also devoid of support equipment necessary for liquid-propellant propulsion: an SRM needs no turbopumps, plumbing, control valves, pressurant gas, or other dead weight, thus increasing its mass ratio.

However, two specific disadvantages plague SRMs, which to date have remained unresolved: the solid-propellant motor is almost guaranteed to have a lower specific impulse than a mass-comparable liquid-propellant engine (SRMs have, on average, an I_{sp} of 200–300 s), and SRMs cannot be throttled (although their thrust-time curves can be tailored, by changing the shape of their grain, to provide a varying thrust profile to match their intended mission). Hence, once an SRM is ignited, it will continue to burn (and produce thrust) until all of its fuel is consumed. This means that SRMs cannot be test-fired before their actual use; thus, their reliability must be established by analysis and similitude, which becomes quite an important consideration when designing a human-rated launch vehicle.

The solid propellant in an SRM is typically a rubbery combination of fuel, oxidizer, and a binding agent—typically aluminum, ammonium perchlorate, and a binder such as polybutadiene acrylonitrile (PBAN) are used. (Black powder is still used for smaller rockets.) Unlike for liquid propellants, the choice of propellants for an SRM is a considerably narrower field: most manufacturers will have their own optimized blends of ingredients, from which their SRMs are made. These ingredients are mixed and then cast into a pressure vessel to create the final form of the fuel. Once the fuel has solidified, it is operational.

4.4.1 Basic Configuration

From the standpoint of physics, a solid-propellant motor is identical to a liquid-propellant engine, in that both types create a hot gas via combustion and expel it as a high-speed exhaust through a nozzle. The performance parameters given in Table 4.3 are calculated in the exact same fashion. A basic layout of a generic SRM is shown in Fig. 4.8. The requisite hot gas is produced by igniting a surface of the fuel block, known as the charge or grain. This grain is almost always bonded to the inside wall of the casing, in order to prevent the flame front from migrating along the wall of the casing, which would prematurely (and unevenly) ignite other parts of the grain and possibly cause an explosion. There is often a layer of thermal insulation to protect the inside of the motor casing from the hot

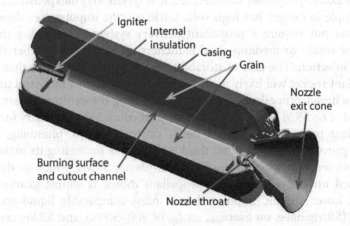

Fig. 4.8 Basic arrangement of a solid-propellant rocket motor. Source: Northrop-Grumman.

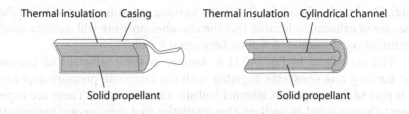

Thermal insulation   Casing   Thermal insulation   Cylindrical channel

Solid propellant   Solid propellant

Fig. 4.9 Two primary types of SRMs: left: end burner; right: core burner.

burning propellant. The charge contains both fuel and oxidizer, as well as the aforementioned binder material, meaning that SRM ignition is, for the most part, a "light and leave" affair. The exhaust velocity of an SRM is not high (on the order of 2,700 m/s for the most advanced types), which contributes to its low specific impulse.

4.4.2 SRM Types and Burn Rates

There are two primary types of SRMs, as shown in Fig. 4.9: the end burner and the core burner. The *thrust period* is the primary difference between them. An end burner will have a thrust period that is low-power and relatively long (a "cigarette burn"), whereas a core burner will have a higher-thrust, shorter period (an "inside-out burn").

One of the most important parameters in the performance of an SRM is its thrust stability. For a liquid-propellant engine, it is reasonable to assume that the chamber pressure remains relatively constant throughout its operation and is determinable based on the mass flow rate of propellant through the injectors. For a solid-propellant motor, the mass flow rate is determined by the rate at which the grain is consumed; not surprisingly, this parameter is known as the *burn rate*, and for solid propellant, is itself a function of the chamber pressure.

Because of this, stable combustion in an SRM cannot be automatically assumed, because the mass flow rate of the propellant increases with chamber pressure. This is given by

$$\dot{m} = aP_c^{\beta} \tag{4.31}$$

where the parameter β (burn rate pressure exponent) controls the stability of combustion. This equation is commonly known as *de Saint Robert's law*. It has already been shown earlier in the chapter that the mass flow rate out of the combustion chamber depends on the pressure within it. Thus, for $\beta > 1$, the combustion gas from the burning grain increases with pressure at a higher rate than the rate of exhaust, which could result in an uncontrolled rise in combustion rate and chamber pressure (and, similarly, a small decrease in chamber pressure could result in a catastrophic drop in burn rate). On the other hand, for $\beta < 1$, the burning rate will always be lower than the rate of exhaust, indicating that the chamber pressure will stabilize after a perturbation. Typically, β will be between 0.4 and 0.7.

The variable a in Eq. (4.31) is known as the coefficient of pressure (or burning rate constant); together with the burn rate pressure exponent, it is part of the propellant internal ballistic characteristics. These are experimentally measured propellant characteristics that vary among propellants. Equation (4.31) describes the *nonerosive burn rate*, and it should thus stand to reason that if nonerosive burning exists, so too does erosive burning. The total burn rate for an SRM, then, is a summation of these two burn rates. For a solid-propellant motor, the conditions at the upper and lower extremes of the charge are different: at the top of the charge, the newly formed exhaust gas is fairly stagnant, whereas at the bottom of the charge the gas is moving very fast. Near the nozzle, this results in a high recession (the rate at which the propellant is consumed and moves away from the nozzle), which if left unchecked can cause a burn-through of the nozzle or a failure of the casing before the upper portion of the propellant is consumed. This phenomenon can be mitigated by having an increasing cross-sectional area near the nozzle. For a constant flow rate, an increase in cross-sectional area causes a decrease in velocity, and thus erosive burning can be forestalled.

4.4.3 Thrust Profile and Grain Shape

The rate at which the grain is combusted directly influences the chamber pressure, and hence the thrust. The chamber pressure depends on two parameters: the area of the burning surface and the recession rate of the grain. The mass flow rate of propellant depends on the volume of grain consumed per second. Although SRMs cannot be throttled, their grain can be shaped to produce different levels of thrust for different periods of time. This is accomplished by inserting mandrels into the motor casing during the solid propellant casting process (Fig. 4.10), which produce cutouts in the cross-section of the grain. This changes the total burn area, which in turn produces varying *thrust profiles*, some examples of which are shown in Fig. 4.11.

The mass flow rate of an SRM is entirely dependent on the burn area A_b, the burn rate b_r, and the density of the propellant ρ

$$\dot{m} = \rho A_b b_r \tag{4.32}$$

Thus, the relationship between mass flow rate (and hence thrust) and burn area is easily seen.

The simplest thrust profile is that of the end burner, whose grain is illustrated on the left side of Fig. 4.9; this type of grain will produce a nearly constant thrust level until all of the propellant is consumed. For this profile, the burning area is limited to the cross-sectional area of the casing,

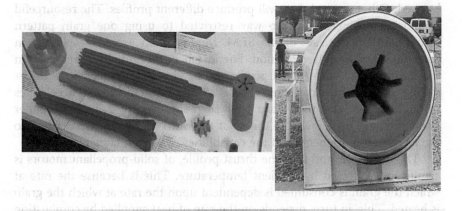

Fig. 4.10 Mandrels used during casting process of solid propellant.

and the burning rim of the grain will be in contact with the inner wall of the casing; because active cooling of the casing wall is not an option with solid-propellant motors, this type of grain is typically only used for short, low-thrust periods.

The hollow cylindrical grain (#1 in Fig. 4.11) begins burning along the inner surface of the cutout, which presents the advantage of the unburned grain insulating the motor casing from the flame front. The burn area of the grain can also be much larger than for a solid cylindrical grain, improving thrust performance. A grain with a cylindrical cutout will produce thrust levels that increase with time, which may be handy for certain applications (e.g., a solid-propellant missile can be fired from a wingtip without forcing the nose of the aircraft to yaw off-target). For a constant thrust, a secondary grain cylinder or a cog-shaped cutout (#2 and #3 in Fig. 4.11, respectively) can

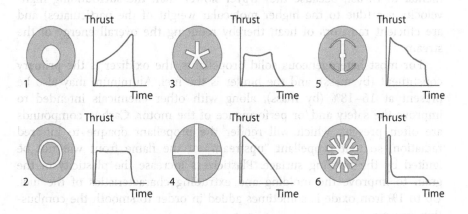

Fig. 4.11 Examples of grain shapes and accompanying thrust profiles. Source: http://www.braeunig.us/

be used. Other grain shapes will produce different profiles. The resourceful propulsion engineer is in no way restricted to using one grain pattern throughout the entirety of the SRM. Mixing grain profiles, especially in large boosters, is quite common. For instance, the initial segments of an SRM can have a star-shaped profile (#6 in Fig. 4.11) to maximize thrust during liftoff while reducing thrust as the launch vehicle accelerates through max-q, and then have later segments feature a double-cylinder grain profile. This allows the thrust of the SRM to be tuned according to the mission requirements and the capabilities of a given launch vehicle.

The burn rate, and thus the thrust profile, of solid-propellant motors is significantly affected by ambient temperature. This is because the rate at which the grain is consumed is dependent upon the rate at which the grain is heated. This, in turn, depends on the rate of heat supplied by combustion and the temperature of the grain itself: if the grain is very cold, more heat must be supplied before the grain can combust. The grain itself is very heavy, and is therefore a good insulator. Variations in burn rate by a factor of two due to variations in temperature between $-15°C$ and $20°C$ have been reported [6]. This will affect the thrust profile of the SRM, which could actually cause the mission to fail if the motors do not produce sufficient thrust.

4.4.4 SRM Propellant Additives

In modern propellants, metallic powders are often added to aid in combustion and increase the combustion temperature, thereby increasing the energy released during launch. Aluminum is commonly used, and because the resultant aluminum oxide found in the exhaust stream is *refractory* (i.e., retains its strength at elevated temperatures), it remains solid as it is being ejected from the motor. Solid particles in the exhaust stream are detrimental to thrust, because they travel slower than the surrounding high-velocity gas (due to the higher molecular weight of the particulates) and are efficient radiators of heat, thereby reducing the overall energy of the stream.

For most heterogeneous solid propellants, the oxidizer is the primary constituent (by mass), and the binder is the fuel. Aluminum may also be present at 16–18% (by mass), along with other chemicals intended to improve the safety and/or performance of the motor. Carbon compounds are often present, which will render the propellant opaque to infrared radiation, so that propellant "upstream" of the flame front will not be ignited by the burning surface. Plasticizers increase the plasticity of the grain to improve the molding and extruding characteristics of the fuel. Up to 1% iron oxide is sometimes added in order to smooth the combustion process.

In addition to improving the performance of the motor, several other additives can also be added to increase the stability of the fuel, which affects its storage qualities and mechanical strength. The mechanical strength of the fuel is a very important property: the fuel must be resistant to cracking during handling, because cracks will increase the total surface area of the fuel exposed to a flame front. This can lead to uneven burning, or even an explosion. The fuel not undergoing combustion must also withstand the launch loads, including accelerations of 10 g or more, without failing.

4.4.5 SRM Exhaust Toxicity

When firing a rocket engine near the ground, the majority of the exhaust is dispersed across a wide area surrounding the launch site. Unlike liquid-propellant engines, the combustion products of SRMs usually contain particulate matter, as well as approximately 12% hydrogen chloride by mass (which the chemically minded reader will recognize as a gas that forms hydrochloric acid when in contact with atmospheric humidity). Other products of combustion can include aluminum oxide, carbon monoxide, carbon dioxide, water vapor, nitrogen, and hydrogen. The combined molecular weight of the gaseous exhaust products is approximately 25 g/mol. By comparison, the molecular weight of aluminum oxide is 102 g/mol. Note that this two-phase flow is actually more beneficial than the alternative: if the aluminum oxide was also in a gaseous state, the molecular weight of the exhaust would be very high, producing a slow-moving exhaust stream that would degrade the performance of the motor.

Because of exhaust toxicity, it is therefore important that the SRM exhaust stream be channeled via ductwork as appropriate. Obviously, this is not possible once the rocket is in flight, and atmospheric dispersal must be relied upon. Most solid-propellant launch sites will require a thorough cleaning after launch, which can take several days.

4.5 Liquid-Propellant Engines

A liquid-propellant rocket engine is, simply, a rocket engine that uses liquid propellants. These engines may be *monopropellant, bipropellant,* or of the more exotic *tripropellant* variety, depending on the number of propellants the engine consumes to produce thrust. Although any number of propellants is theoretically possible, the vast majority of liquid engines in use today will be of the mono- or bipropellant variety. These propellants are typically fed to the thrust chamber using pumps or pressurant gas, where they are then mixed and combusted to produce thrust. A monopropellant is unique in that it contains both a combustible material and an oxidizer in a single substance; monopropellants may be multiple substances or a single substance with the desired properties, such as hydrogen peroxide or hydrazine.

Literally thousands of combinations of liquid propellants have been tried over the course of history, but in the end only three practical types of liquid propellant are used: cryogenic, semicryogenic, and hypergolic. A cryogenic propellant, as the name implies, is one that requires very low temperatures in order to liquefy (e.g., for liquid oxygen, this is around –193°C or 90 K). A semicryogenic fuel (e.g., kerolox, a mixture of kerosene fuel and liquid oxygen oxidizer) is one in which at least one component is storable (i.e., liquid at ambient temperature). Hypergolic propellants are storable chemicals that auto-ignite upon contact with one another.

The primary advantage of a liquid engine over a solid motor is that a liquid engine will typically have a much higher specific impulse than a solid motor, with the attendant increases in overall vehicle performance. Also, many liquid engines can be throttled by changing the propellant mass flow rate, which provides designers more flexibility in how a launch vehicle (or spacecraft) delivers the thrust necessary for its mission. Liquid propellants typically have a higher energy density than their solid-propellant counterparts, which means that the volume of the propellant tanks may be relatively low; the structural weight of the propellant tanks can also be reduced by using pumps and compressors to deliver the propellants to the combustion chamber, which allows the propellants to be kept under low pressure to minimize the tank wall thickness. This arrangement improves the mass ratio for the rocket. Alternatively, an inert gas (e.g., helium) at high pressure can be used in lieu of pumps to force the propellants into the combustion chamber; these engines will have a lower mass ratio, but are usually more reliable [7]. If high-pressure gas is used in lieu of pumps, the propellant storage tanks must sustain the pressure provided by the high-pressure gas, resulting in more wall thickness and more massive tanks.

The primary disadvantages of liquid engines arise due to two factors: propellant storability and all-up mass of the propellant subsystems. Cryogenic propellants must be kept cold to be usable, which in turn requires insulation and venting. Propellant boil-off is a very real issue, which is governed by the amount of thermal leakage of the tanks as well as the percentage of propellants present (partially filled tanks experience faster boil-off). By the square-cube law, the smaller the tank, the faster the boil-off. Boil-off is also dependent upon temperature: the low temperatures required to store liquid hydrogen (–253°C or 20 K) mean it can experience a boil-off of 0.13% per day, whereas the higher temperature of liquid oxygen can limit the boil-off to 0.26–0.48% of tank capacity per day. This can become a very real concern when a launch must be continually delayed. By contrast, hypergolic propellants are typically easily storable, at the cost of requiring very specific handling procedures (and the associated constraints on the vehicle lifecycle), due to the fact that many hypergolic propellants have a high toxicity and carcinogenicity. The liquid engine architecture also requires that propellants be transported from one location to another (e.g., from the propellant tanks to the combustion chamber) via pumps or pressurant gas, which increases the inert weight of the propulsion

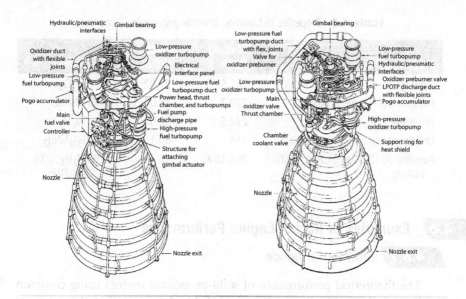

Fig. 4.12 Two views of the Space Shuttle main engine. Source: [7].

system. Illustrations of liquid-propellant engines, such as the Space Shuttle main engine shown in Fig. 4.12, offer a hint at just how complex these engine architectures can become.

The mass of the propellant subsystems must also be carefully taken into account during the design phase. Although the propellant mass is an important parameter, equally important is the mass of the propellant tanks: they must not only be strong enough to store the propellants and survive the launch loads, but also be sufficiently lightweight to maintain a favorable mass ratio. With cryogenic propellants, this problem is compounded, because the tanks must also be insulated. Additional tank mass is required for pressure-fed systems. Properties of common liquid propellants are given in Tables 4.4 and 4.5.

Table 4.4 Properties of Common Cryogenic Propellants

Propellant	Boiling Point, K	Mass, g/mol	Formula	I_{sp} with LOx	O/F, kg/kg_{LOx}
LOx	90	32	O_2	–	–
RP-1	460+	175	$CH_{1.97}$	250	0.43
LH_2	20.4	2.01	H_2	360+	0.15
Methane	112	16.04	CH_4	268	0.41
Propane	231	44.1	C_3H_8	260	0.34

Table 4.5 Properties of Common Storable (Hypergolic) Propellants

Propellant	Melting Point, K	Mass, g/mol	Formula
Dinitrogen tetroxide (DNTO or NTO)	261.5	92.02	N_2O_4
Monomethyl hydrazine (MMH)	220.7	46.08	CH_3NHNH_2
Hydrazine	274.5	32.05	N_2H_4
Unsymmetrical dimethyl hydrazine	216	60.1	$(CH_3)_2NNH_2$
Aerozine 50 (50% Hydrazine – 50% UDMH)	266.15 K	92.14	50% N_2H_4, 50% $(CH_3)_2NNH_2$

4.6 Examples of Rocket Engine Performance

4.6.1 SRM Performance

The theoretical performance of solid-propellant motors using common oxidizer–fuel combinations is given in Table 4.6. Note that these figures are for a chamber pressure of 1,000 psia (6,895 kPa) at sea level, with a perfectly expanded nozzle and frozen combustion. For comparison, the performance characteristics of several large SRMs, including sea-level I_{sp} and burn time t_b, are given in Table 4.7. Note that for large SRMs, mass ratios of 8–10 are not uncommon, reinforcing the idea that rocket motors must maximize the total propellant mass carried for each pound of inert structure.

Table 4.6 Theoretical SRM Performance

Oxidizer	Fuel	ρ, g/cm^3	T, K	c^*, m/s	MW, g/mol	I_{sp}, s
NH_4NO_3	11% binder, 7% additives	1.51	1,282	1,209	20.1	192
NH_4ClO_4	18% polymer, 4–20% Al	1.69	2,816	1,590	25.0	262
NH_4ClO_4	12% binder, 20% Al	1.74	3,371	1,577	29.3	266

Table 4.7 Performance of Sample Large SRMs

Motor	Gross Mass, kg	Empty Mass, kg	T, kN	I_{sp}, s	t_b, s
Space Shuttle SRB	590,000	88,000	11,520	235	123
Delta II GEM-46	19,327	2,282	628.3	273	75
Atlas V	40,824	4,000	1,270	245	94
Delta IV GEM-60	33,978	3,849	826.6	243	90
Ariane V EAP	278,330	38,200	6,470	250	130
Minuteman M-55	23,077	2,292	792	237	60

Table 4.8   Performance of Several Kick Motors

Motor	Total Impulse, N	Gross Mass, kg	ζ_p	$\overline{T}$, N	I_{sp}, s
STAR-13B	1.16E5	47	0.88	7,015	286
STAR-30BP	1.46E6	543	0.94	26,511	292
STAR-30C	1.65E6	628	0.95	31,760	285
STAR-30E	1.78E6	667	0.94	35,185	289
STAR-37F	3.02E6	1,149	0.94	44,086	291
STAR-48A	6.78E6	2,559	0.95	79,623	284
IUS SRM-2	8.11E6	2,995	0.91	80,157	304
LEASAT PKM	9.26E6	3,658	0.91	157,356	285
IUS SRM-1	2.81E7	10,374	0.94	198,435	296

The performance figures for several "kick" motors is given in Table 4.8. These are more commonly referred to as *apogee kick motors (AKMs)*, and they are regularly used on spacecraft intended for geostationary orbit. This allows a launch vehicle to place a spacecraft in an elliptical orbit with a nonzero inclination, at which point the spacecraft must generate its own thrust to provide the Δv required to reach geostationary orbit. An AKM thus "kicks" the speed of the spacecraft up to to the necessary orbital speed.

4.6.2   Liquid Engine Performance

The theoretical performance of some common liquid fuels and oxidizers is listed in Table 4.9. Note that any fuel listed in Table 4.9 with a percent sign refer to fuels by mass percent composition; also, the fuel "Aerozine 50" refers to a fuel that is equal parts UDMH and hydrazine. The two sets of values listed in Table 4.9 are for frozen (upper) or shifting (lower) equilibrium at an assumed chamber pressure of 1,000 psia (6,895 kPa) and a sea-level ambient pressure (14.7 psia, 101.4 kPa) at the nozzle exit. These figures also assume optimum expansion ratio, adiabatic combustion, and isentropic expansion. Specific gravities are provided at their boiling point or at 68° F/1 atm (20° C/101.4 kPa), whichever is lower.

It is instructive to compare the theories presented so far with real-world examples of liquid-propellant engines. Perhaps the most famous liquid-propellant engine is the Rocketdyne F-1, which powered the first stage of the Saturn V launch vehicle. Its performance specifications are given in Table 4.10, and the engine is shown in Fig. 4.13. In the table, "SA5xx" refers to the xx-th mission of the Saturn V LV.

The specifications for the Rocketdyne J-2, which powered second stage of the Saturn IB and the second and third stages of the Saturn V, are given in

Table 4.9 Theoretical Performance of Liquid-Propellant Engines

Oxidizer	Fuel	Mass f	Vol. f	S.G.	T_c, K	c^*, m/s	MW, kg/kmol	I_{sp}, s
LOx	75% ethanol	1.30	0.98	1.00	2,904	1,641	23.4	267
		1.43	1.08	1.01	2,957	1,670	24.1	279
	Hydrazine	0.74	0.66	1.06	3,027	1,871	18.3	301
		0.90	0.80	1.07	3,127	1,992	19.3	313
	Hydrogen	3.40	0.21	0.26	2,416	2,428	8.9	388
		4.02	0.25	0.28	2,724	2,432	10.0	391
	RP-1	2.24	1.59	1.01	3,282	1,774	21.9	286
		2.56	1.82	1.02	3,399	1,804	23.3	300
	UDMH	1.39	0.96	0.96	3,171	1,835	19.8	295
		1.65	1.14	0.96	3,321	1,864	21.3	310
NTO	Hydrazine	1.08	0.75	1.20	2,857	1,765	19.5	283
		1.34	0.93	1.22	2,977	1,782	20.9	292
	Aerozine 50	1.62	1.02	1.18	2,957	1,731	21.0	278
		2.00	1.24	1.21	3,088	1,745	22.6	288

Table 4.11, and the engine is shown in Fig. 4.14. In the table, "SA2xx" refers to the xx-th mission of the Saturn IB LV, and "SA5xx" refers to the xx-th mission of the Saturn V LV.

The Rocketdyne RS-68 is used as the first-stage engine for the Delta IV launch vehicle, and currently is the largest hydrogen-fueled rocket engine in the world. Unlike many other liquid-propellant engines, the RS-68 has an ablative nozzle lining; this nozzle lining burns away during flight to dissipate heat. Although this solution is heavier than a conventional tube-wall

Table 4.10 Rocketdyne F-1 Specifications

	Vehicle Effectivity	
	SA501-SA503	SA504 and Subsequent
Thrust (sea level), MN	6.672	6.77
Thrust duration, s	150	165
I_{sp}, s	260	263
Engine dry mass, kg	8,353	8,391
Engine mass (burnout), kg	9,115	9,153
Exit/throat area ratio	16	16
Propellants	LOx/RP-1	LOx/RP-1
f	2.27% ± 2%	2.27% ± 2%

Source: NASA.

Fig. 4.13   Rocketdyne F-1 engine on display at Kennedy Space Center. Source: NASA.

Table 4.11   Rocketdyne J-2 Specifications

	Vehicle Effectivity		
	SA201-SA203	SA204-SA207 SA501-SA502	SA208 and Subsequent SA504 and Subsequent
Thrust (altitude), kN	889.6	1,000.9	1,023
Thrust duration, s	500	500	500
I_{sp}, s	418	419	421
Engine dry mass, kg	1,579	1,579	1,584
Engine mass (burnout), kg	1,637	1,637	1,642
Exit/throat area ratio	27.5	27.5	27.5
Propellants	LOx/LH$_2$	LOx/LH$_2$	LOx/LH$_2$
f	5.00 ± 2%	5.50 ± 2%	5.50 ± 2%

Source: NASA.

Fig. 4.14 Rocketdyne J-2 during test firing. Source: NASA.

regeneratively-cooled nozzle, it reduces part count and is easier to manufacture, thus lowering the engine's cost. The specifications for the RS-68 are given in Table 4.12, and the engine is shown in Fig. 4.15.

Table 4.12 Rocketdyne RS-68 Specifications

	Vehicle Effectivity	
	RS-68	RS-68A
Thrust (sea level), kN	2,936	3,136
I_{sp}, s	410	412
Engine dry mass, kg	6,604	6,745
Exit/throat area ratio	21.5	21.5
Propellants	LOx/LH$_2$	LOx/LH$_2$

Source: Aerojet Rocketdyne.

Fig. 4.15   Rocketdyne RS-68 during test firing. Source: NASA.

A summary of performance characteristics of common liquid-propellant rocket engines, compiled from various sources, is given in Table 4.13.

4.7   Rocket Engine Power Cycles

It may correctly be surmised from Table 4.13 that the power cycle will have an impact on an engine's performance, either directly (via combustion) or indirectly (via feed systems and support hardware). The term *power cycle* is a useful method for categorizing rocket engines, and refers to how power is derived to drive pumps that feed propellants to the combustion chamber(s). This control and feeding of propellants to the combustion chamber is accomplished through the use of various plumbing, pump, valve, and control components known as the powerhead. Generally speaking, the most common types of power cycles are as follows:

- Gas generator cycle
- Staged combustion cycle

Table 4.13   Performance Characteristics of Selected Liquid-propellant Engines

Motor	Thrust, kN		I_{sp}, s		P_c, MPa	f	A/A^*	Geometry		Ox/Fuel	Cycle
	SL	Vac	SL	Vac				L, m	D, m		
MA-5A Sust	269	378	220	309	5.07	2.27	25:1	2.46	1.22	LOx/RP-1	Gas gen
MA-5A Boost	1,910	2,104	265	296	4.96	2.25	8:1	2.57	1.22	LOx/RP-1	Gas gen
RS-27A	890	1,054	255	302	4.83	2.25	12:1	3.78	1.44	LOx/RP-1	Gas gen
RD-170	7,255	1,909	307	337	24.13	2.6	36:1	4.01	1.45	LOx/RP-1	Staged
RD-180	3,827	4,150	311	338	25.51	2.72	36.8:1	3.58	1.45	LOx/RP-1	Staged
Merlin 1V	N/A	411	N/A	342	N/A	N/A	N/A	N/A	N/A	LOx/RP-1	Gas gen
Merlin 1A	319	N/A	270	300	N/A	N/A	N/A	N/A	N/A	LOx/RP-1	Gas Gen
Merlin 1C	614.7	N/A	275	304	N/A	N/A	N/A	N/A	N/A	LOx/RP-1	Gas gen
LR87-AJ-11	1,984	2,353	254	302	5.70	2	15:1	3.84	1.50	N_2O_4/A-50	Gas gen
LR91-AJ-11	N/A	445	N/A	314	5.70	1.86	49.2:1	2.79	1.63	N_2O_4/A-50	Gas gen
RD-253	1,472	1,637	285	316	14.70	2.67	26.2:1	2.72	1.43	N_2O_4/UDMH	Staged
RD-0210	N/A	588	N/A	327	14.69	2.65	78.2:1	2.30	1.46	N_2O_4/UDMH	Staged
HM-60	885	1,140	334	431	11.00	5.3	45:1	2.87	1.75	LOx/LH_2	Gas gen
RL10A-4-1	N/A	99.2	N/A	451	4.21	5.5	84:1	1.78	1.17	LOx/LH_2	Expander
RL10B-2	N/A	110.3	N/A	467	4.44	6.0	285:1	3.05	2.14	LOx/LH_2	Expander
RS-68	2,891	N/A	365	410	9.72	6.0	12.5:1	N/A	N/A	LOx/LH_2	Gas gen
SSME 109%	1,818	2,279	380	453	22.48	6.0	77.5:1	4.27	2.44	LOx/LH_2	Staged

- Expander cycle
- Open expander cycle
- Pressure-fed cycle
- Electric pump-fed

Each power cycle is discussed briefly in the following sections.

4.7.1 Gas Generator Cycle

The gas generator cycle is shown schematically in Fig. 4.16. In this power cycle, a small amount (approximately 3–7%) of fuel and oxidizer is diverted from the main propellant flow into a preburner, which powers a turbine that drives the fuel and oxidizer pumps, which in turn feed propellants into the combustion chamber. A gas generator engine will typically burn propellants at a suboptimal mixture ratio (the engine will run oxidizer-rich) in order to keep the turbine blade temperature low. Unused exhaust from the preburner is either dumped or sent into the nozzle. Because there will always be some

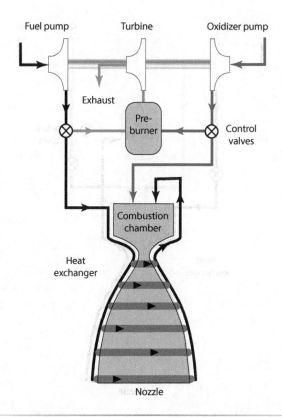

Fig. 4.16 The gas generator power cycle.

amount of propellant exhausted from the power turbine, the gas generator cycle is also referred to as an *open cycle*.

The key advantage of the gas generator cycle is that the power turbine does not need to inject the turbopump exhaust against the backpressure of combustion, simplifying the design of the engine's plumbing and turbine, which may result in a lighter and less expensive engine. The disadvantage of this power cycle is, of course, the lost efficiency stemming from the diverted propellant used to drive the power turbine; this results in gas generator engines operating at a lower specific impulse than engines utilizing other power cycles. The RS-68 engine is an example of the gas generator cycle.

4.7.2 Staged Combustion Cycle

The staged combustion cycle, also known as a *topping cycle*, is one where the propellants are combusted in stages. This power cycle is often used for high-power applications, such as the RD-180 or NK-33 engines. A preburner

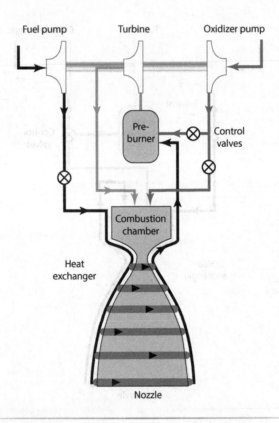

Fig. 4.17   A fuel-rich staged combustion power cycle.

generates the working fluid for a power turbine (which drives the propellant pumps) as either a fuel- or oxidizer-rich gas mixture. Unlike the gas generator cycle, the staged combustion cycle injects the exhausted gas mixture from the power turbine into the combustion chamber, where the gas mixture burns again along with the second propellant. The staged combustion cycle is shown schematically in Fig. 4.17; as is usually the case, one or more propellants can be used as a coolant, simultaneously cooling the engine and preheating the propellant.

The primary advantage of this "closed" type of power cycle is obvious: all of the propellant is routed through the combustion chamber. Staged combustion engines typically use high-power turbopumps, resulting in high chamber pressures and allowing the use of large nozzle area ratios to improve the low-altitude performance of the engine. These advantages come at a cost, however, and staged combustion cycles usually require exotic materials for plumbing and valves to carry the hot gas mixture, along with complicated control systems. Engines using oxygen-rich staged combustion (such as the NK-33), in particular, require advanced materials due to the corrosive oxidizer-rich gas mixture; by contrast, the primary concern of a fuel-rich staged combustion engine is the use of a noncoking fuel, such as liquid hydrogen.

A variant of the staged combustion cycle is known as the full-flow staged combustion cycle (FFSC; see Fig. 4.18). An FFSC uses both an oxidizer- *and* fuel-rich preburner, which permits the power turbine to be powered by both propellants. (This arrangement also results in full gasification of both propellants, leading FFSC engines to occasionally be referred to as "gas-gas" engines.) This also allows a unique advantage over regular staged combustion: an FFSC engine will have the fuel and oxidizer pumps driven by two separate power turbines, thus permitting control over the flow of one propellant independently of the other. Both turbines will usually operate at lower temperatures and pressures in this configuration, which has a direct impact on both engine life and reliability.

4.7.3 Expander Cycle

An expander cycle is similar to a staged combustion cycle, with the key difference being that an expander cycle does not use a preburner. The heat in the combustion chamber cooling jacket vaporizes the fuel, which then passes through the power turbine and injects into the combustion chamber along with the oxidizer. Expander cycles work well with propellants that have a low boiling point and are easily vaporized, such as liquid hydrogen and methane. The expander cycle is shown schematically in Fig. 4.19.

Although the propellants will burn at or near the optimal mixture ratio in the combustion chamber (and no flow is dumped overboard), the expander cycle is limited by the amount of heat able to be transferred to the fuel in the cooling jacket, which in turn limits the available power to the turbine.

Therefore, the expander cycle is usually best employed in small and midsize engines such as the RL-10, RL-60, LE-5A/B, and Vinci engine.

This thrust limitation in conventional closed expander cycle engines is due to the square-cube law: as the nozzle size increases (with increasing thrust levels), the surface area from whence heat is extracted to vaporize the fuel increases as the square of the nozzle radius; however, the volume of fuel that must be vaporized increases as the cube of the nozzle radius. There exists, therefore, a practical limit (of about 300 kN of thrust) beyond which a conventional closed expander cycle (such as that shown in Fig. 4.19) will not be able to vaporize sufficient fuel to drive the power turbine.

To bypass this limitation, several expander cycle engines utilize a dual-expander cycle (i.e., both the fuel and oxidizer are vaporized in their cooling jackets and are then used to drive their respective, independent power turbines), a closed split-expander cycle (i.e., propellants are fed through a secondary pump stage prior to injection into the combustion chamber, in order to recover some of the pressure losses induced by

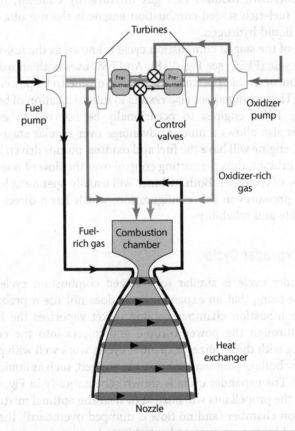

Fig. 4.18 The full-flow staged combustion cycle.

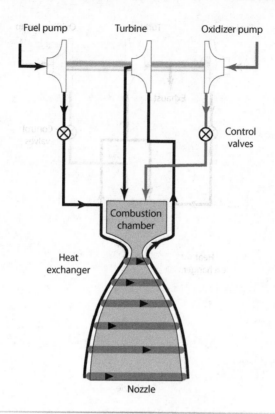

Fig. 4.19   The closed expander cycle.

circulating through the cooling jacket), and even gas generator–augmented expander cycles (different from a conventional gas generator cycle in that, in this case, the combustion products from the gas generator are pumped through a heat exchanger and then exhausted overboard, rather than directly driving the power turbine).

A notable variation on the expander cycle is the open expander cycle (or "bleed" expander), wherein only a small portion of the fuel is used to drive the power turbine, and the working fluid is exhausted overboard (to ambient pressure) after passing through the turbine. This increases the turbine pressure ratio, increasing the power output of the engine via higher chamber pressures than are typically available with the closed expander cycle at the cost of decreased efficiency (lower specific impulse) due to the overboard flow. This is shown schematically in Fig. 4.20. The expander cycle has a number of advantages over other configurations. First and foremost, the operating temperature of the turbine in an expander cycle is very low relative to a gas generator or staged combustion cycle; after the gasification of the fuel, the working fluid is usually near room temperature and will do little damage to the turbine. Secondly, an expander cycle can be more tolerant

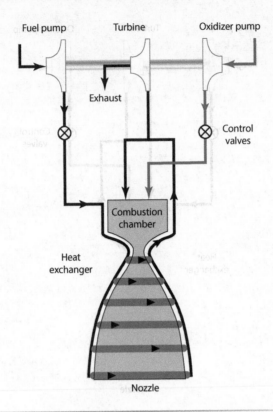

Fig. 4.20 The open expander cycle.

of fuel contamination due to the wide channels used in the nozzle cooling jacket; by contrast, gas generator cycles are very sensitive to blockage, and even a small hot spot can cause an "unscheduled rapid disassembly" of the engine. This also leads to the characteristic of inherent safety: because the nozzle is used as a heat exchanger, an expander cycle engine is immune to unintended feedback problems caused by malfunctioning valves.

4.7.4 Electric Pump-Fed Cycle

Note that it is not necessarily a requirement to use a preburner to drive a power turbine, which in turn drives the propellant pumps. The propellant pumps can instead be driven via electric motors, which are then powered by an onboard generator or battery. This arrangement is known as an electric pump-fed cycle. (The word *electric* is specifically used to differentiate this configuration from others that use turbine-driven pumps, such as the gas generator cycle and staged combustion cycle.) Typically, a electric motor of this type can increase propellant pressure up to 1,500 to 2,900 psi at the combustion chamber. Additionally, electric pumps can have operating efficiencies on the order of 95%, as opposed to approximately 50% for the power turbine on a typical gas generator cycle.

The main advantage of this cycle is its potential simplicity: a great deal of the complexity arising from the various plumbing, valves, turbines, pumps, preburners, feed lines, return lines, and so forth inherent in many other cycles can simply be done away with. To appreciate the magnitude of this idea of simplified propellant feeds, consider a conventional peroxide pump-fed alcolox (LOx/alcohol) rocket engine, as shown in Fig. 4.21. The attraction of an electric pump-fed system becomes readily apparent.

It should be noted that the simplicity of an electric pump-fed configuration is only a *potential* advantage: an electric pump-fed architecture will, by its very nature, require some manner of electric power to operate. Because propellant pumps are typically high-powered turbomachinery, it stands to reason that the electric pumps will need a significant battery bank, generator, or some other electric source to be effective, which may increase the inert mass of the launch vehicle. The Electron two-stage launch vehicle (developed by New Zealand Rocket Lab), designed to loft CubeSats for commercial customers, is one of the few well-funded, large-scale projects using electric pump-fed engines. The Rutherford engine powering the Electron is the first flight-ready engine to use the electric pump-fed cycle, and can provide 5,000 lb$_f$ of (vacuum) thrust at a specific impulse of 327 s. Interestingly, the Electron jettisons several battery modules during launch, as they are expended. The electric pump-fed cycle is shown schematically in Fig. 4.22.

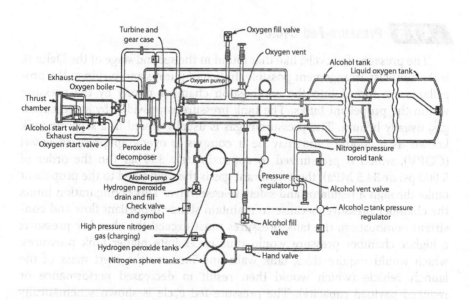

Fig. 4.21 Schematic of a LOx/alcohol engine. Source: Ordway et al. [8].

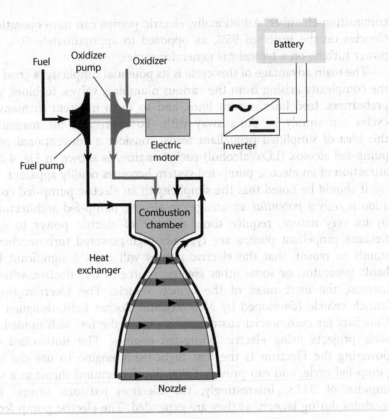

Fig. 4.22 The electric pump-fed combustion cycle.

4.7.5 Pressure-Fed Cycle

The pressure-fed cycle, like that found in the second stage of the Delta II, is the simplest arrangement possible. In lieu of pumps and turbines, the propellants are forced into the combustion chamber purely by the pressure within the propellant tanks. The tank pressure is provided by a pressurant gas, usually helium. The pressurant gas is usually stored in a separate tank known as a "bottle" and may be a composite overwrap pressure vessel (COPV), which is pressurized to approximately 350 bar (on the order of 5,000 psi or 34.5 MPa); the pressurant gas is then plumbed to the propellant tanks through a regulator and safety release valve. This configuration limits the chamber pressure, because to maintain steady propellant flow and consistent combustion, the tank pressures must exceed the chamber pressure; a higher chamber pressure would thus necessitate higher tank pressures, which would require thick tank walls and increase the inert mass of the launch vehicle (which would then result in decreased performance or reduced payload capacity). The pressure-fed cycle is shown schematically in Fig. 4.23.

The pressure-fed cycle may not produce the highest thrust levels, but it *is* reliable due to the greatly reduced parts count and low inherent complexity. An example of this simplicity is seen in Fig. 4.24. One specific type of pressure-fed system, known as a *blowdown* system, can have as few as seven components (in addition to the required plumbing): the engine, two propellant tanks, two fill valves for the pressurant gas, and two fill valves for the propellants. This creates a unique design problem: during extended burns, excessive cooling of the pressurant gas (due to adiabatic expansion) must be avoided. The pressurant gas will not necessarily liquefy, but the excessive cold could freeze the propellants, reduce tank pressure, and damage components that are not rated for cryogenic use.

Due to the simplicity and reliability of pressure-fed engines they are used almost exclusively by spacecraft attitude control systems, maneuvering thrusters, and landers. The reaction control system (RCS) and orbital maneuvering system (OMS) of the Space Shuttle orbiter were pressure-fed designs, as were the RCS and service propulsion system engines on the Apollo command/service module. Other examples include the SuperDraco and Draco RCS engines on the Dragon V2; the RCS, ascent, and descent engines on the Apollo lunar module; the Armadillo Aerospace Pixel

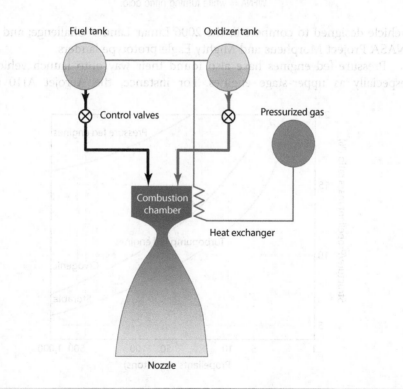

Fig. 4.23   The pressure-fed combustion cycle.

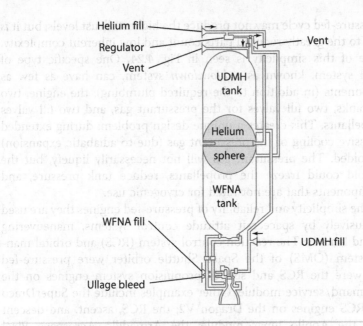

Fig. 4.24 Hypergolic pressure-fed engine. Source: Riley and Sailor [9].
WFNA = white fuming nitric acid.

vehicle designed to compete in the 2006 Lunar Lander Challenge; and the NASA Project Morpheus and Mighty Eagle prototype landers.

Pressure-fed engines have also found their way onto launch vehicles, especially as upper-stage engines. For instance, the Aerojet AJ10 and

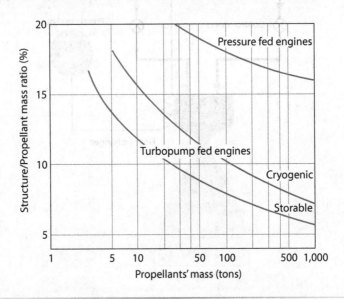

Fig. 4.25 Mass ratio of common power cycles.

TRWTR-201 have both been used on Delta launch vehicles; the SpaceX Kestrel engine has been used on the Falcon 1 upper stage and the CONAE Tronador II upper stage. Pressure-fed systems have even been used on rocket-powered aircraft, such as the XCOR EZ-Rocket.

It is apparent that the selection of power cycle is not an arbitrary decision. As shown in Fig. 4.25, the amount of launch vehicle all-up mass devoted to structural elements is strongly tied to not only the propellant type, but also the power cycle used. Although the data in Fig. 4.25 represent trends rather than absolutes, this is a useful figure for the early stages of the design process.

4.8 Aerospike Engines

With a conventional, fixed nozzle, the expansion of exhaust gases is ideal at only a single ambient pressure. This means that the rocket engine will be operating in a nonoptimal state at all but one specific altitude. Turning a conventional nozzle "inside out," however, would allow the exhaust gas to be ideally expanded at all altitudes. This type of engine known as an *aerospike*, was to be explored in the X-33 VentureStar. A typical aerospike, known as a linear aerospike, is shown in Fig. 4.26.

The effects of altitude on nozzle performance are shown pictorially in Fig. 4.27. From Fig. 4.27a, it is readily seen that as a launch vehicle ascends, the exhaust plume will expand past the boundaries of the nozzle and assume an increasingly random motion, degrading performance. An aerospike avoids this by directing exhaust along the contour of a wedge-shaped protrusion (the "spike" portion of the aerospike name), which

Fig. 4.26 Linear aerospike engine.

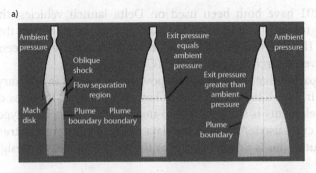

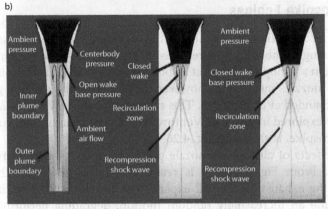

Fig. 4.27 Comparison of a) conventional and b) aerospike nozzle performance as a function of altitude. Source: AerospaceWeb.

forms one side of a "virtual nozzle," with the ambient pressure forming the second side (the "aero" portion of the aerospike name).

At low altitudes, the ambient pressure compresses the exhaust plume against the spike. As the vehicle ascends, the ambient pressure decreases, and the exhaust plume is free to expand; this means that the nozzle behaves as if it contained an "altitude compensator," in that the size of the virtual nozzle automatically changes as the ambient pressure changes. Note, however, that an aerospike nozzle is not a panacea: as seen in the middle figure of Fig. 4.27b, there will still exist only one altitude where an aerospike will operate at maximum efficiency.

Several versions of the aerospike exist, although the linear configuration of the X-33's XRS-2200 engine—in reality a collection of Rocketdyne J-2S hardware outfitted to a custom spike (see Fig. 4.28)—is arguably the most well-known. The linear aerospike has the primary advantage of allowing several smaller engines to be "stacked" along the spike to attain the desired thrust levels, while providing steering inputs by individually throttling the engines. Another type of aerospike nozzle is known as the *toroidal aerospike*,

Fig. 4.28   Test-firing of XRS-2200 linear aerospike.

shown in Fig. 4.29. For this configuration, exhaust is blown from a ring along the outer edge of the nozzle, and the central spike section operates in a fashion similar to a linear aerospike. This aerospike configuration was

Fig. 4.29   Toroidal aerospike nozzle.

flight-tested by a joint industry/academic team (Garvey Spacecraft Corporation and California State University Long Beach) in Sept. 2003.

Despite the cancellation of the X-33 program, aerospikes remain an area of active research. Further successes came in 2004, with the successful tests of solid-propellant aerospikes conducted at NASA Dryden (now Armstrong) Flight Research Center. The rockets used in these flight tests reached altitudes of 26,000 ft and speeds of approximately Mach 1.5.

4.9 Hybrid Rockets

A hybrid rocket engine is one that uses propellants in at least two different phases; typically, one propellant is solid and the other is either a liquid or a gas, but other combinations have been used. A hybrid rocket is shown schematically in Fig. 4.30.

The simplest form of a hybrid rocket is a pressure vessel containing the oxidizer, combustion channels along the fuel, and an ignition source. Combustion occurs almost entirely along a diffusion flame adjacent to the surface of the solid propellant. Commonly, the liquid propellant is the oxidizer and the solid propellant is the fuel. This permits the use of solid fuels such as hydroxyl-terminated polybutadiene (HTPB) and paraffin wax, into which fuel additives such as metal hydrides can be integrated. Other common fuels include polymers, cross-linked rubbers, or liquefying fuels. Depending on the propellants used, a hybrid rocket can avoid some of the disadvantages of solid-propellant motors (e.g., lack of means to control burn once initiated), while also avoiding some of the disadvantages (e.g., plumbing and mechanical complexity) of liquid-propellant engines. Hybrid rockets have one unique advantage over other configurations: because the propellants are two different states of matter, it is difficult for them to mix intimately, and thus the failure modes of a hybrid rocket tend to be more benign. Hybrids can also be throttled, although the specific impulse of hybrid rockets tends to be between 250 and 300 s; modern hybrids circumvent this issue through the use of metallized fuels and have demonstrated specific impulses on the order of 400 s.

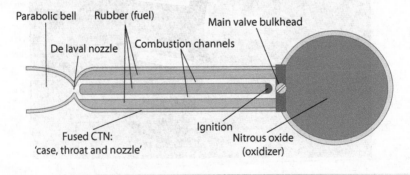

Fig. 4.30 A hybrid rocket engine. Source: Scaled Composites.

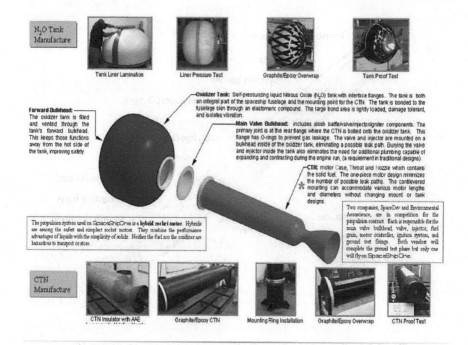

Fig. 4.31   Details of the SpaceShipOne hybrid rocket. Source: Scaled Composites.

Hybrids also exhibit some unique disadvantages, the most critical of which is known as an oxidizer–fuel shift (O–F shift), caused by the rate of fuel production to oxidizer flow changing as the solid grain regresses. Additionally, the use of fuels with a low regression rate (the speed at which the solid fuel recedes during combustion) will usually require a fuel grain with multiple combustion channels, which can contain structural deficiencies.

Designing a large (orbital booster–scale) hybrid rocket would also present the need to use turbomachinery to pressurize the oxidizer (or suffer increased vehicle weight from a thick-walled tank), which must obviously be powered in some fashion. In a liquid-propellant engine, this is done via one of the power cycle methods discussed previously. In a hybrid, such a pump would need an external power source or a monopropellant (such as hydrogen peroxide or nitromethane) that can be used to drive a power turbine; however, these oxidizers are less efficient than liquid oxygen, which cannot be used to power a turbopump. A secondary fuel would likely be needed to power the turbomachinery, which in turn would require its own tank and plumbing and decrease the vehicle's performance.

Successes have been had in the field of hybrid rocket research. For instance, the rocket that powered SpaceShipOne was a hybrid powered by nitrous oxide and HTPB. A breakaway view of this rocket is shown in Fig. 4.31. Hybrid rockets have even been considered for use as the strap-on boosters for the Space Shuttle, as shown in Fig. 4.32. For more information

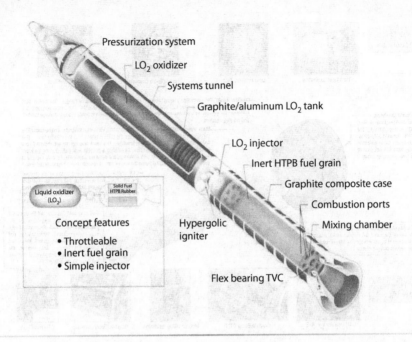

Fig. 4.32 Concept of a Space Shuttle hybrid rocket strap-on. Source: [6].

on the Space Shuttle SRB study and on different types of hybrid rockets, the reader is referred to Zurawski and Rapp [10].

Another advantage of hybrids is in the cost of propellant. Consider the costs shown in Table 4.14 reproduced with data from [11] and adjusted to 2020 constant dollars. It is immediately apparent that the propellants most likely to be used in a hybrid rocket are, far and away, also the least expensive. Although the performance of hybrids may not necessarily exceed that of other types of rockets, and as we show in Chapter 17 that the cost of propellants for a launch vehicle is almost negligible compared to other costs, the long-standing desire to drive down the launch cost per kilogram of payload may drive future launch vehicles increasingly toward hybrid rockets. This

Table 4.14 Cost of Commonly Used Propellants, 2020 US Dollars

Propellant Name	Cost, $/kg
Liquid hydrogen (LH$_2$)	8.61
Liquid oxygen (LOx)	0.25
Kerosene/RP-1	3.85
Dinitrogen tetroxide (N$_2$O$_4$, "NTO")	105.81
Hydrazine (N$_2$H$_4$, also MMH, UDMH, Aerozine 50)	408.84
Solid propellant	10.82

is especially likely in the commercial world, where "cash is king," and innovation is often required to directly influence the bottom line.

4.10 Example Problems

1. A spacecraft has the properties listed in the table below. Calculate Δv, exhaust velocity, total impulse, and propellant mass flow rate.

Item	Value
M_0	3,200 kg
ζ	0.25
Burn time	3.35 s
Δp (change in momentum)	73,500 kg m/s
I_{sp}	275 s

2. A liquid oxygen–liquid hydrogen (hydrolox) motor produces 44.5 kN of thrust while operating at a chamber pressure of 6.895 MPa and a mixture ratio of 3.40. The exhaust products have a ratio of specific heats of 1.26 and a temperature of 2,689 K. Assume the motor operates at an altitude where the freestream pressure is a constant 9,500 Pa, and that the actual specific impulse is 96% of the theoretical value. Calculate:
   a) The speed of the exhaust stream
   b) Propellant mass and mass flow rate
   c) Nozzle exit area for optimal expansion
   d) Total propellant mass for 3 min. of burn time
3. A rocket engine uses RP-1 propellant to produce 9,000 N of thrust with an exit velocity of 1,250 m/s and will have a burn time of 67 s. Helium gas ($R = 2,077$ J/kg-K; $\gamma = 1.66$) is used as a pressurant and is pressurized to 14 MPa. The ambient temperature is 297 K. Calculate the mass of pressurant gas needed to pressurize the RP-1 tank to 3 MPa. Assume 1.2% residual propellant and 3% residual pressurant.
4. A propellant combination has a molecular mass of 23 kg/kmol and a specific heats ratio $\gamma = 1.24$. The combustion chamber temperature and pressure are 3,400 K and 6,900 kPa with $P_e = 101.3$ kPa (1 atm). The mass flow = 100 kg/s and the engine's exit diameter D_{Exit} is 0.62 m.
   a) Find the exhaust velocity v_e (m/s) at sea level.
   b) Find the *effective* exhaust velocity c (including pressure effects), the I_{sp}, and the thrust T at sea level.
   c) Find the exhaust velocity v_e (m/s) at altitude 9,150 m, where $P_{atm} = 37.88$ kPa.
   d) Find the *effective* exhaust velocity c, the I_{sp}, and the thrust T at 9,150 m altitude.
   e) Find the *effective* exhaust velocity c, the I_{sp}, and the thrust T in space.

    f) What is the theoretical maximum value of v_e and I_{sp}?

    g) Does thrust typically *increase* or *decrease* as altitude in an atmosphere increases? Explain.

5. A launch vehicle has a liftoff T/W of 1.5, a propellant mass fraction $\zeta_p = 0.9$, and $I_{sp} = 260$ s.

    a) Find the time to burnout t_b in seconds.

    b) Find the T/W at burnout, assuming constant thrust.

6. The S-II (second) step of the *Saturn V* had a very small structural mass fraction, because it was the last step designed of the three, and had to meet very stringent mass requirements. Since it had such an efficient structure, it is attractive to consider it alone as a high-performance vehicle when Δv performance is calculated with the rocket equation. In actual operation, the S-II's five J-2 engines burned together until such time when the center engine was shut down to limit peak acceleration.

    Assume that the S-II had the following characteristics:

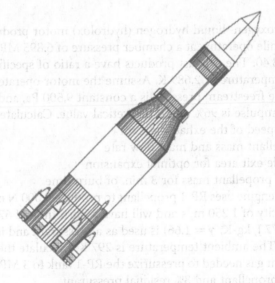

    Initial mass $m_0 = 480$ T (T = tonne = 1,000 kg), inert mass $m_s = 36$ T, $T = 5,100$ kN total, J-2 engine vacuum $I_{sp-vac} = 425$ s. Use $g_0 = 9.80665$ m/s^2.

    a) Calculate the S-II's burn time t_b in seconds, assuming all five engines burned all the way to burnout and ignoring the effect of exhaust pressure.

    b) Determine the maximum payload (kg) that could be delivered to a velocity of 9,150 m/s using this step as a single-stage launch vehicle.

    c) Comment on your results for the amount of thrust available at sea level. Extra Credit: determine $I_{sp-sea-level}$ (s), and calculate the sea level thrust value.

d) Determine liftoff T/W ratio. Ignoring the fact that the J-2 engines were optimized for high-altitude operation, could the S-II work as a launch vehicle? Comment on the situation, and explain why or why not.

References

[1] Kuo, K. K., *Principles of Combustion*, Wiley-Interscience, New York, 2005.
[2] Kuo, K. K., and Acharya, R., *Fundamentals of Turbulent and Multi-Phase Combustion*, Wiley, New York, 2012.
[3] Turns, S. R., *An Introduction to Combustion: Concepts and Applications*, McGraw-Hill, New York, 2011.
[4] Sellers, J., Astore, W., Giffen, R., and Larson, W., *Understanding Space: An Introduction to Astronautics*, Pearson, Upper Saddle River, NJ, 2007.
[5] Ley, W., Wittman, K., and Hallmann, W., *Handbook of Space Technology*. AIAA/John Wiley & Sons, Reston, VA, 2009.
[6] Turner, M. J. L., *Rocket and Spacecraft Propulsion*, Praxis, Chichester, UK, 2009.
[7] Sutton, G. P., and Biblarz, O., *Rocket Propulsion Elements*, John Wiley & Sons, New York, 2001.
[8] Ordway, F., Gardner, J., and Sharpe, M., *Basic Aeronautics*, Prentice-Hall, Englewood Cliffs, NJ, 1962.
[9] Riley, F., and Sailor, J., *Space Systems Engineering*, McGraw-Hill Inc., New York, 1962.
[10] Zurawski, R. L., and Rapp, D. C., "Analysis of Quasi-Hybrid Solid Rocket Booster Concepts for Advanced Earth-to-Orbit Vehicles." NASA Technical Paper TP-2751, August 1987.
[11] Koelle, D. E., "Economics of Fully Reusable Launch Systems (SSTO vs. TSTO Vehicles)," 47th IAF Congress, Beijing, 7–11 Oct. 1996.

Chapter 5

Launch Vehicle Performance and Staging

oving forward in the design process, we show the design roadmap in Fig. 5.1 with the highlighted signpost indicating the nature of the information to be covered in this chapter: the performance of launch vehicles based on their descriptive parameters, such as burnout mass, propellant mass, specific impulse, and payload to be carried.

We will begin with performance parameters and their definitions. Using the rocket equation, we'll look at how to predict different masses, staging, single-stage-to-orbit vehicles, and staging analysis.

5.1 The Three Categories of Launch Vehicle Mass

A generic launch vehicle is shown schematically in Fig. 5.2. The figure shows the major components including payload, structure, tankage, electronics, engines, and so forth.

All of the components in a launch vehicle can be categorized by whether they remain after firing or how they're used before the engines are shut down or burned out. These three categories are very simple:

1. *Propellants:* These are "thrown out the back" by the vehicles engine(s) in order to provide forward thrust. We will call the propellant mass m_p. Note that this term includes all the *propulsive* mass that is ejected. The jettisoning of parts such as payload fairings is not included.

2. *Payload:* All the other components of the vehicle exist *to deliver the payload to the correct velocity and altitude.* Payload mass will be defined as m_{PL} (*uppercase* subscript *PL* to distinguish it from lowercase *p* for propellants).

3. *"Structure" Mass:* This includes everything besides the ejected propellant and payload that doesn't burn and isn't part of the payload: "structure" mass $= m_s$. We write the word "structure" in quotes because in reality, the so-called "structure" mass also includes the mass of engines, plumbing, pumps, tanks, guidance, GNC electronics, unused propellants (margin), and so forth. It is the mass after usable propellants are consumed WITHOUT any payload. The final mass is the structure mass m_s + the payload mass m_{PL}. These definitions are shown pictorially in Fig. 5.3.

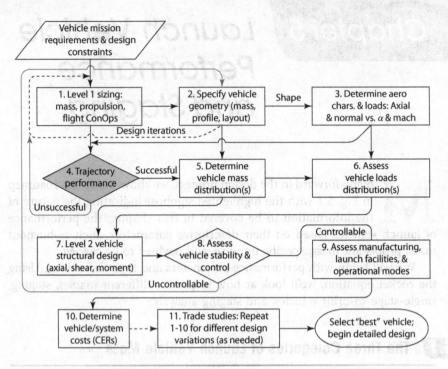

Fig. 5.1 This chapter will look at trajectory analysis using the design data generated in the previous chapter.

The abbreviations listed are not used universally. The subscript p could be used to represent propellant or payload; the subscript f could be used for final or fuel. (Some people use the incorrect term *fuel* to represent the propellants, both the fuel and the oxidizer.) The subscript i may be used for initial or inert, having the opposite meanings. Some use E or e for

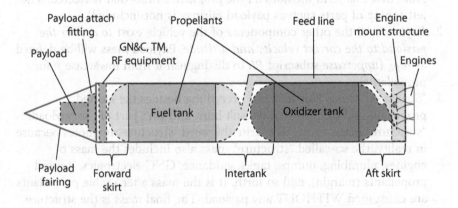

Fig. 5.2 The entire mass of the launch vehicle may be split into payload mass (front), final mass (payload attach fitting, equipment, and engine mount structure), and propellant mass (tanks).

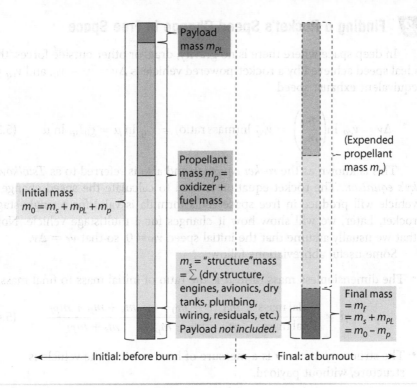

Fig. 5.3 The entire mass of the launch vehicle can be split into payload mass, final mass, and propellant mass.

empty. To be consistent, we will use m_p, m_{PL}, and m_s, the three abbreviations used in the list, and caution the reader to be sure to understand the definitions that other works might use. Also, we will use the subscript "0" to indicate *initial* or *starting* value, and the subscript "f" to indicate *final* or *ending* value.

Mathematically speaking, here are the interrelationships among the three variables:

The *launch mass* is the vehicle's mass at time $t = 0$

$$\text{Launch mass} = m(t = 0) = m_0 = m_s + m_p + m_{PL} \tag{5.1}$$

The launch mass may also be referred to as gross liftoff mass (GLOM).

The *final mass* is the vehicle's mass at final time $t = t_f$

$$m(t_f) = m_f$$

$$m_f = m_0 - m_p = m_s + m_{PL} \tag{5.2}$$

It is important to not confuse final mass m_f with structure mass m_s (or inert or empty mass), because the final mass must also include the payload mass.

5.2 Finding a Rocket's Speed Change in Free Space

In deep space where there is no gravity, drag, or other outside forces, the ideal speed achieved by a rocket-powered vehicle is $\Delta v = v_f - v_0$, and $v_{eq} =$ equivalent exhaust speed

$$\Delta v = v_{eq} \ln\left(\frac{m_0}{m_f}\right) = v_{eq} \ln(\text{mass ratio}) = v_{eq} \ln \mu = g_0 I_{sp} \ln \mu \quad (5.3)$$

This is known as the *rocket equation* and also is referred to as *Tsiolkovsky's equation*. The rocket equation is used to calculate the speed change a vehicle will produce in free space. This formula is valid for a single-stage rocket. Later, we will show how it changes for a multistage vehicle. Note that we usually assume that the initial speed $v_0 = 0$, so that $v_f = \Delta v$.

Some useful abbreviations follow:

- The dimensionless mass ratio μ is the ratio of initial mass to final mass.

$$\mu \equiv \frac{\text{initial mass}}{\text{final mass}} = \frac{m_0}{m_f} = \frac{m_0}{m_0 - m_p} = \frac{m_s + m_p + m_{PL}}{m_s + m_{PL}} \quad (5.4)$$

- The structural factor σ is a measure of how much of the vehicle is structure, without payload.

$$\sigma = \frac{m_s}{m_s + m_p} = \frac{m_s}{m_0 - m_p} \quad (5.5)$$

- The payload ratio π describes how much of the vehicle is payload.

$$\pi = \frac{m_{PL}}{m_s + m_p + m_{PL}} \quad (5.6)$$

- The payload fraction π_{PL} tells what fraction of *initial mass* m_{01} is payload, and is usually small.

$$\pi_{PL} = \frac{m_{PL}}{m_{01}} = \frac{m_{PL}}{m_s + m_p + m_{PL}} \ll 1 \quad (5.7)$$

The reciprocal of the mass ratio is $\mu^{-1} = m_f/m_0$. That fraction can be written as

$$\frac{m_f}{m_0} = \frac{m_s + m_{PL}}{m_s + m_p + m_{PL}} = \frac{(m_s + m_{PL} + m_p) - m_p}{m_0} = 1 - \frac{m_p}{m_0} = 1 - \frac{m_s + m_p}{m_0} \frac{m_p}{m_s + m_p}$$

$$= 1 - \left[\frac{(m_s + m_p + m_{PL}) - m_{PL}}{m_0}\right]\left[\frac{(m_p + m_s) - m_s}{m_s + m_p}\right] = 1 - (1 - \pi)(1 - \sigma)$$

$$= 1 - (1 - \pi - \sigma + \sigma\pi) = \sigma + (1 - \sigma)\pi \quad (5.8)$$

We can use these definitions to simplify the rocket equation. Recall the rocket equation

$$v_f = v_{eq} \ln \mu = -c \ln(\mu^{-1}) \qquad (5.9)$$

so

$$v_f = -c \ln[\sigma + (1 - \sigma)\pi] \qquad (5.10)$$

The maximum speed $v_{f\text{max}}$ occurs with *zero* payload mass, or $\pi = 0$ (not practically useful but a limiting case).

$$v_{f\text{max}} = -c \ln \sigma \qquad (5.11)$$

5.3   Burnout Speed

We can use the rocket equation to generate a family of curves relating the final speed to the payload ratio, as shown in Fig. 5.4. This illustrates the fact that the lower the structural ratio, the higher the normalized speed. This is the basis of the extreme emphasis on launch vehicle inert mass to be as small as possible.

5.4   Single-Stage-to-Orbit

During the 1980s and 1990s, the "holy grail" for launch vehicles used to be *single-stage-to-orbit (SSTO)*. This construction would need only a single stage to reach orbit, and no staging would be necessary. In addition, many thought

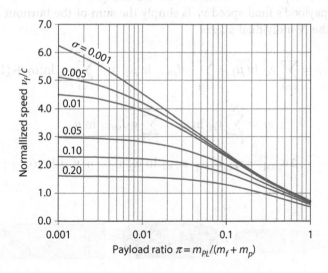

Fig. 5.4   The relationship between the payload ratio and the vehicle speed change normalized by exhaust speed c.

it possible to recover and reuse the single-stage vehicle over and over. Let's consider the required performance for SSTO.

We have seen that the burnout speed is

$$v_f = c \ln \mu = c \ln[(1 + \pi)/(\sigma + \pi)] \tag{5.12}$$

For $\sigma = 0.08$, a reasonable lower limit, and $\pi = 0.05$ (a 5% payload ratio), we find that $v_f = 2.09\ c$, or $c = v_f/2.09$. To achieve a final speed of 9.4 km/s [typical for low Earth orbit (LEO) including the losses due to gravity and drag, as shown in Chapter 3], we find that $c \geq 4.5$ km/s, or $I_{sp} = 459$ s required, which turns out to be a few percent greater than the best vacuum performance we can achieve with hydrolox propellants LOx/LH$_2$! When one also considers the reduction in engine performance for a vacuum engine in the atmosphere, it appears that SSTO vehicles with reasonable payload fractions are just not practical at this time. So, what can we do in order to achieve the required speed with losses? The answer is *staging*.

5.5 Staging

A vehicle has a burnout speed that is determined by the rocket equation. If we were to *stack* two, three, or more rockets end-to-end, we could calculate the speed by simply adding up the speed that each part of the rocket produces, keeping in mind that the lowest part has the remaining part of the rocket as payload, working our way down from the top to the bottom. The stacking of rockets is called *staging*, and it allows us to achieve higher Δv. A three-step stack of rockets is shown in Fig. 5.5.

The payload's final speed v_f is simply the sum of the burnout speeds of each of the N individual stages

$$v_{total} = v_f = \sum_{i=1}^{N} v_{ei} \ln \mu_i = \sum_{i=1}^{N} g_0 I_{sp-i} \ln \mu_i = - \sum_{i=1}^{N} v_{ei} \ln[\sigma_i + (1 - \sigma_i)\pi_i]$$

$$v_f = \sum_{i=1}^{N} v_{ei} \ln \mu_i = \sum_{i=1}^{N} g_0 I_{sp-i} \ln \mu_i \tag{5.13}$$

$$\Delta v_{total} = v_{e1} \ln \mu_1 + v_{e2} \ln \mu_2 + \cdots + v_{eN} \ln \mu_N$$

Fig. 5.5 Higher speeds are achieved by stacking rockets and firing them sequentially, then dropping off the empty steps. Three-step Saturn V shown.

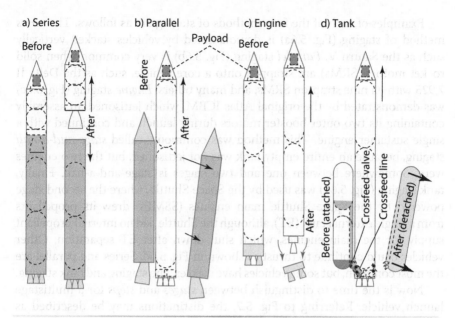

Fig. 5.6   Four different types of launch vehicle staging. The prestaging configuration is shown on the left side of each of the diagrams, and the poststaging configuration is on the right. a) for *series* staging, the lower step of the rocket fires first, then is detached, and then the upper step fires. b) in *parallel* staging, multiple parts of the rocket burn simultaneously, or in parallel. c) for *engine* staging, heavy rocket engines are jettisoned partway through the burn, and the remainder of the vehicle continues firing. d) for *tank* staging, an external tank provides propellant to a separate part of the rocket via crossfeed lines and valves. Once depleted, the valves are closed and the external tank is separated.

or

$$v_f = -\sum_{k=1}^{N} v_{eq-k}\ \ln[\sigma_k + (1-\sigma_k)\pi_k] \qquad (5.14)$$

The definitions of the variables are step k's structural factor

$$\sigma_k = \frac{m_{fk}}{m_{fk} + m_{pk}} \qquad (5.15)$$

and payload ratio, k-th step

$$\pi_k = \frac{m_{0,k+1}}{m_{0,k}} \qquad (5.16)$$

5.5.1   Types of Launch Vehicle Staging

One way to stage rockets is to place them end-to-end, as mentioned previously, and to discard the spent part as soon as it is empty, in order to increase the subsequent mass ratio. This process is known as *series* staging. However, there are three other methods that have been used for staging, as shown in Fig. 5.6, where the parts to be jettisoned are shaded gray.

Examples of each of the four methods of staging are as follows. The *series* method of staging (Fig. 5.6a) is demonstrated by vehicles stacked vertically such as the Saturn V. *Parallel* staging (Fig. 5.6b) is very common when solid rocket motors (SRMs) are strapped onto a core vehicle, such as the Delta II 7,925 with its nine strap-on SRMs, and many others. *Engine* staging (Fig. 5.6c) was demonstrated by the original Atlas ICBM, which jettisoned an assembly containing its two outer booster motors during launch and continued with a single sustainer engine. This method was commonly called *stage-and-a-half* staging, because an entire empty tank was not jettisoned, but the two engines were. Somewhere between one and two stages is stage-and-a-half. Finally, tank staging (Fig. 5.6d) was used by the Space Shuttle, where the second stage powered by the Space Shuttle main engines (SSMEs) drew its propellants from a single external tank (ET), although the Shuttle had no internal propellant supply for the main engines, which shut down after ET separation. Other vehicles could continue to thrust as shown in Fig. 5.6d. Series and parallel are the most common, but some vehicles have used engine staging and tank staging.

Now is the time to distinguish between *stages* and *steps* for a multistage launch vehicle. Referring to Fig. 5.7, the distinctions may be described as follows:

* A *stage* refers to the *entire* portion of the rocket at a given time.
* A step refers to a *propulsive* piece of the rocket that is going to separate or be jettisoned at some later time. Non-propulsive parts such as payload fairings are not considered steps.

The public uses both interchangeably and often *incorrectly* at times.

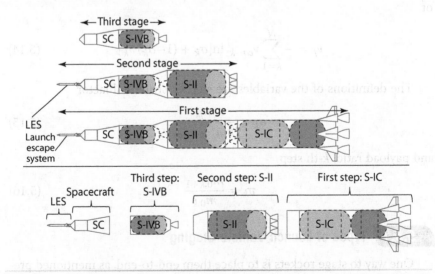

Fig. 5.7 The three assemblies above the horizontal line are the Saturn V's three *stages*. The vehicle's *steps* or individual assemblies are below the horizontal line. Note that in the top figure, the LES is not included in the third stage, because it is jettisoned during the second-stage burn.

When identifying and designing parts of launch vehicles, it is handy to use the vehicle's steps to be sure that the proper portion of the vehicle is being looked at. To make all this perfectly clear, please refer to Fig 5.7, which shows the three *stages* (top of figure) as well as the three *steps* (bottom of figure) of the Saturn V.

5.6 Calculation of Speed Supplied by a Multistage Rocket

Looking at the Saturn V in Fig. 5.7, we will demonstrate how to calculate its delivered speed v_f. The speed delivered by each *stage* of the Saturn is calculated via the rocket equation, where we include each step i's equivalent exhaust speed c_i or v_{eq-i} (or equivalently its specific impulse I_{spi}) and its mass ratio, which is the ratio of the stage's loaded mass to the stage's empty mass. The payload mass m_{PL}, propellant masses m_{pi} and final masses m_{si} must be known or specified. Alternatively, the initial masses m_{0i} and final masses m_{fi} must be known.

To calculate the speed change caused by the *Saturn V*'s first stage burn, we assemble the mass ratio using the loaded and empty masses as shown in Fig. 5.8a. The variables used in the figure are defined as follows:

i = number of stages/steps
m_{0i} = **stage** i initial mass
m_{fi} = **stage** i final mass
m_{si} = **step** i structure mass
m_{pi} = **step** i propellant mass
m_{PL} = **payload** mass

The speed provided by the first stage of the Saturn V is calculated as

$$\Delta v_1 = g_0 I_{sp1} \ln\left(\frac{m_{01}}{m_{f1}}\right)$$
$$= g_0 I_{sp1} \ln\left(\frac{m_{s1} + m_{p1} + m_{s2} + m_{p2} + m_{s3} + m_{p3} + m_{PL}}{m_{s1} + 0 + m_{s2} + m_{p2} + m_{s3} + m_{p3} + m_{PL}}\right) \quad (5.17)$$

Note that the mass ratio includes *all* of the mass that is located above the portion of the vehicle that is firing.

Continuing for the Saturn V's second stage, we assemble the mass ratio now using the configuration of loaded and empty masses shown in Fig. 5.8b.

The speed provided by the second stage of the Saturn V is calculated as

$$\Delta v_2 = g_0 I_{sp2} \ln\left(\frac{m_{02}}{m_{f2}}\right) = g_0 I_{sp2} \ln\left(\frac{m_{s2} + m_{p2} + m_{s3} + m_{p3} + m_{PL}}{m_{s2} + 0 + m_{s3} + m_{p3} + m_{PL}}\right)$$
$$(5.18)$$

Finally, for the Saturn V's third stage, we assemble the mass ratio now using the configuration of loaded and empty masses shown in Fig. 5.9.

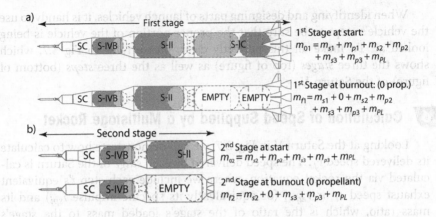

Fig. 5.8 a) The loaded (first step) and empty configurations for the Saturn V's first stage to be used for the Stage 1 Δv. b) The loaded (second step) and empty configurations of the second stage of the Saturn V.

3rd Stage at start:
$$m_{03} = m_{s3} + m_{p3} + m_{PL}$$

3rd Stage at burnout (0 propellant)
$$m_{f3} = m_{s3} + 0 + m_{PL}$$

Fig. 5.9 The loaded (third step) and empty configurations of the third stage of the Saturn V.

The speed provided by the third stage of the Saturn V is calculated as

$$\Delta v_3 = g_0 \, I_{sp3} \ln\left(\frac{m_{03}}{m_{f3}}\right) = g_0 \, I_{sp3} \ln\left(\frac{m_{s3} + m_{p3} + m_{PL}}{m_{s3} + 0 + m_{PL}}\right) \qquad (5.19)$$

The ideal speed provided by the Saturn V is simply the sum of the Δvs provided by the three steps, or $v_f = \Delta v_1 + \Delta v_2 + \Delta v_3$. Note that this is the *ideal* speed and does not include gravity, drag, propulsion, or steering losses. Those may be calculated once the actual trajectory is known, or estimated based on the vehicle's liftoff T/W_0 ratio, using the methods in Chapter 3.

Example 5.1

For the three steps of the Saturn V, we have the following approximate values (T = metric tonne):

- Step 1: $m_{p1} = 2,080$ T, $m_{s1} = 240$ T, $I_{sp-SL} = 263$ s
- Step 2: $m_{p2} = 450$ T, $m_{s2} = 36$ T, $I_{sp} = 419$ s
- Step 3: $m_{p3} = 108$ T, $m_{s3} = 11$ T, $I_{sp} = 426$ s
- Payload: $m_{PL} = 50$ T

(Continued)

Example 5.1 (Continued)

Stage 1:

$$
\begin{aligned}
\Delta v_1 &= g_0\, I_{sp1} \ln\left(\frac{m_{01}}{m_{f1}}\right)\\
&= g_0\, I_{sp1} \ln\left(\frac{m_{s1}+m_{p1}+m_{s2}+m_{p2}+m_{s3}+m_{p3}+m_{PL}}{m_{s1}+0+m_{s2}+m_{p2}+m_{s3}+m_{p3}+m_{PL}}\right)\\
&= 9.81 \text{ m/s}^2 \times 263 \text{ s} \times \ln\left(\frac{240+2{,}080+36+450+11+108+50}{240+0+36+450+11+108+50}\right)\\
&= 3{,}099 \text{ m/s}
\end{aligned}
$$

Stage 2:

$$
\begin{aligned}
\Delta v_2 &= g_0\, I_{sp2} \ln\left(\frac{m_{02}}{m_{f2}}\right) = g_0\, I_{sp2} \ln\left(\frac{m_{s2}+m_{p2}+m_{s3}+m_{p3}+m_{PL}}{m_{s2}+0+m_{s3}+m_{p3}+m_{PL}}\right)\\
&= 9.81\frac{\text{m}}{\text{s}^2} \times 419 \text{ s} \times \ln\left(\frac{36+450+11+108+50}{36+0+11+108+50}\right)\\
&= 4{,}775 \text{ m/s}
\end{aligned}
$$

Stage 3:

$$
\begin{aligned}
\Delta v_3 &= g_0\, I_{sp3} \ln\left(\frac{m_{03}}{m_{f3}}\right) = g_0\, I_{sp3} \ln\left(\frac{m_{s3}+m_{p3}+m_{PL}}{m_{s3}+0+m_{PL}}\right)\\
&= 9.81\frac{\text{m}}{\text{s}^2} \times 426 \text{ s} \times \ln\left(\frac{11+108+50}{11+0+50}\right) = 4{,}259 \text{ ms}
\end{aligned}
$$

Final speed $v_f = 3{,}099 + 4{,}775 + 4{,}259 \text{ m/s} = \mathbf{12{,}133 \text{ m/s}}$.

Even using the sea-level specific impulse for the entire first-stage burn, which is very conservative, the final speed is much greater than typical orbital speed ($\sim$8,000 m/s). The reason this speed is so great is not just to overcome gravity, drag, and steering losses (from Table 3.3, these are 1,534, 40, and 243 m/s, respectively), but also because the Saturn V injected its payload into a translunar orbit, requiring significantly more speed than just orbiting.

One last interesting fact: the calculated ideal speed may not even be feasible! The rocket equation does not incorporate thrust values, yet it is used to calculate speed changes. It's entirely possible that a "good performing" launch vehicle as determined by the staged rocket equation might not even be able to get off the ground without proper sizing of its engines. In addition to the speed calculation, one must also select one or more engines to assure that the thrust-to-weight ratio is greater than one, at least for the first stage. Upper stages commonly have T/W ratios that are less than one. Thus, in mathematical terms, we can say that performance as indicated by the rocket equation is necessary, but not sufficient to indicate feasibility.

5.7 Payload Ratio

The *overall* payload ratio is defined as $\pi_{PL} = m_{PL}/m_{01}$. We can do some algebraic tricks and expand the expression to make this more interesting.

$$\pi = \frac{m_{PL}}{m_{01}} = \left(\frac{m_{PL}}{m_{0N}}\right)\left(\frac{m_{0N}}{m_{0,N-1}}\right)\left(\frac{m_{0,N-1}}{m_{0,N-2}}\right)\left(\cdots\right)\left(\frac{m_{02}}{m_{01}}\right) = \prod_{k=1}^{N} \pi_k \quad (5.20)$$

This relation indicates that a small overall payload ratio may be generated as the *product* of several step payload ratios. Because the payload ratios are multiplied, they need only be "reasonably" small to still attain a good overall payload ratio. This process relaxes the design (in terms of minimizing mass) of individual steps somewhat.

There was a scheme in the 1990s where consideration was given to having inexpensive steps with moderate (meaning lower-cost) mass ratios built by commercial companies, say Chicago Bridge and Iron, instead of very expensive aerospace companies who charge "an arm and a leg" for a very high-grade structure. Some called a vehicle assembled from "less than aerospace grade components" the "big dumb" booster [1]. The idea, unfortunately, never caught on.

5.8 Unrestricted Staging

Let us consider the design of vehicles with N steps that may have different exhaust speeds c_i (or I_{sp-i}) and different structural ratios σ_i. With variations in those parameters, we would expect to have different mass ratios μ_i also. Clearly, this description refers to many existing launch vehicles that use different propellants, engines, constructions, etc. One might ask: how do we determine what relative size to make each step? A possible answer to these questions is to use mathematics to numerically optimize the mass ratios of the rocket having differing step parameters c_i (or I_{sp-i}) and σ_i.

The material presented in this section is based on section 8.2 of [2]. We note that the speed provided by a multi-stage rocket, ignoring gravity and drag losses, is:

$$v_f = \sum_{i=1}^{N} c_i \ln \mu_i \quad (5.21)$$

Since v_f is specified by our mission, we can specify different characteristic velocities (or specific impulses) and choose different mass ratios for the various steps of the rocket. But what is the *best* set of mass ratios to use?

One way to approach this is to minimize the ratio of m_{01}/m_{PL} (minimize the GLOM needed for a given payload mass) or to maximize its inverse m_{PL}/m_{01} (maximize the amount of payload for a given GLOM). To do this minimization, we need an expression for m_{01}/m_{PL} in terms of all the step mass ratios μ_i.

Recall that the initial mass of the "next higher" $(i+1)$ stage $m_{0,i+1}$ is equal to the initial mass of the "current" (i) stage mass m_{0i} minus the propellant consumed by the previous step (m_{pi}) and the previous step's empty mass (m_{si}). Thus, the initial mass of the second stage is $m_{02} = m_{01} - m_{p1} - m_{s1}$. With this observation, we can write the ratio m_{01}/m_{PL} as:

$$\frac{m_{01}}{m_{PL}} = \left(\frac{m_{01}}{m_{01} - m_{p1} - m_{s1}}\right) \times \left(\frac{m_{02}}{m_{02} - m_{p2} - m_{s2}}\right) \times \cdots \times \left(\frac{m_{0N}}{m_{PL}}\right)$$

(5.22)

Now remembering the definitions of the mass ratio as

$$\mu_i = \frac{m_{0i}}{m_{0i} - m_{pi}}$$

(5.23)

And the structural ratio σ_i defined as

$$\sigma_i = \frac{m_{si}}{m_{si} + m_{pi}}$$

(5.24)

We can rewrite one of the factors in parenthesis above as follows:

$$\frac{m_{0i}}{m_{0i} - m_{pi} - m_{si}} = \left(\frac{m_{0i}}{m_{0i} - m_{pi}}\right)\left(\frac{m_{pi}}{m_{si} + m_{pi}}\right)$$

$$\times \left[\frac{(m_{0i} - m_{pi})(m_{si} + m_{pi})}{m_{pi}(m_{0i} - m_{pi} - m_{si})}\right] = \frac{\mu_i(1 - \sigma_i)}{1 - \mu_i\sigma_i}$$

(5.25)

With this, m_{01}/m_{PL}, the reciprocal of the overall payload ratio can be re-written as

$$\frac{m_{01}}{m_{PL}} = \left[\frac{\mu_1(1 - \sigma_1)}{1 - \mu_1\sigma_1}\right] \times \left[\frac{\mu_2(1 - \sigma_2)}{1 - \mu_2\sigma_2}\right] \times \cdots \times \left[\frac{\mu_N(1 - \sigma_N)}{1 - \mu_N\sigma_N}\right]$$

(5.26)

To minimize the above ratio, it is convenient to take its natural logarithm, making the expression into a summation rather than a product:

$$\ln\frac{m_{01}}{m_{PL}} = \sum_{i=1}^{N} \ln\left(\frac{\mu_i(1 - \sigma_i)}{1 - \mu_i\sigma_i}\right) = \sum_{i=1}^{N} [\ln\mu_i + \ln(1 - \sigma_i) - \ln(1 - \mu_i\sigma_i)]$$

(5.27)

Now we are ready to find the minimum of this expression, subject to the constraint that

$$v_f = \sum_{i=1}^{N} v_{eq-i} \ln \mu_i \qquad (5.28)$$

In mathematics, we say that a function may have one extremum (maximum or minimum) or multiple extrema (maxima and/or minima).

We desire to find the minimum of payload ratio function $z = \ln(m_{01}/m_{PL}) = f(\sigma, \mu)$, subject to the constraint

$$g(\sigma, \mu) = \sum_{i=1}^{N} v_{eq-i} \ln \mu_i - v_f = 0 \qquad (5.29)$$

where g is the expression for the burnout speed rearranged to equal zero. One mathematical way to do this is to use the method of Lagrange multipliers. Here we form the function $h = f + Lg$, where we have added the constraint equation g multiplied by the variable L, a *Lagrange multiplier*, to f. This is permitted because we are simply adding $L \times 0 = 0$ to the above equation, which does not change it. We then find the extrema by simultaneously solving $\partial h/\partial \mu = 0$ and $\partial h/\partial L = 0$. In these equations, the steps' c_i are determined by the chemistry of their selected propellants, and their σ_i are selected by the analyst.

For our problem, we now define the function $h(\sigma, \mu)$ to be minimized as

$$h(\mu, \sigma) = \ln\frac{m_{01}}{m_{PL}} = \sum_{i=1}^{N} \left\{ [\ln \mu_i + \ln(1 - \sigma_i) - \ln(1 - \mu_i \sigma_i)] + L[c_i \ln \mu_i - v_f] \right\}$$

$$(5.30)$$

We differentiate this equation with respect to μ_i, so we can find the optimum values of μ_i. This differentiation yields N equations of the form

$$\frac{1}{\mu_i} + \frac{\sigma_i}{1 - \mu_i \sigma_i} + L\frac{c_i}{\mu_i} = 0 \qquad (5.31)$$

Which can be solved to get

$$\mu_i = \frac{1 + Lc_i}{Lc_i \sigma_i} \qquad (5.32)$$

Next, we substitute the expressions for μ_i back into the burnout speed equation $v_f = \sum_{i=1}^{N} c_i \ln \mu_i$ to get

$$v_f = \sum_{i=1}^{N} c_i \ln\frac{1 + Lc_i}{Lc_i \sigma_i}, \quad \text{or} \quad v_f - \sum_{i=1}^{N} c_i \ln\frac{1 + Lc_i}{Lc_i \sigma_i} = 0 \qquad (5.33)$$

With c_i, s_i, and v_f known or specified, the v_f equation is numerically solved for L. The resulting value is substituted back into the μ_i equation above to find the mass ratio of each step.

Example: Two-Stage Optimized Launch Vehicle

These formulae may be applied to the simplest multi-stage launch vehicle, a *two-stage* vehicle. After calculation, we find

$$\mu_1 = \frac{1 + Lc_1}{Lc_1\sigma_1}, \quad \mu_2 = \frac{1 + Lc_2}{Lc_2\sigma_2}, \quad \text{and} \quad v_f = c_1 \ln \mu_1 + c_2 \ln \mu_2 \quad (5.34)$$

Substituting μ_1 and μ_2 into the v_f expression, we find

$$v_f = c_1 \ln \frac{1 + Lc_1}{Lc_1\sigma_1} + c_2 \ln \frac{1 + Lc_2}{Lc_2\sigma_2} \quad (5.35)$$

This expression may be solved numerically for the value of L that satisfies it, using the selected structural ratios and the respective exhaust speeds of the two steps. Then, the resulting value of L is used to obtain the step mass ratios μ_1 and μ_2, using the two expressions above, and the vehicle is completely specified.

Example: Special Case Launch Vehicle With Identical Specific Impulse for All Steps

For this special case, the $c_i = c$ are equal for all steps, and the above equation becomes

$$\frac{v_f}{c} = N\ln \frac{1 + Lc}{Lc} - \sum_{i=1}^{N} \ln \sigma_i \quad (5.36)$$

Solving, we find

$$\frac{1 + Lc}{Lc} = \exp\left[\frac{1}{N}\left(\frac{v_f}{c} + \sum_{i=1}^{N} \ln \sigma_i\right)\right] \quad (5.37)$$

Since

$$\mu_i \sigma_i = \frac{1 + Lc}{Lc} \quad (5.38)$$

the mass ratio for step i is

$$\mu_i = \frac{1}{\sigma_i}\exp\left[\frac{1}{N}\left(\frac{v_f}{c} + \sum_{i=1}^{N} \ln \sigma_i\right)\right] \quad (5.39)$$

If we make the additional assumption that each step has the same structural ratio σ, it can be shown that for minimum mass, each stage provides $1/N$ of the total Δv. So, for a two-stage vehicle, each step would provide one-half of the total speed; for a three-stage vehicle, each stage would provide one-third of the total speed, etc.

(Continued)

Example: Special Case Launch Vehicle With Identical Specific Impulse for All Steps *(Continued)*

This situation is called *restricted staging*, which refers to the construction of a launch vehicle that has a simple (but unrealistic) assumption: all steps are *similar*, meaning that they all have same specific impulse I_{sp-i}, same payload ratio π_i, same structural ratio σ_i, and, consequently, same mass ratio μ_i.

Reference [2] shows that the normalized final or burnout speed v_f for a rocket with N similar steps (restricted staging) in free space is

$$\frac{v_f}{v_e} = N \ln\left[\frac{1}{\pi^{1/N}(1-\sigma)+\sigma}\right]$$

For very large $N (N \to \infty)$,   (5.40)

$$\frac{v_f}{v_e} = (1-\sigma)\ln\frac{1}{\pi}$$

The results of the normalized speed ratios produced by the upper relation is shown in Figure 5.10 as a bar graph for one to nine stages. Also included is the result produced by the lower expression for an *infinite* number of stages ($N = \infty$, clearly an impossible situation). For these results, it's assumed that steps have the same specific impulse I_{sp} or exhaust speed c, fixed structural ratio $\sigma = 0.10$, and fixed payload ratio $\pi = 0.05$.

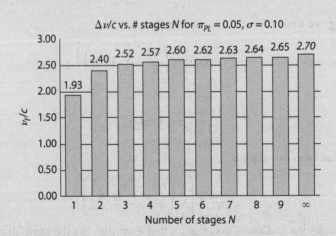

Fig. 5.10   Burnout speed Δv normalized by exhaust speed c vs number of stages N with fixed values of specific impulse, structural ratio, and payload ratio. There is little benefit to having more than three stages.

(Continued)

Example: Special Case Launch Vehicle With Identical Specific Impulse for All Steps (Continued)

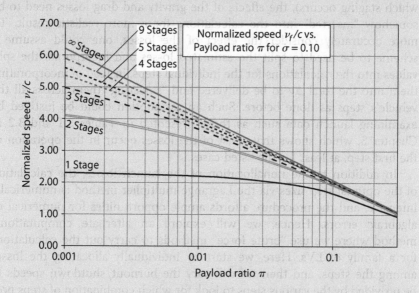

Fig. 5.11 By normalizing speeds to exit speed c, we can see that the effects of more than two steps are minimal except for very low payload fractions for restricted staging.

A glance at Figure 5.10 shows that multi-step launch vehicle can achieve higher burnout speeds than single-step vehicles, but there is very little benefit to having more than three steps in a launch vehicle. Figure 5.11 plots a family of curves relating normalized speed to payload ratio π for different numbers of steps. Again, it can be seen that for practical purposes, there's not much improvement after three steps ($N > 3$).

Note: although we have calculated the "optimum" step parameters that will minimize liftoff mass, *this staging optimization process does not consider gravity, drag, and propulsion losses.* The results will likely be different when the losses are included. If we were to add some assumed losses to the ideal v_f so as to obtain a vehicle whose burnout speed would result in the correct value *after* losses, the result might produce a vehicle that is not optimal. This is because simply adding the assumed losses "smears" them across the entire LV, while the *actual* losses are NOT evenly distributed among the steps. Nevertheless, restricted staging is still a useful tool to study rocket performance.

Later in this chapter, we will describe a "brute force" family-of-launch-vehicle analysis that assigns the losses to individual stages of the vehicle and possibly provides a more realistic "optimal" design. Additional information on the results of trajectory optimization carried out on realistic vehicles is provided in Chapter 7.

5.9 Pitfalls of the Lagrangian "Optimization" Procedure

As mentioned previously, when one uses the Lagrangian method to calculate the steps' mass properties (which in turn determine the speeds at which staging occurs), the effects of the gravity and drag losses need to be somehow "inserted" into the calculations for a more realistic result. To more accurately consider the effect of the losses, one could assume a scheme to be used to split the estimated losses and incorporate the split values into the calculations for the individual steps, instead of incorporating them into the total Δv to be delivered and "smearing" them across all the vehicle's steps as done before. Such an assumption could be justified by examining launch data such as those presented in Tables 3.1 and 3.2 in Chapter 3, which shows that most of the losses occur in the operation of the first step, at least for the listed cases.

In addition to the nonallocation of the launch losses, the calculation of the optimal step sizes via the Lagrange multiplier method is numerically intensive, and the procedure affords ample opportunities for numerical or algebraic errors. Hence, we will explore an alternate computational method where we use "brute force" methods to carry out the calculations for a family of LVs. Here, we start by individually allocating the losses among the steps, and then we will vary the burnout/shutdown speeds to be provided by the various steps to look for which combination of steps provides the smallest liftoff mass.

We will pursue this idea by understanding that the total *burnout (BO)* or shutdown speed that the vehicle must deliver to the payload (the orbit speed) is equal to the vehicle's *designed* speed capability Δv_{design} (what it has to be designed to provide via the rocket equation) less the *losses* Δv_{loss} that occur during ascent (due to gravity, drag, and steering) $\pm$ the speed of the launch site Δv_{launch_site} (whose magnitude depends on the launch site's latitude and the LV's launch azimuth angle, as described in Chapter 3), or

$$\Delta v_{design} = \Delta v_{burnout} + \Delta v_{loss} \tag{5.41}$$

Here, in the definition of Δv_{design}, we have already included the gain or loss associated with Δv_{launch_site}. (In other words, we have subtracted the component of the launch site speed that is parallel with our launch azimuth.) In addition, we state that the burnout or shutdown speed is attained by the sequential firing of a total of N steps

$$\Delta v_{burnout} = \Delta v_{burnout\,step1} + \Delta v_{burnout\,step2} + \cdots + \Delta v_{burnout\,stepN}$$

$$= \sum_{i=1}^{N} (\Delta v_{burnout\,step\,i}) \tag{5.42}$$

We also break the losses into those produced by climbing against gravity and aerodynamic drag.

$$\Delta v_{loss} = \Delta v_{grav\,loss} + \Delta v_{drag\,loss} \tag{5.43}$$

Now, we can allocate the contributions provided by each step of the vehicle, provided that the sum of the allocations provides the needed burnout/shutdown speed after losses. (It's assumed that the required burnout/shutdown *altitude* is already implicitly incorporated into the losses.) For the ith step of an LV, such allocations might look like

$$\Delta v_{design\,step\,i} = \alpha_i \,\Delta v_{burnout} + \beta_i \,\Delta v_{loss} \tag{5.44}$$

and the coefficients α_i and β_i are defined such that

$$\alpha_i, \beta_i \geq 0$$

$$\sum \alpha_i = 1 \tag{5.45}$$

$$\sum \beta_i = 1$$

This set of definitions for α_i and β_i will ensure that $\Delta v_{design} = \Delta v_{burnout} + \Delta v_{loss}$.

To illustrate this process, let's consider $N = 2$ for a two-step vehicle [i.e., two-stage-to-orbit (TSTO)]. With the previous definitions, we assume that we will carry out the calculations needed for a vehicle where the first step would provide a speed $\Delta v_{burnout\,step1}$ (*after* losses), calculated as a fraction α_1 of the total burnout speed needed, and the second step would provide a speed $\Delta v_{burnout\,step2}$, the remainder of the speed needed, or $(1 - \alpha_1)$ fraction of the total speed. It would be also assumed that the losses "absorbed" by the step would be determined by the fraction β_1, such that

$$\Delta v_{design\,step\,1} = \alpha_1 \,\Delta v_{burnout} + \beta_1 \,\Delta v_{loss} \tag{5.46}$$

Similarly, the second step would provide

$$\Delta v_{design\,step\,2} = (1 - \alpha_1)\Delta v_{burnout} + (1 - \beta_1)\Delta v_{loss} \tag{5.47}$$

With these considerations, the two steps would provide the proper value of

$$\Delta v_{design} = \Delta v_{design\,step\,1} + \Delta v_{design\,step\,2}$$

$$= [\alpha_1 + (1 - \alpha_1)]\Delta v_{burnout} + [\beta_1 + (1 - \beta_1)]\Delta v_{loss} \tag{5.48}$$

$$= \Delta v_{burnout} + \Delta v_{loss}$$

to achieve the desired orbit *with* losses. The value of α_1 would be varied from (say) 0.1 to 0.9. The value of β_1 allocates the fraction of the losses to each stage. To make the following more concise, we'll drop the subscripts from α_i and β_i without any loss of generality.

If the combination steps were to be sized so the launch losses were in the same proportion as that of the speed provided (in other words, the losses are smeared over the entire launch vehicle), one would choose $\beta = \alpha$. Then the speeds provided by the two steps would be determined by the family of calculations obtained by varying α

$$\Delta v_{\text{design step 1|smeared}} = \alpha(\Delta v_{\text{ideal}} + \Delta v_{\text{loss}}) \tag{5.49}$$

and

$$\Delta v_{\text{design step 2|smeared}} = (1 - \alpha)(\Delta v_{\text{ideal}} + \Delta v_{\text{loss}}) \tag{5.50}$$

If instead, the combination were to be sized so the first step took care of *all* of the launch losses, one would choose $\beta = 1$. The assumption here would be that all of the drag and most of the gravity loss occurs during the firing of the first stage, and so the losses should all be lumped there. Then, we would determine the speeds provided by the two steps by a second family of calculations:

$$\Delta v_{\text{design step 1|}\beta=1} = \alpha \Delta v_{\text{ideal}} + \Delta v_{\text{loss}} \tag{5.51}$$

and

$$\Delta v_{\text{design step 2|}\beta=1} = (1 - \alpha)\Delta v_{\text{ideal}} \tag{5.52}$$

Now, this procedure would not be complete without also modifying the concept of the staging speed. The design speed predicted by the rocket equation does not consider external losses. Hence, the speed at staging would be calculated by taking the stage's design speed and subtracting the fraction of the losses assigned to that step. Hence, the staging speed, or the speed at the burnout of step 1, is calculated from

$$\Delta v_{\text{staging}} = \Delta v_{\text{design step 1}} - \beta \Delta v_{\text{loss}} \tag{5.53}$$

Similarly, the speed change from start to burnout of the second step would be

$$\Delta v_{\text{burnout 2}} = \Delta v_{\text{design step 2}} - (1 - \beta)\Delta v_{\text{loss}} \tag{5.54}$$

Hence, the total speed achieved would be

$$\Delta v_{staging} + \Delta v_{burnout\,2} = (\Delta v_{design\,step\,1} - \beta \Delta v_{loss})$$

$$+ [\Delta v_{design\,step\,2} - (1 - \beta)\Delta v_{loss}]$$

$$= \Delta v_{design\,step\,1} + \Delta v_{design\,step\,2} - [\beta + (1 - \beta)]\Delta v_{loss}$$

$$= \Delta v_{ideal} - \Delta v_{loss} \qquad (5.55)$$

as expected.

Example: Gross Mass vs Staging Speed for Families of TSTO LVs with Differing Propellants

We'll now take a look at the liftoff masses of families of $N = 2$ two-stage-to-orbit (TSTO) vehicles using combinations of different propellants: kerolox, hydrolox, and solids. To do this, we will need to know what values to use for the specific impulses of the different combinations in their first steps, which fly from sea level to a near-vacuum.

Normally, engine data are presented as either sea-level or vacuum specific impulses, or both. However, the first-stage flight begins at or near sea level and ends in the upper atmosphere in near-vacuum conditions, so the question arises as to what value should be used for the I_{sp}. Obviously, the effective value would be somewhere in between its sea-level and vacuum values.

Leondes [3] suggests using a value two thirds of the way between sea level and vacuum conditions as the effective I_{sp} value

$$I_{sp\,eff} = 0.95[I_{sp-SL-theo} + 2/3(I_{sp-vac-theo} - I_{sp-SL-theo})] \qquad (5.56)$$

If we assume that commonly available I_{sp} values are delivered and not theoretical values, we may drop the *theo* subscripts and the 0.95 multiplier to yield, after a bit of algebra,

$$I_{sp\,eff} = (I_{sp-SL} + 2 I_{sp-vac})/3 \qquad (5.57)$$

or

$$c_{eff} = (c_{SL} + 2 c_{vac})/3 \qquad (5.58)$$

These formulations account for the increase in engine performance between launch elevation and the vacuum of orbit. As a note, Leondes also provides information on how to approximate the effective specific impulse with varying chamber pressures and expansion ratios.

We will use the Falcon 9 and Space Shuttle vehicles as sources of propulsion information. The kerolox engine performance is based on the Merlin 1D and Merlin Vac engines from SpaceX. The hydrolox performance and the solid data are for the SSME and the Shuttle's SRB motors, respectively. These LV propellants' characteristics are given in Table 5.1.

For all members of the families, we assume all first steps have a structural ratio $\sigma = 0.06$, and all second steps have a structural ratio $\sigma = 0.08$, values that represent the approximate state of the art. All vehicles carry the same payload mass of $m_{PL} = 1,000$ kg = 1 T.

The example's orbit destination is at an altitude requiring a circular orbit speed of 7.8 km/s. The losses associated with the ascent are identical for all members of the three families, such that the total loss is 1.15 km/s. Such an assumption may be unrealistic if the families do not have similar T/W_0 values. In particular, a hydrolox first step could conceivably have a lower T/W_0 value than a kerolox step would. Finally, the example's launch site provides an eastward speed of 0.35 km/s. With all these assumptions and an eastward trajectory, the LV must be designed to produce a required $\Delta v_{design} = v_{orbit} + v_{loss} - v_{launch \; site} = 7.8 + 1.15 - 0.35$ km/s = 8.6 km/s over its two steps' operation.

An example of a family of solid/kerolox LVs calculated following this procedure is shown in Table 5.2. The parameter used to vary the ratio between the first and second step Δv, α, is shown in the second line. The table uses a value of the parameter $\beta = 1.0$, meaning that all losses are applied during the operation of the vehicle's first step and can be seen in the "stage 2's loss" row containing all zeros. The Δvs per step are shown in the corresponding lines, and the mass ratio μ comes from the rocket equation: $\mu = \exp(\Delta v/(g_0 \, I_{sp}))$. Once known, μ can be used to calculate the propellant m_p, structure m_s, and initial mass m_0 using the equations listed at the end of this chapter.

The results of this procedure are shown graphically in Figure 5.12, which plots the variation in liftoff masses for the three families. Two cases are shown for each family: one case where the losses are assumed to be spread over both steps' burns, and one where the losses are assumed to occur

Table 5.1 Propulsion System Parameters for Example LV Families

	Step 1 Propellants	I_{sp1} (s)	Step 2 Propellants	I_{sp2} (s)	Notes
A	RP-1/LOx	291	RP-1/LOx	304	I_{sp1} effective of M1D SL and M1D vac; I_{sp2} M1 vac
B	RP-1/LOx	291	LH_2/LOx	450	I_{sp1} effective of M1D SL and M1D vac; I_{sp2} SSME vac
C	LH_2/LOx	424	LH_2/LOx	450	I_{sp1} effective of SSME SL and vac; I_{sp2} SSME vac
D	Solid	258	RP-1/LOx	340	I_{sp1} effective of Shuttle SRB SL and vac; I_{sp2} M1 vac

Table 5.2 Calculation of Speeds and Masses for LV Family with Varying Staging Speed Fractions

Family of TSTOs		rocket engine/motor information m/s				g_0 9.80665 m/s²			v_{ideal} 7,450
σ_1 0.06		I_{sp1} 258.1 s		c_1 2.531		Payload m_{PL} 1 kg			= v_{orbit} − $v_{launch\,site}$
σ_2 0.08		I_{sp2} 340 s		c_2 3.334					v_{req} 8,600 m/s
						v_{orbit} 7,800 m/s			
						$v_{launch\,site}$ 350 m/s			
						v_{loss} 1,150 m/s			

Losses in 1st stage only

Case	1	2	3	4	5	6	7	8	9	10	11	12	13	14	15	16	17
step 1 %Δv (= α)	85%	80%	75%	70%	65%	60%	55%	50%	45%	40%	35%	30%	25%	20%	15%	10%	5%
step 2 %Δv	15%	20%	25%	30%	35%	40%	45%	50%	55%	60%	65%	70%	75%	80%	85%	90%	95%
required Δv_2	1,118	1,490	1,863	2,235	2,608	2,980	3,353	3,725	4,098	4,470	4,843	5,215	5,588	5,960	6,333	6,705	7,078
stage 2's loss	0	0	0	0	0	0	0	0	0	0	0	0	0	0	0	0	0
mass ratio μ_2	1.398	1.563	1.748	1.955	2.186	2.444	2.733	3.056	3.417	3.821	4.273	4.778	5.343	5.975	6.681	7.470	8.353
m_{p2}	0.413	0.594	0.803	1.045	1.329	1.661	2.056	2.526	3.095	3.789	4.651	5.742	7.156	9.049	11.693	15.616	21.987
m_{s2}	0.037	0.054	0.073	0.095	0.120	0.150	0.186	0.229	0.280	0.343	0.421	0.520	0.648	0.819	1.058	1.413	1.990
m_{02}	1.450	1.647	1.875	2.140	2.449	2.812	3.242	3.755	4.375	5.132	6.072	7.261	8.804	10.868	13.751	18.029	24.978
required Δv_1	7,483	7,110	6,738	6,365	5,993	5,620	5,248	4,875	4,503	4,130	3,758	3,385	3,013	2,640	2,268	1,895	1,523
stage 1's loss	1,150	1,150	1,150	1,150	1,150	1,150	1,150	1,150	1,150	1,150	1,150	1,150	1,150	1,150	1,150	1,150	1,150
staging speed	6,333	5,960	5,588	5,215	4,843	4,470	4,098	3,725	3,333	2,980	2,608	2,235	1,863	1,490	1,118	745	372
mass ratio μ_1	19.231	16.599	14.327	12.366	10.674	9.213	7.952	6.864	5.924	5.113	4.414	3.809	3.288	2.838	2.450	2.114	1.825
m_{p1}	−161.56	5933.85	167.34	88.60	61.93	48.54	40.51	35.19	31.42	28.63	26.50	24.86	23.59	22.63	21.97	21.63	21.75
m_{s1}	−10.312	378.756	10.682	5.655	3.953	3.098	2.586	2.246	2.005	1.827	1.692	1.587	1.506	1.445	1.402	1.381	1.388
m_{01}	−171.25	6289.88	177.38	93.92	65.64	51.45	42.94	37.30	33.30	30.34	28.09	26.35	25.00	23.99	23.29	22.93	23.06
m_{00} (with PL)	−169.80	6291.52	179.26	96.06	68.09	54.26	46.19	41.06	37.68	35.48	34.16	33.61	33.81	34.86	37.04	40.96	48.03

during the first step only ($\beta = 1$). The various family curves are calculated by varying α, which determines the speed at which staging occurs with a fixed burnout speed. The effect of assigning all the launch losses to occur during the first step's operation is to reduce the staging speed for minimum mass.

The lowest point on the curve indicates a minimum-mass case and a possible preferred staging speed. The solid curves show all losses in the first step, whereas the dashed curves assume losses spread over the entire launch. Note that the smeared and nonsmeared cases have almost the same liftoff mass even though the staging speeds are quite different.

In Fig. 5.12, we see that the more realistic scheme of including all losses in the first step's operation reduces the preferred staging speed (shifts it to the left), but does not significantly affect the vehicles' gross liftoff mass. In addition, it can be seen that the preferred staging speeds for RP/RP ($3.0\sim3.35$ km/s), solid/RP ($2.6\sim3.0$ km/s), and LH_2/LH_2 ($3.3\sim3.7$ km/s) are significantly higher than that of RP/LH_2 ($1.1\sim2.6$ km/s). This is probably because the RP/LH_2's second step, with its much higher I_{sp}, is far more efficient than its RP first step, so a larger second step accomplishing more of the burn is a more mass-efficient way to orbit.

The minimum launch mass for a 1-T payload is 9.5 T for LH_2/LH_2, 13.4 T for RP/LH_2, 27.3 T for RP/RP, and 33.6 T for solid/RP. Particularly noteworthy is that the use of hydrolox second steps, or both first and second steps, has the potential to reduce overall liftoff mass by a factor of ~2 to 3 over solid/RP and RP/RP. However, this may not necessarily indicate an overall cost savings, because dealing with LH_2 and its considerably larger tank sizes (due to its low density) and insulation (due to its extremely low temperature) can be more expensive than the denser but less-costly RP-1 (kerosene).

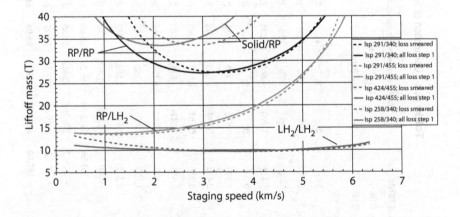

Fig. 5.12 Families of launch vehicles using solid, kerolox, and hydrolox first steps along with kerolox or hydrolox second steps, with varying speeds being delivered by steps 1 and 2 and losses attributed to both steps equally, or only to the lower step. A payload mass of 1 T is assumed for all cases.

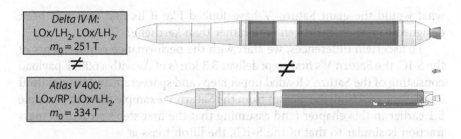

Fig. 5.13   The A5 and D4 with similar performance but different relative sizes. The larger D4 has 83 T, or about 25% less mass than the smaller A5, because the D4's LH_2 is much less dense than the A5's RP-1 and requires more volume. Source: Richard Kruse, HistoricSpacecraft.com.

The procedure of generating a family of launch vehicles as given previously provides a practical alternative to the method of optimization using Lagrange multipliers and is the basis of a homework assignment at the end of this chapter. A more realistic analysis would be an actual numerical trajectory optimization, as described in Chapter 6.

Let's see if these results compare to some actual launch vehicles. Figure 5.13 shows the Atlas V and Delta IV launch vehicles sold by ULA. The Atlas V (A5) has a LOx/RP first step and a LOx/LH_2 second stage, similar to the plot in Fig. 5.12. The Delta IV (D4) has two LOx/LH_2 stages, also similar to Fig. 5.12.

It may be seen that the Atlas V (no strap-ons) appears to be much smaller than the Delta IV model, although both vehicles carry the same payload to the same orbits. However, the Atlas has a liftoff mass of 334 T, whereas the comparable Delta has a liftoff mass of "only" 251 T, despite its larger appearance. This really demonstrates the difference between kerosene and hydrogen as a fuel: hydrogen is so much less dense that its airframe is larger in size, but smaller in overall mass—25% in this case. The family-of-launch-vehicle analysis shows a mass reduction of 29% for the same combinations of propellants, not very far from the actual vehicles considered.

We repeat that although the all-hydrolox Delta IV booster has less mass than the kerolox-hydrolox Atlas V launcher due to its higher-energy propellants, it may be more costly overall due to its increased bulk along with issues associated with handling the extremely cold temperatures of its liquid hydrogen.

5.10 All-Hydrogen Saturn V?

We've learned how LH_2/LOx (hydrolox) propulsion, with its higher specific impulse, has significantly better mass efficiency than RP-1/LOx (kerolox). It is used for the Delta IV, Ariane 5, Space Shuttle, Centaur upper steps, and several other vehicles. So, the question might be asked:

what would the giant Saturn V have looked like if its first step had been designed to run on hydrogen fuel, rather than kerosene?

To ascertain differences, we start with the performance requirements of the S-IC, the Saturn V's first step: deliver 3.3 km/s of Δv with a 608-T payload consisting of the Saturn's loaded upper steps and spacecraft, and have a liftoff $T/W_0 > 1.16$. Using the masses from the Saturn V example given in Example 5.1 earlier in this chapter (and assuming that the first step's structural mass fraction is similar to that of the S-IC), the liftoff mass is

$$m_{01} = (2,320 + 486 + 119 + 50)\,\text{T} = 2,975\,\text{T} \tag{5.59}$$

and the burnout mass is

$$m_{f1} = (240 + 486 + 119 + 50)\,\text{T} = 895\,\text{T} \tag{5.60}$$

With a burnout speed $\Delta v_1 = 3.3$ km/s, we find that the effective value of the F-1's specific impulse is

$$I_{sp1\_F-1_{eff}} = \frac{\Delta v}{g_0 \ln\left(\dfrac{m_{01}}{m_{f1}}\right)} = \frac{3,300\,\text{m/s}}{9.80665\,\dfrac{\text{m}}{\text{s}^2}\ln\left(\dfrac{2,975}{895}\right)} = 280\,\text{s} \tag{5.61}$$

The F-1 engine's sea level I_{sp-SL} and vacuum I_{sp-vac} were 265 s and 304 s, respectively, so the fraction that the effective I_{sp} value of 280 s lies between the sea level and vacuum values is

$$\frac{280 - 265}{304 - 265} = \frac{15}{39} = 0.38 \tag{5.62}$$

or 38% of the way between its SL and vacuum values.

The only large hydrogen engine that has been built was the Aerojet M-1 engine with thrust measuring 5,336 kN (1.2 million lbf) and whose I_{sp-SL} and I_{sp-vac} were 310 s and 428 s, respectively. This large engine, shown in Fig. 5.14a, was tested but did not make it into service. It may also be the physically largest rocket engine ever; note the size of the figure (the author) for scale, and the large exit area compared to the F-1 in 5.14b. Its higher specific impulses relative to the F-1 indicate more efficiency at both liftoff and at altitude, so we would expect a reduction in the required propellant mass. What would happen if the M-1 replaced the F-1 engines on the Saturn V?

To determine the propellant mass of the LH$_2$/LOx first step, we need an effective value for the M-1's specific impulse between sea level (SL) and vacuum. Assuming that the M-1 engine characteristics are such that its specific impulse varies similar to the F-1's, we'll use the same effective

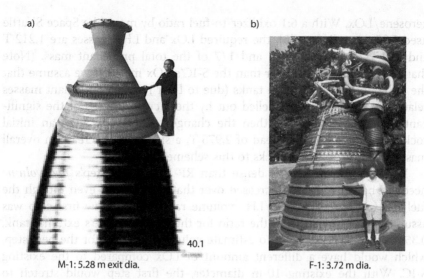

M-1: 5.28 m exit dia.   40.1   F-1: 3.72 m dia.

Fig. 5.14 The Aerojet M-1 and the Rocketdyne F-1 large engines, to scale. Sources: a) [4], with actual photo of M-1 engine at Evergreen Museum from author. b) author.

value relative to the SL and vacuum values for the M-1 specific impulse

$$I_{sp\_M\text{-}1eff} = I_{sp\_M\text{-}1SL} + 0.38\left(I_{sp\_M\text{-}1Vac} - I_{sp\_M\text{-}1SL}\right)$$

$$= I_{sp\_M\text{-}1SL}\left[1 + 0.38\left(\frac{I_{sp\_M\text{-}1vac}}{I_{sp\_M\text{-}1SL}} - 1\right)\right]$$

$$= 310\ \text{s}\left[1 + 0.38\left(\frac{428\ \text{s}}{310\ \text{s}} - 1\right)\right]$$

$$I_{sp\_M\text{-}1eff} = 355\ \text{s}. \tag{5.63}$$

Now we use this value to calculate the propellant mass for the M-1. One key assumption is that the empty first-step structure mass is about the same as the S-IC mass, considering that the hydrogen propellant tank (larger volume but smaller mass) combined with the smaller required volume of LOx (due to the different fuel:ox ratio) and required engines would total roughly the same mass as the existing tank and engine masses. Using the formula for propellant mass at the end of this chapter, the M-1's propellant mass is

$$m_{p\_M\text{-}1} = m_{f1}\left[e^{\left(\frac{\Delta v}{g_0 I_{sp\_M\text{-}1}}\right)} - 1\right] = 895\ \text{T}\left[e^{\left(\frac{3300\ \text{m/s}}{9.80665\ \text{m/s}^2\ \times 355\ \text{s}}\right)} - 1\right]$$

$$= 1{,}414\ \text{T}. \tag{5.64}$$

The replacement of the F-1 engines with the M-1s gives a first step that reqatres 1,414 T of LH$_2$/LOx propellants, replacing 2,080 T of kerosene/LOx. With a 6:1 oxidizer-to-fuel ratio by mass (the Space Shuttle used 5.95 according to [5], the required LOx and LH_2 masses are 1,212 T and 202 T respectively, 6/7 and 1/7 of the total propellant mass. (Note that the LOx mass is smaller than the S-IC's LOx mass.) If we assume that the reduction in mass of the tanks (due to their reduced propellant masses relative to the S-IC) is cancelled out by the increased mass of the significantly-larger M-1 engines, then the change to hydrolox gives an initial rocket mass of 2,310 T instead of 2,975 T, a significant decrease in overall mass. Are there any drawbacks to this scheme? Well, yes.

Because LH_2 is far less dense than RP-1, the first step's *fuel volume* needs to be substantially increased over that of the S-IC, even though the fuel *mass* is less. The LOx/LH_2 volume ratio for the new first step was assumed to be the same as the ratio for the Space Shuttle's external tank, 0.351. This value was used to estimate volumes required for the new step, which would have a different amount of LOx compared to the existing S-IC. With the existing 10-m diameter, the first step would stretch to 77 m from 42 m, with the hydrogen first step becoming very long and skinny. This "skinny" version would also probably have clearance issues for the mounting of its significantly larger M-1 engines, as well as gimbaling. Increasing the diameter of the first step to 13 m (from 10 m) would result in a 55-m first-step length and more room for the engines. These configurations are depicted in Fig. 5.15.

If the as-built Saturn V (top) had been designed with a high-efficiency LH_2/LOx first step rather than the kerolox S-IC, it might have looked something like the second and third figures, Saturn V-H10 (LH_2 first step, 10-m diameter) and Saturn V-H13 (LH_2 first step, 13-m diameter). The first step's volume would be increased due to the reduced density of liquid hydrogen, but the liftoff mass drops from 2,970 to 2,593 T. Six hydrolox M-1 engines would replace the S-IC's five F-1s.

In both the 10- and 13-m diameter first steps at the bottom of the figure, the total LH_2 carried would be about 3.5 times that of the original Saturn V, which would also increase the ground LH_2 storage requirements by the same factor. The drag penalty of the 13-m-diameter first step wouldn't be excessive, even with the larger diameter. Referring to Fig. 3.10 in Chapter 3, if the burnout angle is similar, with the improved specific impulse of 428 s, the value of K_D becomes about 0.9×10^6, a reduction of about 25%. If we assume a drag coefficient increase of 10% due to the larger base diameter, reference area scaling by $1.3^2 = 1.69$, and the 22% reduction of liftoff mass, the original 46 m/s drag loss (Table 3.2) would be multiplied by $(0.75)(1.1)(1.69)/0.78$, or about 1.78, increasing its drag loss to 82 m/s, a relatively small change.

To meet or exceed the same initial thrust-to-weight ratio requirement, six 1.2-million-lb-thrust M-1s (5.34 MN) would be needed; the six engines would generate $T/W_0 = 1.26$ at liftoff. During the ascent, as the vehicle's mass decreased and specific impulse/thrust increased, one or two of them

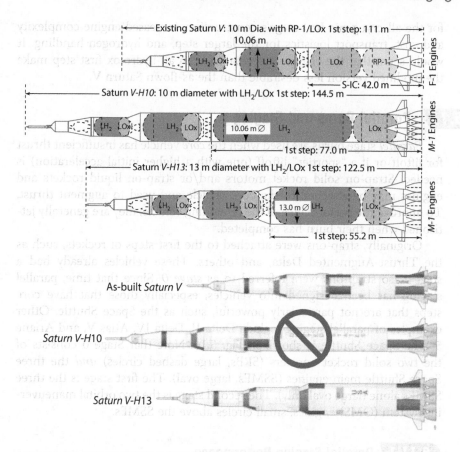

Fig. 5.15   Saturn V, Saturn V-H10, and Saturn V-H13. Bottom: Artist conceptions of how the hydrolox LV might be painted. The 144-m-long H10 model seems impractical. Source: Artist's conceptions courtesy Richard Kruse, HistoricSpacecraft.com.

would likely be shut off over the course of the burn to limit acceleration, with the remainder left running until shutdown, as the center engine was shut down on the Saturn V. With a similar T/W_0 profile to the existing Saturn, the hydrolox Saturn's ascent could be expected to have similar gravity losses.

Note that both hydrolox designs were laid out with the LOx tank below the LH_2 tank. We will see later that for gust-induced loading, it's preferable to place the heavier tank on the *bottom* of the step. Another reason to orient them as specified is the hydraulic losses associated with the very long ducting needed to provide LOx to the engines, if the LOx tank was in front. Finally, there may be issues where the LOx flowing through the LH_2 propellant might freeze. These layout issues are described in more detail in Chapters 7 and 14.

Bear in mind that even though these hydrolox rockets would have less total mass, they probably would cost more because of the changes required for the all-cryogenic first stage's construction, increased engine complexity and cost, transport logistics for the larger step, and hydrogen handling. It would seem that the changes necessary for an all-hydrolox first step make the modified version less desirable than the as-flown Saturn V.

5.11 Parallel Burns and Staging

Parallel stage burns are used when the *core* vehicle has insufficient thrust for liftoff or if a "sportier" liftoff (one with a higher initial acceleration) is needed. Strap-on solid rocket motors and/or strap-on liquid rockets and engines may be attached to the core vehicle and used to augment thrust. To improve the performance of the remaining vehicle, they are generally jettisoned when their burn has completed.

Originally, strap-ons were attached to the first steps of rockets, such as the Thrust-Augmented Delta, and others. These vehicles already had a stage 1, so strap-ons were referred to as *stage 0*. Since that time, parallel staging has been designed into vehicles, especially those that have core steps that are not particularly powerful, such as the Space Shuttle. Other examples of parallel staging include Delta II, Delta IV, Atlas V, and Ariane 5. The Space Shuttle is shown in Fig. 5.16. Note that Stage 0 consists of the two solid rocket boosters (SRBs, large dashed circles) *and* the three Space Shuttle main engines (SSMEs, large oval). The first stage is the three SSMEs alone (large oval only). The second stage is the two orbital maneuvering system (OMS) engines, small circles above the SSMEs.

5.11.1 Parallel Staging Performance

Parallel burns of engines/motors of different sizes, thrusts, or specific impulses are accounted for using the *effective* weighted exhaust speed

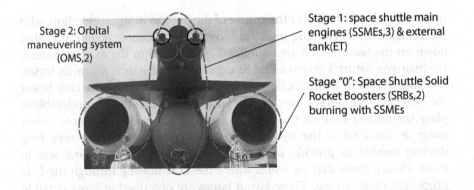

Stage 2: Orbital maneuvering system (OMS,2)

Stage 1: space shuttle main engines (SSMEs,3) & external tank(ET)

Stage "0": Space Shuttle Solid Rocket Boosters (SRBs,2) burning with SSMEs

Fig. 5.16 A bottom view of the Space Shuttle stack and its three different types of firing during operation.

Fig. 5.17 The Delta IVH utilizes parallel staging to increase its payload capability.
Source: Courtesy ULA.

while all burn. The total thrust may be calculated by adding the contributions
of each of the components

$$T_{tot} = v_{e0}\dot{m}_0 + v_{e1}\dot{m}_1 = v_{e0-ave}(\dot{m}_0 + \dot{m}_1) \qquad (5.65)$$

Here v_{e0-ave} is the mass-averaged exit speed of the parallel steps

$$v_{e0-ave} = \frac{v_{e0}\dot{m}_0 + v_{e1}\dot{m}_1}{\dot{m}_0 + \dot{m}_1} \qquad (5.66)$$

The remaining steps individually act as before, with a single exhaust
speed/specific impulse.

In many cases, if the core and its strap-ons have identical performance,
the core vehicle will often launch with its engine(s) throttled back so as to
not consume as much propellant. When this is done, the core can fire substantially longer, rather than burning out when the strap-ons burn out.
This is the case with the Delta IV Heavy, which is shown in Fig. 5.17. This
vehicle is essentially three very similar common booster core vehicles
strapped together. (The three are not identical, because the center one has
to sustain loads from the strap-ons in addition to its own loads.) The
bright areas of exhaust are not the same size, and the core (center) engine
can be seen to be smaller, developing less thrust.

5.11.2 Parallel Staging Procedures

Vehicles use parallel staging to enhance performance. The added thrust
of the additional propulsion systems attached allows larger payloads to be
flown without redesigning the core vehicle. In some cases, strap-ons are

solid motors (Space Shuttle, Delta II and IV, Atlas V, Ariane 5, etc.); in other cases, the strap-ons are powered by liquid engines (Delta IV Heavy, Falcon 9 Heavy, some Long March models). In all cases, the idea is for the strap-ons to burn and separate while the "inner" vehicle continues to fire.

It's common to see both stage 0 and stage 1 used for the augmented core vehicle; the reader should be sure to understand which terminology is being used.

The operation of a parallel stage system is shown in Fig. 5.18, which separates the mission into three phases: a "boosted" or augmented core vehicle (stage 0 or 1), the core vehicle operating without augmentation (stage 1 or 2), and the core's upper stage firing (stage 2 or 3). In all cases, the lowest stage number refers to the parallel burn of *both* the core and the strap-ons, and the mass-averaged parallel-burn equations given previously must be used for analysis.

In the figure, stage 1 (left) has a core vehicle with strap-ons firing in parallel. The core is operating at a lower throttle level so as to have propellant remaining for stage 2 (center). Here, the empty strap-ons are jettisoned, and the core's first step continues firing until empty. When empty, the first step of the core is jettisoned and stage 3 (right) begins. Here, the upper stage of the core fires until shut down.

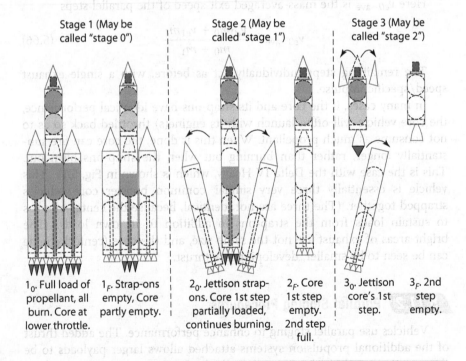

Stage 1 (May be called "stage 0")

Stage 2 (May be called "stage 1")

Stage 3 (May be called "stage 2")

1_0. Full load of propellant, all burn. Core at lower throttle.

1_F. Strap-ons empty, Core partly empty.

2_0. Jettison strap-ons. Core 1st step partially loaded, continues burning.

2_F. Core 1st step empty. 2nd step full.

3_0. Jettison core's 1st step.

3_F. 2nd step empty.

Fig. 5.18 A three-stage rocket with a parallel burn.

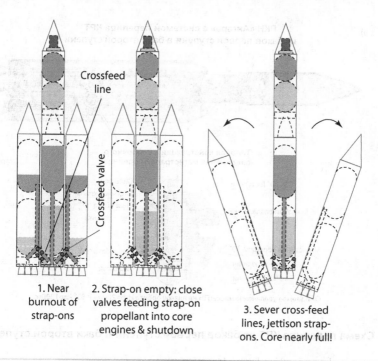

1. Near burnout of strap-ons

2. Strap-on empty: close valves feeding strap-on propellant into core engines & shutdown

3. Sever cross-feed lines, jettison strap-ons. Core nearly full!

Fig. 5.19 Crossfeed lines feed the core's engines with propellants from the strap-ons.

5.11.3 Parallel Stage Enhanced Performance

Imagine that instead of having a partially empty core vehicle when the strap-ons are separated, instead you had a more-loaded vehicle. One would guess that it would have better overall performance. One means to accomplish this feat is called *crossfeeding*. The idea is illustrated in Fig. 5.19.

The staging process would be like this: parallel burning is accomplished using cross-fed propellant lines from the strap-ons' propellant tanks to the center vehicle. The propellants in the strap-ons would be supplied simultaneously to both the strap-ons' engines and the core vehicle's engines. When the strap-ons are empty, the strap-on engines shut down, valves on either side of the crossfeed close, and the strap-ons are jettisoned. This would occur in a manner similar to the means used to jettison the booster section of the Atlas ICBM (see Chapter 2 and/or Chapter 7 for more details). This scheme, shown in Fig. 5.19, eliminates the need for heavy, high-capacity pumps.

Figure 5.20 shows that Russia seemed to be considering a similar scheme for its Angara launch vehicle. The Russian design does not have crossfeed lines. Instead, as can be seen in Fig. 5.20b, it appears to use differences in internal pressure between the strap-ons and the center (higher pressure in the strap-ons) to transfer strap-on propellant directly across to the center vehicle. Note that in Fig. 5.20b, there seem to be

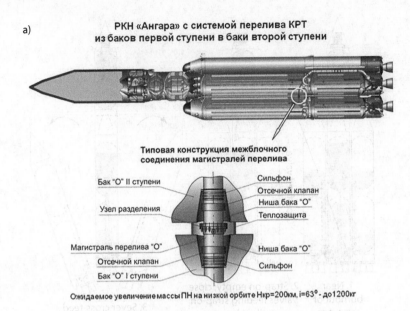

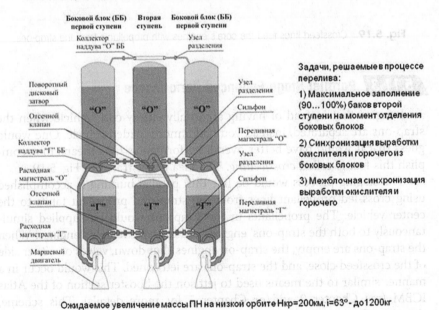

two pressure vessels, each above vehicle's two center tanks, that would provide increased pressure to the strap-ons. Also, note that the center vehicle has consumed less propellant than its outside neighbors. At the time of this writing, we are unaware of the status of this interesting Russian concept.

5.12 Launch Vehicle Design Sensitivities

In this book, we've mentioned how several launch vehicles have undergone an evolution where they were changed to increase their launch mass capabilities. Steps were lengthened, adding propellant (and mass). This raises the question, where is the best place to add propellant, or subtract mass, or add a higher-performance engine? Lower step, upper step? This section is based on material found in Wiesel [6].

We can use the rocket equation to answer this question. We can mathematically calculate its derivative, holding the final speed v_f constant, to calculate rates of change of payload with respect to changes in the mass of a structure, increased propellant volume, and so forth. The results are called *tradeoff ratios* or *design sensitivities*.

Let's assume we are able to calculate these derivatives with constant final speed. A *positive* tradeoff ratio indicates the change will cause the payload mass to increase; a *negative* ratio indicates the payload will be reduced. The magnitude of these ratios helps us to determine the best places to (1) add propellant, and/or (2) improve engine performance (improved v_{eq} or I_{sp}), and/or (3) reduce inert mass. Note that either of the first two changes may also be accompanied by the third (i.e., a tank stretched to contain additional propellant requires additional volume, which may also increase its inert mass).

As an example, consider the evolution of the Delta vehicle shown in Fig. 5.21. The GTO payload capability (shown on the vertical scale) increased from about 54 kg to over 1,800 kg. This change in payload was accomplished by making the changes noted on the figure: improved-performance main engine, stretched propellant tank, changing metallic strap-on SRM cases to graphite-epoxy, and more. The locations for these changes were determined by analysis of tradeoff ratios, which provide a numerical value to the changes being considered. Table 2.3 may be used to see how some of the improvements were made to improve the vehicle's performance.

5.12.1 Tradeoff Ratio Calculation

Part of this section is based on Section 7.6 of [6]. We start by defining the masses involved. The *initial* mass of kth step = (payload mass) + (propellant mass kth step) + (structure mass kth step) + (mass of steps above k) can be written as follows:

$$m_{0k} = m_{PL} + m_{pk} + m_{sk} + \sum_{j=k+1}^{N} (m_{sj} + m_{pj}) \qquad (5.67)$$

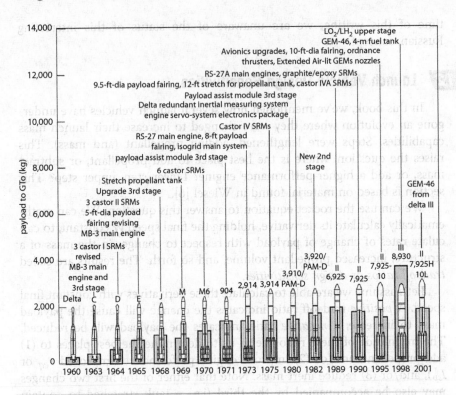

Fig. 5.21 The Delta rockets' evolution dramatically increased payload capability, shown by the length of the gray bar on the bottom of the graph. Source: Wambolt [7].

The *final* mass of step k = (payload mass) + (0 propellant) + (structure mass, kth step) + (mass of steps above k)

$$m_{fk} = m_{PL} + 0 \cdot m_{pk} + m_{sk} + \sum_{j=k+1}^{N} (m_{sj} + m_{pj}) \qquad (5.68)$$

We wish to calculate the rates of change of payload mass m_{PL} while holding the final speed v_f constant due to changes in:

- *Inert* mass m_{sk} of kth step (tradeoff ratio = $\partial m_{PL}/\partial m_{sk}$, dimensionless)
- *Propellant* mass m_{pk} of kth step (tradeoff ratio = $\partial m_{PL}/\partial m_{pk}$, dimensionless)
- *Specific impulse* $I_{sp,k}$ of kth step = $\partial m_{PL}/\partial I_{sp,k}$, units kg/s, *or*
- *Effective exhaust speed* $v_{eq,k}$ of kth step = $\partial m_{PL}/\partial v_{eq,k}$, units kg/[m/s]

The derivatives are assumed to behave *linearly*, with the assumption that changes are "small" in an engineering sense (several orders of magnitude less than their associated quantity).

How Tradeoffs/Sensitivity Derivatives Are Used

The change in payload mass due to a change in one of the three variables listed in the previous section is found by multiplying the derivative by the proposed input change. For example, the rate of change in payload Δm_{PL} due to a mass change Δm_{sk} is calculated by assuming that for "small" Δs, the derivative may be approximated by the quotient of the finite values

$$\frac{\partial m_{PL}}{\partial m_{sk}} \approx \frac{\Delta m_{PL}}{\Delta m_{sk}} \tag{5.69}$$

which can be rearranged to provide the change in payload Δm_{PL} as

$$\Delta m_{PL} \approx \frac{\partial m_{PL}}{\partial m_{sk}} \Delta m_{sk} \tag{5.70}$$

If the tradeoff ratio is *negative*, then the payload mass *decreases*, and vice versa

Inert Mass Tradeoff Ratio

We wish to calculate $\partial m_{PL}/\partial m_{sk}$ while holding v_f (burnout speed) constant, meaning that the total differential $dv_f = 0$.

$$dv_f = 0 = \sum_{i=1}^{N} v_{eq,i} \frac{m_{fi}}{m_{0i}} \left(\frac{1}{m_{fi}} - \frac{m_{0i}}{m_{fi}^2} \right) dm_{PL} + \sum_{j=1}^{k} v_{eq,j} \frac{m_{fj}}{m_{0j}} \left(\frac{1}{m_{fj}} - \frac{m_{0j}}{m_{fj}^2} \right) dm_{sk} \tag{5.71}$$

Rearrange to get the mass tradeoff ratio, which tells us the change in payload due to change in inert mass.

$$\left. \frac{\partial m_{PL}}{\partial m_{sk}} \right|_{dv_f=0} = - \frac{\sum_{j=1}^{k} v_{eq,j} \left(\frac{1}{m_{0j}} - \frac{1}{m_{fj}} \right)}{\sum_{i=1}^{N} v_{eq,i} \left(\frac{1}{m_{0i}} - \frac{1}{m_{fi}} \right)} \tag{5.72}$$

Consider the top or last step: for this step, $k = N$, and we find that

$$\frac{\partial m_{PL}}{\partial m_{sN}} = -1 \tag{5.73}$$

There is a 1:1 ratio between added inert mass and lost payload mass. In other words, this means that any mass added to the final step *subtracts an equal amount directly from the payload mass*, which is what one would expect.

5.12.4 Propellant Mass Tradeoff Ratio

Now we take the derivative with respect to m_{pk}, the *propellant* mass. The result tells us the change in payload due to a change in propellant mass.

$$dv_f = 0 = \sum_{i=1}^{N} v_{eq,i}\left(\frac{1}{m_{0i}} - \frac{1}{m_{fi}}\right)dm_{PL} + \sum_{j=1}^{k-1} v_{eq,j}\left(\frac{1}{m_{0j}} - \frac{1}{m_{fj}}\right)dm_{pk}$$

$$+ v_{eq,k}\frac{dm_{pk}}{m_{0k}} \tag{5.74}$$

Then, rearranging we get the following tradeoff ratio:

$$\left.\frac{\partial m_{PL}}{\partial m_{pk}}\right|_{dv_f=0} = -\frac{\sum_{j=1}^{k-1} v_{eq,j}\left(\dfrac{1}{m_{0j}} - \dfrac{1}{m_{fj}}\right) + \dfrac{v_{eq,k}}{m_{0k}}}{\sum_{i=1}^{N} v_{eq,i}\left(\dfrac{1}{m_{0i}} - \dfrac{1}{m_{fi}}\right)} \tag{5.75}$$

We would expect the propellant tradeoff ratio to always be positive, because adding propellant will always increase the payload; however, that's not the end of the analysis. Adding propellant means adding tank volume, which in turn requires additional structure that *adds* mass. We dealt with the added mass tradeoff expression in the previous section. So the proper analysis would weigh (no pun intended) the benefit of the *addition* of propellant against the penalty of the increased tank mass needed.

5.12.5 Rocket Specific Impulse Tradeoff

Consider the replacement of an older rocket engine with another having a better specific impulse or exit speed. This can often be done for zero or very small mass increase. You would expect an increase in payload mass for *any* increase in specific impulse or exhaust speed. The derivatives or trades are of the form $\partial m_{PL}/\partial I_{sp-k}$, kg payload per second of I_{sp} improvement.

Recall the definitions of burnout speed and initial and final masses:

$$v_f = \sum_{k=1}^{N} v_{eq,k} \ln\frac{m_{0k}}{m_{fk}} \tag{5.76}$$

$$m_{0k} = m_{PL} + m_{pk} + m_{sk} + \sum_{j=k+1}^{N}(m_{sj} + m_{pj}) \tag{5.77}$$

$$m_{fk} = m_{PL} + 0 \cdot m_{pk} + m_{sk} + \sum_{j=k+1}^{N}(m_{sj} + m_{pj}) \tag{5.78}$$

Take the total differential of v_f allowing changes only in v_{ek} and m_{PL}.

$$dv_f = 0 = \ln\left(\frac{m_{0k}}{m_{fk}}\right)\partial v_{eq,k} + \sum_{j=1}^{N} v_{ej}\frac{m_{fj}}{m_{0j}}\left(\frac{1}{m_{fj}} - \frac{m_{0j}}{m_{fj}^2}\right)\partial m_{PL} \qquad (5.79)$$

Now solve for the trade ratio, which supplies the change in payload due to change in specific impulse.

$$\left.\frac{\partial m_{PL}}{\partial v_{eq,k}}\right|_{dv_f=0} = \frac{-\ln\left(\frac{m_{0k}}{m_{fk}}\right)}{\sum_{j=1}^{N} v_{eq,j}\left(\frac{1}{m_{0j}} - \frac{1}{m_{fj}}\right)} \qquad (5.80)$$

or by dividing both sides by g_0,

$$\left.\frac{\partial m_{PL}}{\partial I_{sp,k}}\right|_{dv_f=0} = \frac{-\ln\left(\frac{m_{0k}}{m_{fk}}\right)}{\sum_{j=1}^{N} I_{sp,j}\left(\frac{1}{m_{0j}} - \frac{1}{m_{fj}}\right)} \qquad (5.81)$$

The values of $\partial m_{PL}/\partial v_{eq,k}$ or $\partial m_{PL}/\partial I_{sp,k}$ may be used to calculate specific impulse tradeoffs for any rocket stage.

The underlying assumption in calculating changes from tradeoff ratios is that the changes are small. This allows us to treat the derivatives as linear near their nominal points. The changes in payload mass due to added mass, added propellant, and added specific impulse are calculated from

Added mass: $\Delta m_{PL} \approx \dfrac{\partial m_{PL}}{\partial m_{sk}}\Delta m_{sk}$

Added propellant: $\Delta m_{PL} \approx \dfrac{\partial m_{PL}}{\partial m_{sk}}\Delta m_{pk}$

Added specific impulse: $\Delta m_{PL} \approx \dfrac{\partial m_{PL}}{\partial I_{sp,k}}\Delta I_{sp,k}$

Added exhaust speed: $\Delta m_{PL} \approx \dfrac{\partial m_{PL}}{\partial v_{eq,k}}\Delta v_{eq,k}$

5.12.6   Improved Saturn IB Performance by Adding Propellant

Table 5.3 provides the three derivatives for the two steps of the Saturn IB. Notice that the upper step's mass trade is 1:1, as expected. Comparing this to the lower step's mass trade, if one *has* to add mass, it's better to add to the first step (loss of about 1 kg payload for 9 kg added mass) vs 1:1 for adding to the upper step.

Table 5.3 Saturn IB Mass Tradeoff Ratios

k Stage No.	$V_{eq,k}$ Equiv. Exit Speed (m/s)	m_{0k} Initial Mass (kg)	m_{sk} Final Mass (kg)	$\dfrac{\partial m_{PL}}{\partial m_{sk}}$ (kg/kg)	$\dfrac{\partial m_{PL}}{\partial m_{pk}}$ (kg/kg)	$\dfrac{\partial m_{PL}}{\partial I_{sp,k}}$ (kg/s)
0	2,568	588,000	180,000	−0.1102	+0.0486	+129.3
1	4,126	144,000	38,000	−1.000	+0.2088	+145.5

Note: m_{PL} = 21,000 kg payload mass (held constant)
   $V_{eq,k}$ = exhaust speed of kth stage in m/s
   $k = 1$ = Step 1 = S-IB stage (LOx/RP-1)
   $k = 2$ = Step 2 = S-IVB stage (LOx/LH$_2$)

Suppose that we wish to improve the payload capability of the Saturn IB by adding propellant to one of its steps. The tradeoff ratios in the sixth column of the table indicate which step would provide the *most* benefit. There is always a benefit from propellant addition; its tradeoffs are *always positive*. To improve performance by adding propellant, one gets more than *four times* more payload by adding to the second step vs the first step: $\partial m_{PL}/\partial m_{p2} \div \partial m_{PL}/\partial m_{p1} = 0.2088 \div 0.0486 = 4.30$).

However, one also must consider the addition of mass (added tank volume) to carry the extra propellant (added mass = negative tradeoff ratio). Hence, the *net benefit* is the difference of the payload mass added due to the additional propellant less the payload mass lost by the added tank mass, the difference of the two effects.

5.12.7 Space Shuttle Tradeoff Ratios

As another example, let's take a look at the Space Shuttle's tradeoff ratios in Table 5.4. Because it was a parallel-staged launch vehicle, the first stage was stage 0. At liftoff, both the solid rocket boosters (SRBs) and the main engines were burning. Stage 1 was the core vehicle, main engines, and external tank (no SRBs). Stage 2 was the orbital maneuvering system (OMS, rhymes with "homes"). The OMS was used to circularize the Shuttle's orbit after it had jettisoned the ET. (The ET was jettisoned with 99% of orbital speed, to ensure that it entered the atmosphere and did not stay in orbit.)

5.12.7.1 A Practical Application of Tradeoff Ratios: Space Shuttle Application 1

The Space Shuttle provides an example of practical benefits that can be estimated from tradeoff ratios. Two Shuttle launches are shown in Fig. 5.22. A careful examination of the two photos will reveal a difference between the two vehicles.

Table 5.4   Space Shuttle Tradeoff Ratios

k Stage No.	$v_{eq.k}$ Equiv. Exit Speed (m/s)	m_{0k} Initial Mass (kg)	m_{sk} Final Mass (kg)	$\dfrac{\partial m_{PL}}{\partial m_{sk}}$ (kg/kg)	$\dfrac{\partial m_{PL}}{\partial m_{pk}}$ (kg/kg)	$\dfrac{\partial m_{PL}}{\partial I_{sp,k}}$ (kg/s)
0	3,060	2,015,000	834,000	−0.078	+0.055	+275
1	4,459	675,500	143,300	−0.964	+0.161	+620
2	3,067	108,300	104,700	−1.000	+0.059	—

Source: Wiesel [6].
Note: m_{PL} = payload mass = 29,500 kg (held constant)
$k = 0$ = Stage 0 = SRBs + SSMEs + ET
$k = 1$ = Stage 1 = SSMEs + ET (after SRB jettison)
$k = 2$ = Stage 2 = OMS only.

The noticeable difference is the shuttle on the left (the first shuttle launch, STS-1) has an external tank that's white. In contrast, the shuttle on the right has an orange ET. Why the difference? Take a look at the fifth column of Table 5.4, showing payload loss due to added mass. Note that for stage 1 (SSMEs + ET), 1 kg of mass added reduces the payload by

Fig. 5.22   Use of tradeoff ratios on the NASA Space Shuttle. The shuttle on the left has a painted ET, while the one on the right *does not* have the 270 kg of paint, producing a payload increase of 260 kg. Source: NASA.

0.964 kg. And what is the mass of the added white paint? 270 kg (595 lb)! By omitting that white paint, the Shuttle can carry an extra 260 kg (573 lb) of payload, the amount determined by multiplying the $k = 1$ trade ratio by the elimination of the paint mass: payload change $= (-0.964) \times (-270 \text{ kg}) = +260 \text{ kg}$.

Suppose you wished to improve the Shuttle's payload capacity by adding propellant. What location would provide the best benefit? This time, look at column 6 of Table 5.4. It can be seen that from the tradeoff ratios, the best place is in the ET, $k = 1$ stage, LOx/LH_2. Here, 6.21 kg of propellants will raise the payload mass by 1 kg ($6.21 = 1 \div 0.161$, the reciprocal of the tradeoff ratio). If you were to add propellant to either stage 0 or stage 2, the added propellant mass would only be about one third as effective compared to stage 1.

We must emphasize that the designer must also consider the additional tank structure mass needed to accommodate propellant volume, which will *subtract* from the additional payload capacity. The problems at the end of this chapter address the calculation of added payload mass due to added propellant vs the loss of payload due to the structure required to contain the added propellants.

A historical note: some of the older rockets, such as Delta, added strap-ons because the then too-heavy core vehicle couldn't take off on its own thrust, and it was too costly to modify the design to incorporate an engine of the thrust needed. Strap-ons are a relatively easy way to fix the thrust-to-weight problem.

5.12.7.2 Shuttle Application 2: Carrying Out Missions to ISS

The Space Shuttle launched from Kennedy Space Center (KSC), located at 28.5 deg latitude. However, missions to the ISS required an orbit with a 51.6-deg inclination. This increased inclination cost the shuttle 5,900 kg (13,000 lb) of payload. What could be done? The Shuttles were all built, but because the external tanks were used only once, there was an opportunity to produce an ET with less mass. And as we saw with the white paint, the benefit of reducing stage 1 mass was 0.964:1, almost a one-to-one ratio.

The new "super lightweight" design for the ET provided a total mass savings of 3,629 kg (13,000 lb), from the following modifications to the original design:

- Switching to higher strength 2195 Al-Li alloy saved 2,268 kg (5,000 lb).
- A revised waffle-grid (orthogrid) design saved 1,134 kg (2,500 lb) by eliminating older T stiffeners.
- Reducing the thickness of sprayed-on foam insulation (SOFI) saved 227 kg (500 lb).

The result of the ET's "diet" made it possible for the Shuttle to deliver 15,200 kg (33,510 lb) of payload to the ISS's orbit. These data were found in *Aviation Week* [8] and Jenkins [9].

5.12.7.3 Shuttle Application 3: High-Performance SRBs

In the 1980s, the U.S. Air Force planned to operate a Space Shuttle from Vandenberg Air Force Base in California in order to launch military and classified payloads to high-inclination orbits. Better vehicle performance was needed to make up for the loss of payload capacity due to the higher energy needed for these orbits.

One solution had the SRB manufacturer Thiokol (now part of Northrop Grumman) provide the SRBs in lower-mass *graphite-epoxy composite* SRB cases, known as filament wound case solid rocket motors (FWC-SRM). As seen in Fig. 5.23, each graphite-epoxy SRB case had 12,500 kg less mass than equivalent size steel casings [9].

Looking at the tradeoff ratios, two SRMs provided a benefit of $-0.078 \times (-25{,}000 \text{ kg}) = +1{,}954 \text{ kg}$ payload gain from reduced mass. Ultimately, however, the idea was not pursued. After the *Challenger* disaster, concern about fatigue issues arising from repeated uses of the casings (remember that the Shuttle's SRBs were used to fly multiple missions) and its effect on safety made them impractical.

The *best* way to improve the Shuttle's payload capability is to remove the Shuttle entirely! Because much of the Shuttle exists to support the crew, it would be possible to instead utilize a cargo container without wings and life support systems, thus saving huge amounts of mass. This concept, shown in Fig. 5.24, was called Shuttle-C where the *C* stood for *cargo*. With this substitution, the entire mass formerly needed by the orbiter (wings, pressure vessel, human accommodations including life support system, etc.) could be used as payload instead, increasing payload capacity to 100 T or more. Like the composite SRBs, however, this concept was never pursued.

Fig. 5.23 Shuttle FWC-SRM casing developed to increase the Shuttle's payload when flying out of VAFB. Note black color of graphite. Source: NASA/Morton Thiokol.

Fig. 5.24 The Shuttle-C concept would be able to orbit approximately four to five *times* the payload the Shuttle could orbit. The concept never reached the hardware stage. Source: NASA.

5.12.8 Another Method to Calculate Tradeoff Ratios

The four formulae (eqs. 5.72, 5.75, 5.80, and 5.81) given previously to calculate tradeoff ratios are numerically complex, and there is a significant chance of calculation error. Another way to calculate tradeoffs is to create a spreadsheet or computer code that will calculate a vehicle's Δv given step mass ratios, structural fractions, and specific impulses. This tool may be used to change the LV's appropriate final mass, propellant mass, or specific impulse by one "unit" (either kg or seconds). The result of this one-unit change will be a slight change of the final Δv. Because we want to maintain the *exact* Δv, we can use the goal seek function of a spreadsheet program (and the equivalent in computer code) and observe the resulting new payload value. This brute force approach will provide identical values to those in the tradeoff tables given earlier—as it *should*! In some cases, however, you may run into roundoff issues, and the speed change ends up being too small to accurately calculate. In this case, you need to change the input by (say) 5 or 10 units rather than a single unit. This will make the change 5 or 10 times larger, and so the change has to be divided by 5 or 10 to yield the appropriate tradeoff ratio.

5.13 **Some Useful Results: Determining Component Mass Values**

For a given Δv requirement and specific impulse, the usable propellant mass m_p may be calculated given the equivalent exhaust speed c or v_{eq} or the specific impulse I_{sp}, along with either the initial mass m_0 or the burnout mass $m_f = m_s + m_{PL}$ = step final mass + payload. The propellant mass as it relates to initial or final mass may be calculated from

$$m_p = m_0 \left[1 - e^{\left(-\frac{\Delta v}{g_0 I_{sp}} \right)} \right] = m_0 \left[1 - e^{\left(-\frac{\Delta v}{v_{eq}} \right)} \right] \tag{5.82}$$

$$m_p = m_f \left[e^{\left(+\frac{\Delta v}{g_0 I_{sp}} \right)} - 1 \right] = m_f \left[e^{\left(+\frac{\Delta v}{v_{eq}} \right)} - 1 \right] \tag{5.83}$$

In some circumstances, the structure mass, propellant mass, and gross mass values are needed for sizing purposes. In this case, we do not know the initial or final masses, but we do know the required Δv, selected specific impulse I_{sp} (or equivalent exhaust speed v_{eq}), assumed step mass fraction σ, and payload mass m_{PL}. The mass ratio μ is calculated from

$$\mu = \frac{m_0}{m_f} = e^{\left(\frac{\Delta v}{g_0 I_{sp}} \right)} = e^{\left(\frac{\Delta v}{v_{eq}} \right)} \tag{5.84}$$

Then, the following equations may be used to calculate the necessary propellant mass m_p, the necessary gross mass m_0, and the final or burnout mass m_f

$$m_p = m_{PL} \frac{(\mu - 1)(1 - \sigma)}{1 - \mu \sigma} \tag{5.85}$$

or payload mass given propellant mass m_p, mass ratio μ, and structural ratio σ

$$m_{PL} = m_p \frac{1 - \mu \sigma}{(\mu - 1)(1 - \sigma)} \tag{5.86}$$

or initial mass given payload mass, mass ratio μ, and structural ratio σ

$$m_0 = m_{PL} \frac{\mu(1 - \sigma)}{1 - \mu \sigma} \tag{5.87}$$

or final mass given payload mass, mass ratio μ, and structural ratio σ

$$m_f = m_{PL} \frac{1 - \sigma}{1 - \mu \sigma} \tag{5.88}$$

or structural ratio σ given payload mass, propellant mass, and mass ratio μ

$$\sigma = \frac{m_{PL}(\mu - 1) - m_p}{m_{PL}(\mu - 1) - \mu m_p} \qquad (5.89)$$

The structure mass is calculated from rearranging the defining equation

$$\sigma = \frac{m_s}{m_p + m_s} \qquad (5.90)$$

to get

$$m_s = m_p \frac{\sigma}{(1 - \sigma)} \qquad (5.91)$$

Note that the structure mass *is not* the simple expression ($\sigma\, m_p$) as one might (incorrectly) assume. These formulae allow us to immediately begin sizing a vehicle needing to contain the needed propellants.

Sizing a multistep vehicle uses these same equations starting at the top of the vehicle, step k. Once the initial mass m_{0k} of the upper stage (loaded step k + payload m_{PL}), that m_{0k} becomes the payload m_{PL-k-1} of the next lower step, and the calculations are repeated until the lowest step is completed.

5.14   Summary

In this chapter, we have discussed launch vehicle performance parameters, the rocket equation, handling different types of staging, determination of masses of multistep LVs to minimize liftoff mass, and design sensitivities. In the next chapter, we will apply these performance values to the calculation of launch vehicle trajectories.

References

[1] Office of Technical Assessment, "Big Dumb Boosters—A Low-Cost Space Transportation Option?" NTIS Order #PB89-155196, International Security & Commerce Program, Office of Technical Assessment, Congress of the United States, Washington, DC, 1989.

[2] Thomson, W. T., Introduction to Space Dynamics, Dover Publications Inc., New York, 1986.

[3] Leondes, C. T., and Vance, R. W., *Lunar Missions and Explorations*. John Wiley & Sons, New York, 1964.

[4] Dankhoff, W., "The M-1 Rocket Engine Project." NASA Technical Memorandum TM X-50854, 1963.

[5] Rockwell International, "Space Shuttle Transportation System Press Information," unnumbered, March 1982.

[6] Wiesel, W. E., *Spaceflight Dynamics*, Aphelion Press, Beavercreek, OH, 2012.

[7] Wambolt, J. F., "Medium Launch Vehicles for Satellite Delivery," Aerospace Corp., Crosslink Magazine, Vol. 4, No. 1, Winter 2002–03, pp. 26–31.

[8] Taverna, M. A. and Morring, F. Jr. "Working Together: Shuttle advanced cooperation with Spacelab, space stations," *Aviation Week*, 6 Dec. 2010, pp. 64–65.

[9] Jenkins, D. "The History of Developing the National Space Transportation System: The Beginning through STA-75". Published by Dennis R. Jenkins, Cape Canaveral, FL. 1997.

[10] Space Exploration Technologies, "Falcon 9 Launch Vehicle Payload User's Guide, Rev 1". SCM-2008-010 Rev.1, 2009.

5.15 Exercises

Assignment: Launch Vehicle Performance Problems

Please create a summary sheet that gives the values requested below, and any requested explanations of comments. Also include spreadsheet calculations that are clearly labeled and show payload mass m_{PL} and total Δv highlighted in yellow. Use $g_0 = 9.80665$ m/s^2 for all calculations. *Provide no more than four significant digits for your results: round numbers as necessary (but please, no scientific notation!). Also, please use the numbers provided in the problems for your calculations, even though they may not be what is listed currently in references.*

1. Space Shuttle Performance Calculation

    Components of the U.S. Space Shuttle had the following characteristics (astronautix.com):

    * Shuttle SRB (each): *Gross Mass:* 589,670 kg; *Empty Mass:* 86,183 kg. *Thrust (vac):* 11,520 kN. $I_{SP-SRB-SL} = 237$ sec; $I_{SP-SRB-vac} = 269$ sec. *Burn time:* 124 sec. Quantity: 2.
    * External Tank (ET). *Gross Mass:* 750,975 kg. *Empty Mass:* 29,930 kg.
    * Shuttle Orbiter with 3 SSMEs & 2 OMS engines, dry (no propellants): *Mass:* 77,658 kg dry (not including OMS propellant). *Thrust (vac):* 6,834 kN. $I_{SP-SSME-SL} = 363$ sec; $I_{SP-SSME-vac} = 455$ sec. *Burn time:* 480 sec.
    * Orbital Maneuvering System (OMS): *dry mass:* 3,600 kg and *is included in Orbiter mass given above. Thrust (vac):* 27 kN. $I_{SP-OMS-vac} = 313$ sec. OMS propellant mass = 21,660 kg. OMS circularizes orbit after SSME shutdown & ET separation.
    * Desired orbit 278 km at 28.50°, total Δv delivered by SRBs and SSMEs: 9086 m/s from 1st and 2nd stages; 144 m/s from OMS firing. Motion of launch site already included.
        a) Stage 1 has two SRBs (red) burning in parallel with the three SSMEs (blue) on the Orbiter, all attached to the ET (see figure). Set up a spreadsheet that will calculate the Δv_1 this stage provides at the end of the SRBs' 124 sec burn time, including a value for the payload that you can change to achieve a certain total Δv (start with a value $m_{PL} = 10,000$ kg). It is necessary to determine how much propellant mass the Orbiter/SSME

loses during the 124 sec. Since you have the gross and empty mass of the ET and its burn time, you can calculate mass flow by dividing ET propellant mass by the total burn time (you may assume constant mass flow and neglect throttling). For the overall specific impulse of the parallel burn I_{SP1}, calculate the mass-averaged exit speed using the relation on Chapter 5 p. 23, of the text (don't forget the total SRB mass flow is 2 × one SRB's mass flow). Since Stage 1 launches from sea level and ends at high altitude, use the average specific impulse $I_{sp-eff} = \frac{1}{3} I_{sp-SL} + \frac{2}{3} I_{sp-vac}$ for the mass-averaged exit speed calculation (both the SRBs and SSMEs). The initial stage 1 mass $m_{10} = 2 \times m_{0\ SRB}$ (loaded) $+ m_{Orbiter+SSMEs+OMS} + m_{0ET}$ (loaded) $+ m_{pOMS} + m_{PL}$. The final stage 1 mass $m_{1f} = 2 \times m_{sSRB}$ (empty) $+ m_{Orbiter+SSMEs+OMS} + m_{sET}$ (empty) $+ m_{pET}$ (at $t = 124$ s) $+ m_{pOMS} + m_{PL}$. Use these to find the mass ratio for the 1st stage, and then calculate the Δv_1 stage 1 provides at the end of the SRBs' 124 sec burn time.

b) Now continue your spreadsheet with the addition of Stage 2 calculations. Stage 2's starting mass is found by jettisoning the empty SRB masses, and continuing the ascent (see figure to right). You need to determine the initial stage 2 mass $m_{20} = m_{Orbiter+SSMEs+OMS} + m_{sET}$ (empty) $+ m_{pET}$ (at $t = 124$ s) $+ m_{pOMS} + m_{PL}$. The final stage 2 mass $m_{2f} = m_{Orbiter+SSMEs+OMS} + m_{sET}$ (empty) $+ m_{pOMS} + m_{PL}$. Use these to find the mass ratio for the second stage, and then use the mass ratio along with the value of $I_{SP-SSME-vac}$ (since this stage all occurs essentially in vacuum) to calculate the Δv_2 stage 2 provides at the end of the SSMEs' 480 sec burn time.

c) Now, iterate through parts 1a and 1b above, entering different payload masses until the total speed reached is **9,086 m/s** (this number is the final design speed and already includes earth rotation gain and ascent losses, so *don't change it*). It is the Δv the shuttle must provide before the OMS system fires. Excel®'s "goal seek" function is a quick way to determine the payload.

EXPLAIN: What is the resulting mass of the payload? How does the mass compare to the "official" payload value of 29,500 kg? Explain any agreements or differences in a few sentences.

d) Now add stage 3, which is the OMS system burn after the ET is jettisoned (OMS engines are on the Orbiter tail above the SSMEs). Determine the final mass of the shuttle after the OMS system provides the specified 144 m/s. You know the initial mass of stage 3

$m_{30} = m_{Orbiter+SSMEs+OMS} + m_{pOMS} + m_{PL}$. The final stage 2 mass $m_{3f} = m_{Orbiter+SSMEs+OMS} + m_{pOMS}$ (remaining after 144 m/s Δv) + m_{PL} can be determined by finding the m_{3f} that produces the specified Δv. How much OMS propellant is consumed for the stage 3 burn, and how much of the original mass remains?

e) Now assume that there is no OMS system mass, but $m_{p-OMS} = $ 14,700 kg of propellant must be on board for orbital attitude controls and plane changes only (this propellant can be considered payload, as it remains in the orbiter when orbit is reached). This means subtract 3,600 kg from $(m_{Orbiter+SSMEs+OMS})$ to get $(m_{Orbiter+SSMEs})$. Repeat steps 1a, 1b, & 1c with these new values, and find the payload mass where the total speed $\Delta v_1 + \Delta v_2 = \textbf{9,230 m/s}$, the required Orbiter design speed (do NOT repeat 1d, as there is no OMS system, and only stages 1 & 2 for this configuration). *This is equivalent to going to orbit with the ET attached, and not using the OMS at all.* What is the resulting NO-OMS payload? How does it compare to the result you obtained in 1c above? Explain any agreements or differences in a few sentences.

Notes:

(A) Be sure to include the proper number of SRBs and SSMEs to produce the needed thrust values.

(B) The numbers or specifications provided by different sources on launch vehicles are often different and disagree with one another. Discrepancies can be due to different vehicle models, manufacturing techniques, trajectories, etc. *Please use the given values, so that your answer may be directly compared to the answer obtained by the instructor, "apples to apples."*

(C) Some the parameters provided above are not needed to solve this problem, but they are included so you can get a "feel" for their values. If in doubt, do the calculations the simplest way possible.

2. Heavy-Lift Launch Vehicle Performance Analysis

Consider a space shuttle-derived vehicle using existing shuttle parts. This vehicle consists of four (*TWO pairs*) solid rocket boosters (SRBs) burning together as the first stage ONLY (no main engines), and an external tank (by itself, no orbiter) with *two* SSMEs as the second stage (no parallel burn, SRBs are jettisoned). Assume that the main engines and their required systems mass adds an additional 12,000 kg over the ET masses given. Assume $I_{sp-eff} = \frac{1}{3}I_{sp-SL} + \frac{2}{3}I_{sp-vac}$ for the SRBs. Use data from the problem above as needed to create a spreadsheet to make Δv calculations and iteration easier.

a) How much payload can this vehicle provide a $\Delta v = \textbf{9,450 m/s}$?

b) How much advantage in payload mass does this derivative configuration provide over the conventional space shuttle system? Discuss.

As a minimum, your spreadsheet should contain the following information (please display four significant digits in all cells):

item	Series, 4 SRBS
m_{PL}	kg
Mass, begin stage 0	kg
Mass, end stage 0	kg
I_{sp0}	s
Δv_0	m/s
Mass, begin stage 1	kg
Mass, end stage 1	kg
Δv_1	m/s
total Δv	9.450 km/s

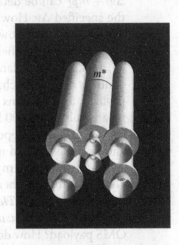

3. Falcon Heavy Performance Analysis

In April 2011, SpaceX proposed a new version of its Falcon LV, the *Falcon 9 Heavy* or *F9H* (right). According to the news release, "Falcon Heavy will be the first rocket in history to feature propellant cross-feed from the side boosters to the center core. Cross-feeding leaves the center core still carrying the majority of its propellant after the side boosters separate, giving Falcon Heavy performance comparable to that of a 3-stage rocket." The release says that *F9H's* payload mass is 53 T (T = tonne = 1,000 kg).

According to [10], the SL thrust of the center vehicle 1st step = 5,000 kN; the 2nd step is 445 kN. On the 1st step, all nine

An artist's rendering of the Falcon Heavy. Space Exploration Technologies hopes to hold a demonstration launch by the end of 2012. (Space Exploration Technologies / April 4, 2011)

Merlin 1C (M1C) engines burn for 155.5 s, then two shut down and the remaining seven burn to 174.2 s. The single 2nd step Merlin 1 vacuum (M1V) engine burns from 179.2 to 475.9 s.

From www.spacelaunchreport.com/falcon9.html (visited 04/2011), the 1st step $I_{sp1SL} = 266$ s and $I_{sp1vac} = 304$ s, the 1st step $m_{1f} = 14.73$ T; the 2nd step $I_{sp2} = 340$ s, and the 2nd step $m_{2f} = 3.0$ T.

a) Determine the propellant masses for the first and second steps of the *core* vehicle only. Use the sea-level thrust and I_{sp} to determine the mass flow of nine M1C engines, divide by nine to get the mass flow of one engine, and then use the one-engine mass flow value and the number of engine-seconds that the first stage consumes to calculate the total first step propellant mass. Repeat with the second step thrust and specific impulse, and burn time to get the second step propellant mass. Use these propellant masses for all of the calculations below, and assume all strap-ons have the same propellant mass as the core vehicle. Also assume that the strap-on empty mass is the same as the empty mass of step 1 of the core vehicle (this would be assuming that whatever structure separates the first and second steps of the two-stage vehicle has the same mass as the nose cones placed on top of the two strap-ons).

For the following, assume all three first-step parts (core and two strap-ons) are identical except for variations in their I_{sp}s. Assume the ground-launched steps' I_{sp} are $I_{sp-eff} = \frac{1}{3} I_{sp-SL} + \frac{2}{3} I_{sp-vac}$ during their entire burn. For items b, c, and d, determine 1) how much payload this vehicle can accelerate to the target speed for vehicle burnout, 9150 m/s, and 2) show the Δv each stage provides. Iteration will be necessary.

b) *Cross-flow analysis*: for 'stage 0,' assume that core and strap-ons burn at equal rates with effective I_{sp-eff} for first burn. After $\frac{2}{3}$ of the *all-nine-engine* burn time, the two strap-ons' propellant is instantaneously transferred into the core, filling it completely, and the strap-ons separate. Neglect the mass of the transfer system. The stage 1's now fully-loaded core continues burning all nine engines with $I_{sp-1vac}$ until it consumes all propellant, then stage 2 fires normally.

c) *Non-cross-flow, series burn*: calculate the payload this vehicle could accelerate to the same speed if 0) only the outer two segments burn simultaneously to burnout (stage 0 with I_{sp-eff}), are dropped, then 1) the center segment burns (1st stage, $I_{sp-1vac}$), then 2) the 2nd stage fires ($I_{sp-2vac}$).

d) *Non-cross-flow, parallel burn*: calculate the payload mass the F9H vehicle could accelerate to the same speed if there is NO transfer of propellants, assuming *all three segments* burn simultaneously. Use I_{sp-eff} for first stage burn, vacuum I_{sp} for second stage burn.

e) Summarize payload mass information and Δvs per stage for configurations in letters b, c, and d neatly in a table. How do your masses compare to SpaceX value? If you show a different value, explain in a few words some possible causes of any difference.

4. Assume that you need to increase the Space Shuttle's
payload by 2,000 kg, and you have decided to increase
the fuel volume of the external tank (ET) based on its
higher tradeoff ratio in Table 5.4, where $\partial m_{PL}/$
$\partial m_p = +0.161$ and $\partial m_{PL}/\partial m_s = -0.964$. Note that
these values already incorporate the effect of small
changes on the lower stages.

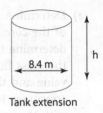

Tank extension

a) Use the appropriate ratio to determine the mass of
propellant (kg) needed to provide this payload increase only, ignoring
the possible structural weight increase.

b) With $\rho_{LH2} = 70.97$ kg/m^3, $\rho_{LOx} = 1,137.48$ kg/m^3,
$m_{LH2} = 106,261$ kg, $m_{LOx} = 629,340$ kg, and a LOx:LH$_2$ mixture ratio
of 5.922 to 1.0 by mass, determine the average density $\rho_{propellant}$ and
Vol$_{propellant}$, the volume of both propellants (fuel and oxidizer)
needed (in m^3).

c) The ET has to be lengthened to carry the extra propellant, but
you cannot change its diameter due to its existing fabrication
process. With the inner diameter of the ET at 8.4 m, determine the
added length h (in m) of the ET necessary to provide the
needed volume. Assume that you are just extending the
cylindrical "barrel" section of the tank, and that the top and bottom
bulkheads of each of the propellant tanks would be unchanged by
this addition.

d) There is no free lunch! (Or is it "free *launch*"?) To carry this propellant,
you need to include structural (tank) mass for the added tank length, so
the mass of the tank will also increase, which will *reduce* the payload
(as predicted by the payload/added mass trade ratio). Using the
geometry of the ET, determine the cylinder side area of the tank
addition. Assuming the tank's area density $\sigma_A = 10$ kg/m^2, calculate
the mass of the added tank walls (kg).

e) Use the tradeoff to determine the payload loss due to the added wall
mass (kg).

f) Make an Excel spreadsheet like the one below, and use the Goal
Seek function to find the actual fuel addition needed to carry
2,000 kg extra payload, *including* the penalty produced by the added
mass of the tank extension (change the mass in the first line to make
the net payload after losses = 2,000 kg). Summarize by providing the
following:
  (i) Net propellant mass needed (kg)
  (ii) Resulting new payload mass before subtracting added tank
  mass (kg)
  (iii) Cylinder volume added (m^3)
  (iv) Cylinder height (m)
  (v) Cylinder outer surface area (m^2)
  (vi) Added cylinder mass (kg)

(vii) Payload loss due to added cylinder mass (kg)

(viii) Net payload added after losses (kg)

(ix) The added fuel there simply to offset the added tank mass (kg)

5. The Shuttle's orbital maneuvering system (OMS) is used to provide the remaining ~1% of speed needed for the injection of the Shuttle into orbit. Shuttle information is provided to the right, and results in a payload of 19,561 kg to 9.45 km/s if done correctly.

Use the Shuttle information to calculate the missing term for $\partial m_{PL}/\partial I_{sp2}$ in the last column of the table below. The result is the increase in payload mass per second of specific impulse increase for OMS. How does your number compare with stage 0's SRB $\partial m_{PL}/\partial I_{sp0}$ and stage 1's SSME ($\partial m_{PL}/\partial I_{sp1}$) values in the table? Comment on your result. Is it worth a big effort to increase the I_{sp} of this hypergolic system?

Note: This problem can be done either analytically using the formulae given in the text or by brute force where the I_{sp} is changed by 1 s, and the new payload is calculated while the Δv is held to be the same.

Shuttle SRB (each)	
Loaded	589,670 kg
Empty	86,183 kg
Thrust	11,520 kN
I_{sp} vac	269.0 s
I_{sp} SL	237.0 s
Burn time	124.0 s
Burn rate	4060.4 kg/s
External Tank (ET)	
Loaded	750,975 kg
Empty	29,930 kg
Propellant	721,045 kg
Shuttle with SSMEs & Payload	
Mass	99,318 kg
Thrust	6834.0 kN
I_{sp} vac	455.0 s
I_{sp} SL	363 s
Burn time	480 s
Burn rate	1502.2 kg/s
Shuttle/OMS	
Loaded	25,200 kg
Propellant	3,600 kg
I_{sp}	316.0 s
Burn time	1250.0 s

k (stage no.)	v_{ek} (m/s)	m_{0k} (kg) (initial mass)	m_{fk} (kg) (final mass)	$\frac{\partial m_{PL}}{\partial m_{fk}}$ (kg/kg)	$\frac{\partial m_{PL}}{\partial m_{pk}}$ (kg/kg)	$\frac{\partial m_{PL}}{\partial I_{sp,k}}$ (kg/s)
0	3,060	2,015,000	834,000	−0.078	+0.055	
1	4,459	675,500	143,300	−0.964	+0.161	
2	3,067	108,300	104,700	−1.000	+0.059	

6. Problem 2 above considers a Space Shuttle–derivative vehicle consisting of four (two pairs) solid rocket boosters and an external tank. Assume that this vehicle will carry m_{total} = 166 T. Part of mass m_{total} must be allotted for a payload shroud m_{shroud}, so the actual useful mass $m_{useful} = m_{total} - m_{shroud}$. Assume that the payload is electronics whose average density is $\rho = 320$ kg/m^3 inside of an 8.4-m inner diameter shroud of areal density $\sigma = 10$ kg/m^2.

You may want to follow a procedure something like the following: first, estimate the volume of the payload Vol$_{payload}$ in m^3. Next, estimate the height of a cylinder h to provide the calculated volume. (Part of the volume is supplied by the 8.4-m-diameter hemispheric nose cone.) Now calculate the fairing's surface area, and use the area to get its mass using the given areal density. (Assume that the shroud has a top hemisphere and cylindrical sides, but no bottom: the tank dome underneath serves as the bottom.) A spreadsheet is a handy way to iterate and find the shroud mass m_{shroud} as accurately as you can. Summarize by giving the following parameters:

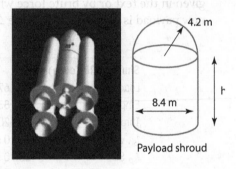

Payload shroud

a) Net payload mass m_{useful} (does not include shroud mass, kg)
b) Net payload volume for 7a, Vol$_{payload}$ (m^3)
c) Cylinder part (see figure) of shroud height h (m, not including hemisphere)
d) Shroud outer surface area (hemisphere + cylinder, m^2)
e) Shroud mass m_{shroud} (hemisphere + cylinder, kg)

7. **Design Task:** The Air Force needs to launch a spacecraft into orbit for communication and surveillance purposes. Your assignment is to design the smallest possible launch vehicle to carry out this mission. Size a family of two-stage-to-orbit (TSTO) SLVs to do the following task with varying lower- and upper-step Δvs for a set of propellants based on your last name.
   Design requirements:
• Payload = 1,000 kg including mounting and separation hardware.

- Orbit required: 300-km circular orbit, best inclination for launch from either ETR or WTR; may directly ascend, or use transfer orbit and circularize.
- Use step propellants as specified in the table below, based on the first letters of your last name.

C = cryogenic hydrolox = LOx/LH_2
H = hypergolic
M = metalox = $LOx/methane$
R = kerolox = $LOx/kerosene$
S = solid

Last Name	A	B	C	DE	F	G	H	IJK	L	M	N	O	P	Q	R	S	T	UV	WX	YZ
1st step	H	H	S	S	S	H	H	H	C	C	C	R	R	R	R	M	M	M	M	M
2nd step	S	H	C	R	M	C	R	M	C	M	R	H	C	R	M	R	H	C	S	M

Your assignment is to do the following:

Using your specific propellants, you will create two "families" of SLVs consisting of two groups of vehicles whose first and second stages supply varying fractions of Δv_{total}, such that $\Delta v_{total} = \Delta v_{stage1} + \Delta v_{stage2}$.

a) Based on existing launch vehicle mass properties and specific impulses (you can get ideas from online information sources such as astronautix.com), it will be up to you to select "reasonable" values for each step's parameters. This is a matter of engineering judgment along with selecting any improvements due to technical advancements. Provide your required propellant(s), specific impulses I_{sp-i}, structural mass fractions σ_i, and brief explanations as a table, with rows for your LV's first and second steps as shown in the example table below:

Step No.	Propellant	I_{sp}	Struct. Mass Fraction σ	Notes/Explanation
1	Kerolox	255	0.0855	I_{sp-1} from V-2 second step. Structural mass fraction from Saturn V S-IC with 5% credit for Al-Li
2	Hydrolox	435	0.09	J-2 engine, S-IVB step mass fraction

The two structural mass fractions that you select should be based on existing vehicles, although you can assume a small technological improvement, especially if you are looking at older vehicles as references. For example, for a LOx/kerosene first stage, you might want to look at the Saturn V S-IC first step as a reference. But you might think about what improvement you could achieve with a slightly lighter metal (aluminum-lithium, i.e., Atlas V or Falcon 9) or composite construction. If you take advantage of such improvements, write a sentence or two in the explanation column justifying your assumptions.

For propulsion, please select a value for I_{sp} "effective" for the first stage ("effective" means the integrated value from liftoff to burnout), and use I_{sp} vacuum for the second stage. If you cannot locate an effective value for the first stage, then use an average of SL and vacuum I_{sp}s.

b) Make a table with rows for first and second steps, and design a family of *nine* different launch vehicles to carry the given payload, whose first and second stage Δvs are split as follows:

	Vehicle #1	2	3	4	5	6	7	8	9
Stage 1 Δv_1	30%	35%	40%	45%	50%	55%	60%	65%	70%
Stage 2 Δv_2	70%	65%	60%	55%	50%	45%	40%	35%	30%

Use a spreadsheet or another method to specify/determine (some are inputs, some outputs) the following parameters for each vehicle: m_{01}, m_{s1}, m_{p1}, σ_1, I_{sp1}, Δv_1; m_{02}, m_{s2}, m_{p2}, σ_2, I_{sp2}, Δv_2, gross liftoff mass (GLOM) or m_0, and finally Δv_{total}, making sure that Δv_{total} is met and is the same value for all cases, and the mass of each step is minimized.

Use the same structural mass fraction σ_1 for all vehicles' lower steps, and the same structural mass fraction σ_2 for all vehicles' upper steps. (The upper and lower steps' values may be different.) Similarly, use the same specific impulse I_{sp-1} for all vehicles' lower steps (for first steps, use effective I_{sp}), and the same specific impulse I_{sp-2} for all vehicles' upper steps. (The upper and lower steps' values do not need to be the same.) You may find it convenient to use eqs. (5.80)–(5.85) instead of using a solver routine.

Provide all of these values in a neat summary table similar to Table 5.2, with one column of 12 rows for each of the designs. Please display no more than five significant digits. You may use metric tonnes (T) or kg units for mass.

Note 1: Some ratios may end up unfeasible, meaning that it's physically impossible for the combination to provide the required Δv. These may produce a computation error.

Note 2: The Δv ratios in your table may not encompass the minimum GLOM. You may need to add another column on the right or left in order to capture the minimum mass.

c) Use your table to construct a plot showing the GLOM of the different designs similar to the TSTO Example in the text Section 5.9.5, Fig. 5.12. *Only show those designs that have positive Δv. Do not attempt to plot any designs that will not work.* Your plot should resemble that shown in Fig. 5.12.

d) Find the minimum GLOM by numerically adjusting the percent Δv per stage (to the nearest percent), which may lie between the ones in your table. Insert a *highlighted* separate column for the minimum GLOM case in your table, if the optimum percentages are not already included in (or outside: your minimum may fall outside the required values) cases 1–9. Make a plot showing GLOM vs design number. Highlight or otherwise show which of the vehicles provides the required Δv for the lowest GLOM. If there is an existing vehicle with the same propellants, see if your design has similar numbers or values (in what variables?).

e) Now consider that the gravity and drag losses are likely to occur mostly in the first step's burn. As a gross approximation, assume that the first step is completely responsible for dealing with the losses, and the second step does not provide for any losses. Make a set of calculations for the second family of vehicles similar to the first family, so that the first and second stages supply varying fractions of Δv_{total}. Here $\Delta v_{total} = \Delta v_{stage1} + \Delta v_{stage2} + \Delta v_{grav\ loss} + \Delta v_{drag\ loss} = \Delta v_{ideal} + \Delta v_{loss}$, and $\Delta v_{loss} = \Delta v_{grav\ loss} + \Delta v_{drag\ loss}$. Split family 2's stage Δvs as follows, using your design values:

	Vehicle 1	2	3	4	5	6	7	8	Vehicle 9
Stage 1 Δv_1	(30% of Δv_{ideal}) + Δv_{loss}	(35% of Δv_{ideal}) + Δv_{loss}	(40% Δv_{ideal}) + Δv_{loss}	(45% Δv_{ideal}) + Δv_{loss}	(50% Δv_{ideal}) + Δv_{loss}	(55% Δv_{ideal}) + Δv_{loss}	(60% Δv_{ideal}) + Δv_{loss}	(65% Δv_{ideal}) + Δv_{loss}	(70% of Δv_{ideal}) + Δv_{loss}
Stage 2 Δv_2	70% of Δv_{ideal}	65% of Δv_{ideal}	60% of Δv_{ideal}	55% of Δv_{ideal}	50% of Δv_{ideal}	45% of Δv_{ideal}	40% of Δv_{ideal}	35% of Δv_{ideal}	30% of Δv_{ideal}

As before, each vehicle will utilize identical structural fractions and chemistry for its first steps, and identical structural fractions and chemistry (which may be different from the first step) for its second steps.

For example, suppose $\Delta v_{total} = \Delta v_{ideal} + \Delta v_{loss} = 10$ km/s, and so $\Delta v_{ideal} = 9$ km/s and $\Delta v_{loss} = 1$ km/s (your numbers will be different). For this example set of numbers, the table would look like:

	Vehicle 1	Vehicle 2	Vehicle 3
Stage 1 Δv_1	0.3 $\Delta v_{ideal} + \Delta v_{loss} =$ 3.7 km/s	0.35 $\Delta v_{ideal} + \Delta v_{loss} =$ 4.15 km/s	0.4$\Delta v_{ideal} + \Delta v_{loss} =$ 4.6 km/s
Stage 2 Δv_2	0.7 $\Delta v_{ideal} = 6.3$ km/s	0.65 $\Delta v_{ideal} = 5.85$ km/s	0.6 $\Delta v_{ideal} = 5.4$ km/s

Repeat the steps given above for your family 2, and provide a similar table. Plot the family 2 GLOM curve on the same window as family 1. Compare the results from family 1 and family 2, and comment about how the incorporation of losses into the first step affects the GLOM of the vehicle. Which would be more appropriate for design purposes?

Chapter 6   *Ascent Trajectory Analysis and Optimization*

Moving forward, the design roadmap in Fig. 6.1 shows the high-lighted signpost indicating the information to be covered in this chapter: the modeling and analysis of launch vehicle trajectories under various conditions and environments.

It is instructive to begin with a very simple set of assumptions, so that some simple behaviors and trends can be observed. We will continue adding assumptions in order to approach the situation for typical or realistic launch and flight.

Starting with vertical flight in a vacuum, we add inclined flight (vacuum), and finally an atmosphere to provide aerodynamic forces and torques. We'll simulate a launch trajectory using numerical integration. We will discuss

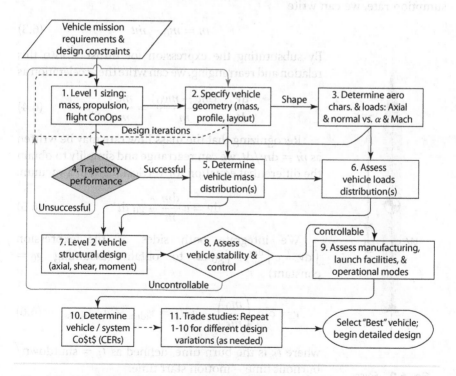

Fig. 6.1 This chapter will look at trajectory analysis and optimization using the design data generated in Chapter 5.

optimal trajectories, and conclude with a number of examples of trajectories of existing LVs.

6.1 Vertical Flight in Gravity, No Atmosphere

Consider the motion of a launch vehicle (LV) in an environment where gravity exists, but aerodynamic forces do not (in other words, a vacuum environment, such as the moon). The LV's thrust axis points vertically upwards; gravity's attraction g is vertically downwards and appears at the vehicle's center of mass (CM), illustrated by the black-and-white circular symbol in Fig. 6.2.

T, the vehicle's thrust, may be calculated as a function of propellant mass flow $\dot{m}$ and exhaust speed c via equation 4.14a, $I_{sp} = T/(g_0 \dot{m}_p)$

$$T = \dot{m}c \qquad (6.1)$$

Using Newton's law, the sum of the applied forces = mass × acceleration. In our situation, the vehicle's acceleration may be written

$$a = (T - W)/m \qquad (6.2)$$

where m is the vehicle's instantaneous mass. By knowing propellant consumption rate, we can write

$$m = m_0 - \dot{m}t \qquad (6.3)$$

Fig. 6.2 Forces acting on a rocket in vertical flight.

By substituting the expression for thrust T into this relation and rearranging, we can write the acceleration as

$$\frac{dv}{dt} = \frac{(\dot{m}c - mg_0)}{m} = \frac{\dot{m}c}{m} - g_0 \qquad (6.4)$$

Recognizing that the mass flow rate may be written as $\dot{m} = dm/dt$, we can rearrange and simplify to obtain the differential equation for vertical flight in a vacuum.

$$dv = c\frac{dm}{m} - g_0\,dt \qquad (6.5)$$

We integrate both sides of the expression $\int dv = c\int \frac{dm}{m} - \int g_0\,dt$ to obtain (assuming $g_0 =$ constant)

$$v = c\ln\left(\frac{m_0}{m_f}\right) - g_0\,t_b = v_{ideal} - v_{grav-loss} \qquad (6.6)$$

where t_b is the burn time, defined as $t_b =$ shutdown/burnout time − motion start time.

We now can calculate the speed the vertical vehicle obtains based on only three parameters: exhaust speed, mass ratio, and burn time. We can gain some more insights by noticing that Eq. (6.6) is composed of two terms. The first term, $v_{ideal} = c \ln (m_f/m)$, can be seen to be identical to Tsiolkovsky's equation, which provides the rocket's speed in a gravity-free and drag-free environment (an "ideal" environment).

The second term, $v_{grav-loss} = g_0\, t_b$, which is *subtracted* from the first term, is the product of the vehicle's burn time and (constant) gravity. This operation also produces a number with units of speed. Because it is *subtracted* from the ideal speed, this term represents a loss that *reduces* the effective speed of the vehicle, due to the fact that part of the vehicle's thrust is not used to accelerate, and instead is used to overcome the force created by the acceleration of gravity (which we commonly call *weight*). It is formally referred to as the vehicle's *gravity loss*. Hence, we can refer to the speed calculation as [*actual speed delivered*] = [*ideal speed from rocket equation*] − [*gravity loss*]. In reality, we need to consider the variation of gravity with altitude, which is done later in this chapter. There is more discussion about gravity loss and numerical values in Chapter 3.

It's interesting to notice that these two terms, v_{ideal} and $v_{grav-loss}$, in the speed equation of Eq. (6.6) have very different behaviors. The first term is the "ideal speed" as calculated by the rocket equation. It is *independent* of thrust time history. The second term, the gravity loss, is *not* independent of thrust history. In fact, one needs to obtain the vehicle's burn time t_b and perform the calculation.

One might also conclude that gravity loss could be minimized by considering the effect of burn time t_b. The burn time is directly related to the thrust required and the mass flow rate: a *long* acceleration, with *low* thrust and *lower* mass flow rates produces a larger gravity loss than a second vehicle with a rapid burn time, meaning a less lengthy period of acceleration, higher accelerations, and larger mass flow rates.

What do these conclusions have to do with the design of a launch vehicle? Quickly exhausting propellants minimizes losses, but also produces high acceleration, which might be at a level too high for delicate on-board equipment or human flight. One of the ways to contrast designs against each other is to use the vehicles' *thrust-to-weight ratio T/W*, a dimensionless parameter. The most commonly used thrust-to-weight ratio is the initial or liftoff thrust-to-weight ratio

$$\frac{T}{W_0} = \frac{T}{m_0 g_0} = \frac{\dot{m}c}{m_0\, g_0} \tag{6.7}$$

Here the subscript 0 is used to indicate conditions at time zero, otherwise known as liftoff.

6.1.1 Gravity Loss

The gravity loss term of the vertical speed equation may be written incorporating T/W_0 as a parameter—this implicitly includes the burn time. The

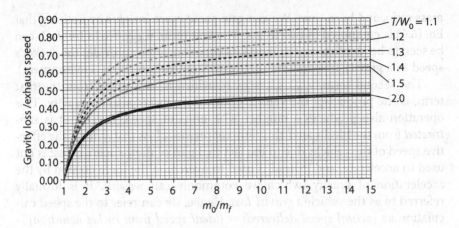

Fig. 6.3 Nondimensional gravity loss for vertical climbs having varying values of T/W_0.

vertical gravity loss term is calculated as

$$v_{\text{grav-loss}} = c\left(1 - \frac{m_f}{m_0}\right)\left(\frac{1}{T/W_0}\right) = c\left(1 - \frac{1}{\mu}\right)\left(\frac{1}{T/W_0}\right) \quad (6.8)$$

To get an order-of-magnitude assessment, assume an engine with an exhaust speed of 3.5 km/s, a mass ratio of 10, and a thrust-to-weight ratio of 1.4; this results in a loss of about 2.25 km/s. Remember that this calculation is for a vertical trajectory; a typical launch ascent transitions to horizontal near burnout, so its gravity losses will be reduced as burnout is approached.

Our simple analysis has provided some insights related to vertical flight. Let's consider these results in more detail.

Figure 6.3 shows the normalized gravity loss $v_{\text{grav-loss}}/c$ [the gravity loss from Eq. (6.8) divided by the engine's effective exhaust speed] for a number of mass ratios m_0/m_f and liftoff thrust-to-weight ratios T/W_0. It shows that the gravity loss *increases* as the liftoff thrust-to-weight ratio decreases, regardless of the mass ratio. This makes sense physically, because the lower the T/W_0 ratio, the longer it takes to burn the propellant, and thus the longer period where the vehicle is exposed to gravity. This would encourage larger T/W_0 ratios to minimize the gravity loss, which potentially conflicts with maximum acceleration limits for carbon-based inhabitants and sensitive silicon-based electronics.

6.1.2 Burnout Altitude

To calculate burnout altitude, we integrate the speed equation in Eq. (6.6) to get $s(t)$, the distance traveled during the rocket's burn time.

$$s = c \int_0^t \ln \frac{m_0}{m} \, dt - g \int_0^t t \, dt$$

$$s = c \frac{m_0}{\dot{m}} \left[1 - \frac{m}{m_0} \left(\ln \frac{m_0}{m} - 1 \right) \right] - \frac{1}{2} g t^2$$

$m = m_0 - \dot{m}t$, so $t = \frac{m_0}{\dot{m}} \left(1 - \frac{m}{m_0} \right)$. Then

$$s = c \frac{m_0}{\dot{m}} \left[1 - \frac{m}{m_0} \left(\ln \frac{m_0}{m} - 1 \right) \right] - \frac{1}{2} g \left(\frac{m_0}{\dot{m}} \right)^2 \left(1 - \frac{m}{m_0} \right)^2$$

$$h_{bo} = c \frac{m_0}{\dot{m}} \left[1 - \frac{m_f}{m_0} \left(\ln \frac{m_0}{m_f} - 1 \right) \right] - \frac{1}{2} g \left(\frac{m_0}{\dot{m}} \right)^2 \left(1 - \frac{m_f}{m_0} \right)^2$$

This result is only valid during powered flight, where $0 \leq t \leq t_b$. Post-burnout coasting is treated in the following section.

6.1.3 Coast After Burnout

After burnout, the rocket coasts upward, and the only force acting on the vehicle is its own weight, which produces downward (negative) deceleration. The general equations for this case, with the subscript bo representing burnout, may be seen as follows (here $t_b \leq t \leq t_{impact}$):

$$v = v_{bo} - g_0(t - t_{bo}) = c \ln \left(\frac{m_0}{m_f} \right) - g_0(t - t_{bo}) \qquad (6.9)$$

$$h = h_{bo} + v_{bo}(t - t_{bo}) - \frac{1}{2} g_0(t - t_{bo})^2 \qquad (6.10)$$

One interesting event for vertical flight might be the vehicle's maximum altitude, which occurs when $v = 0$. If we denote this time as t_{max-h}, one can see that

$$t_{max-h} = \left(\frac{c}{g_0} \right) \ln \left(\frac{m_0}{m_f} \right) = \left(\frac{c}{g_0} \right) \ln \mu \qquad (6.11)$$

where $\mu = m_0/m_f$.

Inserting the value of t_{max-h} into the height equation in Eq. (6.10) produces the maximum height h_{max}

$$h_{max} = \frac{1}{2} \left(\frac{c^2}{g_0} \right) \ln^2 \mu - \left(\frac{cm_0}{\dot{m}} \right) \frac{\mu \ln \mu - (\mu - 1)}{\mu} \qquad (6.12)$$

One last observation to end this section: one may inspect Eq. (6.12) to determine how to maximize altitude. The first term has parameters that are fixed that we can't really adjust: c, g_0, and μ. The second term has mass flow $\dot{m}$ in its denominator, which is a variable that we can control.

This implies that the altitude approaches its maximum value as $\dot{m} \to \infty$, in other words, all propellant consumed instantaneously. This type of launch is commonly referred to as a *mortar* or *cannon*, a device that burns all its propellant at the beginning of the projectile's flight. This produces the maximum possible altitude, because there is no need to elevate propellant that has not yet been consumed as a conventional rocket does. The reader is encouraged to consider the structural implication of space launches consuming their entire propellant mass instantaneously.

6.1.4 Summary

The previous equations apply to vertical flight in *vacuum*. The values for burnout speed and altitude are reduced due to gravity pulling backwards on the vehicle. We can calculate the gravity loss for a vertical mission, but in general it will be very costly in terms of performance loss.

Now, consider that the primary *orbital* requirement for a space launcher is to achieve very large *horizontal* speed (not vertical). Intuition suggests that we would like to transition to horizontal flight as soon as possible. We will have to do something else to achieve this speed.

6.2 Inclined Flight in Gravity, No Atmosphere

Most space launch vehicles begin their flight in a vertical attitude, but they also need to be approximately horizontal at burnout or when injected into orbit. The horizontal flight portion needs to be defined, and the combination of horizontal and vertical motion analyzed. This section will consider what happens if the flight path is *inclined* relative to vertical.

We will find that a non-vertical pitch angle provides speed in the horizontal direction. We will also see that gravity affects the flight by turning it away from vertical. In the atmosphere, inclined flight will produce large structural loads, but not out of the atmosphere—but there can be significant gravity losses.

6.2.1 Equations for Inclined Flight in Gravity, No Atmosphere

We now replace the previous vertical equations with those appropriate for planar motion components z (the vertical direction) and x (the horizontal direction), as illustrated in Fig. 6.4. The angle θ represents a third degree of freedom, so that motion of a rigid body in a plane is said to have three degrees of freedom (3 DoF). Speeds v_x and v_z are the horizontal and vertical components of the velocity vector $\mathbf{v}$

$$v_z = c \sin \theta \ln\left(\frac{m_0}{m}\right) - gt$$

$$v_x = c \cos \theta \ln\left(\frac{m_0}{m}\right)$$

The v_z equation is similar in appearance to the previous section's Eq. (6.6), with a sine term added to determine the vertical component of the velocity vector for the vehicle's direction angle θ. The familiar $-gt$ term is also included, so we would expect the vertical speed to behave similarly to the speed in vertical flight (previous section). The x equation has a *cosine* term but no gravitational terms, so we may expect to see that motion in the x-direction (horizontal) does not experience gravity losses.

If the vehicle flies with a constant pitch angle $\theta = \theta_0$, we can calculate the magnitude of the resulting total speed by taking the root sum square (RSS) of the two component equations.

$$v = \left(v_z^2 + v_x^2\right)^{\frac{1}{2}} = [c^2 \ln^2\left(\frac{m_0}{m}\right)$$

$$- 2cg_0 t \sin \theta_0 \ln\left(\frac{m_0}{m}\right) + g_0^2 t^2]^{\frac{1}{2}}$$

Figure 6.5 provides curves showing vehicle speed as a function of mass ratio and inclination angle. Choosing a mass ratio $\mu = 10$ (achievable in practical engineering efforts), the vertical speed gain including gravity loss is only about 67% of horizontal gain. This suggests that to minimize gravity loss, *we want our vehicle to transition to horizontal flight as soon as possible.*

A "real" launch does not fly at a constant angle, so the previous equations are of limited utility and will not be considered further. For vertical launch and horizontal burnout or shutdown, the vehicle's pitch angle must somehow transition from 90 deg to 0 deg. This means we would have to calculate increments of speed and position as the angle changes continuously with time. This will be carried out in trajectory simulations described later in this chapter.

Many space launchers having upper steps with low thrust-to-weight ratios may utilize a technique called *lofting*, where the vehicle is boosted up to a high enough altitude that it can tilt below a 90-deg angle so as to accelerate horizontally while *falling* vertically. This is not a bad thing to do

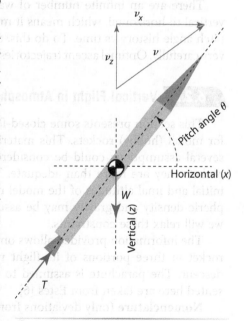

Fig. 6.4 Inclined flight in a vacuum. If the LV's pitch angle θ is less than 90°, part of the thrust accelerates the LV horizontally, and the vertical component of thrust works against gravity. This reduces the amount of thrust available to accelerate.

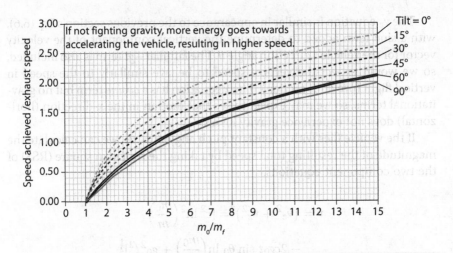

Fig. 6.5 Velocity gain vs pitch angle for vehicle with $c = 3,000$ m/s, varying mass ratios.

as long as the altitude is sufficient for the vehicle to accelerate up to the desired orbital speed during the "falling" time. The technique of lofting will be discussed later in this chapter.

There are an infinite number of ways to carry out the transition from vertical to horizontal, which means it may be possible to find an "optimum" pitch angle history vs time. To do this, the word *optimum* has to be defined very carefully. Optimal ascent trajectories will be discussed later in this chapter.

6.2.2 Vertical Flight in Atmosphere, with Gravity

This section presents some closed-form results that have been obtained for model (hobby) rockets. This material has limited applicability because several assumptions could be considered overly restrictive, but for model rockets they are more than adequate. The main assumptions are that the initial and final altitudes of the model rocket are small enough that atmospheric density and gravity may be assumed constant. In the next section, we will relax these constraints.

The information provided allows one to predict performance of a model rocket in three portions of its flight regime: boost, coast, and parachute descent. The parachute is assumed to open at apogee. The formulae presented here are taken from Estes [6].

Nomenclature (only deviations from the previous section are listed):

C_D = drag coefficient, where $D = C_D A\frac{1}{2}\rho V^2$
A = rocket's reference area, πR^2
D = aerodynamic drag of rocket
ρ = air density
$()_B$ = burnout
$()_C$ = coasting (no thrust)

$K = \frac{1}{2}C_D A\rho$ (a convenient constant)

S = altitude

S_B = burnout altitude (altitude at the time engine burns out)

S_{C0} = coast altitude (altitude gained between burnout and peak)

$S_{C\text{-general}}$ = altitude gained after arbitrary coast time

S_{max} = maximum possible altitude (burnout altitude plus coast to zero velocity altitude)

T = *average* thrust of the rocket engine; T = motor's total impulse ÷ its burn time.

t_B = burnout time

t_{C0} = coast time to apogee (time it takes to slow down to zero velocity)

V = velocity

V_B = burnout velocity (velocity obtained at the time the engine burns out)

W = rocket weight

6.2.3 Thrusting Equations

The equation of motion of a rocket under thrust, and *not* neglecting weight and drag, is

$$T - C_D A \frac{1}{2}\rho V^2 - W = m\frac{dV}{dt} \tag{6.13}$$

This can be separated into a differential equation as we did before:

$$t_B = m\frac{1}{\sqrt{K(T-W)}}\tanh^{-1}\left(V_B\sqrt{\frac{K}{(T-W)}}\right) \tag{6.14}$$

so we now have the time to burnout as a function of the burnout velocity.

We invert the equation to get velocity as a function of time

$$V_B = \sqrt{\frac{W}{C_D A \frac{1}{2}\rho}\left(\frac{T}{W}-1\right)} \cdot \tanh\left[g \cdot t_B\sqrt{\left(\frac{T}{W}-1\right)\frac{C_D A \frac{1}{2}\rho}{W}}\right] \tag{6.15}$$

Now we have burnout velocity as a function of burn time, but we need burnout altitude. This can be obtained by a second integration. We'll skip the details and provide the result

$$S_B = \frac{1}{g}\frac{W}{C_D A \frac{1}{2}\rho}\ln\left(\cosh\left(g\sqrt{\left(\frac{T}{W}-1\right)\frac{C_D A \frac{1}{2}\rho}{W}} \cdot t_B\right)\right) \tag{6.16}$$

6.2.4 Coasting Equations

The equation of motion of a rocket under no thrust, including weight and drag, is

$$- C_D A \frac{1}{2}\rho V^2 - W = m\frac{dv}{dt} \tag{6.17}$$

Without proof, the equations obtained from solving Equation 8 are:

Coast altitude (altitude when $V = 0$)

$$S_{CO} = \frac{1}{2g}\frac{W}{C_D A \frac{1}{2}\rho}\ln\left(1 + \frac{C_D A \frac{1}{2}\rho V_B^2}{W}\right) \tag{6.18}$$

Coast time (time to coast to zero velocity)

$$t_{CO} = \frac{1}{g}\sqrt{\frac{W}{C_D A \frac{1}{2}\rho}}\tan^{-1}\sqrt{\frac{C_D A \frac{1}{2}\rho V_B^2}{W}} \tag{6.19}$$

The maximum altitude that the rocket can achieve occurs when it reaches the apex of the trajectory with zero velocity, and can be calculated from

$$S_{\max} = S_B + S_{CO} \tag{6.20}$$

Because the rocket parachute will be ejected at a fixed time after burnout (either before or after the apex of the flight), the actual height will be less.

It is unlikely that the parachute will be ejected precisely at the top of the flight. The equation for the rocket's altitude after a fixed amount of *coast* time t, such as the model rocket engine's delay time, is

$$S_{C\text{-general}} = S_{c0} + \frac{W}{C_D A \frac{1}{2}\rho g}\ln\left(\cos\left[g\sqrt{\frac{C_D A \frac{1}{2}\rho}{W}}(t_c - t)\right]\right) \tag{6.21}$$

While this may seem like an oversimplification, these equations form the basis of more general, higher-fidelity flight equations. The previous information was used as the basis for a freshman astronautics class project where the students built solid-propellant model rockets and then attempted to predict the parameters of their rocket's flight.

6.3 General Flight with Gravity, Atmosphere Effects

We are now ready to look at flight dynamics for space launch vehicles in a more realistic way, where the atmospheric loads and the loads from motion through a moving atmosphere are incorporated, and we also allow variable flight path angle. This analysis has the potential to add many details that are pertinent to the flight of rockets and therefore their mechanical design:

- Aerodynamic forces and torques as a result of motion through the atmosphere
- Bending moments and shear forces due to flight winds (gusts, wind shears, jet streams) superimposed onto the quasi-static forces previously described

As was stated earlier, most LVs begin their flight in a vertical attitude, but they also need to be approximately horizontal at burnout or when injected into orbit. For now, we will not worry about the details of how we will obtain the needed trajectory time history, and we will see that there are some interesting factors to consider.

For preliminary design and analysis purposes, we make the following assumptions:

- The trajectory is confined to a plane passing through the center of the attracting body. All aerodynamic, gravitational, and inertial forces considered are in the plane of the trajectory, and no yawing motions are allowed. Gravitational attraction is assumed to vary as $1/r^2$ from the center of the planet.
- The ascent is assumed to be over a "flat planet," assuming that the altitude gained and downrange distances are very small fractions of the radius of the planet from which we're launching. Equivalently, the curvature of the planet's surface at these scales is negligible. We will apply a correction term to properly account for the curvature of a planet's surface.
- The flight path angle γ is a variable that can be controlled as needed to generate a specified trajectory.
- The planet is not rotating and is a sphere with uniform density. We will add the eastward speed of the launch site to any orbital motions. Perturbations to the flight path from the moon, the sun, and nearby planets are assumed negligible.
- The vehicle may be considered either a point mass or a rigid body when stability and control aspects are considered. Flexible vehicles are discussed in Chapter 12.
- An average specific impulse (the effective value between launch and vacuum) may be used to model the behavior of engines(s) during boost from low in the atmosphere to vacuum. This can be refined if details of the engines (exhaust pressure, exit area) are known.
- All aerodynamic forces and torques, along with any corresponding dimensionless coefficients, will have linear behavior, an assumption that is usually true at small angles of attack. The drag coefficient is assumed constant regardless of Mach number. (This may be refined later with detailed knowledge of the vehicle's C_D.)
- The behaviors of density and pressure properties of the atmosphere may be approximated as simple exponentials using an atmospheric scale height.

This set of assumptions will allow us to develop a relatively simple set of equations of motion that may be used for trajectory simulation and calculations.

6.3.1 Launch Vehicle Boost Trajectory Coordinate System

Normally, we would like the origin to be inertial and not accelerating, as a flying vehicle might. However, the choice of the moving vehicle's CM as origin makes it particularly easy to incorporate all of the applied loads on the vehicle into the equations of motion. This configuration is shown in Fig. 6.6. Note that the velocity vector is always tangent to the trajectory curve.

In general, analyses of statics, stability, and rigid-body dynamics of vehicle motion will require two types of mass properties:

1. Location of center of mass (CM), as described
2. Values of mass moments of inertia (MoIs) in a specified coordinate system

Because CM changes with time, it may be easier to select a fixed reference location and determine the instantaneous CM offset as a function of time. This may be easier to manipulate because it requires only a simple set of offsets to translate the properties about the new frame.

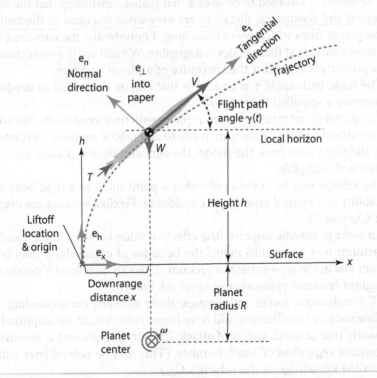

Fig. 6.6 Local vertical, local horizontal (LVLH) coordinates, also referred to as altitude and downrange coordinates.

6.3.2 Equations of Motion

Referring to Fig. 6.6, we can assemble the following description:

Our vehicle has velocity $\mathbf{v} = v\mathbf{e_t}$, where $\mathbf{e_t}$ is the *unit vector* in the tangential velocity direction by definition. The unit vector $\mathbf{e_n}$ is normal to (perpendicular to) velocity.

We use the symbol γ for flight-path angle; this angle is measured with respect to the local horizon.

The vector derivative rule gives the inertial acceleration $\mathbf{a}$ of the center of mass, and is readily found in most dynamics textbooks (e.g., [7]):

$$\mathbf{a} = \frac{d\mathbf{v}}{dt} = \left.\frac{d\mathbf{v}}{dt}\right|_{\text{in local frame}} + \vec{\omega} \times \mathbf{v}$$

This relation allows us to convert any quantity measured on the vehicle to an absolute quantity with respect to inertial space, as long as we know the instantaneous rotation rate $\boldsymbol{\omega}$ at each time point. The rate ω is easily measured by rate sensors or gyros mounted on board the moving vehicle. This formulation of measurements on a moving platform is very useful, because the sensors located on the platform may be used for all trajectory calculations.

In Fig. 6.6, the value of $\vec{\omega} = \dot{\gamma}\mathbf{u_\perp}$, where $\mathbf{u_\perp}$ is a unit vector pointing into the paper. Using this,

$$\vec{\omega} \times \vec{v} = \dot{\gamma}\mathbf{u_\perp} \times v\mathbf{u_t} = -\dot{\gamma}v\mathbf{u_n}$$

The absolute acceleration $\mathbf{a} = \dot{v}\mathbf{u_t} - \dot{\gamma}v\mathbf{u_n} = \mathbf{a_t} + \mathbf{a_n} = a_t\mathbf{u_t} + a_n\mathbf{u_n}$. Comparing, we find that the acceleration in the tangential or *velocity* direction is

$$a_t = \dot{v} = \frac{dv}{dt}$$

and the normal acceleration that is *perpendicular to* the velocity direction is

$$a_n = -\dot{\gamma}v = -v\frac{d\gamma}{dt}$$

One last step remains before we perform the assembly: to account for the planet's curvature using "flat surface" local vertical, local horizontal (LVLH) coordinates, we need to add a *centrifugal force* term. (Remember, centrifugal force is simply the centripetal acceleration term located on the "wrong" side of the equation.) The added term corresponds to the centripetal acceleration of the vehicle's coordinate system around the circular planet (recall centripetal acceleration $= v^2/r$).

$$a_n = -v\frac{d\gamma}{dt} + \frac{v^2}{R_e + h}\cos\gamma$$

where R_e is the planet's radius and h is the height about the planet's surface. For a detailed derivation of the centrifugal term of the normal acceleration, the reader is referred to Sec. 1.7 of [25].

6.3.3 Forces and Torques on a Launch Vehicle Due to Aerodynamic, Thrust, and Steering Forces

Now we have accelerations in the LVLH system. We need to obtain the forces on the vehicle, then use Newton's law: $\sum$ forces = mass $\times$ acceleration, in both the tangential and normal directions. So, let's look at the forces acting on a launch vehicle.

As shown in the previous section, we have to identify which forces and their associated directions would be assigned to the normal and tangential directions. We can do this by referring to Fig. 6.7.

The reader should note that this figure contains some new information, particularly several forces and torques that we need to consider. The primary change is that the vehicle's velocity vector is no longer parallel to its long axis. The angular difference between the two directions is known as the *angle of attack* α (AoA) and will ultimately be the source of aerodynamic forces that produce internal shear forces, bending moments, and torques and will become design drivers for the vehicle's structure.

The vehicle's rocket engine(s) produce(s) thrust whose magnitude depends on the mass flow $\dot{m}$ exhaust speed v_e, exhaust pressure P_e, and ambient pressure P_∞.

$$T = \dot{m}v_e + A_e(P_e - P_\infty).$$

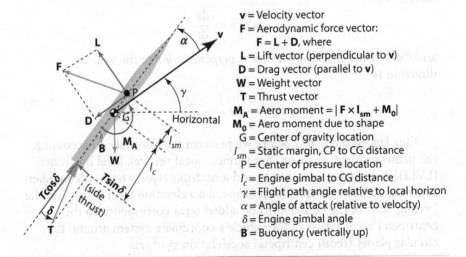

$\mathbf{v}$ = Velocity vector
$\mathbf{F}$ = Aerodynamic force vector:
 $\mathbf{F} = \mathbf{L} + \mathbf{D}$, where
 $\mathbf{L}$ = Lift vector (perpendicular to $\mathbf{v}$)
 $\mathbf{D}$ = Drag vector (parallel to $\mathbf{v}$)
$\mathbf{W}$ = Weight vector
$\mathbf{T}$ = Thrust vector
M_A = Aero moment = $|\mathbf{F} \times \mathbf{l}_{sm} + M_0|$
M_0 = Aero moment due to shape
G = Center of gravity location
l_{sm} = Static margin, CP to CG distance
P = Center of pressure location
l_c = Engine gimbal to CG distance
γ = Flight path angle relative to local horizon
α = Angle of attack (relative to velocity)
δ = Engine gimbal angle
B = Buoyancy (vertically up)

Fig. 6.7 Aerodynamic (lift, drag), thrust, and gravitational forces (weight and buoyancy), and torques acting on a launch vehicle in the atmosphere.

Later we shall show how to obtain the ambient pressure. Alternatively, the thrust may be calculated from

$$T = \dot{m}c = \dot{m}v_{eq}$$

if the effective exhaust speed c or v_{eq} is known.

The weight force $\mathbf{W}$ ($\mathbf{W} = m\mathbf{g}$) pulling the vehicle back towards the ground acts at the vehicle's center of mass (CM). The CM's location G is at the tail-end of the gravitational force $\mathbf{W}$ vector arrow, and is shown in the figure with the ◉ symbol. Notice that the magnitude of the vehicle's weight W changes for two reasons: first, the weight changes rapidly due to propellant consumption, $m = m_0 - \dot{m}t$. However, we also have to consider the reduction in the magnitude of g, the acceleration due to gravity, due to the changing distance from the Earth's center. To account for the latter effect, we write

$$g(h) = \frac{g_0}{\left(1 + \dfrac{h}{R}\right)^2},$$

where g_0 is the gravitational acceleration on the surface of the planet ($h = 0$).

We show the generalized aerodynamic force $\mathbf{F}$ assumed to be generated by the vehicle's flight at an AoA and effective at a location P that we call the *center of pressure (CP)*, forward of the CM location G. The aerodynamic force $\mathbf{F}$ may be resolved into two perpendicular components: $\mathbf{L}$, a lift force whose direction is perpendicular to the speed vector $\mathbf{v}$, and a drag force $\mathbf{D}$ in the direction opposite to the speed vector. These lift and drag forces correspond to the same quantities in aircraft flight, perpendicular and antiparallel to the velocity vector's direction. Generally speaking, if the vehicle flies at zero angle of attack, the lift force becomes negligible, and the drag force points down the long axis of the vehicle. The calculation of the vehicle's lift and drag forces will be shown later in this chapter.

An aerodynamic moment $\mathbf{M_A}$ is produced by the crossproduct of the force $\mathbf{F}$ and the vector $\mathbf{l_{sm}}$: $\mathbf{M_A} = \mathbf{F} \times \mathbf{l_{sm}}$. If the vehicle is not symmetrical, there may be an additional moment $\mathbf{M_0}$ produced at zero AoA that must be added to M produced by the force $\mathbf{F}$.

The vehicle shown in Fig. 6.7 may be characterized as *unstable*, because a disturbance from zero angle of attack produces a moment that tends to *increase* the AoA rather than reduce it as a stable vehicle would. An unstable vehicle will try to turn away from a horizontal gust of wind. Contrast this with a *stable* vehicle, whose CP lies *behind* its CM. This vehicle, if disturbed by a gust of wind, would exhibit a stabilizing force that would turn it *towards* the wind. Most sounding rockets have stabilizing fins and do not have active steering systems. Most space launchers have no fins and are unstable, so some sort of stabilizing system is required to keep the vehicle from diverging from its trajectory.

Another specialization of Fig. 6.7 indicates that the thrust force vector may be gimbaled, or rotated an angle δ back and forth. This rotation produces a side force, which is used as an effector or actuator producing a rotational torque M within a *thrust vector control (TVC)* system. The inclusion of a TVC system usually means that the vehicle is unstable without a control system. The vehicle in the figure is therefore able to fly at a nonzero angle of attack, which we will assume to be small and in linear ranges for force computations.

Figure 6.7 shows the vehicle's *buoyancy force* as the vector **B**. This the force produced by the vehicle's parts displacing their equivalent volume of atmosphere, just as a piece of wood floats in water because it displaces a volume of water. As might be expected, for very small or dense parts, the effect is small and negligible; however, for larger parts such as a low-density tank, the effect is not so small. As an example, consider the retired Space Shuttle's external tank (ET) with a volume of 2,200 m^3 (77,700 ft^3). We can calculate that the sea level buoyancy of the tank in air produces an effective lifting force of 26,000 N or ~6,000 lb$_f$! Author Edberg was once told that an investigation was made into a performance deviation noticed during several launches of the Space Shuttle, where it was ultimately determined that when the missing buoyancy force was properly included in the simulations, they agreed *exactly* with the observed behavior. Note that the buoyancy force would be applied to the vehicle's center of buoyancy, just as we apply weight forces to the center of mass and aero forces to the center of pressure.

We group the forces into those that are largely parallel to the vehicle's long axis and those that are perpendicular to its long axis. The forces will be assembled into the equations at the appropriate places.

6.3.4 Launch Vehicle Force (and Torque) Balance

Considering the forces in Fig. 6.7, finding the tangential and normal components of the forces, and comparing to the tangential and normal accelerations, we find the following:

A force balance in vehicle coordinates yields, respectively, the following tangential and normal equations:

$$ma_t = m\frac{dv}{dt} = T\cos(\alpha + \delta) - D - mg\sin\gamma \qquad (6.22)$$

$$ma_n = mv\frac{d\gamma}{dt} + \frac{mv^2\cos\gamma}{R_e + h} = L - mg\cos\gamma + T\sin(\alpha + \delta) \qquad (6.23)$$

A third, rigid-body rotation equation may be generated from summing moments in the vehicle's *pitch* plane (trajectory plane) about the CM

$$M - l_{CG}T\sin\delta + l_{ACL}\cos\alpha = I_{pitch}\frac{\partial^2\theta}{\partial t^2} \qquad (6.24)$$

This pitch-plane equation would be solved for engine gimbal angle δ vs time to achieve a desired pitch angle profile or for static trim calculation.

If we assume the vehicle is a point mass, then we are basically assuming that the rocket is implicitly steered properly to achieve the desired trajectory and that the dynamics are not of concern at this time.

If α and δ are small ($\ll 1$), we can neglect side thrust (which basically means that you don't have to worry about its small side force contribution). The normal and tangential equations become

$$ma_t = m\frac{dv}{dt} = T - D - mg\sin\gamma \tag{6.25}$$

$$mv\frac{d\gamma}{dt} = L - mg\cos\gamma - \frac{mv^2\cos\gamma}{R_e + h} \tag{6.26}$$

These are the 2-DoF flight trajectory equations for motion in a plane. Additional equations exist for yaw motion, as well as the rotation of the vehicle on its long axis (roll); these will be discussed in Chapter 12.

Note: properly executed, a trajectory simulation would require six degrees of freedom (6 DoF): three equations for translation and three for rotations. This, of course, can get very complicated, so we assume that we are only operating in the pitch plane, and therefore only need two translations—vertical and downrange—and one rotation—pitch. A more accurate simulation would also include the Earth's curvature and its rotational speed, making things much more complex. For a proper treatment of a realistic launch simulation, please refer to Chapter 9 of Ashley [1].

6.4 Aerodynamics of Launch Vehicles

The LV equations of motion incorporate a number of aerodynamic forces. We will assume that all aerodynamic forces F (units F = force) may be converted to a nondimensional coefficient by the following relation: $C_F = F/(q\ S_{ref})$. In this standard definition, C_F = nondimensional force coefficient, $q = \frac{1}{2}\rho v^2$ is the dynamic pressure (units F/L^2), and S_{ref} is a reference area (units L^2). The reference area is based on the largest cross-sectional area along the length of the vehicle. The determination of the force coefficient and the atmospheric density are described in the following sections.

General torques or moments M may also be nondimensionalized. In this case, we write $C_M = M/(qL\ S_{ref})$. Here the moment is in units of $[F \cdot L]$, q and S_{ref} are as before, and the length normalization L is usually the vertical height of the vehicle. (Sometimes maximum outer diameter D is used instead, so carefully read any references that discuss scaling factors.) One comment is that a vehicle's dimensions may change during the design process, so be sure to either (1) maintain the original reference dimension, even if it changes; or (2) keep the dimension up-to-date with endless revisions by making sure that *every single document* is kept updated for every change, ever. Maintaining a single reference area is a lot less work.

6.4.1 Assessment of Launch Vehicle Drag

There are several ways to determine the drag of a launch vehicle. Unfortunately, some of these methods are more difficult, expensive, or time-consuming compared to others, and there are particular problems assessing drag around $M = 1$ to 1.5 or so.

One tried-and-true method is to do wind tunnel (WT) testing. As long as the WT model is built with great care to simulate each and every protruding part, then there is a chance to get a useful value of drag, with more caveats:

* The model testing must occur at nearly the same Reynolds number (RN) as the flight vehicle so that the test conditions match those of flight as much as possible. Note that RN = VL/v, where V = speed, L = characteristic length, and v = kinematic viscosity of atmosphere. If a scale model is to be tested, a high-pressure WT facility may be needed to maintain RN.
* Best results are obtained when the WT model includes a simulation of the exhaust plume of the vehicle, which also has an effect on vehicle drag.
* Off-nominal drag measurements (for instance, AoA varying between ±10 deg) must also be made to account for behavior in wind gusts and the like.

Wind tunnel test plots of drag coefficients measured for the V-2, Vanguard, and Saturn C-1 are shown in Figs. 6.8, 6.9, and 6.10, respectively.

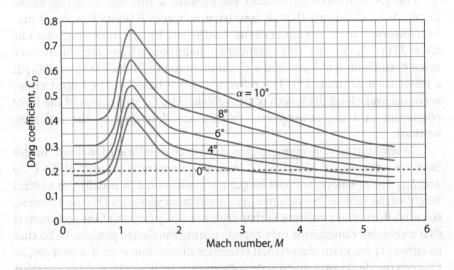

Fig. 6.8   V-2 C_D vs Mach and α, based on cross-sectional area, with no exhaust plume. Source: [19].

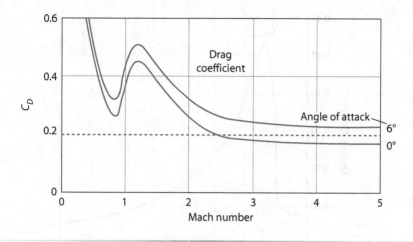

Fig. 6.9 Vanguard C_D vs Mach and AoA of 0 deg and 6 deg. Source: [9].

Information on other vehicles may be found by library and/or online searching, or by resorting to a drag prediction program such as Missile DATCOM [22]. The generic drag coefficient plot in Fig. 6.11 comes from Dergarabedian [5].

There is a rule of thumb regarding missile C_D values. If you look at the plots for Figs. 6.8 to 6.10, the minimum drag coefficient tends to be around 0.2 for much of the flight envelope.

As far as actual incorporation of these drag values, ideally one would create a "drag table" relating the speed or Mach number to generate the

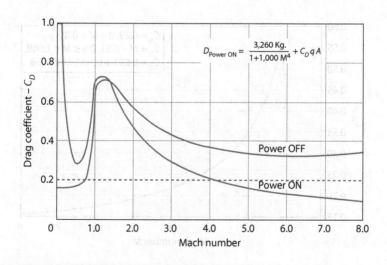

Fig. 6.10 Saturn C-1 C_D vs Mach, $\alpha = 0$, power on and power off. Source: [12].

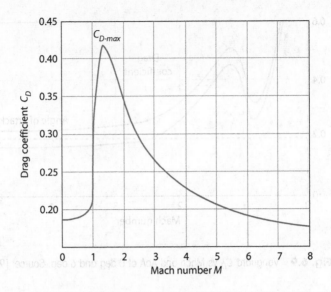

Fig. 6.11 Generic drag coefficient curve for launch vehicles. Source: [5].

C_D at varying Mach numbers. Such analytical descriptive equations for a Titan II vehicle are shown in Fig. 6.12, taken from Linshue [10].

6.4.2 Assessment of Launch Vehicle Lift

Similar to the assessment of drag, there are several ways to determine the *lift* a launch vehicle may generate. Wind tunnel testing is commonly used as

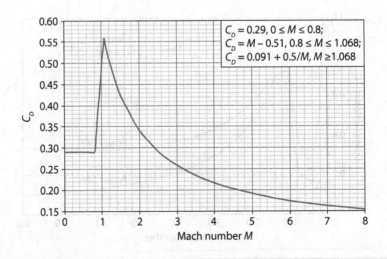

$$C_D = 0.29, 0 \le M \le 0.8;$$
$$C_D = M - 0.51, 0.8 \le M \le 1.068;$$
$$C_D = 0.091 + 0.5/M, M \ge 1.068$$

Fig. 6.12 Titan II drag coefficient curve and analytical description equations. Source: [10].

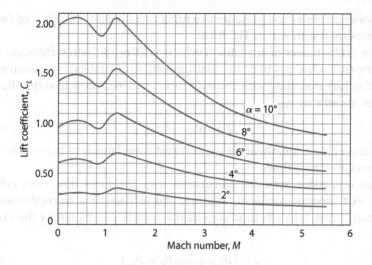

Fig. 6.13   V-2 C_L vs Mach and α, based on cross-sectional area, with no exhaust plume.
Source: [19].

for lift assessment, along with the same caveats. A wind tunnel lift coefficient plot for the German V-2 is shown in Fig. 6.13. As with the drag coefficients, the values of these coefficients reach a maximum slightly above Mach $M = 1$.

6.4.3   Ascent Aerodynamic Forces, Propulsion Models, and Gravity

We are now at the stage where we can create a simulation of a launch ascent. We have seen that there are no closed-form solutions for this problem, so it must be calculated numerically. In this section, we will assemble the equations governing each of the vehicle's parameters needed to simulate the ascent.

The vehicle's drag force D is calculated via the usual definition of C_D

$$D = C_D S_{\text{ref}} \frac{1}{2} \rho v^2 \qquad (6.27)$$

We have previously seen how to calculate or obtain the drag coefficient C_D; however, we need to use the instantaneous values of the atmospheric density ρ, and v, its speed. S_{ref} is already defined.

For these calculations, it is expedient to use a simple exponential atmosphere density model. The model assumes air density ρ decreases exponentially with altitude h

$$\rho(h) = \rho_0 e^{-h/h_0} \qquad (6.28)$$

ρ_0 is the density at $h = 0$ (1.225 kg/m^3 on Earth), and h_0 is the scale height, approximately 7.64 km on Earth. With these values, one may calculate the atmospheric density and pressure as a function of height $h(t)$. Plots of density and pressure are shown in Fig. 6.14.

The prime mover of any launch vehicle is its propulsion system, generating a thrust T. One way to do calculations with the engine(s) is to assume that it (or they) are operating with constant thrust and mass flow $\dot{m}$ at the effective exhaust speed c or v_{eq}.

$$T = \dot{m}c = \dot{m}v_{eq} \tag{6.29}$$

In this calculation, the increase in thrust from launch to burnout is averaged over the burn time.

Another, more accurate, calculation uses the mass flow rate, exhaust speed, exit pressure, ambient pressure, and exhaust area to recalculate the (varying) thrust for each time step as the vehicle climbs out of the atmosphere.

$$T = \dot{m}v_e + A_e(P_e - P_\infty). \tag{6.30}$$

P_e, the engine's exhaust pressure, is fixed. The ambient pressure P_∞ is also calculated assuming the ambient pressure drops exponentially with height.

$$P_\infty(h) = P_0 e^{-h/h_0} \tag{6.31}$$

P_0 is the pressure at altitude $h = 0$ (101.325 kPa or 101,325 N/m^2 on Earth). The latter thrust calculation may be difficult to carry out if the engines' parameters are not known.

Most launch vehicles are designed to be strong in their axial direction (long and skinny) in order to withstand acceleration from thrust. At the same time, many are more flexible in bending, which saves mass and improves the vehicles' performance. In Chapter 9 we will carry out a loads analysis for an in-flight vehicle; for now, please accept that flight through wind shear conditions leads to body-bending modes and increased dry mass to reduce bending stresses. Therefore, we normally design vehicles to fly through the atmosphere with as small an angle of attack as possible to

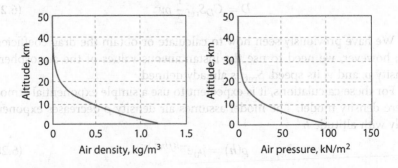

Fig. 6.14 Exponential behavior of Earth atmosphere; scale height $h_0 \approx 7.64$ km.

minimize lift, which minimizes body-bending moments and their associated mass increases.

Let us assume that any lift forces L and side thrusts due to engine gimbaling are negligible. With this assumption, the final result needed for trajectory simulation is a set of ordinary differential equations, as shown in Table 6.1. Many variables depend on speed v and/or atmospheric density (which depends on altitude h); drag depends on Mach number (see the drag coefficient plots earlier in the chapter). The flight path angle γ may be calculated via the perpendicular force equation (second line), causing a "gravity turn"— this means that the trajectory turns automatically and no outside steering or control of the flight path angle is needed. Alternately, the flight path angle γ may be explicitly prescribed (*explicit guidance*), in which case the equation in the second line in the table is not used.

Before we employ these equations for simulation, there are several related concepts that have to be discussed. One such topic considers the efficiency of a launch process, and we shall learn how to compute such things as gravity, drag, and propulsion losses and view them graphically. Then we'll consider how to control a vehicle's flight path angle.

Table 6.1   The Set of Ordinary Differential Equations That Can Be Numerically Integrated to Yield Trajectories

$$\frac{dv}{dt} = \frac{T}{m} - \frac{D}{m} - g\sin\gamma$$

$$\frac{d\gamma}{dt} = -\left(\frac{g}{v} - \frac{v}{R+h}\right)\cos\gamma$$

$$\frac{dh}{dt} = v\sin\gamma$$

$$\frac{dx}{dt} = \frac{R_E}{R_E + h}v\cos\gamma$$

$m(t) = m_0 - \dot{m}t$; mass flow $\dot{m} = \dfrac{T}{g_0 I_{sp}}$

$D = C_D S_{ref}\dfrac{1}{2}\rho v^2$; C_D is a function of Mach no.

$T = \dot{m}v_e + (P_e - P_\infty)A_e = \dot{m}c$

$P_\infty(h) = P_0 e^{-h/h_0}$ [$P_0 = P(h=0)$ = surface pressure 101,325 Pa for Earth]

$\rho(h) = \rho_0 e^{-h/h_0}$ [$\rho_0 = \rho(h=0)$ = surface density 1.225 kg/m^3 for Earth]

h_0 = scale height [approximately 7.64 km for Earth]

$g(h) = \dfrac{g_0}{(1 + h/R_E)^2}$, R_E = planet radius, g_0 = surface gravity = 9.80665 m/s^2 for Earth

6.4.4 Speed or Energy Losses During Launch

For the simple case of the vertical launch presented earlier in this chapter, we showed that the actual speed achieved was less than the ideal max speed, and the difference had to do with how long the burn time was. We called this a *gravity loss* and showed how it reduced system efficiency. Let's see if we can come up with a general way to calculate gravity losses during flight.

Let's begin with the tangential speed equation with the centrifugal term omitted.

$$\dot{v} = \frac{T \cos \delta}{m} - \frac{D}{m} - g \sin \gamma, \quad D = C_D A \frac{1}{2} \rho v^2 \qquad (6.32)$$

Now assume exponential atmosphere, with scale height h_0.

Density: $\rho = \rho_0 e^{-h/h_0}$, $\rho_0 = \rho(0) = $ density when $h = 0$

Pressure: $P_\infty = P_0 e^{-h/h_0}$, $P_0 = P(0) = $ pressure when $h = 0$ (6.33)

Also, recall that the definition of power (work per unit time) is

$$P_{\text{delivered}} = \int_0^{t_f} \mathbf{T} \cdot \mathbf{v} dt = \int_0^{t_f} Tv \cos \theta \, dt \qquad (6.34)$$

where $\theta = $ angle between thrust and velocity.

Now, we can rewrite the parallel thrust component as

$$T \cos \delta = T - T(1 - \cos \delta) \qquad (6.35)$$

Using the trigonometric identity $1 - \cos \delta = 2 \sin^2 \delta/2$ and the definition of thrust $T = \dot{m}c$, we can formally integrate to get

$$v_f = \underbrace{c \ln \frac{m_0}{m_f}}_{\substack{\text{Ideal speed} \\ \text{(Rocket eq.)}}} - \underbrace{2T \int_0^{t_f} \frac{\sin^2 \delta/2}{m} dt}_{\substack{\text{Steering} \\ \text{Loss}}} - \underbrace{\int_0^{t_f} \frac{D}{m} dt}_{\substack{\text{Drag} \\ \text{loss}}} - \underbrace{\int_0^{t_f} g \sin \gamma dt}_{\substack{\text{Gravity} \\ \text{Loss}}}$$

This result indicates that the delivered final speed v_f (also referred to as Δv) is the ideal speed (from the rocket equation) with the following items subtracted: gimbaled engine steering loss, drag loss, and gravity loss. This suggests that we fly a trajectory that minimizes the sum of the three losses. Some typical values for these losses were provided in Chapter 3.

The *steering loss* may be attributed to the reduction of power transfer suffered when an engine is producing thrust that is not aligned with the velocity vector. (Recall that instantaneous thrust power $P = \mathbf{T} \cdot \mathbf{v} = Tv \cos \theta$, so any angle between the two produces less effective power than if they are parallel with $\theta = 0$). The closer the angle θ is to zero, the higher the propulsive efficiency of the step is.

One example of a steering loss has to do with the TVC system as shown in Fig. 6.7. In the figure, it can be seen that due to gimbaling of the engine, the thrust line does not line up with the velocity vector and maximize the power transfer. This loss may also be generated by a *lofted trajectory*, which will be discussed later in this chapter. During portions of lofting trajectories, the vehicle is pitched upward so that a portion of the thrust works against gravity, while the remainder

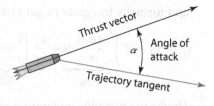

Fig. 6.15 Steering losses occur when the thrust line is not parallel to the velocity vector, which is the tangent line to the trajectory. This occurs when launch vehicles loft.

accelerates the vehicle horizontally. Referring to Fig. 6.15, the larger the angle between the thrust line and the velocity vector, the larger the "cost," or steering loss, which depends on the cosine of the angle between the thrust vector and the velocity vector.

The *drag loss* has to do with the work done by the LV in forcing it through a draggy atmosphere. Recall that instantaneous thrust power $P = \mathbf{T} \cdot \mathbf{v} = Tv \cos \theta$ is reduced by the drag power $P_{drag} = \mathbf{D} \cdot \mathbf{v} = Dv$ because the drag force is always parallel to the velocity vector. Obviously the power needed to overcome drag reduces the power available for other things, which subtracts from the thrust available, reducing the delivered acceleration.

The *gravity loss* is similar to the one found for a vertical launch (and in fact, becomes identical to that term if $\gamma = 0$ deg). This loss, similar to a drag loss, indicates that part of the LV's thrust is consumed, increasing the potential energy associated with the mass of the rocket and its propellants. The longer the period of time that the thrust is fighting gravity, the larger the gravity loss.

6.4.5 Changing the Flight Path Angle γ

It's instructive to consider the LV's flight path angle γ in a similar fashion. Let's see if we can generically determine the contributors to the change of flight path angle during flight. Let's start by considering the following "normal" equation of motion:

$$\dot{\gamma} = \frac{T \sin \delta}{mv} + \frac{L}{mv} + \left(-\frac{g}{v} + \frac{v}{R+h}\right) \cos \gamma$$

$$L = C_L S \frac{1}{2} \rho v^2 = \kappa_L \rho v^2,$$

where $\kappa_L = \frac{1}{2} C_L S$, a constant.

and formally integrate to get (assuming exponential atmosphere)

$$\gamma_f = \gamma_0 + \underbrace{T\int_0^{t_f}\frac{\sin\delta}{mv}dt}_{\text{steering}} + \underbrace{\frac{\kappa_L}{h_0}\int_0^{t_f}ve^{h/h_0}dt}_{\text{lift}} - \underbrace{\int_0^{t_f}\frac{g}{v}\cos\delta\,dt}_{\text{gravity}} + \underbrace{\int_0^{t_f}\frac{v\cos\gamma}{R+h}dt}_{\text{centripetal}}$$

Unlike the previous integration of the speed equation, these four terms—steering, lift, gravity, and centripetal—are not actual losses. They are simply the *changes* in the flight path angle due to those terms. The final flight path angle, γ_f, is equal to the initial angle (γ_0, often 90 degrees) with the addition of the four parts shown. (Presumably their sum would be negative and produce a final flight path angle that approaches 0 deg.) The second (steering) term is the LV's active control [thrust vector control (TVC)].

The lift term, third indicates that positive lift can increase the flight path angle. Turning things the other way, negative lift can be used to reduce the flight path angle without using TVC. Most vehicles are symmetric and produce little lift, but the Space Shuttle's wing could be used to produce the negative lift needed—if the Shuttle were to turn upside down during launch. This is actually done with the Shuttle, whose dramatic roll maneuver just after liftoff placed the Shuttle in a heads-down attitude. Generally speaking, lift can be used for this purpose, but one must be extremely careful to not exceed allowable loads, which is more challenging the faster the LV travels through the atmosphere. This procedure is discussed in some detail later in this chapter. For the rest of this text, we will assume that lift will not be used to make changes to the flight path angle.

The fourth term, gravity, automatically reduces the flight path angle without any active steering needed. This effect from gravity is the origin of a gravity turn trajectory, which occurs naturally and is discussed later in this chapter, section 6.4.7.

6.4.6 Events During Liftoff and Ascent

Here is a short list of maneuvers and events that occur during a typical launch. Some of these maneuvers are illustrated in Fig. 6.16.

1. *Vertical climb*: The LV gains altitude vertically to clear its tower or other nearby structures before making the maneuvers following. Note: Occasionally an LV will make a small maneuver immediately after liftoff so as to increase clearance from nearby structures, as described in Chapter 14.

2. *Roll maneuver*: The vehicle rotates along its long axis so that its pitch heading is in the proper azimuth for the desired orbital plane (not shown in figure). One vehicle with a rather spectacular roll maneuver is the Space Shuttle.

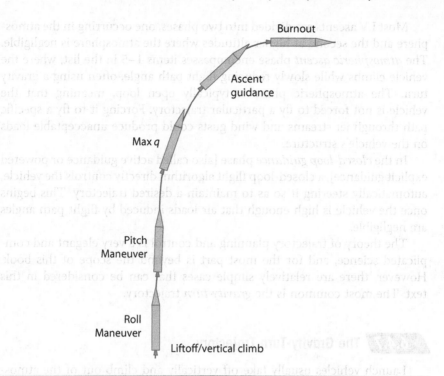

Fig. 6.16   Some of the maneuvers that occur during launch.

3. *Pitch maneuver or program*: The LV's attitude control method (most often TVC) "nudges" or "kicks" the LV slightly away from vertical to initiate a gradual trajectory turn. This may be a gravity turn (see the next section) where the force of gravity naturally turns the LV towards horizontal. At the same time, the LV is flying directly pointed into its velocity vector, which minimizes the angle of attack and thus the air loads on the LV.

4. *Max q (maximum dynamic pressure $q = \rho v^2/2$)*: This is the time along the trajectory when the LV experiences the largest magnitude air loads, because they are defined by force coefficients multiplied by q and a reference area.

5. *Wind shear, max air, max-qα*: Here the LV flies into a horizontal gust, which makes an effective angle of attack α (not in figure). This, along with max-q, can be the defining aerodynamic loads condition. Interestingly, the wind gusts are often jet streams that often occur at max-q or max-$q\alpha$.

6. *Ascent guidance*: The ascent is automatically guided onto a desired trajectory. This is often called powered explicit guidance (PEG).

7. *Burnout or shutdown*: The vehicle is injected into its orbit, and there is no more propulsion.

Most LV ascents are divided into two phases, one occurring in the atmosphere and the second at higher altitudes where the atmosphere is negligible. The *atmospheric ascent* phase encompasses items 1–5 in the list, where the vehicle climbs while slowly reducing flight path angle, often using a gravity turn. The atmospheric phase is typically open loop, meaning that the vehicle is not forced to fly a particular trajectory. Forcing it to fly a specific path through jet streams and wind gusts could produce unacceptable loads on the vehicle's structure.

In the *closed-loop guidance* phase [also called active guidance or powered explicit guidance], a closed-loop flight algorithm directly controls the vehicle, automatically steering it so as to maintain a desired trajectory. This begins once the vehicle is high enough that air loads induced by flight path angles are negligible.

The theory of trajectory planning and control is a very elegant and complicated science, and for the most part is beyond the scope of this book. However, there are relatively simple cases that can be considered in this text. The most common is the *gravity-turn* trajectory.

6.4.7 The Gravity-Turn Trajectory

Launch vehicles usually take off vertically and climb out of the atmosphere nearly vertically while building up speed. However, at orbit injection they must be flying approximately horizontal. When in the thin upper atmosphere, the trajectory slowly "bends over" to build up horizontal speed. One way to gradually transition from vertical to horizontal flight is called a *gravity-turn* trajectory.

A gravity-turn trajectory is useful because it happens automatically once the initial pitch rate is provided, and no thrusters besides the engines themselves are needed. This also means that the LV spends most of its turn at zero angle of attack, which reduces the air loads to which the vehicle is subjected.

At liftoff, the LV is vertical, and its flight path angle $\gamma = 90.00$ deg exactly. We can rearrange the normal acceleration equation to obtain the pitch *rate*

$$\dot{\gamma} = -\left(\frac{g}{v} - \frac{v}{R+h}\right)\cos\gamma \tag{6.36}$$

By inspection we can see that the pitch rate $\dot{\gamma}$ is zero unless $\gamma \neq 90$ deg. Therefore, the LV continues climbing vertically. If we wish to initiate a gravity turn, we need to give it a little nudge so that γ is no longer 90 deg. We usually wait a few seconds until the vehicle climbs above ground objects such as launch towers, lightning posts, and the like. When clear, a pitch maneuver, pitch program, or pitch kick produces an initial flight path angle γ_0 a bit less than 90 deg to initiate pitch rate. This will cause the pitch angle γ to continue to decrease ("fall over") naturally at a rate dictated by Eq. (6.36). Selecting a few variables as an example: $h = 2$ km above Earth, $\gamma = 85$ deg, $v = 110$ m/s. With these, $\dot{\gamma} = -0.00777$ rad/s $= -0.44$ deg/s.

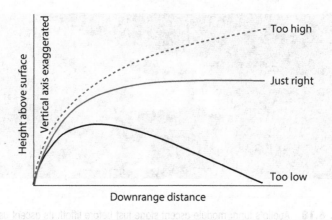

Fig. 6.17   Gravity turn trajectories are *very* sensitive to the initial kick angle, so we use the method of shooting to find the "Goldilocks" kick angle—not too high, not too low, *just right!*

Remember that no automatic pitch rotation or falling over will happen unless the pitch angle $\gamma \neq 90$ deg. Once initiated, the LV continues turning on its own and flies with an angle of attack $\alpha \approx 0$ in the atmosphere, which minimizes air loads. The small angle of attack is a beneficial side effect of a gravity turn trajectory.

As the LV's speed increases, the coefficient of $\cos \gamma$ in the pitch angle equation decreases, and the rate of change of the flight path angle tends towards zero as LV approaches orbital speed; it is exactly zero at orbital speed. This works well because we want the LV flying approximately horizontally ($\gamma \approx 0$) when orbital speed is reached (for a circular orbit injection). A family of gravity-turn trajectories is shown in Fig. 6.17.

Because the gravity turn occurs over a relatively long time period, it is very sensitive to small changes in the initial kick angle. That makes it a bit difficult to find the desired value of γ_0. However, one can use the method of "shooting" to converge on the answer. Similar to shooting a gun, one adjusts γ_0 to be larger or smaller depending on whether the terminal state is too high (>0 deg) or too low (<0 deg).

Gravity turns were used successfully for Apollo's two-astronaut lunar module in its ascent from the moon to rendezvous with the command module. The still photo in Fig. 6.18 is from a video of Apollo 17's lunar module launch and ascent. During the video, you can hear the astronauts' "Pitchover!" callout. This is in reaction to the pitch kick executed by the control system.

6.4.8 Other Methods of Guidance

There is nothing sacred about guidance. The only considerations for guidance algorithms are structural (minimize angles of attack during flight in atmosphere), efficiency (it may not be obvious whether a selected guidance scheme is optimal in terms of payload delivery), and ease of use. The LM

Fig. 6.18 Apollo's lunar module ascent stage just before liftoff. Its ascent used a gravity-turn trajectory. Source: NASA.

ascent problem at the end of this chapter can be carried out with a gravity-turn trajectory but is somewhat challenging. Instead, the reader could use a guidance program that doesn't happen automatically (as does the gravity turn) but instead needs to be scheduled for the details of the ascent.

Another means of solution would be to use a *constant flight path angle rate (CFPAR)* trajectory for the LM problem. It might be applied to the second portion of the ascent after the gravity turn algorithm had commenced. With some trial and error, it was possible to create a change in flight path angle that transitioned to zero at the time orbital speed was expected to be achieved. You can find more information on CFPAR and constant pitch rate algorithms in Walter [23].

Another experiment that can be considered is to *throttle* the engine, in addition to adjusting the flight path angle. Again, this requires a bit of iteration and experimentation, but can be done. Using the methods mentioned in the previous examples—gravity turn, CFPAR, and throttling—will give some insights into just how difficult it is to achieve required performance, (i.e., burnout speed and altitude).

We will now mention some other steering laws. One common law is the *linear tangent* steering law. This control law is described in Perkins [17] and works to make the tangent of the flight path angle change linearly from the point at which insertion guidance starts until circular orbit is achieved. In equation form, the pitch angle $\theta(t)$ is found from

$$\tan\theta = \left(1 - \frac{t}{t_f}\right)\tan\theta_0 \qquad (6.37)$$

where θ_0 is the initial pitch angle, and t_f is the duration of the boost. It can be shown that the linear tangent law is optimal for orbital insertion of a launch vehicle [20]. For details on its application to the Saturn launch vehicles, refer to NASA [14].

Another steering law is known as *powered explicit guidance (PEG)*. Descriptions of this law are available from Orbiterwiki (http://www.orbiter wiki.org/wiki/Powered_Explicit_Guidance) and Kerbal (https://forum.- kerbalspaceprogram.com/index.php?/topic/142213-pegas-powered-explicit- guidance-ascent-system-devlog).

Detailed considerations of linear tangent guidance and PEG laws are beyond the scope of this book; for detailed information, please consult the References and the recommendations for further reading at the end of this chapter.

6.4.9 The General Ascent Problem

Generally speaking, each SLV launch needs to start with a set of specified initial conditions and finish with a second (terminal) set of conditions that specifies the orbit or trajectory in which the SLV/payload must be placed.

Initial conditions (time $t = 0$) may be specified as follows:

$v(0) = v_0 = 0$ (or the launch site/carrier vehicle speed)
$h(0) = h_0 =$ elevation of launch site (or carrier vehicle altitude h_0)
$\gamma(0) = \gamma_0 = 90$ deg usually ($\gamma < 90$ deg if launched from a tilted rail, or $\gamma \approx 0$ if air-launched)

Burnout, shutdown, or final (t_f) conditions are the desired orbit's speed v_f, flight path angle γ_f, and altitude h_f

$$v(t_f) = v_f = \sqrt{\frac{\mu}{a_f\left(1 - e_f^2\right)}}\sqrt{1 + 2e_f \cos\theta_f + e_f^2} \qquad (6.38)$$

$$\cos\gamma(t_f) = \cos\gamma_f = \frac{1 + e_f \cos\theta_f}{\sqrt{1 + 2e_f \cos\theta_f + e_f^2}} \qquad (6.39)$$

$$h(t_f) = h_f = r_f - R = \frac{a_f\left(1 - e_f^2\right)}{1 + e_f \cos\theta_f} - R \qquad (6.40)$$

Here a_f, e_f, and θ_f are the desired orbit's semimajor axis, eccentricity, and true anomaly, respectively. We can control thrust and flight path angle in order to fly to the desired terminal conditions. The time of flight t_f is not restricted, but it seems clear that the time of flight should be minimized to reduce gravity losses.

6.5 Getting to Orbit

As we have seen, getting to orbit involves attaining the proper amount of kinetic energy (speed) and potential energy (altitude) after losses. Because the energy comes from chemical propellants, we naturally want to minimize the energy needed so as to launch with the least-mass launcher. For minimum energy, we wish to launch to the lowest possible orbital altitude in order to minimize the vehicle's Δv requirements. This is not compatible with orbits having higher required altitudes.

To efficiently reach higher altitude orbits, we recommend launching to a low orbit altitude while adding sufficient speed at burnout/shutdown to allow a coast to the desired higher altitude, then a burn to circularize the orbit at the higher altitude. For still higher orbits, the same procedure applies, but in this case the vehicle injects into an elliptical transfer orbit with apoapsis at the desired orbital altitude. Once the vehicle reaches apoapsis altitude, a second kick burn is used to circularize the orbit. These injection schemes are shown in Fig. 6.19. To minimize the launcher's Δv requirements, the vehicle should inject into the lowest possible orbital altitude. Depending on the payload's requirements, this may be done in one of the ways shown in the figure. Left: lower orbits (~200 km) burn directly to injection. Center: medium orbits (~200–400 km) are achieved by coasting after shutdown, then circularizing with a second burn. Right: higher orbits (>400 km) use a technique similar to medium orbits, except an elliptical Hohmann transfer orbit is used to coast to apoapsis, which becomes the altitude of the final orbit after a kick burn.

It is common, but not universal, for the payload to separate from the launch vehicle after injection into the transfer orbit and to carry a propulsion system adequate to provide its own kick burn Δv. However, if the launcher has enough excess performance, it may provide the apoapsis circularization burn *before* the payload separates. This will leave the LV's upper step in a possibly undesirable high orbit, unless it is deliberately maneuvered elsewhere.

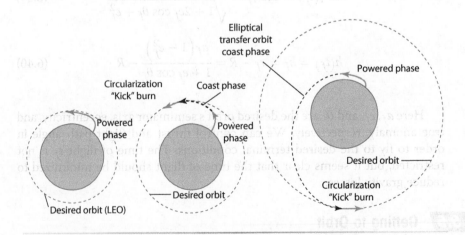

Fig. 6.19 Ways to launch to the desired orbit using the minimum possible energy.

6.6 Launch Vehicle Trajectory Simulation

The calculation of a trajectory uses a method called *numerical integration* with these equations: you start with a set of initial conditions for the rocket launch, including altitude of launch pad, initial mass, propulsion specs (thrust, mass flow, specific impulse), reference area, drag coefficient, and so on. You basically make a spreadsheet with one column corresponding to each engineering variable's value at time $t = 0$. Then, you increment time to the next time period (say 1 second, or $\Delta t = t_i - t_{i-1} = 1$ s) and recalculate. Your program or spreadsheet will march forward in time and update all the parameters at each time step. At some point, your vehicle might run out of propellants, and so the burnout parameters must match the requirements of the particular mission. If the requirements are met, then the engine may be commanded to shut down (stop producing thrust).

6.6.1 Notes on Numerical Integration

In case you're not familiar with numerical integration, the idea is to use a variable and its rate of change to calculate subsequent values. For example, we say that acceleration is the rate of change of velocity, or $a = dv/dt$. We assume that changes are small enough (if our subsequent value is only a short time after the previous one) that we can approximate

$$a = \dot{v} = \frac{dv}{dt} \approx \frac{\Delta v}{\Delta t} = \frac{v(t_2) - v(t_1)}{t_2 - t_1}. \tag{6.41}$$

Then, one may calculate

$$v(t_2) \overset{\text{def}}{=} v_2 = v_1 + \frac{\Delta v}{\Delta t} \Delta t \tag{6.42}$$

if Δv is the speed change ($v_2 - v_1$), and $\Delta t = (t_2 - t_1)$ is the time interval. This procedure is known as Euler or second-order integration. The process is continued by marching forward in time until the desired final conditions are reached, which are often given as a required burnout speed and elevation. This procedure is illustrated in Fig. 6.20.

Euler integration is less accurate than higher-order methods such as the fourth-order Runge-Kutta integration method, which is known as RK4 in MATLAB®.

When things don't work, problems that come up tend to be associated with using the *proper, consistent set of dimensions* for your

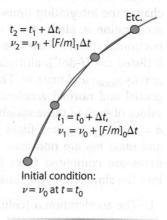

Etc.

$t_2 = t_1 + \Delta t,$
$v_2 = v_1 + [F/m]_1 \Delta t$

$t_1 = t_0 + \Delta t,$
$v_1 = v_0 + [F/m]_0 \Delta t$

Initial condition:
$v = v_0$ at $t = t_0$

Fig. 6.20 The calculation of updated speed v_1 based on previous speed v_0 and acceleration ($a = dv/dt = F/m$).

◇	A	B	C	D	E	F	G	H	I	J	K	L	M
1	EARTH & LAUNCH SITE DATA				TRAJECTORY REQUIREMENT			LV Configuration Data			Propulsion data		
2	μ_{earth}	4E+14	m³/s²		h_{final}	200	km	C_D	0.2		T	3.3E+07	N
3	g_0	9.8067	m/s²		V_c	7766.74	m/s	R_{rocket}	5	m	I_{sp}	450	s
4	R_{earth}	6378000	m		V_{box}	7357.81	m/s	m_i	3.00E+06	kg	fuel flow	7477.9	kg/s
5	h_0	6000	m					m_f	1.00E+05	kg			
6	L_0	28.45	°	γ_{pitch}	0.00100	°	(guess)	S_{ref}	78.5398	m²			
7	ρ_0	1.225	kg/m³	time step	1	s							
8													
9	Time t (s)	Accel. a (v-dot) m/s²	γ-dot (rad/s)	Velocity V (m/s)	Flight path γ (rad)	X-dot Horiz. Velocity (m/s)	X Down-range Dist (m)	H-dot Vert. Veloc (m/s)	Altitude H (m)	ρ Atm. Dens. (kg/m³)	Drag D (N)	g_{local} (m/s²)	mass M (kg)
10	0	1.19	0	0.0	1.571	0	0	0	0	1.225	0	9.807	3000000
11	1	1.22	0.00000	1.2	1.571	7.3E-17	0	0	1.1934	1	14	9.807	2992522
12	2	1.25	0.00000	2.4	1.571	1.2E-15	0	0	2.4142	4	56	9.807	2985044

Fig. 6.21   Part of a spreadsheet made to do trajectory simulations.

calculations, and that *angles are in the proper format*: either degrees or radians.

A final comment: your vehicle only has to burn out with a speed equal to the required orbital speed minus the speed of the launch location. For example, if your required orbital speed is 8.0 km/s to the east, and the launch site has a speed of 0.5 km/s east based on its latitude, your vehicle only needs to provide an additional 7.5 km/s.

Let's suppose you wish to simulate a gravity-turn trajectory, where the vehicle rises vertically and then is given a kick to initiate the pitchover for the gravity turn. Figure 6.21 shows the first few lines of an Excel® spreadsheet developed to do a trajectory analysis of a single-stage-to-orbit hydrolox vehicle using a gravity turn.

The vehicle parameters and orbital requirements are contained in the cells in rows 2-7. Row 9 contains text "headers" labeling the contents of the columns underneath. Time begins at zero in cell A10 and increments by one timestep (cell E7) for each row below (this allows one to easily change the integration timestep). The other columns contain vehicle states: acceleration a, pitch rate $\dot{\gamma}$ (listed as "γ-dot"), speed v, flight-path angle γ, horizontal speed $\dot{x}$ (listed as "X-dot"), downrange distance x, vertical speed $\dot{h}$ (listed as "H-dot"), altitude H, atmospheric density ρ, drag force D, local gravity g_{local}, and mass m. These are all the variables needed to calculate tangential and normal accelerations in the equations of flight. Known initial values of each of these variables are inserted into line 10, where acceleration $a = \frac{T-D}{m} - g_{local}\sin\gamma$, flight path angle $\gamma = 90°$, density ρ_0, local gravity g_0, and mass m_0 are non-zero. The remaining states are zero. The unknown items are computed from left-to-right using the initial conditions input into the simulation equations:

1. The acceleration a (column B) = $\sum$(forces) ÷ mass, or

$$a\,(\text{col.B}) = [T(\text{L2}) - D(\text{col.K})]/m(\text{col.M}) - g_{local}(\text{col.L}) * sin(\gamma(\text{col.E}))$$

For the first timestep, $\gamma_0 = 90°$ and $v_0 = 0$, so $D = 0$, and acceleration a_0 should just be $T/m_0 - g_0$.

2. Next, we calculate the pitch rate $\dot\gamma$ (column C) using the trajectory equation:

$$\dot\gamma \text{ (col. C)} = -\{[g_{local} \text{ (col. L)}/v\text{(col. D)}] - v\text{(col. D)}/[R\text{(cell B4))}$$
$$+ h\text{(col. D)}]\} * \cos(\gamma\text{(col. E))}.$$

for the first timestep, the pitch rate should be zero because the initial flight path angle (column E) is vertical: $\gamma_0 = 90° = \pi/2$ radians, and cos $\gamma_0 = 0$. Notice also that the pitch rate should be zero when orbit is achieved at the end of the time integration, since the two terms in the parenthesis become equal. This may not happen if the launch site speed is significantly less than the required orbital speed, since the pitch rate equation does not take initial speeds into account.

Note 1: MS Excel® does not recognize square brackets "[x]" or curved brackets "{x}" in its formulae. The ones in the formula above, provided for clarity, must be changed to parenthesis "(x)" to execute properly.

Note 2: in MS Excel®, it is possible to name cells in its "Insert" → "Name" → "Define Name" menu commands. Hence, the cell with mass flow (L4) may be named "mdot", and that name will appear in all formulae using it. In addition, it's possible to name an entire column which will also appear in equations. In this spreadsheet, some columns are named "gamma", "g_local", "V", etc. This makes troubleshooting quite a bit simpler as the equations can resemble their symbolic forms.

Note 3: MS Excel® carries out trigonometric function calculations using *radians*, where π rad = 180°. If you need to use degrees, be sure to convert to and from radians as needed, either by multiplying by pi()/180 or using the built-in degrees() and radians() functions (yes, Excel® uses the term pi() to enter the value of π, 3.14159..., in calculations.

Note 4: if you need to refer to a particular cell location that doesn't change, you can use dollar signs to "freeze" it. For example, to refer to gravity at the surface, you would use the designation "B3" rather than just "B3". Of course, it's preferred to name the variable as in Note 2 above.

3. Next we calculate the speed v (column D) using the approximate that acceleration $a \approx \Delta v/\Delta t$; Δt is the time step (stored in E7), so the new speed v is
New v = old v (0 at start) + acceleration (col. B) * Δt (cell E7)

4. The flight path angle γ (column E) is calculated using
New γ = previous γ (col. E) + $\dot\gamma$ (col. C) * Δt (cell E7)

Note that once γ is less than 90°, $\dot\gamma$ is negative, and the flight path angle will decrease since a negative value is being added each timestep.

5. Next, compute the horizontal speed $\dot{x}$ (column F):
   $\dot{x} = v$ (column D) $*$ cos γ (column E) $* R_{earth}/(R_{earth} + h$(col. I)).
6. The downrange distance x (column G) is found using the horizontal speed and the time step:
   New distance x = previous x + horizontal speed (col. F) $* \Delta t$ (cell E7)
7. Compute the vertical speed $\dot{h}$ (column H):
   $\dot{h} = v$ (column D) $*$ sin γ (column E).
8. The altitude h (column I) is found similarly:
   New h = old $h + \dot{h} * \Delta t$ (cell E7).
9. The local atmosphere density ρ (column J) is found using the altitude h (column I) in the exponential atmosphere equation, with ρ_0 stored in cell B7, and h_0 in cell B5:

$$\rho = \rho_0 * \exp(-h/h_0).$$

10. The drag D (column K) is calculated using the usual definition $D = C_D S \rho v^2/2$. The speed v and density ρ come from columns D and J respectively, and constants C_D and S are stored in cells I2 and I6 respectively.
11. For g_{local} (column L), we use the altitude h (column I) in the g_{local} gravity equation:

$$g_{local} = g_0 /(1 + h/R)^{\wedge}2$$

12. Now, we end by calculating the new mass (column M) at the end of the time step:
   New m = old $m - \Delta t$ (cell E7) $* \dot{m}$ (cell L4).

Note that if mass flow is a positive number, the amount of propellant change $\dot{m}\Delta t$ must be subtracted; if a negative number (mass is being reduced), $\dot{m}\Delta t$ must be *added*.

Once the spreadsheet's Step 12 is completed, there is enough numerical information to calculate the next row below, so its contents are inserted by repeating steps 1-12 above. This continues until the simulation is completed (a combination of desired speed, altitude, and flight path angle is achieved).

However, there is one special "trick" needed. As we saw in step 2, $\dot{\gamma}$ will remain 0 and the LV will continue to fly straight up unless we somehow induce a *non-zero* pitch rate which will initiate a gravity turn at some time during the flight. The simplest way to do this is to impose a pitch kick γ_{kick} (a step change in flight path angle) into the flight path angle column: $\gamma_{new} = \gamma_{old} - \gamma_{kick}$. The new γ will cause a non-zero cosine and will introduce a pitch rate, and thus begin the vehicle's automatic pitch over. In Fig. 6.21, the pitch kick is stored in cell E6. After inducing the pitch kick, we continue the integration. As shown in Fig. 6.17, the value of γ_{kick} is changed to get the desired trajectory by the method of "shooting," or trial and error.

The process of marching forwards in time continues until orbit is achieved OR propellants are gone OR the vehicle impacts the surface ($h < 0$, "too low" in Fig. 6-17). If the latter occurs during a gravity-turn simulation, the initial kick angle γ_{kick} must be reduced. In some cases, the value of γ_{kick} can be very small, much smaller than a degree.

Note that the same techniques would be used for a simulation carried out in a different software package, such as MATLAB®. Such a code may be easier to program for this simulation, and should yield identical results.

It may become apparent that no combination of pitch kick and initiation time converges to the desired terminal conditions. In this case, it may be that less performance is needed. One way to carry this out is to throttle the engine, or reduce its thrust. This is often done partway through a flight, but the amount of thrust reduction and the time it begins may not be apparent. Trial and error is again called for.

Figure 6.22 shows the variables produced by an Excel spreadsheet simulation. This plot shows a successful boost to orbit, as seen by a (slightly low) burnout altitude of 108 km, a burnout speed of about 7.4 km/s, a smooth transition of flight path angle from 90 to 0 deg, and a reasonable-looking plot of q with max-q occurring about 90 s into the flight. Notice also that the burnout speed of 7.4 km/s (cell F4) is about 300 m/s less than the needed orbital speed (cell F3). This is because we have subtracted the speed of the launch site, whose latitude is given in cell B6.

There are a few things to look for in the results of a trajectory's numerical integration. Observing Fig. 6.22, the reader should notice that for a successful trajectory, the following should be true:

* For reasonable liftoff performance, a $T/W_0 \sim 1.3$ to 1.5 is recommended for decent liftoff acceleration and smaller gravity loss.

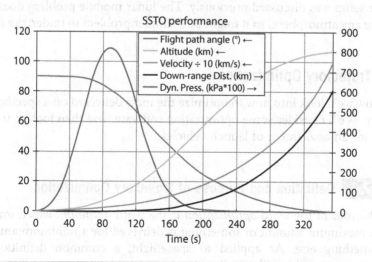

Fig. 6.22   Results of a spreadsheet made to do trajectory simulations.

- The flight path angle for a gravity turn trajectory changes smoothly from ~90 deg to 0 deg (or required orbit γ_f) at burnout/shutdown.
- Max-q should occur somewhere between 50 and 90 s after liftoff. Higher T/W_0 vehicles tend to reach it earlier.
- For most launches, time to orbit should be less than ~600 s. (Higher T/W_0 reduces time to orbit.)
- Altitude climbs smoothly from 0 (or drop altitude, if air drop) to orbital altitude h_f.
- If injecting into a circular orbit, the slope of the altitude curve should approach zero at burnout or shutdown.
- Speed increases smoothly from 0 (or drop speed, if air drop) to orbit speed v_{BO}, except for staging events.
- Acceleration level changes appropriately during throttling, staging, and jettison events.
- For due-east launches in a flat-Earth simulation: actual speed achieved (v_{BO}) should be less than orbital speed v_f by the amount v_{ROT}, the earth's rotational speed contribution at the launch site ($v_f = v_{BO} + v_{ROT}$). For other azimuths, achieved speed should incorporate launch site speed appropriately.
- Calculated gravity loss is reasonable (~1.5 km/s).
- Calculated drag loss is reasonable (~100–200 m/s).

A downrange distance of about 600 km is much less than R_{earth}, so the flat-Earth approximation we introduced earlier seems to be valid.

The reader is encouraged to carry out numerical simulations as directed in the problems at the end of the chapter. In particular, there is an Apollo lunar module ascent problem as experienced by the Apollo astronauts as they began their return from the moon to the Earth, as well as an SSTO problem whose setup was discussed previously. The lunar module problem does not involve any atmosphere, so it might be a simpler problem to undertake first.

6.7 Trajectory Optimization

Now we'll look into how to optimize the mass delivered on a specific trajectory. We'll consider some optimization software, and then look at trajectories for an assortment of launch vehicles.

6.7.1 Definition and Purpose of Trajectory Optimization

The root of the word *optimization* is the word *optimum*, which implies that a maximum amount of something is delivered for a minimum amount of something else. As applied to spaceflight, a common definition of *optimum* is something like "deliver the largest payload mass possible using minimum fuel" or "deliver the largest payload possible in less than 90 days" or something similar. Occasionally it may involve minimum cost or some other parameters.

Optimization is particularly useful for space launch vehicles. We know that they (usually) launch from the ground vertically and end up in orbit flying horizontally, but there is no upfront way to guess how that path from 90 deg to 0 deg should be executed. Does the vehicle climb vertically out of the atmosphere, and then turn horizontally to accelerate horizontally? This might produce a trajectory with a large gravity loss. On the other hand, maybe the vehicle should pitch over as soon as possible so as to start developing horizontal speed to get to orbit. This would involve "carving" through the atmosphere, minimizing the gravity loss but probably increasing the drag loss, and possibly overstressing the vehicle due to air loads and/or aerodynamic heating in the denser lower atmosphere. So it's far from clear.

Before we tackle launch optimization, let's take a look at the first optimization problem, solved by Isaac Newton hundreds of years ago.

6.7.2 The First Optimization Problem: The Brachistochrone or Shortest Time Problem

In the time of Newton in the 17th century, differential calculus was just being invented and used to solve various problems. One problem introduced by Johann Bernoulli was: "What shape would result in a minimum-time slide from top to bottom?" A more interesting formulation of the problem is "Find the optimal shape for a frictionless playground sliding board that would allow a youngster to slide from top to bottom in minimum time." It is understood that the object/child starts with zero speed, and gravity is vertical. Obviously this is a problem that cannot be solved by inspection; some options are shown in Fig. 6.23.

The four curves in Fig. 6.23 were chosen to be possible solutions based on some intuition. Curve A, the straight line, is the shortest length between start and finish. Curve B starts out vertical to accelerate faster and then transitions to horizontal. Curve C accelerates the object the maximum possible, then makes a (frictionless) turn and continues horizontally with the same speed. Finally, Curve D is a compromise between the shortest distance A and the smooth transition from high acceleration to horizontal of Curve B. You're probably wondering what the answer is!

The method of solution for this problem ultimately created a new branch of mathematics called the *calculus of variations*. The problem was solved by several mathematicians of the time, including Jakob and Johann Bernoulli, and Isaac Newton, who solved the problem in a single afternoon—that could itself be called a *minimum time* solution! When Newton's unsigned solution was received later, it has been said that Johann Bernoulli was so impressed with its cleverness that he has been quoted as saying, "I recognize the lion by his paw."

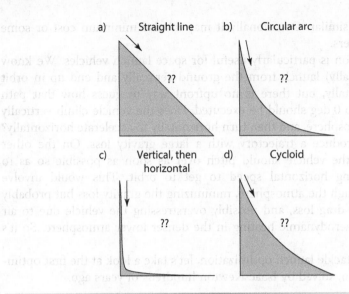

Fig. 6.23 Some possible shapes for the minimum time problem.

Exposés of this problem and its solution can be found in many mathematics history books. It turns out that the fastest path is the shape of a *cycloid* (D in Fig. 6.23), which is the curve traced out by a fixed point on a rotating wheel of a moving vehicle relative to the nonmoving ground. Even though this path is longer than a straight line (the shortest path), the elapsed time is less, because it has the right combination of steepness for acceleration and shortness to reduce transit time. A straight line would be a reasonable guess but takes about 10% longer! The numerical data associated with the different trajectories are shown in Fig. 6.24.

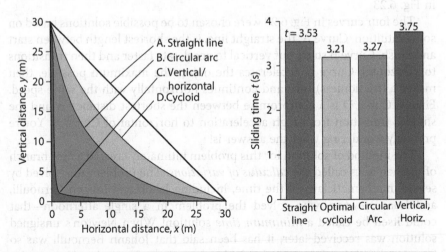

Fig. 6.24 A selection of possible path shapes for the minimum time problem.

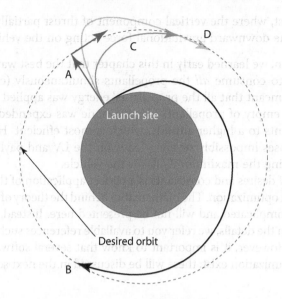

Fig. 6.25   Which of trajectories A, B, C, or D is the optimal trajectory to orbit? See the text!

This new field invented by Newton, the calculus of variations, would have many applications in the following years. Although we'd love to mention others, we'll only consider space launch vehicle applications here. Let's start by looking at the problem that has been mentioned several times when gravity loss and drag loss were calculated. We found the following (refer to Fig. 6.25):

* To limit drag losses, fly the launch vehicle straight up, and then turn right and accelerate horizontally, as shown in trajectory A in Fig. 6.25. The vertical ascent produces a large gravity loss. Gravity loss increases with the amount of time that the vehicle climbs in a vertical direction, suggesting that the launch vehicle should turn towards horizontal as soon as possible.
* Turn the launch vehicle towards horizontal as soon as possible, as shown in Trajectory B. However, turning too early towards horizontal exposes the launch vehicle to excessive air loads and possible extreme aerodynamic heating—but less gravity loss.
* An optimal trajectory or "best" method for a vehicle with a moderate T/W ratio is a compromise somewhere between A and B, as shown by trajectory C: its early portion is near-vertical for small drag loss, followed by a shallow climb to minimize gravity loss, transitioning to horizontal for orbit injection.
* If the vehicle has a lower T/W ratio (underpowered), a lofted trajectory may be more desirable, as shown in curve D. During a lofted trajectory, the vehicle climbs above the desired orbital altitude to allow for more time for the vehicle to accelerate horizontally while "falling down" vertically. The vehicle may be flying at a significant angle of attack during the later portion

of this boost, where the vertical component of thrust partially or fully counters the downward gravitational force acting on the vehicle.

In addition, we learned early in this chapter that the best way to launch a payload was to consume *all* the propellants simultaneously (e.g., a mortar shell), which meant that all the propellants' energy was applied to accelerating the (now empty of propellant) LV, and none was expended raising leftover propellants to a higher altitude, which is most efficient. However, this method imposes impossibly severe g-loads on the LV and payload, another constraint being the maximum T/W for the vehicle.

This set of desires and constraints is a perfect application of the calculus of variations and optimization. The mathematics behind the theory of optimization is relatively complicated and will not be presented here. Instead, if the reader is interested in the details, we refer you to available references such as Patterson & Rao [16]. However, it is important to know that several software codes for trajectory optimization exist; these will be discussed in the next section.

6.7.3 Optimization Software

Trajectory designers use optimization software to develop optimized LV trajectories. There are several traditional optimization codes that have been used in the aerospace industry for a long time.

* *Optimal Trajectories by Implicit Simulation (OTIS)*: Explicit or implicit, considered to be more versatile (NASA Glenn & Boeing supported) [24]
* *Program to Optimize Simulated Trajectories (POST-II)*: Explicit time-step integration, a method of shooting (NASA Langley) [3].
* *Simulation and Optimization Rocket Trajectories (SORT)*: Developed for Space Shuttle trajectories by Lockheed Martin

These programs are International Traffic in Arms Regulations (ITAR) controlled and are not available to the public or to anyone not working in the U.S. government or one of its contractors. However, there are alternative optimization programs. The author's favorite is GPOPS-II, a MATLAB® program that is a generic optimization "machine" that can be used for many different optimization problems, as opposed to POST, OTIS, and SORT, which are specifically tailored to vehicle trajectory problems. It uses the well-known SNOPT optimization subroutine by Phil Gill, U.C. San Diego. A license for GPOPS-II may be purchased online for a nominal amount, and the documentation for the program includes working examples containing the inputs necessary to analyze a multistage rocket. Some of the plots you will see in this chapter were generated by GPOPS running on the author's laptop.

6.8 Some Examples of Launch Profiles and Trajectories

This portion of the chapter contains some example launch profiles, trajectories, and explanations. Some of these launches are governed by such constraints as thrust or g limits, maximum dynamic pressure, staging delay, payload fairing jettison (when aerodynamic heat levels are low enough), engine loss, crew abort scenarios, and the like. The optimization of the launch cannot take place until the constraints are given in numerical form.

The launch profiles you will see in the following pages contain multiple engineering parameters plotted against time since liftoff. They contain a very large amount of information that can be used to get a really good idea of what the SLV is capable of. The only tricky part is to make sure that you are reading the desired parameter on the correct scale. Some of these plots have four or more different scales, so double-check to see if you are reading the right scale.

The launch parameters that are usually shown include some of the following: acceleration (m/s^2 or multiples of g_0), dynamic pressure (q in Pa or N/m^2 or atmospheres), speed (m/s or km/h), altitude, downrange distance, and pitch/flight path angle

6.8.1 LV Example: Delta II Launch to Mars Transfer Escape

Figure 6.26 is an illustration for the boost profile for the Mars Exploration Rover (MER) on board a Boeing Delta II. This Delta II has nine strap-on solid rocket motors that are consumed at the beginning of the flight and then jettisoned. Interestingly, only six SRMs are utilized at liftoff; the remaining three are used after the first six burn out.

In the profile, each major event is called out with its mission elapsed time, altitude, and velocity, inertial (VI). The three plots shown in

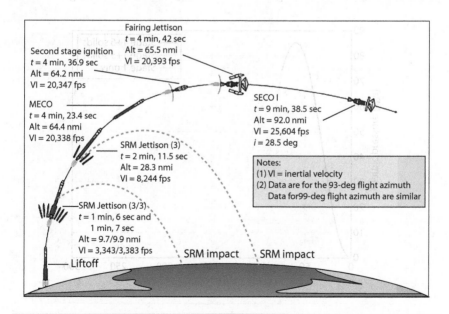

Fig. 6.26 A launch of a Delta II carrying MER-B to orbit and then on to Mars. Source: [2].

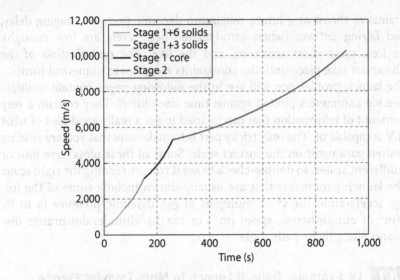

Fig. 6.27 The speed profile of the launch of a Delta II. One of the optimization program's requirements is that the vehicle speed must be continuous across the change of a phase, such as staging. GPOPS output.

Figs. 6.27, 6.28, and 6.29 were generated from GPOPS and show time histories of a simulated Delta II's speed, dynamic pressure, and altitude, respectively. Figure 6.29 shows an unusual trajectory called *lofting*. Lofting involves the trajectory first climbing and then falling while accelerating to orbital

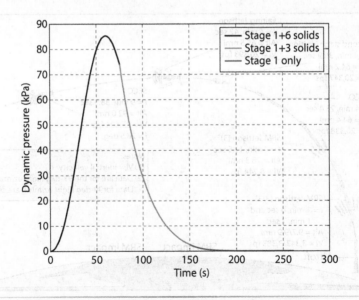

Fig. 6.28 The variation of the dynamic pressure q during the first step burn of a Delta II. Max-q is approximately 84 kPa and occurs around $t = +60$ s. GPOPS output.

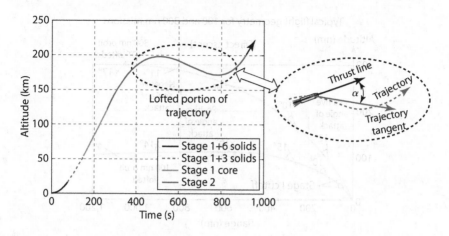

Fig. 6.29   The variation of the Delta II's altitude vs time. Positive slope at burnout shown by the arrow on the right indicates entry to climbing orbit or escape—in this case, to Mars. The circled portion is the lofting part of the trajectory; the vehicle climbs to ~200 km, then descends ~30 km while continuing to accelerate towards burnout. Typically the thrust line is not horizontal during this time, leading to a propulsion loss. GPOPS output.

speed. It is nonintuitive that losing altitude is a way to improve performance. Lofting will be described in more detail in the following section.

6.8.2 What Is Lofting, and Why Does It Occur?

Lofting typically occurs when an upper step is underpowered compared to other steps. A better way to describe this situation might be to say that the step is underaccelerated, which is a symptom of a second step engine with lower thrust. Because of the lower thrust level, the step has to burn longer, and during a horizontal burn, will fall towards the planet it orbits until it finally does reach orbital speed. Because of this, it is lofted to a higher altitude so that it may fall vertically while accelerating horizontally to reach orbital speed.

To reduce the falling that occurs, the lofted step may be pitched up above the local horizon, so that a portion of the step's thrust works against gravity and reduces the altitude that would otherwise be lost, as discussed earlier in this chapter in Section 6.2 on inclined flight. However, because the thrust vector **T** and velocity vector **v** are not parallel, the power being delivered goes as an integral

$$P_{\text{delivered}} = \int_0^{t_f} \mathbf{T} \cdot \mathbf{v}\, dt = \int_0^{t_f} Tv \cos\theta\, dt$$

where θ = the angle between the thrust line and velocity vector. The closer the angle θ is to zero, the smaller the *propulsion loss* and the higher the efficiency of the step.

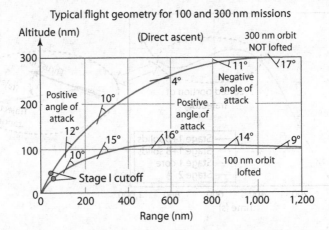

Fig. 6.30 Angles of attack in Saturn C-1 trajectories to 100- and 300-nm orbits. Source: [8].

One example of the angles of attack that have been considered is from Jean [9] presented at the NASA-Industry Apollo Technical Conference in Washington DC. Commenting about Saturn C-1 ascent trajectories for 100- and 300-nm orbits:

> It is interesting to note that rather large angles of attack are encountered during the S-IV burning. This is due to the low thrust-to-weight ratio of this stage, which is 0.6 *g* at ignition. The optimum program in the 100-nm case encounters angles of attack up to 16-deg magnitude, and in the 300-nm case, 17 deg near cutoff of the terminal stage.

The trajectories are shown in Fig. 6.30, and it's clear that the 100-nm case is a lofted trajectory. In contrast, the 300-nm trajectory starts at a positive angle of attack, perhaps to increase the climb rate, and ends with a negative angle of attack, to cancel out the vertical velocity and burnout at 0-deg trajectory. This is one of very few cases of angles of attack being shown on a trajectory plot.

The following comments about lofting are from *Air & Space Magazine*, May 2009 (http://www.airspacemag.com/space-exploration/Is-It-Safe.html):

> Rockets that have relatively under-powered second steps and fly a *'lofted trajectory'* where the first stage shoots them very high and they actually start descending before climbing to orbit. If astronauts abort near the high point, their capsule could 'plummet straight down' and 'belly flop' on the atmosphere with extreme gs. Elon Musk says: "This was one of the main reasons given by NASA for not using those vehicles for manned space-flight."
>
> *Falcon 9* was designed with a second stage about four times as powerful as that of an *Atlas* or *Delta*, for a more slanted, softer trajectory. The fuel weight adds cost, but if astronauts abort, the flight path will catapult *Dragon* horizontally, slicing more gradually into the atmosphere.

There are several reasons to loft, and they all are the result of optimization programs:

* Lofting can deliver more payload to orbit.
* Lofting reduces max-q and max-$q\alpha$ aerodynamic load levels, because LV gets out of the atmosphere faster.
* Lofting can provide extra altitude margin in an engine-out situation (vehicle may lose altitude while still accelerating horizontally).

Note that any changes to an already-optimized lofting trajectory will occur at a cost of loss of performance. The discussion of Shuttle trajectories in the following section will shed more light on this.

6.8.3   The Space Shuttle Ascent: Complex, Many Constraints, Nonoptimal Trajectories

NASA's Space Shuttle was a very complicated launch system. Its design features, including parallel staging (orbiter next to tanks rather than in front of), orbiter airframe with wings and tail, human-carrying, fragile thermal protection systems, and external tank (ET) disposal, led to a number of constraints that made the trajectory planning interesting. These features and their advantages and disadvantages are listed in Table 6.2.

The following is an edited excerpt from NASA [15]:

> NASA's challenge was to put wings on a vehicle that would survive the atmospheric heating that occurred during re-entry into Earth's atmosphere. The addition of wings resulted in a much-enhanced vehicle with a lift-to-drag ratio that allowed many abort options and a greater cross-range capability, affording more return-to-Earth opportunities. This Orbiter capability did, however, create a unique ascent flight design challenge. The launch configuration was no longer a smooth profiled rocket. The vehicle required new and complex aerodynamic and structural load relief capabilities during ascent.

Table 6.2   Advantages and Disadvantages of Space Shuttle Features

Space Shuttle Feature	Why Feature Was Incorporated	Effect on Launch Trajectory
Airframe with wings and tail	Crossrange performance during entry, runway landing	Very restricted angle of attack and sideslip angles during ascent
External tank (ET)	Allows reuse of main engines	ET disposal must occur in a safe area
Fragile thermal insulation foam on ET	Reduced cryogenic propellant loss due to heat transfer	Restricted dynamic pressure during launch
Human crew	Advanced on-orbit capabilities	Acceleration <3 g requires engine throttling and SRB thrust profiling

The Space Shuttle ascent flight design optimized payload to orbit while operating in a constrained environment. The Orbiter trajectory needed to restrict wing and tail structural loading during max-*q* [Fig. 6.31] and provide acceptable first stage performance. This was achieved by flying a precise angle of attack and sideslip profile and by throttling the space shuttle main engines (SSMEs) to limit dynamic pressure. The Solid Rocket Boosters (SRBs) had a built-in thrust tailoring design that worked with the SSMEs to minimize the maximum dynamic pressure the vehicle would encounter.

During the first stage of ascent, the vehicle's angle of attack and dynamic pressure produced a lift force from the wings and vehicle structural loading. First-stage guidance and control algorithms ensured that the angle of attack and sideslip did not vary significantly and resulted in flying through a desired "keyhole," [Fig. 6.32] defined by the product of dynamic pressure and angle of attack which maintained the desired loading on the vehicle tail.

Because day-of-launch winds aloft significantly altered vehicle angle of attack and sideslip during ascent, balloon measurements were taken before liftoff near the launch site. Based on these wind measurements, Orbiter guidance parameters were biased and updated via telemetry.

Just after liftoff and clearing the tower, a roll maneuver was initiated to achieve the desired orbital inclination and put the vehicle in a heads-down attitude during ascent.

Vehicle performance was maximized during second stage by a linear steering law called "powered explicit guidance." This steering law guided the vehicle to orbital insertion and provided abort capability to downrange abort sites or return to launch site. Ascent performance was maintained. If one main engine failed, an intact abort could be achieved to a safe landing site. Such aborts allow the Orbiter and crew to either fly at a lower-than-planned orbit or land.

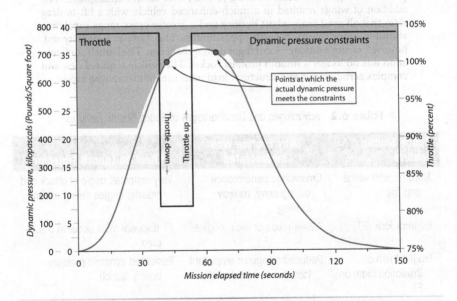

Fig. 6.31 To meet the shuttle's dynamic pressure constraints during ascent, the main engines were throttled down. The goal was to maximize performance by getting as close as possible to the constraints.

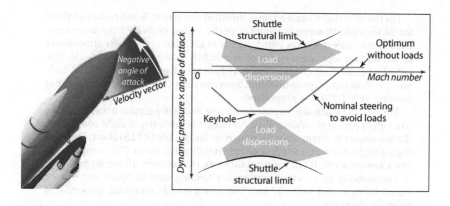

Fig. 6.32 To avoid load dispersions due to variations in thrust and atmosphere, the shuttle had to "fly through a keyhole." This meant deviating from the optimum angle of attack at certain Mach numbers.

Ascent flight design was also constrained to dispose the ET in safe waters: the Indian or Pacific oceans, or in a location where tank debris was not an issue. After MECO and ET separation, the remaining main engine fuel and oxidizer were dumped. This event provided some additional performance capability.

After the shuttle became operational, ascent performance was improved to accommodate heavy payloads. Many guidance and targeting algorithm additions provided more payload capability. For example, standard targets were replaced by direct targets, resulting in one Orbital Maneuvering System maneuver instead of two, saving propellant and resulting in more payload.

In some cases, the act of departing from an optimal (in one sense) path to a less-optimal one can provide some useful insights into how a system behaves. The Space Shuttle provided a few examples of this. When it was operational, the Space Shuttle flew more *depressed* (lower or less steep) trajectory than some expendable launch vehicles. The depressed trajectories allow for less difficult abort trajectories following a premature engine shutdown or failure.

The first shuttle flight after Challenger followed an even safer "abort-shaped" trajectory, but the loss of performance (in terms of payload to orbit) was too high. All subsequent flights reverted to the standard shuttle launch trajectory.

From Harwood in Spaceflight Now [18], slightly edited:

To ease stress on External Tank insulation, NASA redesigned the ascent profile, changed the main engine throttle settings, and lofted shuttle trajectory. This so-called 'Low-q ascent profile' was not new, having been used in the past for a 1999 Hubble servicing mission.

The normally used 'high-q' profile kept the shuttle main engines at a higher throttle setting when climbing out of denser atmosphere

One previous high-q mission had the main engines throttled down to 72% power at 38 seconds and held before throttling back up to 104.5% at 52 seconds. During throttled-down phase, shuttle was subjected to max-$q = 695$ psf.

For the next flight: main engines throttled down to 67% and remained there for 24 seconds, 10 seconds longer than the high-q profile. At the same time, the shuttle's trajectory was lofted slightly to get it out of the lower atmosphere faster. Booster separation occurred at a slightly lower speed and altitude. The net effect of all these changes was to reduce aerodynamic loads on LOx feed-line, ice-frost ramps, pressurization lines and cable tray by 7% *at the expense of burning up 1,180 lbs. of propellant.*

This low-q profile also exposed the crew to slightly greater abort risks due to a longer period of vulnerability. An engine failure early in flight could force the astronauts to attempt a risky return-to-launch site (RTLS) abort. During a high-q flight, the RTLS window of vulnerability was about 155 seconds long. For a low-q ascent, RTLS vulnerability was 165 seconds, 10 sec longer.

Vulnerability to aborts to emergency landing sites in Spain and France would be extended some 17 sec. The low-q profile increased crew risk of abort by about 2%.

Pilot Kelly, who monitored Discovery's main engines during the climb to space, said " . . . as we approached max-q, when we got the maximum force on the vehicle, we usually throttle the main engines down so we can get through that heavier, thicker part of the atmosphere at a lower speed and accelerate more slowly until we're out of the atmosphere and there are no air loads on the vehicle. Some flights we've been flying a high-q profile so we don't throttle down as much, we go faster, put more loads on the vehicle."

"Because our flight had some additional ascent performance, we could carry an extra 1,500 lbs. to orbit. If we just gave up about 1,000 lbs. of the excess, we could throttle down sooner, stay throttled down longer, and throttle back up later, which would reduce loads on the tank by around 7 to 10%."

"Imagine you roll down your car window at 40 MPH and you stick your hand out – you're going to feel a certain force on your hand. If you accelerate to 60 MPH and stick your hand out, you can tell there's a big difference. So doing this low-q means less force on the tank and a little bit more margin to deal with the uncertainties."

" . . . It was almost transparent for us. The only thing for me that was different, instead of the engines throttling back to 71%, they throttle back to 67% – that's it. It changed abort boundaries, it changed the profile a little bit, and it was a little bit more lofted. We've flown a lot more flights at low-q than we have at high-q."

6.9 Some Typical Launch Trajectories

In the following sections, we'll present figures showing the trajectory parameters during ascent for a number of launch vehicles including the Space Shuttle, Saturn V, Japan's M-class, and Pegasus. We will comment on any unusual traces that appear, including a very unique "mixture-change" optimization that was executed on the Saturn V to increase its payload to the moon.

6.9.1 Shuttle STS-122 Ascent Trajectory

Figure 6.33 provides a plot of the parameters associated with the STS-122 shuttle launch on 7 Feb. 2008. It shows plots of acceleration, speed, altitude, and range for the ~8.5-min boost period. The throttle-down of the main engines may be seen clearly, and it's also clear that this Shuttle flight flew a lofted trajectory. See the notes in the previous section for more discussion of the Shuttle's trajectory and constraints.

6.9.2 Saturn V Ascent Trajectory

In the 1960s, the United States was in a "space race" with the Soviet Union to be the first to land a human on the moon and return safely. The Apollo spacecraft to travel to and from the moon's surface was very complex and massive. So massive, in fact, that its launch vehicle, the Saturn V, did not have enough payload capability. Because of the many Δvs needed to carry out this mission, weight and performance were at a premium. In fact, Logsdon [11] says that each additional pound of payload carried to the moon "was worth about $2,000." These 1960s dollars are worth about 8 times the value of a 2020 dollar.

Because there is extensive documentation available on the Saturn V, we'll take a look at some of the parameters during launch. Let's start with Fig. 6.34. In the figure, you see that the altitude vs time is about what you'd expect it to be, until about 7 min elapsed time, where there are a few "wiggles" in the plot.

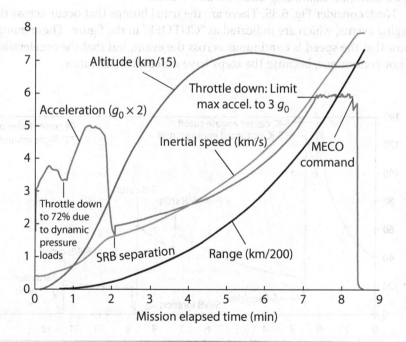

Fig. 6.33 Ascent profile for STS-122. Notice that the inertial speed of the shuttle is not zero due to Earth's rotational speed. Be careful doing unit conversions when reading values from curves. Source: [23]. MECO = Main engine cutoff.

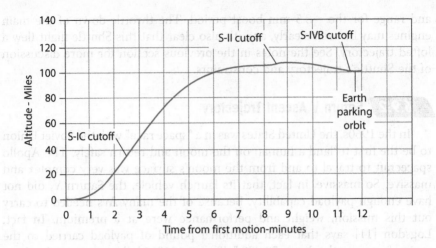

Fig. 6.34   Saturn V altitude, liftoff to Earth parking orbit. Even the mighty Saturn V flew a
lofted trajectory. Source: [13].

We also see that the trajectory is lofted, because the cutoff altitude is less than
some of the intermediate altitudes. We will learn the source of these wiggles
as we continue examining other Saturn V data.

Next, consider Fig. 6.35. There are the usual bumps that occur across the
staging events, which are indicated as "CUTOFF" in the figure. These bumps
show that the speed is continuous across the event, but that the acceleration
is not continuous, because the steps have different T/W ratios.

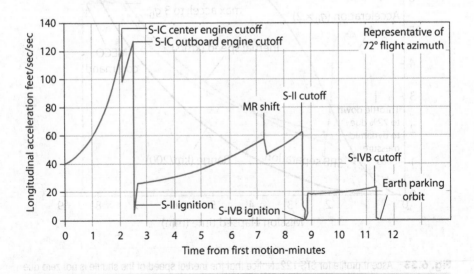

Fig. 6.35   Saturn V longitudinal acceleration during boost to parking orbit. Source: [13].

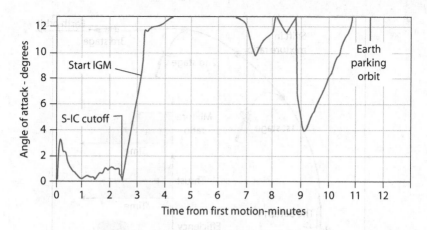

Fig. 6.36 Saturn V/Apollo VIII angle of attack, liftoff to parking orbit. There is a significant AoA, so we can conclude that this is a lofted trajectory. Source: [13]. IGM = Iterative Guidance Mode (closed-loop guidance mode).

Next, consider Fig. 6.36. This is a plot of the angle of attack (AoA) of the Saturn V during launch. As one would expect, the AoA is small until the first step S-IC cuts off. Then, the AoA rises steadily up to between 10 deg and 14 deg during the S-II burn. For the S-IVB's burn, the AoA rose steadily from 4 deg up to about 14 deg at cutoff. These angles of attack confirm that the Saturn is flying a lofted trajectory; however, there is even more to look at in terms of optimization for the Saturn V flight.

6.9.3 Saturn V Second Step (S-II) Variable Mixture-Ratio Scheme

We've already mentioned the need for high performance of the Saturn V in carrying out its mission to return astronauts and lunar samples to Earth. During the development of the Saturn V for the Apollo moon landings, many simulations were run. North American (NA) was responsible for the S-II, which had five J-2 engines that were designed to burn their propellants at a constant steady-state mixture ratio of 5:1 by mass (5 kg of LOx for every kg of LH_2). NA's Monte Carlo simulations showed that random variations in payload delivered were due to statistical variations in the step's mix ratio (mass $LOx:LH_2$). The question then was asked: If something causes our payload to vary, what can we do to accentuate that effect in our favor?

After considerable study, it was determined that the mixture ratio could be changed from the nominal higher-thrust mix ratio of 5.5:1 to a higher-I_{sp} mix ratio of 4.5:1 during boost, as shown in Fig. 6.37. This minor change resulted in a *1,225 kg (2,700 lb$_m$) payload increase*!

Why? When at nominal mix ratio, the S-II uses more propellant earlier in the burn, and therefore doesn't have to haul as much propellant later in the

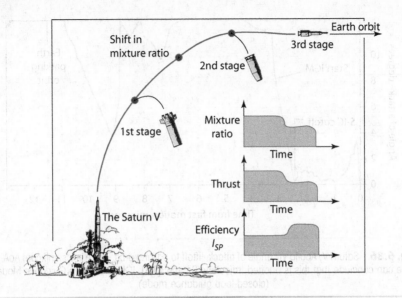

Fig. 6.37 The five J-2 engines on the S-II second step of the Saturn V. North American, contractor of the step, showed that if the mix ratio shifted, the rocket could carry an extra 1,225 kg to the moon. Source: [11].

burn, which amounts to a *more efficient* payload delivery. The beauty of this discovery? *No changes were needed to any hardware!* Just command the software to change the position of the five J-2 engines' mixture valves during the S-II burn. This scheme is very simple to implement—the best kind of modification.

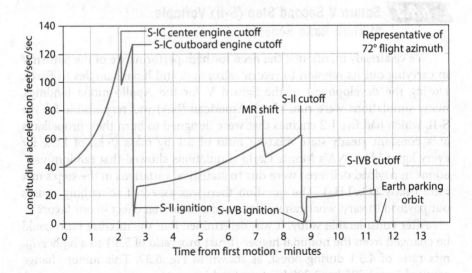

Fig. 6.38 Saturn V longitudinal acceleration, liftoff to parking orbit. Note that center engine cutoffs limit acceleration approaching burnout of S-IC and S-II steps. Source: [13]. MR = Mix ratio.

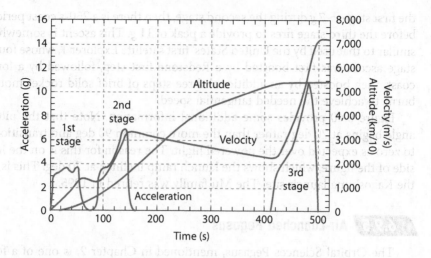

Fig. 6.39   Mu-3-S-2: speed, acceleration, and altitude vs time. Note the ~310-s coast period between second-step burnout and third-step ignition. Source: [21].

Figure 6.38 shows the physical result of the mix ratio change, which occurs around 7 min 15 s on the plot. The change results in an abrupt reduction in thrust, but with more efficiency. The end result is a larger payload mass.

6.9.4 Mu-3-S-2 (Japan)

Japan's Mu family of rockets, sometimes called M-Class, consists of three-step solid-fueled orbital vehicles. Their characteristics may be understood by examining Figs. 6.39 and 6.40. Acceleration reaches ~4 g during

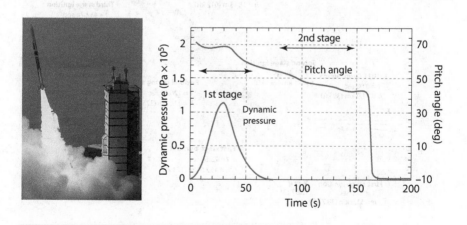

Fig. 6.40   Mu-3-S-2: Left: vehicle launching from 70-deg launch rail. Right: dynamic pressure and pitch angle plot starting at 70 deg. Source: [21].

the first stage, ~7 g during the second stage, then there is a 250-s coast period before the third stage fires to provide a peak of 11 g. This ascent is somewhat similar to that used by the United States' first satellite Explorer 1, whose four-stage ascent was first boosted by a Redstone first step, followed by a long coast, then boosted by an additional three steps of brief solid rocket motor burns to achieve the needed tangential speed.

Figure 6.40 provides some additional information. Note that the pitch angle begins at 70 deg rather than the more common 90 deg and transitions to zero as expected over the powered flight. The reason for this is on the left side of the figure, which shows the launch ramp inclined at 70 deg. This is at the Kagoshima launch site. The Mu family was retired in 1995.

6.9.5 Air-Launched Pegasus

The Orbital Sciences Pegasus, mentioned in Chapter 2, is one of a few space launch vehicles that are air-launched. Figure 6.41 depicts the launch profile as the vehicle is dropped from its carrier aircraft through orbital insertion. Except for the horizontal release and a first step that has a wing and articulated tail surfaces, the profile is not much different than a conventional ground-launched LV.

Speed, altitude, and flight path angle are shown in Fig. 6.42. As expected, with a horizontal launch, the flight path angle begins at ~0 deg and increases

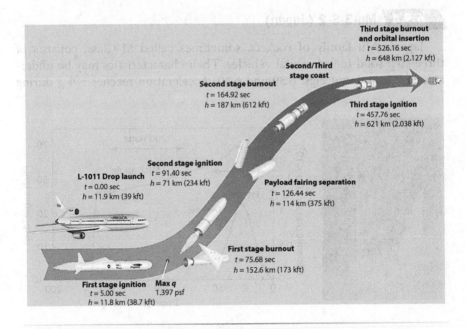

Fig. 6.41 Pegasus small launch vehicle is air-dropped from a converted L-1011 passenger aircraft. Source: Northrop Grumman Corp.

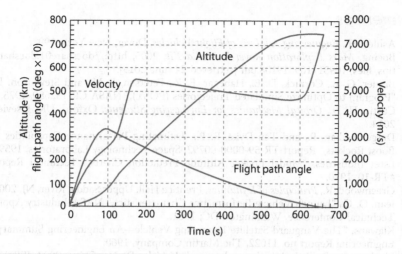

Fig. 6.42 Air-launched Pegasus: altitude, velocity, and flight path angle (starts at 0 deg).
Source: [21].

to a maximum value of ~34 deg before tapering down back to 0 deg at orbit injection. Interestingly, the speed decreases between about 160 s and 600 s, implying a coast to higher altitude before the third step fires to provide the speed needed to complete the ascent.

6.10 Conclusion

Let's conclude by considering parameters that have been allowed to vary during a flight and their ramifications:

* Climbing vertically produces gravity loss, so try to get horizontal as soon as possible to minimize this.
* A shallow climb produces significant drag loss, higher aerodynamic loads, or even overheating, so don't go shallow or level off too soon.
* The combination of a quick climb out of atmosphere with a shallow climb after appears to provide optimality in many cases.
* In some cases, it makes sense (in an optimality aspect) to loft the vehicle to a higher than desired burnout altitude, then fly at a positive angle of attack accelerating while also descending to a desired burnout altitude. Such a trajectory produces a significant steering loss due to the thrust line not being parallel with velocity.
* Mix ratio shift, as used by the Saturn V LV's second step, provides a more efficient blend of higher thrust level up front and higher efficiency (higher I_{sp}) in the later portion, thus increasing payload significantly.

References

[1] Ashley, H., *Engineering Analysis of Flight Vehicles*, Dover, New York, 1992.

[2] Boeing, *Mars Exploration Rover-B Media Kit*, 2005, http://docshare01.docshare. tips/files/10931/109319108.pdf [retrieved 24 June 2019].

[3] Brauer, G. L., Cornick, D. E., Habegar, E. R., Petersen, F. M., and Stevenson, R., "Program to Optimize Simulated Trajectories (POST)," NASA CR-132689, 1975.

[4] Curtis, H. D., *Orbital Mechanics for Engineering Students*, Oxford, UK, Elsevier, 2014.

[5] Dergarabedian, P., and Ten Dyke, R. P., "Estimating Performance Capabilities of Boost Rockets," Report TR-59-0000-00792, Space Technology Laboratories, 1959.

[6] Estes Industries, "Model Rocket Altitude Prediction Charts," Technical Report #TR-10, 1970.

[7] Greenwood, R., *Principles of Dynamics*, Prentice Hall, Upper Saddle River, NJ, 2002.

[8] Jean, O. C., "Launch Vehicle Performance Characteristics," NASA-Industry Apollo Technical Conference, Washington DC, 1961.

[9] Klawans, "The Vanguard Satellite Launching Vehicle—An Engineering Summary," Engineering Report no. 11022, The Martin Company, 1960.

[10] Linshue, H., *Ballistic Missiles and Launch Vehicles Design*, China, 2002. ISBN 7-81077-187-6.

[11] Logsdon, T., *Orbital Mechanics: Theory and Applications*, Wiley, New York, 1998.

[12] NASA, "The Apollo 'A'/Saturn C-1 Launch Vehicle System," NASA TM-X-69174, MSFC report MPR-M-SAT-61-5, 1961.

[13] NASA, "Saturn V Flight Manual SA-503," NASA TM-X-72151 or MSFC-MAN-503, 1968.

[14] NASA, "Description and Performance of the Saturn Launch Vehicle's Navigation, Guidance, and Control System," NASA TN D-5869, July 1970.

[15] NASA, "Ascent Flight Design," NASA SP-2010-3049, 2010.

[16] Patterson, M. A., and Rao, A. V., "GPOPS-II: A MATLAB Software for Solving Multiple-Phase Optimal Control Problems Using hp-Adaptive Gaussian Quadrature Collocation Methods and Sparse Nonlinear Programming," *ACM Transactions on Mathematical Software*, Vol. 41, No. 1, Oct. 2014, doi:10.1145/2558904.

[17] Perkins, F. M., "Derivation of Linear-Tangent Steering Laws," Air Force Report #SSD-TR-66-211, 1966, doi:10.21236/ad0643209.

[18] Harwood, W., "Unprecedented Coverage of Slightly Gentler Ascent," Spaceflight Now, http://www.spaceflightnow.com/shuttle/sts121/060629preview/part2.html [retrieved May 2018].

[19] Sutton, G. P., and Biblarz, O., *Rocket Propulsion Elements*, 8th ed., John Wiley & Sons, Hoboken, NJ, 2010.

[20] Thomson, W. T., *Introduction to Space Dynamics*, John Wiley & Sons, New York, 1961.

[21] Turner, M. J., *Rocket and Spacecraft Propulsion—Principles, Practices, and New Developments*, Springer Praxis, Berlin, 2009.

[22] Vukelich, S. R., and Bruns, K. D., *Missile DATCOM*, U.S. Dept. of Commerce, National Technical Information Service, Springfield, VA, 1988.

[23] Walter, U., *Astronautics*, 3rd ed., Fig. 6.13, Wiley-VCH, Weinheim, 2012.

[24] Brauer, G. L., Cornick, D. E., and Stevenson, R., "Capabilities and Applications of the Program to Optimize Simulated Trajectories," NASA CR-2770, Feb. 1977.

[25] Wiesel, W. E., *Spaceflight Dynamics*, Aphelion Press, Beavercreek, OH, 2012.

Further Reading

Boeing, "Mars Exploration Rover-B Media Kit," https://www.scribd.com/document/109319108/Mars-Exploration-Rover-B accessed February 25, 2020

Cherry, G., "A General, Explicit, Optimizing Guidance Law for Rocket-Propelled Space-flight," Astrodynamics Guidance and Control Conference, 1964, doi:10.2514/6.1964-638.

Falangas, E., *Performance Evaluation and Design of Flight Vehicle Control Systems*, Wiley-IEEE Press, Piscataway, NJ, 2015.

Greensite, A. L., *Analysis and Design of Space Vehicle Flight Control Systems*, Spartan, New York, 1970.

Kim, C. H., *Commercial Satellite Launch Vehicle Attitude Control Systems Design and Analysis*, CHK, Fountain Valley, CA, 2007.

Seifert, H. S., Space Technology, Wiley, New York, 1959.

Online Simulation Software

ZOOM: Conceptual Design and Analysis of Rockets and Their Missions, http://trajectorysolution.com/ZOOM%20Program.html [retrieved 23 June 2019].

The Orbital Launch Simulator and Trajectory Visualisation Software, https://www2.flightclub.io [retrieved 23 June 2019].

Launch Vehicle Performance Calculator, http://www.silverbirdastronautics.com/LVperform.html [retrieved 23 June 2019].

Trajectory Data

Launch ascent profile data for Space Shuttle (and lots of other space data): Space Place Downloads, http://www.cbsnews.com/network/news/space/home/flightdata/downloads.html [retrieved 31 July 2019].

6.11 Exercises

1. Apollo Lunar Module Ascent Simulation

Simulate a trajectory for the Apollo *Lunar Module*'s (LM) upper stage ascent from the lunar surface into orbit. The LM flew a powered trajectory to its shutdown ('final') height $h_f = 30$ km altitude, where its velocity vector was horizontal. Its speed had to be higher than needed for circular orbit in order to enter an *elliptical Hohmann transfer orbit* that coasted halfway around the moon, where it rendezvoused with the Apollo *Command Module* (CM) at its 250-km circular orbit. Note: please use the following *exact* values for your orbital calculations: $\mu_{moon} = 4902.8$ km^3/s^2, $r_{moon} = 1740$ km.

a) Assuming that the LM's liftoff location is the moon's equator, determine the launch site speed (m/s).

b) Calculate v_p, the shutdown speed of the LM (m/s) required at the perilune (periapsis) of the elliptical transfer orbit.

c) Calculate v_a, the speed at apolune (apoapsis) of the transfer ellipse (km/s).

d) Calculate v_f, the CM's circular orbit speed (km/s).

e) Calculate the $\Delta v = v_f - v_a$ (m/s) required for an apoapsis "kick burn" to match speed with the CM at rendezvous. (Hopefully there is propellant left for this second burn!)

f) What is the orbital period of the CM circular orbit (sec)?

g) Calculate the required coast time (seconds) between shutdown at periapsis of transfer, and apoapsis. In order to assure a rendezvous, the LM and CM must arrive at the rendezvous point at the same time. Because the CM is in a higher orbit, its speed is lower than the LM's transfer orbit, so the CM must be *ahead* of the LM when it launches. Relative to the Moon's center, how much past vertically above the LM must the CM be at LM shutdown? Cite your answer as 1) a time differential *and* 2) an angle differential (in degrees) of the CM ahead of the LM shutdown point. This sets the "launch window" for the LM takeoff.

h) Can we up-front specify the time it takes for the powered ascent with only the above parameters? How do we tell the LM pilots exactly when to liftoff? Explain your answer.

i) **Flight simulation portion**: simulate the powered ascent of the LM. You may use MATLAB® or Excel® or another method for your calculations, and either use Euler or a 4th-order Runge-Kutta integration (see comments below). You should use the trajectory equations given in the text.

   LM Ascent Stage specs: total mass = 4,780 kg, propellant mass = 2,375 kg, ascent engine thrust = 15.57 kN, propellant: NTO/Aerozine-50, I_{sp} = 311 s (from http://www.braeunig.us/space/specs/lm.htm).

   Assume that the LM lifts off vertically and climbs above 100 m, where it changes pitch angle γ_{pitch} to begin a gravity-turn trajectory. Provide the appropriate value of γ_{pitch} to achieve the following *three* conditions at shutdown: (1) $\gamma_f = 0 \pm 3°$ (nearly horizontal at shutdown); (2) shutdown speed $v_f = v_p + 1\% -0\%$ (v_p is the proper shutdown perilune speed for required ascent ellipse; and (3) 25 km $\leq h_f \leq$ 35 km (a $\pm$ **5 km** tolerance on shutdown height). Your trajectory *must* use a gravity turn to achieve these conditions; other than the pitch kick, you may not use a pitch program or prescribe any steering vs. time. You MAY use a throttle reduction (fraction of full thrust capability) if you find it useful (specify the time and throttle percentage used if you do). There is *no* restriction on the downrange distance at shutdown: you only need to provide the proper altitude, speed, and flight path angle. The closer you get to the nominal values, the better, but some tolerances are included so you don't have to hit them exactly. You should find that you don't need to burn all the propellant to meet these shutdown conditions; use a logical operation to shut off your engine when the proper speed is reached. The expression that provides this in Excel® is

"IF(logical_test,[value_if_true], [value_if_false])"; you can create a similar expression in MATLAB®. *Be sure to continue your integration 10-20 seconds after shutdown* to show explicitly the speed and flight path angle (note speed should drop and altitude should increase after the engine shutdown occurs, because the LM should be decelerating as it is coasting uphill on a transfer orbit, eventually reaching apolune altitude of 250 km.

j) Once you have met or come close to the required shutdown conditions, make plots of speed $v(t)$, altitude $h(t)$, flight path angle $\gamma(t)$, and acceleration $dv(t)/dt$ vs. time, all on a single graph; similar to the (non-compliant) example plot below. Dual vertical scales make this easier, and can be done in MS Excel®. *Be sure to show each variable in the legend, and use solid, dashed, dotted, double lines, etc. to differentiate the different variables.* TURN OFF 'SYMBOLS' (dots, squares, triangles, etc.) to make the curves' lines thin (not broad). Hand label curves if necessary. Show the time where shutdown occurs on your plot, and continue plots for at least 20 seconds.

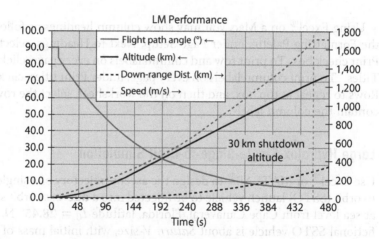

Sample multi-plot of LM ascent. Note that the above is a FAILED trajectory: shutdown occurs at about 450 s, the shutdown altitude is too high (70 km instead of 30), and the flight path angle is not zero at shutdown. Note that it's ok for the slope of the altitude curve to not be horizontal at shutdown; this is to be expected since the LM continues to coast upwards after shutdown. To try to fix this incorrect ascent trajectory, one should increase the pitch angle kick after 100 m climb to increase the rate at which the pitch angle decreases (remember that this pitch rate decrease is determined by the trajectory equations).

k) On another graph, plot altitude vs. downrange distance.

l) If you get close to reaching the desired 30 km altitude, adjust your shutdown time to provide the actual speed needed to reach the specified rendezvous altitude with the CM.

m) Your assignment should contain upload your plots and your spreadsheet or program's time history, which should contain a column for each of the following items (you may need more for intermediate calculations): elapsed time t (s), acceleration a (m/s^2), flight path angle rate $d\gamma/dt$ (rad/s), speed v (km/s), flight path angle γ (deg), altitude h (km), horizontal speed (km/s), downrange distance x (km), vertical speed (km/s), local g (m/s^2), mass m (kg). **Highlight the time of shutdown.** Be sure there are column headings on EACH page output, if you output more than a single page.

Time t (s)	a Accel. (m/s^2)	$\dot{\gamma}$ (rad/s)	v Speed (m/s)	γ Flight path angle (rad)	$\dot{x}$ Horiz. Speed (m/s)	x Down-range Dist. (m)	$\dot{h}$ Vert. Speed (m/s)	Altitude H (m)	g_{local} (m/s^2)	m mass (kg)
0.0	1.64	0	0.0	1.571	0	0	0	0	1.619	4,780

Using Excel® on a Mac, you may show column headings as follows: on the Formatting Palette, under Page Setup, Next to Headings, select the Print check box. To print row and column labels on every page, click "Print Titles." To print column labels on every page, under Print titles, click in the Rows to repeat at top box, and then on the worksheet, select the rows that contain the column labels.

2. Large Hydrolox Single-Stage-to-Orbit Simulation

Use Excel® or MATLAB® to design the ascent trajectory of a single-stage to orbit (SSTO) launch vehicle to be injected into a **due-east** orbit starting at sea level from Cape Canaveral, Florida, latitude $L_0 = 28.45°$ N. The fictional SSTO vehicle is about *Saturn V*-size, with initial mass of $m_0 = 3E6$ **kg** and a dry mass of $m_f = 1E5$ **kg**. Assume the SSTO carries five high-performance hydrolox M-1 engines providing a total constant thrust value of $T = 3.3E7$ **N** and $I_{sp} = 450$ **s**. The vehicle is to be placed into a 28.45° inclination circular orbit.

You will be integrating the equations of motion given in Table 6.1. You may wish to modify the LM program by adding drag, atmospheric density ρ, reference area S, and changing the planetary radius and gravity parameters ($g_0 = 9.80665$ m/s^2 and $R_{earth} = 6,378$ km). For aerodynamic drag, use $D = C_D S \frac{1}{2}\rho v^2$. Assume the vehicle's drag coefficient $C_D = 0.2$ is constant, and the reference area S is that created by its **10 m** base diameter. Assume that the vehicle produces no normal forces, only drag forces, and ignore thrust changes due to ambient pressure. You will need to calculate the atmospheric density ρ as your vehicle climbs: assume a simple exponential atmosphere defined by $\rho(h) = \rho_0 \exp(-h/h_0)$ with a scale height $h_0 = 7,194$ m, and $\rho_0 = 1.225$ kg/m^3.

The SSTO's mass will decrease rapidly during ascent due to propellant consumption (calculate $\dot{m} = dm/dt$ using the definition of thrust T and I_{sp}). Its weight will decrease due to mass flow AND the distance from the earth's center decreasing during climb, $g = g_0/(1 + h/R_{earth})^2$.

Your spreadsheet should carry out one integration cycle for each second of flight, until achieving the proper speed and height (at which point you will shut down the engines). Start with a vertical climb with $\gamma = 90°$. After a vertical climb to about 1 km, try a pitch maneuver of $\gamma_{pitch} = 0.01°$ to the east (new $\gamma = 89.99°$) to begin a gravity–turn trajectory, where the LV slowly "tips over" during ascent. The idea is to have it tip over to $0°$ at shutdown. You will have to "adjust" the value of the "pitch kick" γ_{pitch} using the "method of shooting" to achieve the following three conditions at shutdown: (1) $\gamma_f = 0 \pm 1°$ (nearly circular orbit); (2) $v_f = [\mu_{earth}/(R_{earth} + h_{final})]^{\frac{1}{2}} - v_{LS} + 1\% -0\%$, the proper shutdown speed including launch site speed; and (3) $h_f = 200 \pm 10$ km. Your trajectory must use a gravity turn to achieve these conditions; other than the pitch kick, you are not allowed to use a pitch program or prescribe any steering vs. time. You ARE allowed to use reduced throttle (fraction of full thrust capability) if you find it useful (specify the time and throttle percentage used if you do). There is *no* restriction on the downrange distance at shutdown: you only need to provide the proper altitude, speed, and flight path angle. Tolerances are included so you can "get close" and finish in a reasonable time, and understand how hard trajectory planning actually is!

Your spreadsheet should contain a column for each of the following items (you may use more if you like): elapsed time t (s), altitude h (km), density ρ (kg/m^3), speed v (km/s), downrange distance x (km), flight path angle γ (deg), acceleration dv/dt (units of m/s^2), mass m (kg), propellant consumed (kg), weight W (N), and drag D (N). A spreadsheet showing possible headings (not including gravity and drag losses) is shown in Fig. 6.21 in the text. Once you have met the shutdown conditions:

1. Make a plot showing v, h, γ, acceleration a, and dynamic pressure q vs. time. A sample plot follows. Show time where shutdown occurs, and continue plots for at least 20 seconds. Fig. 6.22 in the text shows a (failed) SSTO simulation. Note that its achieved orbital altitude is 90 km short of the required 200 km value.

2. Show the time t_{q-max}, speed v_{q-max}, and altitude h_{q-max} where max q ($= \frac{1}{2}\rho v^2$) occurs during ascent. Using h_{q-max}, calculate the approximate Mach number M_{q-max}.

3. On another graph, plot altitude vs. downrange distance. Show point where shutdown occurs, and continue plot for at least 20 seconds.

4. Determine the gravity loss and drag loss for your trajectory and compare to known values.

Provide a copy of your spreadsheet, including column headings on EACH sheet, or any code written.

Sample (failed) SSTO simulation. Note altitude is far short (~90 km) of required 200 km value.

Here are possible headings on a spreadsheet (gravity and drag losses not shown):

Scoring: simulations for problems 1 & 2 will be scored using the following rubric for your assignment's grading. Please provide the requested values for your work by filling in the blanks. LM simulations will not be graded on atmospheric effects that are absent (shaded areas for drag loss, max q, etc.).

Student Name (Last, First) _____ Date _____

Item	LM value	Score	SSTO value	Score	Max
1. Shutdown: speed v (km/s)					5
Shutdown: Altitude H (km) $dH/dt = 0$ (horizontal)?					2 + 3
Shutdown: Flight path angle γ (deg) $d\gamma/dt \sim 0$?					2 + 3
2. Gravity loss: calculated by sim (m/s)					5
0.80 × cons. energy (m/s)					2
3. Drag loss: calculated by sim (m/s)*	(omit)				5
Fig. 3.13 est. (m/s)*	(omit)				2
4. Sec 3.3.9 comb. Loss m/s	(omit)				4
5. Max-q value (Pa)*	(omit)				5
Occurs at Flight time (s)*	(omit)				1
Speed (m/s)*	(omit)				1
Mach no.*	(omit)				1
Altitude (km)*	(omit)				1
6. Time to orbit or shutdown (time in s)					5
7. Peak acceleration (g)					1

(Continued)

Item	LM value	Score	SSTO value	Score	Max
8. Plots: speed v	Refer to student plot				5
Altitude h	"				5
Flight path angle γ	"				5
Dynamic Pressure q^*	" (omit)				5
Acceleration a	"				5
Altitude h vs. downrange distance x	"				5
Penalties?					(As needed)
$*$ = omit for LM sim	**Total Scores**		LM = 54 max		78 max

3. Launch Vehicle Trajectory Optimization Using GPOPS-II Software

Obtain a copy of GPOPS-II (available at http://www.gpops2.com/index.html) and examine the sample "Launch Vehicle Ascent Problem" that is provided with the program. Get GPOPS-II running properly (you will need a computer that runs MATLAB®) so you can match the results provided in the sample problem.

Then, choose your favorite launch vehicle and modify the LV Ascent Problem's inputs to match your vehicle. You can find useful engineering data (specific impulses, propellant masses, dry masses, etc.) on the launch vehicle manufacturer's website or on astronautix.com. Run GPOPS-II with your modified inputs, and see how close you come to the "official" performance given in such places as the LV's payload planner's guide or user's guide or equivalent.

If you model a complex vehicle such as the Space Shuttle, be sure to include the maximum 3-g_0 acceleration as a constraint, as well as a value for max-q. You will have fun figuring out what each of the subroutines in GPOPS-II does!

Chapter 7

Space Launch Vehicle Structures and Layout

his chapter will serve as an introduction to the layouts, geometry, and practices used in structures in launch vehicles. We will provide a number of historical examples of different practices used in laying out a space launch vehicle design. The information contained in this chapter will be helpful for you to make design decisions when you begin the layout of your own vehicle, starting in the next chapter.

Here's what you will find in this chapter:

* Structural examples of selected launch vehicles including layouts, components, and details
* Launch vehicle structural types and vocabulary: interstages, intertanks, skirts, payload attach fitting/launch vehicle adapter, payload fairing
* Structural materials: metals, composites and their properties, selection criteria

7.1   The Thor IRBM

Douglas Missile and Space (now part of Boeing) originated the Thor missile in the 1950s. The Thor was an intermediate-range ballistic missile (IRBM) that relied on its *isogrid* tank walls to maintain its structural integrity. Isogrid refers to a metal skin that has equilateral triangles machined into its surface, which removes perhaps 90% of its mass without changing the structural characteristics much. (Isogrid will be discussed in more detail later in this chapter.) The internal arrangement of a Thor missile is shown in Fig. 7.1.

Thor's internal propellant line passed through the liquid oxygen tank situated underneath, allowing a more direct path between the tank and the engine. This pass-through scheme also required two penetrations in the LOx tank below and caused the fuel to be chilled by the LOx within the tank it passed through. Most likely, the engineers considered the weight savings and straightness of the internal fuel line against the simplicity of an external

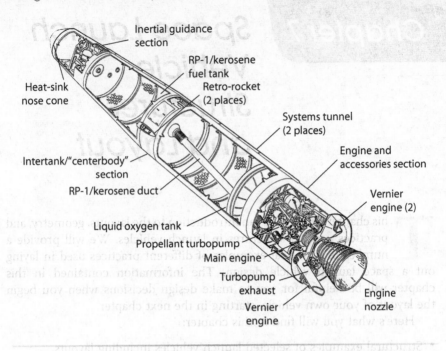

Inertial guidance section

RP-1/kerosene fuel tank

Heat-sink nose cone

Retro-rocket (2 places)

Systems tunnel (2 places)

Engine and accessories section

Intertank/"centerbody" section

RP-1/kerosene duct

Vernier engine (2)

Liquid oxygen tank

Propellant turbopump

Main engine

Turbopump exhaust

Vernier engine

Engine nozzle

Fig. 7.1 The Thor missile's inboard profile. Source: U.S. Air Force Space & Missile Command (SMC).

line. In engineering, there are usually many ways to solve problems successfully; the trick is to find the "best" solution.

During the 1950s and 1960s, engineers realized that the Thor might be a good booster with an upper step attached. There were many variations of the design, including longer propellant tanks, strap-on solid rocket motors, and so forth. One urban legend is that so many changes were made to Thors and their upper steps that it was decided to call the resulting vehicles *Delta*, which is the Greek letter engineers use to represent the concept of change.

Now we'll look at one of the descendants of Thor, the Delta II. Because the Delta II has samples of most of the features used in launch vehicles and uses three different types of propellants, we'll spend some time looking at its details.

7.2 The Delta II: Evolved from Thor

The Delta II (1962–2019) orbital launcher traces its origin back to the original Thor missile. The Delta II-7925 model is shown in Fig. 7.2, and the design evolution history is shown in Fig. 5.21, Chapter 5. The tapered upper section of the Thor was lengthened and turned into a constant diameter cylinder, resulting in a cylindrical vehicle. The effect of the enlargement and lengthening of the tanks is that the existing Rocketdyne (now Aerojet

Rocketdyne) RS-27A rocket engine was no longer adequate to lift the rocket off of its pad, so the design was improved by attaching a number of solid rocket motors (three, four, six, or nine, depending on performance requirements). These so-called "thrust augmentation solids" have graphite-epoxy casings and are jettisoned at their burnout to reduce vehicle mass and increase performance. When nine SRMs are used, six are lit for liftoff, and the remaining three are ignited when the first six have burned out and separated.

The Delta II's first step is shown in the bottom left of Fig. 7.2 and close up in Fig. 7.3. Its rear end features a RS-27A LOx/kerosene rocket engine, above which are oxidizer and fuel tanks. In between the two tanks is an *intertank* section, or "centerbody" as labeled in the figure. The main engine gimbals for pitch and yaw control, and there are two small vernier engines on the bottom used for roll control and attitude control (not visible). On the top of the first step is a structure called the *interstage*, which connects the first step to the second. The interstage is mostly empty but serves to house the exhaust nozzle of the second step's rocket engine. Again, after separation, the interstage remains with the first step to reduce mass.

The Delta II's second step is shown in the middle of Fig. 7.2 and in detail in Fig. 7.4. It is powered by an Aerojet (now Aerojet Rocketdyne) AJ10-118K

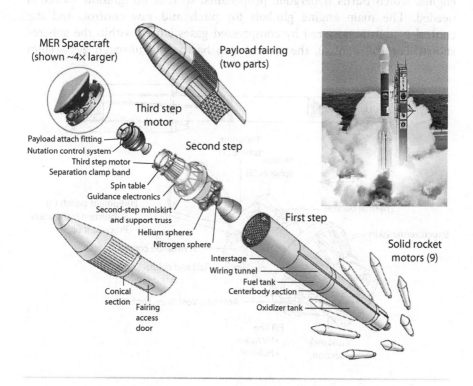

Fig. 7.2   The major features of the Delta II 7925 design. Source: [9].

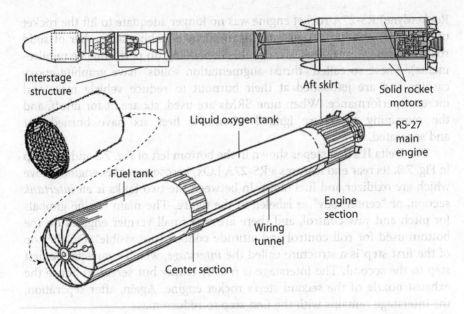

Fig. 7.3   Details of the Delta II first step. Source: [5].

engine, which burns hypergolic propellants, so that no ignition source is needed. The main engine gimbals for pitch and yaw control, and the control system is powered by compressed gases stored within the spheres shown. For roll control, the second step has a cold nitrogen gas system

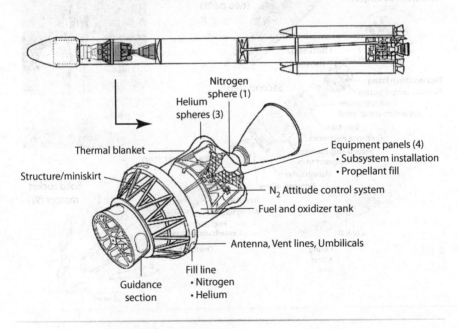

Fig. 7.4   Delta II second step features. Source: [5].

that doubles as a three-axis attitude control system once the second-step engine shuts down. The attitude control system, the redundant inertial flight control assembly (RIFCA), uses ring laser gyros and accelerometers to perform navigation, guidance, and sequencing functions. The "redundant" part of the name indicates that there is a second, backup system in case one malfunctions.

Notice the extreme efforts that were made to reduce mass. The entire second- and third-step assemblies as well as the payload and its fairing are supported by a small portion of the outer rocket body called the *miniskirt* along with the support truss, which is the set of rods connecting the miniskirt to the second- and third-step assemblies. The first step's interstage below the miniskirt is jettisoned at staging, and the payload fairing is jettisoned after a portion of the second-step burn.

At the top end of Delta II's second step, we see a *spin table* that is used to spin up the third step prior to its firing. The spin table is a mechanical bearing that allows the parts above to spin relative to the parts below. There are two reasons for spinning the third step: first, the spin provides stabilization and tends to keep the third step's spin pointing in the same direction, which is especially helpful because the third step does not have an active control system. Second, because the third step is a solid rocket motor, its spin tends to average out any thrust variations that there happen to be. (Note that some Deltas are only two-step and do not fly with a third step at all.)

Also at the top of the second step is the *third-step motor separation clamp band*. This clamp band or Marmon clamp tightly clamps the third step to the second step until the band is released after spin-up. (Clamp bands will be discussed in some detail in Chapter 11.) When the band is released, four springs (labeled "spring actuator" in Fig. 7.5) gently push the two steps apart, after which the third-step solid rocket motor is commanded to fire. The label "separation plane" refers to the plane where the two pieces separate.

The Delta II third step is on the top left of Fig. 7.2, and is also shown in much more detail in Fig. 7.5. Just above the spin table are two spin rockets, which fire to raise the spin rate to 70 RPM (revolutions per minute). Once the SRM has finished firing, a *yo-yo despinner* (not shown) releases two weights on cables that separate outwards from the spin table to absorb most of the rotational energy, dropping the spin rate from 70 to 12 RPM, which keeps the spacecraft spin-stabilized but not rotating so fast that it would take too much propellant to turn the spacecraft.

Above the two spin rockets is a spin table separation clamp assembly. A single thruster is used as a *nutation control system thruster (NCS)* and labeled NCS Thruster in Fig. 7.5. Nutation is a kind of wobble that occurs in some spinning systems, and it's generally not desired, thus the NCS to get rid of it.

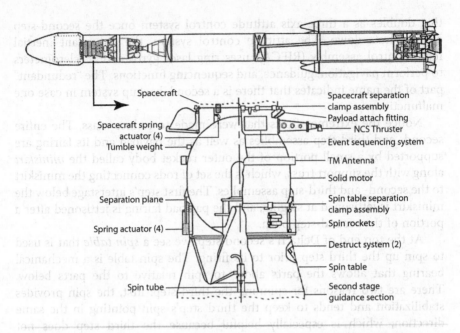

Spacecraft

Spacecraft separation
clamp assembly

Spacecraft spring
actuator (4)

Payload attach fitting

NCS Thruster

Tumble weight

Event sequencing system

TM Antenna

Solid motor

Separation plane

Spin table separation
clamp assembly

Spring actuator (4)

Spin rockets

Destruct system (2)

Spin table

Spin tube

Second stage
guidance section

Fig. 7.5 Delta II third-step details, along with the top of the second step. Source: [5].

Above the SRM is the spacecraft separation clamp assembly, which releases the spacecraft from the third step; springs are used to separate the two. Once the spacecraft has separated from the launch vehicle (LV), the LV's job is done.

7.3 Atlas Takes Tank Structure Principle to Extremes

Why not pressure-stabilize the entire structure like a balloon and get rid of most internal structure also? General Dynamics Convair (GDC), now part of Lockheed Martin, introduced Atlas as the United States' first intercontinental ballistic missile (ICBM). It was designed as a pressure-stabilized structure that did resemble a stainless-steel balloon. To keep it from collapsing, an internal pressure of about 5 psi had to be maintained always, or the vehicle had to be stored between stretching fixtures. Figure 7.6 shows a number of Atlas missiles being fabricated. Inside the tanks, the vehicle had minimal structure between the forward bulkhead and the corrugated engine compartment in the rear, thus minimizing mass.

GDC's Atlas was also unique in that it incorporated so-called stage-and-a-half operation. At the time of Atlas's design, engineers were not sure that they could reliably air start a rocket engine (have it start at a high altitude after a lower step burned out). To avoid the risk of a failed air start, and to meet the performance requirements without the traditional

Fig. 7.6 The Convair/General Dynamics (now Lockheed Martin) three-engine Atlas production line, San Diego, California. Source: https://www.flickr.com/photos/sdasmarchives/16983828850/in/photostream/.

staging scheme, they designed Atlas so that it would drop two of its three engines during boost to get rid of mass, and the single engine and remaining airframe flew to orbit. Because *something* was jettisoned that did not contain empty propellant tanks, it was considered a half stage. The jettisoning of the two engines and support structure improved the vehicle's mass ratio enough to gain the desired performance.

The portion of the Atlas that would separate is the corrugated portion on the right of in Fig. 7.7, and is shown after separation in Fig. 7.8. The latter

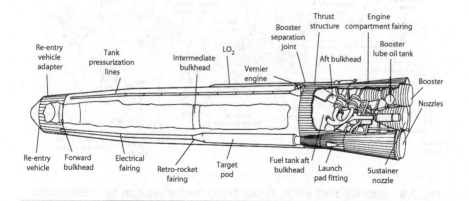

Fig. 7.7 Atlas series B missile cutaway. Source: Courtesy nasaspaceflight.com.

figure shows the rails (*jettison track*) that were used to guide the bottom portion away from the upper portion in a controlled manner.

Another Atlas engineering concept to consider is that early in the flight when there are three gimbaled engines, rotations can be controlled in all three axes: all engines up/down for pitch, all engines left/right for yaw, and outer engines down/up and up/down for roll. Once the two booster engines separated, the remaining (center) engine could only control pitch and yaw. So, the Convair engineers incorporated two *vernier* engines whose differential pointing could be used to control roll. The verniers are located above the booster engine module's separation plane, so they stay on the vehicle after separation.

The Atlas design also had an external liquid oxygen line running down the body. In addition, what is labeled "intermediate bulkhead" in Fig. 7.7 appears to be a single common bulkhead, which eliminates a portion of the body that would not be carrying propellant.

GDC, now part of Lockheed Martin, also provides a pressure-stabilized upper step known as the Centaur—the first hydrogen-fueled upper step, which was introduced in the early 1960s. It is also a pressurized steel balloon, and in some cases was too flexible during the launch, when it was enclosed by the payload fairing. GDC solved this problem by placing a forward load reactor (FLR), a structure made to center the Centaur step on the inside of its payload fairing to keep it from vibrating and causing damage. The Centaur and a diagram of the FLR may be seen in Fig. 7.9. (The vehicle in the photo is the Centaur intended to fly inside the Space Shuttle cargo bay, which allowed the Centaur to be "squashed" axially. The result was called the "short-fat-squat" Centaur.)

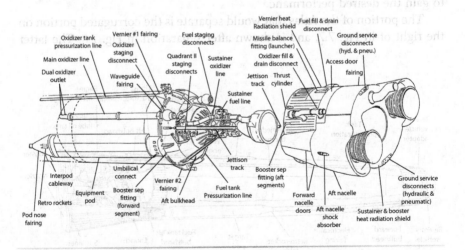

Fig. 7.8 Stage-and-a-half vehicle. Partway through the first step burn, two outside booster engines are jettisoned. The center (*sustainer*) engine continues burning all the way to orbit. Source: Courtesy nasaspaceflight.com.

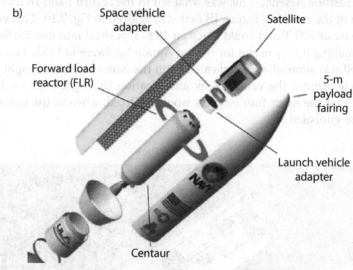

Fig. 7.9 a) The GDC Centaur upper step was like its lower-step Atlas booster, also pressure-stabilized. b) The FLR ring used to keep the flexible Centaur from being damaged by vibration forcing it into the payload fairing.
(The bottom version of Centaur is an entirely different version than the top photo). Sources: photo: Edberg; diagram: [24].

7.4 The Mighty Saturns

In the 1960s, the United States embarked on an ambitious program to land humans on the moon by the end of the decade, which was deemed to be 31 Dec. 1969. ("deemed to be" because a decade starts with the year 1 and ends with the year 10, so the *actual* end of the decade was 31 Dec. 1970.) The requirement of landing humans on the moon meant that heavy-lift launch vehicles needed to be developed, and this process was begun with the Saturn series of launch vehicles.

The work began at Redstone Arsenal near Huntsville, Alabama, where Wernher von Braun was directing the U.S. Army's missile program. That group had developed the Redstone and Jupiter vehicles, and was considering how to scale up to a heavier launch vehicle. Their idea was to use an extended Jupiter tank surrounded by eight Redstone tanks. This configuration is known as *parallel* tanking, because all the tanks are next to each other. This cluster of tanks sits on top of a thrust structure holding eight Rocketdyne H-1 rocket engines. (As far as the authors know, the only other successful LVs having eight or more engines would be the British Black Arrow and the Falcon 9 from SpaceX.)

During the period of Saturn I development, the civilian entity NASA was created, and von Braun and his group were transferred to NASA and set up a new center called the Marshall Space Flight Center (MSFC) on the grounds of the Redstone Arsenal. This was what led to the Saturn I and IB models; a diagram of the resulting Saturn IB first step is given in Fig. 7.10. This vehicle had a mass of 590 T and could place an 18.6-T payload into low Earth orbit (LEO), making it very useful for testing Apollo hardware in LEO. The structure itself was somewhat complicated, with the issues of locating eight tanks in a circle around the central tank and locating the eight engines. It was decided that the inner four engines would be fixed, whereas the outer four would be gimbaled.

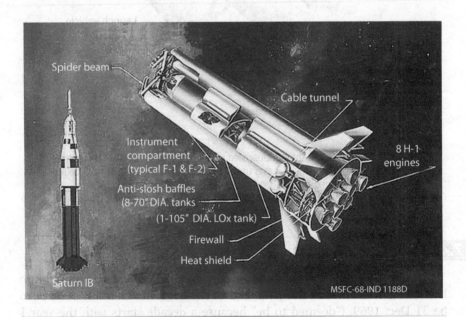

Fig. 7.10   Saturn IB First Step (designated S-IB). Six Redstone-diameter RP-1 tanks and two Redstone-size LOx tanks surround one Jupiter-diameter LOx tank. Spider beams, with associated structure, hold it all together. Source: NASA.

Fig. 7.11 Saturn I front end. The spider beam is in the center of the photo and connects the eight outer tanks to the center tank. Source: Edberg.

Grouping these nine tanks required some complex structures. A photo of the so-called "spider beam" (see label at top left of Fig. 7.10) and nearby structure is provided in Fig. 7.11. We will see later in this chapter that this configuration has poor weight efficiency.

There is a lot of flexibility in the design process. In general, there may be several ways to solve a particular set of requirements. This leaves the designer the freedom to choose the internal arrangement.

One example of this idea is the geometry and plumbing of tanks. The designer may choose from a number of arrangements, as shown in Fig. 7.12. Here we see tandem tanks with external plumbing and internal plumbing, tanks with common bulkheads, clustered parallel tanks, and a toroidal (doughnut-shaped) tank arrangement. All of these have been used for rocket designs in the past, and each has its advantages and disadvantages.

The first two—tandem tanks with separate bulkheads and external or internal piping—are probably the most frequently used. Using tandem tanks with a common bulkhead can reduce overall length and dry mass, but with the complexity of the engineering of the common bulkhead itself, which must provide both separation and insulation. Multiple parallel tanks have been used in several vehicles, the most famous of which are the Saturn I, Saturn IB, and Proton. Toroidal tanks have been utilized in several vehicles' upper steps, where overall length may be at a premium.

Toroidal tanks, as shown to the right of Fig. 7.12, are utilized when length is at a premium. Note that there is room inside the center of the tanks for equipment that would otherwise have to be in front of, or behind, the tanks. Also note that the toroids must be inclined slightly so that the contained propellants may be removed from a single low point. Toroidal tanks have been used on the upper steps of some Russian rockets, as well as the terminal guidance steps of U.S. ICBMs.

Figure 7.13 provides the masses of some of these design concepts. Figure 7.13a, with tandem tanks and common bulkhead, is taken as the reference version with standard mass of 100%. The next design (Fig. 7.13b), with its integrated helium bottle, gives the lowest mass (93%), but there may be issues with piping routes. The integrated helium bottle adds another level of complexity, as well as difficulty with inspections. The third design (Fig. 7.13c), with concentric tanks at 118%, provides simpler propulsion duct routing, but it's not clear how the inner tank could be inspected on its outer face. Finally, the fourth design (Fig. 7.13d), multitanks, is a cluster of tanks mounted together in parallel. This can be attractive if there are existing tanks available with the right size or diameter. This configuration is relatively easy to fabricate and handle, but has the highest mass compared to the standard design at top: 162%.

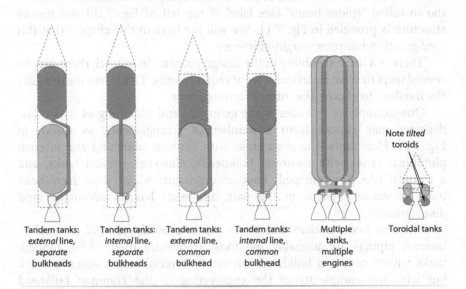

| Tandem tanks: external line, separate bulkheads | Tandem tanks: internal line, separate bulkheads | Tandem tanks: external line, common bulkhead | Tandem tanks: internal line, common bulkhead | Multiple tanks, multiple engines | Toroidal tanks |

Fig. 7.12 Different possibilities of internal tank arrangements. Note that the toroidal (doughnut-shaped) tanks are slightly inclined so the propellant can be drained from the lowest point in the tanks.

a) Tandem propellant tanks with common bulkhead and separate helium bottles—tankage weight, 100%

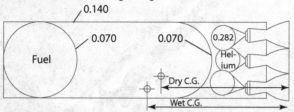

b) Tandem propellant tanks with integrated helium bottle in the middle—tankage weight, 93%

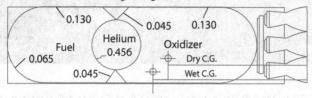

c) Concentric propellant tanks with integrated helium bottle at aft end—tankage weight, 118%

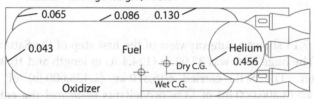

d) Multiple propellant tanks and helium bottle in cluster—tankage weight, 162%

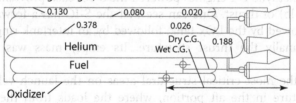

Fig. 7.13 The masses of different configurations are compared to the baseline at top, common bulkhead. Numbers provided appear to be tank wall thicknesses in inches; they are not specified by the author. Source: [11].

7.5 The Saturn V

The large booster work continued at MSFC, where an even larger LV was being developed to take the Apollo lunar-landing spacecraft and its crew to the moon and back. That vehicle was the Saturn V, a vehicle taller than a 36-story building. Because of the wealth of data available on the Saturn V, we have used and will continue use it as an example for design calculations.

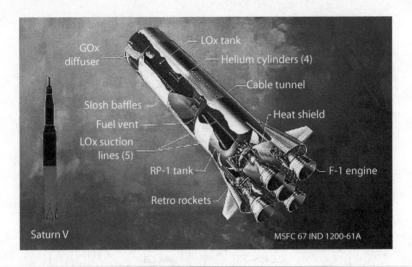

Fig. 7.14   Saturn V's S-IC step provided 6 million lb of thrust to deliver the rest of the vehicle to about 40 miles altitude. Propellants were LOx and RP-1. Source: NASA-MSFC.

Figure 7.14 shows a cutaway view of the first step of the Saturn V, called the S-IC. The huge S-IC was 37.92 m (124.4 ft) in length and 10.06 m (33 ft) in diameter. The S-IC carried 646,370 kg (1,425,000 lb) of RP-1 and 1,499,580 kg (3,306,000 lb) of LOx propellants to propel the entire Saturn V stack up about 40 miles in about 2.5 min.

Five Rocketdyne F-1 engines powered the S-IC, each providing 6.67 MN (1.5 million lb) of thrust. One can see that the S-IC has a forward skirt at the top, followed by the LOx tank, followed by an intertank ring, the RP-1 tank, and finally the thrust structure. Its empty mass was 130,410 kg (287,500 lb).

One of the most critical structural areas on the launch vehicle is the thrust structure in the aft portion, where the loads from the engine or engines are transmitted to the rest of the vehicle. This area also supports the entire weight of the loaded launch vehicle until it begins flight.

Generally speaking, the engines' thrust produces a concentrated "point" load that must be beamed out to the body of the LV. Because of the thin-walled nature of the tanks and other body pieces, it's imperative to have the loads uniformly spread out around its circumference. In addition, the lower structure must also accommodate propellant feed lines, wiring, and the thrust vector control (TVC) actuation system. Some different arrangements are shown in Fig. 7.15.

Most launch vehicles use a TVC system for stability and control. So, we find that the engine has to be mounted firmly in the three translations, but must be free to rotate in pitch and yaw in order to control the vehicle. (The engine is also restrained in the roll axis.) Hydraulic actuators are used

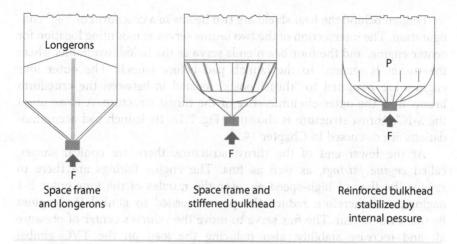

Longerons

P

F

F

F

Space frame
and longerons

Space frame and
stiffened bulkhead

Reinforced bulkhead
stabilized by
internal pessure

Fig. 7.15 Several ways to beam out or distribute the thrust load to the rest of the launch vehicle structure. Source: [20].

to rotate or gimbal the engine to provide side thrust. Figure 7.16 illustrates some of these concepts.

The structural arrangement of the S-IC step is shown in Fig. 7.17. This illustration provides some detail as to what each of the parts is made of and what its design requirements might be. For example, we see that the top ring is a nonpressurized forward skirt with circular rings and longitudinal stiffeners. The intertank section has similar construction, as does the outer portion of the thrust structure. These parts are not pressurized, so in order for them to withstand the imposed loads, *longitudinal stiffeners allow each part to take the required compressive loads*, while the *rings help with buckling stability*. This is true for *all* nonpressurized rings that are not sandwich construction.

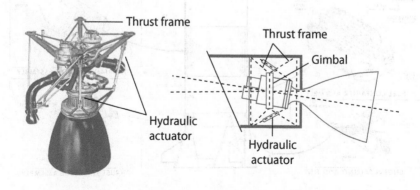

Thrust frame

Thrust frame

Gimbal

Hydraulic
actuator

Hydraulic
actuator

Fig. 7.16 How the engine can be gimbaled inside the thrust structure.

Hidden behind the heat shield are two beams in a cruciform or " + " configuration. The intersection of the two beams serves as mounting location for center engine, and the four beam ends serve as the holddown points where the rocket is secured to the launch pad before launch. The outer four engines are mounted to "thrust posts" located in-between the cruciform beams near the outer circumference of the thrust structure. A close-up of the S-IC's thrust structure is shown in Fig. 7.18. Its launch pad accommodations are discussed in Chapter 14.

At the lower end of the thrust structure, there are conical shapes, called *engine fairings*, as well as fins. The engine fairings are there to reduce the flow of high-speed air over the nozzles of the four outer F-1 engines, and therefore reduce the power needed to gimbal the engines by the TVC system. The fins serve to move the vehicle's center of pressure aft and increase stability, also reducing the load on the TVC gimbal system.

Above the thrust structure, one finds the serial (one on top of the other) oxidizer and fuel tanks, with oxidizer (LOx) above the fuel (RP-1). Both of these structures have ellipsoidal domes (bulkheads) at either end; in addition, there are a number of rings inside of both tanks. The rings are installed internally both for compressive strength and to dissipate *sloshing*, which is a side-to-side motion of the contained fluids that occurs when the tank is partially empty. The sloshing can cause instabilities, which will be discussed in Chapter 12.

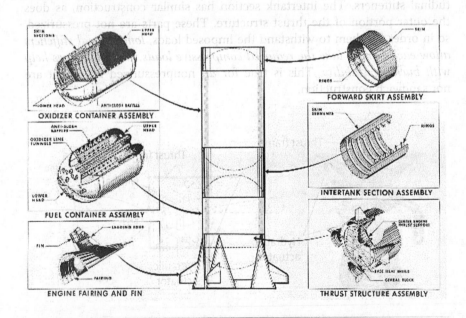

Fig. 7.17 Saturn V S-IC structural arrangement. Source: NASA.

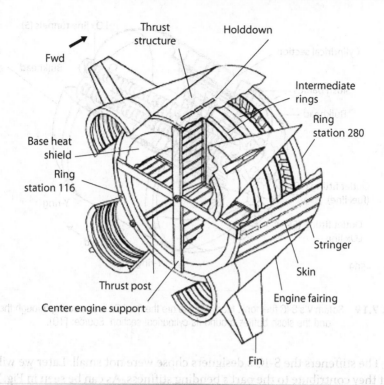

Fig. 7.18   The Saturn V's thrust structure has a skin-stringer cylindrical frame along with crossbeams to mount the center engine, and to hold the rocket on to the launch pad. The four outer engines were attached to thrust posts near the outer perimeter of the thrust structure. Engine fairings protect the outer four engines from the slipstream, and fins provide added stability while flying in the atmosphere. Source: NASA.

Above the thrust structure is the fuel tank, whose construction is more complicated because five oxidizer lines pass directly through the tank to supply the LOx to the engines underneath. This means that there were an additional 10 tank penetrations for these lines, each of which is 43–53 cm (17–21 in.) in diameter. And remember, these feed lines had to be able to withstand the vibrations imposed at launch, and also expand and contract so they were not damaged by the large temperature difference between the RP-1 and LOx. The layout of the fuel tank may be seen in Fig. 7.19.

The tank is all-welded 2219 aluminum alloy. There are small pieces labeled "Y-Ring" at the top and bottom of the cylindrical section. These machined metal Y-shaped rings are used to provide attachments from the domes to the cylindrical sections above and below the tank. A photo of a Y-ring is shown in Fig. 7.20.

Between the two tanks is an intertank section, which is a conventional nonpressurized cylinder with longitudinal stiffeners and internal rings. Figure 7.21 shows a drawing of the intertank structure on the left and the external surface of the intertank structure on the right.

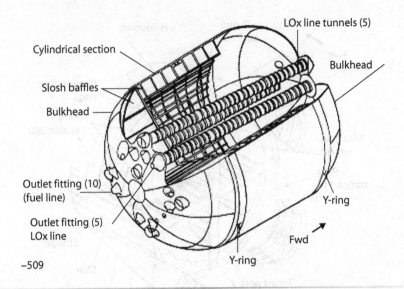

Fig. 7.19 Saturn V's S-IC fuel tank. Most notable are the five LOx lines passing through the tank, and the slosh baffles around its cylindrical section. Source: [10].

The stiffeners the S-IC's designers chose were not small. Later we will see that they contribute to the part's bending stiffness. As can be seen in Fig. 7.22, the stiffeners were 8.84 cm (3.48 in.) tall and spaced at 29-cm (11.40-in.) intervals. The designers chose an elegant arrangement to attach the stiffeners on the intertank to the Y-ring, as can be seen in Fig. 7.22b.

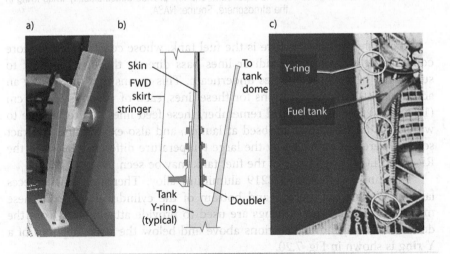

Fig. 7.20 a) Metal Y-rings on the tanks' tops and bottoms were used to join the tank to the structures above and below them. b) How the Y-ring's straight side was attached to the outer skin of the rocket and the curved part connected to the dome. c) The stiffeners on the inside of the tank wall. Sources: photos: Edberg; drawing: [12].

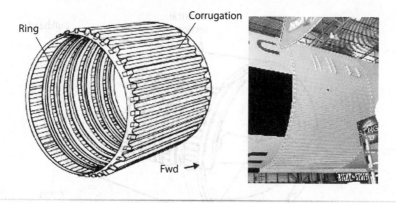

Fig. 7.21   The Saturn V's S-IC intertank area, showing the internal rings and external stiffeners.
Sources: left: [10]; right: Edberg.

Above the S-IC's intertank is the LOx tank, as shown in Fig. 7.23. It is similar in construction to the fuel tank below it. The LOx tank is an all-welded 2219 aluminum-alloy structure.

Finally, we come to the last major piece of structure on the S-IC: the forward skirt. The forward skirt connects to the top of the LOx tank with a Y-ring, and above it is the interstage separating the S-IC from the S-II. Its construction is similar to the intertank described previously. A drawing of the forward skirt is shown in Fig. 7.24.

Now we'll move on to the second step of the Saturn V, the S-II step. Like the S-IC first step, the S-II had five engines, but these were lower-thrust Rocketdyne J-2 engines. Unlike the first step, the S-II propellants were

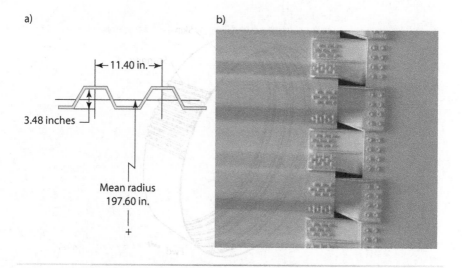

Fig. 7.22   The geometry and attachment details of the Saturn V's interstage stiffeners.
Sources: drawing: [11]; photo: Edberg.

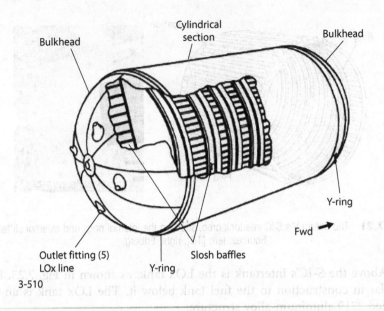

Fig. 7.23 The construction of the S-IC's LOx tank. Source: [10].

liquid hydrogen and liquid oxygen, so we should expect to see both similarities and differences between the two steps, because of their different propellants. The inboard profile of the S-II step can be seen in Fig. 7.25.

There are a number of differences between the S-IC and the S-II. The S-II step carried cryogenic propellants, namely liquid hydrogen and liquid

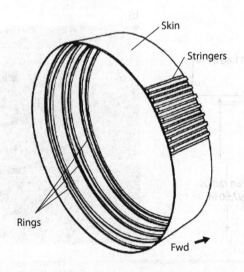

Fig. 7.24 The S-IC's forward skirt connects the S-IC step to the S-II step above it. Source: [10].

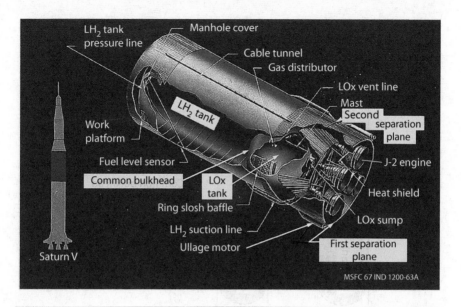

Fig. 7.25   Saturn V second step, the S-II. This step also has serial tanks and a common bulkhead.
Source: NASA MSFC.

oxygen. These propellants are known for their higher specific impulse; however, the drawback is that the LH_2 tank is very large because the fuel is not very dense.

In the S-II, there is no intertank structure connecting the two propellant tanks. Instead, there is one *common bulkhead*, which is between the upper LH_2 tank and the lower LOx tank. This bulkhead design was selected because it saved a lot of mass due to the elimination of (1) a bulkhead and (2) an intertank structure between the fuel and oxidizer tanks. The common bulkhead for the S-II had to provide insulation between a $-252°C$ ($-422°F$) temperature for LH_2 and $-182°C$ ($-295°F$) for LOx, a very large $70°C$ ($127°F$) difference. Because of its low temperature, the LH_2 tank was covered with external insulation. The bonding of the insulation caused many problems; finding an adhesive that worked at such low temperatures was difficult.

Let's look at the characteristics of the common bulkhead configuration. Referring to Fig. 7.26, a common bulkhead saves length (an extra dome and external ring structure are eliminated), which also saves mass, although the common bulkhead may be somewhat heavier than the two single bulkheads it replaces due to the insulation it must carry if the propellants are stored at different temperatures. This construction was used in the second step of the Saturn V, the S-II, where a common bulkhead saved 3,600 kg (8,000 lb) over a conventional two-bulkhead design. A section of the S-II's common bulkhead is shown in Fig. 7.27. The construction is approximately 15 cm-thick (5.9 in.-thick) fiberglass honeycomb between two aluminum sheets.

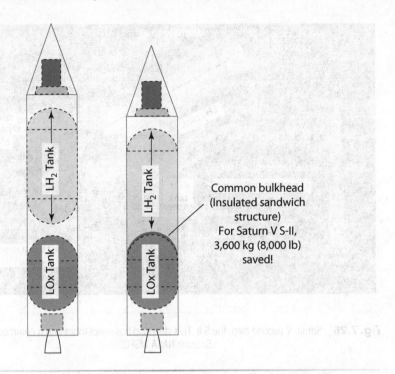

Fig. 7.26 Conventional layout (two separate tanks) vs common bulkhead (tanks share a dome, and no more cylinder between). Note length savings, weight savings (3,600 kg on Saturn V S-II step), and one less dome to worry about–albeit the remaining common bulkhead is MORE complicated than a regular dome.

Fig. 7.27 The S-II's common bulkhead had to withstand a very large temperature difference of 70°C from LH$_2$ to LOx. Construction is aluminum face sheets, fiberglass honeycomb. Source: Edberg.

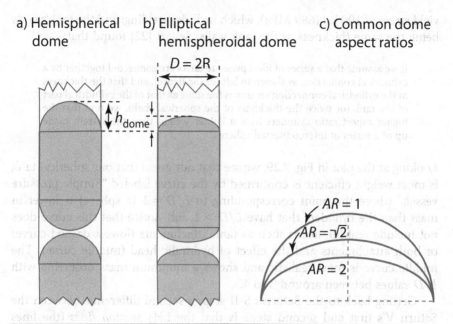

Fig. 7.28   Tank dome geometries. a) hemispherical domes, b) Elliptical hemispheroidal dome shape, and c) dome shape vs. aspect ratio of hemispherical ($AR = 1$) and two elliptical hemispheroidal domes.

7.6   Another Way to Save Mass: Tank Dome Shapes

Besides the hemispherical dome shape, another shape that is commonly used is the Elliptical hemispheroidal (EH) dome, as shown on the right side of Fig. 7.28. Notice that the EH dome allows the two tanks to be closer together, which can reduce the rocket's mass due to the shorter length needed for the intertank structure.

7.7   Spherical vs Cylindrical Tanks: Which Have Less Mass?

We are often told that a spherical tank (not a cylindrical tank with hemispherical domes, but a fully spherical tank) is the lowest-mass configuration, because it has the minimum surface area for a given volume, and its stresses when pressurized are equal in all directions, minimizing the material required to build them. Although this is true for an individual tank in microgravity, it's not necessarily true for a series of tanks that is experiencing multiple-g loading. What effect does the required hardware for mounting have? What about the effect of *hydraulic head* (the pressure that varies with depth in a fluid due to acceleration)?

The plot in Fig. 7.29 will help to answer these questions. With assumptions of aluminum walls; 30-psi head (207 kPa); a load factor $n = 2.5\ g_0$; a yield stress of 100 ksi (689 MPa), which includes welding; and the tank's ends being the same thickness as the tank walls, Seifert [22] found that

> If we assume that a series of ideal pressure tanks are connected together by a cylindrical connection as shown to left [of the figure], and that the thickness of the cylindrical connection section is the same as that of the cylindrical part of the tank (or twice the thickness of the spherical shell), we find that the higher aspect ratio cylinders have a lighter weight than the system made up of a series of interconnected spheres.

Looking at the plot in Fig. 7.29, we see that our guess that our spherical tank is most weight efficient is confirmed by the curve labeled "Simple pressure vessel," where the point corresponding to $L/D = 1$ (a sphere) is lower in mass than the cylinders that have $L/D > 1$. But notice that this curve does not include real-life needs such as tank attachments (lowest dashed curve) or both attachments and the effect of hydraulic head (middle curve). The middle curve is most realistic and shows a minimum mass occurring with L/D values between around 3 to 4.

Getting back to the Saturn's S-II step, a second difference between the Saturn V's first and second steps is that the LH_2 *suction lines* (the lines that feed hydrogen to the engines) go outside and around the lower LOx tank, rather than through it.

Because the S-II's engines extend past the bottom end of the LOx tank, volume must be provided for them above the S-IC. This space is inside the *interstage ring* that extends between the first separation plane and the second separation plane shown in Fig. 7.25, and on the lower right of Fig. 7.32. Figure 7.30 shows still frames from a famous movie of the Saturn V's staging event.

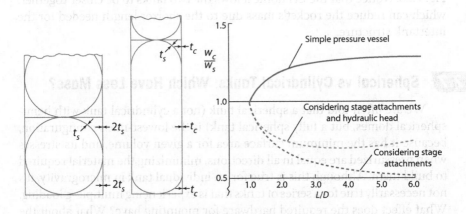

Fig. 7.29   For aluminum tanks having yield at 689 MPa (100 ksi) with 207-kPa (30-psi) head and load factor of 2.5, cylinders$_c$ have a lower mass than a system made up of a series of interconnected spheres$_s$. A cylinder L/D ratio between 3 and 4 seems about the minimum mass. Source: [22].

Fig. 7.30 Saturn V's S-IC/S-II interstage ring being jettisoned. Source: NASA.

The S-II step has a similar structural crossbeam system whose four ends and intersection are used to mount its five engines, but unlike the S-IC, it has a truncated-cone shape to accommodate the bottom end of the LOx tank. This thrust structure can be seen in Fig. 7.31, and the truncated-cone shape is visible on the right side of Fig. 7.32. Details of the S-II's construction are shown in Fig 7.33.

The Saturn V's uppermost step is the S-IVB. The layout of this step is shown in Fig. 7.34. Like the S-II, the S-IV carried hydrolox propellants and also featured a common bulkhead to separate its serial tanks; however,

Fig. 7.31 Saturn V S-II step J-2 engines and thrust structure. Source: Edberg.

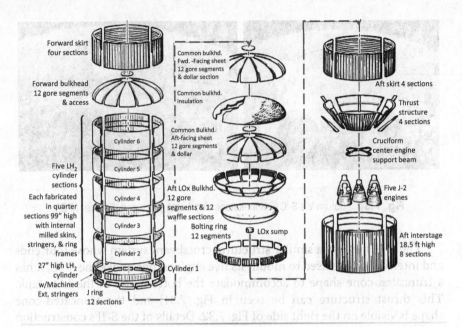

Fig. 7.32 Saturn V S-II step major components. Source: NASA.

it had some differences as well. Its designers opted for internal insulation that allowed a warmer adhesive bond compared to external insulation used on the S-II. Also, because the step had a single J-2 engine, it required an auxiliary

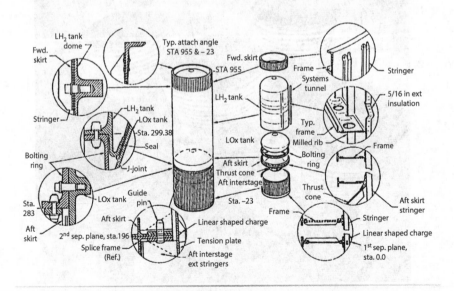

Fig. 7.33 Saturn V S-II step structural details. The external surface of the LH$_2$ tank was covered with 8-mm-thick (5/16-in.-thick) insulation. Source: NASA.

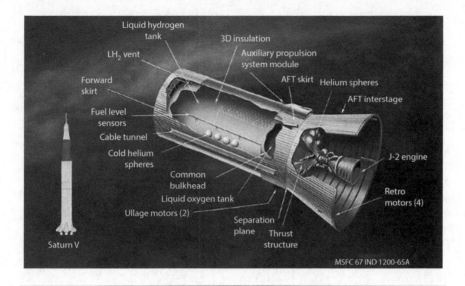

Fig. 7.34   The Saturn V third step, S-IVB, which used hydrolox propellants as did the S-II. It also had a common bulkhead. Source: NASA MSFC.

propulsion system (APS) for both attitude control and restart capability. Helium spheres were carried in the liquid hydrogen tank, so that more helium could be stored at the low temperature of LH_2.

One of the notable features of the top of the S-II step was the large conical interstage skirt used to house the S-IVB's single J-2 engine, as can be seen on the right side of Fig. 7.34. The interstage was mostly empty space, so it seems like there might be a more efficient way to pack things.

Earlier, we saw that one way to reduce a vehicle's length is to use elliptical bulkheads or a common bulkhead. Another issue that can occur is that the upper step's rocket engine nozzle can be quite large, which could mean a lot of dead space inside of the interstage structure, not to mention the excess mass for the structure enclosing this mostly dead space.

This problem has been solved using the so-called *telescoping* or *crank down* nozzle, as shown in Fig. 7.35. The figure shows that the RL-10's nozzle has two parts that can be collapsed (left), and then the outer portion is driven downwards with ball-screws until it's in solid contact with the upper portion (right). This allows a length reduction of about 1.44 m, which results in a shorter LV with less mass.

Another consideration for steps with liquid propellants of differing density (e.g., hydrolox) is the fore-and-aft orientation of the two propellants. As in the case of the Saturn V's S-II and S-IVB steps, it's usually preferable to put the LH_2 tank *forward* of the heavier LOx tank, which offers the advantage that the side loads produced by wind gusts are smaller, meaning it can reduce mass. However, having the lighter LH_2 tank forward of the heavier LOx tank

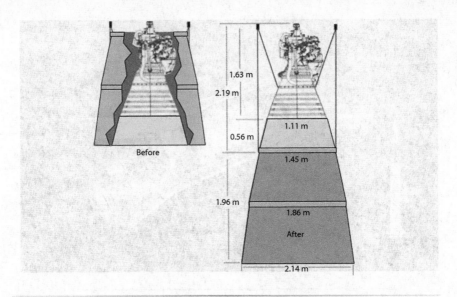

Fig. 7.35   RL-10 engine with deployable nozzle stowed (L) and deployed (R). Stowing nozzle reduces needed storage length by almost 2 m. Credit: Norbert Brügge.

moves the center of gravity (CG) aft, and it increases CG travel with respect to the bottom of the rocket, both of which can reduce stability. For this reason, both the Ariane V and the Space Shuttle are notable exceptions and have the LOx tank forward, because they have controllability issues. We will discuss the tank arrangement in more detail in Chapters 9 and 12.

Figure 7.36 shows a cross-section of the Ariane V first step. Like the Saturn V's S-II and S-IVB steps, the Ariane V uses hydrolox propellants

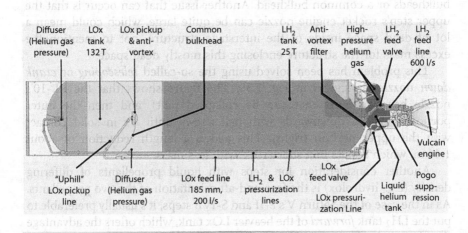

Fig. 7.36   Some of the internal parts of the hydrolox-powered Ariane V first step. Source: [23].

and a common bulkhead (bottom of LOx tank). The LOx tank is forward for this design; note that the common bulkhead is concave upwards, rather than concave downwards as shown earlier in Fig. 7.26. This makes an interesting LOx pickup system that moves the propellant "uphill" from the center of the tank, exits the tank, and continues down the outside of the body. Pressurization lines pass parallel to the LOx duct. High-pressure helium bottles are placed near the engines, so the heat from the engines may be used to warm up the helium for pressurization.

7.8 The Space Shuttle

The U.S. Space Shuttle consisted of an orbiter, an external tank (ET), and two large solid rocket boosters (SRBs). The configuration of the Space Shuttle is shown in Fig. 7.37. The orbiter is the winged vehicle. The ET is the large structure between the two SRBs. The ET carried all of the liquid propellants used by the Space Shuttle main engines (SSMEs).

The ET, shown in Fig. 7.38, was literally the backbone of the vehicle during boost and had to absorb loads applied by both SRBs and the orbiter's main engines. The external tank actually consists of two tanks, one each for LOx (forward) and LH_2 (larger aft tank) separated by an intertank structure. The ET tank structures consist of mainly preformed aluminum elements along with integrally machined longitudinal stringers and ring frames. Unlike the common bulkheads seen on the Saturn V and Ariane V hydrolox

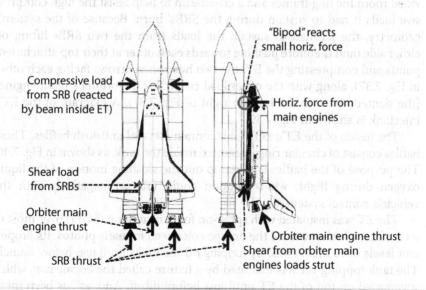

Fig. 7.37 Plan (left) and side (right) views of the Space Shuttle stacked for launch. The arrows indicate thrust forces that occur during launch and reactions on the ET. Source: [15].

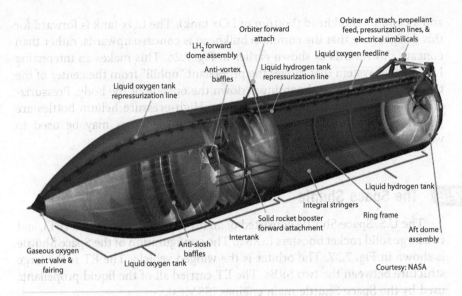

Orbiter forward attach

LH$_2$ forward dome assembly

Anti-vortex baffles

Liquid oxygen tank repressurization line

Orbiter aft attach, propellant feed, pressurization lines, & electrical umbilicals

Liquid oxygen feedline

Liquid hydrogen tank repressurization line

Liquid hydrogen tank

Integral stringers

Solid rocket booster forward attachment

Ring frame

Aft dome assembly

Intertank

Anti-slosh baffles

Gaseous oxygen vent valve & fairing

Liquid oxygen tank

Courtesy: NASA

Fig. 7.38 Shuttle external tank (ET) construction. Source: [18].

steps, the ET featured two separate bulkheads between the propellant tanks. The intertank was constructed similar to other nonpressurized intertanks shown earlier and featured a conventional semimonocoque structure.

The empty volume between the two bulkheads inside the intertank provided room for ring frames and a crossbeam to help resist the high compressive loads it had to sustain during the SRBs' burn. Because of the system's geometry, the ET had to sustain the loads from the two SRBs lifting on either side (and therefore pushing towards each other at their top attachment points and compressing the ET, the two horizontal arrows facing each other in Fig. 7.37), along with the offset load caused by the orbiter's main engines (the slanted arrow in the bottom right of Fig. 7.37). A diagram of the ET's intertank is shown in Fig. 7.39.

The inside of the ET's LOx tank contains several antislosh baffles. These baffles consist of circular rings mounted inside the tank, as shown in Fig. 7.40. The purpose of the baffles is to damp out the sloshing motion of the liquid oxygen during flight, whose motion could interact negatively with the vehicle's control system.

The ET was insulated with spray-on foam insulation (SOFI) over most of its surface, which provided the orange color seen in many photos. Its propellant levels were maintained by topping off the tanks until just before launch. The tank topping-off was executed by a fixture called the *beanie cap*, which was placed on top of the ET until just before liftoff. And, as has been mentioned previously, each ET was discarded after launch even though each achieved about 99% of orbital speed.

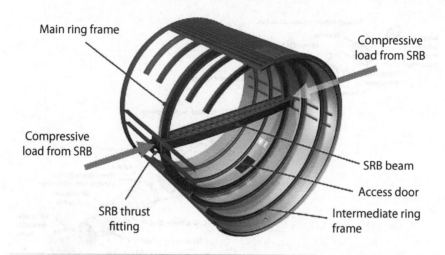

Fig. 7.39 The ET's intertank structure features a crossbeam (SRB beam) to sustain the SRB compressive loads. Compare to Fig. 7.21 showing the Saturn V's S-IC intertank, which does not have any external side forces applied and requires no crossbeam. Source: Ed LeBoutillier.

The Space Shuttle's SRB strap-ons comprise an 11-piece structure, as shown in Fig. 7.41. The large, cylindrical portion containing propellants was assembled with propellant-loaded cylinders made of 13-mm-thick, high-strength D6AC steel. The SRB's fore and aft external covers, nose cap, frustum, and aft skirt were conventional skin-stringer construction.

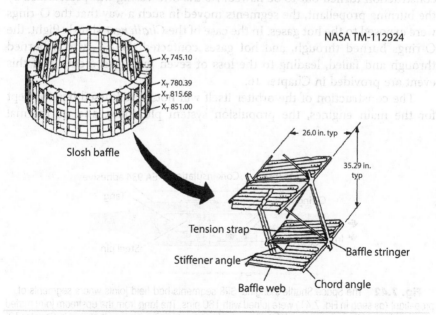

Fig. 7.40 The slosh baffle rings inside the ET LOx tank. Source: [17].

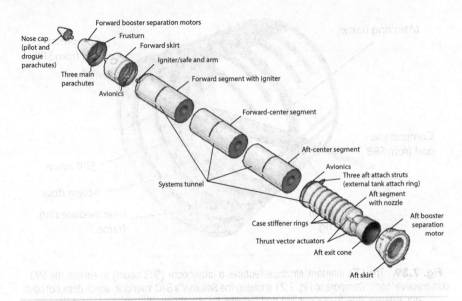

Fig. 7.41   Shuttle SRB details. Source: [18].

The propellant segments were stacked and then joined with steel pins, as shown in Fig. 7.42. The stacking process is described in Chapter 13. The segments' junctions were wrapped with fiberglass and sealed with O-rings. This construction turned out to be flawed. As the SRB casing was pressurized by the burning propellant, the segments moved in such a way that the O-rings were exposed to the hot gases. In the case of the *Challenger* shuttle flight, the O-rings burned through, and hot gases contacted the ET, which burned through and failed, leading to the loss of seven lives. More details of this event are provided in Chapter 16.

The construction of the orbiter itself will not be discussed here. Except for the main engines, the propulsion system plumbing, and its thermal

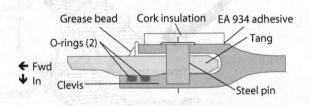

Fig. 7.42   The Space Shuttle's original SRB segments had field joints where segments of propellant (as seen in Fig. 7.41) were joined with 180 pins. The tang from the upstream joint mated with the clevis of the downstream joint. The actual segment joining process is shown in Chapter 13. Source: NASA.

protection system, it is more representative of an aircraft than a launch vehicle. More information on the orbiter may be found in many references, such as [16].

The Space Shuttle system was retired in 2012 after making 135 flights. The system was designed to be reusable, although only the SRBs and the orbiter were reused. Modified versions of the SRBs and ET are to become the primary components of NASA's Space Launch System (SLS), which is scheduled to fly in the 2020s.

7.9 Delta IV

The last vehicle we'll consider is the Delta IV. This vehicle has a "heavy" configuration that is shown in Fig. 7.43. The heavy vehicle is assembled from *common booster cores (CBCs)*, the implication being that any CBC could be used on any vehicle. Upon inspection, one can see that the loads on the center part of the vehicle are very different from those in the two side strap-ons, as with the case of the Space Shuttle in Fig. 7.37. Either all of the CBCs are built to withstand the center body's loads (and thus are heavier), or the strap-ons can be built lighter. If the latter is true, than the common booster core isn't so "common."

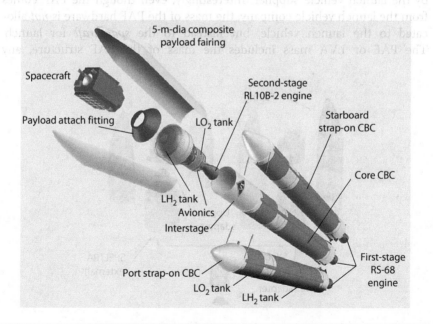

Fig. 7.43   The Delta IVH vehicle's lower step is assembled from three CBCs fastened to each other. Source: [25].

7.10 Other Design Layout Considerations

There are many more considerations that can drive the design and layout of a vehicle's tanks, bulkheads, propellant conditioning, and other factors. For more information, references [26,27,28] are recommended.

7.11 Payload Accommodations

Now it's time to look at the LVs' payload accommodations. We have to ensure that the payload (a spacecraft or satellite) survives the launch process. Besides designing the payload to survive the shock and vibration environment (which will be described in Chapter 11), as the launch provider we need to (1) securely attach the payload to our launch vehicle and (2) protect it from the prelaunch and launch environments as much as possible. The former is solved by a piece of attachment hardware. The latter is solved by an enclosure that provides protection to the payload. Samples of both of these parts are shown in Fig. 7.44.

7.11.1 Payload Attach Fitting

As its name suggests, the payload attach fitting (PAF) [or alternately launch vehicle adapter (LVA)] is a part that is supplied in almost all cases by the launch vehicle supplier. Interestingly, even though the PAF comes from the launch vehicle company, the mass of the PAF hardware is *not* allocated to the launch vehicle, but instead to the *spacecraft* for launch. The PAF or LVA mass includes the mass of the PAF structure, any

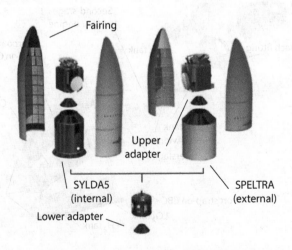

Fig. 7.44 An assortment of payload attach fittings and fairings for dual payloads on the Ariane 5, as described in its user's manual. Source: [8].

separation devices, cabling to and from the LV, and any other equipment needed to mount the spacecraft to the LV. A typical PAF is visible near the bottom of Fig. 7.45.

The spacecraft designer must choose the proper PAF for his or her design. As a rough order-of-magnitude (ROM) estimate, the PAF mass can be estimated from a *mass estimating relationship (MER)* provided in Brown [6]:

$$m_{adapter} = 0.0755$$
$$\times\, SC\, launch\, mass\, (kg)$$
$$+\, 50\, kg$$

This estimate utilizes the "wet" or fully loaded mass of the spacecraft, not its dry mass. Of course, this estimate should be replaced as soon as possible with a more accurate one based on the actual mass and geometry of the payload and its attachment fitting.

Fig. 7.45 Maven spacecraft mounted to a payload attach fitting (circled). Source: NASA (2014).

For existing LVs, the PAF dimensional and mass data may be found in the launch vehicle manufacturer's *payload planner's guide, payload user's guide, mission planner's guide,* or equivalent. Figure 7.46 shows a 6915 PAF in an exploded view of a Delta II and an excerpt from a Boeing Delta II payload planner's guide (PPG). Note that the mass of the 6915 PAF is given as 93 kg (205 lb$_m$) and it includes four "hard points" for explosive bolts to secure and later release the spacecraft. More research provides specific data on the 6915 PAF, including its diameter of 175 cm (69 in.) and its height of 38 cm (15 in.). The PAF remains attached to the LV's upper step after the spacecraft separates. Attachment and separation systems are described in Chapter 11.

Besides dimensional data, each PAF has a certain mass capability, based on the loads that the rocket sees during launch and ascent. As might be expected, because of lateral loading, the higher the center of mass of the spacecraft, the lower the PAF's mass capability. This is shown in Fig. 7.47.

PAFs may be metallic, as shown in the Delta II payload planner's guide, or they may be composite, as shown in the Delta IV PPG, reproduced in Fig. 7.48. The notation shows the diameter in millimeters and inches, and describes the required type of separation system. Both clamp bands and bolted interfaces are provided. (See Chapter 11 for a description of these separation systems.)

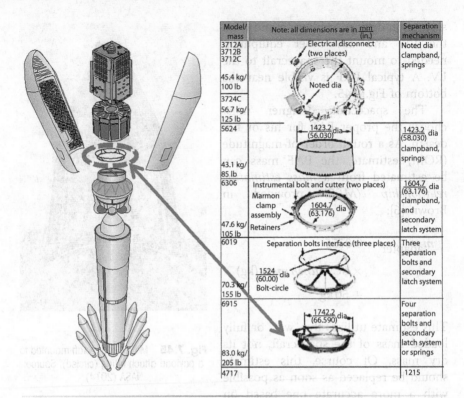

Model/mass	Note: all dimensions are in $\frac{mm}{(in.)}$	Separation mechanism
3712A 3712B 3712C 45.4 kg/ 100 lb	Electrical disconnect (two places) Noted dia	Noted dia clampband, springs
3724C 56.7 kg/ 125 lb 5624 43.1 kg/ 85 lb	$\frac{1423.2}{(56.030)}$ dia	$\frac{1423.2}{(58.030)}$ dia clampband, springs
6306 47.6 kg/ 105 lb	Instrumental bolt and cutter (two places) Marmon clamp assembly $\frac{1604.7}{(63.176)}$ dia Retainers	$\frac{1604.7}{(63.176)}$ dia clampband, and secondary latch system
6019 70.3 kg/ 155 lb	Separation bolts interface (three places) $\frac{1524}{(60.00)}$ dia Bolt-circle	Three separation bolts and secondary latch system
6915 83.0 kg/ 205 lb	$\frac{1742.2}{(66.590)}$ dia	Four separation bolts and secondary latch system or springs
4717		1215

Fig. 7.46 Delta II 6915 payload attach fitting (PAF). Source: [3].

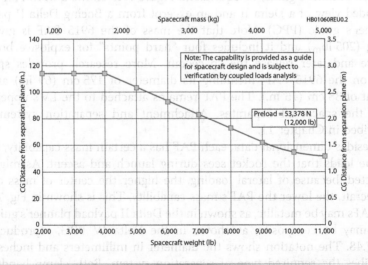

Note: The capability is provided as a guide for spacecraft design and is subject to verification by coupled loads analysis

Preload = 53,378 N (12,000 lb)

HB01060REU0.2

Fig. 7.47 Capability of the 6915 Payload Attach Fitting. The higher the spacecraft's center of gravity is above the separation plane, the smaller the spacecraft mass permitted. Source: [29].

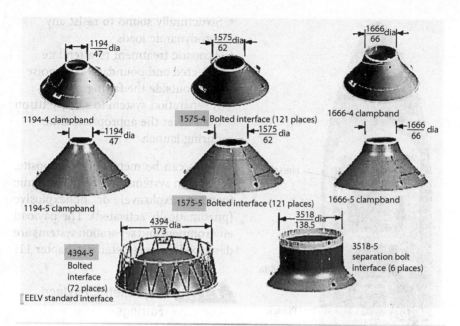

Fig. 7.48   Delta IV PAFs. Source: [4].

Occasionally (or commonly for the Ariane 5), the LV has enough capability to send two spacecraft to orbit at the same time, referred to as a *dual manifest* mission. This means several things, mainly that the two spacecraft have to be released on fairly similar orbits and that a dual payload attach fitting (DPAF) is required. Such a system is shown in Fig. 7.44 at the beginning of this section, and is also shown in Fig. 7.49.

7.11.2   Payload Fairings

The other part of the protection provided by an LV to a spacecraft is the *payload fairing (PLF)* or *shroud*. The PLF is nothing more than an insulated windshield to protect the spacecraft from thermal excursions, acoustic excitation (noise), and hypersonic flow during liftoff and ascent through the atmosphere. A photo of a spacecraft mounted on its PAF and about to be encapsulated by its PLF is shown in Fig. 7.50.

The features found in most fairings are as follows:

• Airtight, so that air conditioning can keep the payload in the proper temperature range and humidity before launch
• Venting during ascent to allow the trapped atmosphere to escape so that pressure can be equalized
• Thermal insulation, and possibly an ablative material treatment, so the aerodynamic heating from supersonic and hypersonic flow during ascent does not significantly raise the internal temperature

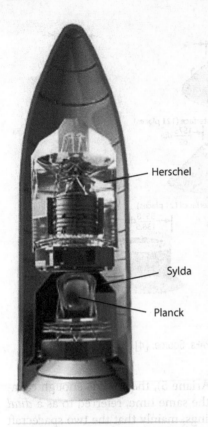

Herschel

Sylda

Planck

Fig. 7.49 Launch arrangement: Ariane 5 DPAF can launch two spacecraft together. Of course, a dual launch means that the forward spacecraft and its adapter need to be ejected before the lower spacecraft can be released. Source: http://www.arianespace.com/corporate-news/the-sylda-5-payload-dispensers-deployment-is-imaged-during-ariane-5s-most-recent-mission-success/.

- Structurally sound to resist any aerodynamic loads
- Acoustic treatment to attenuate reflected and boundary-layer noise from outside the fairing
- A separation system to safely jettison the PLF at the appropriate time during launch

Fairings can be metallic or composite. Separation systems may be pyrotechnic (contain explosive) or nonexplosive (pneumatically activated). The payload environment and separation systems are described in more detail in Chapter 11.

7.11.3 Asymmetric Payload Fairings

There may sometimes be a need to launch a very large or awkwardly shaped payload, such as a large uninhabited aerial vehicle to fly in the atmosphere of Mars, or a large parabolic antenna or optical aperture. The unusually shaped asymmetric fairing shown in Fig. 7.51 was designed to accommodate very large payloads on the Atlas V HLV while maintaining the vehicle's structural requirements and current control authority limits. According to ATA Engineering [1], the design was achieved through the use of computational fluid dynamics (CFD)-based geometric optimization, composite structural tailoring, and novel manufacturing methods and was validated through correlation with subscale wind tunnel testing. Its composite sandwich structure meets or exceeds strength, buckling, flutter, thermal, and acoustic requirements, and does not require significant modifications to existing launch pad integration facilities. Its computer-optimized external shape minimizes the changes in aerodynamic forces associated with substituting an asymmetric fairing for the axisymmetric fairing that a heavy lift vehicle was designed to carry.

ATA [1] indicated that the large PLF's design objective was to minimize a "weighted sum of its pitching moment and axial force at the base of the boat tail, where it attaches to the *Centaur* upper step vehicle. The weighting was biased toward the minimization of the pitching moment since this represented the most significant obstacle in producing a useable design." The analysis was conducted at $M = 1.075$, angle of attack $\alpha = 4$ deg, and a max-q dynamic pressure of 21.5 kPa (450 psf) at altitude of 10.3 km (33,800 ft). The quasi-static load factors at this flight condition were $n_x = 3.0$ and n_z (pitch direction) $= 0.3$. As shown in the figure, the resulting PLF was 10% longer, had 97% more internal volume, and yet had only 36% more mass. Such a fairing, although unorthodox in appearance, would certainly solve some issues with large and awkward payloads. (Note: the heavy version of the Atlas V was never built, but these practices would equally apply to other launchers such as Delta IVH, Falcon 9H, etc.)

Fig. 7.50 This PLF for the OCO-2 spacecraft contains black acoustic treatment inside to attenuate sound exposure to payload. Source: [19].

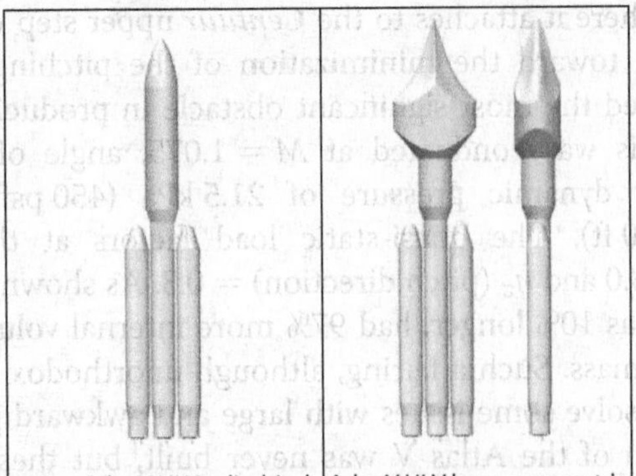

	5.4 m Atlas V HLV cylindrical PLF	Atlas V HLV large asymmetric PLF
Static payload envelope	⌀4.6 m × 12.2 m long	9.3 m w × 10 m h × 4.3 m d
Available payload volume	203 m³	400 m³
Fairing length	26.5 m	29.2 m
Construction	Composite sandwich	Composite sandwich
Mass	4394 kg	5965 kg

Fig. 7.51 Example of a large asymmetric payload fairing mounted on a launch vehicle. Source: http://www.ata-e.com/wp-content/uploads/2018/01/Large-Asymmetric-Fairing-Data-Sheet.pdf, or [1].

7.12 Launch Vehicle Structure Types

So far, we've seen several launch vehicles and their configurations in some detail. Now we will discuss the structures discipline. A variety of structural materials and types of structures are utilized, depending on the environment (loads, thermal, electrical, etc.). Some of the possible structures are:

- Skin-stringer (composite or metal)
- Honeycomb plates (composite and/or metal)
- Machined waffle or isogrid (usually metal, not always)
- Filament wound (usually smaller, high-pressure tanks)
- Composite overwrap pressure vessel (COPV; metal tank or liner wrapped with composite, fiberglass, or aramid)

Examples of these will be presented in the following sections.

7.12.1 Skin and Stringer Construction

The most traditional (meaning the oldest) structural type used for launch vehicles is the so-called *skin and stringer* construction. This construction was (and still is) used for many aircraft. Many of the more established LV suppliers began as offshoots of aircraft companies, so it's not surprising that they

Fig. 7.52 Skin and stringer construction of a Boeing 787. Source: Edberg.

continued with conventional construction. Skin and stringer construction is still used in nonpressurized LV structures such as skirts, intertanks, and interstages. A sample of skin and stringer construction is shown in Fig. 7.52.

Figure 7.52 shows an internal portion of a Boeing 787 fuselage. The skin is the flat surface underneath the stringers, which run from lower left to upper right. The photo also shows parts of two internal ring frames, running from the upper left to the lower right.

7.12.2 Sandwich Construction

Another type of construction is called *sandwich* construction, because it resembles a sandwich in configuration. Figure 7.53 illustrates a sandwich structure consisting of a skin or face sheet bonded to a lightweight core material, which in turn is bonded to another face sheet.

This type of construction may be used with metallic or composite face sheets, and metallic or composite core materials. Common face sheet materials are aluminum or steel and graphite/epoxy. Core materials can be

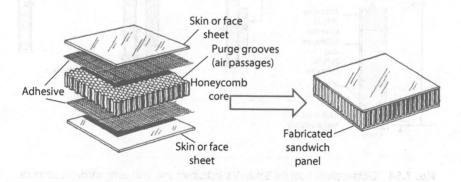

Fig. 7.53 Sandwich or honeycomb construction. Source: NASA.

aluminum, aramid, or phenolic honeycomb such as HexWeb®, flexible core such as Coremat®, or solid foam material such as Rohacell® or Klegecell®.

It is important to remember that when the face sheets are bonded to the core, there may be trapped air within the core material, especially if it is honeycomb. One can imagine what happens when a panel with trapped air is subjected to vacuum, like during a launch. Suffice it to say that now all honeycombs come with perforations to allow the trapped air to bleed out. There was a launch failure in the 1960s that turned out to be caused by a sandwich panel failure due to trapped air.

Sandwich construction results in very lightweight panels, but its nature of being almost empty between the face sheets doesn't work well when the panels need to be joined to other panels or for panels to support concentrated loads for things like equipment mounts. For such a situation, it's necessary to embed either denser core material or hard points into the core material for fasteners. Of course, such a step will increase the mass of the sandwich structure, but it usually is still better than a conventional structure. Figure 7.54 shows the construction used for the Saturn V's instrument unit, which sat above the upper skirt of the Saturn V's S-IVB third step and below the adaptor for the Apollo spacecraft.

Sandwich construction is gradually replacing skin and stringer construction and is used for space launch vehicles for intertanks, interstages, and payload fairings. Many spacecraft use sandwich construction for their structural panels as well.

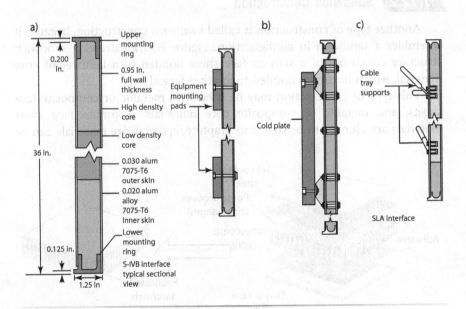

Fig. 7.54 Cross-sections from the Saturn V's instrument unit, built using sandwich structure. Note the upper and lower mounting rings bonded to the core, as well as the hard points inserted into the core for equipment mounting. Source: [13].

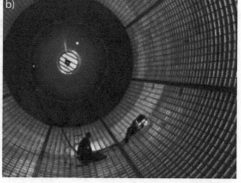

Fig. 7.55 Machined panels whose grid patterns remove most of the mass of the metal sheet while maintaining structural strength. a) Isogrid; b) orthogrid (SLS fuel tank). Sources: Edberg, NASA.

7.12.3 Integrally Machined Stiffeners

A type of metallic construction very commonly used for launch vehicles is called isogrid, orthogrid, waffle, or integrally machined stiffeners. To use this technique, one begins with a sheet of metal (usually aluminum) and machines out the material to provide a triangular, square, or rectangular arrangement of cutouts, as shown in Fig. 7.55a. The flat panel is then roll-formed (Fig. 7.55b), and then welded to other panels to construct the structure as needed. The reader interested in isogrid's technical design and characteristics is referred to [14]. The fabrication of machined stiffeners is discussed in Chapter 13.

7.13 Structural Materials

Just as there are many different types of construction for space launch vehicles, there are also many types of materials used. Table 7.1 provides a partial list of materials used in aerospace.

7.13.1 Metallic Materials

Metals were the traditional structural materials for LVs, although composite materials (see next section) are increasingly being adopted as structural materials. A list of metallic materials and their advantages and disadvantages follows in Table 7.2.

Figure 7.56 shows the materials that were used on the 1960s' design of the Saturn V. Almost every part was metallic, because existing composite materials were relatively heavy and poorly understood.

Table 7.1 Structural Materials

Metallic Materials	Composite Materials
Aluminum	Fiberglass ("glass")
Aluminum-lithium	Aramid (Kevlar, others)
Steel alloys and stainless	Graphite (carbon fiber)
High-temperature alloys	Boron fiber
Titanium	Metal-matrix composites
Beryllium	

7.13.2 Composite Materials

As mentioned previously, it is becoming more and more common for LV structures to contain composite materials. Their advantages are lower mass, meaning a higher strength-to-weight ratio, and somewhat faster fabrication.

Table 7.2 Metal Material Usage Guide

Type	Advantages	Disadvantages	Applications
Aluminum 2014 alloy	Good for cryogenics, good machining, welding, low cost	High temperature limitations, high CTE	Cryogenic tank walls
Aluminum 6xxx series	Higher strength, good welding, machining, low cost	High CTE, galling, stress corrosion issues	Truss structure, skins, stringers, brackets
Al-Lithium e.g. 2195	Lower density & higher modulus compared to Al	Reduced ductility & fracture toughness in short transverse direction	Skins, stringers, face sheets
Titanium	Low density, high strength, good high temps, low CTE & thermal conductivity	Expensive, difficult to machine	Attach fittings for composites, thermal isolators, flexures
Steel e.g. 4130, D6AC	High stiffness, strength, high temps ok, low cost, weldable	Heavy, magnetic, oxidizes if not stainless. Stainless galls easily	Fasteners, threaded parts, bearings & gears
High-Temp Alloys (i.e. Inconel)	High stiffness, strength at high temperatures, oxidation resistance & non-magnetic	Heavy, difficult to machine	Fasteners, high temperature parts
Beryllium	ultra-high stiffness, very low mass	Expensive to machine, *carcinogen*	Ultra stiff, ultra light structures, mirrors

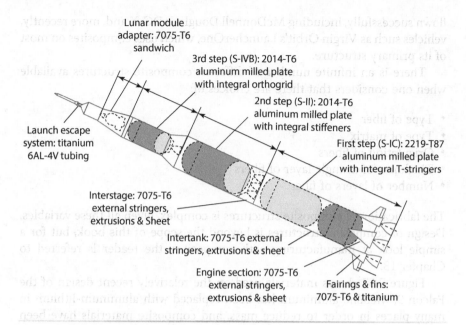

Lunar module
adapter: 7075-T6
sandwich

3rd step (S-IVB): 2014-T6
aluminum milled plate
with integral orthogrid

2nd step (S-II): 2014-T6
aluminum milled plate
with integral stiffeners

Launch escape
system: titanium
6AL-4V tubing

First step (S-IC): 2219-T87
aluminum milled plate
with integral T-stringers

Interstage: 7075-T6
external stringers,
extrusions & Sheet

Intertank: 7075-T6 external
stringers, extrusions & sheet

Engine section: 7075-T6
external stringers,
extrusions & sheet

Fairings & fins:
7075-T6 & titanium

Fig. 7.56 Some of the materials used in the main structural parts of the Saturn V.

Because they are formed to the part's outer mold line or to net shape, there is much less touch labor along with a lower part count.

A composite material consists of a strong fiber material embedded in resin matrix, often epoxy or other plastic resin. There are a number of different composites and abbreviations:

* Graphite-epoxy (GrE), a.k.a. carbon fiber-epoxy, CFRP = carbon fiber reinforced plastic
* HT-CFRP = high *tensile strength*
* HM-CFRP = high *modulus*
* UHM-CFRP = ultra-high modulus
* Fiberglass (FG)
* Aramid fibers/matrix, such as Kevlar®/epoxy
* Exotic composites such as boron/epoxy, metal matrix composites (MMC)

One of the most attractive features of composites is the potential for weight savings. For example, *Aviation Week* [2] reported that "Cryotank ... weight savings over aluminum approached the 35% target set by NASA, according to John Fikes, NASA deputy project manager."

However, composite materials are not without drawbacks. These include high-temperature limitations that may not be present with metals, issues with cyclic fatigue and reusability, their brittleness—composites are often brittle materials that show no yielding—and abrupt failure. Finally, there are microcracking issues with cryogenics that have caused failures, but several have flown successfully, including McDonnell Douglas's DC-X and, more recently, vehicles such as Virgin Orbit's LauncherOne, which uses composites on most of its primary structure.

There is an infinite number of different composite structures available when one considers that there are choices of:

- Type of fiber
- Type of matrix
- Orientation of fibers
- Thickness of a single layer of fibers
- Number of layers of fiber

The fabrication of composite structures is complex, given all these variables. Design of composite structures is beyond the scope of this book, but for a simple look at manufacturing with composites, the reader is referred to Chapter 13.

Figure 7.57 shows materials used on the relatively recent design of the Falcon 9. Metallic aluminum has been replaced with aluminum-lithium in many places in order to reduce mass, and composite materials have been employed as skirts, interstages, and fairings. The Virgin Orbit LauncherOne goes even further with composite tanks for its RP-1 and LOx propellants, indicating that the issues of LOx and composites have been at least partially solved.

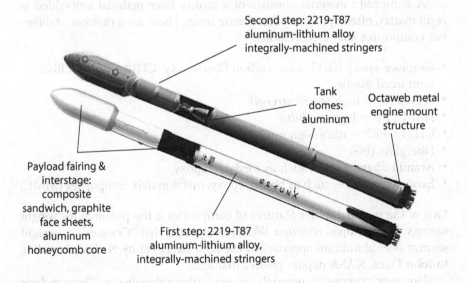

Second step: 2219-T87 aluminum-lithium alloy integrally-machined stringers

Tank domes: aluminum

Octaweb metal engine mount structure

Payload fairing & interstage: composite sandwich, graphite face sheets, aluminum honeycomb core

First step: 2219-T87 aluminum-lithium alloy, integrally-machined stringers

Fig. 7.57 Some of the materials used in the main structural elements of the SpaceX Falcon 9. Source: [30].

Table 7.3 Table of Material Properties, English Units*

Material	Yield Stress Y (ksi,Ave)	Density ρ (lb$_m$/ in^3)	Modulus E (10^6 psi)	Ratio*, Weight to Weight of 2024-T3 Al		
				Tension	Bending	Buckling
Beryllium	70.0	0.069	44.0	0.65	0.67	0.43
High modulus Gr/E	148.0	0.059	30.0	0.26	0.39	0.42
Steel, 4130	190.0	0.289	30.0	1.00	1.70	2.04
Steel, Stainless	185.0	0.286	26.0	1.02	1.71	2.11
High strength Gr/E	178.0	0.056	21.0	0.21	0.34	0.44
Titanium	160.0	0.160	16.5	0.66	1.03	1.38
Aluminum 2024-T3	66.0	0.100	10.5	1.00	1.00	1.00
Aluminum 7075-T6	77.0	0.101	10.4	0.87	0.94	1.01
Alum-Lithium 2090	70.0	0.093	11.5	0.93		
Kevlar/Epoxy	108.0	0.052	10.0	0.32	0.41	0.53
Fiberglass S-2	72.0	0.072	8.6	0.66	0.69	0.77
Magnesium alloy	40.0	0.065	6.5	1.07	0.83	0.76
Laminated plastic	30.0	0.050	2.5	1.10	0.74	0.81
Spruce (wood)	9.4	0.016	1.3	1.10	0.41	0.31

*Less than one is better for these ratios.

The experienced designer will make his or her choice based on requirements, environment, and the following:

* Strength *and* stiffness through the entire range of operating temperatures
* Buckling/elastic stability
* Specific gravity or density
* Machinability, fabrication, and joinability
* Electrochemical and/or sea water corrosion
* Electrical and thermal conductivity
* Other issues

If there is a need for electrical transparency or nonconductivity (e.g., a part that has to be transparent to radio frequency (RF) such as a window to be transparent to an antenna or dish), it's necessary to use either glass or aramid fibers. Conversely, if there is a need for conductivity, graphite or carbon fibers provide some capability, but may need augmentation with a metal mesh.

A variety of room-temperature material properties is provided in Table 7.3. Columns 2–4 provide the values of yield stress, density, and

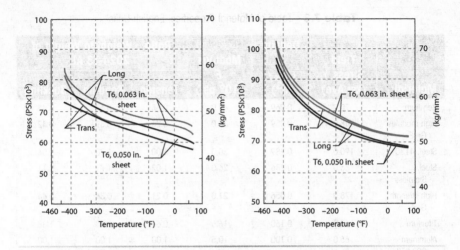

Fig. 7.58 The yield and ultimate tensile strength properties of 2014 aluminum commonly used for propellant tanks in LVs. Note the improvement in properties as temperature (°F) drops, and "improper" SI units used for stress. There may also be a loss of strength at higher temperatures than those shown. Source: Harpoothian [7].

elastic modulus. Columns 5–7 provide the materials' properties normalized to aluminum. This means that if the number is less than 1.0, then the material is superior to aluminum for the particular property evaluated.

The previous information referred to room-temperature properties. LVs, however, are subjected to extremes of temperature, from near-zero cryogenic propellant temperatures to thousands of degrees due to combustion and aerodynamic heating. With this in mind, we point out that the designer must take operating temperatures into account when selecting materials. As an example, we present plots of the yield and ultimate tensile strengths of 2014 aluminum as a function of temperature in Fig. 7.58.

7.13.3 Miracle Materials

Finally, the reader is warned about "miracle" materials. Experienced designers often refer to these when a design clearly requires or specifies a material or property that cannot possibly exist (e.g., a massless material, infinite modulus, or negative density). If one hears colleagues referring to Unobtanium, Balonium, Eludium, Handwavium, Miraculum, or Wishalloy, it is suggested that a careful examination of one's work be made in order to learn what the problem is!

References

[1] Ochinero, T., Deiters, T., Higgins, J., Arritt, B., Blades, E., and Newman, J., *Design and Testing of a Large Composite Asymmetric Payload Fairing*, Presented at the

50th AIAA/ASME/ASCE/AHS/ASC Structures, Structural Dynamics, and Materials Conference, Palm Springs, CA, April 2009.

[2] Frank Morring Jr., "Tank Technology," *Aviation Week*, 25 May–7 June 2015, p. 64.

[3] Boeing, "Delta II Payload Planners Guide," Boeing VDD12453.3 M8BJ, 2006, Fig. 5-1, p. 5-2.

[4] Boeing, "Delta IV Payload Planners Guide," Boeing 06H0233, 2007, Fig. 5-2, p. 5-3.

[5] Boeing, "The Delta II Launch Vehicle," Document VDD12453.3 M88J, undated presentation, pp. 7–9.

[6] Brown, C., *Elements of Spacecraft Design*, AIAA, Reston, VA, 2002.

[7] Harpoothian, E., "The Production of Large Tanks for Cryogenic Fuels," Douglas Paper No. 3155, prepared for presentation at German Society for Rocket Technology and Astronautics, Munich, Germany, 12 Nov. 1964.

[8] European Space Agency, *Ariane 5 User Manual, Issue 4, Revision 0*, 2004, p. 1-7.

[9] Jet Propulsion Laboratory, "Mars Exploration Rover (MER) Project Mission Plan," JPL Report D-19659, 2002.

[10] NASA, *Apollo Systems Descriptions*, Vol. II, NASA TM-X-881, MSFC, 1964.

[11] Huzel, D. K., and Huang, D. H., "Design of Liquid Propellant Rocket Engines," NASA SP-125, 1967.

[12] Bellcomm Inc., "Description of the S-IC Stage Structure – Case 330," NASA CR-153761, 1967.

[13] NASA, "Saturn V Flight Manual SA-503," NASA TM-X-72151, MSFC-MAN-503, 1968.

[14] NASA, "Isogrid Design Handbook," NASA CR-124075-Revision A, 1973.

[15] The Presidential Commission on the Space Shuttle Challenger Accident, "Report to the President," Vol. I, N86-24726, 1986.

[16] NASA, *NSTS 1988 News Reference Manual*, 1988, https://science.ksc.nasa.gov/shuttle/technology/sts-newsref/et.html [retrieved 21 Aug. 2018].

[17] NASA, "Propellant Management in Booster and Upper Stage Propulsion Systems," NASA TM-112924, 1997.

[18] NASA, "Wings in Orbit," NASA SP-2010-3049, 2010.

[19] Kremer, K., "NASA Set to Launch OCO-2 Observatory on July 1 – Sniffer of Carbon Dioxide Greenhouse Gas," *Universe Today*, 29 June 2014, https://www.universeto day.com/112869/nasa-set-to-launch-oco-2-observatory-on-july-1-sniffer-of-carbon-dioxide-greenhouse-gas/ [retrieved 25 April 2019].

[20] Ordway III, F. I., Gardner, J. P., Sharpe Jr., M. R., *Basic Aeronautics—An Introduction to Space Science, Engineering, and Medicine*, Prentice Hall, Englewood Cliffs, NJ, 1962.

[21] Pisacane, V. L., *Fundamentals of Space Systems*, 2nd ed., Oxford University Press, New York, NY, 2005.

[22] Seifert, H., *Space Technology*, John Wiley and Sons, New York, 1959.

[23] Snecma, "Ariane 5 Main Cryogenic Stage Propulsion System," Sep division de snecma, undated.

[24] United Launch Alliance, "Atlas® V—The Future of Space Is Now," 2006.

[25] United Launch Alliance, "Delta IV Payload Planners Guide," 06H0233, September 2007.

[26] Hellebrand, E. A. *Structural Problems of Large Space Boosters*. Conference Preprint 118, presented at the ASCE Structural Engineering Conference, Oct. 1964.

[27] Platt, G. K., Nein, M. E., Vaniman, J. L., and Wood, C. C. *Feed System Problems Associated with Cryogenic Propellant Engines*. Paper 687A, National Aero-Nautical Meeting, Washington DC, April 1963.

[28] McCool, A. A., and McKay, G. H. *Propulsion Development Problems Associated With Large Liquid Rockets*. NASA TM X-53075, Aug. 1963.

[29] United Launch Alliance, "Delta II Payload Planners Guide," Publication 06H0214, 2006

[30] Space Exploration Technologies, "Falcon User's Guide Rev 2," Oct. 2015

Further Reading

McDonnell Douglas Astronautics, "Isogrid Design Handbook," NASA CR-124075, Feb. 1973. Look for the "improved" version, April 2004.

Chapter 8

Sizing, Inboard Profile, Mass Properties

oving forward from our look at possible structures and layout configurations, we show the "design roadmap" as Figure 8.1 with the highlighted signposts indicating the information to be covered in this chapter: sizing, layout, and mass properties. Our goal for this chapter is to develop a quantitative description of our launch vehicle.

In the following pages, we will cover the following topics:

- Inboard profile: where does all the equipment fit?
- Required propellants calculation
- Tanks, tank domes, tank sizing, and volume calculation
- Engine(s), thrust structure
- Intertank structure
- Payload fairing and payload attach fitting
- Mass Properties Estimation (mass, CM location, moments of inertia)

8.1 Inboard Profile

Inboard profile is a term used by the aerospace industry to indicate how all the equipment that is to be carried by a vehicle will be arranged within the vehicle. The designer comes up with an inboard profile that provides geometric information to be used for further calculations, including mass properties, performance, aerodynamic loads, stresses, thermal loads, stability & control, manufacturing processes, and cost estimation.

At this stage in the design, we need to accommodate the following major components:

- Engine(s)
- Thrust structure
- Propellant tank volumes
- Intertank structure
- Interstage structure
- Avionics/astrionics and wire harness locations
- Payload accommodations: attachment and protection
- Payload attach fitting (PAF)
- Payload fairing (PLF)

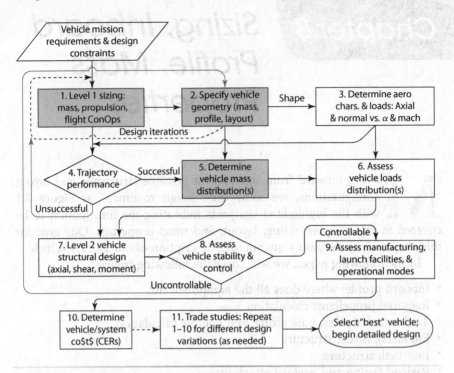

Fig. 8.1 The launch vehicle design process with sizing, inboard profile, and mass estimation.

We will also need to consider the following items, which are discussed in Chapters 11 and 14:

- Hydraulics or other systems for flight control
- Pressurization, pneumatics
- Staging and separation mechanisms
- Ground attach structure and hold-down methods

A simple inboard profile is shown in Fig. 8.2. The major items are labeled, and each needs to be sized geometrically. Internal items, such as the payload, payload attach fitting, avionics, and wiring are also shown. We do not show the propellant loading lines, feed lines, and other features, but they will be discussed in detail in Chapters 13 and 14.

8.1.1 Vehicle Sizing and Layout Process

The vehicle sizing and layout process proceeds as follows (see Fig. 8.3):

1. Payload mass, I_{sp} of the engines, structural fractions σ_i, and Δv requirements are known. The required mass ratios, propellant masses, and structural masses are found using the rocket equations given at the end of Chapter 5. Increase propellant masses for "real-life" effects.

2. Choose liftoff T/W_0 and other steps' T/W ratios. Solids may specify total impulse instead.
3. Choose vehicle diameter(s)
4. Select tank dome geometry(s)—hemispherical or ellipsoidal (solids use hemispherical due to high internal pressures). Calculate tank volumes including real-life effects. Size propellant tanks' cylindrical volumes and lengths using total required volumes.
5. Size the interstage structure, ground attachment system, thrust structures, and intertank rings. Size engine(s): find throat area and other engine dimensions. This process defines the thrust structure, interstage, and ground attachment structure dimensions. For solids, the thrust structure is included in the casing.
6. Size/choose payload attach fitting and payload fairing. Size & locate internal parts: avionics, wiring, etc.

The vehicle sizing and layout process is somewhat more complex than might be understood from the six steps listed above. For instance, the actual propellant masses must also include additional propellant for cooling down and starting the engines, for propellant boiloff, and for residual propellant that can't be extracted from the tanks. The tank volumes must include additional volume for pressurization gas, for self-shrinkage when their temperatures drop from contained cryogenic propellants, and for any internal insulation or other components. Some of these factors are illustrated in Fig. 8.4. Solid rocket motors need additional volume for case insulation, grain cutouts, and ignition systems. We will now discuss the ways to incorporate these additional needs for liquid-propelled LVs.

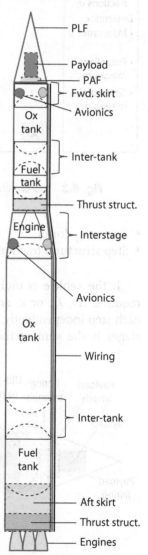

8.2 Vehicle or Step Mass Calculations

The process begins with the calculation of the basic vehicle parameters. We will assume that the following quantities have been provided or selected, for each step:

- Payload mass m_{PL}
- Required Δv

Fig. 8.2   Inboard profile of LV.

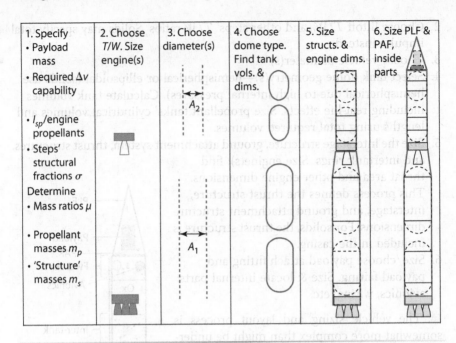

1. Specify	2. Choose	3. Choose	4. Choose	5. Size	6. Size PLF &
• Payload mass	T/W. Size engine(s)	diameter(s)	dome type. Find tank vols. & dims.	structs. & engine dims.	PAF, inside parts
• Required Δv capability					
• I_{sp}/engine propellants					
• Steps' structural fractions σ					
Determine					
• Mass ratios μ					
• Propellant masses m_p					
• 'Structure' masses m_s					

Fig. 8.3 Launch vehicle sizing process in six steps (liquid-propellant shown).

- Engine I_{sp} (or c)
- Step structural fraction σ

If the vehicle is multi-step, it's assumed that each step would have a required Δv, I_{sp} or c, and σ, so that the calculations are carried out for each step independently. Remember that the payload for the kth stage of N stages is the sum of the masses of the fully-loaded steps above the kth

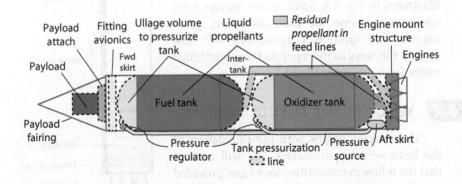

Fig. 8.4 Some of the major components of a liquid-propellant LV. Propellant lines transfer propellant to engines. Tanks have volume at top for pressurized gas ducted to the top of the tanks for ullage pressure.

stage and the payload. So $m_{PL1} = (m_{s2} + m_{p2}) + (m_{s3} + m_{p3}) + \cdots + (m_{sN} + m_{pN}) + m_{PL}$.

Referring to Chapter 5, we utilize Eq. (5.82) to find the mass ratio μ of the uppermost step:

$$\mu = e^{\left(\frac{\Delta v}{g_0 I_{sp}}\right)} = e^{\left(\frac{\Delta v}{c}\right)} \tag{5.82}$$

Next, we calculate the ideal or mission propellant mass m_p required to reach the specified Δv using Eq. (5.83):

$$m_p = m_{PL} \frac{(\mu - 1)(1 - \sigma)}{1 - \mu\sigma} \tag{5.83}$$

The "structure" or inert mass m_s comes from Eq. (5.89):

$$m_s = m_p \frac{\sigma}{1 - \sigma} \tag{5.89}$$

This process is repeated for each of the vehicle's steps, moving down until the first step is reached.

With this information in hand, we are almost ready to begin the sizing process, but first, some additional "real life" factors have to be considered.

8.3 Liquid Propulsion System Real-Life Additions to Mass and Volume

The mass of propellant found above is not the complete amount of propellant necessary. A practical operational liquid propulsion system needs additional propellant mass and tank volume to accommodate the following needs and requirements, as shown in Fig. 8.5.

8.3.1 Liquid Propulsion System "Real-Life" Additions

The liquid propellant mass and tankage volume needs to be modified to provide for the additions listed above:

1L. Propellant *mass* is consumed during the engine or engines' startup procedure. The amount is determined by the propellant flow rates and the time needed to reach desired operating conditions. Flow rates are determined by thrust level and engine specific impulse values.

2L. Propellant *mass* is lost due to unusable and residual propellants, which are those that can't be extracted from the tank and are present in the feed lines when the engine(s) shut(s) down, also shown in Fig. 8.5. Residuals are usually taken as a percentage of tank volume.

3L. Extra propellant tank *volume*, called "ullage," is needed at the top of a tank to provide space for pressurization gas. Refer to Fig. 8.5 for an illustration of ullage volume and pressurization hardware. Ullage is usually defined as a percentage of the tanks' volumes.

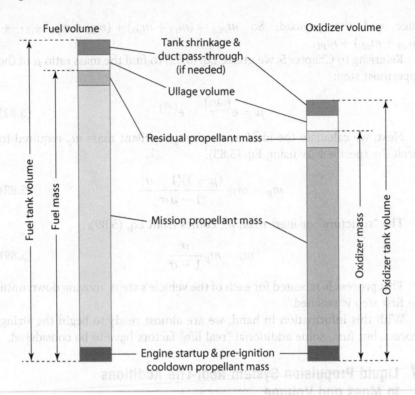

Fig. 8.5 Definitions of masses and volumes for a liquid-propellant launch vehicle. Note: quantities shown are representative only, and are not to scale of any actual values.

4L. Additional tank *volume* will needed if propellant ducts or lines pass through a tank, in order to account for the volume of propellant displaced by the duct. External ducting that passes *outside* of any propellant tanks *does not* subtract volume from the lower tank. These two types of ducts may be seen in Fig. 7.12, and Fig. 7.19 shows the five large internal ducts in the Saturn V's S-IC step. In this chapter, we will assume that all propellant ducting is external as shown in the left-most figure in Fig. 7.12. The design of ducting, and the calculation of its required dimensions requires knowledge of flow rates and ducting loss factors, and is beyond the scope of this book. The interested reader is referred to Ring [11] and other propellant flow references.

For cryogenic propellants, more effects need to be included:

5L. Additional tank internal *volume* must be added to compensate for the volume decrease due to thermal contraction. The shrinkage is estimated by applying the known temperature difference between manufacturing (approximately room temperature) and the operating temperature of the propellant and the tank material's thermal expansion characteristics.

6L. Internal *volume* must be added for internal thermal insulation (if utilized). Currently, most vehicles use external insulation which does not affect the tank internal volume.

7L. Additional *volume* must be added for cryogenic propellant boil-off: at some time during the preflight preparations, there is no more topping off, and the propellants may boil off. More volume may be needed for propellants that will not be present at the desired liftoff time. We will assume that our pad facilities provide top-off services, and this volume will not be calculated for our example.

8L. Additional propellant *mass* is needed to chill down the cryogenic engine(s) before upper-stage starts and restarts. This will depend on the engine chill-down requirements that are typically different for each model of engine, so we will neglect this addition for our example.

The resulting masses and volumes for these effects are conceptually illustrated in Fig. 8.5.

Liquid Propellant Mass Buildup

To properly do the vehicle sizing, we first must calculate the entire mass of propellants needed, including that needed to carry out the mission, and the additional masses listed above in nos. 1L, 2L, and 8L (as needed). The resulting "entire" mass of propellant is used to find the required volume for the tanks using the propellant's density, which will then be increased by adding volume for the effects listed in nos. 3L, 4L, 5L, 6L, and 7L (as needed) above.

Now, we will begin by showing the process used to determine the total propellant mass that is needed for the vehicle. This will include additional propellant mass for the previous steps given for liquid propulsion: 1L, engine startup, and 2L, residual propellants trapped in feed lines and therefore not available for propulsion. Afterwards, we'll continue with the volume calculations.

Startup Liquid Propellant Mass

To calculate the startup propellant mass, we need the mass flow rate, which in turn is determined by the engine thrust and specific impulse. We will specify a thrust-to-weight ratio and use it to calculate mass flow. With an assumed startup time, we have the startup propellant mass $m_{p\text{-startup}}$.

Now, let's select a liftoff thrust-to-weight ratio $\psi = T/W_0 = T/(m_0 g_0)$, so $T = \psi m_0 g_0$. The propellant mass flow $\dot{m}$ is obtained from the thrust equation

$$\dot{m} = \frac{T}{g_0} I_{sp} = \frac{\psi m_0}{I_{sp}} = \frac{\psi m_0 g_0}{c} \tag{8.1}$$

The propellant mass needed for start-up, $m_{p\text{-startup}}$, comes from multiplying the mass flow by the time needed to start up, t_{start}:

$$m_{p\text{-startup}} = \dot{m}t_{\text{start}} = \frac{\psi m_0 t_{\text{start}}}{I_{sp}} = \frac{\psi m_0 g_0 t_{\text{start}}}{c} \qquad (8.2)$$

Residual Liquid Propellant Mass

Next, we calculate the residual propellant mass $m_{p\text{-residual}}$ by a residual fraction method where f_{residual} = residual fraction:

$$m_{p\text{-residual}} = f_{\text{residual}} m_p, \qquad (8.3)$$

Total Liquid Propellant Mass

We can now find the total propellant mass $m_{p\text{-tot}}$ (ignoring boil-off) by summing "ideal" or mission propellant mass m_p, startup mass $m_{p\text{-startup}}$, and residual mass $m_{p\text{-residual}}$:

$$
\begin{aligned}
m_{p\text{-tot}} &= m_p + m_{p\text{-residual}} + m_{p\text{-startup}} \\
&= m_p + f_{\text{residual}} m_p + \dot{m}\, t_{\text{start}} \\
&= m_p(1 + f_{\text{residual}}) + \frac{\psi m_0 t_{\text{start}}}{I_{sp}}.
\end{aligned}
\qquad (8.4)
$$

Individual Liquid Propellant Masses

We now have the total mass of liquid propellant needed for the mission, engine startup, and to account for residual propellant left in the ducts; however, we actually have two propellants: one fuel, and one oxidizer. To size these two tanks, we have to split the total into the individual propellants.

The vehicle's engines run at a fuel:oxidizer mass ratio mixture ratio of f. The mix ratio $f = m_f / m_{ox}$, so $m_f = f\, m_{ox}$. Since the total propellant mass $m_p = m_f + m_{ox}$, we find for Step k:

$$m_f = m_{pk}\left(\frac{f_k}{1+f_k}\right) \qquad (8.5a)$$

$$m_{ox} = m_{pk}\left(\frac{1}{1+f_k}\right) \qquad (8.5b)$$

Here, m_f is the total fuel mass, and m_{ox} is the total oxidizer mass for step k, and f_k is the kth step's mix ratio. These calculations are repeated for the number of steps k.

8.3.2 Needed Liquid Propellant Tank Volumes

With the "total" masses of each of the liquid propellants (fuel and oxidizer) calculated, we can begin the tank sizing process. The "total" masses of propellants will be used to find the required volume for the tanks using the propellant's density. Then, the required volumes will be increased by adding volume for the effects listed in nos. 3L, 4L, 5L, 6L, and 7L (as needed) given in Section 8.3.1 above.

We begin by finding the propellants' volumes, then modify those volumes to account for the additional volumes mentioned above. In our calculations, we will use the abbreviation Vol to represent the volume of a tank, and the subscripts $-ox$ and $-f$ to represent oxidizer and fuel, respectively. We do this to avoid the use of the lowercase letter v so as not to eliminate any confusion between volume and velocity.

"Ideal" Liquid Propellant Tank Volumes

The "ideal" tank volumes are calculated using the definitions of mass, density, and volume: $\rho = m/\text{Vol}$, so

$$\text{Vol}_{\text{prop}-k} = \frac{m_{\text{prop}-k}}{\rho_{\text{prop}-k}} \tag{8.6}$$

The calculation in Eq. (8.7) is repeated for each propellant (prop $= f$ or ox) and each step k.

Needed Liquid Propellant Tank Volumes

The tank volumes found above do not include the necessary added volume for ullage and for tank shrinkage (cryogenic propellant only), nos. 3L and 5L above. We will now carry out this calculation, which will yield the "real" or needed volume $\text{Vol}_{p\text{-all}}$ for each tank.

To do this, we assume that the needed volume for propellant p (f or ox) is multiplied by factors for the needed ullage volume (ullage$_p$) and the shrinkage $shrink_p$: This leads to the relation

$$\text{Vol}_{p\text{-all}} = \text{Vol}_p + \text{Vol}_{p\text{-ullage}} + \text{Vol}_{p\text{-shrinkage}} \tag{8.7}$$

Here $\text{Vol}_{p\text{-ullage}}$ represents the added volume fraction of propellant "p" to account for ullage, and $\text{Vol}_{p\text{-shrinkage}}$ represents the additional volume added to maintain volume after thermal contraction.

For ullage, we use a simple fraction ullage$_p$ to account for the additional space on top of the fuel tank to allow for the pressurization gas's volume:

$$\text{Vol}_{p\text{-ullage}} = \text{Vol}_p \times (\text{ullage}_p). \tag{8.8}$$

If the propellant is cryogenic, we need to add additional volume to account for the tank's shrinkage due to the temperature drop. If the tanks are manufactured at room temperature, when we load the tank with cryogenics, it will experience a temperature change of ΔT. This temperature change ΔT along with the appropriate coefficient of thermal expansion α (also known as CTE, or coefficient of thermal expansion) may be used to calculate dimensional changes as follows.

The shrinkage of the tank may be calculated by observing that each linear dimension of the tank will be multiplied by the factor $(1 + \alpha\Delta T)$ due to a temperature change of ΔT. The volume Vol of a tank is proportional to $R^2 h$, and both of the dimensions R and h change by the same factor $(1 + \alpha\Delta T)$, so the volume changes by the factor cubed, or $(1 + \alpha\Delta T)^3$. Therefore, the cold or "shrunken" volume is calculated from

$$\text{Vol}_{\text{cold}} = \text{Vol}_{\text{room temp}}[1 + 3\alpha\Delta T + 3(\alpha\Delta T)^2) + (\alpha\Delta T)^3]$$
$$\approx \text{Vol}_{\text{room temp}}(1 + 3\alpha\Delta T)$$

after we neglect the quadratic and cubic terms of the result. The change in volume, ΔVol, is obtained by subtracting the original, room-temperature volume from the cold volume:

$$\Delta\text{Vol} = \text{Vol}_{\text{cold}} - \text{Vol}_{\text{room temp}} = \text{Vol}_{\text{room temp}}(1 + 3\alpha\Delta T) - \text{Vol}_{\text{room temp}}$$
$$= (3\alpha\Delta T)\text{Vol}_{\text{room temp}}$$

Then, dividing by the original volume, we find that the fractional change in volume shrinkage$_p$ is

$$\text{shrinkage}_p = \frac{\Delta\text{Vol}}{\text{Vol}} = 3\alpha\Delta T. \tag{8.9}$$

We can use this result to calculate the shrinkage fraction of the propellant tank as its temperature drops.

Combining the ullage volume and the cryogenic shrinkage, the required volume for any propellant tank will be

$$\text{Vol}_{p\text{-all}} = \text{Vol}_p(1 + \text{ullage}_p + \text{shrinkage}_p)$$
$$= \text{Vol}_p(1 + \text{ullage}_p + 3\alpha\Delta T_p) \tag{8.10}$$

To calculate the thermal shrinkage factor, we need two parameters: the thermal expansion coefficient α and the temperature change ΔT for the propellant p.

For common aluminum, the thermal expansion coefficient is $\alpha_{\text{Al}} \approx 23.4 \times 10^{-6}/\text{K}$.

The temperature change is the difference of the boiling temperature of the cryogenic propellant we are considering, and the ambient propellant loading temperature (usually approximately room temperature). Table 8.1 provides a sampling of cryogenic fluid boiling temperatures for some common propellants.

Table 8.1   Boiling Temperatures for Cryogenic Fluids (source NASA TN-D-4701)

Substance	Normal Boiling Temp., K	Normal Boiling Temp., °C	Normal Boiling Temp., °F
Helium	4.6	−267	−451.4
Hydrogen	20.7	−252	−422.4
Nitrogen	77.4	−196	−320.4
Oxygen	90.1	−183	−297.4
Methane	111.7	−161	−258.5
Room Temp.	293	+20	+77

Using these factors, we can compute the shrinkage factors for three cryogenic propellants: LH_2, LOx, and LCH_4 (liquid methane). For an aluminum tank:

$$\text{shrink}_{LH2} = 3\alpha\Delta T = 3(23.4 \times 10^{-6}/\text{K})(20 - 293\text{ K}) = -0.0192, \quad (8.11)$$

so add 1.92% volume to compensate for thermal shrinkage

$$\text{shrink}_{LOx} = 3\alpha\Delta T = 3(23.4 \times 10^{-6}/\text{K})(90 - 293\text{ K}) = -0.0143, \quad (8.12)$$

so add 1.43% to compensate

$$\text{shrink}_{LCH4} = 3\alpha\Delta T = 3(23.4 \times 10^{-6}/\text{K})(111 - 293\text{ K}) = -0.0128,$$
$$(8.13)$$

so add 1.28% to compensate.

8.3.3 Other Liquid Propellant Volume Changes to Consider

Volume Changes Due to Pressurization

Most LVs pressurize their propellant tanks during flight, which will increase geometric tank volume. However, because most propellant loading operations are done with minimal pressurization, we will neglect Δ(Volume) due to tank pressurization.

Liquid Propellant Boil-off

Propellant mass may need to be added to account for the boil-off of cryogenic propellants. While on the pad, the tanks may be topped off until just before liftoff to alleviate this problem. We will assume that the tanks can be topped off, and excess propellants are not needed.

However, there may be situations where the vehicle cannot be topped off, so additional volume must be added to compensate for boil-off. One example is Virgin Orbit's LauncherOne SLV, which is carried under the wing of a Boeing 747. The LauncherOne does not have any in-flight propellant addition capability, so it was necessary to add sufficient volume to the tank(s) to account for lost propellant(s) due to boil-off. Naturally, the lower the temperature of the propellant and the higher the amount of external heating or heat convection, the higher the rate of boil-off. This means that insulation is very likely required.

Cryogenic Upper-Stage Engine Startup

In addition to propellant mass needed for boil-off, for cryogenic engines, propellant mass must be added to account for what is utilized during operations leading to startup. Some missions require that the upper-stage engines be capable of multiple starts. Some LOx/LH$_2$ engines may require a chill-down process, where cold hydrogen propellant is pumped through the engine's plumbing prior to startup. The tanks must accommodate boil-off and the needed volume of cool-down propellant, even though neither provides significant propulsion to the vehicle.

Non-Aluminum, Nonlinear CTE-Material Tanks

To do the volumetric change estimation, two factors must be considered. First, the CTE appropriate for the actual tank materials must be used. Second, in the case of a material whose CTE varies with temperature, the volume change must be determined by mathematically integrating the CTE over the beginning-to-end temperature range. Figure 8.6 shows the CTE for 2014 aluminum; note that its value decreases with temperature. Thus, the percentage shrinkage corrections for this material would depend on the temperature drop.

8.3.4 Tank Sizing

We'll begin by calculating the volume needed to completely contain the propellant mass found previously. Then, we modify the calculation to also include "real world" effects, such as ullage volume and tank shrinkage.

Volume of "Ideal" Tanks

We'll carry out the sizing process with tanks having a cylindrical shape of constant diameter D. We use the name h_{cyl} to indicate the height of the cylindrical portion. Note that there may be different diameters for different steps.

With a diameter specified, we now have to choose the shape of the tank domes (also known as bulkheads). Figure 8.7 shows a cylindrical tank with

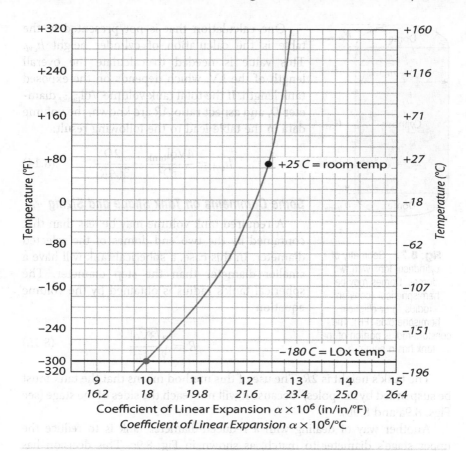

Fig. 8.6 Coefficient of Thermal Expansion for 2014 Aluminum. Courtesy: [11].

non-hemispherical domes, but we are not limited to that geometry. For tanks with moderate pressure requirements, we can use any "round" shape for the domes, such as a hemisphere (one-half of a sphere) or the elliptical hemispheroid (EH) dome shape shown in Fig. 7.28 in Chapter 7. For high-pressure applications such as solid rocket motor cases, a hemispherical dome is required due to the higher stresses generated.

For layout purposes, we provide the following guidelines, which are based on the dome's aspect ratio (AR), defined as $AR = R/h_{\mathrm{dome}}$. For hemispherical domes, $h_{\mathrm{dome}} = R$, so their aspect ratio $AR = 1$. For elliptical hemispheroid (EH) domes, the dome height is calculated by dividing the dome radius by the dome AR: $h_{\mathrm{dome}} = R/(AR)$. A typical EH dome AR is $\sqrt{2} = 1.414214$. Some dome geometries are shown in Fig. 8.8.

The volume of an EH dome is calculated by dividing the hemispherical volume of a dome with the same radius by the aspect ratio: $\mathrm{Vol}_{EP\,\mathrm{dome}} = \frac{2}{3}\pi R^3/(AR)$. This information is contained in Table 8.2.

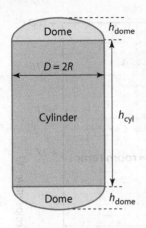

One calculation that is not presented in the table is the calculation of cylinder height h_{cyl}. This value is needed to calculate the overall length of the LV, which depends on the exposed tank length. If the total tank volume Vol_{tank}, diameter D, and aspect ratio AR are known, the volume data in the table lead to the following result:

$$h_{cyl} = \frac{4 Vol_{tank}}{\pi D^2} - \frac{2D}{3(AR)} \qquad (8.14)$$

Some Comments on Tank Shape and Sizing

A required tank volume may be less than that contained by the two end domes of the selected diameter. In this case, a spherical tank will have a smaller diameter than the step diameter. The spherical tank's radius is obtained by the volume equation

$$R = \sqrt[3]{\frac{3 Vol}{4\pi}} \qquad (8.15)$$

The tank's height is $2R$. The use of this method means that the tank must be suspended by its "poles," because it will not reach the sides of the stage (see Figs. 8.9a and 8.9b).

Another way of dealing with a reduced-diameter tank is to reduce the upper stage's diameter to match, as shown in Fig. 8.9c. This decision has other ramifications. First, the tapered adapter can result in a lengthier vehicle. Second, this diameter change will reduce the payload fairing's diameter, unless a second adapter is added to enlarge the diameter underneath the PLF. These adapters or tapered sections of structure are known as skirts.

Earlier, we specified a diameter D for our vehicle, but there is no *prima facie* reason to select any particular diameter. We could use anything from the smallest possible (which would be limited by the fit of the systems inside and the engine mount at the bottom, as well as structural flexibility) to a large value (which would provide excessive aerodynamic drag). The designer is recommended to stay with a slenderness ratio (ratio of length ÷ diameter) of no more than 20. The slenderness ratio is also known as the fineness ratio.

Fig. 8.8 Geometry of tank domes. Domes are characterized by their aspect ratio AR. A hemispherical dome with constant radius R has $AR = 1$. Elliptical hemispheroid domes (EH domes) can have other ARs as shown. A commonly-used EH dome AR is $\sqrt{2}$.

Table 8.2 Geometry of Tanks and Domes

Shape	$AR = R/h_{dome}$	Outer Surface Area A	Internal Volume Vol
Cylinder	—	$\pi D h_{cyl} = 2\pi R h_{cyl}$	$\pi D^2 h_{cyl}/4 = \pi R^2 h_{cyl}$
Hemisphere	1.0	$2\pi R^2$	$\dfrac{2}{3}\pi R^3$
Cyl. + 2 Hemisphere dome	—	$2\pi R h_{cyl} + 4\pi R^2$	$\pi R^2 h_{cyl} + 4\pi R^3/3$
Elliptical Hemispheroid (EH) Dome	1 – 2 (typ.)	$\pi R^2\left[1 + \dfrac{1}{2E(AR^2)}\ln\dfrac{1+E}{1-E}\right]$	$\dfrac{2}{3}\pi R^2 h_{dome} = \dfrac{\frac{2}{3}\pi R^3}{AR}$
$\sqrt{2}$ EH dome ($AR = \sqrt{2}$)	$\sqrt{2} = 1.4142$	$1.6234\ \pi R^2$	$1.48096\ \pi R^3$
Cyl. + 2($\sqrt{2}$EH domes)	—	$2\pi R h_{cyl} + 1.6234\ \pi R^2$	$\pi R^2 h_{cyl} + 1.48096\ \pi R^3$

Aspect Ratio $AR = R/h_{dome}$; Eccentricity $E = \sqrt{1 - h^2/R^2} = \sqrt{1 - 1/AR^2}$

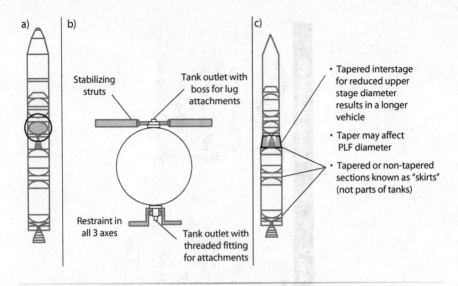

Fig. 8.9   Two ways to accommodate a tank with a smaller diameter: a) suspended tank with polar mount, b) details of a polar mount (Source: [5].) c) tapered skirt (shaded) with reduced upper-stage diameter.

Other Propulsion System Factors Influencing Vehicle Layout

There are many more considerations that can drive the design of a vehicle's tanks and bulkheads (including their location and orientation), propellant delivery system, and other factors. For more information, the works of Hellebrand [7], Platt [10], and NASA [9] are recommended.

8.4   Other Launch Vehicle Components

We now have knowledge of the tanks needed to carry the required propellants. We will need to size the entire vehicle before we can calculate the launch vehicle's other masses, which include the payload fairing (PLF); any skirts, intertanks and/or interstage structures; and the thrust structure that transfers the engine thrust load to the lower portion of the rocket.

Chapter 7 provides several examples of such structures, and Fig. 8.10 shows some guidelines for these parts' sizes in terms of the nearby vehicle diameter. Table 8.20 provides numerical data for these dimensions, and the locations of their mass centers. These guidelines are simply recommendations that may be modified if the designer has better knowledge of the required geometries or systems involved.

An intertank structure separates the bottom dome of the upper tank from the upper dome of the lower tank, to allow for fuel lines, pressurization, and so on. The physical space between the two tanks also provides insulation between propellants that may be at differing temperatures, such as RP-1 at room temperature vs cryogenic propellants. Note that an intertank is not needed when a common bulkhead is used, as discussed in Chapter 7.

An interstage structure is used to connect a lower step to an upper step. It must provide enough length to accommodate the engine bell of the upper step. This interstage is usually attached to the lower step, so as to minimize the upper step's mass. In the case of the Saturn V, the interstage had two separations: first from the lower S-IC step, then from the upper S-II step. This was implemented because of concerns of re-contacting between the lower and upper steps. The Saturn V interstage is shown in Fig. 7.30 in Chapter 7.

One of the most critical structural areas on the launch vehicle is the aft portion, where the loads from the engine or engines are transmitted to the rest of the vehicle. This area also supports the entire weight of the loaded launch vehicle on the launch pad, until it begins flight.

Generally speaking, the engines' thrust produces a concentrated "point" load that must be "beamed out" (distributed) to the body of the LV. Because of the thin-walled nature of the tanks and other body pieces, it's imperative to have the loads uniformly spread out around the perimeter of the aft end of the LV. In addition, the lower structure must also

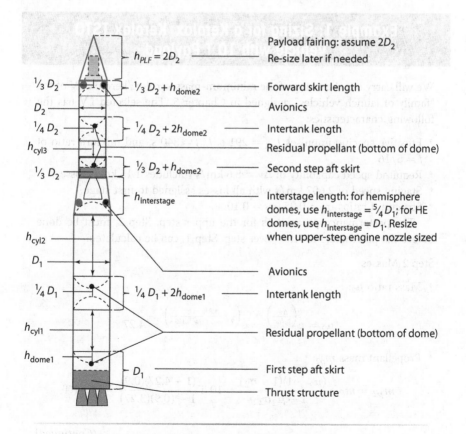

Fig. 8.10 Suggested dimensions to be used to size external structures, such as intertanks, interstage, skirts, etc.

accommodate propellant feed lines, wiring, and the thrust vector control (TVC) actuation system. Several different methods for beaming out the engine's thrust point load are shown in Fig. 7.15 in Chapter 7.

Once all the parts are dimensioned, overall vehicle length is a simple calculation:

Step 1 length: $l_1 = l_{\text{aft skirt1}} + h_{\text{cyl\_ox1}} + h_{\text{intertank1}} + h_{\text{cyl\_f1}} + h_{\text{interstage}}$
Step 2 length: $l_2 = l_{\text{aft skirt2}} + h_{\text{cyl\_ox2}} + h_{\text{intertank2}} + h_{\text{cyl\_f2}} + l_{\text{fwd skirt2}} + h_{\text{PLF}}$
Overall length: $l = l_1 + l_2$

The dimensions shown will also be used to determine the parts' masses, which when totaled provide the vehicle mass, and can be also used to calculate its mass properties, notably its center of mass and moments of inertia.

Our next step will be to use the known masses and dimensions to calculate these components' masses.

In addition, we have to consider some other "hidden" LV components, which include avionics (sometimes called astrionics), wiring, rocket engine(s), and engine steering parts.

Example 1: Sizing for a Kerolox/Kerolox TSTO Vehicle with 10 T Payload

We will carry out the sizing of the minimum-mass kerolox/kerolox LV of the "family of launch vehicles" designed in Chapter 5. The selected LV has the following characteristics:

- Effective specific impulses $I_{sp1} = 291$ s, $I_{sp2} = 340$ s, and mixture ratio of $f = 5/16$
- Required speed capability of $\Delta v = 8.6$ km/s (includes 1.15 km/s losses)
- Staging speed of 2.607 km/s with all losses assigned to first stage
- Structural ratios $\sigma_1 = 0.08$, $\sigma_2 = 0.10$.

The performance parameters for the upper step, Step 2, must be done before the parameters for the lower step, Step 1, can be calculated:

Step 2 Masses

- Mass ratio μ_2:

$$\mu_2 = e^{\left(\frac{\Delta v_2}{g_0 I_{sp2}}\right)} = e^{\left(\frac{4{,}842 \text{ m/s}}{9.80665 \text{ m/s}^2 \times 340 \text{ s}}\right)} = 4.27$$

- Propellant mass m_{p2}:

$$m_{p2} = m_{PL2}\frac{(\mu_2 - 1)(1 - \sigma_2)}{1 - \mu_2\sigma_2} = 10 \text{ T}\frac{(1 - 4.27)(0.9)}{1 - (0.9)(3.27)} = 51.44 \text{ T}$$

(Continued)

Example 1: Sizing for a Kerolox/Kerolox TSTO Vehicle with 10 T Payload *(Continued)*

- 'Structure' mass m_{s2}:

$$m_{s2} = m_{p2} \frac{\sigma_2}{1 - \sigma_2} = 51.44 \text{ T} \left(\frac{0.1}{1 - 0.1} \right) = 5.72 \text{ T}$$

- *Step* 2 loaded + PL = payload for Step 1 =

$$m_{p2} + m_{s2} + m_{PL} = 51.44 + 5.72 + 10 \text{ T} = 67.15 \text{ T}$$

Step 1 Masses

- Mass ratio μ_1:

$$\mu_1 = e^{\left(\frac{\Delta v_1}{g_0 I_{sp1}} \right)} = e^{\left(\frac{3{,}757 \text{ m/s}}{9.80665 \text{ m/s}^2 \times 291 \text{ s}} \right)} = 3.73$$

- Propellant mass m_{p1}:

$$m_{p1} = m_{PL1} \frac{(\mu_1 - 1)(1 - \sigma_1)}{1 - \mu_1 \sigma_1} = 67.15 \text{ T} \frac{(3.73 - 1)(1 - 0.08)}{1 - (0.08)(3.73)} = 240.5 \text{ T}$$

- 'Structure' mass m_{s1}:

$$m_{s1} = m_{p1} \frac{\sigma_1}{1 - \sigma_1} = 240.5 \text{ T} \left(\frac{0.08}{1 - 0.08} \right) = 20.9 \text{ T}$$

- *Stage* 1 initial mass = liftoff mass

$$m_{10} = m_{p1} + m_{s1} + m_{PL1} = 240.5 + 20.9 + 67.15 \text{ T} = 328.6 \text{ T}$$

a) **Propellant Mass calculations**

The masses given above are the mission or "ideal" ones produced by the rocket equation, so we need to augment them for engine startup propellant masses and residual propellant. We'll use the assumptions in Table 8.3 to calculate these additions.

These parameters are used to calculate the LV's fuel and oxidizer masses for both the first and second steps, as shown in Table 8.4.

Next, we add the residual and startup masses whose calculation was shown previously. The total fuel & oxidizer masses (including additions) are:

$$m_{p1f\text{-tot}} = m_{p1\text{-}f}(1 + f_{\text{residual}}) + \text{frac}_{f\text{-}1} t_{\text{startup}} \dot{m}_1$$

$$= (57{,}267 \text{ kg})(1.02) + 0.2381(1 \text{ s})(1{,}467.9 \text{ kg/s})$$

$$= 58{,}762 \text{ kg}$$

(Continued)

Example 1: Sizing for a Kerolox/Kerolox TSTO Vehicle with 10 T Payload *(Continued)*

Table 8.3 Propellant "Real" Mass Additions Calculation Parameters

Item	Step 1	Step 2	Comment
T/W_0	1.3	0.7	Step 1: Reasonable liftoff performance, possible mass growth Step 2: Reduced need for lofting
Mass flow $\dot{m}$	$\dfrac{T_1}{g_0 I_{sp1}}$	$\dfrac{T_2}{g_0 I_{sp2}}$	Definition
Engine start time t_{start}	1 s	1 s	Assumed
Residual propellant fraction $f_{residual}$	0.02	0.02	Assumed
Residual propellant	$f_{residual}$ m_{p1}	$f_{residual}$ m_{p2}	Definition

Table 8.4 Step Fuel & Ox *Mass* Calculations

Item	Step 1	Step 2
f (mix ratio)	$1/3.2 = 5/16^\dagger$	$1/3.2 = 5/16$
Fuel Mass Fraction $frac_f$	$frac_{f\text{-}2} = 5/21 = 0.2381$ $= \dfrac{f_1}{1+f_1}$	$frac_{f\text{-}2} = 5/21 = 0.2381$ $= \dfrac{f_2}{1+f_2}$
Ideal Fuel mass*	$= m_{p1\text{-}tot}\, frac_{f\text{-}1} = 0.2381$ $m_{p1\text{-}tot} = 57{,}267$ kg	$m_{p2\text{-}tot}\, frac_{f\text{-}2} = 0.2381$ $m_{p2\text{-}tot} = 12{,}247$ kg
Oxidizer Mass Fraction $frac_{ox}$	$frac_{ox\text{-}1} = 0.7619 = 16/21$ $= \dfrac{1}{1+f_1}$	$frac_{ox\text{-}2} = 16/21$ $= 0.7619 = \dfrac{1}{1+f_2}$
Ideal Oxidizer Mass*	$= m_{p1\text{-}tot}\, frac_{ox\text{-}1} = 0.7619$ $m_{p1\text{-}tot} = 183{,}254$ kg	$m_{p2\text{-}tot}\, frac_{ox\text{-}2} = 0.7619$ $m_{p2\text{-}tot} = 39{,}191$ kg

*Ideal = before adding startup & residual
†Reasonable kerolox performance. Same for both steps.

(Continued)

Example 1: Sizing for a Kerolox/Kerolox TSTO Vehicle with 10 T Payload (Continued)

$$m_{p1ox\text{-}tot} = m_{p1\text{-}ox}(1 + f_{residual}) + frac_{ox\text{-}1}t_{startup}\,\dot{m}_1$$
$$= (183{,}254 \text{ kg})(1.02) + 0.7619(1 \text{ s})(1{,}467.9 \text{ kg/s})$$
$$= 188{,}038 \text{ kg}$$

$$m_{p2f\text{-}tot} = m_{p1\text{-}f}(1 + f_{residual}) + frac_{f\text{-}2}t_{startup}\,\dot{m}_2$$
$$= (12{,}247 \text{ kg})(1.02) + 0.2381(1 \text{ s})(138.3 \text{ kg/s})$$
$$= 12{,}525 \text{ kg}$$

$$m_{p2ox\text{-}tot} = m_{p1\text{-}ox}(1 + f_{residual}) + frac_{ox\text{-}2}t_{startup}\,\dot{m}_2$$
$$= (39{,}191 \text{ kg})(1.02) + 0.7619(1 \text{ s})(138.3 \text{ kg/s})$$
$$= 40{,}080 \text{ kg}$$

b) **Propellant volume calculations**

Next, we find the propellant-only volumes of the above "total" propellant masses. The propellant-only volumes for the fuel & oxidizer do *not* include the volume to be added to account for ullage and tank shrinkage, and are calculated using the propellant densities $\rho_{RP\text{-}1} = 820 \text{ kg/m}^3$ and $\rho_{LOx} = 1{,}140 \text{ kg/m}^3$.

The propellant-only volumes are calculated using $Vol = m/\rho$:

Step 1's propellant-only volumes
- $Vol_{RP\text{-}1} = m_{RP\text{-}1}/\rho_{RP\text{-}1} = 58{,}762 \text{ kg}/(820 \text{ kg/m}^3) = 71.66 \text{ m}^3$
- $Vol_{LOx} = m_{LOx}/\rho_{LOx} = 188{,}038 \text{ kg}/(1{,}140 \text{ kg/m}^3) = 164.95 \text{ m}^3$

Step 2 propellant-only volumes
- $Vol_{RP\text{-}1} = m_{RP\text{-}1}/\rho_{RP\text{-}1} = 12{,}525 \text{ kg}/(820 \text{ kg/m}^3) = 15.27 \text{ m}^3$
- $Vol_{LOx} = m_{LOx}/\rho_{LOx} = 40{,}080 \text{ kg}/(1{,}140 \text{ kg/m}^3) = 35.16 \text{ m}^3$

The propellant-only volumes above must be increased, because did not include "real world" additions for ullage, shrinkage, etc. All values must be increased to account for ullage volume, and the cryogenic LOx volumes must also be increased for tank shrinkage (RP-1 tanks do not shrink). Using the values of ullage = 0.03 and shrink$_{LOx}$ = 0.0143 provided earlier in Eq. (8.13), we compute the additional volume factors below:

$$Vol_{p\text{-}LOx} = Vol_p\left(1 + ullage_p + shrink_{LOx}\right)$$
$$= Vol_p(1 + 0.03 + 0.0143) = Vol_p(1.0443)$$
$$Vol_{p\text{-}RP\text{-}1} = Vol_p\left(1 + ullage_p + 0\right) = Vol_p(1.03)$$

Next, we calculate the total required volumes including these ullage & shrinkage factors:

(Continued)

Example 1: Sizing for a Kerolox/Kerolox TSTO Vehicle with 10 T Payload *(Continued)*

Step 1 total required propellant volumes
- $Vol_{RP-1} = 71.66 \text{ m}^3 \ (1.03) = 73.81 \text{ m}^3$
- $Vol_{LOx} = 164.95 \text{ m}^3 \ (1.0443) = 172.2 \text{ m}^3$

Step 2 total propellant volumes
- $Vol_{RP-1} = 15.27 \text{ m}^3 \ (1.03) = 15.73 \text{ m}^3$
- $Vol_{LOx} = 35.16 \text{ m}^3 \ (1.0443) = 36.71 \text{ m}^3$

c) **Assumed vehicle diameter**

To continue with tank sizing, we need to select values for the body diameters. For this example, we choose $D_1 = 4$ m and $D_2 = 3$ m for the diameter of the first and second steps respectively. This geometry, providing a slightly smaller-diameter upper step, is shown in Fig. 8.11.

d) **Tank cylinder heights**

Continuing with the sizing analysis, we next need to obtain the dimensions of the propellant tanks. Before we can size the tanks, we must select the tank dome geometry.

In this example, we choose to make the design a bit more compact and use "$\sqrt{2}$" HE domes $(AR = \sqrt{2})$. We could have just as easily chosen hemispherical domes instead.

With that assumption, and the formula for HE-domed tank cylinder height

$$h_{cyl} = \frac{4Vol}{\pi D^2} - \frac{D\sqrt{2}}{3} \qquad (8.16)$$

We find the Step 1 Cylinder heights as
- $h_{RP-1 \text{ step1 cyl}} = 4(73.81 \text{ m}^3)/[\pi(4.0 \text{ m})^2] - \sqrt{2}(4.0 \text{ m})/3 = 3.99 \text{ m}$
- $h_{LOx \text{ step1 cyl}} = 4(172.2 \text{ m}^3)/[\pi(4.0 \text{ m})^2] - \sqrt{2}\ (4.0 \text{ m})/3 = 11.82 \text{ m}$

and the Step 2 Cylinder heights to be
- $h_{RP-1 \text{ step2 cyl}} = (15.73 \text{ m}^3)/[\pi(3.0 \text{ m})^2] - \sqrt{2}(3.0 \text{ m})/3 = 0.81 \text{ m}$
- $h_{LOx \text{ step2 cyl}} = (36.71 \text{ m}^3)/[\pi(3.0 \text{ m})^2] - \sqrt{2}(3.0 \text{ m})/3 = 3.78 \text{ m}$

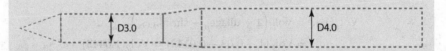

Fig. 8.11 A lower step diameter of 4 m and an upper step diameter of 3 m chosen for example LV.

(Continued)

Example 1: Sizing for a Kerolox/Kerolox TSTO Vehicle with 10 T Payload *(Continued)*

Table 8.5 External Sizes of Example LV Components

Item	Size: $D \times h$ or D_1, D_2, h (m)
1. Payload Fairing	3 m × 6 m (conical)
2. Forward Skirt, 2nd Step	3 m × (1 + 1.061) m = 3 m × 2.061 m
3. Intertank, 2nd Step	3 m × (0.75 + 2 × 1.061) m = 3 m × 2.87 m
4. Aft Skirt, 2nd Step	3 m × (1 + 1.061) m = 3 m × 2.061 m
5. Interstage	4 m, 3 m, 1 m
6. Intertank, 1st Step	4 m × (1 + 2 × 1.414) m = 4 m × 3.83 m
7. Aft Skirt, 1st Step	4 m × 4 m

e) **Remaining external parts**

The remaining external parts are the payload fairing (PLF); forward skirt, intertank, and aft skirt (Step 2), and interstage, intertank, and aft skirt (step 1). Using the recommended dimensions in Fig. 8.8, we have the sizes given in Table 8.5.

The surface areas in the table are calculated using the exposed area formulae given in Table 8.2.

With these data assembled, we can locate the parts end-to-end, with which we arrive at the layout of the external surfaces of our example launch vehicle, as shown in Fig. 8.12.

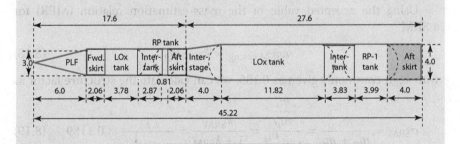

Fig. 8.12 The external layout can be drawn, once all the external parts are sized.

8.5 Solid Propulsion System Sizing

Now we'll discuss the sizing process for solid rocket motors (SRMs). Because solids are quite a bit less complex than liquids—some have no moving parts at all—their design procedures are also a bit simpler. In general, they follow the liquid propellant procedures given above, so we'll omit the similar parts.

Also, remember that solids have both their fuel and oxidizer chemicals mixed together and are typically cast directly into their motor casings, so the propellant "grain" has a single composition and density. Because the cast propellant contains both fuel and oxidizer, there are no intertanks within a solid rocket motor.

Most solids are cast with an opening in the grain to deliver a specified thrust profile. The thrust profile depends on how the burn area progresses as the motor burns, usually from inside to outside.

8.5.1 Solid Propulsion System Initial Sizing

Solid rocket motors may have their performance requirements specified in two different forms: a required Δv similar to the liquid propellant system previously examined, or a required total impulse. We will consider each of these specifications for layout purposes.

8.5.2 Solid Propulsion System With Δv Specified

For a specified Δv, we use the rocket equation along with the specific impulse I_{sp} of the propellant, the payload mass m_{PL}, and the required Δv to determine the mass of propellant required using the rocket equation:

$$\mu = e^{\left(\frac{\Delta v}{g_0 I_{sp}}\right)} = e^{\left(\frac{\Delta v}{c}\right)} \tag{8.17}$$

Using the accepted value of the mass-estimation relation (MER) for a SRM

$$m_{\text{SRM casing}} = k_{\text{SRM}} m_p, \tag{8.18}$$

with k_{SRM} commonly given as 0.135, we can calculate the structure factor as

$$\sigma_{\text{SRM}} = \frac{m_s}{m_p + m_s} = \frac{\dfrac{m_s}{m_p}}{1 + \dfrac{m_s}{m_p}} = \frac{k_{\text{SRM}}}{1 + k_{\text{SRM}}} = \frac{1}{1 + \dfrac{1}{k_{\text{SRM}}}} = 0.1189 \tag{8.19}$$

As before, we can then use the calculated values of μ, σ, and m_{PL} to calculate ideal or mission propellant mass m_p required:

$$m_p = m_{PL} \frac{(\mu - 1)(1 - \sigma_{\text{SRM}})}{1 - \mu \sigma_{\text{SRM}}} \tag{8.20}$$

The resulting propellant mass is used for sizing purposes, as was done previously.

8.5.3 Solid Propulsion System With Specified Total Impulse

Some SRM applications, such as strap-on motors for other vehicles, may specify the motor's total impulse rather than its Δv; for that case, the motor propellant mass may be derived using a different procedure. Given a total impulse requirement and I_{sp} assumption, we can solve for the mass ratio using the following:

First, recall the definitions of total impulse, propellant flow, and thrust as a function of mass flow:

$$I_{tot} = T_{ave} t_{burn} \tag{8.21a}$$

$$m_p = \dot{m}\, t_{burn}, \text{ and} \tag{8.21b}$$

$$T_{ave} = \dot{m}c = \dot{m}\, g_0\, I_{sp} \tag{8.21c}$$

These three relations can be combined to solve for the propellant mass m_p:

$$m_p = \dot{m} t_{burn} = \dot{m}\frac{I_{tot}}{T_{ave}} = \dot{m}\frac{I_{tot}}{\dot{m}c} = \frac{I_{tot}}{c} = \frac{I_{tot}}{g_0 I_{sp}} \tag{8.22}$$

The resulting propellant mass is used for sizing purposes, as was done previously.

8.5.4 Solid Propellant Volume Calculation

With the propellant mass known, we calculate the required volume of propellant as

$$\text{Vol}_{propellant} = m_p / \rho_{propellant} \tag{8.23}$$

Some values for propellant densities and specific impulses may be found in Chapter 4.

Solid Propulsion System Real-Life Additions to Volume

A solid propellant system needs to add volume for the opening in the propellant grain to control thrust profile, internal insulation, and for an ignition system; these are shown in Figs. 8.13a and 8.13b.

For a solid propulsion system, we need to size the motor casing to provide additional volume to accommodate the following:

1S. Internal volume must be added to accommodate motor grain modifications (openings) that are used to adjust the motor's thrust-vs-time profile or its burn rate. This modification usually consists of a longitudinal opening down the center of the grain. Some possible grain openings are provided in Chapter 4, propulsion.

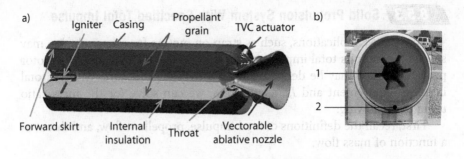

Fig. 8.13 (a) The major parts of a solid motor. Source: Northrop Grumman. (b) A solid motor needs volume for grain openings (1), insulation (2), ignition system not shown here but can be seen in Fig. 8.13a. Source: Edberg.

2S. Internal volume must be provided for motor case insulation. The motor's casing must be insulated from the hot gases of the burning propellant. This internal insulation requires a percentage of the casing's volume.

3S. A small volume is needed for the hardware used to ignite the motor. The volume is small and will be neglected for the first estimation process.

4S. Volume also is needed for the embedded nozzle (if utilized). Embedded nozzle configurations are shown in Chapter 4.

We do not add mass for startup, because we assume the ignition process is instantaneous. We also assume complete combustion with no residual propellant. We assume that the masses of the insulation material and the ignition system are incorporated into the SRM casing mass.

Solid Propulsion System Volume

We'll assume that the openings ("holes" or channels) in the propellant grain for thrust tailoring consume about 10% of the internal volume of the motor casing (see Fig. 8.13b). We also assume that insulation required to protect the casing and the igniter system together add 2% volume. If we neglect any embedded nozzle portion, we find that the volume of the motor casing can be found from

$$\text{Vol}_{\text{casing}} = 1.12 \, \text{Vol}_{\text{propellant}} \qquad (8.24)$$

The resulting casing volume may be used to determine the SRM's dimensions.

For initial design purposes, one may assume that the front and rear domes are hemispheric (because of the high internal pressures of a SRM), and filled with propellant (except for the nozzle itself and the grain's opening). This can be seen in Fig. 8.13a.

There are a few observations regarding that figure: first, the casing is specified to be carbon epoxy. Many other SRMs use steel (often type D6AC), which is heavier but more robust (in terms of mechanical use and damage, and capable of much higher internal temperatures). Internal insulation is shown between the HTPB propellant grain and the internal wall of the casing. Another unique feature of an SRM is that the exhaust nozzle may be internal or submerged (the latter to save space).

For sizing purposes, we only need find the protruding length of the diverging portion of the motor's nozzle; this may be estimated from the engine sizing equations given below, or actuals can be used, if available. The protruding length is added to the motor casing length to find the overall length of the step.

8.6 Comments about Upper Steps and Payload Fairings

8.6.1 Upper Stage Layouts

Depending on the stage's requirements, it may make sense to use a smaller diameter for the upper stage, as shown in Fig. 8.14. The procedure is to follow the previous guidelines regarding adapters, skirts, and the like. Note that with a geometry as shown in the figure, there may be added vibration due to vortex shedding by the PLF. This will be discussed in Chapter 11.

Fig. 8.14 This Taurus has an upper step that is smaller in diameter than the lower step, easily handled with the skirt connecting the two. Note also the increase in diameter for the payload. Source: NASA.

8.6.2 Payload Fairings

Obviously, the PLF needs to be sized so as to accommodate any payload and provide the amount of volume required. Be aware that the PLF structure has to protect the payload from prelaunch to the elevation and speed where the atmospheric heating becomes negligible.

The simplest PLF is a right circular cone with a base diameter approximately the same as the top of the rocket body and a metal tip on its nose to alleviate the heat absorbed at its tip. The PLF can be either smooth curve or cone, but smooth curves probably will need to be made from composite materials. In some cases, the vehicle may need a larger diameter. Figure 8.15 shows some possible fairing shapes.

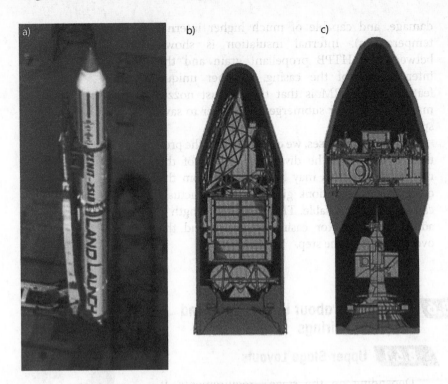

Fig. 8.15 The LandLaunch vehicle has a simple right circular cone for a payload fairing. b) A PLF with a rounded nose. c) A PLF with a bi-conic shape (two cones connected to a cylinder); notably, it also contains two spacecraft that are launched together.

One can find out about an LV supplier's PLF geometry by consulting its payload planning guide or user guides. There is no shame in imitating the PLF of an existing manufacturer. Take a look at a similar vehicle comparable to yours, or design your own. One configuration that might work well is a cylindrical shell with a right circular cone for simplicity, as may be seen on the LandLaunch vehicle in Fig. 8.15a.

To get a feel for the sizes of different PLFs, the dimensions of the PLFs on 12 current or retired launch vehicles are shown in Fig. 8.16. Please note that the plots are not to scale; the width or diameter dimension is magnified compared to the length dimension.

8.7 Mass Estimation Process

At some point in the design process, it is necessary to determine the mass of the tanks, tank insulation (if any), PLF, intertanks, interstage, engines, steering hardware, avionics, wiring, and other parts. The masses of all known components and propellants are summed to provide the first-order mass of the vehicle. The value of this mass with the payload mass added determines whether the vehicle can achieve the desired structural ratio and performance. If it doesn't, it's back to the drawing board.

Ideally, this might be done using a bottom-up method with the known propellant volumes and masses, vehicle diameters, dimensions, loads, and so forth and carrying out a complete stress analysis that would, in turn, allow us to do detail sizing of all the parts and calculate each mass. This is impossible to do unless you have the vehicle's dimensions, trajectory, and loads details.

Another method, much better suited to preliminary design, is to use statistical mass estimating relationships (MERs) to estimate the masses of components. The MERs are mathematical formulae that correlate known dimensions and materials to calculate masses, and are based on physical parameters such as size, area, contained mass, vehicle length, thrust, and many others.

The MERs are based on curve fits to known masses of components of previously-built vehicles or components, and so they may be somewhat conservative because they do not indicate advances in technology that may have occurred since the component was manufactured. For instance, improved materials, such as composite materials instead of metals, improved manufacturing processes, use of miniaturized electronics instead of older configurations, and the like will not be reflected.

In addition, because they are based on statistics and curve fits, there is some scatter in the MER curves. A good curve fit will have a correlation coefficient R^2 (related to scatter) of 0.9 or higher, but even then, the real number could end up higher or lower. This means that at best, the results

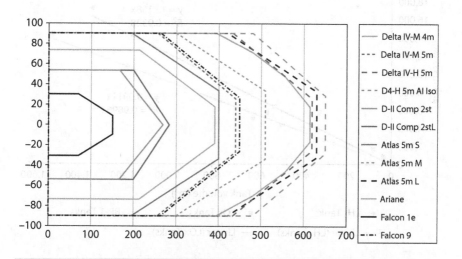

Fig. 8.16 An assortment of PLF dimensions (in in.). Note: not to scale—vertical scale exaggerated.

obtained from MERs should be presented with *no more than three significant digits.*

In addition, be sure that you use the proper input units of the MER. Some MERs use English/Imperial/FPS system, whereas some use SI/metric units.

Next, we'll discuss using MERs to estimate masses of liquid propellant tanks, structure, and other LV components. We will then discuss mass calculations for solid propellant LVs.

8.7.1 Estimation of Liquid Propellant Tank Masses

Figure 8.17 shows an example plot of tank mass vs volume along with curve fits for two types of propellant tanks. Note that most of the data points these curves are based on do not even lie on the fitted curve! MERs derived for liquid mass contained in tanks are given in Tables 8.1 and 8.2.

A list of MERs for tanks of launch vehicles vs contained mass is listed in Tables 8.6 and 8.7, courtesy of Akin [1]. They are split into tanks containing LH_2, LOx, and room temperature propellants. Also included are insulation areal densities (mass of insulation per unit area), and masses of solid motor casings, small liquid tanks, and small pressurized-gas tanks. The MERs in Table 8.6 are more recent and preferred when possible.

We have also included Tables 8.8 and 8.9, so that all of the MER tables will be in a common location for ease of use. These tables will be used in subsequent calculations.

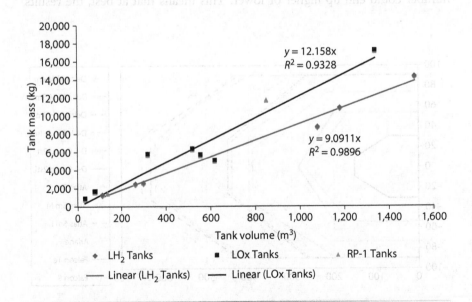

Fig. 8.17 Propellant tank regression data and derived MERs with correlation coefficients R2. Source: [1].

Table 8.6 MERs for Launch Vehicle Tanks vs. Contained Propellant Mass

Tank Mass by Mass Contained (kg)	MER
m_{LH2} tank	$0.128\, m_{LH2}$ for $\rho_{LH2} = 71\ kg/m^3$
m_{LH2} insulation	$2.88\ kg/m^2$
M_{LOX} tank	$0.0107\, m_{Lox}$ (kg) for $\rho_{LOx} = 1140\ kg/m^3$
m_{LOX} insulation	$1.123\ kg/m^2$
$m_{RP\text{-}1}$ tank	$0.0148\, m_{RP\text{-}1}$ (kg) for $\rho_{RP\text{-}1} = 820\ kg/m^3$
$m_{SRM\ casing}$	$0.135\, m_{propellants}$

Source: [1]
Storables liquid at room temp: RP-1/kerosene, N_2O_4, hydrozines; solids.

Table 8.7 MERs for Tanks vs. Contained Volume

Tank Masses by Volume (kg)	MER
$m_{tank\ not\ LH2}$	$12.16\ Vol_{prop}$ (m^3), all except LH_2
$m_{LH2\ tank}$	$9.09\ Vol_{LH2}$ (m^3)
$m_{COPV\ tank}$	$115.3\ Vol_{contents}$ $(m^3) + 3$
$m_{small\ pressurized\ gases\ tank}$	$2(m_{contents})$

Source: [1]
Vol = contained volume; COPV = Composite overwrap pressure vessel, used for high pressures.
Note: Composites could provide 20–30% mass reduction compared to metallic structures;
however, there are issues with the use of composites with cryogenics.

Table 8.8 MERs for Different Structures and Components of LVs

#	Structure Description (mass in kg)	Mass Estimating Relationship (MER)
1	PLFs and shrouds: $m_{fairing}$/area	$13.3\ kg/m^2$
2	Thrust structure mass: $m_{thrust\ structure}$	$2.55 \times 10^{-4}\ T(N)$
3	Payload attach fitting (PAF) or launch vehicle adapter (LVA) mass: m_{PAF}	$0.0755\, m_{PL} + 50\ kg$
4	Avionics mass: $m_{avionics}$	$10\, m_0^{0.361}$
5	Wiring mass: m_{wiring}	$1.058\, m_0^{1/2} L_0^{1/4}$

Source: [1] (lines 1, 2, 4, 5); Brown [3] (line 3).
Note: m_0 = liftoff mass in kg; m_{PL} = payload mass in kg; L_0 = overall length in m, T = liftoff thrust in N.
These MERs tend to overestimate the masses of avionics and wiring for smaller LVs. Designers may need to
find other, more realistic values for these LVs.

Table 8.9 MERs for Different Propulsion Components of LVs

#	Item Description	Estimating Relationship
6	Liquid pump-fed rocket engine mass: $m_{rkt\ engine}$ (kg)	$7.81 \times 10^{-4}\ T(N) + 3.37 \times 10^{-5}$ $T(N)[A_e/A_t]^{1/2} + 59\ kg$
7	Gimbals: $m_{engine\ gimbals}$ (kg)	$237.8\,[T(N)/P_c\ (Pa)]^{0.9375}$
8	Gimbal torque: $Torque_{gimbals}$ (kNm)	$990\,[T(N)/P_c\ (Pa)]^{1.25}$

Note: T = engine thrust (N), P_c = combustion chamber pressure, $A_e/A_t = \varepsilon$ = nozzle expansion ratio.

8.7.2 Cryogenic Tank Insulation Mass

Tanks containing cryogenic propellants such as LH_2, LOx, or other liquified gases need insulation applied to minimize propellant boiloff. The second and fourth lines of Table 8.6 contains MERs that may be used to estimate the mass of insulation applied to tanks. Both of these MERs are based on the surface area of the tank. To utilize them, we simply calculate the tank's surface area and multiply it by the insulation's areal density.

The area of a tank is simply the area of its cylindrical portion added to the area of its two domes, or $A_{tank} = \pi D h_{cyl} + 2A_{dome}$. Areas for different aspect-ratio domes are provided in Table 8.2. Then, we find the insulation mass using the MERs:

$$m_{\text{LOX tank insulation}} = m_{\text{LOX insulation}} A_{\text{LOx tank}} \tag{8.25}$$

$$m_{\text{LH}_2 \text{ tank insulation}} = m_{\text{LH}_2 \text{ insulation}} A_{\text{LH}_2 \text{ tank}} \tag{8.26}$$

Insulation mass values for other propellants or temperatures might reasonably be estimated by assuming the insulation mass varies linearly with the temperature difference.

8.7.3 Masses of Thin-Shelled Structures

We need to calculate the masses of the non-tank structures in Fig. 8.12. Structures such as skirts, intertanks, interstages, and fairings are dealt with by

Table 8.10 Properties of Thin Shells

Shape	z_cCM location	Outer Surface area A (no base, top, no caps)	J_x & J_y	J_z
Thin Cylinder	$h/2$	$2\pi R h$	$m\left(\dfrac{R^2}{2} + \dfrac{h^2}{12}\right)$	mR^2
Thin Cone	$h/3$	$\pi R\sqrt{R^2 + h^2}$	$m\left(\dfrac{R^2}{4} + \dfrac{h^2}{18}\right)$	$m\dfrac{R^2}{2}$
Thin Frustum of Cone	$\dfrac{h}{3}\left(\dfrac{2r+R}{r+R}\right)$	$\pi(r+R)\sqrt{(R-r)^2+h^2}$	$m\left\{\dfrac{R^2+r^2}{4} + \dfrac{h^2}{18}\left[1 + \dfrac{2rR}{(r+R)^2}\right]\right\}$	$m\dfrac{R^2+r^2}{2}$

Open cylindrical shell (no end caps)

Open circular cone lateral surface

Open cone frustum ("skirt") lateral surface

assuming their mass depends on their exposed surface area. Table 8.10. contains formulae for exposed surface area of the three generic shapes (a cone, frustum of a cone, and uniform cylinder) that can be used for dealing with empty tanks, shells, and skirts. Other, more complicated shapes' areas may be calculated by using formulae in Tables 8.10, tabulations of mass for unusual structures, or by direct integration or CAD programs.

We utilize the formulae in the table to calculate masses in the same manner as we did to calculate the mass of insulation earlier. For these objects, we simply calculate the object's surface area, given in the 3rd column of Table 8.10, and multiply it by the areal density of the object. In Table 8.8, we see that line has an entry for the areal density of payload fairings and shrouds. We will use this information to find the mass of the PLF, interstage, skirts, and intertanks. The area of the object is calculated using the area formula, then multiplied by areal density as follows:

$$m_{\text{thin shell object}} = (\text{areal density}) \times A_{\text{thin shell object}} \qquad (8.27)$$

This process is used to calculate the areas and masses of the following types of structures:

* Payload fairing (PLF)
* Forward & aft skirts
* Intertanks
* Interstages

8.7.4 Masses of Other Structures and Components

A few other parts' masses need to be estimated. These are:

* Liquid propellant rocket engine
* Liquid engine thrust structure
* Liquid engine gimbal structure
* Payload attachment
* Avionics
* Electrical wiring
* Solid propellant motor masses

The mass estimating relationships for most of these items are contained in Tables 8.8 and 8.9. Starting at the top of the list, we calculate the masses of the engine(s) and its(their) associated hardware.

8.7.5 Rocket Engine and Thrust Structure Mass Estimation

The masses of propulsion system elements may be found using the MERs provided in Table 8.9. In general, these relationships use thrust level and chamber pressure as inputs, so it may be necessary to make some assumptions about these parameters, usually based on existing engines. The last item in the table is a "TER": a torque-estimation relationship for TVC actuation. Its output is used to size hydraulic or electric servos or actuators.

Engine mass MERs come from plots such as that shown in Fig. 8.18, with engine mass vs thrust level. Clearly there is a correlation between the two variables, and the scatter is relatively small.

Engine Mass

The mass of the engine depends on the thrust and engine expansion ratio, as given by MER #6 in Table 8.9:

$$m_{engine}(kg) = T(N)\left(7.81 \times 10^{-4} + 3.37 \times 10^{-5}\sqrt{\frac{A_e}{A_t}}\right) + 59 \text{ kg}$$

To utilize the engine MER, we'll need to assume a thrust level. For the first step, it's convenient to assume a needed liftoff thrust-to-weight ratio $T/(mg_0) = \psi$ based on a constraint such as limiting shutdown acceleration, or operation with one engine inoperative (OEI). We also must choose an engine expansion ratio; for first steps, it should be something that provides reasonably good sea level to vacuum performance, such as $A_e/A_t = 20$ or so.

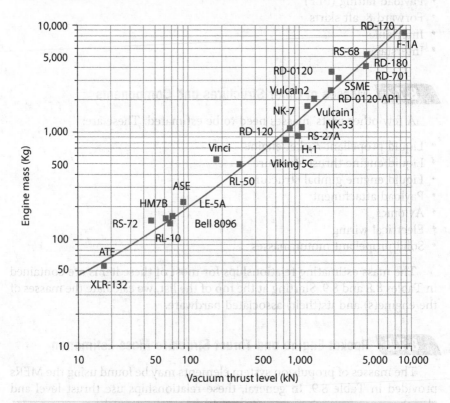

Fig. 8.18 Liquid rocket engine dry mass vs. vacuum thrust. Courtesy: TransCostSystems [6].

For the second and higher steps, we might choose a psi value to reduce the need for lofting or for preferred abort circumstances. We would also choose a much larger expansion ratio, perhaps 100 or larger.

Thrust Structure

As might be expected, the mass of the engine thrust structure depends on the amount of thrust, as given by MER #6 in Table 8.9, where T (N) is total thrust being generated (per engine × number of engines):

$$m_{\text{thrust structure}}(\text{kg}) = 2.55 \times 10^{-4} T(\text{N})$$

Gimbal Mass

Most launch vehicles use a TVC system to gimbal the engine in two axes (pitch and yaw) for stability and control. The frame of the engine must be mounted firmly in the three translations, but must be free to rotate in pitch and yaw in order to control the vehicle. (The engine is also restrained in the roll axis.) Hydraulic actuators are used to rotate or 'gimbal' the engine to provide side thrust. Figure 7.16 in Chapter 7 illustrates some of these concepts.

The mass of the gimbal for the engine depends on the thrust and combustion chamber pressure, as given by the MER in Table 8.9:

$$m_{\text{gimbals}}(\text{kg}) = 237.8 \left[\frac{T(\text{N})}{P_c(\text{Pa})}\right]^{0.9375}$$

Here, T (N) is the connected engine's thrust in newtons, and P_c is combustion chamber pressure in pascals. The total gimbal mass for multiple engines (quantity n) is slightly modified, because of exponential nature of the MER:

$$m_{\text{gimbals } n \text{ engines}}(\text{kg}) = n_{\text{engines}} \times 237.8 \left[\frac{T(\text{N})/n_{\text{engines}}}{P_c(\text{Pa})}\right]^{0.9375}$$

Of course, the mass of a single gimbal would be the total gimbal mass ÷ n:

$$m_{1 \text{ gimbal}}(\text{kg}) = 237.8 \left[\frac{T(\text{N})/n_{\text{engines}}}{P_c(\text{Pa})}\right]^{0.9375}$$

8.7.6 Mass of Other Items

Payload Attach Fitting/Launch Vehicle Adapter

Near the top of the LV, we have the payload attach fitting (PAF), also known as the launch vehicle adapter (LVA). The PAF mass may be estimated from Item #3 in Table 8.8. This item's mass depends on the mass of the payload being attached to it; the relation is (in kg)

$$m_{PAF}(kg) = m_{LVA}(kg) = 0.0755 m_{PL} + 50.$$

This relation comes from Brown [3] and is a statistical fit of existing PAFs.

The reader should note that it is much better to use data from an actual PAF if possible, since it will provide "real" numbers associated with mass. Actual PAF data may be found in LV manufacturers' payload planners guides, mission planners guides, mission user guides, and other documentation with similar names. (Note that the PAF mass is charged to the payload, so the actual payload mass must be reduced to accommodate the PAF.)

Avionics Mass

The avionics mass depends on the gross liftoff mass of the LV and is estimated as follows, using MER #4 in Table 8.8:

$$m_{avionics}(kg) = 10[m_0(kg)]^{0.361}$$

It seems reasonable to assume that the avionics mass is split evenly among the steps at this preliminary stage.

When applied to smaller LVs, the avionics MER presented here provides masses that seem unreasonably high, probably because of improvements in electronics that have occurred since the MER was originated. *In general the designer is encouraged to use their judgement in "knocking down" the avionics mass (and any other quantity that seems "off") as long as they have a reasonable justification for doing so.*

Electrical Wiring

The electrical wiring mass is dependent on the vehicle's overall length. Thus, it is necessary to have a good dimensional layout before doing any mass estimation. Wiring masses are estimated as follows, using MER #5 in Table 8.8:

$$m_{wiring}(kg) = \left(1.058\sqrt{m_0(kg)}\right)\left(\sqrt[4]{length(m)}\right)$$

Since the wiring mass depends on the LV's overall length, it seems reasonable to assume the step masses could be calculated by splitting as a ratio of the step length fraction:

$$m_{wiring\,1} = m_{wiring}(l_1/l)$$

$$m_{wiring\,2} = m_{wiring}(l_2/l)$$

Summary

We have provided a means to estimate the masses of all the components of a launch vehicle. In the next section, we will provide examples of calculations utilizing most of these MERs, where we calculate the dry mass of the two-stage liquid propellant launch vehicle example that we sized previously.

Example 2: Mass Calculations for a Kerolox/Kerolox TSTO Vehicle with 10 T Payload

In Section 8.1.1 Example 1, we determined the dimensions for the example vehicle, which are summarized below:

Total fuel & oxidizer masses (including additions) are:
Step 1: $m_{p1f\text{-}tot} = 58,762$ kg
$m_{p1ox\text{-}tot} = 188,038$ kg
Step 2: $m_{p2f\text{-}tot} = 12,525$ kg
$m_{p2ox\text{-}tot} = 40,080$ kg

Total required propellant volumes
Step 1: $\text{Vol}_{RP\text{-}1} = 73.81$ m^3
$\text{Vol}_{LOx} = 172.2$ m^3
Step 2: $\text{Vol}_{RP\text{-}1} = 15.73$ m^3
$\text{Vol}_{LOx} = 36.71$ m^3

Assumed vehicle diameters
Step 1: $D_1 = 4$ m
Step 2: $D_2 = 3$ m
Tanks: $\sqrt{2}$ domes, all steps

With the above data, we can begin mass calculations.

Tank masses: calculate with tank mass MERs lines 3 and 5, Table 8.6:

$m_{ox\ tank\,1} = 0.0107\, m_{p\text{-}ox1}(kg) = 0.0107(188,038\,kg) = 2,012\,kg$
$m_{f\ tank\,1} = 0.128\, m_{p\text{-}f1}(kg) = 0.128(58,762\,kg) = 870\,kg$
$m_{ox\ tank\,2} = 0.0107\, m_{p\text{-}ox1}(kg) = 0.0107(40,080\,kg) = 495\,kg$
$m_{f\ tank\,2} = 0.128\, m_{p\text{-}f1}(kg) = 0.128(12,525\,kg) = 185\,kg$

Estimating Cryotank Insulation Mass
We need the surface areas of the tanks to estimate insulation mass, and tank wall thicknesses. The tank area formula is:

$$A_{tank} = \pi D h_{cyl} + 2A_{dome} = \pi D h_{cyl} + 3.2468\pi R^2 \quad \text{for } \sqrt{2}\text{ domes}$$

Step 1 Tank Areas (subscript "f" is fuel, and subscript "ox" is oxidizer)

$$A_{f\ tank\,1} = \pi(4.0\,m)(3.99\,m) + 3.2468\pi(2.0\,m)^2 = 90.9\,m^2$$
$$A_{ox\ tank\,1} = \pi(4.0\,m)(11.82\,m) + 3.2468\pi(2.0\,m)^2 = 189.3\,m^2$$

(Continued)

Example 2: Mass Calculations for a Kerolox/Kerolox TSTO Vehicle with 10 T Payload (Continued)

Step 2 Tank Areas

$$A_{f\,\text{tank}2} = \pi(3.0\,\text{m})(0.81\,\text{m}) + 3.2468\,\pi(1.5\,\text{m})^2 = 30.6\,\text{m}^2$$

$$A_{\text{ox tank}2} = \pi(3.0\,\text{m})(3.78\,\text{m}) + 3.2468\,\pi(1.5\,\text{m})^2 = 58.6\,\text{m}^2$$

The insulation mass is the product of the tank surface area and the insulation's areal density. From MER tables: $m_{\text{LOX insulation}}(\text{kg}) = 1.123\ \text{kg/m}^2$. For our example LV:

$$m_{f\,\text{insul}1} = 1.123\,\text{kg/m}^2\,(189.3\,\text{m}^2) = 212.6\,\text{kg}$$
$$m_{f\,\text{insul}2} = 1.123\,\text{kg/m}^2\,(58.6\,\text{m}^2) = 65.8\,\text{kg}$$

No insulation is needed for fuel (RP-1) tanks.

Mass Estimation for Other Structures

Fairings & Shrouds: calculate with MER lines 1 from Table 8.8, which gives $m_{\text{fairing}}/\text{area} = 13.3\ \text{kg/m}^2$. Using the area data presented in Table 8.5, we find the masses of the external portions of the example LV (excepting the propellant tanks and insulation found previously) as listed below:

Payload Fairing: $m_{PLF} = A_{PLF} \times$ (Areal density)
$$= 29.1\ \text{m}^2 \times 13.3\ \text{kg/m}^2 = 388\ \text{kg}$$

Forward Skirt 2: $m_{\text{FwdSkirt}2} = A_{\text{FwdSkirt}2} \times$ (Areal density)
$$= 19.42\ \text{m}^2 \times 13.3\ \text{kg/m}^2 = 258\ \text{kg}$$

Intertank 2: $m_{\text{IntTank}2} = A_{\text{IntTank}2} \times$ (Areal density)
$$= 27.06\ \text{m}^2 \times 13.3\ \text{kg/m}^2 = 360\ \text{kg}$$

Aft Skirt 2: $m_{\text{AftSkirt}2} = A_{\text{AftSkirt}2} \times$ (Areal density)
$$= 19.42\ \text{m}^2 \times 13.3\ \text{kg/m}^2 = 258\ \text{kg}$$

Interstage: $m_{\text{Intstg}} = A_{\text{Intstg}} \times$ (Areal density)
$$= 44.3\ \text{m}^2 \times 13.3\ \text{kg/m}^2 = 590\ \text{kg}$$

Intertank 1: $m_{\text{IntTank}1} = A_{\text{IntTank}1} \times$ (Areal density)
$$= 48.1\ \text{m}^2 \times 13.3\ \text{kg/m}^2 = 640\ \text{kg}$$

Aft Skirt 1: $m_{\text{AftSkirt}1} = A_{\text{AftSkirt}1} \times$ (Areal density)
$$= 50.3\ \text{m}^2 \times 13.3\ \text{kg/m}^2 = 669\ \text{kg}$$

Engine Masses

To utilize the engine mass MER #6 in Table 8.9, we'll make some assumptions. For Step 1, we will assume that we need a liftoff $T_1/(mg_0) = \psi_1 = 1.3$ for lower peak acceleration at engine shutdown. To ensure that our vehicle will have positive vertical acceleration even with one engine inoperative (OEI), we will utilize six individual engines. For this situation, $T_{\text{OEI}}/mg_0 = \psi_1(n-1)/n = 1.3(5)/6 = 1.08$, which is larger than one (and will continue to increase

(Continued)

Example 2: Mass Calculations for a Kerolox/Kerolox TSTO Vehicle with 10 T Payload *(Continued)*

as propellant mass is consumed). Finally, we choose to have a step 1 engine expansion ratio $A_{e1}/A_{t1} = 20$, which provides reasonably good vacuum performance as recommended by Leondes [8].

With this set of assumptions, we can compute engine mass parameters. Recalling the liftoff mass of 328,590 kg, the individual Step 1 engine thrust is

$$T_1 = g_0 m_0(\psi_0)/n_{\text{engines}} = (9.80 \text{ m/s}^2)(328,590 \text{ kg})(1.3)/6 = 698.2 \text{ kN}.$$

Using this in the engine mass MER, the individual Step 1 engine mass is

$$m_{\text{engine1}}(\text{kg}) = 698,200(7.81 \times 10^{-4} + 3.37 \times 10^{-5}\sqrt{20}) + 59 \text{ kg} = 710 \text{ kg}$$

A similar set of calculations can be done for the second step's engine. For Step 2, we will assume that we need an initial $T/m_{20}g_0 = \psi_2 = 0.7$ to reduce need for lofting. We will utilize one engine with an expansion ratio of $A_{e2}/A_{t2} = 100$. With these assumptions and recalling stage 2's initial mass of 67,154 kg, the Step 2 engine thrust: $T_2 = g_0 m_2 \psi_2$ = $(9.80 \text{ m/s}^2)(67,154 \text{ kg})(0.7) = 461 \text{ kN}$. This leads to the step 2 engine mass:

$$m_{\text{engine 2}}(\text{kg}) = 461,000(7.81 \times 10^{-4} + 3.37 \times 10^{-5}\sqrt{100}) + 59 \text{ kg} = 574 \text{ kg}$$

Thrust Structure Masses

The masses of the engine thrust structures are given by MER #6 in Table 8.9, where T (N) is total thrust being generated (per engine × number of engines):

$$m_{\text{thrust structure}}(\text{kg}) = 2.55 \times 10^{-4}T(\text{N})$$

For our example, the first step thrust structure mass (for all six engines) is calculated as

$$m_{\text{thrust struct 1}} = 2.55 \times 10^{-4}(6 \times 698,200) = 1,068 \text{ kg}$$

The second step's thrust structure mass:

$$m_{\text{thrust struct 2}} = 2.55 \times 10^{-4}(461,000) = 118 \text{ kg}$$

Gimbal Masses

The mass of the gimbal for the engine depends on the thrust and combustion chamber pressure, as given by MER #7 in Table 8.9:

$$m_{\text{gimbal}}(\text{kg}) = 237.8\left[\frac{T(\text{N})}{P_c(\text{Pa})}\right]^{0.9375}$$

(Continued)

Example 2: Mass Calculations for a Kerolox/Kerolox TSTO Vehicle with 10 T Payload (Continued)

For Step 1, the gimbal mass is the sum of the gimbal masses of each of the six engines:

$$m_{\text{gimbals1}}(\text{kg}) = 6 \times 237.8 \left[\frac{4{,}189{,}100/6}{7{,}093{,}000}\right]^{0.9375} = 162 \text{ kg}$$

For Step 2, the gimbal mass is for the single step 2 engine:

$$m_{\text{gimbal2}}(\text{kg}) = 237.8 \left[\frac{461{,}000}{7{,}093{,}000}\right]^{0.9375} = 18 \text{ kg}$$

Payload Attach Fitting/Launch Vehicle Adapter Mass

The PAF mass may be estimated from Item #3 in Table 8.8:

$$m_{\text{PAF}}(\text{kg}) = m_{\text{LVA}}(\text{kg}) = 0.0755 m_{PL} + 50.$$

For our example vehicle, the payload is 10,000 kg, so the PAF mass is

$$m_{\text{PAF}} = m_{\text{LVA}} = 0.0755 \times 10{,}000 + 50 \text{ kg} = 805 \text{ kg}.$$

Avionics mass

Avionics mass is estimated as follows, using MER #4 in Table 8.8:

$$m_{\text{avionics}}(\text{kg}) = 10[m_0(\text{kg})]^{0.361} = 10[328{,}600]^{0.361} = 981 \text{ kg}$$

We assume a split of 50-50 upper & lower steps at this preliminary stage, so

$$m_{\text{avionics 1}} = m_{\text{avionics 2}} = 490 \text{ kg}.$$

Wiring Mass

Wiring mass is estimated as follows, using MER #5 in Table 8.8:

$$m_{\text{wiring}}(\text{kg}) = \left(1.058\sqrt{m_0(\text{kg})}\right)\left(\sqrt[4]{\text{length}(\text{m})}\right)$$

$$= \left(1.058\sqrt{328{,}600}\right)\left(\sqrt[4]{45.2}\right) = 1{,}573 \text{ kg}$$

Since the wiring mass depends on the LV's overall length, step wiring masses are assumed to be split as a ratio of their step's length fraction:

$$m_{\text{wiring 1}} = m_{\text{wiring}}(l_1/l) = 961 \text{ kg}$$

$$m_{\text{wiring 2}} = m_{\text{wiring}}(l_2/l) = 612 \text{ kg}$$

(Continued)

Example 2: Mass Calculations for a Kerolox/Kerolox TSTO Vehicle with 10 T Payload (Continued)

Mass Summary

Finally, we list all of these computed masses in Table 8.11. Note that the table provide data for the items in both steps of the LV, as well as information specific to the payload. Also, it is very important to include the masses of the residual propellants as part of the total inert mass, since they are not consumed during the LV's flight.

Table 8.11 shows that we have achieved positive margins in the 'structure' or inert masses of both steps, 20% and 2.2% for the first and second steps, respectively. Note that the positive margins allow some weight growth so the structural fractions can increase, yet the vehicle can still perform its mission.

Table 8.11 Mass Summary for Example TSTO Launch Vehicle

Step 1 Item	Mass (kg)	Step 2 Item	Mass (kg)
Est. 'struct' Mass	20,915	Est. 'struct' Mass	5,715
		Payload Fairing PLF	388
		PAF, $h = D/4$	805
Avionics 1 (50%)	490	Avionics 2 (50%)	490
Wiring 1 (by length ratio)	961	Wiring 2 (by length ratio)	612
Interstage	590	Forward Skirt 2	258
Ox Tank 1	2,012	Ox Tank 2	429
Ox Insulation 1	213	Ox Insulation 2	66
Intertank 1	640	Intertank 2	360
Fuel Tank 1	870	Fuel Tank 2	185
Aft Skirt 1	669	Aft Skirt 2	258
Thrust Structure 1	1,068	Thrust Structure 2	118
Gimbals 1	162	Gimbals 2	18
Engines 1	4,257	Engines 2	574
Residual propellant 1	4,810	Residual propellant 2	1,029
Total 'Structure' Mass 1	**16,742**	**Total 'Structure' Mass 2**	**5,590**
Step 1 Design Margin	*4,173 kg 20%*	*Step 2 design margin*	*125 kg 2.2%*
Total Inert Mass no PL	**22,332**		

8.7.7 **Engine Dimensioning Calculations**

In the initial sizing process we've just carried out, it's not always possible to find an already-existing rocket engine the needed performance. Therefore, it may be necessary to produce a "rubber" engine that is scaled from existing engines and their geometry. We will now provide a way to size an engine for a vehicle. The engine-sizing material presented is courtesy of Straw [12].

To start, we need values for thrust, chamber pressure, throat area, and specific impulse to size our engine. Engine thrust is determined using the liftoff thrust-to-weight ratio and the liftoff mass, $T_{SL} = \psi m_0 g_0$. We then calculate the engine's throat area:

$$A_t = \frac{c_{del} T_{vac}}{P_c g_0 I_{sp\text{-}vac}} \qquad (8.28)$$

where the variables are as follows:

A_t = throat area
c_{del} = delivered c (exit velocity)
T_{vac} = vacuum thrust
P_c = chamber pressure
$I_{sp\text{-}vac}$ = delivered specific impulse
ε_{nozzle} = nozzle expansion ratio = A_e/A_t
A_e = exit area = $\varepsilon_{nozzle} A_t$

Recall also that $T_{SL} = T_{vac} - P_\infty A_e$, where:

T_{SL} = sea level thrust
P_∞ = ambient pressure

These can all be solved for an expression providing T_{vac} knowing T_{SL}:

$$T_{vac} = \frac{T_{SL}}{1 - \dfrac{P_\infty c_{del} \varepsilon_{nozzle}}{P_c g_0 I_{sp\text{-}vac\text{-}del}}} \qquad (8.29)$$

It will be helpful to have some typical engine parameters to continue; a list of some parameters is provided in Table 8.12.

We need to provide dimensions for the combustion chamber diameter and length, divergent section length, diameter of throat, and nozzle length. The geometry of our engine will be defined as shown in Fig. 8.19.

Now, we can go through the engine sizing routine. First, we calculate the nozzle length. In practice, we can assume that the nozzle length is 80% of the

Table 8.12 Typical Rocket Engine Parameters

Item	LOx/LH₂ 1st Stage	LOx/LH₂ 2nd Stage	LOx/RP-1 1st Stage
Chamber Pressure MPa (psi)	8.274 (1,200)	17.24 (2,500)	20.68 (3,000)
Mixture Ratio r	6.0	6.0	2.47
Expansion Ratio n	40	150	40
Theoretical c, m/s (ft/s)	2,312 (7,585)	2,325 (7,629)	1,825 (5,988)
Delivered c, m/s (ft/s)	2,266 (7,433)	2,279 (7,476)	1,789 (5,868)
Theoretical I_{sp} (s)	452.3	474.8	357.6
Delivered I_{sp} (s)	430	456	336

Source: Straw [12]

ideal 15° cone angle:

$$L_{\text{nozzle}} = \frac{(D_{\text{exit}} - D_{\text{throat}})(\%\text{length})}{2 \tan 15°} \quad (8.30)$$

The combustion chamber length is calculated from:

$$L_{\text{comb chamber}} = \sqrt[3]{\frac{4(L^*)A_{\text{throat}}(\text{LthRatio}_{\text{chamber}})}{\pi}} \quad (8.31)$$

Straw suggests using the following: set $L^* = 0.75$ m = characteristic chamber length, and LthRatio $= 0.9 = \frac{L_{\text{comb chamber}}}{D_{\text{chamber}}}$.

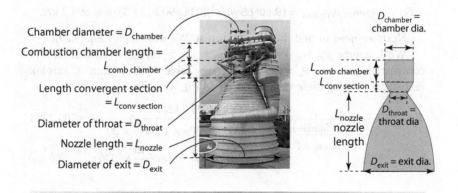

Chamber diameter = D_{chamber}

Combustion chamber length = $L_{\text{comb chamber}}$

Length convergent section = $L_{\text{conv section}}$

Diameter of throat = D_{throat}

Nozzle length = L_{nozzle}

Diameter of exit = D_{exit}

D_{chamber} = chamber dia.

$L_{\text{comb chamber}}$
$L_{\text{conv section}}$

L_{nozzle} nozzle length

D_{throat} = throat dia.

D_{exit} = exit dia.

Fig. 8.19 Definitions of the dimensions of different parts of an engine. Source: NASA (photo).

The chamber diameter is defined by

$$D_{chamber} = \frac{L_{comb\ chamber}}{LthRatio_{chamber}}$$

(8.32)

The convergent section length is

$$L_{conv\ section} = \frac{D_{chamber} - D_{throat}}{2\tan\alpha_{conv\ section}}$$

(8.33)

Here

L_{nozzle} = length of nozzle,
$L_{conv\ section}$ = length of convergent section,
$D_{chamber}$ = diameter of chamber,
D_{throat} = diameter of throat,
$\alpha_{conv\ section}$ = convergent section angle (30° suggested).

These lengths and dimensions are just estimates and should be replaced as soon as a suitable exiting engine is found (use its actual dimensions, if available). If at all possible, the designer should attempt to locate an existing engine/motor with performance that is close to meeting the requirements, rather than creating their own.

Example 3: Engine Dimensions for Example Two-Step Vehicle

First, use the liftoff mass, 328,590 kg, to determine the sea level thrust required. From earlier,

$$T_{SL} = \psi g_0 m_0 / n_{engines} = (9.80665\ m/s^2)(328{,}590\ kg)(1.3)/6 = 698.2\ kN.$$

Next, we need to find the vacuum thrust. We will employ Eq. (8.29).

We'll assume $P_\infty = P_{SL} = 101.3$ kPa, $\varepsilon_{nozzle} = 20$, $c_{del} = 1{,}789$ m/s, and combustion pressure $P_c = 20.68$ MPa (see Table 8.12, column 3, kerolox first stage). Putting these values into Eq. (8.29), we have:

$$T_{vac} = \frac{T_{SL}}{1 - \dfrac{P_\infty c_{del}\varepsilon_{nozzle}}{P_c g_0 I_{sp\text{-}vac\text{-}del}}} = \frac{698.2\ kN}{1 - \dfrac{101.325\ kPa \times 1789\dfrac{m}{s} \times 20}{20{,}680\ kPa \times 9.80665\dfrac{m}{s^2} \times 336\ s}} = 705.2\ kN$$

(Continued)

Example 3: Engine Dimensions for Example Two-Step Vehicle (Continued)

We then calculate the engine's throat area:

$$A_t = \frac{c_{\text{del}} T_{\text{vac}}}{P_c g_0 I_{sp\text{-vac}}} = \frac{1{,}789 \frac{\text{m}}{\text{s}} \times 705.2 \text{ kN}}{20{,}680 \text{ kPa} \times 9.80665 \frac{\text{m}}{\text{s}^2} \times 336 \frac{\text{m}}{\text{s}}} = 0.0183 \text{ m}^2$$

The throat diameter is $D_{\text{throat}} = \sqrt{\frac{4A_t}{\pi}} = 0.153$ m.

Assuming that nozzle length is 80% of the 15° cone angle length, the nozzle length is

$$L_{\text{nozzle}} = \frac{(D_{\text{exit}} - D_{\text{throat}})(\%\text{length})}{2 \tan 15°} = \frac{D_{\text{throat}}(\sqrt{\varepsilon} - 1)(\%\text{length})}{2 \tan 15°}$$

$$= \frac{0.153 \text{ m} \left(\sqrt{20} - 1\right)(0.80)}{2 \tan 15°} = 0.79 \text{ m}$$

Nozzle diameter calculation:

$$D_{\text{exit}} = \sqrt{\varepsilon} \, D_{\text{throat}} = \sqrt{20} \times 0.153 \text{ m} = 0.68 \text{ m}$$

The combustion chamber length is calculated using the suggested values:

$$L_{\text{comb chamber}} = \sqrt[3]{\frac{4(L^*)A_{\text{throat}}(\text{LthRatio}_{\text{chamber}})}{\pi}}$$

$$= \sqrt[3]{\frac{4(0.75 \text{ m})0.0183 \text{ m}^2(0.9)}{\pi}} = 0.251 \text{ m}$$

The chamber diameter is

$$D_{\text{chamber}} = \frac{L_{\text{comb chamber}}}{\text{LthRatio}_{\text{chamber}}} = \frac{0.252 \text{ m}}{0.9} = 0.278 \text{ m}$$

The convergent section length:

$$L_{\text{conv section}} = \frac{D_{\text{chamber}} - D_{\text{throat}}}{2 \tan \alpha_{\text{conv section}}} = \frac{0.279 \text{ m} - 0.154 \text{ m}}{2 \tan 30°} = 0.109 \text{ m}$$

Overall length = $L_{\text{comb chamber}} + L_{\text{conv section}} + L_{\text{nozzle}} = 0.251 + 0.109 + 0.79$ m = 1.15 m.

The results of this engine sizing process are a set of dimensions for the proposed engine(s), which may be used to make a drawing of our LV with approximately-sized engine or engines.

Repeating the calculations with the Step 2 engine, we have the following results, which are used for a scaled drawing of the vehicle, and a summary table:

(Continued)

Example 3: Engine Dimensions for Example Two-Step Vehicle *(Continued)*

Dimension	Step 1 engine	Step 2 engine
Thrust (kN)	698.2 (ea.)	461
Expansion	20:1	100:1
Chamber Dia. (m)	0.278	0.242
Comb. Chamber length (m)	0.251	0.217
Conv. Section length (m)	0.109	0.102
Throat Diameter (m)	0.153	0.123
Nozzle Length (m)	0.79	1.87
Exit Diameter (m)	0.68	1.23
Overall Length (m)	1.15	2.19

Scaled sketches of the engines on the lower and upper steps are shown in Figure 8.20. Note that the right-hand engines in the two steps' drawings appear more realistic by curving the straight nozzle lines on the left figures slightly as shown.

First Step Multi-Engine Arrangement

With six engines and a 4-m diameter, there are two possible arrangements, as shown in Fig. 8.21. The $5+1$ center arrangement may be preferable because it provides more side-to-side space for TVC motion, but it also

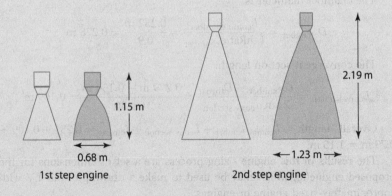

1st step engine   2nd step engine

Fig. 8.20 The result of the scaling process given in the text is a set of realistically-sized engines. Curved lines have been used to make the nozzles appear more realistic.

(Continued)

Example 3: Engine Dimensions for Example Two-Step Vehicle (Continued)

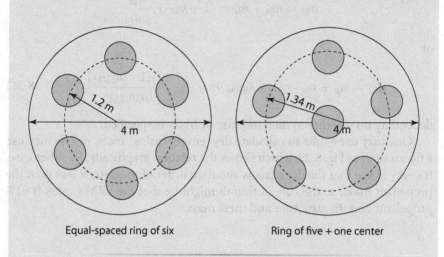

Equal-spaced ring of six   Ring of five + one center

Fig. 8.21   The 4-m-diameter vehicle can accommodate the six engines in a hexagonal arrangement (left) or a pentagonal arrangement with a center engine (right). The moment of inertia calculations done in this chapter used the left arrangement. The right configuration's MoI would be slightly different, since one engine is on the centerline.

would suffer from a more complicated thrust structure, because the center engine's thrust has to be beamed out to the outer radius of the rocket. With either arrangement, no base fairing is needed to reduce the aerodynamic loads on the TVC actuators. However, with hydrolox engines having a larger exit diameter, there might be clearance problems. It would also be possible to slightly enlarge the mounting footprint of such engines and add a base fairing to accommodate the larger footprint. The addition of such a fairing could affect the vehicle's aerodynamic drag slightly.

8.7.7.1 Solid Propulsion Motor Mass Estimation

A common mass estimating relationship (MER) for solid rocket motors (SRMs) is

$$m_{\text{SRM casing}} = k_{\text{SRM}} m_p \qquad (8.34)$$

where the variable k_{SRM} is usually taken as 0.135, which leads to a value for $\sigma_{\text{SRM}} = 0.1189$ as given previously in eq. (8.19). The "casing" mass is usually interpreted in a manner similar to the "structure" mass for a liquid step, and includes all inert items such as wiring, hydraulics, nozzle, ignition system, etc. With that value, we can calculate the step's loaded or 'wet' mass m_0 as:

$$m_0 = m_p + m_s = (1 + k_{SRM})\frac{I_{tot}}{g_0 I_{sp}}$$

or

$$m_0 = m_p + m_s = (1 + k_{SRM})m_{PL}\frac{(\mu - 1)(1 - \sigma_{SRM})}{1 - \mu\sigma_{SRM}} \quad (8.35)$$

depending on whether a) total impulse or b) Δv is specified.

One may use MERs to calculate dry (empty casing) mass, or one may use a figure such as Fig. 8.22, which shows the relation graphically. In either case, it's easy to see that the dry mass is about an order of magnitude less than the propellant mass, so the rule of thumb might be that the SRM's mass is 91% propellant and 9% structure and inert mass.

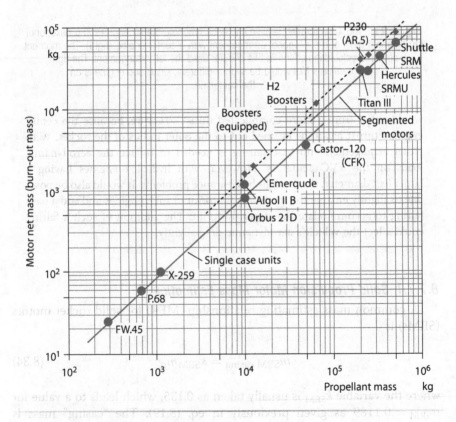

Fig. 8.22 Solid rocket burnout mass vs. propellant mass. Source: [6].

A solid-propellant LV step is a monolithic structure and does not require any intertank structure or propellant feed lines. It will, however, require interstages or skirts on the forward and aft portions for vehicle stacking purposes.

8.7.7.2 Mass Estimation for More Complex Shapes

In many cases, the vehicle we are sizing is not simple "cigar tubes" with engines on the bottom. The Space Shuttle is one example, as are some strap-on liquid and solid boosters, and the infamous X-33. Rohrschneider [4] provides a method to estimate masses for these noncylindrical shapes.

Rohrschneider body mass estimation works as follows:

$$M_{\text{body}} = K_b A_{\text{body}} N_z^{1/3} + K_f V_f + K_{\text{ox}} V_{\text{ox}}^{1.1} + K_t \left(N_{\text{eng}} T_{\text{vac}} \right) \qquad (8.36)$$

where
  M_{body} = Total mass of body group, in pounds, so use English units
  A_{body} = surface area of vehicle body
  Nz = Ultimate load factor = 1.5×2.5 = (factor of safety × limit load)

The body construction constants K are determined from:

  K_b = 272 — composite structure, no TPS

  = 3.20 — aluminum structure, no composites, no TPS

  = 3.40 — hot metallic Ti/Rene honeycomb, no TPS required

  = 4.43 — mold line tankage tank, body structure cryogenic insulation

Values for the constants K_f, K_{ox}, and other parameters are chosen from the items found in Table 8.13, and have to do with the configuration being studied.

Other parameters needed for the Rohrschneider method are:

  K_t = 0.0030: aluminum thrust structure;

    = 0.0024: composite thrust structure

  N_{eng} = Number of main engines on stage

  T_{vac} = Vacuum thrust per main engine

  V_f = Total fuel volume

  V_{ox} = Total oxidizer volume

The reference provides MERs for the following items. **Bold-face** items are those required for expendable LVs:

Table 8.13 Body Shapes and Associated Constants for Rohrschneider's Weight Estimation Method

Source	Propellant	K LB/FT²	Vol FT²	Ullage Pressure	Integral or non-Integral	Material	Geometry	Comments
Shuttle E/T	LH₂	.5918	53,515	36	Integral	AL2219		Does not include insulation
EN-155*	LH₂	.8430	60,387	30	Integral	INC 718		Honeycomb Sandwich added honeycomb for thermal protection
EN-178*	LH₂	.5760	41,646	20	Integral	AL2219		Isogrid includes 4,364 LB insulation
Shuttle E/T	LOx	.6458	19,609	38	Integral	AL2219		Does not include insulation
EN-155*	LOx	.7660	18,355	20	Non Integral	AL2219		Polymide honeycomb for insulation and structural stabilization
EN-178	LOx	.5160	21,841	15	Integral	AL2219		Isogrid includes 1,704 LB insulation
S-1C	LOx	.804	47,250	5	Integral	AL2219		Conventional skin stiffened construction W/O insulation
EN-155	JP-5	.7000	4,819	5	Non Integral	AL2219		
	JP-5	.28					N/A	Penalty for dry-wet wing
S-1C	RP-1	.867	30,000		Integral	AL2219		

Source: [1, 4]
Table showing tank constants from (ref. 1).
*EN designates in-house LARC study vehicles.

Tank Description (spanning: Integral or non-Integral, Material, Geometry, Comments)

1. Wing	15. Personnel Equipment
2. Tail	16. **Dry Weight Margin**
3. Body (part given in Eq. (8.36) above)	17. Crew & Gear
4. TPS	18. **Payload Provisions**
5. Landing Gear	19. **Cargo (up and down)**
6. **Main Propulsion**	20. **Residual Propellants**
7. RCS	21. OMS/RCS Reserve Propellants
8. OMS	22. RCS Entry Propellants
9. **Primary Power**	23. OMS/RCS On-Orbit Propellant
10. **Electrical Conversion & Distribution**	24. Cargo Discharged
11. **Hydraulics**	25. **Ascent Reserve Propellants**
12. **Surface Control & Actuators**	26. **In-flight Losses & Vents**
13. **Avionics**	27. **Ascent Propellants**
14. ECLSS	28. **Startup Losses**
ECLSS = Environmental Control & Life Support Systems	RCS = reaction control system
TPS = thermal protection system	OMS = orbital maneuvering system.

Finally, in Table 8.14, we provide a checklist of mass items to be included in any design. Note that this chapter considered only the items in bold; a proper weight estimate would include all of the items in the table.

8.7.7.3 Mass Properties

Accurate *mass properties*, particularly mass and moments of inertia (MoIs) of a vehicle are required for performance, stability analysis, control system design, manufacturing, and several other needs. In order to calculate these properties, it's necessary to keep track of the *location* and *mass* of each and every piece of the vehicle. In this section, we will cover

* Coordinate systems
* Center of mass (CM) of individual items, and assemblies
* Moments of Inertia of individual items about their own mass centers
    * Symmetry
    * Effects of offset CMs of individual items
* Moments of Inertia of assemblies

8.7.7.4 Coordinate Systems

A coordinate system is needed to position and orient each of the parts of the launch vehicle. This system will be used for several distinct aspects of the design process. A typical coordinate system is shown in Fig. 8.23. These are so-called "airplane" coordinates:

* The long axis (forward direction on long body) is the axial direction and usually is close to the velocity vector. We will define this as the x-axis, also known as the roll axis.
* Rotation in the pitch direction is the y-axis.
* Finally, yaw rotation corresponds to the z-axis.

Table 8.14　A Collection of Items to Be Included in Mass Estimating

Structure	Equipment
Tankage	Control elements
Antislosh/antivortex parts	Telemetry
Cryogenic insulation	Environmental control provisions
Forward interstage	Power supply
Aft Skirt	Electrical network
Thrust structure	Range-safety and destruct systems
Base heat protection	Miscellaneous
Separation systems	Residuals
Miscellaneous	Propellant trapped in engines
Propulsion System/Accessories	Propellant trapped in lines
Engines/accessories	Propellant trapped in tankage
TVC: actuators, high-pressure bottles, hydraulic fluid	Gaseous oxygen and hydrogen
	Thrust-decay propellant
Propellant distribution system	Propellant utilization residual
Pressurization system	Miscellaneous
Fill/drain systems	Expanded Items
Retrorockets and attachments	Ullage rocket cases/propellant
Ullage rocket installation Miscellaneous	Propellant for chill-down and start/ restart

Mass estimation technical papers may be found in the Additional Reference section at the end of this chapter.

The origin of the coordinate system (where the three axes meet) may be located anywhere, but two common locations are the center of mass (CM), or a position related the LV's rear engine mount.

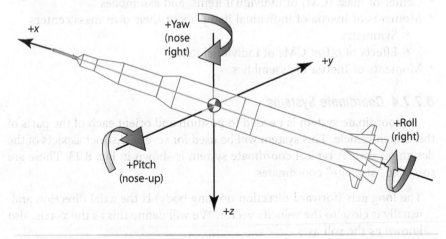

Fig. 8.23　One commonly used coordinate system mimics that used for aircraft.

8.7.7.5 Mass Properties Calculations

The statics, stability analyses, and rigid-body dynamics and control of vehicle motion require two types of mass properties information: the location of the CM, and the values of mass MoIs in the same coordinate system as the CM.

Usually a body-centered coordinate system is used, such as the CM. Because the vehicle's center of mass changes with time, it may be more convenient to use a geometrically fixed reference location, called the moments reference center (MRC). In this case, the CM offset is factored in as a function of time.

In order to calculate the vehicle's center of mass, the following procedure is used:

1. Choose a reference location P. The location of engine *gimbal block* is commonly used, as this location tends to stay fixed regardless of the type or number of engine(s) used. A reference point defined by the engine bell would be less satisfactory, since it would depend on the particular engine's geometry. For convenience, we recommend use of aft end of the aft skirt as a reference plane.

2. Next, determine the magnitudes and locations of all (quantity n) on-board masses relative to the reference location. It is necessary to obtain the x-, y-, and z- coordinates of the CM of each element i. Table 8.15 provides numerical data for item dimensions and the locations of their mass centers.

    It's recommended to tabulate all this information in a spreadsheet, with columns containing its mass, and at least the x and z distance (or

Table 8.15   Dimensions and locations of mass centers for various LV components.

Item	Suggested Height	CM Loc. (above base)
Payload Fairing	$2D$	$\frac{1}{3}D$
Intertank struct.	$\frac{1}{4}D + 2h_{dome}$ ($\frac{1}{4}D$ between domes)	$\frac{1}{2}h_{intertank}$
Forward skirts; Step 2 Aft skirt	$\frac{1}{3}D + h_{dome}$ ($\frac{1}{3}D$ space below/above dome)	$\frac{1}{2}h_{skirt}$ (non-tapered) Tapered use cone frustum.
Residual propellant	Bottom of lower dome	$(h_{intertank}$ or $h_{skirt}) - h_{dome}$
Interstage	D_{lower} (until upper step nozzle sized)	$\frac{1}{2}h_{interstage}$ (non-tapered) Tapered use cone frustum.
Step 1 Aft skirt	D_{lower}	$\frac{1}{2}h_{skirt}$
Thrust structure	$\frac{1}{2}D$	$\frac{1}{2}h_{thrust\ struct}$
Engine	Point mass, no height	$\frac{1}{8}D$ above bottom of thrust structure
Avionics, lower	Thin ring, no height	$\frac{1}{3}D$ above interstage base
Avionics, upper	Thin ring, no height	$\frac{2}{3}D$ above fwd. skirt base
Wiring	Length of step	$\frac{1}{2}$ of l_{step} above step bottom

longitudinal and radial distance) from the reference point to the mass centers of each item. A full mass properties analysis does not assume symmetry and instead tabulates the three coordinates of the mass center of each item (x, y, and z).

3. Sum the masses = total mass $M = \sum(m_i)$. This is done quite easily in any spreadsheet program.

4. Sum the "mass moments" $= \sum(m_i)(l_i)$, where l_i are the x-coordinates of the item's CM. Again, let the computer do the work.

    NOTE: the "mass moments" mentioned here, with dimensions of mass $\times$ distance or mL, are NOT the same as the "force moments" created by loads that are torques, i.e. units of force $\times$ distance or FL. Do not confuse the two terms.

5. With the sum of the masses and the mass moments, the location of the CM relative to P is calculated from $x_{CM} = \sum(m_i \, l_i) \div \sum(m_i) = \sum(m_i \, l_i) \div M$

6. If needed, repeat steps 1–5 above for the y- and z-directions as well.

Example 4: Calculation of Center of Mass of Example Kerolox/Kerolox TSTO Vehicle

We will now determine the center of mass of the example design that we considered earlier in the chapter. We will use the bottom of the aft skirt as the reference location, and assemble a table of masses and mass locations. The rocket is configured as shown in Fig. 8.24 which was created using the location data generated in the chapter for convenience.

Element or item CM locations are found using mass properties tables, such as Table 8.10, or by inspection. For instance, all cylinders (both thin, like a tank, and thick) have a CM location that is one half of their height above the base. All thin cones have a CM location that is one third of their

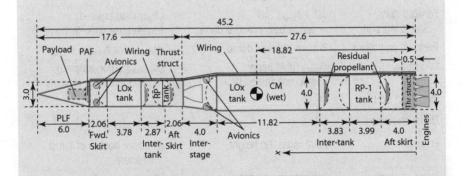

Fig. 8.24 The layout of the example TSTO vehicle is shown above.

(Continued)

Example 4: Calculation of Center of Mass of Example Kerolox/Kerolox TSTO Vehicle *(Continued)*

height above the base. A thin hemispheric dome has a CM that is one-half of its radius above its base. These values are provided in Table 8.16, in the second column labeled "z_c CM location." Data like these can be found in many engineering handbooks and texts.

Table 8.16   Properties of Thin Domes and Rings

Shape	z_C CM Location	J_x & J_y	J_z	Material Volume (Vol)
Thin Hemisphere	$R/2$	mR^2	$\frac{2}{3}mR^2$	$2\pi R^2 t$
Elliptical Hemispheroid (EH)	$\dfrac{2\pi H\left(R^3 - H^3\right)}{3E^2 RS*}$	Use mR^2	Use $\dfrac{2mH}{3R - H}$ $\left(R^2 - \dfrac{3}{4}RH + \dfrac{3}{20}H^2\right)$	$St*$
Thin ring torus	0 (center)	$mR^2/2$	mR^2	~ 0

z_C = CM location above base, mass $m = \rho \times$ Vol, z = axis of symmetry.
Note that this coordinate system is different from the vehicle coordinates given earlier.
†Value given for spherical segment, not elliptical hemispheroid.

For elliptical hemispheroidal shell, Eccentricity $E = \sqrt{1 - \dfrac{H^2}{R^2}}$. *Surface area $S = \pi\left(R^2 + \dfrac{H^2}{2E}\ln\dfrac{1+E}{1-E}\right)$.

$AR = R/h_{dome}$,

so $E = \sqrt{1 - \dfrac{1}{AR^2}}$

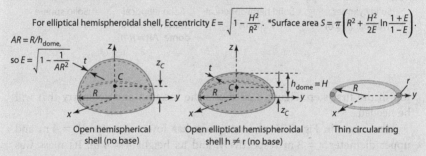

| Open hemispherical shell (no base) | Open elliptical hemispheroidal shell h ≠ r (no base) | Thin circular ring |

Additional calculations must also be done with solids, where a volume is completely filled up with mass. For solid objects, use the information contained in Table 8.17. Note that the thin bar shown is a special case of a solid cylinder with radius R approaching 0. The thin bar inertia calculation is to be used for long, "skinny" objects whose length $L >>$ radius R, such as electrical wiring and thin ducting.

We will need to assemble the mass data from each of the parts of the design into a table to do the center of mass calculation for our vehicle. Now we'll calculate the mass and CM location for the interstage located at the

(Continued)

Example 4: Calculation of Center of Mass of Example Kerolox/Kerolox TSTO Vehicle *(Continued)*

Table 8.17   Properties of Solids

Shape	z_c CM Location	J_x & J_y	J_z	Volume (Vol)
Solid Cylinder	$\frac{1}{2}h$	$m(R^2/4 + h^2/12)$	$\frac{1}{2}mR^2$	$\pi R^2 h$
Thin Bar (cyl. w/ $R = 0$, $t \ll L$)	$\frac{1}{2}h$	$mh^2/12$	~ 0	~ 0
Solid Hemisphere	$\frac{3}{8}R$	$0.259\ mR^2$	$\frac{2}{5}mR^2$	$\frac{2}{3}\pi R^3$
Solid $\frac{1}{2}$ Ellipsoid $h_{dome} \neq R$	$\frac{3}{8}h$	$m(R^2/5 + h^2)$	$\frac{2}{5}mR^2$	$\frac{2}{3}\pi R^2 h$
Solid Sphere	0 (center)	$\frac{2}{5}mR^2$	$\frac{2}{5}mR^2$	$\frac{4}{3}\pi R^3$

z_c = CM location above base, mass $m = r \times$ Vol, z = axis of symmetry. Note that this coordinate system is different from the vehicle coordinates given earlier

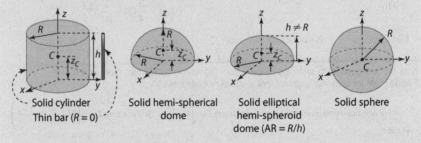

Solid cylinder
Thin bar ($R = 0$)   Solid hemi-spherical dome   Solid elliptical hemi-spheroid dome (AR = R/h)   Solid sphere

top of the first step. Figure 8.25 shows the location and geometry that will be needed.

As shown in Fig. 8.25, the interstage has lower diameter $R = 4$ m and upper diameter $r = 3$ m respectively, and its height $h = 4$ m. Its mass was found to be 590 kg earlier.

The interstage is assumed to be the frustum of circular cone whose properties may be found in the bottom row of Table 8.10. The CM location of the interstage above its base (the x axis, pointing in the 'upwards' direction, is measured towards the left in the figure) is calculated using the formula for z_c, the height of interstage's CM above its bottom, in the table. In this case,

$$z_c = \frac{h}{3}\left(\frac{2r + R}{r + R}\right) = \frac{4\ \text{m}}{3}\left(\frac{2 \times 3\ \text{m} + 4\ \text{m}}{3\ \text{m} + 4\ \text{m}}\right) = 1.90\ \text{m}$$

above the interstage's bottom. Using the dimensions in Fig. 8.25, we find $h_{\text{interstage}}$, the height of bottom of the interstage is $h_{\text{interstage}} = h_{\text{aft skirt}} +$

(Continued)

Example 4: Calculation of Center of Mass of Example Kerolox/Kerolox TSTO Vehicle *(Continued)*

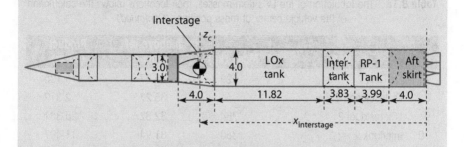

Fig. 8.25 The external layout dimensions are used to locate each of the parts of the vehicle relative to the selected reference location (bottom of the aft skirt). The interstage is highlighted.

$h_{RP\text{-}1\,cyl} + h_{intertank1} + h_{LOx\,cyl} = 23.64$ m from the reference point, and the total distance from the reference point to the CM of the interstage is $x_{interstage} = h_{interstage} + z_c = 25.54$ m $+ 1.90$ m $= 25.54$ m.

The last quantity needed is the product of the mass m and the distance from reference x. For the interstage, the product is $m_{interstage} \times x_{interstage} = 590$ kg $\times 25.54$ m $= 15{,}056$ kg·m.

This process is repeated for each of the items in the LV. The mass and location data for each of the 33 items in the LV are tabulated in Table 8.18. Line 17 of the table includes the interstage mass (column 2), location data (column 3), and the mass moment (column 4) calculated in the previous paragraph. The remaining items in the table were calculated in the same fashion as the interstage shown above: the item's CM location relative to its base was found using Table 8.10, its distance from the reference found using the

Table 8.18 The tabulation of the LV's item masses, their locations allows the calculation of the vehicle center of mass position.

	Item	Mass (kg)	h above ref (m)	Product (kg-m)
1	Payload Fairing $h = 2D$	388	41.22	15,977
2	Payload no-PAF, $h = D/2$	10,000	42.97	429,670
3	Payload Attach Fitting	805	39.59	31,872
4	Forward Skirt 2	258	38.19	9,864
5	Avionics	490	38.19	18,724
6	Wiring	612	36.43	22,276

(Continued)

Example 4: Calculation of Center of Mass of Example Kerolox/Kerolox TSTO Vehicle *(Continued)*

Table 8.18 The tabulation of the LV's item masses, their locations allows the calculation of the vehicle center of mass position. *(Continued)*

	Item	Mass (kg)	*h* above ref (m)	Product (kg-m)
7	Ox Tank 2	429	35.27	15,125
8	Ox Insulation 2	66	35.27	2,319
9	LOx residual 2	784	32.32	25,331
10	Intertank 2	360	31.94	11,497
11	Fuel Tank 2	185	30.10	5,580
12	Fuel residual 2	245	28.63	7,014
13	Aft Skirt 2	258	28.66	7,404
14	Thrust Structure 2	118	28.66	3,370
15	Gimbals 2	18	28.15	516
16	Engines 2	574	27.89	16,021
17	Interstage	590	25.54	15,056
18	Avionics 1 (1/2)	490	25.05	12,282
19	Wiring 1 (by length ratio)	961	13.82	13,280
20	Ox Tank 1	2,012	17.73	35,664
21	Ox Insulation 1	213	17.73	3,768
22	Ox residual 1	3,665	11.15	40,865
23	Intertank 1	640	9.90	6,336
24	Fuel Tank 1	870	5.99	5,213
25	Fuel Residual 1	1,145	4.58	5,245
26	Aft Skirt 1	669	2.00	1,337
27	Thrust Structure 1	1,068	2.00	2,136
28	Gimbals 1	162	1.00	162
29	Engines 1	4,257	0.50	2,129
30	Propellant 2/1?Yes = 1 No = 0	0	0	
31	RP-1 fuel 2	12,280	30.10	369,641
32	LOx oxidizer 2	39,296	35.27	1,385,869
33	RP-1 fuel 1	57,616	5.99	345,354
34	LOx oxidizer 1	184,373	17.73	3,268,077
35	**Totals**	**325,898**	**kg**	6,134,970
36	**CM Location above ref.**	**18.82**	m	

(Continued)

Example 4: Calculation of Center of Mass of Example Kerolox/Kerolox TSTO Vehicle (Continued)

layout drawing in Fig. 8.24, and the mass moment found by multiplying mass by location distance.

The summation of column 2 of the table is the vehicle's mass, 325,898 kg. If the summed product at the bottom of column 4, (6,134,970 kg·m), is divided by the mass, the result is the CM position relative to the reference point, in this case, 18.82 m above the engine gimbal block. This result is also shown in Fig. 8.24.

Finally, we show the mass breakdown for an Atlas V SLV in Fig. 8.26. Note that structures and engines on both steps are a significant contributor to mass: 83% for the first step and 59% for the second step. For this reason, we typically want to choose an engine with as high a thrust-to-weight ratio as possible, to minimize the step's empty mass and improve performance.

8.7.7.6 Some Comments About Engine Selection

When selecting an existing engine or engines, it is worthwhile to consider the engine thrust-to-weight ratio. This is simply because of the rocket

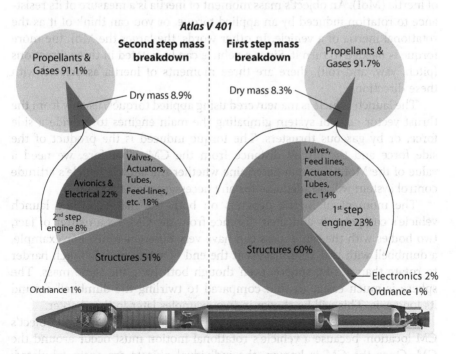

Fig. 8.26   Weight breakdown for the Atlas V 401 LV. Courtesy ULA.

Table 8.19 Thrust-to-Engine Weight Ratios for a Few Engines

Engine	Thrust-to-Weight Ratio
F-1	94
Merlin 1D	180
Raptor	~200

equation, and if the step's empty mass is minimized, the burnout speed is maximized. This is particularly important, because the engine mass can be a large fraction of the step dry mass. As shown in Fig. 8.26, the engines on the Atlas V first and second steps—most significant on the first step at almost 25%—make up a large fraction of the dry mass of the steps. A small change in the thrust-to-*engine* weight of *just the engine* (not to be confused with the stage's $T/W = \psi$ ratio) can have a noticeable influence on the burnout speed. A few values of engine thrust-to-weight ratios are given in Table 8.19. These values back up SpaceX's claim that the Merlin 1D has the highest thrust-to-*engine* weight ratio available.

8.7.7.7 Moment of Inertia Calculations

The second part of the vehicle's mass property calculation is the moment of inertia (MoI). An object's mass moment of inertia is a measure of its resistance to rotation induced by an applied torque, or you can think of it as the rotational inertia of a vehicle. In other words, the larger the MoI, the more torque is needed to turn it. Since a vehicle can be rotated in three directions (pitch, yaw, and roll), there are three moments of inertia associated with these directions.

The launch vehicle is maneuvered using applied torques, usually from the thrust vector control system gimbaling the main engines to provide a side force, or by gaseous thrusters. The torque induced is the product of the side force and the force's distance from the CM. Therefore, we need a value of the MoI so we can determine whether the launch vehicle's attitude control system will be adequate for any necessary maneuvering.

The moment of inertia depends on both the masses of the launch vehicle's components and their distance from the CM. As a matter of fact, two bodies with the same mass can have very different MoIs. For example, a dumbbell with two 1-kg masses at the end of a 1-m stick is much harder to rotate than a 2-kg sphere, even though both have the same mass. The sphere is much easier to spin compared to twirling the dumbbell around its long axis. This will be shown in two examples later in the chapter.

Moment of inertia calculations are based on first knowing the object's CM location, because a vehicle's rotational motion must occur around the CM. Once the CM is known, the individual objects are again tabulated; however, for this calculation, there are two factors that affect the MoI: the object's MoI around its own CM, and the object's distance from the CM. The former (MoI about own CM) may be looked up in a handbook. The latter (effect of distance of object from the assembly's CM) is calculated through the *parallel axis theorem*, which states that the moment of inertia of a body separated from a reference point is equal to the body's MoI about its own CM plus the object's mass multiplied by the separation distance squared. In equation form,

$$J_i = J_{0i} + m_i x_i^2 \tag{8.37}$$

where J_i is object i's total MoI, J_{0i} is the object's MoI about its own CM, and $m_i x_i^2$ is the contribution from the object of mass m_i being offset a distance x_i from the assembled body's CM.

We will assume that the vehicle has an *axis of symmetry*, so that the only two inertias are about the symmetry axis and perpendicular to the symmetry axis. This is true for many launch vehicles, but not all. Notable exceptions are the Space Shuttle, the Delta IVH, Falcon Heavy, and Ariane 5. Objects that do not have an axis of symmetry can be analyzed with similar methods, but we will not cover the details here.

With the assumption of an axis of symmetry, the only two measurements needed for each subcomponent i are x_i, its x-direction or lengthwise separation from the assembly's CM, and r_i, its radial distance from that axis. We will calculate MoIs for both roll (about the long axis) and pitch (perpendicular to the long axis).

Because most launch vehicles are long and skinny, their pitch and yaw MoIs are typically at least an order of magnitude larger than their roll inertia. This provides a "sanity check" opportunity: the value of $J_{roll} <<$ J_{pitch} or J_{yaw}. If not, there is a problem with the calculation, and it needs to be checked.

The following explains how to calculate an assembled body's MoI:

1. Determine the assembly's CM location (as done previously). The CM calculation spreadsheet is a good starting point; it may be easily extended to carry out this MoI calculation.
2. Add two columns to the CM calculation to store the values of L_i and R_i, the axial and radial offset of part i from the CM, for all parts. This can be done by subtracting the CM distance from the reference from the object's distance from the reference.
3. Use each part's mass and dimensions and the MoI tables to determine J_{0xi} and J_{0yi}, the MoIs in roll and pitch, respectively (meaning two more columns).
4. Calculate $m_i L_i^2$ and $m_i R_i^2$ (the parallel axis contributions) for each part, two more columns.
5. Add the results of step 3 and step 4 to determine the part's inertias in roll

$$J_{xi} = J_{0xi} + m_i R_i^2$$

and in pitch

$$J_{yi} = J_{0yi} + m_i L_i^2$$

Total is J for each axis.

6. The vehicle's inertias are calculated by summation of step 5 for roll

$$J_{roll} = J_x = \sum (J_{0i} + m_i R_i^2)$$

and for pitch/yaw

$$J_{pitch} = J_y = \sum (J_{0yi} + m_i L_i^2)$$

One may find, by examining the entries to the MoI table, that usually $J_{0i} << m_i x_i^2$. This means that in many cases, it's not necessary to calculate J_{0i}!

Let's look at a pair of examples to illustrate the process of MoI calculation. We'll use the earlier reference to a sphere vs a dumbbell, both having the same mass.

Example 5: MoI Calculation of 2-kg, 14.1-cm Sphere in Center of 1-kg, 1-m Rod

We will demonstrate the calculation of the MoI of the sphere and rod assembled together, as shown in Fig. 8.27. A 2-kg sphere (part #1) is joined to the center of a 1-kg rod that's 1 m long (part #2). For this simple example there are only two elements that go into the calculation.

We start by creating a table to calculate the center of mass of the assembly, then we will add columns to accommodate offsets from the CM, parallel axis calculation, and total MoI. Table 8.20 provides the framework for the calculation. The first two columns are item number and mass, which are self-explanatory. The third column is the dimension of the object, the radius of the sphere or length of the rod. Column 5 is the object's distance

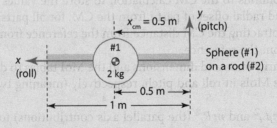

Fig. 8.27 Example 5: Calculation of MoI of rod and single sphere.

(Continued)

Example 5: Mol Calculation of 2-kg, 14.1-cm Sphere in Center of 1-kg, 1-m Rod *(Continued)*

Table 8.20 Moment of Inertia Calculations for Example 1

Item	Mass m_i (kg)	R_i or L_i (m)	J_0 pitch i (kg m^2)	L_i (m)	$m_i L_i$ (kg m)	x_i (m)	$m_i x_i^2$	$J_{0\ \text{roll}\ i}$ (kg m^2)
1	2.0	0.141	0.0159	0.50	1.0	0.000	0.0000	0.0159
2	1.0	1.000	0.0833	0.50	0.5	0.000	0.0000	0.0000
Σ	3.0		0.0992		1.5		0.0000	0.0159

$$CM = \sum m_i L_i / \sum m_i = 0.5000\,\text{m}$$
$$J_y(\text{pitch}) = \sum J_{0i} + \sum m_i x_i^2 = 0.0992\,\text{kg m}^2$$
$$J_x(\text{roll}) = \sum J_{0\,\text{roll}\,i} = 0.0159\,\text{kg m}^3$$

from the reference, part of the CG calculation. Column 6 is the moment of the object, the product of its mass and distance from the reference, which is also part of the CG calculation. At this point, we have enough information to calculate the CM, which is shown as the first line in the highlighted box.

Now we consider the MoI terms. Column 4 is the object's MoI about its own CM, and is calculated using Tables 8.11 and 8.12, presented earlier. For the sphere, it's the numerical value of $2/5\ mr^2$; for the rod, it's $mL^2/12$. These values are also the MoIs about the roll axis in column 9, because they are not offset radially from the x-axis.

The remaining calculations are the contributions from the sphere and rod being offset from the CM. Column 7 gives those offsets ($=0$), and column 8 presents the mx^2 contribution (also $= 0$ for this example). With that, we can calculate the pitch MoI as given in the second and third lines inside the highlighted box.

Now, let's move on to Example 2.

Example 6: Mol Calculation of 2-kg Dumbbell With 1-m Rod

We now calculate the MoI of the rod with a sphere at each end, or a "dumbbell," as shown in Fig. 8.28. Two 1-kg spheres (parts #1 and #2) are joined to each tip of a 1-kg rod that's 1-m long (part #3), for a total of three elements. The arrangement is shown in Fig. 8.28.

We assemble a table for this example, as shown in Table 8.21. As before, the CM must be calculated before the MoI calculation can take place.

(Continued)

Example 6: Mol Calculation of 2-kg Dumbbell With 1-m Rod (Continued)

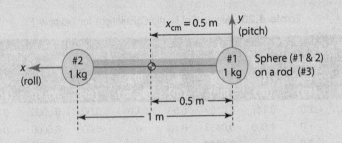

Fig. 8.28 Two-sphere dumbbell with same mass but different Mol from Example 5.

Table 8.21 Mol Calculation for Dumbbell

Item	Mass m_i (kg)	R_i or L_i (m)	$J_{0\ pitch\ i}$ (kg m^2)	L_i (m)	$m_i L_i$ (kg m)	x_i (m)	$m_i x_i^2$	$J_{0\ roll\ i}$ (kg m^2)
1	1.0	0.100	0.0040	0.00	0.0	−0.500	0.2500	0.0040
2	1.0	0.100	0.0040	1.00	1.0	0.500	0.2500	0.0040
3	1.0	1.000	0.0833	0.50	0.5	0.000	0.0000	0.0000
Σ	3.0		0.0913		1.5		0.5000	0.0080

$$CM = \sum m_i L_i / \sum m_i = 0.5000 \text{ m}$$
$$J_y(\text{pitch}) = \sum J_{0i} + \sum m_i x_i^2 = 0.5913 \text{ kg m}^2$$
$$J_x(\text{roll}) = \sum J_{0\ roll\ i} = 0.0080 \text{ kg m}^2$$

The calculations for Example 6 are quite similar to those in Example 5, with the difference being that the two smaller spheres are offset to the two ends of the rod. This is shown by the nonzero numbers in column 5. The offsets in column 7 are also nonzero, contributing to the nonzero moments in column 8. Column 9, as before, is equal to column 4 because nothing is offset from the x-axis.

The end result: the rod + centered sphere (example 5) has a pitch MoI of 0.099 kg·m^2, whereas the dumbbell (example 6) has a pitch MoI of 0.591 kg·m^2—*six times higher!* Remember that they have the same mass, so the locations of the masses are critically important to the MoI's magnitude.

Another result confirms the statement earlier: the value of J_{roll} should be much less than J_{pitch} or J_{yaw}. For example 5, we see a ratio of 0.0159:0.0992, or 1:6.2, as expected. For example 6, it's even larger: 0.591:0.008, or 1:73!

We claimed earlier that usually $J_{0i} << m_i x_i^2$. If you consider Example 6, the sphere's J_0 is 0.004 kg·m², whereas the $m_i x_i^2$ is 0.25 kg·m², some 62 times larger. This confirms that in many cases, it's not necessary to calculate J_{oi}.

8.8 Calculation of Tank or Shell Thicknesses

Launch vehicles are primarily thin-walled structures, mainly cylinders and domes. To do an MoI calculation for a launch vehicle, we need the thicknesses t of the different parts so that we may calculate the part's MoI. If we are to assume that a given shell structure has constant thickness, we can say

$$m_{\text{shell}} = \rho(\text{Vol}_{\text{shell material}}) = (\rho A_{\text{shell}} t) \qquad (8.38)$$

and therefore thickness

$$t = m_{\text{shell}}/(\rho A_{\text{shell}}) \qquad (8.39)$$

We use the mass and area from the CM calculations, along with the material density, to find thickness.

Important Note: The calculated thicknesses from this simple analysis *are not* the "proper" ones that have been designed to sustain all the vehicle's loads. Actual thicknesses must be determined by the analysis of flight loads, which allows a proper stress analysis to size all the parts. This process will be discussed in Chapter 10.

Example 7: Calculation: Moment of Inertia Contribution of Example LV Interstage

As shown in Fig. 8.29, the interstage has lower diameter $R_1 = 4$ m and upper diameter $R_2 = 3$ m respectively, a height $h = 4$ m, and mass $m_{\text{interstage}} = 590$ kg. The interstage's CM height above its bottom is $z_c = 1.90$ m.

The interstage, assumed to be the frustum of circular cone, has properties in Table 8.10. To find its contribution to the vehicle's MoI, we need 1) its mass, 2) its MoI around its own CM, and 3) its CM offset position from LV's CM.

(Continued)

Example 7: Calculation: Moment of Inertia Contribution of Example LV Interstage *(Continued)*

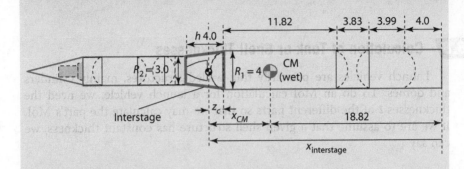

Fig. 8.29 The external layout dimensions are used to locate the interstage's location relative to the center of mass, labeled "CM (wet)". The interstage is highlighted.

Using the formula in the table, the interstage's moment around its own CM is

$$J_{0-\text{pitch intrstg}} = m_{\text{interstage}} \left\{ \frac{R_1^2 + R_2^2}{4} + \frac{h^2}{18} \left[1 + \frac{2R_1 R_2}{(R_1 + R_2)^2} \right] \right\}$$

$$= 1{,}702 \text{ kg·m}^2$$

Next, we calculate the distance from the interstage's CM to the vehicle CM:

$$x_{\text{CM}} = \text{centroid to CM distance} = h_{\text{interstage}} - x_{\text{CM}} = 25.54 - 18.82 \text{ m} = 6.91 \text{ m}$$

With x_{CM} determined, we can calculate the parallel axis contribution to the MoI:

$$m_{\text{interstage}} \, x_{\text{CM}}^2 = 590 \text{ kg} \, (6.91 \text{ m})^2 = 26{,}600 \text{ kg·m}^2$$

The full MoI is the sum of the interstage's MoI about its own CM and the product $m_{\text{intrstg}} \, x_{\text{CM}}^2$:

$$J_{\text{pitch interstage}} = J_{0\text{-pitch interstage}} + m_{\text{interstage}} \cdot x_{\text{CM}}^2$$

$$= 1{,}702 + 28{,}300 \text{ kg} \cdot \text{m}^2 = 28{,}300 \text{ kg} \cdot \text{m}^2.$$

This is the contribution of the interstage to the vehicle's pitch and yaw moments of inertia.

Next, we calculate the interstage's roll moment of inertia contribution. Since the interstage CM is located along the LV's roll axis, there is no parallel axis contribution, and we need only compute $J_{0-\text{roll intrstg}}$, the roll MoI around its own CM. This formula may also be found in Table 8.10:

$$J_{0\text{-roll interstage}} = m_{\text{interstage}} \frac{R_1^2 + R_2^2}{2} = 1{,}842 \text{ kg·m}^2.$$

(Continued)

Example 7: Calculation: Moment of Inertia Contribution of Example LV Interstage *(Continued)*

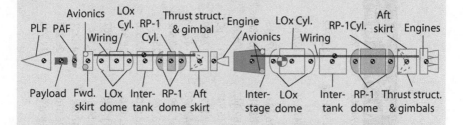

Fig. 8.30 The vehicle's mass properties (total mass, CM location, and total MoI) are calculated by adding up the mass and MoI contributions of each of the LV's components shown above. Note that residual propellants are not shown, but are included in calculations.

The values are the parts of the mass properties of the interstage and are part of the ingredients for the mass property calculation of the example TSTO design. The symbolic breakdown of the entire vehicle is shown in Fig. 8.30.

All of the design's parts are shown in Fig. 8.30, and a tabulation of the design's mass properties, including all of the parts and propellants, is presented in Table 8.22. The values of the interstage's contributions calculated above may be found on Line 17 of Table 8.22.

Notice that there are some "on-off" switches incorporated into the spreadsheet in line 30, where a value of 1 includes the propellant for step 1 and/or step 2 independently. This is convenient for locating the CM in three cases: dry, fully-loaded, and empty step 1 + loaded step 2 (just before staging). More measures would have to be taken if only the step 1's MoIs were needed (for example, maneuvering while flying back to recovery site), or the fully-loaded or empty step 2 only, needed for its control system design.

Also note that we have compared the sum of the J_0 inertias (inertias around own CM) to the total inertia of the LV. Line 35 shows that these two numbers have a very small ratio of 0.42% for the loaded TSTO LV. This confirms the statement that the J_0 inertia terms are essentially negligible for long items like LVs.

We had mentioned earlier that none of the vehicles with a line of symmetry had parts that were offset radially. Of course, that's not strictly true. For example, the LV's wiring is typically carried in channels on the outside of the body. Also, other things such as engine clusters, high-pressure gas bottles, and the like may also be offset from the centerline. We show how to deal with these situations below.

Table 8.22 Mass Properties Listing for the Example TSTO LV

	Item	Mass (kg)	h above ref (m)	Product (kg-m)	Thickness (m)*	Dist from CM (m)	J_0 (kg m²)	mx_{CM}^2 (kg m²)	$J_{pitch/yaw}$ (kg m²)	J_{roll} (kg m²)
1	Payload Fairing h = 2D	388	41.22	15,977	0.0049	22.39	993	194,357	195,350	872
2	Payload no-PAF, h = D/2	10,000	42.97	429,670	—	24.14	10,833	5,828,438	5,839,271	22,500
3	Payload Attach Fitting	805	39.59	31,872	—	20.77	5,501	347,176	352,677	1,811
4	Forward Skirt 2	258	38.19	9,864	0.0049	19.36	382	96,833	97,215	581
5	Avionics	490	38.19	18,724	—	19.36	552	183,812	184,364	1,103
6	Wiring	612	36.43	22,276	—	17.60	15,755	189,447	205,201	1,376
7	Ox Tank 2	429	35.27	15,125	0.0043	16.44	1,566	115,941	117,507	1,038
8	Ox Insulation 2	66	35.27	2,319	—	16.44	151	17,779	17,930	100
9	LOx residual 2	784	32.32	25,331	—	13.49	0	142,690	142,690	0
10	Intertank 2	360	31.94	11,497	0.0049	13.12	652	61,929	62,582	810
11	Fuel Tank 2	185	30.10	5,580	0.0022	11.28	368	23,569	23,937	157
12	Fuel residual 2	245	28.63	7,014	—	9.81	0	23,570	23,570	0
13	Aft Skirt 2	258	28.66	7,404	0.0049	9.84	382	25,010	25,392	581
14	Thrust Structure 2	118	28.66	3,370	—	9.84	0	11,382	11,382	132
15	Gimbals 2	18	28.15	516	—	9.32	0	1,594	1,594	0
16	Engines 2	574	27.89	16,021	—	9.07	0	47,221	47,221	0
17	Interstage	590	25.54	15,056	0.0049	6.71	1,702	26,576	28,278	1,842
18	Avionics 1 (1/2)	490	25.05	12,282	—	6.22	981	18,993	19,973	1,961

(Continued)

Table 8.22 Mass Properties Listing for the Example TSTO LV (Continued)

	Item	Mass (kg)	h above ref (m)	Product (kg-m)	Thickness (m)*	Dist from CM (m)	J_0 (kg m^2)	mx_{CM}^2 (kg m^2)	$J_{pitch/yaw}$ (kg m^2)	J_{roll} (kg m^2)
19	Wiring 1 (by length ratio)	961	13.82	13,280	–	–5.01	61,165	24,102	85,267	3,845
20	Ox Tank 1	2,012	17.73	35,664	0.0039	–1.10	23,261	2,432	25,693	7,261
21	Ox Insulation 1	213	17.73	3,768	–	–1.10	2,458	257	2,715	698
22	Ox residual 1	3,665	11.15	40,865	–	–7.68	0	215,894	215,894	0
23	Intertank 1	640	9.90	6,336	0.0049	–8.92	2,061	50,940	53,001	2,559
24	Fuel Tank 1	870	5.99	5,213	0.0035	–12.83	1,998	143,174	145,172	2,180
25	Fuel Residual 1	1,145	4.58	5,245	–	–14.24	0	232,412	232,412	0
26	Aft Skirt 1	669	2.00	1,337	0.0049	–16.82	2,228	189,244	191,472	2,674
27	Thrust Structure 1	1,068	2.00	2,136	–	–16.82	0	302,383	302,383	2,136
28	Gimbals 1	162	1.00	162	–	–17.82	0	51,581	51,581	0
29	Engines 1	4,257	0.50	2,129	–	–18.32	0	1,429,495	1,429,495	6,370
30	Propellant 2/1? Yes = 1 No = 0	1	1	residuals not included, they're above						
31	RP-1 fuel 2	12,280	30.10	369,641	–	11.28	0	1,561,377	1,561,377	0
32	LOx oxidizer 2	39,296	35.27	1,385,869	–	16.44	0	10,623,694	10,623,694	0
33	RP-1 fuel 1	57,616	5.99	345,354	–	–12.83	0	9,485,336	9,485,336	0
34	LOx oxidizer 1	184,373	17.73	3,268,077	–	–1.10	0	222,851	222,851	0
35	**Totals**	**325,898**	kg	6,134,970			132,988	3.19E + 07	**3.20E + 07**	**62,588**
36	**CM Location above ref.**	**18.82**	m				J_0/J_{pitch}	0.42%	kg m^2	kg m^2
							uniform LV	5.553E + 7	**Pitch Mol**	**Roll Mol**

*Thickness without insulation

Example 8: Avionics MoI Calculation

Looking at Fig. 8.24, the avionics have been drawn so that they are spread around the circumference of the top of the forward skirt of the second step (diameter 3 m), and the bottom of the interstage (diameter 4 m). Because they appear to be shaped like a doughnut, we can use the MoI values for a thin ring to calculate the avionics MoI. From Table 8.16, we see that the CM-centered MoI for pitch is $\frac{1}{2} mr^2$, and for roll the MoI is mr^2. We'll assume that R, the radius of the ring, is either 3 or 4 m depending on which avionics location we're considering, and the avionics mass is split in half for each step, so that $m_{avionics1} = m_{avionics2} = 490$ kg

Avionics MoI Calculation

The general expression for pitch inertia is $J_{pitch\_avionics} = J_{0\text{-}pitch\_avionics} + mh_{pitch\_avionics}^2$.

For step 1 in pitch, $h_{avionics1} = 6.22$ m, so

$$J_{pitch\_avionics\,1} = J_{0\text{-}pitch\_avionics\,1} + m_{avionics\,1} h_{avionics\,1}^2$$

$$= \frac{1}{2} \times 490 \text{ kg} \times (2 \text{ m})^2 + 490 \text{ kg} \times (6.22 \text{ m})^2$$

$$= 981 + 18{,}993 \text{ kg} \cdot \text{m}^2 = 19{,}973 \text{ kg} \cdot \text{m}^2.$$

The step 1 value for roll inertia for its avionics ring is

$$J_{roll\_avionics\,1} = 490 \text{ kg} \times (2 \text{ m})^2 = 1{,}961 \text{ kg} \cdot \text{m}^2.$$

These values can be found in row 18 of the Table 8.22.
For step 2 in pitch, $h_{avionics2} = 19.36$ m, so

$$J_{pitch\_avionics\,1} = J_{0\text{-}pitch\_avionics} + m_{avionics\,1} h_{avionics\,1}^2$$

$$= \frac{1}{2} \times 490 \text{ kg} \times (1.5 \text{ m})^2 + 490 \text{ kg} \times (19.36 \text{ m})^2$$

$$= 552 + 183{,}812 \text{ kg} \cdot \text{m}^2 = 184{,}364 \text{ kg} \cdot \text{m}^2.$$

The step 2 value for roll inertia for its avionics ring is

$$J_{roll\_avionics\,1} = 490 \text{ kg} \times (1.5 \text{ m})^2 = 1{,}103 \text{ kg} \cdot \text{m}^2.$$

These values can be found in row 5 of the Table 8.22.
Note that the arrangement of six engines, their gimbals, and their support structure may also be considered a uniform ring as an approximation. This was executed for these three categories in the table for rows 27–29.

Example 9: Wiring MoI Calculation

We assume that the wiring is essentially a very thin cylinder (meaning zero for J_{0-Roll}) radially separated from the centerline by the radius of the vehicle. We also assume that the wiring is radially split into two point-mass bundles separate by $180°$ on the outside of the tank, each having one-half of the mass, and located at a radius R from the centerline (see Fig. 8.31).

For the longitudinal direction, the wiring is assumed to be split mass-wise by the fraction of the length of the step it's mounted on divided by the overall length. The total wiring mass is 1,573 kg. The LV has a length of 45.2 m, with step 1's length $l_1 = 27.6$ m, and step 2's length $l_2 = 17.6$ m, the masses are $m_{\text{wiring } 1} = m_{\text{wiring}} (l_1/l) = 961$ kg, and $m_{\text{wiring } 2} = m_{\text{wiring}} (l_2/l) = 612$ kg.

Longitudinally (in pitch/yaw), the wire is assumed to be a long, skinny rod, and the CM-centered MoI is $I_0 = mh^2/12$ (see Table 8.17, "thin bar").

The general expression for pitch inertia is $J_{\text{pitch\_wiring}} = J_{0-\text{pitch\_wiring}} + mh_{\text{wiring}}^2$.

For step 1, $m_{\text{wiring1}} = 961$ kg; in pitch, $h_{\text{wiring1}} = -5.01$ m (it is below the CM), so

$$J_{\text{pitch\_wiring 1}} = J_{0-\text{pitch\_wiring 1}} + m_{\text{wiring 1}}h_{\text{wiring 1}}^2$$
$$= 1/12 \times 961 \text{ kg} \times (27.6 \text{ m})^2 + 961 \text{ kg} \times (-5.01 \text{ m})^2$$
$$= 61{,}165 + 24{,}102 \text{ kg} \cdot \text{m}^2 = 85{,}267 \text{ kg} \cdot \text{m}^2.$$

This is one of the very few cases where the mr^2 term is smaller than the I_0 term, due to the fact that the long and skinny wire is less than its own length away from the CM.

The step 1 value for roll inertia for its avionics ring is

$$J_{\text{roll\_wiring 1}} = J_{0\_\text{roll\_wiring 1}} + 2\left(\frac{1}{2}m_{\text{wiring 1}}R_{\text{wiring 1}}^2\right)$$
$$= \sim 0 + m_{\text{wiring 1}}R_{\text{wiring 1}}^2 = 961 \text{ kg}(2.0 \text{ m})^2$$
$$= 3{,}845 \text{ kg} \cdot \text{m}^2$$

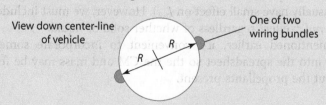

Fig. 8.31 A top view of how the wiring of the launch vehicle is split between two utility ducts separated by $180°$.

(Continued)

Example 9: Wiring MoI Calculation (Continued)

These values can be found in row 19 of the Table 8.22.

For step 2, $m_{wiring2} = 612$ kg; in pitch, $h_{wiring2} = 17.6$ m (happens to be same as length!), so

$$J_{pitch\_wiring\,2} = J_{0-\,pitch\_wiring\,2} + m_{wiring\,2}h_{wiring\,2}^2$$

$$= 1/12 \times 612\,kg \times (17.6\,m)^2 + 612\,kg \times (17.6\,m)^2$$

$$= 15,755 + 189,447\,kg \cdot m^2 = 205,201\,kg \cdot m^2.$$

The step 2 value for roll inertia for its avionics ring is

$$J_{roll\_wiring\,2} = J_{0\_roll\_wiring\,2} + 2\left(\frac{1}{2}m_{wiring\,2}R_{wiring\,2}^2\right)$$

$$= \sim 0 + m_{wiring\,1}R_{wiring\,1}^2 = 612\,kg\,(1.5\,m)^2$$

$$= 1,376\,kg \cdot m^2$$

These values can be found in row 6 of the Table 8.22.

8.8.1 A Vehicle Loaded with Propellants

The table shows non-zero MoIs for propellants. Obviously, for the bulk of the launch period, we need to include them also (and in varying quantities/levels as propellants are consumed).

Solid propellants are solid, so it's relatively simple to add the appropriate shapes to the MoI calculations (cylinders, domes, etc.). One would use the solid moments of inertia in Table 8.23 and add them to the mass properties table (Table 8.22).

Liquid propellants are different, in that it's often possible to neglect J_0 for liquids, especially with respect to rolling rotation around the centerline, because most liquids have low viscosity and do not act as a rigid body (and the J_{0i}s usually have small effect on J_{tot}). However, we must include the propellants' $m_i L_i^2$ effect regardless of whether solid or liquid.

As mentioned earlier, it's convenient to incorporate some "on-off" switches into the spreadsheet so the the CM and mass may be found with or without the propellants present.

8.9 Launch Vehicle Symmetry

So far, many of the vehicles we have considered are symmetric in two axes: in other words, the pitch and yaw moments of inertia are equal;

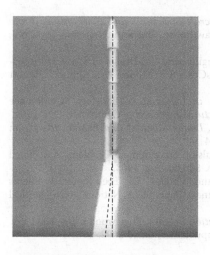

Fig. 8.32 This Atlas V is flying with a single strap-on solid rocket booster. The thrust line is *tilted* relative to the centerline, which creates an angle of attack. Source: Frame from YouTube video (https://youtu.be/dME5amgp0Ug).

however, some vehicles are not. For example, vehicles such as the Falcon Heavy, Delta IVH, Ariane 5, and Space Shuttle all have pitch inertias that do not equal their yaw inertias. This is also true for the large asymmetric payload fairings [2] described in the previous chapter. One may deal with such a situation by recording all three coordinates (x, y, and z positions instead of axial and radial positions) of the location of each component mass and computing inertias as before. This will produce three distinct inertias (pitch, yaw, and roll) rather than two (pitch = yaw, and roll) as before.

There are vehicles that have offset or nonsymmetric thrust lines. A prime example of this is the Space Shuttle, whose SSMEs are located on the aft end of the orbiter, which is not along the vehicle's center of mass. This means that the engines have to be canted, or tilted, so that their thrust line passes through the vehicle's CM. The result of this is easily seen during liftoff: the vehicle not only rises vertically, but also translates forward toward the Shuttle's belly.

In some situations, the vehicle's CM is offset from the centerline, meaning that the thrust line is inclined relative to the vehicle's axis, so the vehicle must fly at an angle of attack. One example of this is the Atlas V launch of 20 April 2006 carrying the Astra 1KR with a single strap-on, shown in Fig. 8.32. Such a vehicle may be seen launching in a YouTube video at https://youtu.be/dME5amgp0Ug. It is easy to see that the exhaust stream is not parallel to the long axis of the vehicle. Evidently, the increased drag and aerodynamic loads of this configuration were not a concern for this particular launch.

References

[1] Akin, D., "Mass Estimating Relations," lecture notes, Course ENAE 483/788D, University of Maryland, 2011.
[2] ATA Engineering, "Large Asymmetric Payload Fairing, Maximizing Payload Volume for Your Launch Vehicle," http://www.ata-e.com/wp-content/uploads/2018/01/Large-Asymmetric-Fairing-Data-Sheet.pdf, 2010 [retrieved 27 April 2019].
[3] Brown, C., *Elements of Spacecraft Design*, AIAA, Reston, VA, 2002.
[4] Rohrschneider, R., "Development of a Mass-Estimating Relationship Database for LV Conceptual Design, AE8900 Report, Georgia Institute of Technology, 2002.

[5] Sarafin, T. P., and Larson, W. J., *Spacecraft Structures and Mechanisms: From Concept to Launch*, Microcosm Press, Hawthorne, CA; Springer, New York, NY, 2007.

[6] TransCostSystems, "Handbook of Cost Engineering and Design of Space Transportation Systems Revision 4 with TransCost 8.2 Model Description," Report TCS-TR-200, Ottobrunn, Germany.

[7] Hellebrand, E. A., "Structural Problems of Large Space Boosters." Conference *Preprint 118, presented at the ASCE Structural Engineering Conference*, Oct. 1964.

[8] Leondes, C. T., and Vance, R. W. (Eds.), *Lunar Missions and Explorations*, John Wiley and Sons, Inc., New York, NY, 1964.

[9] McCool, A. A., and McKay, G. H., "Propulsion Development Problems Associated With Large Liquid Rockets." NASA TM X-53075, Aug. 1963.

[10] Platt, G. K., Nein, M. E., Vaniman, J. L., and Wood, C. C., "Feed System Problems Associated with Cryogenic Propellant Engines." Paper 687A, National Aero-Nautical Meeting, Washington DC, April 1963.

[11] Ring, E., *Rocket Propellant and Pressurization Systems*, Englewood Cliffs, NJ, 1964.

[12] Straw, A. D., *"Launch Vehicle Design"*. Unpublished, dated 2002.

Additional references

Heineman, W., Jr., "Fundamental Techniques of Weight Estimating and Forecasting for Advanced Manned Spacecraft and Space Stations," NASA TN-D-6349, 1971.

Heineman, W., Jr., "Mass Estimation and Forecasting for Aerospace Vehicles Based on Historical Data," NASA JSC-26098, 1994.

MacConochie, I. O., and Klich, P. J., "Techniques for the Determination of Mass Properties of Earth-to-Orbit Transportation Systems," NASA TM-78661, 1978.

Myers, J., "Handbook of Equations for Mass and Area Properties of Various Geometrical Shapes," NAVWEPS report 7827, April 1962, http://www.dtic.mil/dtic/tr/fulltext/u2/274936.pdf [retrieved Feb. 2018], doi:10.21236/ad0274936.

NASA, "WAATS—A Computer Program for Weights Analysis of Advanced Transportation Systems," NASA CR-2420, 1974.

Ordway, F. I., Gardner, J. P., and Sharpe, M. R., *Basic Astronautics: An Introduction to Space Science, Engineering and Medicine*, Prentice Hall, Englewood Cliffs, NJ, 1962.

8.10 Exercises: Sizing, Inboard Profile, and Mass Properties of TSTO LV

Provide no more than four significant digits for your calculations: round numbers as necessary.

Use the mass property details from *your* lowest GLOM Family 2 (losses in 1st step only) TSTO LV design from the previous homework. These parameters will be used to determine your vehicle's geometry, as specified below. **Note: items 1 & 2 can be done in any order.**

1. **Select a body diameter**: there's no "right answer" here, unless you are assembling your vehicle from existing rocket steps. Diameters range from 2 to 5 meters, or more. You don't want to choose too small a diameter or the vehicle may be very flexible. It should "look right" also. One suggestion would be to consider imitating an existing LV with similar performance.

2. **Specify engine performance.** Engine choice will depend on thrust levels, as per the suggestion in Ch. 8. Choose a T/W ratio, remembering that thrust must be greater than liftoff weight! With your known gross liftoff mass, you can calculate the required thrust level. You may be able to split the required thrust among several engines, which may make it possible to use an existing engine, which is best. You will also need to select your upper step's T/W ratio. These are used to calculate the mass flow, which determines the mass of propellants needed for startup.

3. **Create a spreadsheet to calculate tank volumes** using selected diameter to calculate, following the procedures outlined in Chapter 8 of the text. You will need to select dome geometry (hemisphere, EH, etc.) and compute the lengths of the cylindrical sections for each propellant (for now, it's ok to assume inner diameter ≈ outer diameter). For solids there will be only one combustion chamber per step, and don't forget to account for the "empty" internal volume down the motor's center. Of course, you will have to carry out the calculations for each step. State all assumptions, and provide a detailed list of tank specs, including step number (1 or 2), dimensions, dome type, description (i.e. LOx tank), etc. You may assume the engines take 1 second to start up, unless you know differently from actual engine data.

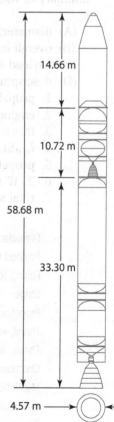

4. **Size intertank, interstage, and thrust structure** enclosures. Use the rules of thumb (fractions of diameter) given in Ch. 8. Provide a list or table of step, item, diameter(s), and length (don't forget to add the height(s) of the dome(s) on the side(s) of the "empty space" length).

5. **Size your payload fairing.** A simple conical fairing is easiest, but you can get fancy if you want like the one shown below! Be sure that your fairing will encompass your payload. As a rule of thumb, you may wish to use an average electronic box density of 320 kg/m³ as given in an earlier assignment. (If you know your payload's actual density, use that value.)

6. **Physically size your engine(s).** Whether you will be using existing engines or you're providing your own engine specifications, give critical dimensions on a simple drawing, following the instructions on engine dimensioning. Existing engines, if used, have existing dimensions! You can also scale an existing engine, by assuming thrust is proportional to exhaust area (i.e. 2 × required thrust would have 2 × exhaust area, or $\sqrt{2}$ × exhaust *diameter*, all

dimensions scale by $\sqrt{2}$). If you do this, be sure to *specify your existing engine and the scale factor you used.*

For your (upper) step 2, you have more freedom, as its T/W ratio may be < 1, or anything (very low T/W ratios require lofting and may reduce performance). Look at upper steps currently in use to perhaps use similar T/W ratios.

7. **Provide a *scale* drawing of your vehicle.** Give locations and dimensions on your drawing for
   (A) payload attach fitting,
   (B) sketch propellant feed lines,
   (C) engine compartment, including engine mounting,
   (D) avionics and wiring.

Use care when drawing: a hand sketch is OK but **must be to scale** on graph paper and should fill the long direction of an 8.5 x 11 sheet of paper. Use a straightedge and compass to make neat, straight lines and circular circles. The figures in the book were done using PowerPoint$^{®}$'s drawing shapes. **Summarize your vehicle.** Provide a technical summary of your vehicle as shown in the table below. It should include

   (A) diameter,
   (B) overall length,
   (C) payload & mission,
   (D) descriptions of each step, including
      1. propellants,
      2. engines,
      3. thrust, SL & vacuum,
      4. I_{sp}, SL & vacuum,
      5. propellant masses,
      6. T/W information (at startup and at burnout),
      7. total step mass, etc., similar to the table shown.

Diameter = 4.57 m	Length = 58.68 m	
Payload to ISS	9,072 kg	
Fairing (drop at stage 1 burnout)	2,994 kg	
Stage	1	2
Propellant	LOx/LH$_2$	LOx/LH$_2$
Thrust, vacuum	3,233 kN	352 kN
Thrust, SL	2,377 kN	NA
Chamber press	8,274 kPa	17,237 kPa
Mix Ratio	6.0:1	6.0:1
I_{sp} Vacuum	430 s	456 s
Expansion ratio	40:1	150:1

Step propellant ratio	0.8801	0.8455
Propellant mass	129.9 T	22.68 T
Step mass w/prop	147.61 T	26.82 T
Stage GLOM	186.5 T	35.896 T
T/W	1.3 – 5.83	1.0 – 2.72
Δv	5,002 m/s	4,446 m/s

8. **Mass properties, part 1: Center of Mass (CM) calculations**.

Assume a coordinate system where x = axial/thrust direction, y = lateral (+ out the right side, pitch axis), and z = perpendicular to x and y (yaw axis). Following Example 2, go through the vehicle and use Excel® to make a table showing all the components (payload, tanks, interstages, intertanks, engines, etc.). Calculate the (empty) mass of each component following the procedures. Total up the masses to get the total dry mass, as shown in line 4 of the table below:

	Item	Mass (kg)
1	Payload fairing	m_1
2	Payload + PAF	m_2
3	...	...
$i+1$	**Total Dry Mass**	$\Sigma\, m_i$

Add a fourth column to the above table for item locations. Use your scale drawing to determine the location of the centroid or center of mass of each part relative to the *bottom of the aft skirt* as a reference ("station 0.0"). You can easily calculate these distances with a ruler and the drawing's scale factor, or by using an object with known size in your drawing program. Keep all dimensions in meters, and positive is above the bottom of the thrust structure. Items not shown on the drawing, such as wiring, should have their CM located at the midpoint of the vehicle or step (wiring should be split into lengths equal to the step lengths). All others should have locations based on their actual position.

	Item	Mass (kg)	Distance from ref. (m)
1	Payload fairing	m_1	L_1
2	Payload + PAF	m_2	L_2
3	...	...	...

"Dry" Vehicle Mass Statement: Add a fifth column to your table titled "Moment (kg·m)." Populate this column with the product of mass and distance for each item. The first part of the table should include all the dry LV components (no propellants). Use the tabulated masses and locations to calculate the dry CM of the vehicle relative to the reference location. This is done by calculating the total moment ($\sum$ individual moments) and dividing by the total mass ($\sum$ individual masses), as shown in lines 1-5 above. Your table should be formatted something like the following:

Launch Vehicle Mass Statement

	Item	Mass (kg)	Distance from ref. (m)	Moment (kg·m)
1	Payload fairing	m_1	L_1	$m_1 \cdot L_1$
2	Payload + PAF	m_2	L_2	$m_2 \cdot L_2$
3	...			...
4		$m_{dry} = \Sigma\, m_i$		$\Sigma(m_i\, L_i)$
5			Dry CM position	$\Sigma(m_i\, L_i)/\Sigma\, m_i$

Mass statement, "wet" (loaded with propellant) vehicle. Insert additional lines to incorporate the full propellant (only) mass and location associated with each tank. The propellant tank's loaded mass is located at the tank's centroid. Re-calculate the CM position. This is shown below with addition of lines 6-7 (one line for each mass of propellant, yours will have more than two). The calculation of the WET (loaded) CM is shown in line 8. You may find it handy to set up your spreadsheet so you can "turn the fuel mass on and off" by changing a designated "Propellant, Yes or No" cell from 1 to 0, or vice versa.

Launch Vehicle Mass Statement

	Item	Mass (kg)	Distance from ref. (m)	Moment (kg·m)
1	Payload fairing	m_1	L_1	$m_1\, L_1$
2	Payload + PAF	m_2	L_2	$m_2\, L_2$
3		...	...	...
4		$m_{dry} = \Sigma\, m_i$		$\Sigma(m_i\, L_i)$
5			Dry CM position	$\Sigma(m_i\, L_i)/\Sigma\, m_i$
6	Propellant, Tank 1	$m_{prop\text{-}tank1}$	L_{tank1}	$m_{prop\text{-}tank1}\, L_{tank1}$
7	...	...	...	...
8		$m_{tot} = \Sigma\, m_{all}$	Wet CM position	$\Sigma(m_{all}\, L_{all})/\Sigma\, m_{all}$

Mass statement summary.

1. Provide a copy of your DRY (all tanks empty) assembled Mass statement per above.

2. Provide a copy of your WET (all tanks full) assembled Mass statement per above.

3. Provide a table with dry and wet (liftoff) masses and CM locations.

9. **Mass properties, part 2:** *Moments of Inertia in Pitch and Roll*

Referring to the mass properties spreadsheets, add new columns to your Excel® CM calculation spreadsheet to match those shown in the text, except: **to save time, YOU MAY OMIT THE J_0 column, and therefore the $[J_{pitch}, J_{yaw}]$ column = $[m·xCM^2]$ column.** Specifically, you will need to add a column for *lengthwise* distance from CM, a column for $m·x_{CM}^2$, and a column for ROLL Moment of inertia of item around own centroid J_{roll}. Your final spreadsheet should look like this:

Launch Vehicle Mass Statement ($J_{0-pitch\ \&\ yaw}$ omitted)

No.	Item	Mass (kg)	$L =$ Distance from ref. (m)	Moment (kg·m)	$x_{CM} =$ Dist. from CM (m)	J_{pitch}, $J_{yaw} =$ $m·x_{CM}^2$ (kg·m²)	J_{Roll} (kg·m²)
1	PLF	m_1	L_1	$m_1\ L_1$	$x_{CM\text{-}1}$	$m_1\ x_{CM\text{-}1}^2$	$J_{Roll\text{-}1}$
2	Payload	m_2	L_2	$m_2\ L_2$	$x_{CM\text{-}2}$	$m_2\ x_{CM\text{-}2}^2$	$J_{Roll\text{-}2}$
3	...	...	...	...	...	...	...
4		$m_{dry} =$ $\Sigma\ m_i$		$\Sigma(m_i\ L_i)$	Dry Pitch & Roll Mols →	$\Sigma(m_i\ x_{CM\text{-}i}^2)$	$\Sigma\ J_{Roll\text{-}i}$
5			$l_{CM\text{-}dry} =$ Dry CM position	$\Sigma(m_i\ L_i)/$ $\Sigma\ m_i$			
6	Propellant, Tank 1	m_{prop1}	L_{tank1}	$m_{prop1}·L_{tank1}$			
7		...	...	...	...	...	...
8		$m_{tot} =$ $\Sigma\ m_{all}$	$l_{CM\text{-}wet} =$ Wet CM position	$\Sigma(m_{all}·L_{all})/$ $\Sigma\ m_{all}$	Wet Pitch & Roll Mols	$\Sigma(m_{all}·x_{CM\text{-}all}^2)$	$\Sigma\ J_{Roll\text{-}all}$

Following the methods in the text, use these added columns to determine the $[J_{pitch}, J_{yaw}]$ **moment of inertia** ($= m·x_{CM}^2$), and roll (J_{roll}) inertias for your vehicle in the following cases:

(A) dry (no propellants)

(B) fully-loaded first and second steps

(C) first step only, dry. In a later assignment, you will use this to take a look at trying to recover your vehicle's first step under rocket power, similar to what SpaceX is now doing with the first step of the *Falcon 9*.

2. Provide a copy of your WET (wet tanks, full) assembled Mass statement per above.

3. Provide a table with dry answer (unfueled) masses and CM locations

9. Mass properties, part 2. Moments of inertia in Pitch and Roll

Referring to the mass properties spreadsheet, add new columns to your Excel™ CM calculation Spreadsheet to match those shown in the text, except to save time, YOU MAY OMIT THE I_b column, and therefore the I_{pitch}-A_{xx} column = [m·CM²] column. Specifically, you will need to add a column for lengthwise distance from CM, a column for ΔCM, and a column for ROLL Moment of inertia of item around own centroid A_{xx}. Your final spreadsheet should look like this:

Launch Vehicle Mass Statement (Q_0 = quiet or low, unfueled)

Following the methods in the text, use these added columns to determine the I_{pitch}-A_{xx} moment of inertia (= [m·ΔCM²]) and roll (I_{roll}) inertias for your vehicle in the following cases:

(A) dry (no propellants)

(B) fully-loaded first and second stages

(C) first step only. In a later assignment, you will use this to take a look at trying to recover your vehicle's first step under rocket power, similar to what SpaceX is now doing with the first step of the Falcon 9.

Chapter 9

Surface and Launch Environments, Launch and Flight Loads Analysis

Moving forward from sizing and mass properties, we show the design roadmap in Fig. 9.1 with the highlighted signposts indicating the information to be covered in this chapter: aerodynamic loads are determined from the vehicle's environments and its aerodynamic

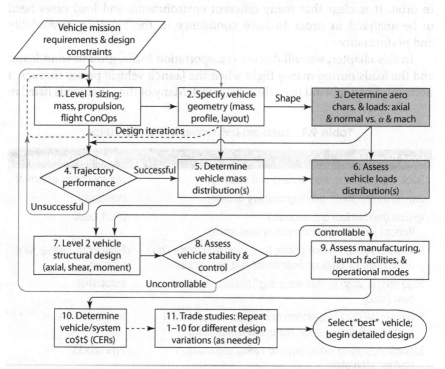

Fig. 9.1 This chapter will look at ground and flight environments, vehicle aerodynamics and loads estimation.

characteristics. We will use the vehicle's geometry, as obtained in Chapter 8, to determine its aerodynamics.

In the following pages, we'll cover the following:

- Loads: transportation, ground winds, and flight loads
- Rigid body acceleration and load factors
- Loads, moments, and shear forces
- Ground winds: load calculation on Saturn V
- Flight winds
- Aerodynamic coefficients and flight loads calculation: Saturn V at max-q
- Loads due to propellants and structure
- Axial, bending, and shear loads
- Load relief during ascent: passive and active

9.1 Launch Vehicle Load Cases

A launch vehicle is exposed to many types of loads, basically from the time the vehicle has been completed until the vehicle delivers its payload (and sometimes longer, if the vehicle is reused). Table 9.1 lists some of these events for a generic vehicle program, from transportation to delivery to orbit. It is clear that many different environments and load cases need to be analyzed in order to have confidence in the vehicle's survivability and performance.

In this chapter, we will discuss transportation loads, ground wind loads, and the loads during max-q flight when the launch vehicle passes through a horizontal gust of wind (e.g., the jet stream). Many of the other loads listed in

Table 9.1 Loads and Environments for a Launch Vehicle

Load Description	Vehicle Condition
Truck, train, air, and sea handling and stacking	Transportation, assembly
Ground winds (shear, bending moment at base)	Prelaunch
Ignition (mechanical and acoustic) Release jerk (vehicle acceleration axial loads)	Liftoff loads
Max-q and max-$q\alpha$ (loads arising from dynamic pressure and angle of attack produce both lateral and axial loads)	Flight loads: max-q, wind shear
Pogo and/or slosh or "tail wags dog" (liquid); resonant burn (solid)	Instabilities
Shutdown/MECO (deceleration axial loads, lateral loads if multiengine); upper-stage startups and shutdowns	Mechanical events
Separation/staging loads; payload fairing separation; payload separation	Pyro shocks
Acoustic, thermal effects	All conditions

MECO = main engine cutoff

Table 9.1 are mechanical, pyrotechnic, or thermal, and will be discussed in other chapters.

9.1.1 Transportation Loads

Space launch vehicles are transported to their launch sites in several ways: by land (truck or train), by sea (barges or ships), or by air. These methods of transportation provide static, quasi-static, and dynamic (transient, periodic, random, or combinations) loads. The loads are usually defined at the vehicle's attachment points.

Figure 9.2 provides a generic schematic of transportation systems, handling systems, and loads. The letter *A* represents loads input *to* a vehicle *from* the vehicle's handling fixture. Moving downwards, letter *B* shows the loads

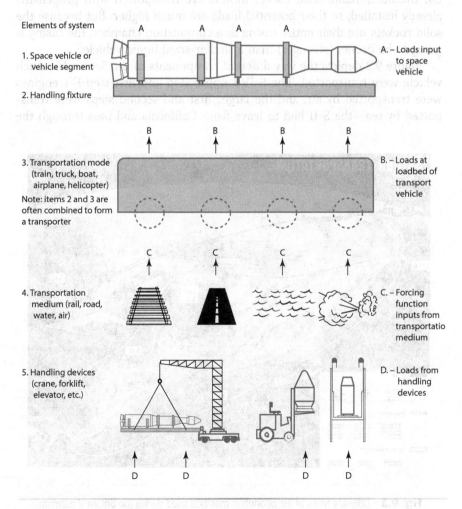

Fig. 9.2 Vehicle transportation systems, handling systems, and loads. Source: [16].

being applied *from* the transport vehicle (truck, train, boat, or aircraft) *to* the handling fixture. Below that, letter *C* indicates different forcing functions provided by different modes of transportation: rail, road, water, and/or air. Letter *D* indicates inputs from handling devices, such as cranes, forklifts, elevators, and the like.

The rocket, or its parts, is typically suspended or cradled as shown at the top of Fig. 9.2. The key point to remember is that the loads imposed by transportation systems are usually *point* loads, meaning they are concentrated in a small area (as opposed to *distributed* loads). The structural designer may need to beef up the structure in these regions to prevent damage from being inflicted on the part. However, most liquid-propellant rockets are transported in an unloaded or empty condition, which will help somewhat to reduce loads, because those loads depend on the mass being accelerated. On the other hand, solid rocket motors are transported with propellants already installed, so their potential loads are much higher. But because the solid rockets use their entire casing as a combustion chamber, the casing is going to be beefier (stronger) than a similar-sized liquid vehicle.

Figure 9.3 depicts the way different components of the Saturn V launch vehicle were transported. The S-IVB upper step and first-step F-1 engines were transported by air, and the larger first and second steps were transported by sea—the S-II had to leave from California and pass through the

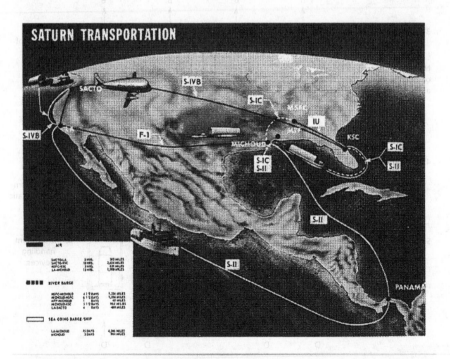

Fig. 9.3   Different types of transportation methods used during the Saturn V program.
Source: NASA [11].

Fig. 9.4 The Super Guppy has transported many large flight vehicles. Source: Frame from NASA video (2014).

Panama Canal. Each of these modes of transportation provides different environments to the part being transported, as shown in Fig. 9.2. Figure 9.3 shows to what great lengths—literally—rocket programs would go to transport their handiwork.

Air transport is certainly the fastest way to transport launch vehicle components and assemblies, although it is likely more expensive. Figure 9.4 shows the famous "Super Guppy" with its vast cargo area exposed, so one can see just how large an item it could carry. In addition to the Apollo/Saturn program, it was also used to carry International Space Station truss assemblies.

Outlandishly large vehicles need another method for air shipment. Figure 9.5 shows Space Shuttle *Atlantis* on top of a Boeing 747 transport aircraft whose vertical tail area had to be augmented with two additional vertical surfaces, one on each side of the horizontal tail, for proper stability and control.

Many vehicles are transported by ground. Some travel by truck, others by train. Figure 9.6 shows a SpaceX Falcon 9 on a trailer behind a truck; Fig. 9.7 shows a Space Shuttle solid rocket booster segment being transported on a railroad car. The sign on the container, "DO NOT HUMP," is not meant to be humorous. Instead, rail humping refers to the way two railroad cars are normally slammed together to ensure the couplers lock and the two cars are linked. Another way of joining cars has to be used, because it's not a good idea to impose mechanical shocks (from the humping) on the loaded propellants.

Both air and ground transportation methods are limited by the size of the transported object, and sea transport becomes attractive for the larger elements. Figure 9.8 shows a ship containing a Russian Soyuz launch

Fig. 9.5 If it won't fit *inside* the airplane, how about putting it on the outside? Space Shuttle *Atlantis* mounted on a modified Boeing 747 transport plane. Source: NASA photo EC98-44740-2.

vehicle on its way to Kourou, French Guiana, in South America, ESA's primary launch site.

Table 9.2 lists typical loads that occur during different types of transportation, in terms of applied acceleration measured in units of g_0, the standard gravity's acceleration at the Earth's surface: $g_0 = 9.80665$ m/s^2 (sometimes referred to as gs or $gees$). A casual look at the table shows loads as high as 3 g_0, quite a bit larger than one might expect.

For more details on transportation and handling loads, the reader is referred to [16].

Fig. 9.6 A SpaceX Falcon 9 being transported by truck down the road.

Fig. 9.7   Space Shuttle SRB segment on railroad car. Source: NASA.

For design purposes, dynamic loads may be specified as a *load factor n* that represents the inertial force as a number of g_0. For example, consider the rocket in Fig. 9.9. If it were in free space with no gravity, its acceleration $a = T/m$. For a given acceleration a, the load factor is of opposite sign.

Fig. 9.8   Russian rocket on the way to ESA launch site at Kourou. Can you guess which rocket it is? Source: Arianespace.

Table 9.2 Typical Transportation Loads

Operation		Applied Accel. (x-, y-, z-axes, g)			Remarks
Clean Room Handling	Dolly	±1.0	±0.75	−1 ± 0.5	Any orientation
	Movement	±0.2	±0.2	−1 ± 0.5	
	Vertical hoist	±0.2	±0.2	−1 ± 0.5	
	LV Mate/demate	±0.5	±0.5	−2/+0	
Container	Hoisting	±0.5	±0.5	−1 ± 0.5	SC horizontal
Road	Quasi-static	±2	±2	−3/+1	40 km/h top speed
Air	Take-off	−1.5	±0.1	−2.5/+1.5	
	Vertical gusts	0	±1.5	−2.5	
	Lateral gusts	0	±1.5	−1.0	
	Landing	±1.5	±1.5	−2.5	
Barge/ship	Slamming	0.0	0.0	−1.8/+0.2	
	Waves	±0.3	±0.5	−1.6/+0.4	
Any transportation	Continuous vibration	±0.1	±0.1	±0.1	Below 10 Hz, not including gravity
Transport	Shock load	±2	±2	±3	

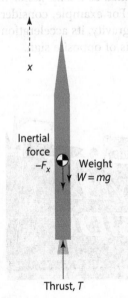

Fig. 9.9 Rocket in flight.

Typically we write the load factor with a subscript indicating its direction; in this case, we would use the load factor $n_x = -(F/m)/g_0 =$ no gravity acceleration in g_0 units.

9.1.1.1 Load Factor Example

Observe the rocket shown in Fig. 9.10. Its mass is 20 T (T = metric ton = 1,000 kg), and its thrust $F = 10^6$ N. Its no-gravity acceleration is

$$a = T/m = (10^6 \text{ N})/(20,000 \text{ kg}) = 50 \text{ m/s}^2$$
$$= (50 \text{ m/s}^2)/(9.80665 \text{ m/s}^2) = 5.1 g_0$$

(9.1)

Most rockets don't operate where there is no gravity. If we include gravity, the vehicle's *actual* acceleration would be $(n_x - 1) g_0 = 4.1 g_0$.

The load factor is opposite sign

$$n_x = -5.1 (\text{units} = g_0) \qquad (9.2)$$

and the inertial force is

$$F_x = n_x mg$$

$$= -5.1(20\text{E}3 \text{ kg})(9.80665 \text{ m/s}^2)$$

$$= -10^6 \text{ N} \qquad (9.3)$$

which is equal and opposite to the thrust. Conclusion: the load factor determines *internal* loading, regardless of external gravity.

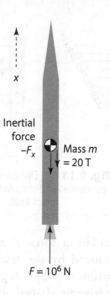

Inertial force $-F_x$ Mass m = 20 T

$F = 10^6$ N

Fig. 9.10 Load factor example.

9.1.2 Calculation of Loads

We need to calculate the loads on a launch vehicle, as well as on the parts inside the launch vehicle. We start with Newton's law $F = ma$ and rearrange it to read $F - ma = 0$. What we call the "inertial force" ma is equal and opposite to the applied force F. This is known as D'Alembert's principle. As Fig. 9.11 shows, the force is also equal to nW, where W is the weight and n is the load factor.

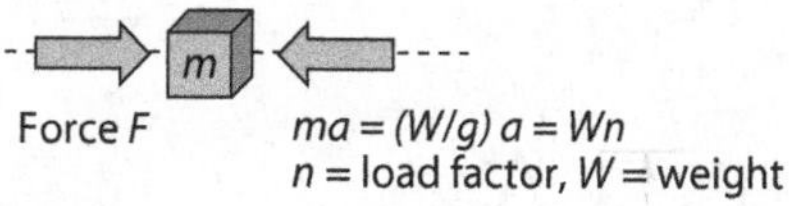

We can use this principle to look at loads applied to parts inside a vehicle (like propellant tanks, boxes, payload, etc.), as shown in Fig. 9.12. Looking at the right side of the figure, we know that the sum of the forces = zero. Hence, if we take positive up,

$$(+)F_B - F_1 - F_2 = 0 \qquad (9.5)$$

We make an imaginary cut, labeled A-A, to see the force on m_1. The load above A-A is

$$F_1 = n g m_1 \qquad (9.6)$$

The load above B-B is

$$F_B = F_1 + F_2 = n g (m_1 + m_2) \qquad (9.7)$$

which happens to be the total of the loads above B-B.

Force F m $ma = (W/g)\,a = Wn$
n = load factor, W = weight

Fig. 9.11 We consider an inertial force that is equal and opposite to the applied force.

Internal masses (i.e., filled propellant tanks, boxes, payload, etc.)

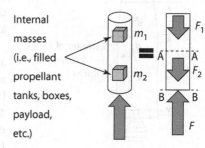

m_1 F_1
A --- A
m_2 F_2
B   B
F

Fig. 9.12 Load factor example.

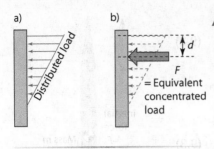

a)   b)

F
= Equivalent
concentrated
load

Fig. 9.13 a) Distributed load; b) conversion of distributed load into point load.

9.1.3 Distributed Loads

In many circumstances, the loads that are applied to an LV are *distributed loads*, spread over a length or area, as depicted in Fig. 9.13a; for example, this could be horizontal drag force produced by a crosswind whose speed increases with height above ground.

If we are not concerned about the local effect of the distributed load, and instead wish to calculate forces above and below the distributed load, we can calculate a force F as shown in Fig. 9.13b, which is the effective force produced by the triangular area. We can determine the magnitude and location of F by integrating over the distributed load to get the area under the triangle-shaped area. It turns out that the effective force acts at the *centroid* of the distributed load. The centroid of the area is shown in Fig. 9.13b as the distance d from the top of the distributed load. The combination of F and d will produce the same effect as the distributed load.

If we are dealing with a constant distributed force applied to a shape, we can analyze the loads in a similar manner, and we need to determine the locations of the centroids of different shapes.

As may be seen in Fig. 9.14, there are really only three different shapes associated with LVs: cones, cylinders (constant diameter), and skirts of trapezoidal shape. As shown in the figure, the top of the rocket is a cone whose bottom diameter is d_1. Underneath the cone is a cylindrical section of diameter d_1. Next comes a skirt whose top diameter is d_1 and bottom diameter is d_3. Finally, there is a bottom cylinder of radius d_3. The side areas and centroids of the cone, cylinder, and skirt or trapezoid are given in Table 9.3.

Now we will use these values to analyze loads created by distributed aerodynamic loads. Figure 9.15 shows two distributed forces (the triangular areas on the right) along with their equivalent concentrated forces F_1 and F_2, and the resultant single force F_{tot}.

As we did before, we will assume that forces act at the centroid of the

l_1

d_1

d_1 = diameter of cone & upper cylinder

l_2

l_3

d_3

diameter of skirt & lower cylinder

l_4

Fig. 9.14 Shapes on an LV.

Table 9.3 Side Areas and Centroid Locations for Several Shapes

Shape	Side Area	Centroid Above Bottom
Cone	$A_i = 1/2\, d_i\, l_i$	$1/3\, l_i$ above bottom
Cylinder segments	$A_i = d_i\, l_i$	$1/2\, l_i$ above bottom
Skirt (trapezoid shape)	$A_3 = 1/2\,(d_1 + d_3)l_3$	$h_3 = \dfrac{l_3(d_1/d_3 + 2)}{3(d_1/d_3 + 1)}$

distributed load. If the part AABB is in equilibrium, we can state that the sum of all horizontal forces is zero. With right $(\rightarrow) =$ positive $(+)$, we can write

$$\Sigma F_{\text{horizontal}} = F_{\text{tot}} - F_1 + F_2 = 0 \qquad (9.8)$$

or

$$F_{\text{tot}} = F_1 - F_2 \qquad (9.9)$$

The next things to consider are the torques or moments that the distributed forces or the concentrated forces produce. A torque or moment is simply a twisting action that results from a force offset from a reference point. Referring again to Fig. 9.15, assume that ends A-A and B-B are *free* (no forces or torques are applied at those locations). Thus

$$M_{A\text{-}A} = M_{B\text{-}B} = 0$$

$$F_{C\text{-}C} = \text{shear force at C-C} = -F_2 \qquad (9.10)$$

Why is the shear at C-C $= -F_2$? Because, if there were no horizontal force at C-C, the bottom of the part CCBB would be pulled to the right by $+F_2$. Hence $F_{C\text{-}C} =$ Shear at C-C $= -F_2$.

Now let's calculate the bending moment *above* C-C. Above C-C, we see that if we take counterclockwise (CCW) as positive $(+)$, the only contributions are from F_{tot} *and* F_1. The moments they produce are

$$M_{C\text{-}C} = -F_{\text{tot}} \times d_{\text{tot}} + F_1 \times d_1 \qquad (9.11)$$

Similarly, the moment *below* C-C is $(\text{CCW} = +)$

$$M_{C\text{-}C} = F_2 \times d_2 \qquad (9.12)$$

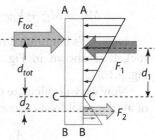

Fig. 9.15 Distributed loads resolve into concentrated forces.

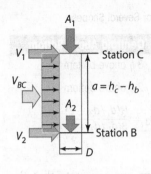

Fig. 9.16 Axial, shear, and bending load calculation.

Because the LV is exposed to gravity, there are vertical or axial forces to consider. In addition, there may be horizontal loads. This situation is illustrated in Fig. 9.16. You might think of this situation that A_1 and A_2 are forces due to equipment at those locations, and the horizontal load is the drag force produced by a horizontal wind.

The key to solving this sort of problem is to *start at the top, and work downwards.*

The top is the station C. At that location,

$$\text{Axial load } A_C = A_1$$

$$\text{Shear load } V_C = V_1$$

$$\text{Bending moment } M_C = 0$$

Now consider the situation at Station B.

$$\text{Axial load } A_B = A_C + A_2 = A_1 + A_2$$

V_{BC} is the equivalent drag force due to the right-arrowed running load between B and C, and is calculated with the usual aerodynamic definitions: drag load V_{BC} = dynamic pressure × drag coefficient × side area, or

$$V_{BC} = q\, C_D(aD) \tag{9.13}$$

The shear load at B, V_B, is the sum of the horizontal forces above B.

$$V_B = V_C + V_{BC} + V_2 = V_1 + V_{BC} + V_2 \tag{9.14}$$

The bending moment at B, M_B, is the sum of the moments above B.

$$M_B = M_C + V_{BC}(a/2) + V_C(a) = V_1(a) + V_{BC}(a/2) \tag{9.15}$$

Now, let's continue down past Station B to include a lower portion of the rocket, as shown in Fig. 9.17 (which can be considered an extension of Fig. 9.16).

Now we will calculate the loads at Station A. The axial load A_A is the sum of all the axial loads above Station A.

$$A_A = A_B + A_3 = A_1 + A_2 + A_3 \tag{9.16}$$

The shear load V_A is the sum of the shear loads above Station A.

$$V_A = V_B + V_{AB} - V_3 = V_1 + V_2 + V_{AB} - V_3 \tag{9.17}$$

The aerodynamic drag load V_{AB} is

$$V_{AB} = q\,C_D(aD) \qquad (9.18)$$

The bending moment is the sum of the moments above Station A.

$$
\begin{aligned}
M_A &= M_B + (V_B)(b) + V_{AB}(b/2) \\
&= (V_1)a + (V_1 + V_2)b + V_{AB}(b/2) \\
&= V_1(a + b) + V_2 b + V_{AB}(b/2)
\end{aligned}
$$

$$(9.19)$$

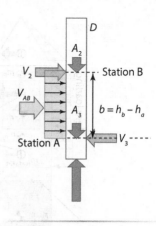

Fig. 9.17 Axial, shear, and bending below Fig. 9.16.

Note: This example used only cylindrical parts. The moment arms will be different when working with cones and skirt shapes.

In this book, we will consider the loads incurred during two different vehicle conditions: on the ground, and in-flight during max-q. Although we have not yet covered the process of load calculation, it is important to understand the sources of the loads, where they are applied to the vehicle, and how the vehicle's structure responds to the loads.

For both ground and flight conditions, the vehicle is exposed to winds: laterally (in the z-direction) while on the ground; at an angle of attack when in flight. In both, aerodynamic forces due to wind loads are applied at the centroid of the surface being analyzed.

Clamped to the launch pad, the vehicle is exposed to horizontal ground winds in a fully-loaded but unpressurized configuration. All axial loads are due to items' weight created by the acceleration of gravity (g_0), and all lateral loads are due to winds. During max-q, the vehicle is exposed to flight accelerating both longitudinally (at rate $n_x\, g_0$) and laterally (at rate $n_z\, g_0$) due to side forces from aerodynamics and engine thrust, and its propellant tanks are pressurized. The terms n_x and n_z are defined as longitudinal and lateral load factors, respectively.

For flight loading, we have to consider the added lateral aerodynamic loads (also called 'normal' forces, as they are perpendicular to the axial direction) as well as inertial loads for all masses. This includes the masses of the LV's structural elements themselves as well as masses attached to or supported by structural pieces. The inertial forces created by accelerating solid masses m_k, whose apparent axial and lateral weights are $m_k\, n_x\, g_0$ and $m_k\, n_z\, g_0$, appear at solid's CM. These loads are applied to and reacted out by LV structure at the lowest connection point of the attached item.

As an example, during flight, the apparent weight a payload $m_{PL}\, n_z\, g_0$ appears laterally at the payload's CM, but is reacted out to the LV at the bottom of the payload attach fitting (PAF). This means that the vertical

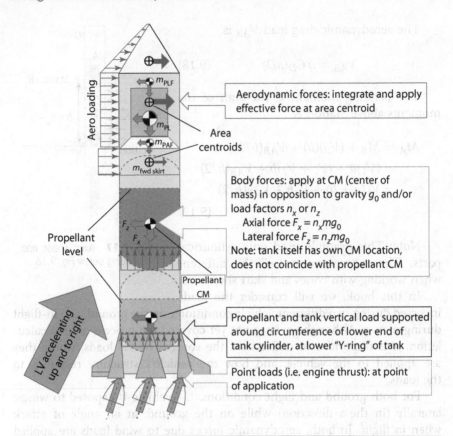

Fig. 9.18 This figure illustrates the locations of application of forces to a launch vehicle. All forces are imposed at shapes' centroids (of area) or centers of mass. Due to the geometry of the vehicle, separations between the force application location and the structural connection also create bending moments which must also be considered for loads analysis.

separation of the payload's CM from the base of the PAF results in both a lateral shear force $V = m_{PL} \, n_z \, g_0$ and a bending moment $M = (m_{PL} \, n_z \, g_0)(h_{CM})$ created by the payload's CM offset h_{CM} that must be applied where the PAF's base attaches to the LV. Additionally, the apparent axial weight of the payload ($m_{PL} \, n_x \, g_0$) is supported axially by the PAF.

Another example describes the inertial loads created by internal masses such as a propellant mass m_p, whose apparent weights $m_p \, n_x \, g_0$ and $m_p \, n_z \, g_0$ appear at propellant's CM, but they are applied laterally to the tank walls at the center of mass of the propellant mass, and axially at the bottom of the cylinder portion of the containing tank where it meets the intertank or skirt supporting structure below.

It is important to point out that each element has to be considered separately. Even though propellant may be contained within a tank or fuel container, the center of mass of the propellant is likely in a different location compared to the CM of the tank. Both must be considered independently. This is particularly important for analyses carried out during flight where propellants have been partially consumed, such as at max-q.

Another comment has to do with the items located around the payload. Since the payload is attached to the PAF, which is in turn attached to the top of the forward skirt along with the PLF, the entire axial load on top of the forward skirt is the apparent weight of the sum $(m_{PL} + m_{PLF} + m_{PAF})$.

Visual representations of these concepts are provided in Fig. 9.18.

9.1.5 Ground Wind Load Calculation

We are now ready to calculate ground wind loads. This is a very important calculation to make, because winds that are too strong at a launch pad could tip over an LV that was not secured to the pad. That situation might look like that shown in Fig. 9.19. Note that this is not an actual Mercury-Redstone vehicle, but a replica that was blown over by hurricane winds in Florida.

When considering the effect of ground wind loads, we need to consider the entire situation. We may experience both steady and unsteady forces parallel to the wind, as well as unsteady cross forces produced by the shedding of vortices behind the cylindrical shape. The situation is illustrated in Fig. 9.20. In addition, the structural loads will get even larger if the vortex shedding frequency coincides with structural bending frequencies. In other words, if the shedding and structural frequencies coincide, a resonant condition will occur with even larger amplitudes.

The drag forces produced by ground winds generate *bending moments* along the vehicle, just like those we saw in the previous examples. Because

Fig. 9.19 The result of horizontal drag forces overcoming ground restraints. This replica was blown down during hurricane Frances. Source: NASA.

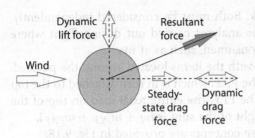

Dynamic lift force

Resultant force

Wind

Steady-state drag force

Dynamic drag force

the rocket may be cantilevered at its bottom (in other words, no other horizontal supports), the maximum bending moment occurs at the bottom of the rocket at its support or mounting points.

9.1.5.1 Ground Winds

Fig. 9.20 The view vertically downward of an LV exposed to steady winds.

We can now calculate the loads on a rocket sitting on its launch pad. This is sometimes called prelaunch loads or ground winds. The winds on the Eastern Test Range (ETR—Cape Canaveral or Kennedy Space Center) are as shown in Table 9.4.

The wind profile in the table may be calculated from the formula

$$v_{ss} = (9.5 \text{ m/s}) h^{0.2} \qquad (9.20)$$

where v_{ss} is in units of m/s, and h is the elevation in meters above the ground. The general formula is

$$\frac{v}{v_{ref}} = \left(\frac{h}{h_{ref}}\right)^{0.2} \qquad (9.21)$$

Remember that winds are a random event. So, statistically speaking, we can never be absolutely sure that a given wind speed won't be exceeded, but we can determine what the chances of a wind speed exceedance might be. If we accept a 99.89% chance that wind would not be exceeded (sometimes referred to as $+3\sigma$), then according to Mackey

Table 9.4 Wind Speed vs Elevation at Eastern Test Range

Height h (m)	Steady-State Wind v_{ss} (m/s)	Peak Wind Speed (m/s)*
3.0	11.8	16.6
9.1	14.8	20.7
18.3	16.9	23.7
30.5	18.8	26.3
61.0	21.6	30.2
91.4	23.4	32.7
121.9	24.7	34.6
152.4	25.9	36.2

*1.4 times steady-state speed. Winds from NASA TN-D-7373
Source: [10].

and Schwartz [10], lateral pressure determined from sum of vortex shedding±max drag load

$$= 1.25 \times v_{ss} + (1.6)^2 \times v_{ss} \text{ for gusts}$$

Thus, the effective wind speed is

$$v_R = [(1.25\,v_{ss})^2 + (2.56\,v_{ss})^2]^{1/2}$$

$$= 2.85\,v_{ss} \tag{9.22}$$

The effective ground wind dynamic pressure p_{ss} is calculated the normal way

$$p_{ss} = q_{ss} = \frac{1}{2}\rho(2.85\,v_{ss})^2 \tag{9.23}$$

For calculations at sea level, it's recommended to use $C_D = 0.77$, and $\rho = 1.2807$ kg/m^3. We see that a factor of almost three times the steady speed is enough to take us to the 99.9% probability region.

Generally speaking, the most critical ground wind load condition is when the LV is exposed to winds when *all its tanks are loaded, but not pressurized.* The nonpressurization means that the allowable stresses in the LV's body are less than they would be when pressurized. Pressurization and its benefits will be discussed in Chapter 10.

9.1.5.2 Simplified Ground Winds Load Calculation Procedure

We now provide the "recipe" for calculating ground wind loads. Referring to Fig. 9.21:

1. a) Determine LV geometry (diameters and lengths of all pieces exposed to wind).
   b) Compute LV cross-sectional areas and centroids using formulae from Table 9.3.
2. a) Determine wind speed v_{ss} vs height using $v_{ss} = (9.5 \text{ m/s})\,h^{0.2}$. The effective speed is $v_R = 2.85\,v_{ss}$.
   b) Calculate dynamic pressure $p_{ss} = 1/2\,\rho v_R^2$ (use $\rho = 1.2807$ kg/m^3).
   c) Calculate the resulting transverse drag load of body i due to wind: $D_i = C_{di}\,p_{ss}\,A_{ref-i}$ (use $C_{di} = 0.77$).
3. Start at the LV nose and numerically sum drag forces to get shear; work down to the base of the LV.
4. Use drag forces acting at centroid heights as moment arms to get moment distribution, again starting at the LV nose and working downward to the LV's base.

Note: This procedure will provide the correct forces and moments at the intersections of the segments n, but *does not provide the variation in between.* To get the *actual* distributed forces and moments, one would have to break the segments down into smaller lengths.

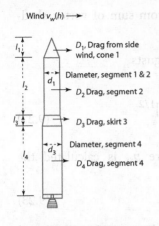

Wind $v_w(h) \rightarrow$

l_1

D_1, Drag from side wind, cone 1

Diameter, segment 1 & 2

d_1

l_2

D_2 Drag, segment 2

l_3

D_3 Drag, skirt 3

Diameter, segment 4

d_3

l_4

D_4 Drag, segment 4

Fig. 9.21 Vehicle broken into segments for analysis.

9.1.5.3 Calculation of Ground Axial Loads

In addition to the ground winds analysis carried out above, we also need to assess the buildup of axial loads while the vehicle is on the launch pad before launching. Every component of the LV structure must be designed to sustain the weight of the structure, fully-loaded propellant tanks, and payload above them, *without the benefit of pressurization*. This is to ensure the safety of the vehicle if a tank has to be emptied while those above are still loaded.

In Chapter 10, we will see that pressurization provides a very significant loads alleviation benefit. It helps to sustain axial loading, and its magnitude also determines the required structural thickness to prevent structural de-stabilization via buckling, which is very important.

The assessment of axial loading follows a process similar to that we showed for shear analysis previously, where the analysis begins at the top of the LV and continues downwards to the vehicle's support location. One can consider the axial load calculation to be simply a "running total" of the weights above selected locations on the LV. It is simply assumed that the weight of a part of the vehicle is sustained by axial forces created by support structure underneath the part, and are reacted out where the support structure connects to the LV body. This idea is shown in Fig. 9.22.

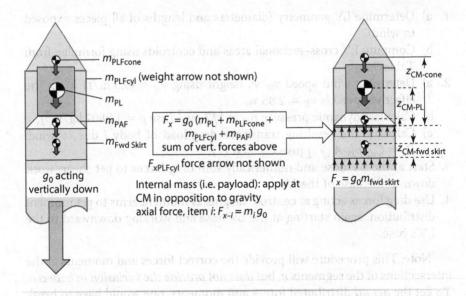

$m_{PLFcone}$

m_{PLFcyl} (weight arrow not shown)

m_{PL}

m_{PAF}

$m_{Fwd\ Skirt}$

g_0 acting vertically down

$$F_x = g_0 (m_{PL} + m_{PLFcone} + m_{PLFcyl} + m_{PAF})$$
sum of vert. forces above

$F_{xPLFcyl}$ force arrow not shown

Internal mass (i.e. payload): apply at CM in opposition to gravity axial force, item i: $F_{x-i} = m_i g_0$

$z_{CM-cone}$

z_{CM-PL}

$z_{CM-fwd\ skirt}$

$F_x = g_0 m_{fwd\ skirt}$

Fig. 9.22 Component weight and axial load forces for LV on ground.

Since the payload is attached to the PAF, which is in turn attached to the top of the forward skirt along with the PLF, the entire axial load on top of the forward skirt is the weight of the sum of the three components, or $F_x = g_0$ $(m_{PL} + m_{PLF} + m_{PAF})$. As the analysis continues towards the LV bottom, the masses of additional components are simply added to the previous running total. Obviously, the axial load at the LV bottom must equal the weight of the LV, or $P_{bottom} = m_0\, g_0 =$ LV weight at $n_x = 1$. We use the traditional symbol for axial loading, a Capital letter "P".

The actual process of calculating axial loading is provided in Section below.

9.1.5.4 Ground Wind Loads Analysis on Saturn V

Now we'll work through the ground winds procedures on a real launch vehicle, the Saturn V moon rocket. In order to do this, we will need its dimensions, which may be found in Fig. 9.23.

There is a great deal of information in Fig. 9.23. The overall height is on the top, and the individual step measurements and tank measurements are provided on the bottom. Fuel and oxidizer propellant tanks are shown fully loaded in dark blue and light blue respectively. Separation planes, where the Saturn steps are separated, are shown in outlined boxes. The diameters of each step, skirt, and payload part are also provided, as are the angles of the tapered skirts. The gimbal locations are also shown; these are needed to determine the torque provided by a gimbaling rocket engine.

9.1.5.5 Calculation of Shear Forces

We will now look at the calculation of shear (horizontal) wind forces, as depicted in Fig. 9.24. We begin at the top.

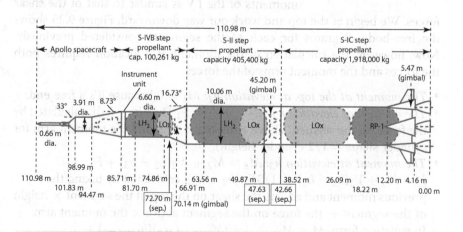

Fig. 9.23 Dimensions in meters and specifications of Saturn V. Source: Drawing by author based on NASA documents.

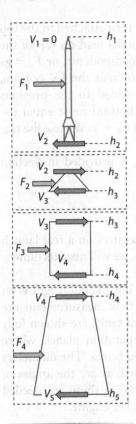

Fig. 9.24 Horizontal drag and internal shear, Saturn V's top.

- *The shear force V_1 at the top of the LV at location h_1:* We know that $V_1 = 0$, because it is a free end.
- *At the bottom of the first segment, at elevation h_2:* Shear $V_2 = F_1$, the force on the segment between h_1 and h_2 above.
- *At the top of the second segment, whose top is at elevation h_2:* There is an equal and opposite shear V_2.
- *At elevation h_3:* Shear $V_3 = V_2 + F_2 = F_1 + F_2$, the previous shear plus the force on the segment between h_2 and h_3 above.
- *At elevation h_4:* Shear $V_4 = V_3 + F_3 = F_1 + F_2 + F_3$, the previous shear + force on the segment above.
- *At elevation h_5:* Shear $V_5 = V_4 + F_4 = F_1 + F_2 + F_3 + F_4$, the previous shear + force on the segment above.

This is a fairly straightforward process, and in a complete analysis, it continues all the way down to the bottom of the LV. It is easy to create a spreadsheet do these calculations, because they can be made for each successive segment of the LV by increasing each term by one.

9.1.5.6 Calculation of Internal Moments

The calculation of the internal bending moments of the LV is similar to that of the shear forces. We begin at the top and work our way downward. Figure 9.25 shows the free-body diagrams for each of the segments considered previously. Now, however, the calculation of the moment at a location requires both the forces and the moment arms of the forces.

- *The moment at the top, at elevation h_1:* $M_1 = 0$ because it's a free end.
- *The moment at elevation h_2:* $M_2 = F_1 l_1 = F_1 [1/2(h_1 - h_2)]$, calculated by the force on the segment above × moment arm (and the moment arm for cylinder shape = 1/2 cylinder height).
- *The moment at elevation h_3:* $M_3 = M_2 + V_2(h_2 - h_3) + F_2 l_2 = M_2 + V_2(h_2 - h_3) + 1/3F_2 (h_2 - h_3)$, calculated by taking the previous moment and adding the shear on the top of the segment × height of the segment + the force on the segment above × the moment arm. In equation form, $M = M_{previous} + (V_{segment\ top})(h_{segment}) + (F_{segment}) \times (shape\ moment\ arm)$. For a triangular shape, the moment arm of the force = 1/3 triangle height.

- *The moment at elevation h_4*: $M_4 = M_3 + V_3(h_3 - h_4) + F_3 \, l_3 = M_3 + V_3(h_3 - h_4) + 1/2 \, F_3 \, (h_3 - h_4)$ is calculated by taking the previous moment and adding the shear on the top of the segment × height of the segment + the force on the segment above × the moment arm.
- *The moment at elevation h_5*: $M_5 = M_4 + V_4(h_4 - h_5) + F_4[l_4(d_4 + 2d_5)/3(d_4 + d_5)]$, calculated from $M = M_{\text{previous}} + (V_{\text{segment top}})(h_{\text{segment}}) + (F_{\text{segment}}) \times$ [trapezoid moment arm].

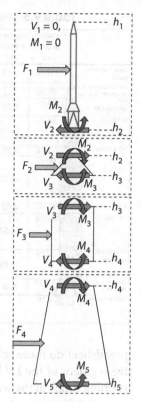

Fig. 9.25 Moments on the top of Saturn V due to horizontal drag forces.

As before, we continue this process to the bottom of the LV. A spreadsheet showing the results of the shear and moment calculation is provided in Table 9.5. Note that we have omitted the effect of the launch escape tower because its forces and moments are negligible.

The shear force and bending moments in Table 9.5 are plotted against the Saturn V outline in Fig. 9.26. As would be expected, both are zero at the top of the rocket, a free end, and both are maximum at the location where the rocket is secured.

9.1.5.7 Calculation of Axial Loads

We will now look at the calculation of axial forces, as depicted in 9.27 below. We begin at the top of the launch vehicle, location h_1. Here, the axial force $P_1 = 0$, because it is a free end.

Moving down to the bottom of the first segment, at elevation h_2, we find that the upwards axial force P_2 must be equal to the weight of the segment above, $W_1 = m_1g_0$. Now, at the top of the segment just underneath this one, we find that there is an equal and opposite force P_2 acting downwards.

Stepping down to elevation h_3: we find that the upwards axial force P_3 must be equal to the weight of the segments above, $P_3 = W_1 + W_2$.

Now, at the top of the segment just underneath this one, we find that there is an equal and opposite force P_3 acting downwards. Including the weight of segment 3, W_3, we find that the upwards axial force $P_4 = P_3 +$ the weight of segment 3, or $P_4 = P_3 + W_3$, or $P_4 = W_1 + W_2 + W_3$.

Moving down to segment 4, a downwards-acting P_4 is joined by the weight of the segment W_4 to be balanced by upwards-facing $P_5 = P_4 + W_4 = W_1 + W_2 + W_3 + W_4$.

As was shown in the shear calculation process, the axial loads calculation is a fairly straightforward process, and in a complete analysis, it continues all the way down to the bottom of the LV. It is easy to augment an existing

Table 9.5 Ground Shear and Moment Calculations for Saturn V LV

	Elevation @ top (m)	Segment length (m)	Segment diameter @ top (m)	Side area (m²)	Wind speed v (m/s)	Pressure (Pa), 2.85 v	Force (kN), $C_D = 0.77$	Shear force (kN)	Moment (kNm)
Launch escape tower	110.98	9.15	0.66	6.0	24.2	3,036	14.1	0	0
Apollo capsule top	101.83	2.84	0.00	5.6	23.9	2,967	12.7	14.1	64.6
Sevice module top	98.99	4.52	3.91	17.7	23.7	2,923	39.8	26.8	116.7
LM Adapter top	94.47	8.76	3.91	46.0	23.4	2,841	100.7	66.6	327.7
S-IV top IU	85.71	13.01	6.60	85.9	22.8	2,698	178.4	167.3	1,390
S-II skirt top	72.70	5.79	6.60	48.2	22.2	2,565	95.3	345.7	4,726
S-II top	66.91	19.28	10.06	194.0	21.3	2,370	353.9	440.9	7,298
S-II bottom	47.63	4.97	10.06	50.0	20.4	2,155	83.0	794.9	19,211
S-IC top	42.66	16.57	10.06	166.7	19.3	1,932	248.0	877.8	23,367
Bottom S-IC LOx tank	26.09	13.89	10.06	139.7	17.1	1,529	164.5	1,125.8	39,967
Bottom S-IC RP-1 tank	12.20	8.04	12.49	107.8	14.5	1,088	90.3	1,290.3	56,747
Bottom fairing	4.16	Includes engine fairing effective side area (bot.) 8.555 m²	Incl. two fin areas, each 8.555 m²	From $v_y = (9.5\ m/s)h^{0.2}$, $h =$ average ht	$q = \rho(K*v)^2/2$	$F = C_D*A*q$	1,380.6	67,860

spreadsheet do these calculations, because they can be made for each successive segment of the LV by increasing each term by one. The analysis is just a "running total" of the weights that are above the top of each segment, and the base axial force is just the weight of the LV.

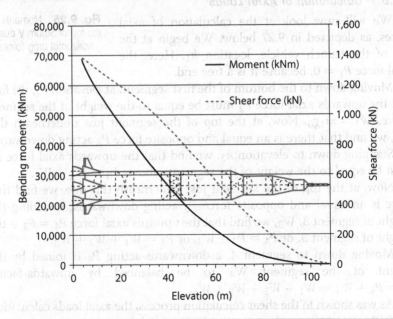

Fig. 9.26 Saturn V ground wind shear and moment distribution.

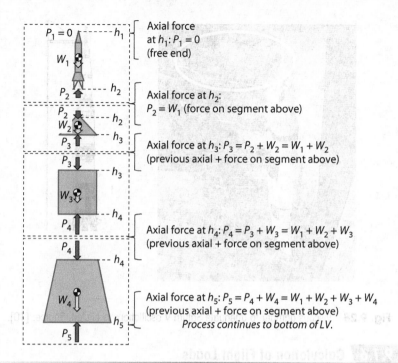

Fig. 9.27   Vertical weight forces and internal axial loads, top of Saturn V.

9.1.5.8 Summary of Ground Loads Calculation

The result of a successful ground loads calculation should be as follows:

- At the top of the vehicle, the shear, moment, and axial forces are all zero
- At the vehicle's ground support (usually bottom of engine section or aft skirt), the shear, moment, axial loads are maximum. The axial load should equal the LV's one-g_0 weight: $P_{bottom} = m_0 \, g_0 = $ LV weight at $n_x = 1$.
- DO NOT support the LV by placing onto the bottom of the engine bell!

9.1.5.9 Notes on Ground Load Calculations

During the Saturn V design process, it was determined that wind loads were underestimated, so NASA's wind criteria were changed. More current wind data may be found in NASA TM-X-53872 [6].

Saturn V was predicted to oscillate in certain wind conditions. To reduce predicted ground loads, Saturn V was connected to a *vibration damper* while on the pad. The damper was removed 4 hours before launch. The damper is shown in Fig. 9.28.

Fig. 9.28 Viscous damper to attenuate Saturn V oscillations is circled. Source: [10].

9.1.6 Calculation of Flight Loads

Now that we've tackled the loads that the LV may experience on the ground before launch, it's time to look at the most critical of loads: the flight loads. These loads come from a variety of sources (see Fig. 9.29):

* Thrust forces producing acceleration loads
* Aerodynamic loads, lateral and axial, proportional to q = dynamic pressure = $1/2\rho v^2$
* Control forces and resulting torques due to engine(s) gimbaling or thrust forces
* Winds (produce angle of attack α that, in turn, creates normal forces and bending moments)
* Vibrations, propellant sloshing within tanks, flexible body vibration
* Thermal loads (not shown)
* Acoustic loads (not shown)

In all these loads, the *design drivers*—the factors that dominate design procedures—are usually:

* Max-q: Maximum dynamic pressure
* Max-$q\alpha$: Maximum (angle of attack × dynamic pressure)
* Maximum acceleration (usually just before burnout)
* Main engine cutoff (MECO)

A photograph of a poster used during the Saturn V S-IVB upper step's design review is shown in Fig. 9.30. Note that the majority of the critical conditions shown are max-$q\alpha$ and engine cutoffs.

We will see that during launch, the LV is compressed axially (by the engine thrust on the bottom and the inertial loads pushing from the top down) and also sees bending loads. This is the most severe part of the life of the LV.

The convention, at least for initial design work, is to assume that the slowly changing loads during launch can be considered static (i.e., g-loading, quasi-steady, quasi-static accelerations). In other words, we can take a snapshot of the vehicle and its loads, and assume they are steady for that phase of flight. The load factors are represented by *quasi-static loads (QSLs)*, the most severe combination of dynamic and steady state acceleration. And, we shall see that both *axial* (compressive) loads and *lateral* loads must be determined (see Fig. 9.31).

One of the driving factors for the design of LVs is horizontal wind or wind gusts, as

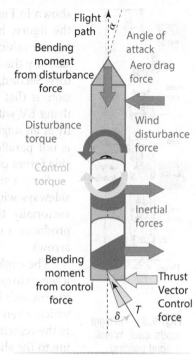

Fig. 9.29 Flight loads. Source: [19].

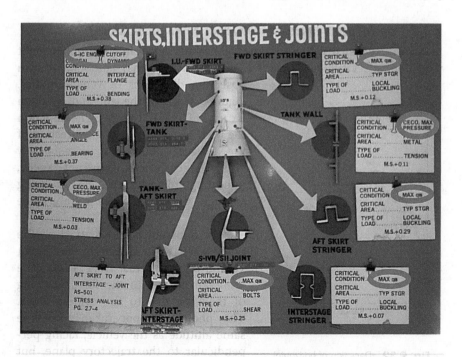

Fig. 9.30 Photo of poster used during the Saturn V's S-IVB design review. Source: author.

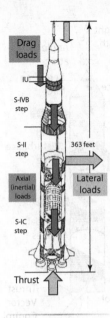

Fig. 9.31 In-flight loads: axial, lateral, and bending.

shown in Fig. 9.32. Horizontal wind gusts, as shown in the figure, have two major effects on the flight of a launch vehicle: first, the LV tends to get "blown to the side" by the winds, which can affect the desired trajectory. Second, and more important to the current discussion, is that the encounter of a more-or-less vertically flying LV with a horizontal wind gust causes the LV to fly at an *angle of attack*, meaning that the relative wind is not parallel to the long axis of the LV and induces side forces on the vehicle.

Such a situation is shown in Fig. 9.33. Note that the sideways wind gust (downward-pointing arrow) added vectorially to the vehicle velocity (horizontal arrow) produces a resultant labeled *relative velocity* (diagonal arrow).

The angle of attack is denoted by Greek letter α, and it serves to generate a varying pressure distribution along the rocket's length (hump-shapes on top of centerline), which when integrated, produce normal (perpendicular to the velocity vector) forces. The normal forces contribute to the shear and bending that occur, and also determine the stability when passing through a gust.

9.1.6.1 Do Launch Vehicles Really Fly at an Angle of Attack?

One can get an insight into some flight events by carefully examining flight video footage. Visual assessment of vehicle attitude may be obtained from observing directions of vehicle bodies. Figure 9.34 shows the launch of a Falcon 9 carrying SES-10 to orbit. The direction the body is pointing and the direction of the exhaust plume are both clearly visible. It is clear that they are not parallel, indicating that the vehicle is flying through the atmosphere at an angle of attack.

The apparent angle of attack in the photo can be measured as about 16 deg, but the actual AoA is less. The reduction is due to the foreshortening of the view due to the camera angle. The photo was not taken from the same altitude as the vehicle, facing perpendicular to the trajectory plane, but instead was taken from below and

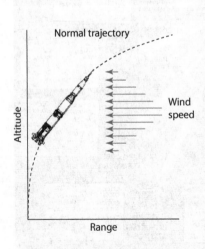

Fig. 9.32 Depiction of horizontal wind gust.

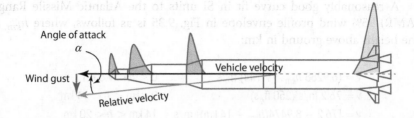

Fig. 9.33 Wind gust induces angle of attack α that produces pressure distribution along body, producing normal forces that, in turn, produce shear and bending loads.

behind the F9. Nevertheless, this indicates a nonzero angle of attack, which would produce significant loading on the vehicle if the dynamic pressure were at a large enough value. In addition, the nonparallel velocity vector and thrust line results in a nonzero steering loss.

A simulation of this launch is available at Flight Club (https://www2. flightclub.io/build/editor/flightprofile). The time history of the vehicle's angle of attack is shown on the right of Fig. 9.34a. At the time of flight shown (1 min, 54 s), the simulation indicates that the vehicle is at an angle of attack of 3.1 deg.

9.1.6.2 Calculation of Angle of Attack

The wind speeds during the entire ascent must be known in order to calculate the angle of attack and the normal forces shown in the figure. Therefore, we need to have a winds-aloft profile, just as we did for the ground wind loads. This is shown in Fig. 9.35.

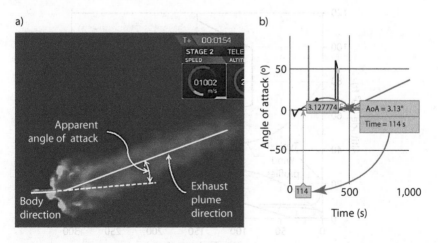

Fig. 9.34 a) Still photo from SpaceX Falcon 9 launch shows the velocity vector direction (indicated by the exhaust plume) is not parallel to the thrust vector direction (assumed to be close to the body direction). Source: SpaceX launch video. b) Simulation shows a 3.1-deg AoA at this time of flight. Source: Flight Club.

A reasonably good curve fit in SI units to the Atlantic Missile Range (AMR) 95% wind profile envelope in Fig. 9.35 is as follows, where h_{km} is the height above ground in km:

$v = (6.9288\ h_{km} + 9.144)$ m/s	$0 < h < 9.6$ km
$v = 76.2$ m/s (250 ft/s)	9.6 km $< h < 14$ km
$v = [76.2 - 8.9474(h_{km} - 14$ km)] m/s	14 km $< h < 20$ km
$v = 24.384$ m/s (80 ft/s)	$h > 20$ km

This set of equations can be of use in trajectory calculations with varying altitudes.

The large wind peaks of ~275 ft/s (~84 m/s) occur between 17.6 and 35 km (25,000 and 50,000 ft) altitude, and these are substantial wind speeds. Coincidentally, this altitude range is also the region where max-q often occurs. In general, we have two design-driving load cases due to horizontal winds:

1. The case where we use the angle of attack when passing through max-q
2. The case of max-$q\alpha$, when the greatest value of the *product* of angle of attack α and q occurs

Until a trajectory analysis is done with the appropriate values inserted, it is not possible to say whether max-q or max-$q\alpha$ is more strenuous on the design. Calculations *should be repeated for max-qα*, because forces $F = C_F q\alpha = C_F\ 1/2\ \rho\ v^2\ \alpha \sim \rho v^2 \alpha$. During ascent, ρ, v, and α are all

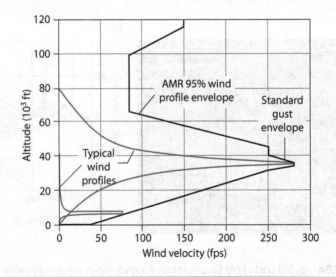

Fig. 9.35 Winds aloft profile showing some typical wind profiles along with a 95% probability envelope. Source: [9].

varying, so their product is also varying and could produce larger forces than the product at max-q.

9.1.6.3 Shear, Bending, and Axial Loads in Flight Due to Wind Shear

The following steps provide the procedure to be used to analyze flight loads at max-q and max-$q\alpha$:

1. *Find max-q information.* Use the launch trajectory to determine max-q. Obtain the following quantities at max-q: time t_{max-q}, altitude h_{max-q}, vehicle acceleration $n_{x\,max-q}$ (in g_0), and speed v_{max-q} from the launch trajectory. Use the vehicle's altitude h_{max-q} to determine atmosphere density ρ_{max-q}, sound speed a_{max-q}, and Mach number M_{max-q}.
2. *Determine partially consumed propellant masses.* Use time after liftoff and known propellant mass flow rates (determined by engine thrust and effective exhaust speed c) to determine the mass of propellant remaining in each tank. The masses will be needed to calculate loads and CM location.
3. *Calculate vehicle center of mass.* Calculate the vehicle's max-q CM location following the procedures given in Chapter 8.
4. *Determine horizontal gust speed; calculate instantaneous angle of attack.* Use the launch site's winds-aloft profile (Fig. 9.35) to determine wind gust speed $v_{w\,max-q}$. Use the gust speed to calculate instantaneous angle of attack: $\alpha_{max-q} = \tan^{-1}(v_{w\,max-q}/v_{max-q})$.
5. *Determine aerodynamic load coefficients and calculate normal and axial loads.* Use Figures 9.38 and 9.39 to find values of C_{N_α} for tapering vehicle segments; use these values to calculate normal loads on each segment. Normal (perpendicular) force $N = (C_{N_\alpha} \cdot \alpha)\, q_{max-q}\, A_{ref}$ for each segment. Note that this reference area may be different from the vehicle's reference area. Axial (drag) force $D = C_d\, q_{max-q}\, A_{ref}$, where $C_d = 0.7$.
6. *Calculate side thrust to trim vehicle.* Use the vehicle geometry (distance between gimbal block and center of mass) to determine the side thrust needed to "trim" the vehicle for moment equilibrium. Calculate the lateral acceleration n_z by summing the normal aerodynamic forces and the side thrust.
7. *Calculate inertia relief forces.* Using the lateral acceleration n_z, calculate lateral *inertia relief* forces ($=m\,n_z$) for every vehicle mass. Lateral forces act at mass centers of solid objects and centroids of partial propellant loads. The centroid of fully loaded tanks can be assumed to be at the tank centroid.
8. *Calculate shear loading.* Starting at the nose and marching downwards, successively add the aerodynamic forces and subtract the *inertia relief* forces for each mass to determine the vehicle's shear loading at each station, similar to the ground winds procedures earlier.

9. *Calculate bending moment distribution.* Start from the nose and use vertical layout information to calculate moment arms and the bending moments resulting from the aerodynamic forces and inertia relief loads, as was done for ground winds. (Ground winds do not have the inertia relief loads, only the aero loads.)

10. *Calculate axial apparent weights.* Using axial acceleration n_x, calculate the *apparent weights* ($W = m\, g_0\, n_x$) of all of the components of the vehicle. Apparent weights act at the lower junction of the weight's support structure (i.e., lower tank bulkhead) and the outer shell of the vehicle.

11. *Calculate axial loading.* Start at the nose and work downward, calculating the axial loading resulting from the apparent weights, similar to the ground winds process.

12. *Repeat analysis for the case of max-qα.* Repeat steps 1–11 at max-qα. It will be necessary to calculate the product of q and α to find its maximum. (The tabular form of the horizontal winds will be helpful.) Determine the worst case of the two.

Earlier we indicated that the worst load case might be max-q or max-$q\alpha$. To determine which, we consider the ascent of a generic LV as shown in Fig. 9.36. The plot shows that at max-$q\alpha$, $\alpha_{max\text{-}q\alpha} \approx 7.7$ deg, $q_{max\text{-}q\alpha} \approx 900$ psf, and $(q\alpha)_{max} \approx 6{,}930$ lbf deg/ft^2, whereas at max-q, $\alpha_{max\text{-}q} \approx 7.25$ deg, $q_{max\text{-}q} \approx 920$ psf, and $(q\alpha)_{max\text{-}q} \approx 6{,}600$ lbf deg/ft^2. Because the aerodynamic forces depend on the product ($\rho \cdot v^2 \cdot \alpha$), it

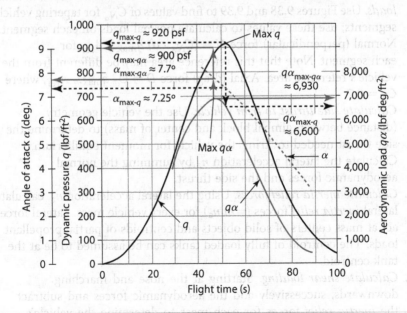

Fig. 9.36 The differences between max-q and max-$q\alpha$. The aerodynamic loads depend on the product $q\alpha$.

Table 9.6 Generic Launch Vehicle Loads
for Max-q and Max-$q\alpha$

Parameter	Max-q	Max-$q\alpha$
Time (s)	~48	~52
q (psf)	**920**	900
α (deg)	7.25	**7.7**
$q\alpha$ (lb$_f$-deg/ft^2)	6,600	**6,930**

appears that the max-$q\alpha$ case is the more strenuous one for this generic vehicle. Load calculations *must be repeated for max-$q\alpha$*, because it's not evident whether max-q or max-$q\alpha$ is more critical—the analyst must check *both*. The generic data are summarized in Table 9.6.

A rule of thumb: during launch, Falangas (in a personal conversation with the author) suggested that the product $q\alpha$ should be less than approximately 168 kPa-deg (3,500 psf-deg). This suggests that the generic vehicle described has too large a value of $q\alpha$ and that the max-$q\alpha$ loads might be too large.

9.1.6.4 Obtaining Aerodynamic Force Data

There are four different methods to get aerodynamic force data: (1) use aerodynamic charts to determine force coefficients, (2) computational fluid dynamics (CFD), (3) wind tunnel tests, or (4) flight data to get pressure distributions. Because we do not have either a math model, a wind tunnel model, or flight test data, we will use method 1. The other three methods are shown in Fig. 9.37.

Figures 9.38, 9.39, and 9.40 show normal force slope coefficients (C_{N_α}) for nose cones and conical flares. These are used to estimate normal force and drag force coefficients for the noncylindrical portions of a vehicle, using

Modeling and simulation   Wind tunnel test   Flight test

Fig. 9.37 Besides hand calculations, there are three ways to obtain aerodynamic force data: CFD, wind tunnel testing, or data from an actual flight test. Source: [5], Boeing, NASA.

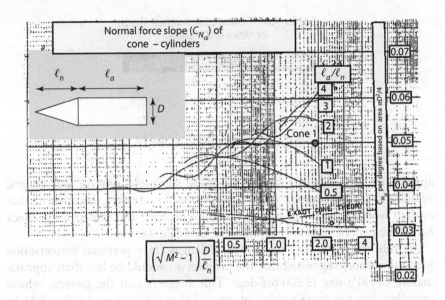

Fig. 9.38 Normal force slope (C_{N_α}) of cone cylinders, per degree, based on reference area $\pi D^2/4$. The labeled dot refers to the nose cone of the Saturn V launch vehicle analyzed later in this chapter. Note that this coefficient's reference area may be different than the vehicle's reference area if the vehicle has multiple body diameters. Source: [8].

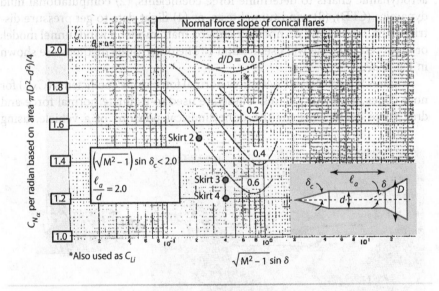

Fig. 9.39 Normal force slope C_{N_α} of conical flares, per radian, based on reference area $\pi(D^2 - d^2)/4$. D and d are the diameters of the cylinder and conical flare as shown. The labeled dots refer to skirts on the Saturn V launch vehicle example analyzed later in the chapter. Note that this coefficient's reference area may be different than the vehicle's reference area if the vehicle has multiple body diameters. Also note that the angles δ and δ_c given here should not be confused with the thrust vector control angle δ used later. Source: [8].

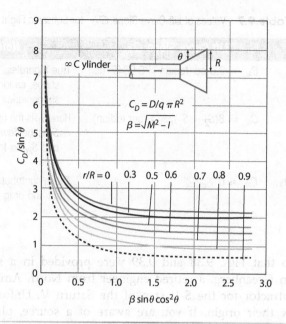

Fig. 9.40 Drag coefficient for conical flares. Source: [20].

the procedure $C_N = C_{N_\alpha} \times \alpha$. C_N is then used in the normal way. It's assumed that the normal force and axial forces of uniform cylindrical portions of the LV are negligible, to first order. Normal forces are usually called *lift* in the airplane world.

In order to use these charts, you will need to get the geometry of your design and then locate the appropriate "point" on the charts. "Point" is in quotes because there's a good chance your particular design won't hit one of the plotted lines, so you must interpolate. You may have noticed dots on Figs. 9.38 and 9.39. These dots represent different cones or skirts on the Saturn V and don't happen to fall on the plotted lines either.

The result you get will be a dimensionless coefficient, C_{N_α}. To obtain the actual normal force itself, you multiply the coefficient by the angle of attack (in *degrees* or *radians* as defined on the plot) due to the crosswind, then multiply by the "based on" area specified on the chart, and finally by q. The result is the actual side force. Be sure you use the correct reference area and radians in your calculations. In addition to the normal force or lift on a segment, one also has to calculate the *drag* of the segment. This is done by using the (constant) value of $C_{di} = 0.7$ in the drag force equation $D_i = C_{di} \, q_{max-q} \, A_{ref-i}$ for the segment. Again, be sure to use the proper reference area for your calculations.

Note that we assume the cylindrical sections of the launch vehicle to have negligible C_{N_α}. This means we only have to worry about places on the LV where the diameter changes.

Table 9.7   Values of Lift Curve Slope C_{N_α} for Subsonic Flight

Vehicle Part	Lift Curve Slope C_{N_α}	Notes
Nose cones	$C_{N_\alpha} = 2.0$ per radian $= 0.0349/\text{deg}$	True regardless of the cone's shape, as long as the shape varies smoothly.
Skirts	$C_{N_\alpha} = 8(S_2 - S_1)/S_{\text{Ref}}$ (per radian)	Here S_1 is the upstream area, S_2 is the downstream area, and S_{Ref} is the reference area.
Cylindrical parts	$C_{N_\alpha} \approx 0$	These contribute negligible lift and drag components.

Source: [2].

Note also that Figs. 9.38 and 9.39 were provided in a short course given by Stan Greenberg, a retired engineer from North American Aviation, the contractor for the S-II step of the Saturn V. Unfortunately, he doesn't know their origin. If you are aware of a source, please contact the author.

These charts are only suitable for supersonic flight. In some instances, the vehicle may be subsonic, with $M < 1.0$. For the subsonic case, use the relations in Table 9.7 to calculate C_{N_α}.

9.1.6.5 Launch Vehicle Pressure Coefficients

The graphs and coefficients you have seen thus far are the result of surface pressure values integrated over the body of the vehicle. Pressure is usually normalized into a *pressure coefficient* C_p: $C_p = (p - p_\infty)/q$. The result of this definition is that a negative C_p represents a local pressure that is below the ambient pressure p_∞, and a positive C_p represents a local pressure that is above ambient.

In Fig. 9.41, it may be seen that large pressure swings occur near the ends of conical sections and diameter changes. In the figure, pressure peaks represent concentrated local pressure loads, and negative peaks occur when flow expands. In any case, Mother Nature integrates the local pressure distribution in order to create the normal force and parallel force (drag) values. To locate the body's center of pressure (CP), integrate along the body length.

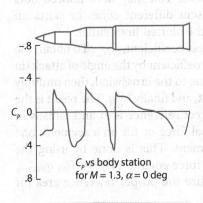

Fig. 9.41   Variation of pressure coefficient along length of launch vehicle. Source: [4].

Note: The two abbreviations CP and C_p are not the same! Don't confuse *center of pressure* (CP) with *pressure coefficient* (C_p).

Figure 9.42 provides the *measured normal force distribution* for a Delta II-6925. This is actually the distributed pressure load at each station, integrated around the circumference. If the plotted data were integrated along the length of the vehicle, the result would be the body's normal force. Notice that the large excursions occur near diameter changes: the nose area (stations 200 to 500) and the area at the front of the solid rocket motors (around station 1,300).

9.1.6.6 Calculation of Aerodynamic Forces and Moments on Vehicle

Our main concerns are the normal or side forces that develop when the vehicle is flying at a nonzero angle of attack, which occurs when the vehicle hits a wind gust. If we know the gust and flight speeds, we can determine the angle of attack, which we can use along with the vehicle's geometry to generate aero forces. At the same time, we can calculate the current propellant quantities and masses in order to get the rocket's internal loads. As we did with ground winds, these loads can be integrated to obtain the shear, axial, and moment distributions. These distributions may be used to determine tank thicknesses and other critical parameters.

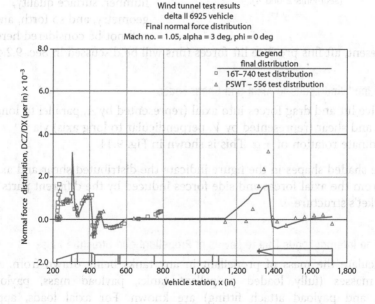

Fig. 9.42   Normal force distribution for a Delta II-6925. Source: McDonnell Douglas.

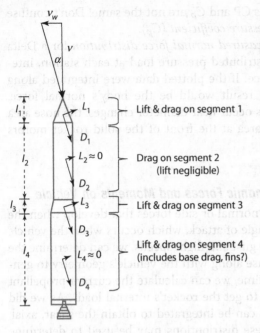

Now we'll demonstrate these steps with a simple LV model, as shown in Fig. 9.43. This model has four segments: conical nose, upper cylinder, skirt, and lower cylinder.

Determine the Air Loads in the Wind Axes

- Calculate lift forces for each segment k ($k = 1, 2, 3, 4$).

$$L_k = C_{Lk} q A_k = C_{N\alpha k} \alpha q A_k$$

Look up $C_{N\alpha}$ on charts.

$$D_k = C_{dk} q A_k \text{ with } C_{dk} = 0.7$$

- Assume constant-diameter cylinders (segments 2 and 4) have negligible lift and drag.
- As an option, we could include skin friction drag on all surfaces. Skin friction depends on the Mach number, surface quality, geometry, and so forth, and will not be considered here.

Fig. 9.43 Lift and drag forces on rocket. Generally, the drag forces on all sections can be neglected, as can the lift forces on the cylinders (segments 2 and 4).

- If present, aft fins produce lift forces (fins will be discussed in Sec. 9.2.4).

Convert the Wind Loads to Axial and Shear Loads

- Resolve lift and drag forces into axial (represented by A, parallel to long axis) and shear (represented by V, perpendicular to long axis) via coordinate rotation of $-\alpha$. This is shown in Fig. 9.44.

The shaded shapes in the figure indicate the distributed shear and axial loads from the axial forces and side forces induced by the different parts of the rocket's structure.

Determine Internal Loads Due to Levels of Propellant and Structure Mass

Calculate the mass of propellant in any tanks being drawn from. All other masses (fully loaded propellant tanks, payload mass, payload fairing, and payload attach fitting) are known. For axial loads, apply the propellant apparent weights ($n_x m g_0$) as concentrated loads at the joint of the aft bulkhead and shell. This procedure is shown pictorially in Fig. 9.45.

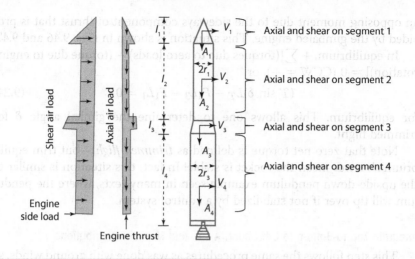

Fig. 9.44 The shear and axial air loads distributed along the length of an LV may be integrated to produce discrete axial and shear forces for hand analysis.

Determine Trimmed Flight Forces

Remember that flying at an angle of attack produces lift forces. In the case of an LV, the lift force is usually in front of the mass center, and therefore is trying to make the vehicle turn. For equilibrium flight, meaning no LV rotation, we gimbal the engine an angle δ. This offset thrust line produces

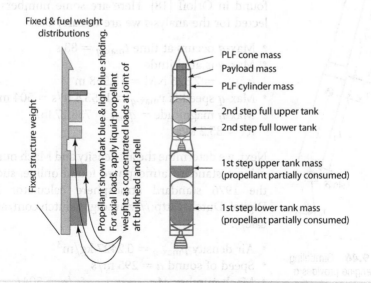

Fig. 9.45 Determination of internal loads due to propellant and structure mass.

an opposing moment due to the sideways component of thrust that is provided by the gimbaled engine. This situation is shown in Figs. 9.46 and 9.47.

In equilibrium, $+\sum[(\text{torques due to aero loads}) + (\text{torque due to engine rotation})] = 0$ (CCW = +), or

$$(T\sin\delta)L_T - F_3L_3 - F_1L_1 = 0 \qquad (9.24)$$

for equilibrium. This allows one to determine the gimbal angle δ for trimmed flight.

Note that zero net torque is defined as *trimmed flight*—but trim equilibrium *does not* mean the rocket is stable! In fact, this situation is similar to the upside-down pendulum example seen in many texts, where the pendulum will tip over if not stabilized by a control system.

Integrate Top-to-Bottom to Get Shear, Axial, and Moment Distributions

This step follows the same procedures as was done with ground winds, so we won't repeat it here.

We will now carry out this procedure on a *very famous* rocket flight!

9.2 Example: Max-*q* Air Load Calculation for Saturn V/Apollo 11 (SA-506)

As an example, to demonstrate the details of the previous load calculation, we will carry out loads analysis on a Saturn V vehicle. To do this,

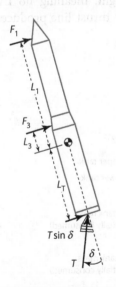

Fig. 9.46 Gimbaling the engine provides a torque to counter the destabilizing aero torques.

we need to collect data at max-*q* for the vehicle. Fortunately, NASA tracked many variables during the Saturn-Apollo program. Many of these data may be found in Orloff [18]. Here are some numbers collected for the analysis we are to carry out:

- Max-*q* occurs at time $t_{\text{max-}q} = 83$ s
- Max-*q* is at altitude $h_{\text{max-}q} = 7.326$ NM $= 13{,}568$ m
- Max-*q* speed is $v_{\text{max-}q} = 1653.4$ ft/s $= 504$ m/s
- Max-*q* magnitude $= q_{\text{max}} = 735.17$ lb/ft^2 $= 35.2$ kPa

Next, we determine the air density and Mach number using a standard atmosphere found online, such as the 1976 standard atmosphere calculator from Digital Dutch (http://www.digitaldutch.com/atmoscalc/):

- Air density $\rho_{\text{max-}q} = 0.2427$ kg/m^3
- Speed of sound $a = 295$ m/s
- Mach number $M_{\text{max-}q} = v_{\text{max-}q}/a = 504/295 = 1.71$

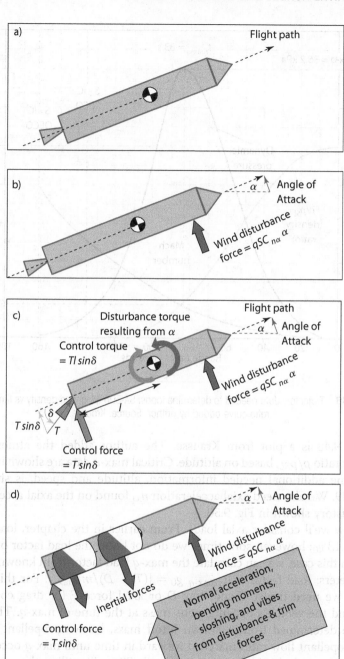

Fig. 9.47 Events during LV flight through a wind disturbance. a) LV in undisturbed air. b) LV encounters wind disturbance, creating an angle of attack α and disturbance torque (pitching upwards). c) LV's control system senses deviation and commands thrust vector control to compensate with opposing moment. d) All forces combine to create normal acceleration, slosh, and vibration. Source: [19].

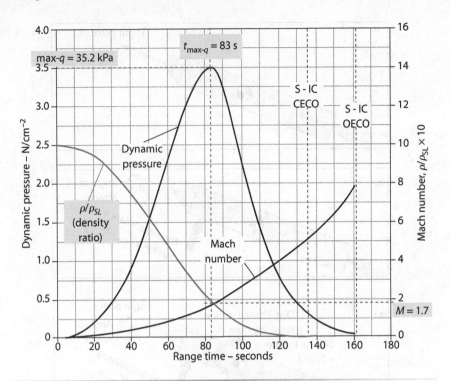

Fig. 9.48 Trajectory data needed to determine loads: SA-506 *M*, *q*, and density vs time. Density ratio curve added by author. Source: Krausse.

Figure 9.48 is a plot from Krausse. The author added the atmosphere's density ratio ρ/ρ_{SL}, based on altitude. Critical max-q data are shown in boxes.

Some additional needed information, altitude and speed, is shown in Fig. 9.49. We also need axial acceleration n_x, found on the axial acceleration time history shown in Fig. 9.50.

Now we'll consider axial loads. From earlier in the chapter, load factor $n_x \approx 2.13\ g_0$; however, sometimes we do not know the load factor or trajectory. In this case, we can calculate the max-q load factor from known vehicle parameters: load factor $= n_{x\,max\text{-}q}\ g_0 = [(T - D)/m]_{max\text{-}q}$. For this calculation, we need thrust T and drag D or axial force F (or drag coefficient C_D), and the vehicle's instantaneous mass at the time of max-q. The latter can be determined from a known liftoff mass, known propellant masses, and propellant flow rates marched forward in time until max-q occurs.

Next, we deal with wind gust speed. We will utilize the worst-case wind gust encountered by *any* Saturn V. As it happens, during max-q on Apollo 9, a wind speed of 250 ft/s = 76.2 m/s was encountered. And note: Apollo 9 is not Apollo 11! So that wind speed happens to be about the maximum value shown on the jet stream chart earlier (Fig. 9.35). The angle of attack that the 76 m/s wind induced was $\alpha = \tan^{-1}(v_w/v) = \tan^{-1}(76.2/504) = 8.6$ deg. We will use this AoA later to calculate aerodynamic loads.

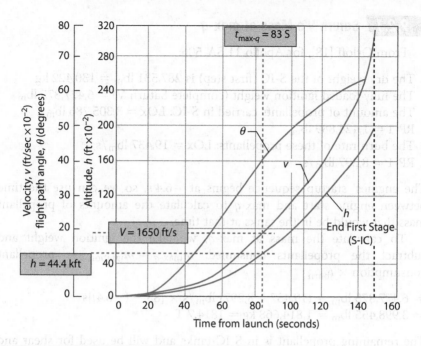

Fig. 9.49 SA-506 trajectory data needed to determine loads: altitude and speed. Source: NASA.

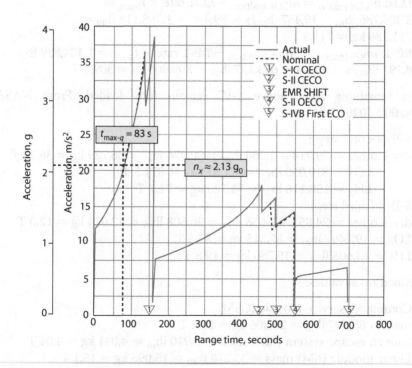

Fig. 9.50 Axial acceleration of SA-506. Source: [12].

9.2.1 Saturn V's Mass at max-q

From Orloff [18], for Apollo 11 SA-506:

- The dry weight of the S-IC (first step) is 287,531 lb_m = 130,422 kg
- The fully loaded ignition weight (complete Saturn V) = 6,477,875 lb_m
- The amount of propellants carried in S-IC: LOx = 3,305,786 lb_m, RP-1 = 1,424,889 lb_m
- The burn rate of these propellants: LOx = 19,437 lb_m/s, RP-1 = 8,297 lb_m/s.

The engines' startup sequence begins at −6.4 s, so we can use the time between engine start and max-q to calculate the amounts of propellant mass that would be in the tanks at that time.

To calculate the mass at max-q, we take the ignition weight and subtract the propellants consumed: $m_{max-q} = m_{ignition}$ − propellant consumption × t_{burn}

$$= 6,477,875 \text{ } lb_m - (19,437 + 8,297) \text{ } lb_m/s \times [83 - (-6.4)]s$$
$$= 3,998,455 \text{ } lb_m = 1,814,668 \text{ kg} = 1814.7 \text{ T}$$

The remaining propellant is in S-IC tanks and will be used for shear and moment calculations).

- LOx: $m_{LOx \text{ } max-q} = m_{LOx \text{ ignition}}$ − LOx rate × t_{burn} = 3,305,786 lb_m − 19,437 lb_m/s × 89.4 s = 1,568,118 lb_m = 711,286 kg = 711.3 T
- RP-1: $m_{RP-1 max-q} = m_{RP-1 ignition}$ − RP-1 rate × t_{burn} = 1,424,889 lb_m − 8,297 lb_m/s × 89.4 s = 683,137 lb_m = 309,866 kg = 309.9 T

The remaining steps above S-IC remain fully loaded. From NASA SP-2000-4029 [18]:

- S-II (second step):
  dry + other = 79,714 + 1,260 lb_m = 80,974 lb_m = 36,729 kg = 36.73 T
    LOx = 819,050 lb_m = 371,515 kg = 371.5 T
    LH$_2$ = 158,116 lb_m = 71,720 kg = 71.7 T
- S-IVB (third step):
  dry + other = 24,852 + 1,656 lb_m = 26,508 lb_m = 12,024 kg = 12.0 T
- LOx = 192,497 lb_m = 87,315 kg = 87.3 T
  LH$_2$ = 43,608 lb_m = 19,780 kg = 19.8 T

Payload mass values:

- Command/service module (CSM)
  mass = 63,507 lb_m = 28,806 kg = 28.8 T
- Launch escape system (LES) mass = 8,910 lb_m = 4,041 kg = 4.04 T
- Lunar module (LM) mass = 33,278 lb_m = 15,095 kg = 15.1 T
- Spacecraft/lunar module (SLM) adapter = 3,951 lb_m = 1,792 kg = 1.79 T
  (add with LM mass to top of S-IVB)

Note: A common assumption is that the LV structural mass is much less than the propellant mass, so it may be neglected for first analysis. Because we have the numerical data on Saturn structures, we will use them.

9.2.2 Acceleration Magnitude During max-q

The sum of the forces $= m_{max-q} \times$ acceleration, so the acceleration load is

$$n_{x\text{-}max\text{-}q}\, g_0 = [(T - D)/m]_{max\text{-}q}$$

To do this calculation, we need the thrust T and drag D or axial force F (or drag coefficient C_D) at max-q. There is an axial force plot available from Marshall Space Flight Center (MSFC) [11]. Note that SA-503 is not the Apollo 11 Saturn V, but we use it to estimate the drag coefficient *only*. As shown in Fig. 9.51, the axial force $F_{max\text{-}q} = 260$ klb$_f$ = 1,156.5 kN. Using $q_{max} = 750$ lb$_f$/ft^2 = 35.91 kPa, and $S_{ref} = 122.5$ m^2, we can calculate the drag coefficient

$$C_D = F_{max\text{-}q}/(q_{max}S_{ref}) = 0.263$$

We will assume C_D for Apollo 11's max-q conditions is the same, even though they were different missions.

From the same reference, we have a thrust vs time plot that we can use for the acceleration calculation, as shown in Fig. 9.52.

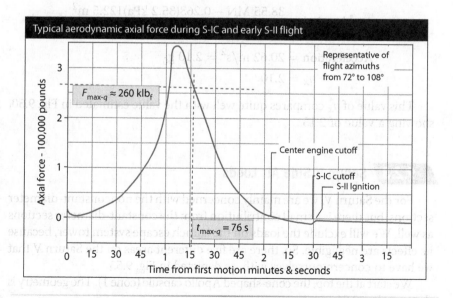

Fig. 9.51 Aerodynamic axial force during S-IC and early S-II flight. Source: [11].

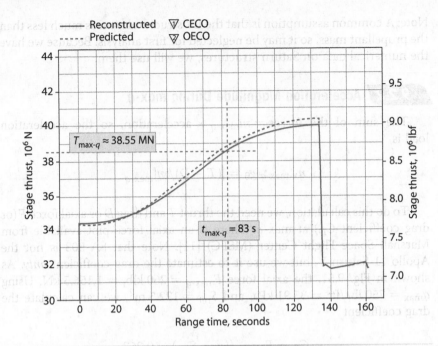

Fig. 9.52 Apollo 11 thrust vs time curve. Source: NASA.

Finally, we can calculate the load factor at max-q:

$$\text{Acceleration} = \frac{T - D}{m} = \frac{T - C_D q S_{ref}}{m}$$

$$\text{Acceleration} = \frac{38.55 \text{ MN} - 0.263(35.2 \text{ kPa})122.5 \text{ m}^2}{18,14,668 \text{ kg}}$$

$$\text{Acceleration} = 20.62 \text{ m/s}^2 = 2.10 \text{ g}_0$$

$$n_x = 2.10$$

This value of n_x compares quite well with the value estimated in Fig. 9.50, showing a value of 2.13.

9.2.3 Saturn V Side Air Loads

For the Saturn V, we are mainly concerned with the nonconstant-diameter sections, but there is a small contribution from the constant-diameter sections as well. We will exclude the loads on the launch escape system tower, because its effects are negligible. So, there are five different areas on the Saturn V that we have to concern ourselves with, as depicted in Fig. 9.53.

We start at the top, the cone-shaped Apollo capsule (cone 1). The geometry is shown in Fig. 9.54.

We determine C_{Li} from the cone/cylinder data given earlier:

$$\frac{D_1}{l_n} = \frac{3.91 \text{ m}}{(101.83 - 98.99) \text{ m}} = \frac{3.91 \text{ m}}{2.84 \text{ m}} = 1.38$$

$$\sqrt{M^2 - 1} = \sqrt{1.71^2 - 1} = 1.387$$

$$\frac{D_1}{l_n}\sqrt{M^2 - 1} = 1.91$$

$$\frac{l_a}{l_n} = \frac{(98.88 - 94.47) \text{ m}}{2.84 \text{ m}} = 1.59$$

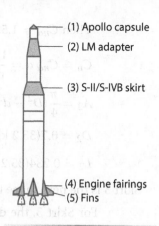

(1) Apollo capsule
(2) LM adapter
(3) S-II/S-IVB skirt
(4) Engine fairings
(5) Fins

Fig. 9.53 To calculate air loads, we consider five areas.

From the cone/cylinder data in Fig. 9.38: $C_{N\alpha} \approx 0.05/\text{deg}$ and is based on *cross-sectional area*. $\alpha = 8.6$ deg, so $C_N = C_{N\alpha}\ \alpha = 0.43$.

The cross-sectional area $A_1 = \pi(3.91 \text{ m})^2/4 = 12.01 \text{ m}^2$, and $D_1 = C_{di}\ q\ A_1 = 0.7(35.2 \text{ kPa})\ (12.01 \text{ m}^2) = 295.86$ kN.

$$L_1 = C_N q A_1 = 0.43(35.2 \text{ kPa})(12.01 \text{m}^2) = 181.74 \text{ kN}$$

Next comes skirt 2, the LM adapter. The geometry is shown in Fig. 9.55.

Saturn V drawing gives $\delta = 8.73$ deg

Then $\sin \delta (\sqrt{M^2 - 1}) = 0.211$,

$$\frac{d}{D} = \frac{3.91 \text{ m}}{6.60 \text{ m}} = 0.59.$$

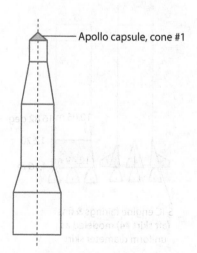

Apollo capsule, cone #1

Fig. 9.54 Apollo capsule geometry for air load calculation.

94.47
8.73°
85.71
3.91
6.60
LM adapter, skirt #2

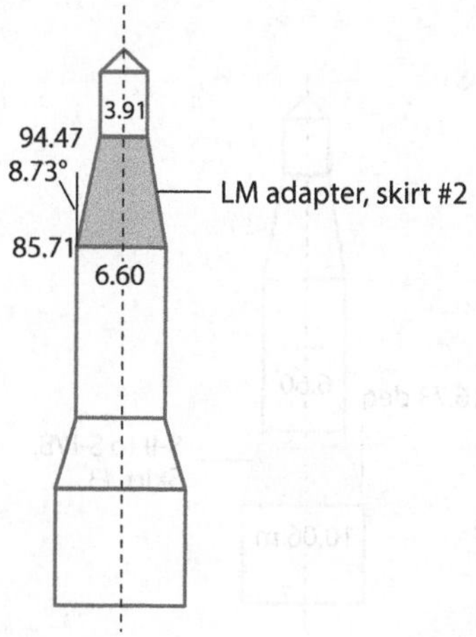

Fig. 9.55 Skirt #2 geometry for air load calculation.

Then $C_{n\alpha} = 1.55/\text{rad}$, and $\alpha = 8.6\,\text{deg}$

$$C_n = C_{n\alpha}\,\alpha = \frac{1.55}{\text{rad}}\left(\frac{\pi\,\text{rad}}{180\,\text{deg}}\right)8.6\,\text{deg} = 0.233$$

$$A_2 = \frac{\pi}{4}(D^2 - d^2) = \frac{\pi}{4}(6.6^2 - 3.91^2)\text{m}^2 = 22.2\,\text{m}^2$$

$$D_2 = 0.7(35.2\,\text{kPa})(22.2\,\text{m}^2) = 547\,\text{kN}$$

$$L_2 = 0.234(35.2\,\text{kPa})(22.2\,\text{m}^2) = 181.8\,\text{kN}$$

Skirt 3 is the next shape to calculate. The geometry is shown in Fig. 9.56.
For Skirt 3, the drawing gives $\delta = 16.73\,\text{deg}$

$$\sin\delta\left(\sqrt{M^2 - 1}\right) = 0.399 \approx 0.4,\ \frac{d}{D} = \frac{6.60\,\text{m}}{10.06\,\text{m}} = 0.66.$$

Then $C_{n\alpha} = 1.3/\text{rad}$. With $\alpha = 8.6\,\text{deg}$, $C_n = 0.235$.

$$A_3 = \pi(10.06^2 - 6.60^2)\text{m}^2/4 = 45.3\,\text{m}^2$$
$$L_3 = 0.235(35.2\,\text{kPa})45.3\,\text{m}^2 = 374.5\,\text{kN}$$
$$D_3 = 0.7(35.2\,\text{kPa})45.3\,\text{m}^2 = 1{,}115.5\,\text{kN}$$

The last area of the Saturn V we have to deal with is the lower portion of the body, where the engines are located. This portion of the LV is fairly complicated, because it has four engine fairings and four fins. This region is shown on the left side of Fig. 9.57. We don't have a simple way to analyze this complex geometry by hand, but we can simplify things by assuming the behavior of the fairings can be "smeared" into an equivalent uniform skirt with the same cross-sectional area, as shown on the right side of Fig. 9.57.

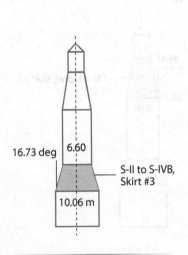

16.73 deg 6.60

S-II to S-IVB, Skirt #3

10.06 m

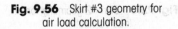

Fig. 9.56 Skirt #3 geometry for air load calculation.

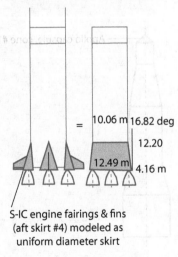

= 10.06 m 16.82 deg

12.20

12.49 m 4.16 m

S-IC engine fairings & fins (aft skirt #4) modeled as uniform diameter skirt

Fig. 9.57 Smearing of fins and fairings into the skirt.

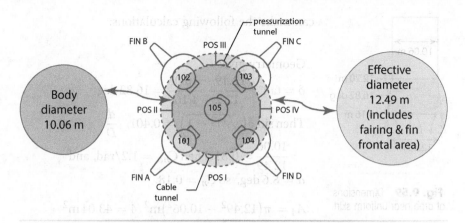

Fig. 9.58 Dimensions of Saturn V base and engine area. Source: NASA.

(We will consider the contribution of the fins later in this section.) Hence, we will have to find the diameter of the uniform skirt; this may be done using the drawing of Saturn V's base shown in Fig. 9.58. This drawing shows the actual body diameter of 10.06 m and the "effective diameter" found by smearing the cross-sectional area of the fins and engine fairings uniformly over the entire circumference.

The effective diameter is calculated as follows: the primary step diameter $D = 10.06$ m, so the step cross-sectional area is

$$\pi D^2/4 = \pi(10.06 \text{ m})^2/4 = 79.49 \text{ m}^2$$

The four curved engine fairings have radius $R = 100$ in. (2.54 m) per NASA [15] the aft lip of these fairings spanned a 180-deg arc. (In other words, their shape is actually a half-circle or semicircle, which isn't properly shown in Fig. 9.55.) Then, the total area of the four fairings $= 4(1/2\pi R^2) = 4\left[1/2\pi(2.54)^2\right] = 40.54^2$.

Now look at the fins. Each fin is 14.4 in. (0.366 m) thick at the root, 4 in. (0.102 m) thick at the tip, and the bottom span is 103 in. (2.62 m). The projected cross-sectional area of the four fins is $4[1/2(0.366 + 0.102 \text{ m}) \times 2.62 \text{ m}] = 2.45 \text{ m}^2$. Fin dimensional data may be found in Barret [1].

We ignore the tiny contributions of the cable tunnel and pressurization tunnel areas. The total step cross-sectional area is the sum of the original body area, the fairing area, and the fins' cross-sectional area, or $A_{\text{effective}} = 79.49 \text{ m}^2 + 40.54 \text{ m}^2 + 2.45 \text{ m}^2 = 122.5 \text{ m}^2$. Using this area, the effective base diameter $D_{\text{effective}} = (4A_{\text{effective}}/\pi)^{1/2} = 12.49\text{m}$. This value is shown in Figs. 9.58 and 9.59, which provide the dimensions of the effective base area.

We can now do the air loads calculation for Skirt 4, the bottom of the S-IC, the first step of the Saturn V. Using the dimensions in Fig. 9.59, we

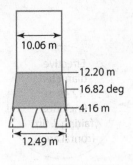

10.06 m

12.20 m

16.82 deg

4.16 m

12.49 m

Fig. 9.59 Dimensions of area near uniform skirt.

can make the following calculations:

Geometry gives

$$\delta = \tan^{-1}\frac{12.49 - 10.06}{12.2 - 4.16} = 16.82 \text{ deg}$$

Then $\sin\delta\left(\sqrt{M^2 - 1}\right) = 0.401$, $\dfrac{d}{D}$

$$= \frac{10.06}{12.49} = 0.81. \text{ Then } C_{n\alpha} = 1.2/\text{rad, and}$$

$\alpha = 8.6$ deg, so $C_n = 0.18$.

$$A_4 = \pi\left(12.49^2 - 10.06^2\right)\text{m}^2/4 = 43.04\,\text{m}^2$$

$$L_4 = 0.18(35.2\,\text{kPa})43.04\,\text{m}^2 = 272.7\,\text{kN}$$

$$D_4 = 0.7(35.2\,\text{kPa})43.04\,\text{m}^2 = 1,060\,\text{kN}$$

9.2.4 Saturn V Supersonic Fin Lift Analysis

Next, we calculate the *sideways* parameters of the fins so we can assess how they act to provide stabilization. To do this, we need normal force coefficient $C_{N\alpha}$, angle of attack α, q_{max} (or speed v, density ρ), span b, sweep angle, and side (planform) area S. The dimensional information may be found in Fig. 9.60. Note that it is traditional to include the shaded area, "submerged" behind the engine fairing, as part of the fin (which we will treat as a wing).

The fin (wing) parameters are:

- Aspect ratio $AR = b_{fin}^2/S_{fin} = 1.463$.
- Taper ratio $\lambda = $ tip chord/root chord $= 0.407 \approx 0.4$.
- $\varepsilon = 90$ deg $- \Lambda_{LE} = 90$ deg $- 30$ deg $= 60$ deg. Note Λ_{LE} is sweep angle of the leading edge.
- Mach no. $M_{max\text{-}q} = 1.71$.
- Parameter $(M_{max\text{-}q}^2 - 1)^{1/2}\tan\varepsilon = 2.402$.
- Parameter $a = \tan\Lambda_{TE}/\tan\Lambda_{LE} = \tan 0$ deg$/\tan 60$ deg $= 0$. Note Λ_{TE} is the sweep of trailing edge.

We can calculate the supersonic lift curve slope $C_{L\alpha}$ using a number of methods. We chose to use the method provided in the handbook published by General Dynamics [7]:

$$C_{L\alpha} = \frac{\partial C_L}{\partial\alpha} = \frac{C_{L\alpha}}{AR}\left(\begin{array}{c}\text{from}\\\text{chart}\end{array}\right) \times AR \times \left(\begin{array}{c}\text{correction}\\\text{factor}\end{array}\right)$$

The fin's lift curve slope is obtained from the plot shown in Fig. 9.61, for a wing with taper 0.4 and $a = 0$. We see that the value of $\dfrac{C_{L\alpha}}{AR} \approx 0.91$

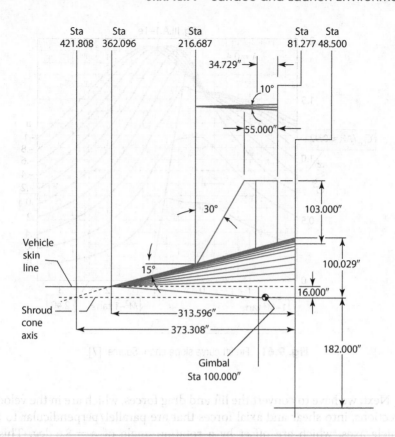

Fig. 9.60 Dimensions of Saturn V's stabilizing fins and engine shrouds. Source: [15].

is obtained when the parameter $\sqrt{M^2 - 1}\tan\varepsilon \approx 2.4$. Then, $C_{L\alpha} = 0.91(1.463)1.04/\text{rad} = 1.38/\text{rad} = 0.0242/\text{deg}$.

The correction factor comes from the lift-curve-slope correlation chart, shown in Fig. 9.62.

The correction factor with $\sqrt{M^2 - 1}\tan\varepsilon = 2.4$ is found to be 1.04. Then,

$$C_{La} = 0.92(1.463)1.04/\text{rad} = 1.38/\text{rad} = 0.0242/\text{deg}$$

The side forces due to the fins' *lift* will be a *stabilizing* force and moment, because they are aft of the CM and trying to turn the vehicle *into* the wind. With fin area $S_{fin} = 8.555 \text{ m}^2$ and the number of fins exposed $= 2$ (assuming the crosswind is perpendicular to two of the four fins as depicted in Fig. 9.63), then the total fin lift $= L_{fin-total} = (\# \text{ fins})(\text{lift/fin}) = (\# \text{ fins})(C_{L\alpha}{\cdot}\alpha)qS_{fin}$, or

$$L_{fin-total} = 2\left[\left(\frac{0.0242}{\text{deg}}\right)8.6\,\text{deg}\right](35.2\,\text{kPa})(8.555\,\text{m}^2)$$

$$L_{fin-total} = L_5 = 125.3\,\text{kN}$$

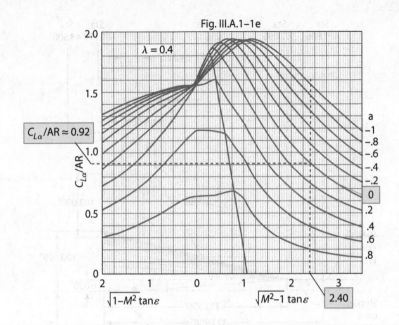

Fig. 9.61 Fin lift curve slope chart. Source: [7].

Next, we have to convert the lift and drag forces, which are in the velocity directions, into shear and axial forces that are parallel/perpendicular to the vehicle axes, which are offset by a rotation angle of $\alpha = 8.6$ deg. This is

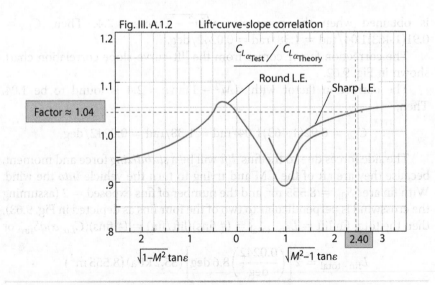

Fig. 9.62 Correction factor for lift-curve-slope. Source: [7].

done with a simple trigonometric rotation

$$V_i = L_i \cos \alpha + D_i \sin \alpha$$
$$A_i = D_i \cos \alpha - L_i \sin \alpha$$

The result of this operation on the five "areas of interest" is shown in Table 9.8.

We need the center of mass location, because all forces act about the vehicle's CM. Its location is also needed for stability assessment. We can use mass properties to calculate (as in Chapter 7), or we can use recorded data. From [3], we find the CM travel vs time shown in Fig. 9.64. At the max-q time of 83 s, the CM is 29 m above the gimbals, or station 34.5 m for our calculations.

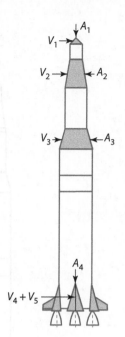

Fig. 9.63 Saturn V points of interest for the example problem.

9.2.5 Lateral Acceleration Due to Air Loads and Engine Gimbaling

Now we must look at all the aerodynamic lateral loads acting on the Saturn V, which create a yawing moment about the CM. As shown in Fig. 9.65, the forces are $V_1 = 223$ kN, $V_2 = 262$ kN, $V_3 = 537$ kN, and $V_4 = 634$ kN. The three forces in front of the CM are destabilizing, and the force in the fin area is stabilizing. In order to *trim* the vehicle, we must add engine gimbal side force F_{gimbal} so that the *sum of all moments* $= 0$. Starting at the top of the vehicle, and with positive moments CCW,

$$(223\,kN)(99.95 - 34.5)m + (262\,kN)(90.46 - 34.5)m + (537\,kN)(70 - 34.5)m$$
$$- (428 + 126\,kN)(34.5 - 8.3)m - F_{gimbal}(34.5 - 5.47)m = 0$$

or

$$F_{gimbal} = 1{,}161\,kN$$

Table 9.8 Wind Forces (Lift and Drag) Converted to Body Forces (Shear and Axial)

Item	Lift L_i	Drag D_i	Shear V_i	Axial A_i
1	181.6	289.0	222.8	258.6
2	181.8	547.0	261.6	513.7
3	374.5	1,115.5	537.1	1,047.0
4	272.7	1,060.0	428.1	1,007.3
5	125.3	11.3	125.6	-7.5

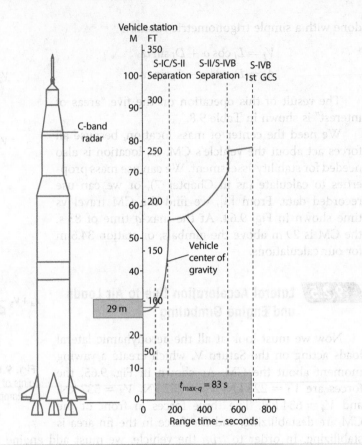

Fig. 9.64 Saturn V CM travel vs time. Source: [3].

The total lateral force F_{lat} on the Saturn V is the sum of all the horizontal forces.

$$F_{lat} = 223 + 262 + 537 + (428 + 126) + 1{,}161\,\text{kN} = 2{,}736\,\text{kN}$$

$$m_{total} = 1{,}814{,}700\,\text{kg}$$

$$W_{total} = m_{total}\,g_0 = 17{,}800\,\text{kN}$$

The lateral load factor n_z is

$$n_z = F_{lat}/W_{total} = 2{,}736\,\text{kN} \div 17{,}800\,\text{kN} = 0.154\,g_0$$

Observing that all of these lateral loads are acting to the right, they are trying to accelerate the vehicle to the right. This means that the acceleration is opposed by the inertial forces from the masses inside the vehicle. In other words, we can define the *net lateral load* at a station as

Net lateral load $=$ air load $-n_z$(weight @ station),
$\quad\quad\quad\quad = $ air load $-n_z(g_0)$(mass @ station)

The Saturn's mass distribution is given in Table 9.9. These values will be used, along with the known propellant masses, to determine the lateral load distribution. We add the dry and other masses in the table to obtain the steps' dry masses. Each step's dry mass is located at the lower end of the cylindrical portion of the lower tank of each assembled step. This is an approximation based on the knowledge that the engines and thrust structure account for most of the dry step's mass.

The net lateral load definition allows us to create a diagram showing all the forces acting on the vehicle at max-q, shown in Fig. 9.66. Each of the *left-facing* forces in the figure are due to the item's mass multiplied by the lateral load factor (in this case, 0.154 g_0). For instance, the top left-facing force of 6.1 kN is from the mass of the launch escape system (LES) of 4 T multiplied by 0.154 g_0. Right-facing loads are aerodynamic loads [except for thrust vector control (TVC) force, bottom left]. Dry step masses are assumed to act at the bottom of the lower tank cylindrical sections.

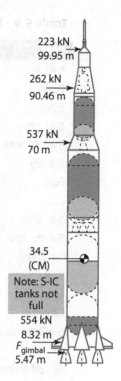

Fig. 9.65 Aerodynamic and gimbal forces on Saturn V.

9.2.6 Inertia Relief

The aerodynamic and engine gimbal forces acting to the *right* on the rocket produce a sideways acceleration to the *right*. The acceleration to the *right* in turn induces *left-pointing* inertial forces due to d'Alembert's principle: $F = ma$ can be written as $F - ma = 0$, or the inertial forces are of the opposite sign of the applied forces. Because the inertial forces are in the opposite direction of the aerodynamic and engine gimbal forces, the overall loads are *reduced* or *relieved*: this is the origin of the term *inertia relief.*

Now we are ready to use the data in Fig. 9.66 to apply all the loads and calculate the shear and bending moments as distributed along the vehicle's axis. The calculations are shown in the spreadsheet in Table 9.10.

Table 9.9 Tabulated Masses of Apollo 11's Saturn V. Source: [18].

Item	klb$_f$	T
S-IC, dry	287.53	130.4
Fuel	1,424.9	646.3
LOx	3,305.8	1,499.5
S-IC, other	6.23	2.8
S-IC dry, total	293.76	133.2
S-II, dry	79.71	36.2
Fuel	158.12	71.7
LOx	819.10	371.5
S-II, other	1.26	0.6
S-II dry, total	80.97	36.7
S-IVB, dry	24.85	11.3
Fuel	43.61	19.8
LOx	192.50	87.3
Other	1.66	0.8
S-IVB dry, total	26.51	12.0
IU	4.28	1.9
SC/LM adapter	3.95	1.8
LM	33.28	15.1
CSM	63.51	28.8
LES	8.91	4.0
Total LV	**6,860.4**	**3,111.8**

The shear load distribution over the vehicle length (column 9) is shown in Fig. 9.67. Similarly, the moment distribution over the vehicle length (last column) is shown in Fig. 9.68.

Figure 9.69 compares the calculated bending moment to the moment from the actual mission. When scaled to maximum values, we see that the shapes of the moment curves are very similar to each other. The dashed line is our hand calculation; the solid line is from MSFC [12].

The hand calculation results are very close in shape to the NASA curve. The NASA curve has smaller amplitude because the vehicle experienced much smaller AoA (1.7 deg compared to our 8.6 deg).

9.2.7 Did the Saturn V *Need* Fins?

The thrust vector control system of the Saturn V was very effective, so one might think that fins weren't needed. After all, the side force that we needed to trim during max-q was $F_{gimbal} = 1,161$ kN. With four F-1

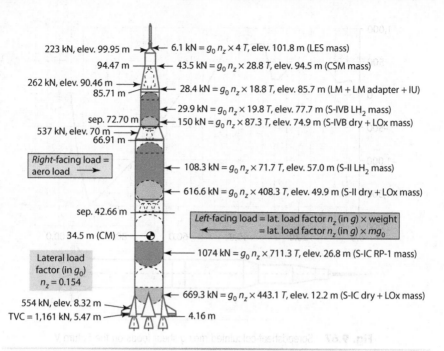

223 kN, elev. 99.95 m →
94.47 m
262 kN, elev. 90.46 m
85.71 m
sep. 72.70 m
537 kN, elev. 70 m
66.91 m

Right-facing load =
aero load →

sep. 42.66 m

34.5 m (CM)

Lateral load
factor (in g_0)
n_z = 0.154

554 kN, elev. 8.32 m
TVC = 1,161 kN, 5.47 m

← 6.1 kN = $g_0 n_z$ × 4 T, elev. 101.8 m (LES mass)
← 43.5 kN = $g_0 n_z$ × 28.8 T, elev. 94.5 m (CSM mass)
← 28.4 kN = $g_0 n_z$ × 18.8 T, elev. 85.7 m (LM + LM adapter + IU)
← 29.9 kN = $g_0 n_z$ × 19.8 T, elev. 77.7 m (S-IVB LH$_2$ mass)
← 150 kN = $g_0 n_z$ × 87.3 T, elev. 74.9 m (S-IVB dry + LOx mass)
← 108.3 kN = $g_0 n_z$ × 71.7 T, elev. 57.0 m (S-II LH$_2$ mass)
← 616.6 kN = $g_0 n_z$ × 408.3 T, elev. 49.9 m (S-II dry + LOx mass)

Left-facing load = lat. load factor n_z (in g) × weight
← = lat. load factor n_z (in g) × mg_0

← 1074 kN = $g_0 n_z$ × 711.3 T, elev. 26.8 m (S-IC RP-1 mass)
← 669.3 kN = $g_0 n_z$ × 443.1 T, elev. 12.2 m (S-IC dry + LOx mass)
4.16 m

Fig. 9.66 Saturn V max-q lateral forces.

Table 9.10 Saturn V Max-q Shear and Bending Analysis

Location	Elev. (m)	Station (m)	Mass (T)	Mass moment ref. Sta. 0.0 (T–m)	Airload (kN)	Airload moment about CG (kNM)	Combined loads (kN)	Shear load (kN)	Bending moment (kNm)
Escape tower bottom	101.8	0.0	4.0	0	0	0	−6	−6	0
CM bottom	99.0	2.8	0.0	0	223	14,381	223	217	0
LM adapter top	94.5	7.4	28.8	212	0	0	−44	173	1,484
LM adapter AC	90.5	11.4	0.0	0	262	14,662	262	435	2,092
S-IVB top	85.7	16.1	18.8	303	0	0	−28	407	4,783
S-IVB LH$_2$ tank	77.7	24.1	19.8	477	0	0	−30	377	7,929
S-IVB LOx tank	74.9	27.0	99.3	2,679	0	0	−150	227	8,957
S-II upper skirt top	72.7	29.1	0.0	0	0	0	0	227	9,285
S-II upper skirt AC	70.0	31.8	0.0	0	537	19,064	537	764	9,898
S-II upper skirt bottom	66.9	34.9	0.0	0	0	0	0	764	13,089
S-II LH$_2$ tank	57.0	44.8	71.7	3,215	0	0	−108	656	20,661
S-II LOx tank	49.9	52.0	408.3	21,213	0	0	−617	39	24,950
S-IC top	42.7	59.2	0.0	0	0	0	0	39	23,008
S-IC RP-1 tank	26.8	75.0	711.3	53,369	0	0	−1,074	−1,035	23,628
S-IC intertank	26.1	75.7	0.0	0	0	0	0	−1,035	22,527
S-IC LOx tank	12.2	89.6	443.1	39,719	0	0	−669	−1,705	8,137
S-IC bottom flare AC	8.3	93.5	0.0	0	554	−14,496	554	−1,143	445
F-1 engine gimbals	5.5	96.4	0.0	0	0	0	1,158	15	−2,024
Total			**1,805**	**121, 189**	**1,576**	**33,610**	**15**		
Masses from NASA SP-2000-4029		Actual	**1,815**			F_{gimbal} =	1085.4	kN	
or calculated using propellant flows		Calc. CG @ stn		67.1	m	Lateral load factor			
Step masses located at bottoms		Calc. CG @ elev		34.7	m	**n_z = 0.154**			
of lower tanks as approximation		Actual. CG @ elev		34.5	m	g_0	9.81		

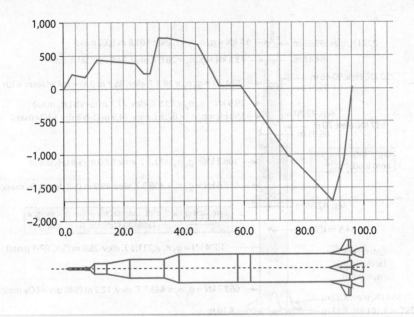

Fig. 9.67   Spreadsheet-calculated max-*q* shear loads on the Saturn V.

engines gimbaled (the center/fifth engine was fixed), that's 290 kN/engine. At that instant, the five engines were producing 38.55 MN of thrust, or 7.71 MN/engine. This results in a gimbal angle of $\tan^{-1}(0.290/7.71) = 2.15$ deg

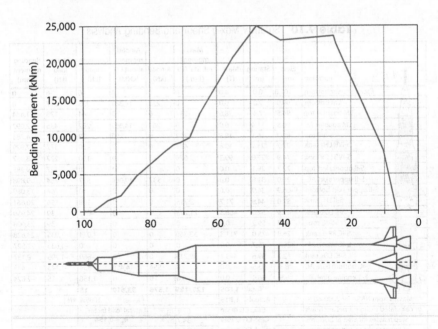

Fig. 9.68   Max-*q* bending loads on the Saturn V.

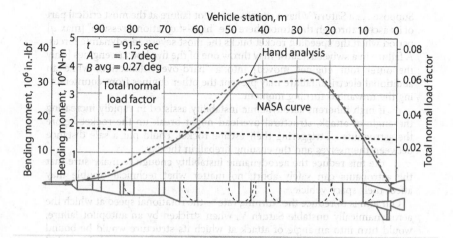

Fig. 9.69 Bending moment: example calculations (dashed line) vs NASA calculations (solid line). Data on top of figure from MSFC [12].

for each of the four engines, much less than their maximum travel ~5 deg. So, why add the mass and cost of fins? To find out the answer, one must refer to an obscure report from the Saturn program development [13], which reads:

> The actual fins were not optimized for aerodynamic properties alone but were designed in conjunction with control output of the swivel engines and the environment that the vehicle would encounter. With four 188 K engines swiveled to 10° control angle, various amounts of wind and shear gradients can be tolerated. A highly unstable vehicle in the high dynamic pressure region of flight is severely restricted unless a very large control torque is available. This might have been designed into the system; however, this would have created another undesirable situation. That is, *in case of shutdown of the engines high divergent rates would exist, a condition which is not desirable for a manned vehicle.* [italics added]

The key phrase is the last one: "a condition which is not desirable for a manned vehicle." The concern was that if one of the gimbaled engine actuators were to fail hard-over or an engine were to fail entirely, the vehicle would "diverge" quickly. The fins on the vehicle had a useful purpose: if one of the engines failed hard-over, especially in the lower atmosphere, the fins would provide a countertorque just strong enough to give the crew time to activate the abort handle before the vehicle self-destructed. The fins also added stability during the period of max-q, where jet stream winds could cause significant wind shear with strong disturbing torques.

Some additional insight can be gained from the master, Wernher von Braun, in a *Popular Science* magazine article [21]:

Suppose ... a Saturn V has a serious autopilot failure at the most critical part of its ascent through the atmosphere-the "high-stagnation-pressure" [max-*q*] period when the speeding rocket bucks the most severe aerodynamic forces. A failure in a swivel actuator may throw one of the five booster engines [only the outer four engines moved] into a "hard-over" deflection, while an additional electrical failure may prevent the other engines from counteracting the unwanted turning moment.

... if high inherent aerodynamic instability assisted in rapidly increasing the angle of attack, structural overload might break up the rocket before the astronauts ... triggering their escape rocket, could put a safe distance between themselves and the ensuing fireball in the sky.

... the fins reduce the aerodynamic instability enough to make sure that the astronauts can safely abort, no matter what technical trouble may affect their space vehicle.

Our aim is to reduce the "turning rate"—the rotational speed at which the aerodynamically unstable Saturn V, when stricken by an autopilot failure, would turn into an angle of attack at which its structure would be bound to fail. One might say that the purpose of the fins is to extend the period of grace that the astronauts have to push the "panic button."

9.2.8 Saturn V Max-*q* Axial Loads

Now we can examine the vehicle's axial loads. Once again, this is done by starting at the top of the vehicle and keeping a running total of the downward-pointing forces, until we reach the bottom. This procedure is shown in Table 9.11. The downward forces are the apparent weights of each of the items under the axial acceleration, which in this case is $n_x = 2.10$. Note that the four axial air loads are shown in the table, associated with the four cones/skirts analyzed earlier. In addition, the propellant masses and structural masses are included as before. We did not include the friction drag forces acting on the cylindrical portions of the vehicle.

A plot of the axial loading during max-*q* is shown in Fig. 9.70. As one would expect, the axial force grows as we move aft on the vehicle, peaking just before the engine gimbals, after which the axial force is zero. The remainder at the bottom of the "Combined Axial Load" column should be zero. The non-zero amount shows the axial load at the engines is about 3.8% higher than the known thrust of 38.55 MN, probably due to errors in propellant consumption, data from incorrect LV, and/or roundoff.

Comparing the axial forces to those from the actual mission, we see that except for scale, the shapes are very close (Fig. 9.71). The dashed line is the hand calculation; the solid line is from NASA. The hand calculation results are reasonably close in shape to the NASA curve.

9.3 Load Curves Rules of Thumb

Here are some general statements about load curves that may be useful as you construct your own curves:

In-flight load curves:

Table 9.11   The Calculation of the Max-q Axial Loads on the Saturn V

#	Location	Elev. (m)	Stn. (m)	Mass (T)	Axial airload (kN)	Combined axial Load (kN)	Running axial load (kN)
1	Esc tower bottom	111.0	0.0	4.0	0	83	83
2	Command module top	101.8	9.2	0.0	295.7	296	379
3	LM adapter top	94.5	16.5	28.8	0	593	972
4	LM adapter AC	90.5	20.5	0.0	514	514	1,486
5	S-IVB top	85.7	25.3	18.8	0	387	1,873
6	S-IVB LH$_2$ tank	77.7	33.3	19.8	0	408	2,281
7	S-IVB LOx tank	74.9	36.1	99.3	0	2,045	4,326
8	S-II upper skirt top	72.7	38.3	0.0	0	0	4,326
9	S-II upper skirt AC	70.0	41.0	0.0	1,047	1,047	5,373
10	S-II upper skirt bottom	66.9	44.1	0.0	0	0	5,373
11	S-II LH$_2$ tank	57.0	54.0	71.7	0	1,477	6,850
12	S-II LOx tank	49.9	61.1	408.2	0	8,407	15,257
13	S-IC top	42.7	68.3	0.0	0	0	15,257
14	S-IC RP-1 tank	26.8	84.2	711.3	0	14,648	29,905
15	S-IC interstage	26.1	84.9	0.0	0	0	29,905
16	S-IC LOx tank	12.2	98.8	443.1	0	9,126	39,031
17	S-IC bottom flare AC	8.3	102.7	0.0	987	987	40,018
18	F-1 engine gimbals	5.5	105.5	0.0	0	−38,550	1,468
	Axial load factor n_x =	2.10	**Total**	**1,805.1**	**2,843.7**	**1,468**	
	Masses from NASA SP-2000-4029 or calculated using propellant flows					Should be ~0	

*Neglects friction drag on cylindrical portions

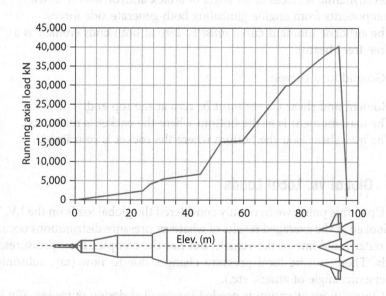

Fig. 9.70   The calculated axial loading during max-q for the Saturn V.

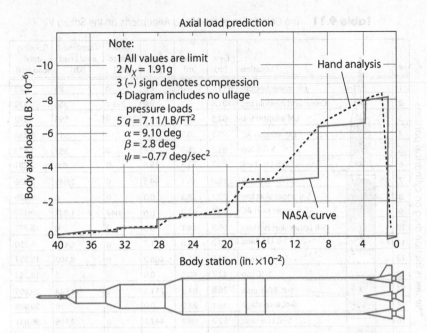

Fig. 9.71 Axial load calculations: example (dashed line) vs. solid. Source: [12].

- Both axial load curve *and* shear curve *do not close*, due to finite loads at the lower end from engine axial thrust and side force from TVC.
- Aerodynamic surfaces at an angle of attack and/or lateral thrust components from engine gimbaling both generate side forces.
- The airborne moment curve *must be zero* at both ends (vehicle is a "free-free" beam).

  Ground load curves:

- Moment and shear curves *must be zero* at the top end.
- The max moment is at the bottom where the rocket is restrained.
- The max shear is at the bottom where the rocket is restrained.

9.4 Global vs. Local Loads

Up to this point, we have only considered the global loads on the LV. Yes, we looked at the averaged results of whatever pressure distributions occur on the outer mold line of the vehicle, but not the complex local pressures and loads. There can be local pressure changes due to flow (say, subsonic vs supersonic, angle of attack, etc.).

Clearly more attention is needed for detailed design purposes. We have complex external flows, pressures, and temperatures, and these need to be

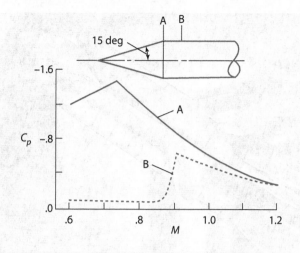

Fig. 9.72 Pressure coefficient vs Mach number for two different locations. Source: [4].

superimposed with internal loads (stresses from pressurization, moments, and shears) to look at the situation of skin loading.

As an example, consider the situation shown in Fig. 9.72. Here we see a conical nose in the front of a uniform cylinder. The local pressure coefficient $C_p = (p - p_\infty)/q$ is plotted vs Mach number for two locations, one at the intersection of the cone and cylinder and one a bit further aft.

The Mach number continuously increases, so it might be thought of as time scale. For corner A, the maximum negative pressure occurs at high subsonic speeds (shown by the solid curve). Placing a vent at A would reduce the pressure there to low values, but at the same time, the net pressure at B would be large. This has a direct application to LVs: payload fairings are vented to release pressure during ascent, but the difference between locations A and B varies quite a bit. This could have a major effect on where to locate pressure vents.

9.5 Real Calculation of Vehicle Loads

Except for very preliminary studies, loads and moments on LVs are not calculated the way we have done in this chapter. The real way to do it is with the use of two very complex and specialized computer programs: computational fluid dynamics (CFD) and finite-element modeling (FEM).

CFD uses the external geometry, angle of attack, and Mach number to provide the pressure distribution over the entire SLV surface. This can be imported into a second software package called FEM, which takes the CFD loads and applies them to a mathematical model of the SLV structure,

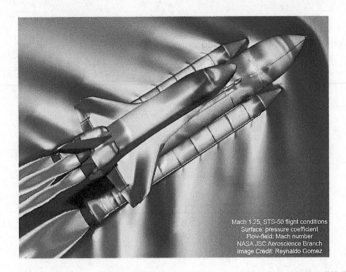

Fig. 9.73 CFD model of Space Shuttle stack during launch. Source: [5].

including material properties, mass distributions, and even temperature distributions.

Figure 9.73 shows the result of running CFD on a computerized model of the Space Shuttle. Influences from both large parts and small protrusions are readily visible.

Fig. 9.74 The Jimsphere sounding balloon used to obtain high-altitude wind data. Source: [17].

9.6 Dealing with High-Altitude Winds

Launch sites have extensive weather measurement facilities whose goal is to measure the local weather and use the information to help ensure launch success. But how do these facilities measure the horizontal wind speed, which we have seen is a major driver of launch loads?

One or more *Jimsphere* sounding balloons are launched in time intervals and tracked with radar. The Jimsphere balloon is shown in Fig. 9.74; they are used at all U.S. launch ranges to obtain high-altitude wind data between the surface and 17 km altitude in about 1 h. The Jimsphere balloons are made with many protruding spikes, because smooth balloons did not ascend stably due to vortex shedding.

The data from the tracking radar are used to do day-of-flight analyses of winds aloft, and along with the LV's computer models, can be used to recommend launching—or no launch—depending on the high-altitude winds situation. Data may also be used to "bias" an LV's guidance program in order to soften the control system so that it can fly through gusts without serious consequences. This introduces the next topic.

9.7 Design Issues for Ascent Phase

As we have seen in Chapter 7, launches can be optimized to deliver the maximum payload to a particular orbit, but the LV must *tightly* follow the guidance commands. This means that the LV will try not to deviate from its optimum path, and it will incur large structural loads from winds, increasing structural weight.

On the other hand, turning a launch vehicle into the wind reduces the angle of attack and therefore reduces the experienced loads, decreasing structural weight. However, turning into the wind incurs drift from the desired trajectory and will cost payload performance. It can also provide increased thermal loads.

Many believe that a balanced response will compromise between no flight path deviation and turning into the wind, and will also consider thermal effects.

One more thing to think about: the gusts that the LV flies through can provide enough energy to excite the vehicle's elastic bending modes, so a satisfactory response to gusts may also require active control of bending modes. We will discuss elastic vibration modes in Chapter 11.

These three schemes—purely flying the desired trajectory regardless of the loads, turning into the wind to reduce aerodynamic loads, and a compromise between the first two—are shown symbolically in Fig. 9.75. The last two lines in the figure indicate the qualitative loss of trajectory accuracy and the loss of performance, both depending on the control method employed by the LV.

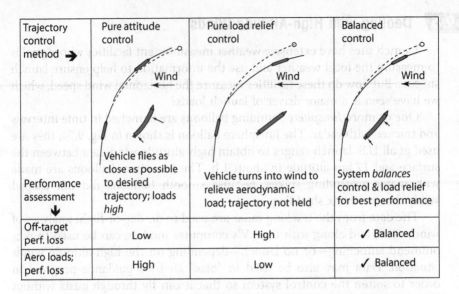

Trajectory control method →	Pure attitude control	Pure load relief control	Balanced control
Performance assessment ↓	Vehicle flies as close as possible to desired trajectory; loads *high*	Vehicle turns into wind to relieve aerodynamic load; trajectory not held	System *balances* control & load relief for best performance
Off-target perf. loss	Low	High	✓ Balanced
Aero loads; perf. loss	High	Low	✓ Balanced

Fig. 9.75 Three possibilities for ascent guidance and the tradeoffs among them. Source: [19].

9.8 Load Relief During Launch

Often during a launch, the commentator will say something like "load relief pitch program." A pitch program is a *passive* method of load relief. It works by doing the following:

First, the wind profile is determined by tracking Jimsphere balloons with radar at the launch site. The now-known wind profile is used to create a virtual headwind or tailwind, and this virtual bias is added to the LV's pitch program. The virtual wind bias in the guidance system effectively reduces angle of attack and reduces loads caused by existing winds. The deviation from the desired ascent can be corrected later in flight.

Another type of load relief can be provided by angle-of-attack sensors or lateral accelerometers. Measurements of this type are used for *active control*, where the LV responds in real-time to the deviations measured (either angle of attack or lateral acceleration, in these two cases). The former was used on the front end of the Saturn V escape tower, which was topped with a device called a *Q-ball*, which measured the differential in dynamic pressures about the Saturn's pitch and yaw axes to compute the angle of attack. (Photos of a Q-ball are provided in Fig. 9.76.) The angles of attack were sensed by the flight computer, which in turn would find a deviation from the desired angles and would send instructions to gimbal the first-step engines firing far below to bring the rocket back to its desired angle of attack.

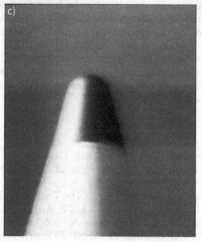

Fig. 9.76   The Saturn V's Q-ball. a) A Q-ball with a transparent housing. Hoses connecting the eight pitot tubes may be seen beneath the top of the nose area. b and c) Apollo 15 Q-ball before and after cover removal. Sources: a) Author photo, Cosmosphere; b and c) frame captures from Apollo 15 liftoff in video Apollo 15 liftoff.mpg (courtesy NASA).

Another way to sense crosswinds would be to measure *lateral accelera-tion* for the same purpose: to minimize AoA and minimize loads. Using lateral acceleration to sense crosswinds could be trickier, because the accel-erometers would also respond to bending vibrations of the LV. Those bending vibrations would have to be filtered out before being sent to the flight computer and control system.

More information on launch vehicle load alleviation may be found in [14].

Endnote

The Saturn V's Q-ball was a critical instrument for safe flight. The following is a quote from Wikipedia explaining how the Q-ball was kept protected from weather until just before launch (http://en.wikipedia.org/wiki/Apollo_%28spacecraft%29).

> A Styrofoam® cover, removed a few seconds before launch, protected the Pitot tubes from being clogged by debris. The cover was split in half vertically and held together by a 2-inch rubber band. A razor blade was positioned behind the rubber band, pinched between the halves of the cover. A wire rope was connected to the top and bottom of the razor blade and to both halves of the cover. The wire rope was routed through a pulley on the hammerhead crane at the top of the launch umbilical tower (LUT) down to a tube on the right side of the 360-foot (110 m) level of the LUT. The wire rope was connected to a cylindrical weight inside a tube. The weight rested on a lever controlled by a pneumatic solenoid valve. When the valve was actuated, 600-PSI nitrogen gas pressure rotated the lever down allowing the weight to drop down the tube. The dropping weight pulled the wire rope, which pulled the blade to cut the rubber band, and the wire rope pulled the halves of the cover away from the launch vehicle.
>
> The apparent over-engineering of this simple system was due to the fact that the launch escape system, which depended on the Q-ball data, was armed five minutes before launch, so retraction of the Q-ball cover was a life-critical part of a possible pad abort.

References

[1] Barret, C., "Review of Our National Heritage of Launch Vehicles Using Aerodynamic Surfaces and Current Use of These by Other Nations," NASA TP-3615, 1996.

[2] Barrowman, J. S., and Barrowman, J. A., "The Theoretical Prediction of the Center of Pressure," R&D Project Report at NARAM-8, Aug. 1966.

[3] Boeing, "Saturn V Postflight Trajectory—AS-506," document no. D5-15560-6, 1969.

[4] Boswinkle, R. W., "Aerodynamic Problems of Launch Vehicles," published in NASA SP-23, 1962.

[5] Buning, P. G., and Gomez, R. J., "20 + Years of Chimera Grid Development for the Space Shuttle," presented at the 10th Symposium on Overset Composite Grid and Solution Technology, Moffett Field, CA, 20–23 Sept. 2010.

[6] Daniels, G. E., Ed., "Terrestrial Environment (Climatic) Criteria Guidelines for Use in Space Vehicle Development," NASA TM-X-53872, 1969 rev.

[7] Smith, C. W., Ed., *Aerospace Handbook*, rev. B, Convair FZA-381-II, General Dynamics Convair Aerospace Division, 1976.

[8] Greenberg, H. S. "Structural Analysis Techniques for Preliminary Design of Launch Vehicles," short course notes, UCLA Extension, 2011.

[9] Heitchue, R., *Space Systems Technology*, Reinhold, New York, NY, 1968.

[10] Mackey, A. C., and Schwartz, R. D., "Apollo Experience Report—The Development of Design-Loads Criteria, Methods, and Operational Procedures for Prelaunch, Lift-off, and Midboost Conditions," NASA TN-D-7373, 1973.

[11] Marshall Space Flight Center, "Saturn V Flight Manual SA-503," MSFC-MAN-503, NASA TM-X-72151, 1968.

[12] Marshall Space Flight Center, "Saturn V Launch Vehicle Flight Evaluation Report-AS-506 Apollo 11 Mission," MSFC report MPR-SAT-FE-69-9, NASA TM-X-62558, 1969.

[13] NASA, "The Apollo A/Saturn C-1 Launch Vehicle System," NASA TM-X-69174, MSFC document MPR-M-SAT-61-5, 17 July 1961.

[14] NASA, "The Alleviation of Aerodynamic Loads on Rigid Space Vehicles," NASA TM X-53397, Feb. 1966.

[15] NASA, "Static Aerodynamic Characteristics of the Aborted Apollo-Saturn V Vehicle," NASA TM-X-53587, MSFC, 1967.

[16] NASA, "Transportation and Handling Loads," NASA SP-8077, Sept. 1971

[17] NASA, "Wind Monitor Environment and Resources Management," NASA Spinoff, 1996, https://spinoff.nasa.gov/spinoff1996/48.html [retrieved 8 May 2018].

[18] Orloff, R. W., "Apollo by the Numbers—A Statistical Reference," NASA SP-2000-4029, 2000, http://history.nasa.gov/SP-4029/SP-4029.htm [retrieved 9 May 2018].

[19] Rakoczy, J., "Fundamentals of Launch Vehicle Flight Control," NASA NESC Webcast Series, 2012.

[20] Stengel, R., "Space Systems Design, MAE 342," class notes, Princeton Univ., 2008.

[21] von Braun, W., "Why Rockets Have Fins," *Popular Science*, Sept. 1964, p. 68.

Further Reading

Bruhn, E. F., Orlando, J. I., and Meyers, J. F., *Missile Structures Analysis and Design*, Library of Congress Card #67-28959, 1967.

Glasgow, R. M., "Static Aerodynamic Characteristics of the Aborted Apollo–Saturn V Vehicle," NASA TM X-53587.

NASA, "The Saturn Launch Vehicles," NASA TM-X-881, 1961.

Krausse, S. C. "Apollo/Saturn V Postflight Trajectory—AS-506," Boeing Document No. D5-15560-6, 1969.

Pedego, G. P., and Orillion, A. C., "Apollo Launch Vehicle Design," NASA TM X-56244.

Seifert, H. S., and Boelter, L. M., *Space Technology*, Wiley, New York, 1959.

Walker, C. E., "Results of Several Experimental Investigations of the Static Aerodynamic Characteristics for the Apollo/Saturn V Launch Vehicle," NASA TM X-53770.

9.10 Exercises

9.10.1 Problem 1: Ground Wind Loads, Shear, Moment, and Axial Load Calculation

You will need the layout of your LV design. *Provide no more than four significant digits for your calculations and answers: round numbers as necessary.* NO SCIENTIFIC NOTATION!

1. Using the dimensions of your vehicle, create a spreadsheet that contains columns for the following data and calculations similar to the ground loads spreadsheet in Chapter 9:

   a) Provide at least a side view of your LV, showing lengths of the two steps, the PLF, and body diameter(s). All lengths should be to scale.

   b) For each major external component of your vehicle, determine 1) projected area and 2) elevation of the centroid (above *ground* level, or AGL). AGL = the centroid height above the bottom of the component + the elevation of the bottom of the component above the ground.

c) Compute the *effective* wind speed at the AGL location of each *centroid* of each major portion of your rocket while standing on the launch pad.

d) Determine the wind loads for each major portion of your rocket, using the drag equation. The forces may be assumed to act at the item's projected area's *centroid*.

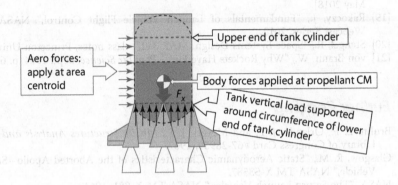

e) Calculate the *shear load* running from the top of the rocket to the bottom due to winds on the pad using the method given in the text. Calculate the shear load at the following locations, starting from the top of the rocket:

1. top of payload fairing,
2. top of 2nd step forward skirt (payload, PLF, and PAF are probably above here),
3. top of 2nd step upper tank cylinder,
4. bottom of 2nd step upper tank cylinder (this is where Step 2's upper tank's propellant weight + tank weight is supported, see vertical up-arrows above)*,
5. top of 2nd step lower tank cylinder*,
6. bottom of 2nd step lower tank cylinder (this is where Step 2's lower tank's propellant + tank weight is supported),
7. top of interstage (all of step 2 is above here),
8. top of 1st step upper tank,
9. bottom of 1st step upper tank cylinder (this is where Step 1's upper tank's propellant + tank weight is supported)*,
10. top of 1st step lower tank cylinder*,
11. bottom of 1st step lower tank cylinder (this is where Step 1's lower tank's propellant + tank weight is supported),
12. bottom of aft skirt or ground support (entire LV is above this).

*If you are using a solid motor or motors, replace items 4 & 5 and/ or 9 & 10 with a calculation at one-half the length of the solid motor casing(s).

*If you have an internal spherical or ellipsoidal tank, replace items 3 & 4, 5 & 6, 8 & 9, and/or 10 & 11 with a single calculation located horizontally at the equator of the tank.

*If using a common bulkhead, replace items 4 & 5 and/or 9 & 10 with a calculation where the bulkhead reaches the tank cylinder.

Provide a table from your spreadsheet with shear calculations.

f) Plot the shear load as a function of the LV length (see below left). If you use MS Excel®, choose the plotting option that connects dots with STRAIGHT lines, not curved ones. The shear load should be zero at the top of the LV, and maximum at the bottom.

Provide a plot of the shear distribution similar to the one shown below.

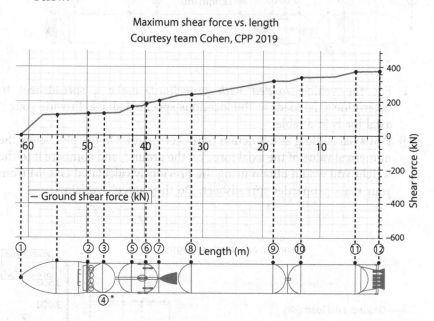

Maximum shear force vs. length
Courtesy team Cohen, CPP 2019

g) Following the procedure in the text, determine the *bending moment* at the same locations. Provide a table from your spreadsheet with moment calculations (this may be the same spreadsheet as done for shear earlier).

h) Provide a plot of the moment distribution as a function of the LV length next to the rocket diagram. This curve should increase "parabolically" from nose to bottom, with peak bending moment at the bottom. It should be zero at the top of the LV, and maximum at the bottom.

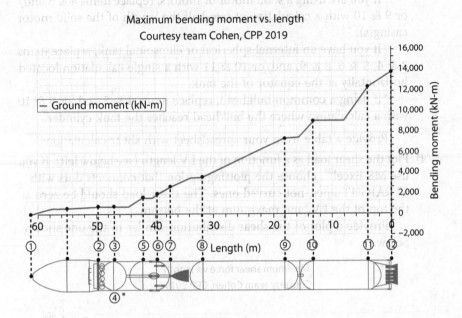

Maximum bending moment vs. length
Courtesy team Cohen, CPP 2019

i) For the vehicle *loaded with propellant,* make a spreadsheet to determine axial loads at the locations specified above. Provide your axial loads in a table.

j) Plot axial loading as a function of rocket length. On the plot, show the numerical value of the axial force on the ground, and compare it to the weight you would obtain using the previously calculated GLOM from your mass properties spreadsheet. Do they agree? Explain.

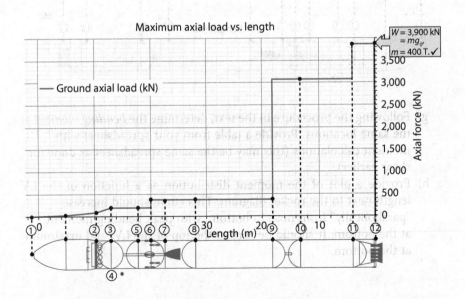

Maximum axial load vs. length

At a minimum, your assignment should contain 1) a side view of your LV to scale, 2, 3, 4) two spreadsheets (shear = 2 + moment = 3 can be combined, but axial = 4 has weights and must be separate), and 5, 6, & 7) the three plots: shear, moment, and axial load distribution. Be sure that the axes and units on the plots can be easily read, and DO NOT provide more than four significant digits.

Problem 2: Launch Vehicle Max-q Flight Loads Calculations

Provide no more than four significant digits for your answers: round numbers as necessary. **NO SCIENTIFIC NOTATION!**

1. a) Fill out a summary table for your LV. Replace all BOLD quantities.:

	Step 1	Step 2
Gross mass (T)	Liftoff mass = **XXX.X (fill in no.)**	Step 2 ignition mass= **XX.X**
Initial Thrust/Weight ratio	**1.XXX**	**0.XXX**
Total thrust (kN)	**X,XXX**	**XXX**
Propellants	**LOx/Gasoline**	**LF/LH$_2$**
Specific Impulse (s)	**3XX**	**4XX**
Structural fraction ?	**0.XX**	**0.XX**
Speed after losses (km/s)	Staging = **X.XXX**	shutdown = **X,XXX**

b) Provide at least a side view of your LV, showing lengths of the two steps, the PLF, and body diameter(s). All lengths should be to scale.

2. Insert your launch vehicle's parameters (thrust, mass, fuel consumption, etc.) into your launch simulation program/spreadsheet from a previous assignment to determine the time at which your LV experiences max-q, and the flight and atmospheric parameters being experienced at that time. Use the same pitch kick angle as was used in that last work. Integrate forwards in time until you have passed the occurrence of max-q (often 80 – 100 s elapsed time; find when it occurs for *your* LV). Determine:

	Quantity	Your Value	
a	$t_{max\text{-}q}$ (s)	**66**	Time from liftoff to max-q
b	$h_{max\text{-}q}$ (km)	**6.66**	Altitude of max-q
c	$v_{max\text{-}q}$ (m/s)	**666**	Vehicle speed at max-q
d	$\rho_{max\text{-}q}$ (kg/m^3)	**0.66**	Standard atmospheric density at max-q
e	$M_{max\text{-}q}$	**1.66**	Mach no. of the vehicle at max-q
f	$m_{max\text{-}q}$ (T)	**666.6**	Mass of the vehicle at max-q
g	$T_{max\text{-}q}$ (kN)	**6566**	Thrust of the vehicle at max-q
h	$n_{max\text{-}q}$	**1.66**	Longitudinal load factor at max-q

You can determine a, the speed of sound for the atmosphere at $h_{\text{max-}q}$ by using on-line atmospheric data tables in the back of aero books. Then, $M_{\text{max-}q} = v_{\text{max-}q}/a$.

Note: if unable to use your simulation, you may assume $t_{\text{max-}q} = 70$ s to calculate your LV's mass, $h_{\text{max-}q} = 13.0$ km and $v_{\text{max-}q} = 500$ m/s to calculate winds, Mach, etc. This will incur a 10-pt penalty.

3. "Max-q" calculations:

a) Show the calculations that determine your LV's *propellant loads* at the max-q time you found in #1. Most likely the 2nd step will not have fired yet, so its tanks remain full, and the 1st step tanks will be *partially empty*. The amount of propellant remaining is determined by subtracting the amount that was consumed up to max-q from the 1st-step tanks; this depends on propellant consumed during the assumed engine startup and known flight time from liftoff to max-q. Use flow rates to determine how much propellant was consumed, and subtract. This is illustrated pictorially in the "determine internal loads" section of Ch. 9.

b) Next, determine your vehicle's *mass* and its *CM location* at the instant of max-q. Revise your previous *Mass Statement* to calculate the mass and CM location by starting with the "dry" vehicle mass statement, and then adding in the masses/locations of the full 2nd step tanks, and *partially-filled* 1st step tanks. Use these data to determine the vehicle CM position at max-q.

   If liquid, the fuel and oxidizer tanks will be partially filled; use the remaining propellant mass in the Mass Statement. If a solid motor, assume that the CM of the reduced amount of propellant has the same location as the fully-loaded SRM: the center of the propellant mass.

c) Determine the jet stream speed using the max-q altitude as input to the *winds aloft profile* chart in Ch. 9 of the text, or you can use the equations under the figure. With the jet stream gust speed, use your max-q parameters from above to determine the relative angle of attack α that the LV, flies using $\alpha = \tan^{-1}(v_{wind}/v)$. Use the max q, α, and Mach information to calculate $C_{N\alpha}$ values for your vehicle's cones, skirts, etc.

d) Use the "Max-$q\alpha$ Airloads Estimation" procedures in the text and your max-q parameters to calculate the lift and drag forces of each segment of your LV. Resolve them into body axes.

e) Determine the engine gimbal angle δ for trim, using your vehicle's thrust value and remembering the side force is $[T \sin \delta]$. Use the required side force magnitude, its location relative to CM, and the vehicle's mass to calculate the vehicle's lateral acceleration. Fill out the following table:

	Quantity	Value
1.	q_{max-q} (kPa)	**36.6**
2.	$v_{wind\ max-q}$ (m/s)	**66**
3.	α_{max-q} (deg)	**6.6**
4.	Req'd. trim force (N)	**666**
5.	δ_{TVC} (deg)	**3.6**
6.	$n_{z\ max-q}$ (g_0)	**0.066**

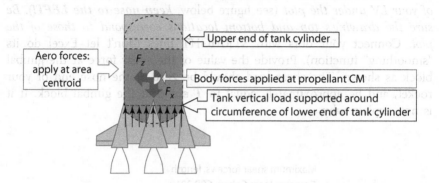

4. Compute the max-*q shear loads* due to angle-of-attack aero loads, inertia relief on masses, and TVC using the text method at the following locations, starting from the top of the rocket:
   1. top of payload fairing,
   2. top of 2nd step forward skirt (payload, PLF, and PAF are probably above here),
   3. top of 2nd step upper tank cylinder,
   4. bottom of 2nd step upper tank cylinder (this is where Step 2's upper tank's propellant weight + tank weight is supported, see vertical up-arrows above)*,
   5. top of 2nd step lower tank cylinder*,
   6. bottom of 2nd step lower tank cylinder (this is where Step 2's lower tank's propellant + tank weight is supported),
   7. top of interstage (all of step 2 is above here),
   8. top of 1st step upper tank,
   9. bottom of 1st step upper tank cylinder (this is where Step 1's upper tank's propellant + tank weight is supported)*,
   10. top of 1st step lower tank cylinder*,
   11. bottom of 1st step lower tank cylinder (this is where Step 1's lower tank's propellant + tank weight is supported),
   12. bottom of aft skirt or ground support (entire LV is above this).

*If you are using a solid motor or motors, replace items 4 & 5 and/or 9 & 10 with a calculation at one-half the length of the solid motor casing(s).

*If you have an internal spherical or ellipsoidal tank, replace items 3 & 4, 5 & 6, 8 & 9, and/or 10 & 11 with a single calculation located horizontally at the equator of the tank.

*If using a common bulkhead, replace items 4 & 5 and/or 9 & 10 with a calculation where the bulkhead reaches the tank cylinder.

Plot the shear loads as a function of LV length and *provide a drawing of your LV under the plot* (see figure below, **keep nose to the LEFT!**). *Be sure the drawing's top and bottom locations correspond to those of the plot.* Connect your dots with STRAIGHT lines (don't let Excel do its "smoothing" function). Provide the value of the shear force at the gimbal block as shown. The shear load should be **zero** at the nose end of your rocket, and be approximately equal to $T \sin \delta$ at the gimbal block. If it is not, explain why.

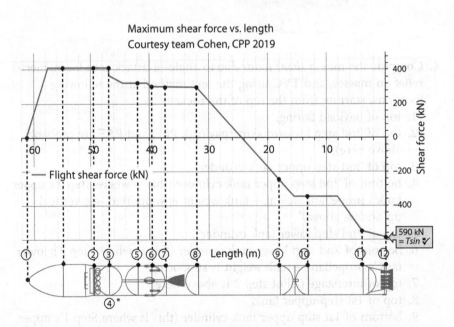

Maximum shear force vs. length
Courtesy team Cohen, CPP 2019

5. Compute the max-q bending loads at the locations specified above, and plot with rocket length. *Provide a drawing of your vehicle under the plot* (see figure below, **keep nose to the LEFT!**). *Be sure the drawing's top and bottom locations correspond to those of the plot.* Connect your dots with STRAIGHT lines. The bending loads should be **zero** at both ends of your rocket.

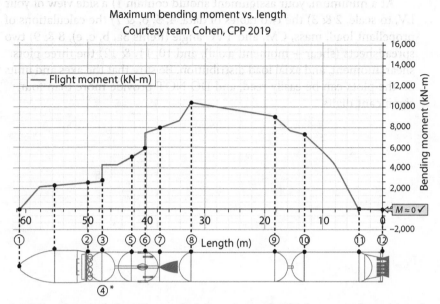

Maximum bending moment vs. length
Courtesy team Cohen, CPP 2019

6. Using the gimbal angle needed for trim and vehicle mass, determine the max-*q* axial ("drag") loads at the locations specified above, and plot with rocket length. *Provide a drawing of your vehicle under the plot* (see figure below, **keep nose to the LEFT!**). *Be sure the drawing's top and bottom locations correspond to those of the plot.* On the plot, provide the value of the axial force at the gimbal block as shown. The axial load should be **zero** at the nose end of your rocket, and be equal to $T\cos\delta \approx T$ at the gimbal block (use your SL value for convenience). Note that the Saturn V's thrust level is about 8 million pounds in the Chapter 9 axial loads calculations. Is your axial load at the gimbal block equal to the thrust? If not, explain why.

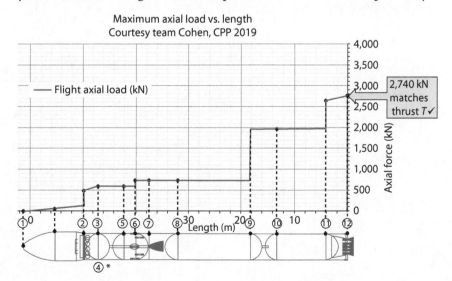

Maximum axial load vs. length
Courtesy team Cohen, CPP 2019

At a minimum, your assignment should contain 1) a side view of your LV, to scale, 2 & 3) the two tables of data, 4, 5, 6, & 7) the calculations of propellant load, mass, CM, and TVC angle (items 3a, b, c, e), 8 & 9) two spreadsheets (shear + moment, axial), and 10, 11, & 12) the three plots: shear, moment, and axial load distribution. Be sure that the axes and units on the plots can be easily read, and DO NOT provide more than four significant digits.

Chapter 10 Launch Vehicle Stress Analysis

Moving forward from ground and flight loads, we show the design roadmap in Fig. 10.1 with the highlighted Signpost indicating the information to be covered in this chapter: the stress analysis portion of vehicle structural design.

In the previous chapter, we calculated the loads that a launch vehicle will be subjected to, both on the ground and in the air. Also, we have done some preliminary sizing and calculations of tank wall thicknesses, but the calculated thicknesses are based on statistical curve fits rather than actual stress calculations. In this chapter, we will unite the loads environments with the structural geometry and material properties to calculate the stresses and to verify that the structures are sufficient to withstand them.

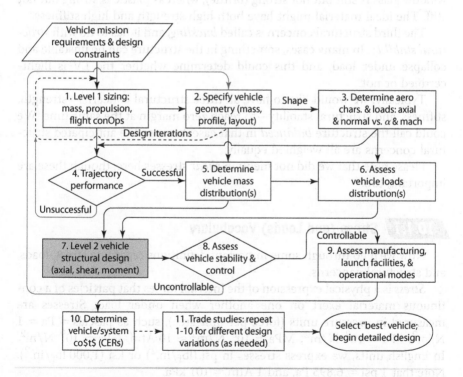

Fig. 10.1 This chapter will look at *stress analysis* using the ground and flight loads generated in the previous chapter.

10.1 Strength and Stress Analysis

In the design of a launch vehicle, mass is at a premium because of the exponential nature of the rocket equation. So, we'd like to make the structure as lightweight as possible. Yet, the launch vehicle (LV) has to survive the ground and launch environments, so it has to be at least somewhat beefy. As a structural designer, you will typically have *three* major structural concerns.

Strength is the property many people would think of first. Strength has to do with a structure's resistance to mechanical failure. The stronger something is, the more force it takes to damage it. But what does "damage" mean? In structures, engineers describe two types of damage: *yielding*, where a permanent deformation occurs (like when you bend a paper clip and it takes a permanent deformation with a new, different shape), and *ultimate*, where the structure actually breaks or fails.

However, there are more concerns than just strength.

The second structural concern is *stiffness*. Stiffness (or flexibility, its opposite) describes a structure's deflection under load. It also dictates a vehicle's natural vibration frequencies that, in turn, can affect its stability. It is important to note that *strength* is not the same as *stiffness*. For instance, window glass is stiff but not strong (brittle), whereas plastic is strong but not stiff. The ideal material might have both high strength and high stiffness.

The third structural concern is called *buckling*, and it has to do with *structural stability*. In many cases, something in the structure can be unstable and collapse under load, and this could determine whether the LV is flight-certified or not.

Typically, we would like to have all three structural concerns—strength, stiffness, and structural stability—to reach zero margin at the same time. We could call the structure *balanced* in that case, because the anticipated structural concerns are all weighted equally.

Please note that we did not mention *shear* stresses here, though these are important, too.

10.1.1 Stress (and Loads) Vocabulary

Now we'll go through some important definitions regarding stress, loads, and structural concerns.

Stress is a physical expression of the internal forces that particles of a continuous material exert on one another when under load. Stresses are measured in pressure units (force per unit area), such as pascals = Pa = 1 N/m^2, kPa = 10^3 N/m^2, MPa = 10^6 $N/m^2 \approx 10$ Atm, GPa = 10^9 N/m^2. In English units, we express stresses in psi ($lb_f/in.^2$) or ksi (1,000 $lb_f/in.^2$). Note that 1 psi = 6,895 Pa, and 1 Atm = 101 kPa.

Engineering stress is usually abbreviated using σ (Greek letter *sigma*). There may be subscripts also; a list of subscripts is given later in the chapter.

10.1.1.1 *Types of Stress*

There are various types of stresses that are dependent on the loading and boundary conditions. Among the most common ones are tension, compression, bending, and shear. Figure 10.2 shows the compressive case of stress.

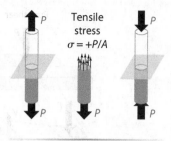

Fig. 10.2 Left to right: tension condition, tensile stress, and compression condition.

Stress numbers can be used to represent pressures in *tension*. Here, *tensile stress* refers to stress caused by stretching loads. To signify tension, we use a positive sign or $+$, according to standard engineering practices.

Another type of stress is called *compression* or *compressive*. This refers to a stress caused by being "squashed" or shortened. To indicate compression, we use a negative sign or $-$. Figure 10.2 shows both of these conditions.

Bending is a situation that creates a combination of tension, compression, and shear in different parts of an object, as shown in Fig. 10.3. This occurs, for example, when a swimmer stands on a diving board. The diving board curves downward, and the top of the board is in tension. The bottom of the board is in compression, and the place in the board where there is no tension or compression is called the *neutral axis*. It turns out that there is shear stress on the neutral axis, as the two "halves" of the beam try to slide past one another.

Shear stress is caused by loads creating a situation where two portions of the material are trying to pass by one another. In Fig. 10.4, we see two types of shearing: pure shear (left) and rivet shear (right). In both cases, one of the materials is being lopped off by another. A third case was described previously, along the neutral axis of a beam in bending.

10.1.1.2 *Stress Subscripts and Material Properties*

The following is a list of common stress terms and their definitions. As mentioned earlier, stress is abbreviated using σ (Greek letter *sigma*), and

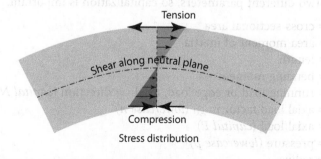

Fig. 10.3 Bending describes a situation where part of the structure is in tension, part is in compression, and part is in shear.

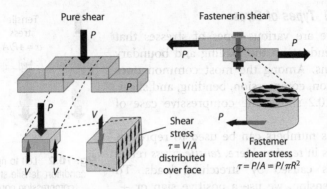

Fig. 10.4 Two examples of shear: pure shear (left) and rivet or fastener shear (right).

the subscript on the stress term refers to the condition. Some of the common subscripts are listed here:

σ_t = tensile stress (tensile stress is defined as positive, or greater than zero)
σ_c = compressive stress (<0)
τ_s = shear stress
σ_u = ultimate (failure) stress
σ_y = yield stress
σ_c = critical stress (buckling)
E = modulus of elasticity (Young's modulus)
G = shear modulus
α = coefficient of thermal expansion (CTE)
ρ = density

10.1.2 Forces, Geometry, and Moments

The stresses in structures result from applied loading and boundary or support conditions. The following is a list of common abbreviations for geometric, structural, load, and safety parameters. Note that some letters are used for two different parameters, so capitalization is important.

A = cross-sectional area
I = area moment of inertia
L = length
M = bending moment
N_x = running load or edge load, axial (x) direction (*capital N*)
n_x = axial load factor in g_0 (*lowercase n*)
P = axial load (*capital P*)
p = pressure (*lowercase p*)
R = radius
t = thickness
FS = factor of safety
MS = margin of safety

The last two items have to do with safety. The *factor of safety* refers to how much stronger a structure is than it needs to be for a planned load. In mathematical form,

$$F.S. = \frac{\text{allowable load}}{\text{design load}} \qquad (10.1)$$

In other words, a structure with $FS = 1.0$ will support the design load and no more; any additional load over the design load will cause failure. A structure with $FS = 1.4$ will fail at 1.4 times the design load. Usually, the design safety factor is specified or provided as a requirement. A collection of safety factors is provided in Table 10.1.

The *margin of safety* definition includes the factor of safety. It is defined as

$$M.S. = F.S. - 1 = \frac{\text{allowable load}}{\text{design load}} - 1 = \frac{\sigma_{\text{allowable}}}{\sigma_{\text{design}}} - 1 \qquad (10.2)$$

In other words, if the design is adequate, then the MS is greater than zero. If the design is deficient, its MS is negative. Finally, in some cases the MS is provided as a percentage rather than a number. In either case, a positive number or percentage is good, whereas a negative number or percentage is not good.

Table 10.1 Typical Factors of Safety

Item	Factors of Safety			
	Limit	Ultimate	Proof	Burst
Flight loads				
Inhabited Vehicle	1.0	1.40	NA*	NA
Uninhabited Vehicle	1.0	1.25	NA	NA
Pressure Loads				
Main liquid propellant tank				
Inhabited Vehicle	1.0	1.40	**	1.40
Uninhabited Vehicle	1.0	1.25	**	1.25
Solid Rocket Motor Casings				
Inhabited Vehicle	1.0	1.25	**	1.25
Uninhabited Vehicle	1.0	1.15	**	1.15
Hydraulic Vessels & Lines	1.0	NA	2.0	4.0

*NA = Not Applicable
** = Unique for Each Design
Source: [4]

10.2 Stress Determination Using External Loads

In the previous chapter, we calculated axial, shear, and bending moment distributions in a LV. In this section, we'll show how to determine the stresses that are caused by the load distributions. The results could be used to size tank wall thicknesses, for example. We will consider the following:

- How external environments and internal payload, structures, and propellants produce axial, shear, and bending loads on a cylindrical structure
- Stress vs strength
- Stress vs elastic stability (buckling)
- Ways to improve elastic stability (structural and other)
- Pressurization benefits

10.2.1 Cylinder Analysis Approach

For our analysis, we will use the *cylinder analysis* approach. The idea is to consider the stresses around perimeters of a simple cylinder whose ends are loaded with axial compression P and bending moment M, as shown in Fig. 10.5.

We will use the usual abbreviation for stress, σ. The axial stress σ_a is calculated as the load divided by the cross-sectional area, or

$$\sigma_a = -P/A \qquad (10.3)$$

where P is axial compressive load, and A is the hollow cylinder wall's cross-sectional area (shown as a thick oval-shaped line).

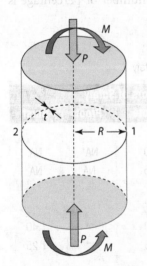

10.2.1.1 Calculation of a Cylinder's Cross-Sectional Area

To calculate stress, we need the cylinder's cross-sectional area A. To calculate A, let the cylinder's outer radius $= R_{out}$, and its inner radius $= R_{in}$. It's easy to see

$$A = \pi\left(R_{out}^2 - R_{in}^2\right) \qquad (10.4)$$

Factoring the term in parentheses,

$$A = \pi(R_{out} - R_{in})(R_{out} + R_{in}) \qquad (10.5)$$

Fig. 10.5 The assumed loads case for the cylinder analysis (the P arrows are pointing through the centerline of the cylinder).

Let's define R as the average radius $R = 1/2(R_{out} + R_{in})$. The wall thickness t is the difference between the outer and inner radii: $t = (R_{out} - R_{in})$. With these definitions,

Eq. (10.5) can be rewritten as

$$A = 2\pi Rt \tag{10.6}$$

which is an exact expression if the cylinder has "thin" walls, defined as $t \ll R$. This expression is identical to saying the area is the product of circumference $2\pi R$ and t, the wall thickness. So, the axial stress is

$$\sigma_a = -P/A = -P/2\pi Rt \tag{10.7}$$

10.2.1.2 Calculation of a Cylinder Cross-Section's Area Moment of Inertia I

We will need the *area* moment of inertia $= I$ in order to calculate stress due to bending. This is entirely different from the *mass* moments of inertia of an object as calculated in Chapter 9. From nearly any engineering handbook or structures textbook,

$$I = \pi\left(R_{out}^4 - R_{in}^4\right)/4$$
$$= (\pi/4)\left(R_{out}^2 - R_{in}^2\right)\left(R_{out}^2 + R_{in}^2\right)$$
$$= (\pi/4)(R_{out} - R_{in})(R_{out} + R_{in})\left(R_{out}^2 + R_{in}^2\right)$$
$$\approx (\pi/4)(t)(2R_{ave})\left(2R_{ave}^2\right)$$
$$I = \pi R^3 t \tag{10.8}$$

where R is the average radius, and the cylinder has "thin" walls, $t \ll R$.

10.2.1.3 Stress Calculation

Most strength of materials textbooks show that the bending stress σ_b in a Bernoulli-Euler beam due to pure bending is

$$\sigma_b = MR/I = MR/\pi R^3 t$$
$$\sigma_b = M/\pi R^2 t \tag{10.9}$$

recalling that $I = \pi R^3 t$ for small t, eq. 10.8.

In the cylinder shown in Fig. 10.5, the axial stress σ_a is negative: the axial load causes uniform compression everywhere along the walls of the cylinder. However, the bending stress σ_b varies among positive, negative, and zero, depending on where on the circumference we look at stresses. In Fig. 10.5, it can be seen that σ_b is negative at point 1 and positive at point 2.

We can also see that the maximum compressive stress occurs at point 1. The magnitude of the maximum stress is the sum of the magnitudes of the axial stress and the bending stress

$$|\sigma_{max1}| = |\sigma_a| + |\sigma_b| = \frac{P}{2\pi Rt} + \frac{M}{\pi R^2 t} \tag{10.10}$$

Remember this peak stress σ_{max1} is compressive and has a negative sign.

The terms *edge load* and *running load* refer to the load averaged over the cylinder's circumference. The edge load N_x for a cylinder of thickness t is defined as

$$N_x = \sigma t = \frac{P}{2\pi R} + \frac{M}{\pi R^2} \tag{10.11}$$

The units of the edge load are force/unit length: either lb_f/in. or N/m.

Example 1a: Propellant Tank Stress Calculations

We now use our newly calculated stress terms in an example.

Given: a cylindrical tank contains a fluid mass $m = 50$ T (T = metric tonne = 1,000 kg).

The tank has geometry as shown in Fig. 10.6: the height of the CM $z_{cm} = 4$ m; its diameter = 2.5 m. Assume the tank has thin walls of thickness $t = 1$ mm $= 10^{-3}$ m, and its average radius $R = 1.25$ m.

The tank's worst-case loading occurs as a combination of axial and lateral loads given as $n_{lateral} = 0.2\ g_0$, $n_{axial} = 2.5\ g_0$.

Determine the maximum stress in the tank at section A-A, which is 4 m below the center of mass of the tank and its fluid. Use a factor of safety of 1.25. Assume that the effective weight of the tank due to the lateral and axial accelerations is concentrated at its CM, and the loads are supported around the tank's circumference. A free-body diagram of the loading is given in Fig. 10.7.

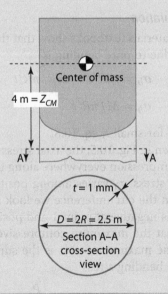

Fig. 10.6   Tank geometry used in example calculation.

(Continued)

Example 1a: Propellant Tank Stress Calculations *(Continued)*

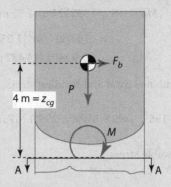

Fig. 10.7   Applied loads on region of interest.

Area A and moment of inertia I calculations:

$$A = 2\pi Rt = 2\pi(1.25 \text{ m})(0.001 \text{ m}) = 0.007854 \text{ m}^2 \qquad (10.12)$$

$$I = \pi R^3 t = \pi(1.25 \text{ m})^3(0.001 \text{ m}) = 0.006136 \text{ m}^4 \qquad (10.13)$$

Propellant tank axial and bending loads and moment calculations:

$$\text{Axial load } P = n_{\text{axial}}\, mg_0$$

$$= 2.5(50 \text{ T})(1{,}000 \text{ kg/t})(9.8 \text{ m/s}^2)$$

$$= 1{,}225 \text{ kN}$$

Horizontal bending force $F_b = n_{\text{lateral}}\, mg_0$

$$= 0.2(50 \text{ T})(1{,}000 \text{ kg/T})(9.8 \text{ m/s}^2) = 98 \text{ kN}$$

$$(10.14)$$

$$\text{Bending moment } M = F_b\, z_{cg}$$

$$= (98 \text{ kN})(4 \text{ m})$$

$$= 392 \text{ kNm} \qquad (10.15)$$

Note: We use the following convention to distinguish concentrated loads from pressure loading: a concentrated *load P* (force units) is capitalized, whereas a *pressure p* (units of force/unit area) is lowercase.

$$\text{Axial stress} = \sigma_a = (P/A)\text{FS}$$

$$= (-1{,}226 \text{ kN}/0.007854 \text{ m}^2)1.25$$

$$= [-156{,}800 \text{ kN/m}^2]1.25$$

$$= -195.1 \text{ MPa} \,(-28{,}295 \text{ psi}) \qquad (10.16)$$

(Continued)

Example 1a: Propellant Tank Stress Calculations *(Continued)*

Bending stress $= \sigma_a = (MR/I)\text{FS} = (392\,\text{kNm})(1.25\,\text{m})/0.006136\,\text{m}^4]1.25$

$$= [79.9\,\text{kN/m}^2]1.25$$
$$= 99.9\,\text{MPa}\,(14{,}487\,\text{psi}) \qquad (10.17)$$

Total stress $\sigma_{\text{tot}} = $ combined axial and bending stresses

$$\sigma_{\text{tot}} = |\sigma_a| + \sigma_b = 195.1 + 99.9\,\text{MPa} = 295.0\,\text{MPa}(42{,}783\,\text{psi}) \qquad (10.18)$$

Now we calculate allowable stress.

Steel:

$$\sigma_{\text{allow-steel}} = 150\,\text{ksi} = 1{,}034\,\text{MPa}$$
$$\sigma_{\text{tot}}/\sigma_{\text{allow-steel}} = 295.0/1{,}034 = 28.5\% \qquad (10.19)$$

Aluminum:

$$\sigma_{\text{allow-Al}} = 65\,\text{ksi} = 448\,\text{MPa}$$
$$\sigma_{\text{tot}}/\sigma_{\text{allow-Al}} = 295.0/448 = 65.8\% \qquad (10.20)$$

These assumed values are typical when one includes the strength loss due to welds.

Shear stress $\sigma_s = $ lateral force $\div$ area.

$$\sigma_s = (F/A)\text{FS} = [98\,\text{kN}/0.07854\,\text{m}^2]1.25$$
$$= [12.48\,\text{MPa}]1.25$$
$$= 15.61\,\text{MPa}\,(2{,}264\,\text{psi}) \qquad (10.21)$$

Minimum Thickness for Strength

The required cylinder thickness $t = $ (edge load or running load) $\div$ (allowable stress), or

$$t = \frac{\dfrac{P}{2\pi R} + \dfrac{M}{\pi R^2}}{\sigma_{\text{allow}}} \qquad (10.22)$$

Using this, we can solve for the required thickness of the steel tank

$$t_{\text{steel}} = \frac{\dfrac{1{,}225\,\text{kN}}{2\pi(1.25\,\text{m})} + \dfrac{392\,\text{kNm}}{\pi(1.25\,\text{m})^2}}{1{,}034\,\text{MPa}} = 0.228\,\text{mm} = 0.009\,\text{in.} \qquad (10.23)$$

This is approximately the thickness of *one* sheet of paper! For the aluminum

(Continued)

Example 1a: Propellant Tank Stress Calculations *(Continued)*

cylinder,

$$t_{Al} = \frac{\dfrac{1{,}225\,\text{kN}}{2\pi(1.25\,\text{m})} + \dfrac{392\,\text{kNm}}{\pi(1.25\,\text{m})^2}}{448\,\text{MPa}} = 0.527\,\text{mm} = 0.0207\,\text{in.} \qquad (10.24)$$

This is about *three* sheets of paper.

These calculations result in thicknesses that are very small and are likely to be unstable under load. The situation where the material is simply too thin is known to cause what is called a *minimum gauge* restriction. Such a structure would have to use thinner materials than are available.

10.2.1.4 Minimum Gauge Issues

The previous example and similar calculations may suggest tank wall thicknesses of ~0.1 or 0.2 mm, which is thinner than a sheet of paper and too thin to consider for a practical structure. In most cases, you may have to choose a larger thickness or *gauge* that you can purchase off-the-shelf.

With space vehicles, sometimes we "jump through a hoop" to get the desired properties, which might be minimum mass. For example, the Titan II missile was manufactured using chemicals to dissolve the skin thickness down to the desired amounts, a process called *chemical milling* or *chem-milling*. This resulted in one of the largest loaded-to-empty mass ratios in the business for its first step.

In general, we suggest utilizing a minimum thickness no less than about ~0.5 mm or about ~0.020 in.

Example 1b: Minimum Thickness for Shear

We computed the shear force $V = F_b = 98$ kN earlier. If we assume that the allowable shear stress is one-half of the compressive allowable, or $\tau_{\text{allow}} = \frac{1}{2}\sigma_{\text{allow}}$, we find that $\tau_{\text{allow-steel}} = 517\,\text{MPa}$, and $\tau_{\text{allow-Al}} = 224\,\text{MPa}$. Now we can determine the needed thickness to withstand shear.

The equation for shear stress is $\tau_{\text{shear}} = \frac{V}{2\pi Rt}$, so we can solve for the minimum thickness $t = \frac{V}{2\pi R\tau_{\text{allow}}}$. For steel thickness: $t_{\text{steel}} = \frac{98\,\text{kN}}{2\pi(1.25\,\text{m})517\,\text{MPa}} = 2.41 \times 10^5\,\text{m} = 0.024\,\text{mm}$; for aluminum structure thickness, $t_{Al} = \frac{98\,\text{kN}}{2\pi(1.25\,\text{m})224\,\text{MPa}} = 5.57 \times 10^5\,\text{m} = 0.056\,\text{mm}$.

These calculations provide thicknesses that are very low, much less than minimum gauge recommendations. For this reason, we will neglect horizontal shear stresses in the remainder of our calculations.

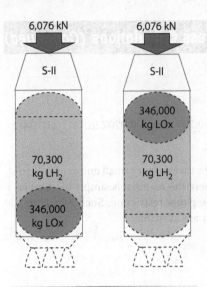

6,076 kN   6,076 kN

S-II   S-II

346,000 kg LOx

70,300 kg LH₂   70,300 kg LH₂

346,000 kg LOx

Fig. 10.8 Placing the heavier LOx tank *forward* (right in figure) increases the bending moment imposed on the step.

10.2.2 A Design Consideration: Relative Tank Position

Let's suppose you are examining the S-II step of the Saturn V, which used hydrolox propellants. The designers were already committed to a common bulkhead to reduce mass, but they had two choices as to the location of the LOx tank vs the LH_2 tank: the LOx tank could be either aft or forward. (Because of the low density of liquid hydrogen, the **loaded** LOx tank is almost five times heavier than the **loaded** LH_2 tank.) These two situations are shown in Fig. 10.8. As you will soon see, the order is very important to the state of stress in the tank cylinders.

- *Case 1: LOx tank aft, S-II as flown*: This arrangement, shown on the left of Fig. 10.8, moves the larger mass aft, and the prelaunch wind loads produced a running load of $N_x = 140.4$ kN/m.
- *Case 2: LOx tank forward, alternate order S-II*: This situation is on the right side of Fig. 10.8. You might guess that having the larger mass forward generally produces more static stability in an aerodynamic sense, and this is in fact true. Structurally, however, we need to consider the prelaunch wind loads, which over the longer length LH_2 tank produced running loads of $N_x = 290.5$ kN/m, more than twice as large as case 1. The result of the higher prelaunch N_x increased the S-II's LH_2 tank wall mass *796 kg* (per Greenberg [3]).

Can you guess why the designers chose case 1?

What effect does the order have on the structure beneath the second step? As shown in Fig. 10.9, we find that for the S-IC, the first step of the

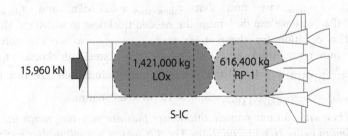

15,960 kN   1,421,000 kg LOx   616,400 kg RP-1

S-IC

Fig. 10.9 The lower step of the Saturn V is unaffected by the tank order in the S-II step above.

Saturn V, the tank mass is insensitive to forward or aft LOx in S-II. This is because the forces applied to the top of S-IC—$F_{above\ S\text{-}II} + F_{S\text{-}II} = 6{,}076$ kN $+ 9{,}884$ kN $= 15{,}960$ kN—are independent of the arrangement of the upper stages: *the total applied axial load is the same regardless of which tank is forward on the S-II.*

10.3   Allowable Stresses Based on Stability (Buckling) Criteria

10.3.1   Critical Stresses and Buckling

A structure, even if adequate in strength and stiffness, can still have instability under load. The elastic stability (or buckling) concerns serve to reduce the allowable stress levels. It is critical that elastic stability be designed into the system so as to prevent instabilities under any circumstances. The so-called "critical" stress σ_c includes stability considerations and may reduce allowable loads below those allowed for strength and stiffness. In the case of a launch vehicle, the stress state includes axial stress from gravity or acceleration, bending stress from ground or flight winds, and shear loads. All of these must be properly taken into account when assessing the flightworthiness of an LV.

LV structures can have instability problems. Figure 10.10 shows the partially collapsed oxidizer tank on a Titan II missile. The loads from the fully loaded second step sitting on top of the empty first step's oxidizer tank exceeded its stability margin, causing its skin to buckle as shown in the photo.

Fig. 10.10   This Titan II's partially collapsed step 1 oxidizer tank was supporting fully loaded step 2 tanks. The chemically-milled tank skins can be seen in the foreground. Source: U.S. Air Force/Maj. Mark Clark [9].

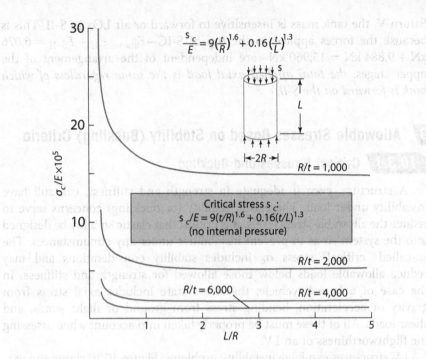

Fig. 10.11 Plots of critical stress vs L/R and radius divided by thickness values R/t. Most values are essentially constant above $L/R > \sim 1.5$ or so. Source: [8].

Guidelines for elastic stability can be found in several places. One guideline is shown in Fig. 10.11, which provides values of critical stress σ_c as functions of the tank's length-to-radius fraction L/R and radius divided by thickness values R/t. It shows that the values for critical stress rapidly converge to steady values once $L/R > 1.5$ or so. The reference provides the unpressurized critical stress σ_c as governed by the following relation:

$$\frac{\sigma_c}{E} = 9\left(\frac{t}{R}\right)^{1.6} + 0.16\left(\frac{t}{L}\right)^{1.3} \tag{10.25}$$

with E being the material's Young's modulus or modulus of elasticity. The plot shows that the critical stress grows with the ratio of R/t dropping, as would be expected. (Lower values of R/t indicate thicker and thus more stable tank walls.)

Let's continue working with the steel and aluminum tanks in the previous example and find the critical buckling stresses in the unpressurized tanks.

Example 2: Critical Stress Calculation (No Internal Pressure)

For the previous examples, we considered steel and aluminum tanks with length $L = 8$ m, $R = 1.25$ m, and wall thicknesses $t_{steel} = 0.228$ mm and $t_{Al} = 0.527$ mm. Also recall that $E_{steel} = 207$ GPa (30×10^6 psi), and $E_{Al} = 69$ GPa (10×10^6 psi). Plugging these values into the critical stress equation, we find that

$$(\sigma_c/E) = 9(t/1.25 \text{ m})^{1.6} + 0.16(t/8 \text{ m})^{1.3} \qquad (10.26)$$

where the thickness and modulus of the desired material, steel or aluminum, is used.

The numbers for steel yield $(\sigma_c/E)_{steel} = 9.58 \times 10^{-6}$ and for aluminum yield $(\sigma_c/E)_{Al} = 3.64 \times 10^{-5}$. Solving for the critical stresses,

$$\sigma_{c\text{-steel}} = (9.58 \times 10^{-6})$$

$$E_{steel} = (9.58 \times 10^{-6})(207 \text{ GPa}) = 1.98 \text{ MPa} = 288 \text{ psi} \qquad (10.27)$$

much less than σ_{allow}; the allowable load is

$$(\sigma A)_{steel} = 1.98 \text{ MPa}(0.00327 \text{ m}^2) = 3,552 \text{ N}$$
$$= 799 \text{ lb}_f(\text{hardly anything!}) \qquad (10.28)$$

For aluminum,

$$\sigma_{c\text{-Al}} = (3.64 \times 10^{-5})E_{Al} = (3.64 \times 10^{-5})(68.95 \text{ GPa})$$
$$= 2.51 \text{ MPa} = 364 \text{ psi} \qquad (10.29)$$

much less than σ_{allow}; the allowable load is

$$(\sigma A)_{Al} = 2.51 \text{ MPa}(0.00755 \text{ m}^2) = 10.4 \text{ kN} = 2,331 \text{ lb}_f(\text{not much}) \qquad (10.30)$$

If this were the lower stage of a rocket, these capabilities would be much too small. Consider that the tank holds 50 T, with a 1-g_0 weight of 490 kN. That weight is orders of magnitude larger than the allowable loads. Therefore, we need to find a way to increase the allowable loads.

10.3.2 Ways to Increase Allowable Critical Axial Stress

Here are seven ways that we can increase the critical axial load or stress for buckling failure in thin-walled cylinders based on elastic stability:

1. Increasing E (higher elastic or Young's modulus)
2. Increasing wall thickness t
3. Using honeycomb/sandwich structure
4. Adding rings to decrease effective length
5. Adding longitudinal stringers
6. Using rings *and* stringers
7. Pressurizing the cylinder

Method 1 could possibly be introduced and reduce total mass—if the replacement material has lower density than the original. Method 2, increasing the wall thickness, seems sure to add mass. We can see that methods 3, 4, 5, and 6 *all add mass* because they add structural materials. Pressurization, method 7, seems the only way that we can improve the situation without adding mass.

10.3.3 Structural Methods to Increase Critical Axial Stress

One way to increase the critical axial load is to increase the modulus of elasticity (also known as Young's modulus). This requires a change to a higher-modulus material, which may not be possible due to material compatibility, fabrication issues, or cost.

A second way to increase the critical axial load is to simply increase the wall thickness, which directly affects the area moment of inertia. The side effect of increasing the thickness is a direct increase of structural mass, which is not desirable.

A third way to increase the critical axial load is to change the cylinder walls from a thin-walled structure to a built-up composite structure, often described as "sandwich" or honeycomb construction, as shown in Fig. 10.12. Here two thin skin sheets are bonded to a hollow (as shown in the figure) or solid lightweight core to produce a lightweight but stiff wall section. Sandwich

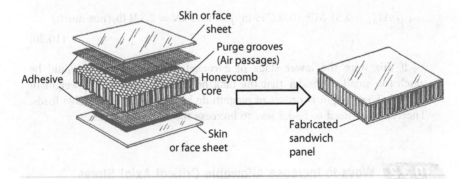

Fig. 10.12 Sandwich or honeycomb construction. Purge grooves allow internal air pressure to vent, preventing delamination of the structure. A delamination event on a Saturn V flight is shown in Fig. 11.46. Source: NASA.

construction is often used for payload fairings, with composite skins and hollow Nomex® honeycomb core or foam core such as Rohacell®.

Note: One thing to consider when using hollow core materials is whether there is a path for air to escape as the vehicle climbs into lower pressure and eventually vacuum. If there is no path, the pressure buildup can *delaminate*, or cause the face sheets to separate from the core material, often with catastrophic results. Most hollow cores have small holes drilled into them to allow for venting; these are labeled "purge grooves" in Fig. 10.12.

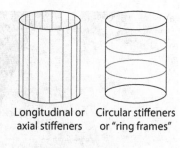

Longitudinal or   Circular stiffeners
axial stiffeners   or "ring frames"

Fig. 10.13   Two arrangements of structural stiffeners.

Yet another way to increase the critical axial load is to add stiffeners to the inside or outside of a structure, as shown in Fig. 10.13. The axial stiffeners can be riveted, bonded, or welded on to provide increased I_x per unit of cross-sectional area (per unit mass), similar to the effect of a sandwich structure. A variety of stiffener cross-sections are used, some of which are shown in Fig. 10.14.

Circular stiffeners or ring frames serve to increase a structure's buckling resistance, whereas axial stiffeners increase a cylinder's effective bending stiffness. Many propellant tanks include parts called *slosh baffles* whose purpose is to damp out sloshing waves of the fluid inside the tank. The slosh baffles, as seen in Chapter 7, could double as circular stiffeners. Figure 10.15 shows both types of stiffeners in action.

The geometric stiffness of a structure that bends about its x-axis is quantified by its *area moment of inertia*. We saw this in the tank stiffness example presented earlier, where we calculated the area moment of the cross-section of a circular tank. The calculation is $I_x = \int_{z_{min}}^{z_{max}} x(z)z^2 \, dz$. Table 10.2 gives the formulae that provide area moments of inertia for several different shapes. Area moments for other shapes may be found in aerospace or mechanical engineering handbooks.

For some structures, one way to remove mass is to machine out or chemically mill out much of the thickness of a metal plate until only a

a)

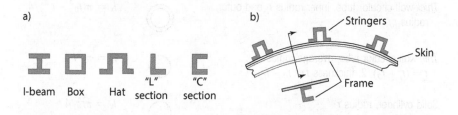

I-beam   Box   Hat   "L" section   "C" section

b)   Stringers

Skin

Frame

Fig. 10.14   a) Stiffener shapes; b) how they are attached to a structure.

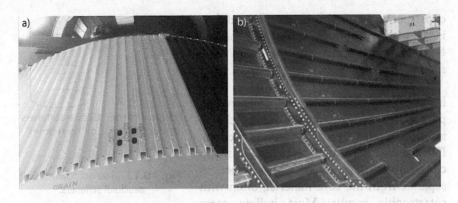

Fig. 10.15 a) Hat sections attached to *outside* of the forward skirt structure of Saturn V S-IC first step. b) Internal *circular ring frames* and *axial* stringers of Boeing 787 graphite/epoxy fuselage.

triangular or rectangular web remains. These types of structures are called *isogrid* and *orthogrid*, respectively. Photos of these structures are shown in Fig. 10.16. Isogrid properties are given in [6].

Finally, we come to pressurization. We will see that pressurization is of great benefit to the design of a launch vehicle.

10.4 Effect of Internal Pressure on Stresses

The stresses in the wall of a cylindrical tank due to pressure inside may be examined by making "cuts" in the tank walls and calculating the forces that they generate. The situation created by internal pressure in a tank is shown in Fig. 10.17. Figure 10.17a shows the effect of cutting the tank in a direction *perpendicular* to its axial or longitudinal axis; Fig. 10.17b shows the effect

Table 10.2 Area Moments of Inertia for Several Shapes

Shape	Illustration	Area Moment of Inertia
Solid rectangle, base width b and height h (x in long direction)		$I_x = bh^3/12$
Thick-wall circular tube, inner radius r_i and outer radius r_o		$I_x = \pi(r_o^4 - r_i^4)/4$
Thin-wall circular tube, average radius $r = (r_i + r_o)/2$, thickness $t = r_o - r_i$		$I_x = \pi r^3 t$
Solid cylinder, radius r		$I_x = \pi r^4/4$

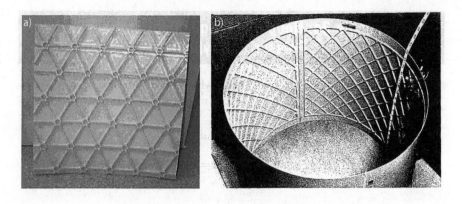

Fig. 10.16   Machining out ~90% of a plate barely affects its longitudinal strength, while saving huge amounts of mass: a) isogrid; b) orthogrid.

of cutting the tank *along* its axial/longitudinal axis. By considering the internal pressure and the forces on the edges needed to preserve equilibrium, it turns out that the axial stress $\sigma_{axial} = pr/2t$. Because the pressure is trying to force the pieces apart, the edges are in tension, so the hoop and axial stresses both have *positive* (+) signs. For both cylindrical and spherical pressurized tanks, the axial stress σ_{axial} is the same value.

We have previously seen that the LV's acceleration produces compressive stresses. We have also seen that any external bending moments produce both compressive and tensile stresses, depending on which side of the LV cylinder we are on. But, pressure *always* produces a tensile load, so it can *increase the amount of allowable axial load!* Thus, we have found a method to deal with the very small critical buckling loads allowed in the examples worked earlier in the chapter.

We didn't discuss the cut along the axis of the cylinder shown in Fig. 10.17b. Here, tensile stresses are again needed for equilibrium. The stress in this situation is called *hoop stress*: $\sigma_{hoop} = pr/t$. The name comes from the metal straps or hoops that are used to hold wooden barrels together. The hoop stress is perpendicular to the axial pressure stress and is twice the magnitude of the axial stress: $\sigma_{hoop} = pr/t = 2\,\sigma_{axial}$.

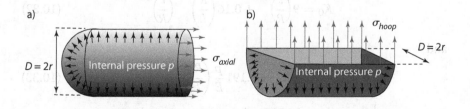

Fig. 10.17   Stresses in spherical-ended tanks caused by internal pressure p: a) axial stress σ_{axial}; b) hoop stress σ_{hoop}.

Example 3a: Hoop and Longitudinal Stresses in Pressurized Cylinders

We continue with the example tank with dimensions length $L = 8$ m (26.25 ft), $t = 1$ mm (0.039 in.), $R = 1.25$ m (4.1 ft), and internal pressure $p = 344.7$ kPa (50 psi).

The longitudinal stress is $\sigma_{axial} = pr/2t = (344.7 \text{ kPa})(1.25 \text{ m})/(2 \times 0.001 \text{ m}) = 215.5$ MPa (31,250 psi).

The hoop stress is $\sigma_{hoop} = pr/t = (344.7 \text{ kPa})(1.25 \text{ m})/0.001 \text{ m} = 431$ MPa (62,500 psi).

10.4.1 Internal Pressure Adds Load Capability

Let's explore the situation caused by internal pressure. As we observed, the positive axial *tensile* stress due to internal pressure $\sigma_{axial} = +pr/2t$ has the potential to cancel out some of the compressive axial stresses and allow the imposition of additional axial load. As a matter of fact, we can see that the internal pressure p allows us to carry an axial load $F_{axial} = \pi R^2 p$ without *any* compressive stress!

For our example, the allowable load $F_{axial} = 4.909p$ (Pa) can be carried, and with 344.8 kPa (50 psi) internal pressure, the cylinder can carry 1.69 MN (380,500 lb$_f$) before the outer shell is subjected to *any* compression. So we have increased the load-carrying capability of a thin-walled tank. What are the critical loads or stresses for a pressurized structure? There is a plot comparable to the unpressurized Fig. 10.11 shown earlier: the corresponding plot is shown in Fig. 10.18.

Referring to Fig. 10.18, the relation for the critical stress is given as

$$\sigma_c = \frac{K_{total}Et}{r} = \frac{(K_0 + K_p)Et}{r} \tag{10.31}$$

where

$$K_0 = 9\left(\frac{t}{R}\right)^{0.6} + 0.16\left(\frac{R}{L}\right)^{1.3}\left(\frac{t}{R}\right)^{0.3} \tag{10.32}$$

and

$$K_p = 0.191\frac{p}{E}\left(\frac{R}{t}\right)^2 \tag{10.33}$$

up to a maximum value of 0.229, when it stays constant.

The variable $K_{total} = K_0 + K_p$. Let's apply these data and examine our example tanks.

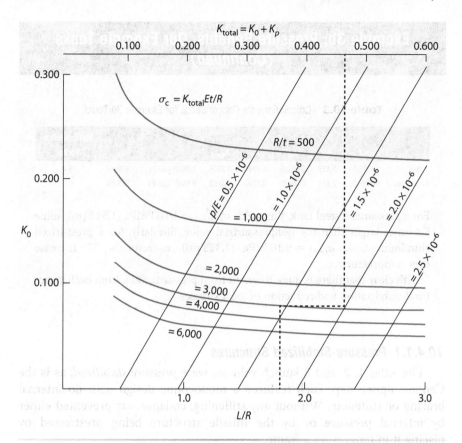

Fig. 10.18 This plot allows one to calculate the critical stress for a pressurized structure. Parameters K are defined in text. Source: [12].

Example 3b: Pressure Benefits, Our Example Tanks

The inputs to the calculation are given in Table 10.3. With a pressure value of 344.75 kPa (50 psi), the appropriate moduli for steel and aluminum, the radius $R = 1.25$ m, and the thicknesses of steel and aluminum are, respectively, 0.228 mm and 0.527 mm, we find

$$K_{\text{p-steel}} = K_{\text{p-Al}} = 0.229 \qquad (10.34)$$

the maximum value.

With $K_{\text{0-steel}} = 0.05249$ and $K_{\text{0-Al}} = 0.08630$,

$$K_{\text{total-steel}} = K_{\text{0-steel}} + K_p = 0.05249 + 0.229 = 0.2815 \qquad (10.35)$$

$$K_{\text{total-Al}} = K_{\text{0-Al}} + K_p = 0.08630 + 0.229 = 0.3153 \qquad (10.36)$$

(Continued)

Example 3b: Pressure Benefits, Our Example Tanks (Continued)

Table 10.3 Calculations for Critical Stress for Example 3b Tanks

	p/E	R/t	K_0	K_p	K_{total}	σ_c	σ_c	Multiplication Factor
Steel	1.667E-06	5,478	0.05249	0.2290	0.2815	1.063E + 07 Pa	1,542 psi	5.4
Aluminum	5.000E-06	2,374	0.08630	0.2290	0.3153	9.158E + 06 Pa	1,328 psi	3.7

For a pressurized steel tank, this yields $\sigma_{c\text{-steel}} = 10.63$ MPa (1,542 psi), some 5.4 times larger than the nonpressurized value. Similarly, for a pressurized aluminum tank, $\sigma_{c\text{-Al}} = 9.16$ MPa (1,328 psi), a factor of 3.7 increase over nonpressurized.

It's clear that there is quite a benefit from pressurization, from both structural stabilization and reduction of axial stresses!

10.4.1.1 Pressure-Stabilized Structures

The Atlas 1, 2, and 3 launch vehicles were *pressure-stabilized*, as is the Centaur upper step. Each featured a monocoque design with no internal bracing or stiffening. Without any stiffening, collapse was prevented either by internal pressure or by the missile structure being prestressed by placing it in tension on a frame.

Atlas used a nominal internal pressure of 34.5 kPa (5 psi) and would collapse if pressure dropped to zero! Figure 10.19 shows the result of a pressure loss in an Atlas-Agena LV. The lower fuel tank was filled with thousands of

Fig. 10.19 This Atlas-Agena LV collapsed when its internal pressure was lost, completely destroying its payload and dumping thousands of gallons of RP-1 on to the launch pad. Source: U.S. Air Force via [2].

gallons of kerosene that spilled onto the pad (see white cloud in the bottom of the fourth photo) when the upper portion of the missile collapsed and the Agena upper stage pivoted to the ground, destroying its payload. There was no fire.

10.4.2 Other Factors to Consider Concerning Pressurization

Levels of internal pressure in the propellant tanks should be based on the *worst* (largest magnitude) of the following four conditions:

1. *Stabilization of the tanks' structure*: As discussed previously, it is necessary that internal pressure be sufficient to produce axial tensile loads that are adequate to carry the applied axial loads and bending moments, taking the buckling stability of the tank into account. Axial loads due to drag and effective weight of structures above the tank are generated forward of the tank and are assumed to act at the tanks' upper cylinder edge. Effective weights are the masses forward of the tank multiplied by the vehicle's acceleration.
2. *Turbopump cavitation*: For pump-fed rocket engines, there must be adequate pressure to prevent cavitation at the turbopumps. The *absolute pressure* in the tank added to the hydrostatic pressure above the pumps must be greater than the sum of the vapor pressure and the net suction head. These pressures are defined as follows:
   * The *gauge pressure* is obtained by subtracting ambient atmospheric pressure, which is dependent on vehicle altitude and atmospheric temperature.
   * The *vapor pressure* of a propellant is a function of its temperature.
   * *Suction head* or *net positive suction head (NPSH)* is determined by turbopump characteristics at the desired flow quantities needed for engine start and running. If needed, additional pressure may be added in order to account for losses due to *flow friction* within the propellant ducting.
   * *Hydrostatic pressure* p_h depends on propellant density ρ, vehicle acceleration n_x, and the top of the propellant's level above the pump intake h: $p_h = \rho g_0 n_x h$. The largest value of hydrostatic pressure occurs when $n_x h$, the product of axial acceleration (increasing with burn time) and propellant level (decreasing with burn time), is maximum, which may occur in the latter portion of a stage's flight, when n_x is increasing due to higher T/W ratio as propellants are consumed.
3. *Propellant boiling*: The tank pressure must be greater than the propellant's vapor pressure (see Point #2, bullet 2) at the propellant's temperature.
4. *Bulkhead reversal*: To prevent the inversion (turning inside out) of a common bulkhead, the pressure on the concave side must be larger than the pressure on the convex side. A difference of at least 35 kPa (5 psi) has

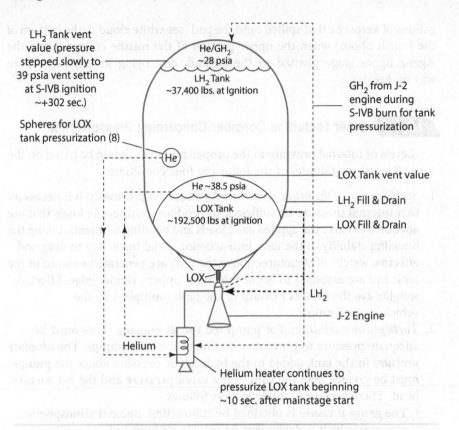

LH$_2$ Tank vent value (pressure stepped slowly to 39 psia vent setting at S-IVB ignition ~+302 sec.)

He/GH$_2$ ~28 psia

LH$_2$ Tank ~37,400 lbs. at Ignition

GH$_2$ from J-2 engine during S-IVB burn for tank pressurization

Spheres for LOX tank pressurization (8)

He

LOX Tank vent value

He ~38.5 psia

LH$_2$ Fill & Drain

LOX Tank ~192,500 lbs at ignition

LOX Fill & Drain

LOX

LH$_2$

J-2 Engine

Helium

Helium heater continues to pressurize LOX tank beginning ~10 sec. after mainstage start

Fig. 10.20 In the S-IVB upper step of Saturn IB AS-204, the lower LOx tank has 72.4 kPa (10.5 psi) greater pressure than the upper LH$_2$ tank to prevent bulkhead reversal. Source: NASA.

been suggested. This pressure differential only applies to tanks with common bulkheads. Figure 10.20 shows the internal pressures in the Apollo 5's S-IVB upper step, with the LOx tank pressure 72.4 kPa (10.5 psi) above that of the LH$_2$ tank located above its common bulkhead [7]. A higher pressure in the tank below will ensure that there is no bulkhead reversal.

Example 4: Hydrostatic Pressure

Assuming our example tank is filled with liquid oxygen of density $\rho = 1{,}141 \text{ kg/m}^3$, we may calculate the hydrostatic pressure p_h on the bottom. We will use the same values for the previous aluminum cylinder examples, $E = 69$ GPa, $P_{\text{axial}} = 1{,}225$ kN, moment $M = 392.3$ kNm, $R = 1.25$ m, and $L = h = 8$ m. We will calculate the hydrostatic pressure for $n_x = 1$ (on the ground), and for flight while accelerating at $n_x = 3.0$:

(Continued)

Example 4: Hydrostatic Pressure *(Continued)*

On the pad, $n_x = 1$, and the pressure at the bottom of the cylinder is $p_h = \rho g_0$
$h = (1{,}141 \text{ kg/m}^3)(9.8 \text{ m/s}^2)(8 \text{ m}) = 89.45 \text{ kPa} = 13 \text{ psi} = 0.88 \text{ atm}$.

In-flight, accelerating with a load factor $n_x = 3.0$, the hydrostatic pressure
$p_h = \rho n_x g_0 \, h = 268 \text{ kPa} = 39 \text{ psi} = 2.65 \text{ atm}$.

The hoop stress for the case of thickness $t = 1$ mm is $\sigma_{\text{hyd-h}} = p_{\text{hyd}}$
$r/t = (268 \text{ kPa})(1.25 \text{ m})/0.001 \text{ m} = 335 \text{ MPa (49 ksi)}$.

How do these values compare to axial and hoop stress due to internal pressure? Previously, an assumed internal pressure value of 3.4 ATM or 50 psi gave axial stress: $\sigma_r = \frac{pR}{2t} = 215.5 \text{ MPa (31.2 ksi)}$, and hoop stress $\sigma_h = \frac{pR}{t} = 431 \text{ MPa (62.5 ksi)}$.

These values are about the same order of magnitude as hydrostatic pressure, so they cannot be neglected or ignored.

Example 5: Calculating Required Thickness for Strength

In the previous Al cylinder examples in this chapter, we selected a wall thickness and then checked the thickness value's stress level to see if it was sufficient. One might ask, what is the minimum thickness for the example? Let's apply the same parameters: $E = 69 \text{ GPa}$, $P_{\text{axial}} = 1{,}225 \text{ kN}$, $M = 392 \text{ kNm}$, $R = 1.25 \text{ m}$, $L = h = 8 \text{ m}$, $n_x = 3.0$, $\rho = 1{,}141 \text{ kg/m}^3$, and FS = 1.25, and determine what the minimum required thickness is.

For the axial or longitudinal direction, we utilize eq. T1:

$$t_{\text{req-axial}} = \frac{FS}{\sigma_{\text{allow}}}\left[\frac{P}{2\pi R} + \frac{M}{\pi R^2} - \frac{R}{2}(p + \rho g_0 n_x h)\right]$$

$$= \frac{1.25}{448 \text{ MPa}}\left[\frac{1.23\text{E}6 \text{ N}}{2\pi \cdot 1.25 \text{ m}} + \frac{392 \text{ kNm}}{\pi(1.25 \text{ m})^2}\right.$$

$$\left. - \frac{1.25 \text{ m}}{2}\left(345 \text{ kPa} + 3.0 \cdot \frac{1{,}141 \text{ kg}}{\text{m}^3} \cdot \frac{9.8 \text{ m}}{\text{s}^2} \cdot 8 \text{ m}\right)\right]$$

$$= 0.00044 \text{ m} = 0.44 \text{ mm}$$

For lateral or hoop direction:

$$t_{\text{req-hoop}} = \frac{FS}{\sigma_{\text{allow}}} R(p + \rho g_0 n_x h)$$

$$= \frac{1.25}{448 \text{ MPa}} 1.25 \text{ m}\left(345 \text{ kPa} + 3.0 \frac{1{,}141 \text{ kg}}{\text{m}^3} \cdot \frac{9.8 \text{ m}}{\text{s}^2} \cdot 8 \text{ m}\right)$$

$$= 0.00214 \text{ m} = 2.14 \text{ mm}$$

(Continued)

These are thicknesses determined by strength (yielding) considerations, but we also must also consider elastic stability or buckling. These will be considered in the next examples.

10.4.2.1 Geysering

The use of cryogenic propellants brings additional considerations. For example, the temperature of LOx cannot exceed −187°C (−297°F), or it will boil into gaseous oxygen, and boiling of LH$_2$ occurs at a much lower temperature. Boiling of propellants cannot be allowed because it can lead to *geysering*, which can begin when a portion of propellant in the propellant line boils. The bubble created by the boiling causes a slug of fluid to rise up into the tank. When the propellant line refills due to gravity and tank ullage pressure, a hydraulic hammer can be developed with enough strength to damage the propellant delivery system.

The occurrence of geysering depends on the vehicle configuration. A large potential heating source occurs in a LOx system whose configuration locates the LOx tank above a nonrefrigerated propellant tank, some examples being the Saturn V's S-IC step and the Space Shuttle's external tank. Both of these vehicles require long LOx suction lines that may pass through a lower, noncryogenic propellant tank, such as RP-1, or may be located on the outer surface of the lower tank.

To prevent geysering, some vehicles (S-IC) utilize a LOx conditioning system that injects helium bubbles into the LOx tank to cool the LOx rapidly and prevent boiling. Other means to prevent geysering are to use a LOx recirculation system or an antigeyser line, as discussed in Fisher [5].

Of course, the simplest method to avoid geysering is to locate the LOx tank at the bottom of the step (if this is permitted by the vehicle dynamics as well as stability and control considerations). Of course, this information applies to any cryogenic propellant, including liquid hydrogen and liquid methane, in addition to LOx.

10.4.2.2 Other Layout Considerations

Besides pressurization and geysering, there are many more considerations that can drive the design of a vehicle's pressurization system, as well as the design of its tanks, bulkheads, propellant conditioning, and other factors. For more information, the works of Fisher [5] and Ring [10], as well as from Hellebrand, Platt, and McCool & McKay in the Further Reading list are recommended.

10.5 Determining the Overall Stress State

To calculate the overall stress state in the tank walls, we must include the following factors:

- Axial stress $\sigma_a = -P/A$ due to compressive loads (axial direction, always $-$)
- Bending stress σ_b (axial direction $\pm$, depends on location)
- Axial stress due to tank pressure p: $\sigma_{axial} = +pr/2t$ (axial direction, always tensile, $+$)
- Hoop stress due to tank pressure p: $\sigma_{hoop} = +pr/t$ (hoop direction, always tensile, $+$)
- Hydrostatic pressure axial stress $\sigma_{axial\text{-}hyd} = +p_h\, R/2t = +\rho g_0 n_x hR/2t$ (always tensile, $+$)
- Hydrostatic pressure hoop stress $\sigma_{hoop\text{-}hyd} = +p_h\, R/t = +\rho g_0 n_x hR/t$ (always tensile, $+$)
- Shear stress $\tau_{shear,} = V/2\pi Rt$, in the shear direction rather than tensile or compressive

With these factors, the magnitude of the maximum axial stress in the tank walls is

$$\sigma_{axial\text{-}max} = |\sigma_a| + |\sigma_b| - |\sigma_l| - |\sigma_h|$$
$$= -P/A - MR/I + (p + \rho g_0 n_x h)R/2t \qquad (10.37)$$

and the magnitude of the maximum tank-wall hoop stress is

$$\sigma_{hoop\text{-}max} = +|\sigma_l| + |\sigma_h| = (p + \rho g_0 n_x h)R/t \qquad (10.38)$$

An additional term is the shear stress τ_{shear}, which is to be combined with the other stresses to determine the overall state of stress. The calculation of the overall state of stress is beyond the scope of this book; the reader is referred to any text on stress analysis or mechanics of materials. The result of all these stress factors is shown in Fig. 10.21.

The tank and hydraulic pressure-induced stresses [the last terms of Eqs. (10.37) and (10.38)] create tensile stresses that usually reduce σ_{max} in the axial direction, and so they *increase* the load-carrying capability. We also have observed that adding tank pressure also benefits our design by providing stabilization to the tank walls. A slight benefit is also obtained from the hydrostatic pressure component, but this component also increases the hoop stress.

Summaries of the ground and flight loads to be analyzed for stress analysis are provided in Tables 10.4 and 10.5 below. These two tables provide summaries of aerodynamic, inertial, and pressure (internal and hydrostatic) loading that occur in these two vehicle states.

For ground loads, all horizontal loads are due to winds, and all vertical loads are due to gravity. *Ground loads DO NOT consider any propellant tank pressurization, or its accompanying benefit, to ensure that the structure remains stable and safe regardless of the state of pressurization of ANY of the*

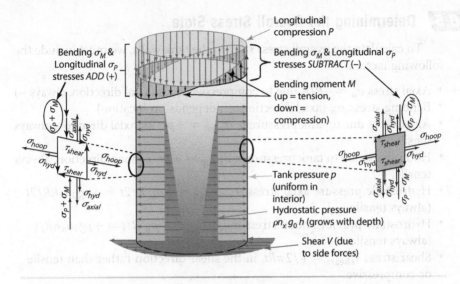

Fig. 10.21 The stresses occurring in a pressurized, accelerating tank. Axial, hoop, and hydrostatic stresses are all tensile (positive). Longitudinal is compressive (negative). Bending is either tensile or compressive depending on the direction of bending. Shear stresses are in a different direction and are labeled with τ rather than σ for tensile or compressive stresses.

tanks on the launch pad. This accommodation is made in case of an emergency requiring depressurization of one or more tanks that might have to occur before propellants are able to be unloaded.

10.5.1 Stress Analysis Summary

To get at least an approximate feel for the state of stress in a LV, one must do the stress calculations for *three cases*: axial or longitudinal (the direction along the "long" axis of the vehicle), lateral hoop (a "sideways" direction,

Table 10.4 A list of ground loads needing stress analysis

Category	Pressurized Structures Analysis Loads: Tanks, Propellant Containers		Unpressurized Structures Loads: Skirts, Intertanks, PLF	
Loads due to	Lateral	Axial	Lateral	Axial
Aero loads: @ centroid	qC_DS_{side}	—	qC_DS_{side}	—
Item's own mass load	—	$m_s g_0$ @ CM	—	$m_s g_0$ @ CM
Inside or attached mass	—	$m_p g_0$ @ CM	—	$m_k g_0$ @ CM
Hydrostatic pressure	—	$\rho g_0 h_{fluid}$ (bottom)	—	$\rho g_0 h_{fluid}$ (bottom)
Ullage pressure?	—	None	—	None

Structure mass m_s is item's empty mass. "Inside" mass m_k is propellant. Attached mass m_k is payload, electronic boxes, etc.

Table 10.5 A list of flight loads needing stress analysis

Category	Pressurized Structures Analysis Loads: Tanks, Propellant Containers		Unpressurized Structures Loads: Skirts, Intertanks, PLF	
Loads due to	**Lateral**	**Axial**	**Lateral**	**Axial**
Aero loads: @ centroid	$qC_N\,S_{ref}$*	$qC_N\,S_{ref}$*	$qC_N\,S_{ref}$*	$qC_N\,S_{ref}$*
Item's own mass load: @ CM	$m_s\,n_z\,g_0$ @ CM	$m_s\,n_x\,g_0$ @ CM	$m_s\,n_z\,g_0$ @ CM	$m_s\,n_x\,g_0$ @ CM
Inside mass or attached	Full‡/part† $m_p\,n_z\,g_0$	Full‡/part† $m_p\,n_x\,g_0$	$m_k\,n_z\,g_0$ @ CM	$m_k\,n_x\,g_0$ @ CM
Pressure?	Flight pressure p	Flight pressure p	—	—
Hydro-static p?	—	$\rho n_x g_0\,h_{fluid}$ (bottom)	—	—

*Transformed from wind to body coordinates.
†partially-loaded bottom step tanks.
‡full upper step tanks.
Structure mass m_s is item's empty mass. "Inside" mass m_k is propellant. Attached mass m_k is payload, electronic boxes, etc.

perpendicular to the axial direction), and buckling or elastic stability (also in the axial direction). The equations for stress existing in these three cases are as follows:

Maximum stress, longitudinal direction of structures:

$$\sigma_{\text{max-vert}} = -|\sigma_a| - |\sigma_b| + \sigma_{\text{axial-pressure}}$$

$$= -|\sigma_a| - |\sigma_b| + \frac{(p+p_h)R}{2t}. \tag{10.39}$$

This is the magnitude of the compressive stresses created by axial and bending forces less the tensile relief created by internal pressure.

Maximum tensile stress in lateral direction of structures:

$$\sigma_{\text{max-horiz}} = \sigma_{\text{hoop}} = \frac{(p+p_h)R}{t}. \tag{10.40}$$

This is the magnitude of the lateral hoop stress created internal pressure from the tank pressurization and any hydrostatic pressure from liquid propellant.

The expression for critical buckling stress for *unpressurized* structures, such as constant-radius skirts, intertanks, and interstages can be re-written from what we have seen earlier in a slightly different form so that it may be later compared with a similar expression that *does* include the effects of internal pressure. The revised form of the unpressurized expression is shown on the right side of eq. (10.41) below:

$$\sigma_c = \frac{P}{A} + \frac{MR}{I} = \left[9\left(\frac{t}{R}\right)^{0.6} + 0.16\left(\frac{R}{L}\right)^{1.3}\left(\frac{t}{R}\right)^{0.3}\right]\frac{Et}{R} \qquad (10.41)$$

The expression for critical buckling stress for *pressurized* structures such as tanks or propellant containers is shown in eq. (10.42):

$$\sigma = \frac{P}{A} + \frac{MR}{I} = \left[9\left(\frac{t}{R}\right)^{0.6} + 0.16\left(\frac{R}{L}\right)^{1.3}\left(\frac{t}{R}\right)^{0.3} + \min\left\langle \begin{matrix} 0.191\frac{p}{E}\left(\frac{R}{t}\right)^2 \\ 0.229 \end{matrix} \right\rangle \right]\frac{Et}{R}$$

$$(10.42)$$

Here it can be seen that the added third term in the square brackets corresponds to the pressurization term. The "min" notation shown in Eq. (10.42) means that the last term within the square brackets will be whichever of the upper expression or the lower express (0.229) is smaller, so it will never exceed 0.229. Also, we have removed the hydrostatic pressure term since it will vanish when the tank is empty.

The practical stress analysis needed means using equations 10.39 & 10.40 along with either 10.41 or 10.42, depending on pressurization. If any of these stresses exceed the allowable value for the part's material, then there will be a negative margin and the LV cannot be expected to survive the expected environment. Practically speaking, *there is more that needs to be done*, a way of saying "back to the drawing board." The parts need to be re-sized.

If any of the LV components stress margins are *negative*, their thicknesses will need to be increased until all margins are above a minimum desired value. Wall thicknesses need to be based on the higher-magnitude loads at the *bottom* of the tank (remember shear, axial load, and hydrostatic pressure tend to increase towards the bottom of the vehicle – as does bending moment when on the ground). *Pressurization* has been shown to decrease wall thickness, as we will see later, when we rearrange these equations to solve for thicknesses.

The equations for minimum thicknesses provided below (eqs. 10.43–10.46) are the thickness equations corresponding to the four stress equations (10.39–10.42) above; this means using equations 10.43 & 10.44 along with either 10.45 or 10.46, depending on whether the item in question is pressurized or not. Eqs. 10.39 and 10.40 could be solved by a trial-and-error approach, but it's easier to just solve them for the minimum thickness required for those items having negative margins, as provided in Eqs. 10.43

and 10.44. The buckling stress equations 10.41 and 10.42 can't be solved in closed form, and must be instead solved numerically.

For axial minimum thickness defined by strength:

$$t_{\text{req-axial}} = \frac{FS}{\sigma_{\text{allow}}}\left[\frac{P}{2\pi R} + \frac{M}{\pi R^2} - \frac{R}{2}(p + \rho g_0 n_x h)\right]. \tag{10.43}$$

For lateral or hoop direction thickness defined by strength:

$$t_{\text{req-hoop}} = \frac{FS}{\sigma_{\text{allow}}}R(p + \rho g_0 n_x h). \tag{10.44}$$

Unpressurized critical buckling thickness t_c is the thickness value of constant-radius skirts, intertanks, interstages, etc. that solves:

$$\frac{P}{2\pi R t_c} + \frac{M}{\pi R^2 t_c} - \left[9\left(\frac{t_c}{R}\right)^{1.6} - 0.16\left(\frac{t_c}{L}\right)^{1.3}\right]\frac{E t_c}{R} = 0. \tag{10.45}$$

Pressurized critical buckling thickness t_c is the thickness value that solves:

$$\frac{P}{2\pi R t_c} + \frac{M}{\pi R^2 t_c}$$
$$- \left[9\left(\frac{t_c}{R}\right)^{0.6} + 0.16\left(\frac{R}{L}\right)^{1.3}\left(\frac{t_c}{R}\right)^{0.3} + \min\left\langle \begin{array}{c} 0.191\frac{p}{E}\left(\frac{R}{t_c}\right)^2 \\ 0.229 \end{array} \right\rangle\right]\frac{E t_c}{R} = 0. \tag{10.46}$$

For Eqs. 10.45 and 10.46, the critical thickness t_c can be easily found numerically using a solver such as Excel®'s "Goal Seek."

Once the necessary minimum thicknesses have been obtained, to make things less complicated one may assume that the wall thickness is constant for each individual tank, intertank, and interstage structure, rather than tapering. Of course, each item will likely have a different thickness depending on its own specific load environment. The reader should also be aware that in actual flight vehicles, it's common practice to vary the thickness along a part's axis so that less mass is needed for portions with reduced loading; this practice is beyond the scope of this book.

Recalling the minimum gauge discussion earlier, the equations 10.43–10.46 may produce unrealistically-low wall thicknesses that need to be replaced with a minimum gauge value, for which we suggest a value of 0.5 mm. To numerically deal with this minimum-gauge issue, it's possible to insert a conditional equation into an analysis that provides the logical equivalent of "if necessary thickness < 0.5 mm, use 0.5 mm." The stress and margin calculations would use the value provided.

Another note: if your design has a tapered interstage or skirt, *do not reduce its thickness.* Equations 10.41 and 10.45 only apply to constant-diameter cylinders, and taper results in a higher than actual buckling load at the reduced-diameter portion. Treatment of these and other geometrical cases are beyond the scope of this book, and can be found in references related to elastic stability in thin structures.

This entire sizing process must be done for all anticipated load cases. In this book, we only have dealt with ground winds and max-q loads, but in reality, there are many other loads cases that also must be analyzed, such as transportation loads, liftoff loads, staging and separation loads, maneuvering, etc.

Once all the thickness re-sizing calculations are completed, each item's required wall thickness will be the *largest* of the values found in the three thickness equations used and a minimum gauge value (0.5 mm suggested), *for all of the loads cases that are analyzed.* Thus, in equation form, the required thickness t_k for item k is

$$t_k = \max\Big[(t_1, t_2, t_3)_{\text{transportation}}, \ (t_1, t_2, t_3)_{\text{ground loads}}, \cdots,$$
$$(t_1, t_2, t_{3 \text{ or } 4})_{\text{flight}}, \ t_{\text{min gauge}}\Big]$$

As soon as the structural thicknesses have been updated, the changes in empty mass that accompany the changes in thickness have to calculated. The calculated values of exposed surface area S_k (i.e. $S_k = \pi D_{cyl\text{-}k} \, h_k$ for a skirt or cylinder of diameter $D_{cyl\text{-}k}$ and height h_k; for a tank, $S_k = S_{cyl\text{-}k} + 2\, S_{dome\text{-}k}$) and the newly-determined thicknesses t_k to find each item's current mass $m_k = \rho_k \, S_k \, t_k$.

In some cases, items will have lower mass, in others, mass will increase. It's important to group the parts by step, i.e. Step 1 parts listed separately from Step 2, so the changes in each step can be determined. As was shown in the tradeoff ratios discussion in Chapter 5, a mass increase in upper steps is much harder to deal with than a mass increase in a lower step. Depending on how much payload mass margin is available, a large enough mass gain can make the mission unfeasible without completely redesigning the LV.

If the inert structural weight is too much, and some of the structure has defaulted to a suggested minimum gauge, it may be possible to use unusual (and expensive) techniques to reduce metal skin thicknesses and lower mass. The *Titan II* ICBM had skins where caustic chemicals were used to dissolve the metal to reduce the skin thickness to the desired amounts! This process is called "chemical milling" or "chem-milling," and is shown in Fig. 10.22 below. A minimum thickness of 0.5 mm is suggested in this text. This minimum-gauge issue can make the "real" mass of a structure end up greater than what we'd like it to be.

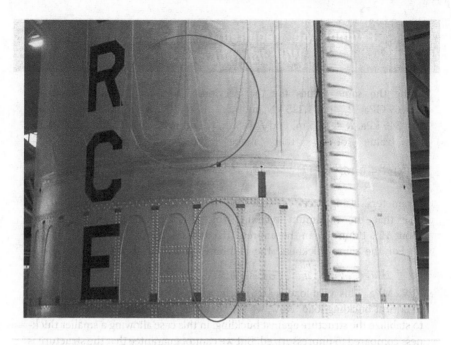

Fig. 10.22   The skins of the Titan II ICBM were chemically-milled to remove unneeded metal and reduce mass where possible. Chemically-milled areas can be seen as the 'rounded' outlines to the right of the letters "R", "C", and "E". Rivets are used to fasten stiffeners where necessary. The systems tunnel is on the right side of the photo. (Photo: Edberg).

Example 6a: Required Thickness: Buckling, *no Internal Pressure*

Using the same values for the previous aluminum cylinder examples, $E = 69$ GPa, $P_{\text{axial}} = 1,225$ kN, moment $M = 392.3$ kNm, $R = 1.25$ m, and $L = 8$ m, we numerically solve eq. (10.45) for the value of t_c that makes the expression $\frac{P}{2\pi R t_c} + \frac{M}{\pi R^2 t_c} - \left[9\left(\frac{t_c}{R}\right)^{1.6} - 0.16\left(\frac{t_c}{L}\right)^{1.3}\right]\frac{Et}{R} = 0$ with the given parameters. We used Excel®'s "goal seek" function to find that the minimum buckling thickness $= 0.00388$ m $= 3.88$ mm.

The combined compressive stress produced by the same axial + bending load applied to a cylinder with the newly-obtained critical thickness of 3.88 m is $\sigma_a + \sigma_b = \frac{P}{2\pi R t_c} + \frac{M}{\pi R^2 t_c} = \frac{1,225\,\text{kN}}{2\pi(1.25\,\text{m})0.00388\,\text{m}} + \frac{392\,\text{kNm}}{\pi(1.25\,\text{m})^2} = 60.8$ MPa, which is only 13.6% of the allowable compressive stress of 448 MPa, *but is 100% of critical buckling load.* This makes it clear that elastic stability may be the driver for many cylinder designs. However, let's look to see if internal pressure will assist in the next example.

Example 6b: Required Thickness: Buckling, With Internal Pressure

Using the same values for the previous aluminum cylinder examples, $E = 69$ GPa, $P_{axial} = 1{,}225$ kN, moment $M = 392.3$ kNm, internal pressure $p = 345$ kPa, $R = 1.25$ m, and $L = 8$ m, we numerically solve eq. (10.46) for the value of t_c that makes the expression $\frac{P}{2\pi R t_c} + \frac{M}{\pi R^2 t_c} -$

$$\left[9\left(\frac{t_c}{R}\right)^{0.6} + 0.16\left(\frac{R}{L}\right)^{1.3}\left(\frac{t_c}{R}\right)^{0.3} + \min\left\langle \begin{matrix} 0.191\frac{p}{E}\left(\frac{R}{t_c}\right)^2 \\ 0.229 \end{matrix} \right\rangle \right] \frac{E t_c}{R} = 0$$ with the given

parameters. Using Excel®'s "goal seek" function, we find for the pressurized case that the minimum buckling thickness $t_c = 0.00329$ m $= 3.29$ mm, about 15% thinner than the unpressurized case.

For the same thickness, the combined mechanical stress is $\sigma_a + \sigma_b -$

$\sigma_p = \frac{P}{2\pi R t_c} + \frac{M}{\pi R^2 t_c} - \frac{pR}{2t_c} = \frac{1{,}225 \text{ kN}}{2\pi(1.25 \text{ m})0.00251 \text{ m}} + \frac{392 \text{ kNm}}{\pi(1.25 \text{ m})^2} - \frac{1.25 \text{ m}}{2}345 \text{ kPa} =$

71.72 MPa. This axial load is 16% of allowable stress $= 448$ MPa, but is 100% of critical buckling load. Now, we see that internal pressurization also assists to stabilize the structure against buckling, in this case allowing a smaller thickness compared to unpressurized. But we cannot guarantee that the structure is always pressurized, especially on the ground.

Example 7: Selecting Required Thickness

The thickness required for a structure subjected to a variety of loads is the largest-magnitude thickness needed to sustain any of the loads. Looking at the examples earlier in the chapter, we have the situation listed in Table 10.6. The worst-case thickness is that from Example 6a, the *unpressurized case*, at **3.88 mm**. Thus, that portion of the structure must be 3.88 mm thick.

Table 10.6 Required Aluminum Cylinder Thicknesses Found in Examples 1–6

Example	Description	Needed Thickness
1a	Strength analysis (axial P & moment M only, unpressurized)	0.527 mm
1b	Strength analysis (shear load V only, unpressurized)	0.056 mm
5a	Strength analysis axial (P, M, p_{hydro}, pressurized) stress	0.44 mm
5b	Strength analysis hoop (p_{hydro}, pressurized) stress	2.14 mm
6a	Buckling analysis (P, M, unpressurized)	**3.88 mm**
6b	Buckling analysis (P, M, pressurized)	3.29 mm

10.5.2   This Is *Just the Start* of Stress Analysis

Note that we have only considered the walls of the tanks. In the *real* world, we have to do *detailed stress analysis* on the entire vehicle, where we must find the stresses in:

* Tank domes
* Intertank-to-tank and interstage-fastened joints
* Welds
* Holes/cutouts for feed lines, pressure lines, etc.
* Stiffeners/suspensions for plumbing
* Tank walls, insulation, thermal gradients, and adhesive bonds that take low/high temperatures
* Many other locations and structures

There is an almost-never-ending list of projects for stress engineers to work on! The stress calculations for these items require detailed knowledge of the shapes of the domes and skirts, and their associated stiffeners (if any) and applied loads, and are beyond the scope of this book. For more detail, the reader should refer to air vehicle structural analysis texts such as Bruhn [1]. Most analyses today are not done by hand, but instead utilize computer modeling techniques.

10.5.3   Finite-Element Modeling

You have been introduced to stress analysis by hand. We have not discussed how to deal with holes, hatches, trusses, fasteners, thermal loading, and a host of other concerns with structures and stress analyses. However, this book is not intended to be a detailed stress manual; it just aims to provide basic knowledge of the "big picture" regarding stress analysis and loads.

These days, much of the detailed stress analysis is done with finite-element modeling (FEM). For this, a mathematical model is made of the launch vehicle by dividing it into thousands of smaller elements whose properties are computed using geometry and material properties. The elements are joined together mathematically, and the appropriate loads are applied. The response is the deformed structure and the stresses associated with the set of loads.

Frequently, the finite-element model (also abbreviated FEM) is generated from computer-aided design (CAD) models generated by vehicle designers. A launch vehicle FEM might look similar to the Vega FEM shown in Fig. 10.23.

10.5.4   FEM Updates

The FEM of the launch vehicle is updated as the design matures and more details are known about its construction, materials, environments,

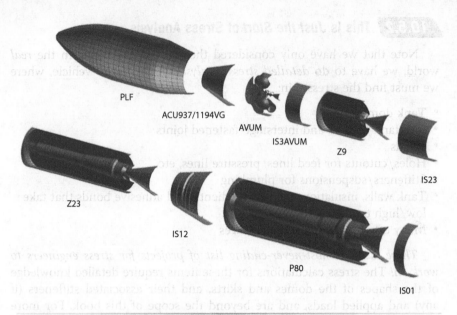

PLF

ACU937/1194VG

AVUM

IS3AVUM

Z9

IS23

Z23

IS12

P80

IS01

Fig. 10.23 This finite-element model of the Vega launcher shows how the LV's major structural components are split into smaller finite elements to be used for detailed stress and thermal analysis. Source: ESA.

and so forth. Once hardware is made, even for prototypes, it is tested, and the FEM is adjusted to agree with the test results in terms of stiffness, natural frequencies, and the like. After flight, the model is again adjusted to agree with measurements, after which the launch vehicle FEM and load environments are well understood. The model may then be used for *coupled loads analysis*, which will be discussed in Chapter 12.

10.6 Summary: Simple Rules for LV Structures

Remember that inert or empty mass is king: to maximize a vehicle's performance, one must minimize mass wherever possible. This entails the following:

- Make all load paths efficient. They should be short, straight, and continuous. An example of a bad structural element would be a bellows, similar to that found on an accordion musical instrument, which has practically zero stiffness.
- Wherever possible, place heavier items lower, remembering the weight savings provided to the S-II step on the Saturn V. However, this may not be feasible due to stability and/or controllability considerations.
- Minimize openings in structures. Cutouts and doors require reinforcements that can be heavy and require additional analyses.

* Some materials have better properties (e.g., higher allowables for yield) at cryogenic temperatures. Take advantage of it! (For example, note the improvement of aluminum's tensile yield properties shown in Chapter 7, Fig. 7.58.)

Be sure to watch out for potential compression or buckling instability! To improve or stabilize these situations, consider:

* Internal pressure, which helps enormously
* Radial and/or circular stringers, especially for unpressurized structures such as intertanks, interstages, and the like
* Sandwich construction, which also helps stabilize unpressurized structures, such as intertanks, interstages, and skirts.

Comment on Structural Design

In some circumstances, such as "small" vehicles, it's not uncommon to come across what's called a "minimum gauge" issue. This is when the calculations suggest that your tank walls (for example) need to be only (say) 0.1 or 0.2 mm thick, which is thinner than a sheet of paper. Clearly, this is too thin to consider for a practical structure.

In a case like this, you may have to go with a larger thickness, referred to as the material *gauge*, that you can purchase from a catalog or off the shelf. However, with space vehicles, it's possible to go to extraordinary measures (usually meaning *costly*!) to get the desired properties. If you take a look at a Titan II missile, chemicals were used to dissolve the skin thickness to the desired amounts. This process is called *chemical milling* or *chem-milling* and is an expensive process. In lieu of such special measures, we recommend using a thickness of 0.5 mm (0.040 in.) as a practical minimum. This minimum-gauge issue can make the "real" mass of a structure result in a greater value than we'd like it to be.

References

[1] Bruhn, E. F., Orlando, J. I., and Meyers, J. F., *Missile Structures Analysis and Design*, Library of Congress Card #67-28959, Tri-State Offset, Cincinnati, OH, 1967.
[2] Day, D. A., "Not a Bang, but a Whimper," The Space Review, 16 March 2009, http://www.thespacereview.com/article/1326/1 [retrieved 10 May 2018].
[3] Greenberg, H. S., "Structural Analysis Techniques for Preliminary Design of Launch Vehicles," short course notes, UCLA Extension, 2011.
[4] Heitchue, R. D., *Space Systems Technology*, Reinhold, New York, NY, 1968.
[5] Fisher, M. F., "Propellant Management in Booster and Upper Stage Propulsion Systems," NASA TM-112924, 1997.
[6] McDonnell Douglas Astronautics, "Isogrid Design Handbook," NASA CR-124075 Rev. A, 1973.
[7] Marshall Space Flight Center, "Technical Information Summary Apollo 5 (AS-204/LM-1) Apollo Saturn IB Vehicle," MSFC R-ASTR-S-67-63, 1967.
[8] Seifert, H. S., and Boelter, L. M., *Space Technology*, Wiley, New York, 1959.

[9] Stumpf, D. K., *Titan II: a History of a Cold War Missile Program*, Univ. of Arkansas Press, Fayetteville, 2000.
[10] Ring, E. *Rocket Propellant and Pressurization Systems*. Prentice-Hall, Inc., Englewood Cliffs, NJ, 1964.

Further Reading

Bruhn, E. F., Orlando, J. I., and Meyers, J. F., *Analysis and Design of Flight Vehicle Structures*, Tri-State Offset, Cincinnati, OH, 1973.
Hellebrand, E. A., "Structural Problems of Large Space Boosters," Conference Preprint 118, presented at the ASCE Structural Engineering Conference, Oct. 1964.
McCool, A. A., and McKay, G. H., "Propulsion Development Problems Associated with Large Liquid Rockets," NASA TM X-53075, Aug. 1963.
Platt, G. K., Nein, M. E., Vaniman, J. L., and Wood, C. C., "Feed System Problems Associated with Cryogenic Propellant Engines," Paper 687A, National Aero-Nautical Meeting, Washington DC, April 1963.

10.7 Exercises

Problem 1: Soda Can Buckling Stress Calculations

Obtain an *unopened* aluminum soda or soft drink can. Try to select one that has no dents or other noticeable flaws in its thin walls. Measure the diameter and length, and try to estimate the wall thickness and internal pressure.

See if you can determine the critical buckling load for *vertical* loads for the soda can. This involves piling weights on top of your can, taking care to center them on the axis of symmetry. (Explain: Why is it important to center the weights?)

You may do the buckling experiment in one or both of two ways:

1. Try this *before* opening the container, so the internal pressure is likely to increase the critical buckling load.
2. Try this with your *opened* can. Note that the stress/strain conditions are different from those of internal pressure. Also note that in the past, the author has been able to support his entire 100-kg mass on an opened can, as long as he climbed up *very carefully*!

Once you've done this, measure the wall thickness (if you have the proper tools). You may expose the can's wall by poking a hole into the side of the can, and then using scissors to cut along the circumference. *Be very careful not to cut yourself: the can edges are very sharp.* If measurement tools are not available, you may be able to find an aluminum can's wall thickness on the Internet. Compare this to your estimate and provide calculations for pressurized and unpressurized buckling loads and stresses. Do your calculations agree with your measurements? Explain why or why not.

Problem 2: LV Stress Calculations

Provide no more than four significant digits for your answers: round numbers as necessary.

2.0. **Introduce** your launch vehicle, some basic information is needed.
   a) Show side view with overall length and diameter(s).
   b) Provide description table with payload mass(es), mission(s), and liftoff mass, liftoff T/W, propellants and I_{sp}s used. For max-q, provide q, v, α, Mach, n_x, n_z (from earlier assignments).
2.1. **Material selection.** Choose material(s) for your LV's structural walls and domes. For liquid steps, use an aluminum (2014-T6 or 2219-T87 suggested) or aluminum-lithium (2195 suggested); for solid motors, use steel from Space Shuttle SRB casings (D6AC suggested) or composites (many use carbon fiber; look into strap-ons for *Delta* or *Atlas*). For intertanks and interstages, you may use whatever material you like, but remember that it's not pressurized, so more material may be required. You can ignore effects of buckling or local crippling for this assignment, but in reality, these factors would also need to be analyzed.
   Provide a table containing the following information to be used for stress analysis, noting that material allowables may *increase* or *decrease* depending on temperature. The table should state the properties and allowables for the appropriate operating temperature(s).
   a) Material(s)
   b) Estimated operating temperature, °C or K. Be sure it includes propellant temperature range(s). Also determine if strength or modulus increases or decreases with high temp.
      Note: if there is a significant change in properties, use two or more lines to capture that information. They may get better or worse!
   c) Density (kg/m^3)
   d) Elastic modulus (MPa)
   e) Yield stress, corresponding to operating temperature(s) in (MPa)
   f) Source of material properties (provide in a format that a fellow colleague could use). For example, note that 2219 aluminum specs may be found at (last ref. 2018 Nov. 17): http://asm.matweb.com/search/SpecificMaterial.asp?bassnum = MA2219T87.
2.2. **Stress calculation, max-q.** Use your calculated max-q shear, bending, and axial load values from the previous assignments along with an assumed value of internal pressure (see below) and your known vehicle diameter(s) to calculate your vehicle's stresses at the load stations specified below, using the methods given in the book. **Use a Factor of Safety FS = 1.25, meaning that the minimum MS is 25% = 0.25. Provide your MS values in percent.**
   Internal pressure. *For liquid tanks*: assume **300 kPa** flight pressurization value. If you have a common bulkhead, make sure you have

35 kPa higher pressure to prevent bulkhead reversal. *Solids*: calculate the maximum chamber pressure based on your required performance, referring to rocket propulsion texts for needed details. You must design to the chamber pressure regardless of ground or max-*q* loads. If you cannot determine this internal pressure, use the maximum pressure inside of the Space Shuttle SRMs or a vehicle similar to the one you are analyzing, and provide the source of your assumed pressure value.

If any of your stress margins are *negative*, you will need to increase the thickness and re-calculate. To make things less complicated, you can assume wall thickness is constant for tank, intertank, and interstage structure, but individual items may have different thicknesses depending on their load environments. Therefore, calculate wall thicknesses based on the loads at the *bottom* of the tank (remember shear, axial load, and hydrostatic pressure tend to increase towards the bottom of the vehicle – as does bending moment when on the ground). *Don't forget to add pressurization*, which may help to decrease wall thicknesses.

Also, note that you need not repeat the tables for any increased wall thicknesses. Instead, just put a conditional equation into your spreadsheet to the effect of "if necessary thickness < 0.5 mm, use 0.5 mm." The stress calculations can continue using the value in the cell.

This analysis should be an extension of the loads spreadsheets you have made in previous homework assignments. You may want to add columns to your loads table showing the location/part description, the selected material, the loads, the stresses and the percent of yield stress or margin of safety at the locations below.

(do not calculate the stress of, or resize the payload fairing):

1. Top of 2nd step forward skirt (payload, PLF, and PAF are probably above here),
2. Top of 2nd step upper tank cylinder,
3. Bottom of 2nd step upper tank cylinder (this is where Step 2's upper tank's propellant weight + tank weight is supported, see vertical up-arrows in figure)*,
4. Top of 2nd step lower tank cylinder*,
5. Bottom of 2nd step lower tank cylinder (where Step 2's lower tank's weight is supported),
6. Top of interstage (all of step 2 is supported above here)†,
7. Top of 1st step upper tank,
8. Bottom of 1st step upper tank cylinder (where Step 1's upper tank's weight is supported)*,
9. Top of 1st step lower tank cylinder*,
10. Bottom of 1st step lower tank cylinder (this is where Step 1's lower tank's propellant + tank weight is supported),
11. Bottom of aft skirt or ground support (all components of LV are supported above this).

*If you are using a solid motor or motors, replace items 3 & 4 and/or 8 & 9 (as needed) with a calculation at one-half the length of the solid motor casing(s).

*If you have an internal spherical or ellipsoidal tank, replace items 2 & 3, 4 & 5, 7 & 8, and/or 9 & 10 with a single calculation located horizontally at the equator of the tank.

*If using a common bulkhead, replace items 3 & 4 and/or 8 & 9 with a calculation where the bulkhead reaches the tank cylinder.

†If you have a tapered interstage or skirt, do not reduce its thickness. The taper results in a higher than actual buckling load at the top, compared to the bottom.

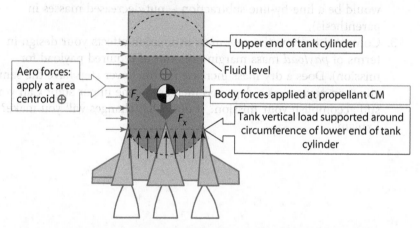

2.3. **Stress calculation, ground winds**. Repeat the calculations done for Problem 2.0 for the situation of ground winds: **fully-loaded, no pressurization**. You will have to compute the weight in each tank and the resulting axial loading as a function of height. All locations should have MS > 25%.

2.4. **Critical stress items.** Since the vehicle has to withstand the worst-case, make a table showing the three items in each of the Problems 2.0 & 2.3 tables that have the lowest safety margin, what the margin is, and identify ground winds or max-q conditions. Your table might look something like this:

Summary of Stress Calculations

Location	Material	Thickness (m)	Load Case	Stress (MPa)	Yield stress (MPa)	MS (%)
1. Interstage	Gr/E	0.003	Max-q	1000.0	1,350	35
2. Lower St1 tank cyl	Al 2014	0.006	Ground	500.0	625	25
3. S1 Aft skirt	Al 6061	0.009	Ground	...	...	25

2.5. **Updated Mass Estimation.** Use your calculated wall thicknesses to estimate the mass of each tank, interstage, or intertank structure. Use these masses to update your mass statement. Provide:

1. Your Mass Statement including columns for the previous AND the *revised* inert and fully-loaded mass values. Provide the mass breakdowns by step (i.e. Step 1 separate from Step 2), so the changes in each step can be seen). A mass increase in the upper step is much harder to deal with than that in a lower step due to the trade-off ratios discussed in Chapter 5.

2. A third column showing the change in mass, plus or minus, from the MER (mass estimating relationship) calculations done previously (this would be a line-by-line subtraction – put decreased masses in parenthesis),

3. Comment on how the new mass estimation affects your design in terms of *payload* mass margin (amount of required payload for mission). Does a dry mass increase for the lower step have the same effect on payload as a dry mass increase of the upper step? Can you still accomplish your mission? If not, what changes will you make? Explain.

Chapter 11

Launch Vehicle and Payload Environments: Vibration, Shock, Acoustic, and Thermal Issues

11.1 Mechanical Loads

hapters 9 and 10 dealt with the mechanical loads associated with *quasi-steady* acceleration. In effect, this analysis assumed that the very quickly changing values of acceleration could be ignored, and focused on the values that slowly varied over time—the quasi-steady levels. However, the actual measured acceleration of a Delta II with nine strap-ons might look like the "fuzzy" trace shown in Fig. 11.1a—the fuzz is a representation of the vibration or dynamic acceleration. If you wanted the quasi-steady acceleration, you would take the average of the values in Fig. 11.1a and obtain a trace like the one in Fig. 11.1b. If you were dealing with a time-varying electrical trace, you would say you were "filtering out the AC" or "filtering out the alternating currents."

If you *subtracted* the quasi-static acceleration (Fig. 11.1a) from the actual measured acceleration on the left side of the same figure, you would obtain the *dynamic acceleration*, the so-called "AC values" of the trace. This would resemble the trace at the top of Fig. 11.2 that is labeled "measured dynamic acceleration." This dynamic acceleration is one of the topics of this chapter.

Now we will take the opposite approach—we will filter out the slowly changing values and concentrate on the quickly changing ones. In electrical terms, we'll filter out the DC and look at the AC. This would yield a trace as shown on the top of Fig. 11.2, which shows the quickly changing accelerations (measured dynamic accelerations) that were omitted from Figure 11.1b.

Figure 11.2 contains a great deal of information. Starting in the lower left and continuing up and over to the right, we see the flight profile of the Delta II. Notice that each event in the flight profile is numbered, from 1 to 13. This is to correlate the numbered events with vibration or shock traces.

Across the top of the figure is a plot of high-frequency vibrations vs time. The top-to-bottom height of the vibration trace indicates its strength. Note that the trace has numbers to indicate where the events in the flight profile occur. Upon inspection, it appears that the waveform near 2 is the widest, except for the up-and-down spikes that occur at places like 1, 4, 6, 7, 8, 10, and so forth. These very narrow but very strong (top-to-bottom width) spikes correspond to *shock* events usually caused by pyrotechnic devices (explosive bolts, etc.) being fired. Upon inspection, it appears 8, the fairing separation transient, is the largest shock event.

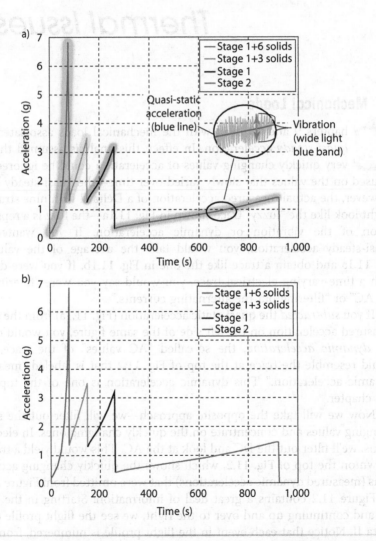

Fig. 11.1 (a) Simulated launch acceleration of a Delta II with nine solid rocket motors *minus* (b) quasi-static launch acceleration = dynamic acceleration. Source: Plot by author using GPOPS.

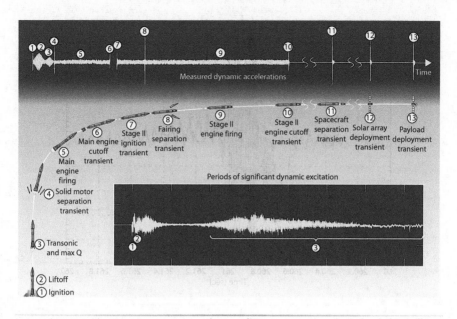

Fig. 11.2   Delta II flight profile, excitation events, and vibration. Source: [1].

Across the bottom right of Fig. 11.2, we see an enlargement of the vibration trace that shows only the first three events in the profile: ignition (1), liftoff (2), and max-q (3). Notice that there is a significant amount of vibration at and during the liftoff event, after which the ride quiets down until the vehicle approaches 3, the transonic and max-q events. This builds up for a while and then slowly tapers down to a relatively quiet (small) trace until 6, where the main engine cuts off and staging, another explosive separation, occurs.

Then there is a short coast period with no measured vibrations, after which the stage 2 engine fires and produces a transient. The ride is relatively quiet until 8, payload fairing separation, occurs. The ride continues relatively quiet until 10, when the second stage cuts off. After 10, the vehicle coasts and various separation maneuvers occur, accompanied by their associated shocks.

From this, one can observe that there are significant vibration and shock events during most of the ascent into orbit and separation from the landing vehicle (LV). We will discuss these events in more detail in the following sections.

11.1.1   Engine Startups and Cutoffs

Transient events occur when engines start up, suddenly putting the structure in compression, or when engines shut down, suddenly releasing the compression. In either case, acceleration suddenly changes from positive to near zero or vice versa, so the structures tend to oscillate for a while

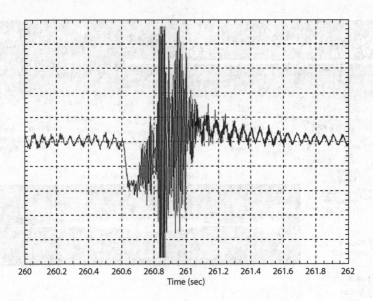

Fig. 11.3 Delta II main engine cutoff transient axial acceleration at guidance section.
Source: [2].

afterwards. Gordon [2] provides some time histories of such an event, which are shown in Fig. 11.3. This shows some high-frequency noise starting about $t = 260.7$ s, followed by a slowly decaying lower-frequency trace at $t = 261.1$ s. The rate of decay of the lower frequency sinusoid depends on the system's damping characteristics.

The behavior of the axial oscillations is understandable when one considers that a launch vehicle is flexible and like a Slinky toy, will axially stretch or contract depending on how the loads are applied or removed. This concept is illustrated in Fig. 11.4, with m_1 as the LV's upper mass and m_2 as its lower one, and a spring in between representing the axial flexibility of the LV body. When under steady thrust (case a), the spring between the two masses is compressed because it must support the apparent weight of m_2. Moving to b, when the engine shuts down, the still-compressed spring suddenly has no apparent weight on top if it, so an instant later in case c, the spring forces the two masses apart. They will continue oscillating towards and away from each other until the stored energy is dissipated by the system's damping or internal energy dissipation.

A trace of a shutdown (or cutoff) accelerations is given in Fig. 11.5. Here a transient event (the sudden reduction of thrust due to a shutdown command) at around $t = 0.25$ s causes the acceleration magnitude to drop from about 7.5 g (typical of shutdown acceleration) to about 0 g. Note that once near the 0-g loading condition, the vehicle, in behavior similar to the springs in Fig. 11.4, suddenly stretches and then contracts, creating a damped sinusoid trace starting about time $t = 0.58$ s.

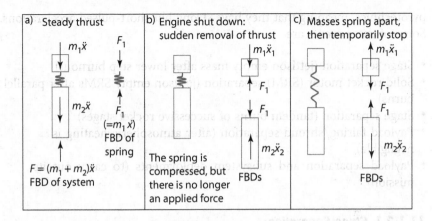

Fig. 11.4 How a flexible body (simulated with springs) under steady thrust can lengthen and then oscillate when the thrust ceases. Source: [3].

It's fairly easy to understand how a single engine shutdown could provide axial excitation. Another possible case of shutdown transients may occur when there are multiple engines running. If all of the engines do not shut down simultaneously, a rotational torque will be generated at the vehicle's bottom due to the engines' differing thrust levels, resulting in significant lateral loads in other areas of the LV.

11.1.2 Separation Events

We've seen that staging, the jettisoning of unneeded mass, is critical to launch vehicle performance. Many of these events are initiated by

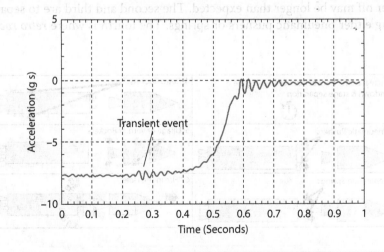

Fig. 11.5 Shutdown transients cause axial ringing in the structure as the loads are suddenly released. Source: [4].

pyrotechnics, meaning that they have strong but short-pulsed accelerations. Some of these events are

* Stage separation (jettison empty mass after lower step burnout)
* Solid rocket motor (SRM) separation (jettison empty SRMs after parallel burns)
* Stage separation (tandem burns of successive rocket stages)
* Payload fairing/shroud separation (after atmospheric heating is negligible)
* Payload separation and subsystem deployments (to carry out its mission)

11.1.2.1 Stage Separation

Many, but not all, launch vehicles use explosively activated separations. Small charges of explosive, also known as *ordnance* or *pyrotechnics*, are used to cut structures apart or to activate devices. Because explosive materials are involved, the induced shock loads can be significant. Other launch vehicles, such as SpaceX's Falcon 9, use a non-pyrotechnic separation system, which has the potential of reducing shock loading, and can be tested without needing replacement.

Stage separation may be executed by either pyrotechnic or non-pyrotechnic separation systems, but it is critical that they operate properly to avoid potential re-contact between a separated lower step (which might still be producing forward thrust after shutdown), and the upper step. Figure 11.6 shows *six* possible stage-separation scenarios for a launch vehicle. The first, not shown, is to increase the time after main engine cutoff (MECO) before separating steps, because the time for the thrust to taper off may be longer than expected. The second and third are to separate using either pneumatic pushers or springs. The fourth is where *retro rockets*

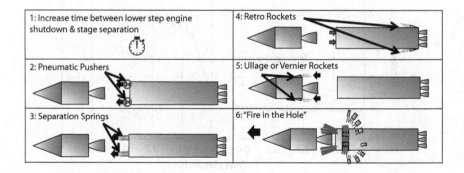

Fig. 11.6 Six different ways to reduce chance of recontact during separation of rocket stages.

Fig. 11.7   Ground separation test indicates considerable energy. Source: http://www.vibrationdata.com/SRS.htm.

thrust the spent steps backwards, and the fifth is small *ullage* or *vernier* engines firing to push the staged pieces apart and also force propellants into the tank bottom to allow the upper stage engine(s) to fire properly. Some lower steps had retro rockets (for example the Saturn V's S-IC), and others, such as the Saturn V's S-II and S-IVB upper steps, had both retro rockets and ullage motors.

The lower right of the drawing shows 6, an alternate method of staging where the upper step begins firing while still attached to the lower step, which is called *fire in the hole*. In this case, vent panels or some other means of relieving internal pressure must be provided. In the figure, these can be seen as small rectangular-shaped objects flying away from the lower step. This method of staging may only be done when there is no risk of the lower step exploding and damaging the upper step when it begins firing. It was used for staging on the Titan II intercontinental ballistic missile (ICBM) and space launch vehicle, except there were no panels to jettison; open *blast ports* (or *exhaust ports*) were provided instead. A video of a Titan II missile using fire-in-the-hole staging is available at https://www.youtube.com/watch?v=PBkZNMM_IUg. A slightly different view is available at https://www.youtube.com/watch?v=AakIvG7Xolc.

Because explosive materials are involved, the events themselves can be *energetic*. Figure 11.7 shows three frames of a ground explosive separation test. The reason for calling such materials *energetics* is obvious. It's also clear that contamination could be an issue.

Figure 11.8 shows two other separation tests—one of a ring used on the Saturn V, and the second a test of the Ares LV separation. In the Saturn movie, the explosive separation provides sufficient blast pressure to tip over the lights in the lower-left corner used to illuminate the test area.

Fig. 11.8 Still frames from videos showing Saturn V and Ares stage separation test. Source: NASA.

11.1.2.2 Pyrotechnic (Explosive) Devices

Explosives activate the separations in the photos. These ordnance devices provide highly reliable methods to make things happen *a single time*. Once they've been fired, they have to be replaced. So there's no way to test an individual device other than actually using it, but many can be tested to develop confidence that each will operate as expected. Different types of ordnance devices are:

* Separation methods, uncontained (as shown)
* Separation methods, contained (breaking and shearing, used for payload fairings)
* Explosive bolts
* Explosive nuts or pyro nuts
* Guillotines
* Pyro valves

(Guillotines and pyro valves will not be discussed in this text.)

The fiery, smoky separations shown in Figs. 11.7 and 11.8 are examples of *uncontained* explosives. By uncontained, we mean that the explosive material is not restricted by anything and can freely expand (and contaminate). This separation is done by *shaped charges* or *linear shaped charges (LSCs)* that are fashioned to *cut* structures for separation. (They are also used to release pressure and render vehicles nonpropulsive for range safety purposes—more on that in Chapter 16.)

Figure 11.9a is a photo of an LSC. These are *chevron* or inverted V-shaped pieces formed by a ductile metal sheath surrounding an explosive core material. The four frames of Fig. 11.9b show the progression of events when the explosive is fired. Explosive initiation causes the chevron to expand until the concave side (bottom side) turns inside out to form a hot metal plume (*jet*). The hot jet extrudes downwards towards the target plate, penetrating it and then severing it. Figure 11.9c shows an actual x-ray of a detonating LSC. The metal jet can be clearly seen punching through the bottom of the metal plate below.

The chevron shape focuses and concentrates the explosive energy into the hot jet firing downwards, making this method much more effective a cutting tool than a plain circular cross-section charge, such as mild detonating cord.

The illustrations of the LSC detonation process show that the process is not very clean, with bits of hot metal and smoke flying around. However, when spacecraft are involved, we have to keep things *very* clean. So, several techniques have been invented to separate structures using explosives, while containing the explosive products. One of these is called *Super*Zip*, invented by Lockheed (now Lockheed Martin).

This item contains the explosive cord within an expandable sleeve that is sized so that the explosive gases do not damage the sleeve and stay contained, but do expand the sleeve. Along with the sleeved explosive cord, the structure itself is grooved on the outside of both sides of the explosive sleeve, so that when the sleeve expands, the metal fractures on both sides. (This is similar to the scoring on the top of a pop-top soda can to ensure the top opens when the pull tab is pulled.) Figure 11.10a shows the operation of Super*Zip in a computer-generated simulation of the device at 0, 40, and 85 μs (microseconds or 10^{-6} seconds) after the initiation of the explosives. Figure 11.10b shows a portion of the Super*Zip intended to separate the Centaur G-Prime stage from its shuttle carrier.

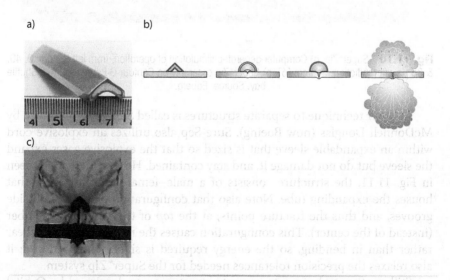

a) b)

c)

Fig. 11.9 a) Linear shaped charge in metal jacket on a centimeter scale. Source: [5]. b) From left to right, four steps of explosive cutting caused by detonation of explosive LSC core. Source: [6]. c) Flash x-ray photo of detonating LSC. High-velocity fragments and molten metal created by the detonation event penetrate through the plate below. Source: [5].

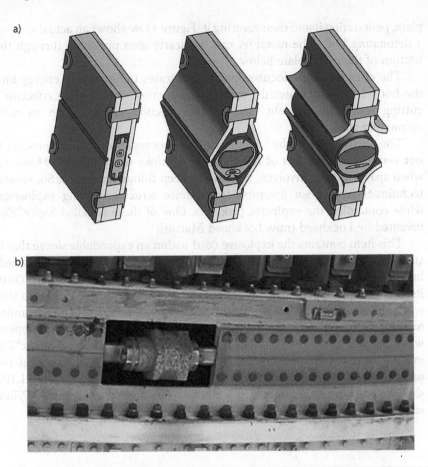

Fig. 11.10 Super*Zip. a) Computer-generated simulation of operation, from left to right: 0, 40, & 85 μs after initiation. Source: [6]. b) Super*Zip assembly on Centaur G-Prime mount in Shuttle bay. Source: Edberg.

Another technique to separate structures is called *Sure-Sep*, developed by McDonnell Douglas (now Boeing). Sure-Sep also utilizes an explosive cord within an expandable sleeve that is sized so that the explosive gases expand the sleeve but do not damage it, and stay contained. However, as can be seen in Fig. 11.11, the structure consists of a male–female joint assembly that houses the expanding tube. Note also that configuration places the outside grooves, and thus the fracture points, at the *top* of the structural member (instead of the center). This configuration causes the material to fail in shear rather than in bending, so the energy required is significantly reduced; it also relaxes the precision tolerances needed for the Super*Zip system.

Still another technique to separate structures was developed by Orbital Sciences (now Northrop Grumman) and Ensign-Bickford Aerospace and Defense, called a *hollow-form frangible joint (HFFJ)*. HFFJ also contains explosive cord within an expandable sleeve but uses a *different* structural

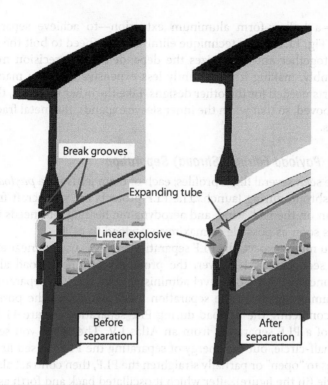

Fig. 11.11   Sure-Sep fails the grooved piece in shear, requiring less energy. Source: [7]

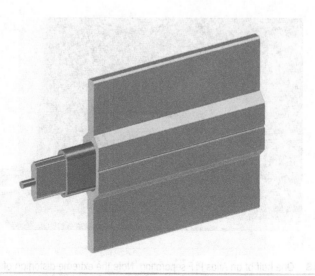

Fig. 11.12   The hollow-form frangible joint uses an aluminum extrusion. Source: [5].

member—a hollow-form aluminum extrusion—to achieve separation, as shown in Fig. 11.12. This technique eliminates the need to bolt the frangible member together and minimizes the dependence on precision machining and assembly, making it significantly less expensive than the machined or forged parts needed for the other designs. Like the other methods, the extrusion is grooved, so that when the inner sleeve expands, the metal fractures on both sides.

11.1.2.3 Payload Fairing (Shroud) Separation

We've seen several flight profiles; each of them jettisons a *payload fairing (PLF)* or shroud during launch. The PLF protects the spacecraft from contamination on the pad, wind, and aerodynamic heating, but needs to be jettisoned as soon as possible to maximize payload to orbit.

As you might expect, the PLF separation involves use of linear explosives as we've seen before. However, the proximity to the payload also raises several concerns: the shock level administered by the PLF separation, possible contamination from the separation mechanism, and the possibility of the PLF contacting the payload during PLF separation. Figure 11.13 shows one half of a PLF separating from an Atlas LV. The shape you see started out as a half-circle, but the energy of separating the PLF caused its two vertical edges to "open" or partially straighten the PLF, then contract almost into a full circle (in the figure), after which it oscillated back and forth as it disappeared below the LV.

A separation system that was developed by McDonnell Douglas (now Boeing) is called the *thrusting joint*. This method also uses explosives to activate, but it operates in a different manner. The explosive material is

Fig. 11.13 One half of an Atlas PLF separating. Note the extreme distortion of the semicircular shape it had before separation. Source: Still frame from video at https://youtu.be/X1sQARtzaEI?t=79.

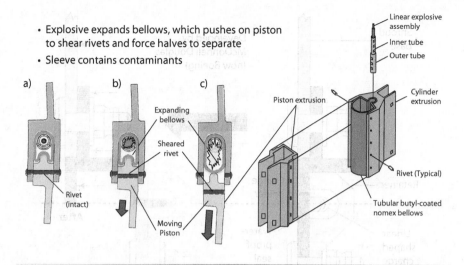

- Explosive expands bellows, which pushes on piston to shear rivets and force halves to separate
- Sleeve contains contaminants

Fig. 11.14 The Boeing thrusting joint separation system expands a bellows to shear rivets that hold the two sides together. Left: (a) before initiation; (b) after initiation, where the bellows has sheared the rivets; (c) bellows continues to expand, thrusting joint apart. Right: exploded view. Source: [8].

contained in a *bellows*, as shown in Fig. 11.14. When the explosive is initiated, the bellows expands to push on the face of the *piston extrusion*. The expanding bellows shears the rivets that had been holding the piston to the other side, causing the two sides to separate. (Five rivets are shown in the figure.) Thus, instead of causing the structural rail on the PLF edge to fail at a groove in the structure, it forces the two sides of the structure apart, shearing rivets.

We also need to consider how to release the bottom of the PLF, because thrusting it upwards is a bad idea, and thrusting it radially is impossible. We therefore want a nonthrusting joint here, such as the *base separation joint* shown in Fig. 11.15. This linear shaped charge cuts around the circumference of the bottom portion of the fairing, propelling it to the right, as shown.

Because the base separation joint shown in Fig. 11.15 employs a LSC, contamination can occur. An alternate method to separate the base of the PLF is shown in Fig. 11.16, which depicts the base joint expanding-tube separation method. Employing a thrusting tube to shear the base of the PLF, this expanding-tube method performs a similar function to the previous method, but contains the explosion in an expanding tube for less potential contamination.

Finally, we need to consider the ordnance needed to release the *entire* PLF. This consists of the base separation mechanism and the thrusting joint mechanisms, both of which are shown in Fig. 11.17. Two redundant detonators on each circuit provide extra reliability. (Either is sufficient to cause separation.)

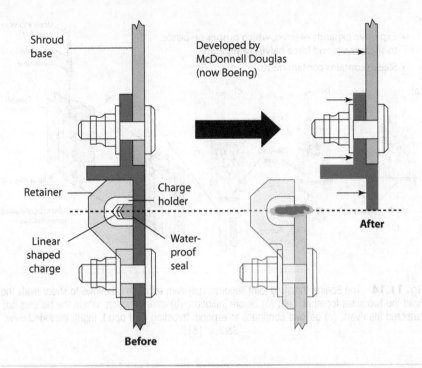

Fig. 11.15 Base separation joint, to be used at the base of the PLF. Vertical thrusting joint along edge of PLF provides force to move PLF to the right. Source: Modified from [8].

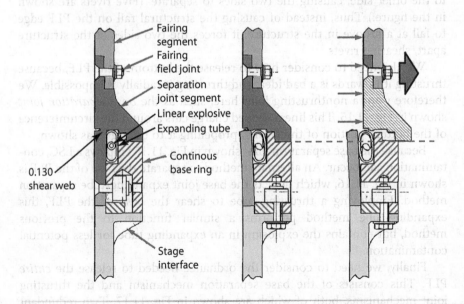

Fig. 11.16 The payload fairing base joint expanding-tube separation method reduces contamination by eliminating the LSC cutter. Source: Modified from [9].

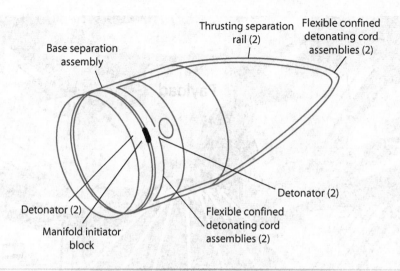

Fig. 11.17 Ordnance required to jettison the PLF. Source: [8].

As far as the payload is concerned, fairing separation is a major shock event, as seen as item 8 in Fig. 11.2 earlier, and in Fig. 11.13. The reason it is felt so strongly by the payload may be because most payload attach fittings (PAFs, also called launch vehicle adapters, or LVAs), look and act like a *funnel* connecting the top of the LV with the bottom of the payload, as may be seen in Fig. 11.18, which shows a fairing separation test for a Boeing Delta IV. The PLF sits on the top of the upper step, shown as a dotted line in the figure. So if there is a shock associated with the fairing release, part of it travels upward and is focused by the PAF (upward converging arrows) into the bottom of the payload (the cylinder in the top center of the photo), and the remainder is passed downwards from the top of the LV (down-pointing arrows).

Sometimes the payload fairing doesn't separate as it is supposed to. Figure 11.19 shows an unseparated PLF on board an Agena upper stage. The Gemini capsule whose nose is in the top foreground was supposed to dock with the Agena, but was unable to because the PLF was in the way. In other cases, the LV may not have enough margin to reach orbit if its PLF fails to separate, and will instead crash into the Earth.

11.1.2.4 Alternatives to Explosive Separation Systems

Explosive-driven systems cannot be tested before they are used... because if you test them, you have to replace them with unexploded ones that have not been tested. So, one has to depend on the statistics of these devices making them reliable enough to use. Another problem with explosive-driven systems is that they create high shock levels for nearby equipment.

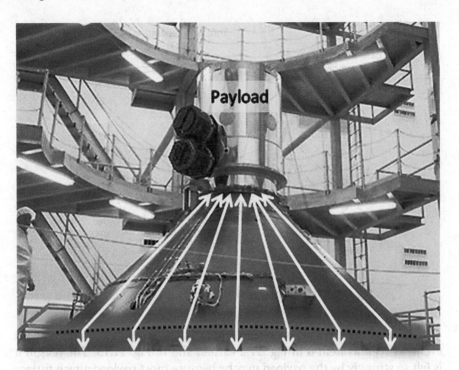

Fig. 11.18 Payload atop a conical composite Delta IV payload attach fitting. The payload fairing sits atop the circular rail on the bottom of the photo. When commanded to separate, the separation shock is funneled upwards to the base of the payload. Credit: Air Force.

Fig. 11.19 The payload fairing on the Agena upper stage did not separate. This vehicle was nicknamed the "angry alligator" by the astronauts in the Gemini capsule that had planned on docking with it. Source: NASA.

Fig. 11.20   SpaceX's Falcon 9 uses nonexplosive stage separation system: couplers clamp the two structures together until release, when pneumatic pushers separate the two parts. Source: [10].

For these reasons, SpaceX utilizes a pneumatic "pnon"-explosive system to separate the Falcon 9's first and second steps. As seen in Fig. 11.20, the Falcon 9 interstage system has a reusable mechanical clamping system. The system has six *separation collets* that work together with a pneumatic long-stroke pusher system to separate the two steps. The beauty of this pneumatic system is that it can be tested before flight, it can be reused, and it provides lower shock levels than a comparable explosive-driven system. However, the price for these advantages is likely a larger mass penalty.

11.1.2.5 Nonexplosive Fairing Separation

SpaceX's Falcon 9 has a nonexplosive payload fairing separation mechanism. According to the Falcon 9 Launch Vehicle *Payload User's Guide* Rev 2 dated 21 Oct. 2015 [10]:

> The two halves of the fairing are fastened by mechanical latches along the fairing vertical seam. To deploy the fairing, a high-pressure helium circuit releases the latches, and four pneumatic pushers facilitate positive-force deployment of the two halves. The use of all-pneumatic separation systems provides a benign shock environment, allows acceptance and preflight testing of the actual separation system hardware, and minimizes debris created during the separation event.

SpaceX attempts to recover their PLFs and reuse them when possible. It is said that the two fairing halves together cost about $6 million, so recovering these could reduce the cost of launching substantially. After separation, jet exhausts have been observed that would be typical of an attitude control system. The mass of the PLF is low and the area is high, so they probably can survive reentry without significant heating.

For recovery, a special ship was observed in Feb. 2018. The ship, carrying something that resembles a very large butterfly net, is shown in Fig. 11.21a. SpaceX owner Elon Musk tweeted about the Falcon 9's PLF: "It [the PLF] has onboard thrusters and a guidance system to bring it through the

a)

b)

Fig. 11.21 (a) SpaceX ship with a built-in large net to catch and reuse its Falcon 9 PLFs. (b) A PLF that landed intact. Reuse could save as much as $6 million. Source: SpaceX.

atmosphere intact, then releases a paraf-
oil and our ship, named Mr. Steven,
with basically a giant catcher's mitt
welded on, tries to catch it."

A floating, apparently intact PLF is
shown in Fig. 11.21b. SpaceX is definitely
revolutionizing the launch industry.

11.1.2.6 PLF Separation Dynamics

Let's pretend we have a vehicle under
rocket power that has acceleration a_v in
the near vacuum of space. A payload
fairing is attached to the top of the
vehicle, each half being mounted on its
own hinge that is specially designed to
release after a certain rotation angle ϕ_1
is passed. This situation is shown in
Fig. 11.22, which includes the vehicle
and one half of its PLF. Each half of the
PLF has mass m and begins to separate.
By "near vacuum," we mean that we can

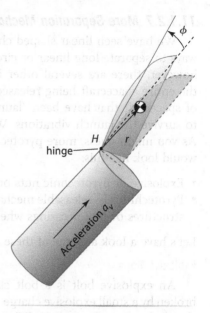

Fig. 11.22 Payload fairing
separation dynamics. Source: After [11].

neglect aerodynamic forces and concentrate on only mechanical forces.

When the PLF halves separate, the half-PLF pivots on the hinge point
shown. We don't want the PLF to contact the rocket's body after it separates.
Let's look at the dynamics of this situation.

During the hinged phase, when $0 < \phi < \phi_1$, the apparent PLF weight is
ma_v, due to the acceleration of the vehicle. That apparent weight creates a
torque about the hinge located at point H. We can write the equation of
motion of the PLF half by summing torques about the hinge H

$$\sum \text{Torques} = ma_v(r \sin \phi) = I_H \ddot{\phi}$$

where I_H is the PLF moment of inertia about its hinge.

One can see that the PLF must reach a certain angle ϕ_0, or the apparent
weight of the PLF will cause it to return to its $\phi = 0$ position. This means
that the segments have to separate with enough rotational speed that they
will reach the angle where the center of mass (CM) is directly above the
hinge H with positive speed. This speed is supplied by the PLF separation
system. After separation ($\phi > \phi_1$), the PLF is in free-fall, with no forces
other than gravity attracting it, so it will continue with whatever linear
and rotational speed its CM had at the moment of release: the PLF will
tumble around its center of mass with the rate $\dot{\phi}_1$ that was the rotation
rate at separation. One can show that the PLF's mass center follows a parabola with respect to the LV's reference axis (see Exercise 2 at the end of
this chapter).

11.1.2.7 More Separation Mechanisms

We have seen linear shaped charges, Super*Zip, and thrusting joints as ways to separate long linear or circumferential parts from a launch vehicle; however, there are several other items that need to be released, typically the entire spacecraft being released from its payload attach fitting, or parts of spacecraft that have been "launch-locked," meaning secured to be able to survive the launch vibrations. What can we use to separate these items? As you might guess, more pyrotechnic devices may be used! A list of them would look like this:

• Explosive or pyrotechnic nuts or bolts (good for point supports)
• Pyrotechnically-releasable mechanical clamp bands (good for circular shell structures or motor casings where you don't want point loads)

Let's have a look at some of these items.

Explosive Bolts

An explosive bolt is a bolt clamping two items together that can be broken by a small explosive charge. A drawing showing how such a bolt operates is shown in Fig. 11.23a. When the charge fires, the shank of the bolt is severed, and the two items that were clamped together are now free to move apart. Figure 11.23b shows an explosive bolt before and after firing.

Explosive Nuts

An explosive nut operates in a similar fashion to an explosive bolt. A small explosive charge "cracks" the nut into several pieces, allowing separation just like the explosive bolt. A drawing of an explosive nut is shown in Fig. 11.24.

Some explosive nuts are called *frangible nuts*. This particular type of nut was used to hold the Space Shuttle onto its launch pad until liftoff was commanded. Each of the Shuttle's two solid rocket boosters had four of these nuts, which can be seen in Fig. 11.25a and b. A special catcher mechanism

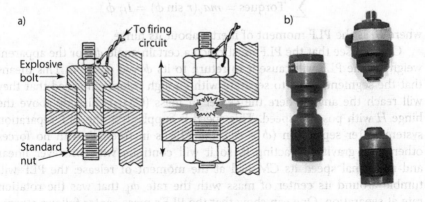

Fig. 11.23   a) How the explosive bolt releases the two items that were clamped together. b) An actual bolt before and after. Sources: [6, 12].

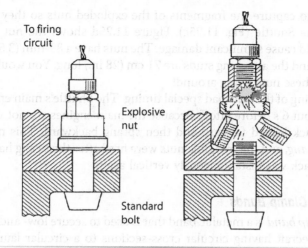

Fig. 11.24 Explosive nut operation for releasing parts. Source: [12].

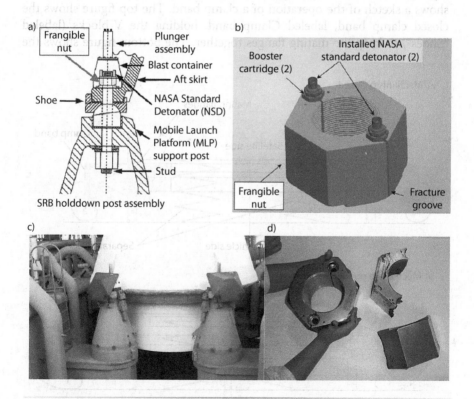

Fig. 11.25 a) The cross-section of the Shuttle's SRB hold-down posts; b) a closeup of the frangible nut along with its explosive detonators; c) the bottom of the SRB showing two of its four hold-downs and the catcher mechanism; d) an intact and a fired Shuttle frangible nut. Source: NASA.

was used to capture the fragments of the exploded nuts so they would not damage the Shuttle (Fig. 11.25c). Figure 11.25d shows the nut fragments, which could cause significant damage. The nuts have a 89-mm (3.5-in.) diameter hole, and the attaching studs are 71 cm (28 in.) long. You would *not* want pieces of these nuts flying around!

The firing of the nuts had special timing. The Shuttle's main engines were ignited about 6 s before liftoff. Because the main engines are not vertical, the Shuttle stack leaned forward and then sprang backward. This motion was called a *twang*, and the frangible nuts were fired after the twang had occurred and the stack was instantaneously vertical again.

11.1.2.8 Clamp Bands

A *clamp band* is a metallic band that is used to secure low- and moderate-mass spacecraft having circular cross-sections to a circular launch vehicle adapter. Clamp bands are also referred to as *V-bands* and *Marmon clamps*. Typically, there are V-blocks on the inside of the clamp band that hold together the bottom of the spacecraft and the top of the LVA. Figure 11.26 shows a sketch of the operation of a clamp band. The top figure shows the closed clamp band, labeled Clamp band, holding the V-blocks (labeled "Shoes") to keep the mating flanges together. The bottom figure shows the

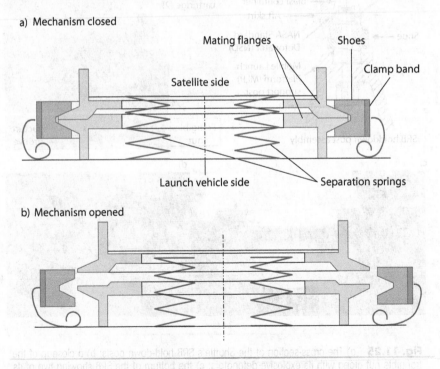

Fig. 11.26 Operation of a clamp band. Top: payload secured by V-blocks (shoes). Bottom: clamp band released, V-blocks retracted by springs, payload pushed away by springs. Source: [13].

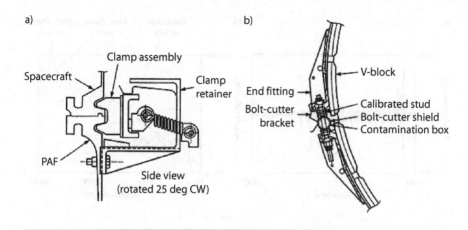

Fig. 11.27   Side/top views of clamp band assembly, showing components. Source: [14].

released clamp band's operation: when the band is released, the V-blocks are pulled away from the mating flanges by springs (often called retractors or extractors), and springs are used to separate the spacecraft from the launch vehicle.

The limitation of a clamp band system is the amount of tension that the clamp band is able to store. When the tension is released by a pyrotechnic bolt cutter, the entire ring releases a large amount of stored elastic energy as a *shock*, some of which goes up to the base of the payload as described in 11.1.2.3 above.

Figure 11.27 shows some details of a clamp band system. The cross-section of the system can be seen in Fig. 11.27a. Figure 11.27b shows details of the tensioning bolt and the pyrotechnic bolt cutter, which releases the tension in the band. Most clamp band systems have two bolt cutters, either one of which can release the band, for redundancy.

Here's an interesting fact about clamp bands: Zeppo Marx, one of the Marx brothers featured in multiple movies in the 20th century, was also active in engineering. He established Marman Products Co. in California in 1941, making clamping devices and straps. During World War II, it produced the Marman clamp, which was used to secure cargo during transport. Currently, the bands are used for spacecraft (http://www.marx-brothers.org/bio graphy/zeppo/inventions.htm).

11.1.3   Pyrotechnic Shocks

The operation of pyrotechnic or explosive devices creates *pyro shocks*, which can be very energetic. Figure 11.28a shows a typical time history of a shock event, and Fig. 11.28b shows what is called the *shock response spectrum (SRS)*.

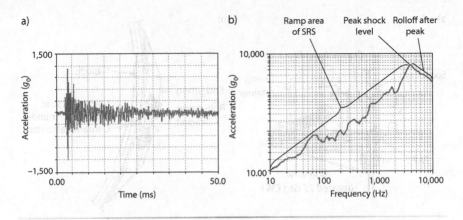

Fig. 11.28 a) The acceleration, measured in *g*s, created by a shock event. Note peaks over 1,000 g_0, but very short duration. b) The SRS indicates how the shock's energy is distributed by frequency. Source: [15].

In the time history in Fig. 11.28a, note the high *g*-levels (1,200 g_0) and the very short durations of the spikes (less than milliseconds). Even though the durations are very short, these levels can damage sensitive equipment. Hardware failures include cracking of parts, dislodging of contaminants, and cracking of solder joints on circuit boards.

The SRS plot in Fig. 11.28b gives an idea of how the energy of the shock is distributed as a function of frequency; however, one must realize that multiple shock events may not provide consistent time histories, so they have to be approached in a statistical manner. The peak levels occur around 3,000–4,000 Hz, and some equipment may be very sensitive to these frequencies. Typically shock levels are measured, averaged, and then an "envelope" is drawn around the SRS to provide some margin in case a shock with more energy appears.

The analysis of shock and other random events is a well-established branch of mathematics, but is beyond the scope of this book. This brief introduction is only intended to provide the reader with some familiarity of the subject and its vocabulary. For more information, the reader is encouraged to study textbooks in the field, such as that written by Wijker [16].

11.2 Acoustic Environment

There is no argument that rockets produce *very high* noise levels; however, we are not concerned at the present with the noise that is perceived by people in the general vicinity. What we *are* concerned with is how the noise produced by the launch vehicle affects the vehicle itself and its payload.

Obviously, the main source of sound in the early stages of liftoff is the rocket engines powering the LV. We start with the ignition process, where the lighting of the engine(s) can produce a pressure wave called *ignition overpressure* (IOP). Next, at and after liftoff, the rocket engine noise reflects off the ground and nearby structure, and is strongly received by the climbing LV.

The second major source of noise for the LV is during climb, where aerodynamic boundary layer noise excites the LV's body and its payload fairing, resulting in two effects: it causes the LV's body panels and PLF to vibrate, and it directly transmits some of the sound to the internal equipment and the payload. This airborne sound case is most apparent during the period of max-*q*, as can be seen as item 3 in Fig. 11.2 presented earlier.

How do we specify sound levels or measure acoustic loads? We start by recognizing that sound is simply an oscillating pressure in air, so we have the opportunity to measure the pressure level that provides the strength of the sound, and its frequency, which will help us with structural vibrations later.

To make things a bit simpler, acoustic loads are measured using a logarithmic scale with units of decibels (dB). The sound pressure level (SPL) is defined *logarithmically*: SPL (dB) $= 20 \log_{10} (P/P_{ref})$ decibels. The value for $P_{ref} = 2.9 \times 10^{-9}$ psi $= 2 \times 10^{-5}$ Pa $= 20$ μPa, which is the accepted value for the threshold of an average human just being able to hear a sound. A listing of the sound pressure levels of different items is presented in Table 11.1.

11.2.1 Ignition Overpressure

Ignition overpressure (IOP) is a significant transient low-frequency pressure event caused by the rapid pressure rise rate of solid rocket motors and/or liquid engines as they begin firing at ignition. In some situations, the designer may stagger the engine ignition times so that the overpressure is not concentrated to a single startup time, if possible.

Table 11.1 Sound Pressure Levels of Different Items

Noise Source	Sound Level
Note: 6 dB ≈ Factor of 2	
Hearing threshold	0
Dishwasher	50–53
Car @ 10 m	60–80
Jackhammer @ 1 m	100
Jet engine @ 100 m	110–140
Jet engine @ 30 m	150
Rocket launch	**165**
Theoretical limit	194

Note: Because the $\log_{10}(2) \approx 0.3010$, $20 \log_{10}(2) \approx 20(0.3010) =$ 6.02 dB ≈ 6 dB. This means that a factor of 2 or doubling is equivalent to a +6-dB change, and a factor of $1/2$ = halving ≈ –6-dB change.

Fig. 11.29 A 5% scale model of SLS undergoes acoustic testing at MSFC. Source: NASA.

IOP is difficult to estimate. One can perform scale model tests and scale up the results until actuals are obtained. Figure 11.29 shows a 5% scale SLS vehicle model with operating scale engines being tested at Marshall Space Flight Center in order to assess potential IOP loads.

11.2.2 High Acoustic Environments 1: Liftoff

During liftoff, the sound emitted by the rocket reflects off of the pad and the ground (see Fig. 11.30). Water spray can suppress noise intensity. Exhaust ducting, especially tunnels, also help by channeling the sound away from the LV. Water noise suppression systems will be discussed in Chapter 14.

11.2.3 High Acoustic Environments 2: Flight

During flight, flow of air makes boundary layer or wind noise that excites the vehicle's structure, its fairing, and all of its internal equipment, including the payload (Fig. 11.31). It often peaks around the time of max-q.

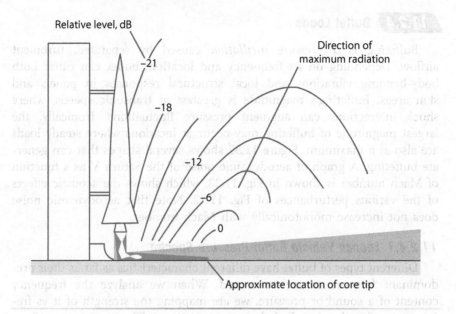

Fig. 11.30   Rocket noise reflects off of pad and ground. Water spray can reduce liftoff noise. Source: [17].

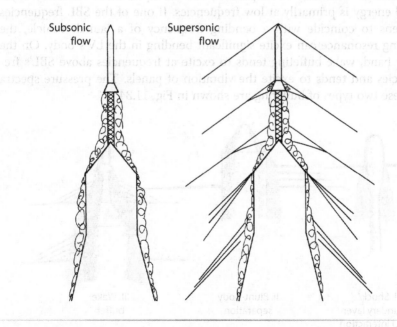

Fig. 11.31   Around max-air, the boundary layer noise, also known as wind noise, creates high levels of sound inside and outside of the vehicle structure. Source: [18].

11.2.4 Buffet Loads

Buffeting is a *pressure oscillation* caused by separated, turbulent airflow. Depending on its frequency and location, buffet can cause both body-bending vibrations and local structural responses in panels and skin areas. Buffeting's magnitude is greatest at transonic speeds, where shock interactions can augment pressure fluctuations. Ironically, the largest magnitude of buffeting may occur at locations where steady loads are also at a maximum. Figure 11.32 shows several shapes that can generate buffeting. A graph of aerodynamic noise of the Saturn V as a function of Mach number is shown in Fig. 11.33, which shows the acoustic effects of the various perturbances of Fig. 11.32. Note that aerodynamic noise does not increase monotonically with Mach number.

11.2.4.1 Launch Vehicle Buffet Pressure Spectra

Different types of buffet have different characteristics as far as their predominant frequencies are concerned. When we analyze the frequency content of a sound or pressure, we are mapping the strength of it vs frequency, a plot that is called the *power spectrum*. The name comes from the original usage of this procedure, because it was used to map the amount of power in electrical noise. A more pertinent name would be the *pressure spectrum* or *SPL spectrum*.

Power spectra show how buffeting's energy is distributed over different frequencies. This can be important; for example, *shock-boundary layer (SBL)* energy is primarily at low frequencies. If one of the SBL frequencies happens to coincide with a bending frequency of a launch vehicle, the ensuing resonance can excite significant bending in the LV's body. On the other hand, wake buffeting tends to excite at frequencies above SBL's frequencies and tends to excite the vibration of panels. The pressure spectra of these two types of buffeting are shown in Fig. 11.34.

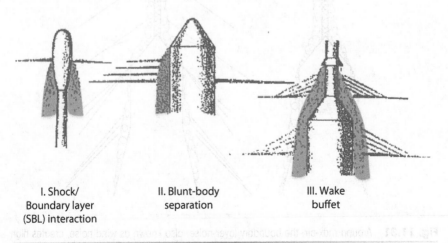

I. Shock/
Boundary layer
(SBL) interaction

II. Blunt-body
separation

III. Wake
buffet

Fig. 11.32 Three different buffet sources. Source: [19].

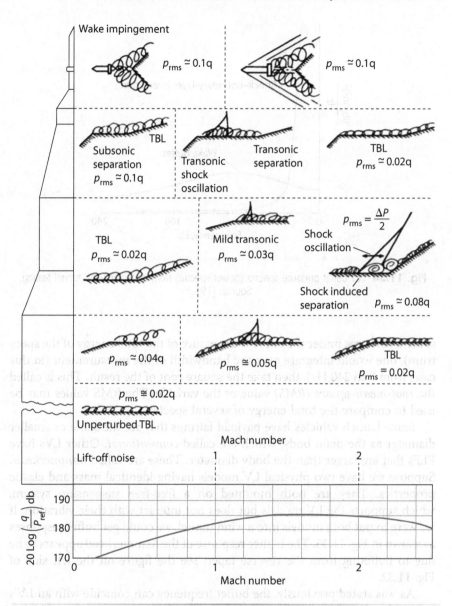

Fig. 11.33 Aerodynamic noise sources as a function of Saturn V vehicle geometry and Mach number. Source: [18]. Note: The letter p represents pressure fluctuation, while $q = \rho v^2/2$ is dynamic pressure. *TBL* is an abbreviation for *turbulent boundary layer*.

The vertical axis units are

$$\Phi = \left(\frac{\text{lbf}}{\text{in}^2}\right)^2 \Big/ \text{Hz}$$

Although these pressure spectrum units seem strange, the reason for them is the following: to compute the area under the curve of these pressure

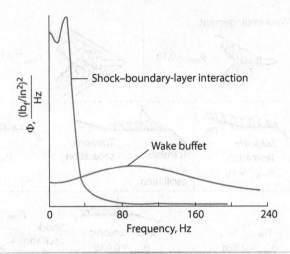

Fig. 11.34 LV buffet pressure spectra (power spectra) from transonic wind tunnel testing. Source: [19].

spectra (the area under the curve is a measure of the total energy of the spectrum), one would integrate over the bandwidth of the measurement (in this case, from 0 to 240 Hz), then take the square root of the result. This is called the *root-mean-square (RMS)* value of the variable. The RMS values may be used to compare the total energy of several spectra.

Some launch vehicles have payload fairings that are the same or smaller diameter as the main body, which are called *conventional*. Other LVs have PLFs that are larger than the body diameter. These are called *hammerhead*. Suppose we have two physical LV models having identical mass and elastic properties. They are both mounted on a free-free suspension system, which supports the LV models but does not interact with their vibration. If we were to put both models into a wind tunnel, we could get buffet responses as shown in Fig. 11.35. The higher response of the hammerhead appears to be due to buffeting from the reverse taper; see the figure on the left side of Fig. 11.32.

As was stated previously, the buffet frequency can coincide with an LV's natural bending frequency, whose narrow-band SPL spectrum forces higher bending responses. As might be expected, the nose shape can have a large effect on the response. Figure 11.36 shows the bending moment power spectra for the two nose shapes we've looked at: hammerhead vs regular.

With the same wind tunnel models and suspension as before, we see that the hammerhead's higher response (the height of the plot) is due to the higher level of buffeting from the reverse taper characteristic of hammerhead noses. These higher peaks usually indicate low structural and aerodynamic damping, which will increase the amplitude to higher levels. Even more importantly, sometimes the first vibration mode shows *negative damping*

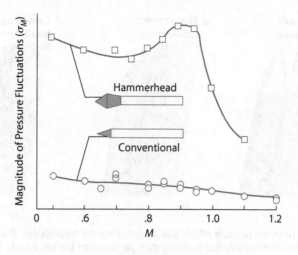

Fig. 11.35   Effect of nose shape on buffet response: conventional nose vs hammerhead.
Source: [19].

at higher Mach numbers: this means that the vibration could gain energy from the flow and could cause sustained oscillation and ultimately could damage the structure.

Sometimes there are issues with the intersections of shapes on the launch vehicle. We saw earlier that on the Space Shuttle, aerodynamic heating was enhanced where some of its shapes were near each other. This is also true for buffeting. Figure 11.37 shows the buffeting effects due to boosters adjacent to the main body.

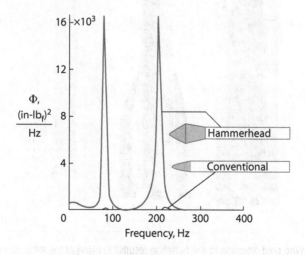

Fig. 11.36   Hammerhead vs regular bending moment power spectra for $M = 0.9$, $a = 0$ deg.
Source: [19].

a) Sharp on-axis booster nose.

b) Sharp canted booster nose.

c) Canted ogive booster nose.

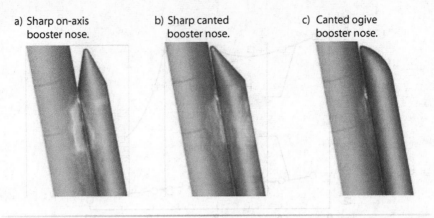

Fig. 11.37 Unsteady pressure effects due to canted booster nose shapes. The shape on the right has considerably less buffeting than the shape on the left. Source: [20].

Fig. 11.38 Who paid attention to the buffeting results? Looking at the strap-on nose shapes, it appears that the Atlas V and Ariane 6 designers did their homework! SLS has retained the SRB nose cone shape from the heritage Shuttle hardware it is utilizing. Illustrations *not to scale*. Sources: NASA, ESA, and Richard Kruse, HistoricSpacecraft.com.

According to the CFD research done by NASA NESC [20]:

> From left: (A) pressure on core stage and solid rocket booster is the base line configuration; (B) sharp canted-nose configuration indicates reduced pressure on core stage but increased pressure on booster, and (C) canted ogive indicates reduced pressure on both booster and core stage. These data were verified by wind tunnel tests.

This would indicate that configuration C was the best in terms of minimizing pressure oscillations.

One would expect that new LV designs would take advantage of this knowledge. Observing the designs in Fig. 11.38, it appears that the Atlas V designers paid lots of attention; the Ariane 6 designers were happy with moderate improvement, whereas the SLS designers apparently did not take advantage of this knowledge. (This is likely because SLS is using heritage Shuttle hardware that already exists.)

11.2.5 Acoustic Suppression

To reduce noise levels at the payload location, passive acoustic insulation is often applied to the inside of the payload fairing. The insulation is bonded or attached to the inside skin. Figure 11.39a shows a treated PLF.

Figure 11.39b shows another means to attenuate noise. This PLF has a number of Helmholtz oscillators on its walls. These oscillators tend to absorb sound waves that have the same frequency as the oscillators, so a

Fig. 11.39 Acoustic attenuation in PLF. a) Twin GRAIL lunar mappers being enclosed with payload fairing atop Delta II rocket with insulated PLF. b) Tuned Helmholtz absorbers on board an Atlas V PLF. Source: NASA.

number of oscillators with different tuned frequencies are needed. The principle is similar to the sound that is heard when one blows gently over a glass bottle.

There is a very high acoustic environment at launch and during flight, particularly around max-q. Acoustic attenuation measures inside the PLF help with the problem. In Chapter 14, we will show the measures taken at the launch pad to reduce the liftoff portion of the acoustic environment.

11.3 Launch Vehicle Thermal Environment

A launch vehicle both *creates* and *is exposed to* a variety of thermal environments, from prelaunch to orbital. The prelaunch environmental conditions are relatively easily controlled, using protective structures, enclosures, and so forth. The launch phase environment, by contrast, is one of the most strenuous. There are a number of heat and cooling sources (Fig. 11.40), some of which may be found in Table 11.2:

- Cryogenic fluids cool tank bulkheads and walls.
- Aerodynamic heating in the atmosphere forces convective heating on parts of the vehicle.
- Material allowables change due to low/high temperatures.
- Thermal gradients cause stresses.
- Radiation from the exhaust plume reaches the bottom of the LV.
- Combustion chamber and nozzle obviously get hot, and recirculated gases from the plume can heat the area more.
- Solar is relatively limited but can cause hot spots.
- Payload is protected with the payload fairing so as not to be subjected to heating during launch.

11.3.1 Base Heating

According to Boswinkle [19], at high altitudes, multiple "engines' exhausts can intersect, and trailing shocks can form. The energy of the air near the jet boundaries is too low to allow the flow to move back through

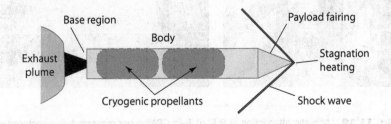

Fig. 11.40 A launch vehicle has a number of sources of heating and cooling.

Table 11.2   Convective, Conductive, and Radiative Heat/Cooling Sources

Region/Mode	Convective	Conductive	Radiative
Payload fairing	Flow	Minimal	Shock, solar
Main body	Flow	Combustion chamber, propellants	Minimal plume, shock, solar
Base region	Recirculating flow, combustion gases	Minimal	Nozzle, plume

the shock waves, so the flow reverses and flows toward the base of the vehicle. It escapes by flowing laterally across the base. The recirculated flow of hot exhaust gases reaches supersonic velocities and can cause damage" to unprotected surfaces. This flow behavior is illustrated in Fig. 11.41.

One way to deal with this problem is to add *base flow deflectors* or *scoops* to guide external airflow into the base region or engine compartment in order to reduce recirculation (and base drag as well). Drawings and photos of scoops on the Saturn V are shown in Fig. 11.42.

As seen in Fig. 11.43, hot gas recirculation is a major problem going into space. Radiation comes from afterburning of the exhaust downstream of the rocket's nozzle(s). Baseburning is a major problem from 5 to 15 km (3 to 10 miles) altitude

The exhaust jet is usually not a problem at low altitudes, unless there is blowback from the flame bucket. At higher altitudes, the jet expands sideways from the nozzle exit, creating backwash, and radiates back into the engine compartment, which can cause surface temperatures above 538°C (1,000°F).

What can be done to minimize base heating? We can add engine shields or full closeouts, as seen in Fig. 11.44. The shields may be supported from a

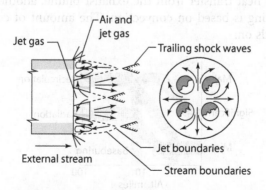

Fig. 11.41   Shock waves and flow reversals can cause heating damage in the base region.
Source: [21].

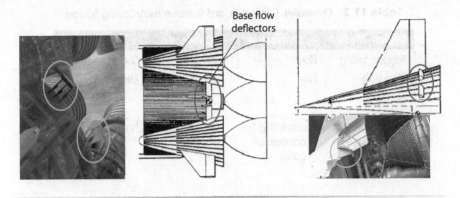

Fig. 11.42 Circles highlight three different kinds of base flow deflectors on Saturn V. More may be seen in Fig. 11.44. Sources: Left and bottom right: Edberg photos. Middle and top right: [22].

stationary center engine or from support structure. Figure 11.44b shows the flexible curtains on the four outer engines of the Saturn IB first step (as well as eight base flow deflectors, circled).

To get a feel about how much variation there is in an exhaust plume, Fig. 11.45 shows the Space Shuttle SRMs' exhaust plume starting at mission elapsed time of 25 s and ending at 125 s in 20-s intervals, corresponding to Mach 0.53, 1.00, 1.66, 2.65, 3.59, and 3.92. It may be seen that the plume starts out approximately the diameter of the SRB aft skirt and gets larger and larger as the Shuttle ascends into the atmosphere. Of course, this is because of the reduced atmospheric pressure with elevation. A larger plume may provide a larger area for radiation to be emitted from, and thus more absorption by the base area of the LV.

11.3.2 Convective Thermal Environment During Boost

Besides the heat transfer from the exhaust plume, another major component of heating is based on convection. The amount of convective heat transfer depends on:

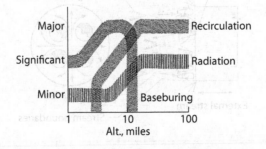

Fig. 11.43 Sources of base heating for clustered nozzles. Source: [19].

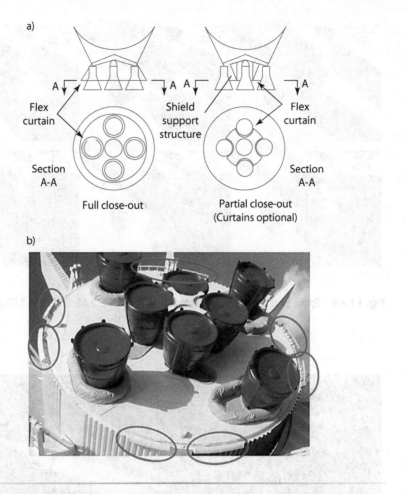

Fig. 11.44 a) Two schemes to help to reduce base heating: flexible curtains or support structure. Source: [23]. b) Saturn IB firewall with penetrations for four outer engines with flexible curtains. Source: Edberg, KSC.

- Vehicle drag, or skin friction coefficient
- Shock structures that are present
- Flow regime and transition points (laminar or turbulent)

We saw in Chapter 6 that the Space Shuttle flew different trajectories based on heating and loading constraints, so we know that trajectory parameters are major contributors to the boost convection. In particular, the speed and altitude profiles are very important. The aerodynamic heating is related to the power generated by the motion through air. Because thrust power $= Tv$, and $T \sim D \sim \rho v^2$, heating $\sim (\rho v^2) v = \rho v^3$.

Other factors that enter into the thermal environment are flight time and internal heat sources such as the rocket engine(s) itself (themselves).

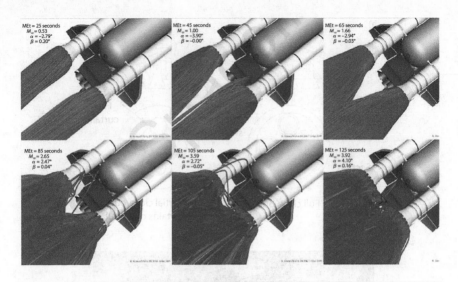

Fig. 11.45 Space Shuttle SRM plume for Mach 0.53, 1.0, 1.66, 2.65, 3.59, and 3.92 (top left to bottom right). Source: [24].

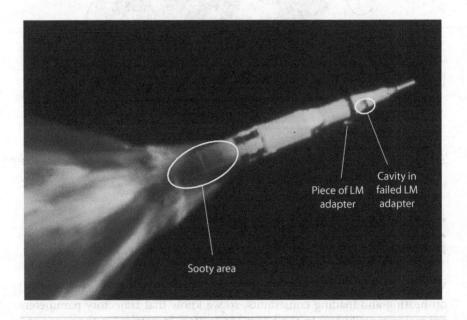

Fig. 11.46 Flow separation occurring around the base of a Saturn V during the launch of Apollo 6 was expected and planned for. A completely unrelated failure is shown by the cavity in the failed portion of the adapter just above the S-IVB step's forward skirt, and the falling piece. This structural failure occurred because expanding gases trapped within the adapter's sandwich structure caused its structure to delaminate. The failure did not cause the loss of the vehicle. Source: NASA.

11.3.3 Saturn V Flow Separation

As seen in Fig 11.46, a situation of *flow separation* or *back flow* was observed during Saturn V launches. This was not a surprise, and although it appears that the lower portion of the vehicle was on fire, there was no danger, because this was an expected event.

At higher altitudes, the lower ambient pressure caused the F-1 engines' plume to expand and create an apparently solid wall as encountered by the air flow near the vehicle. As it passes, the flow near the side of the vehicle encounters this "wall" before it reaches the plume's front, and separates. This phenomenon is shown in Fig. 11.47.

As the plume expands, the region of flow separation, characterized by the gray "sooty" appearance, moves forward. Hot gas from the exhaust recirculates up into the separated flow region from the base area. The sootiness or blackness observed is likely carbon deposits or burned paint.

11.3.4 Thermal Protection Systems

There are special coatings and lightweight insulators available that can be used to help protect launch vehicles, as well as to help keep structure and other components below their maximum allowable temperatures. These can vary from spray-on foam insulation (SOFI) to Nomex Felt Reusable Surface Insulation (FRSI), ablative paints, ceramic tiles, and others. Figure 11.48 shows a Space Shuttle external tank (ET) after separation,

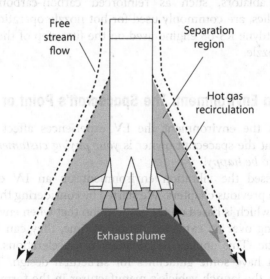

Fig. 11.47 The flow separation that occurs at higher altitudes, causing hot gas to recirculate past the base of the vehicle and up its sides. Figure 11.46 shows this occurring on a Saturn V launch.

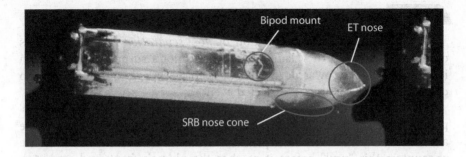

Fig. 11.48   The Space Shuttle's ET was covered with SOFI to reduce cryogenic propellant boil-off and reduce the amount of ice buildup. There were concerns that chunks of ice falling off of the ET during launch could damage the Shuttle's ceramic tiles. Photo shows the ET after separation from shuttle, where dark areas in insulation indicate burning in areas exposed to atmospheric or exhaust plume heating.

where the SOFI insulation has become charred in some areas due to atmospheric heating. The hottest areas appear to be the forward portion near the nose, the area adjacent to the noses of the two SRBs (see Fig. 11.37—could some heating be from the air turbulence?), the rear of the ET near the main engines, and in the vicinity of the bipod mount for the Shuttle. These are all areas where the flow over the bodies can interfere with each other.

An *ablative* material is one that absorbs heat and turns directly into a gas, a process that makes a relatively cool layer to reduce the heat that is absorbed by a structure. Ablatives are not reusable, but are refurbishable. High-density ablators, such as reinforced carbon-carbon (RCC) and carbon phenolics, are commonly used for hot nozzle operation. In particular, the Rocketdyne RS-68 engine used on the first step of the Delta IV has an ablative nozzle.

11.4   Payload Environment: The Spacecraft's Point of View

How does the environment the LV experiences affect the payload? Remember that the spacecraft owner is *your paying customer, so you want the customer to be happy*!

We discussed the vibration environment on an LV earlier in this chapter and in previous chapters. We started by considering the quasi-steady environment, which is based on the assumption that when environments are slowly changing over an extended period of time, they can be considered steady or static. This applies to *g*-loading or accelerations. The payload designer must have some guidelines for structural design. These may be obtained from the launch vehicle's manufacturer in the form of something called the payload planners' guide, payload user guide, user manual, mission planner's guide, or a similar title. This document is where the designer would obtain preliminary design information to start the design process.

All of these documents present something called *design load factors* that are represented by quasi-static loads (QSLs), which are usually the most severe combination of dynamic and steady state acceleration values. These are normally presented as tables or plots of *axial* or *longitudinal* load factors and *lateral* or *transverse* load factors. Note that for design purposes, the highest QSLs are the axial compressive loads, but lateral loads bend the structure and tend to be worse than axial even though their magnitude is less. This is because most spacecraft tend to be long and skinny, so the lateral loads cause bending that is more difficult to handle than the axial compressive loads. These load types are illustrated in Fig. 11.49.

The forces that are exerted on a payload come from the rocket engine's thrust for acceleration to orbital velocity and height, as well as side thrust that is used to trim the vehicle when needed due to aerodynamic forces (see Chapter 9). These forces produce primary loads in the *axial* or thrust direction and in the *lateral* or side-to-side direction.

For initial design, we use load factors in the vehicle's payload planners' guide available from the manufacturer. These levels are preliminary and are based on the LV's past acceleration time histories. Typical values are around 4–8 *g* axial and 1.5–3 *g* lateral. These may be higher values than those found on the QSL plot, because *uncertainty factors* are applied as a margin until things become better understood. "Better understood" means that the stresses are calculated for the joined LV and SC combination.

A typical set of QSLs acting on a spacecraft is provided for the Ariane 5 in Table 11.3. Notice that this table provides static and dynamic loads separately for the longitudinal (axial) direction, and combined static and dynamic loads for the lateral direction. Note that the static axial values are all negative, indicating compression (which is to be expected for a rocket launching). The critical events are listed in the leftmost column.

The Atlas V provides the information in two ways: tabular and graphical. The tabular format is given in Table 11.4, which lists the quasi-static values with either a plus or a minus sign, whereas the dynamic loads are preceded with the plus/minus ($\pm$) symbol, meaning that they can go either direction

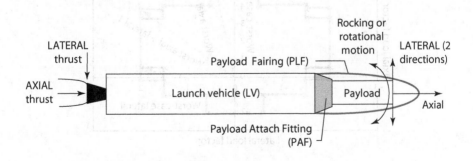

Fig. 11.49 Definitions and directions of forces experienced by a payload.

Table 11.3 Ariane 5 QSL Values

Critical Flight Events	Acceleration (g) Longitudinal Static	Longitudinal Dynamic	Lateral Static + Dynamic
Liftoff	−1.7	±1.5	±2
Max-q	−2.7	±0.5	±2
SRB end of flight	−4.55	±1.45	±1
Main core thrust tail-off	−0.2	±1.4	±0.25
Max tension case: SRB jettison	+2.5		±0.9

Source: ESA.

Table 11.4 Atlas V Design Load Factors

Flight Condition	Lateral (+ = compression, − = tension)	Axial (+ = compression, − = tension)
Liftoff	±2.0 g	+1.8 ±2.0 g
Flight Winds	±0.4 ±1.6 g	+2.8 ±0.5 g
Strap-On Separation	±0.5 g	+3.3 ±0.5 g
BECO (max axial)	±0.5 g	+5.5 ±1.0 g
BECO (max lateral)	±1.5 g	+2.5 ±1.0 g
MECO (max axial)	±0.3 g	+4.5 ±1.0 g
MECO (max lateral)	±0.6 g	±2.0 g

Source: ULA.

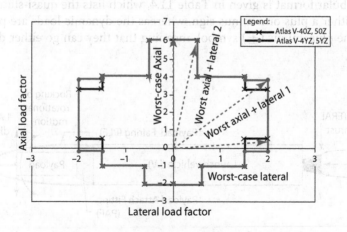

Fig. 11.50 Atlas V design load factors—graphical. Source: ULA.

depending on the way the LV happens to be vibrating. The largest loads are indicated in **bold** type in the table.

The graphical version of the Atlas V QSLs is presented in Fig. 11.50. This is possibly more intuitive, because it physically shows the axial loads (up and down on the graph) and the lateral loads (left to right on the graph). In this case, the worst case loads are indicated by the length of the vector from the origin to the boundary of the plot. Two of these are shown as Worst Axial + Lateral, meaning that the spacecraft has to withstand axial and lateral loading *at the same time.*

11.5   Spacecraft Structure Design Verification Process

How is the design of a spacecraft's structure actually carried out? The process started with static and dynamic load factors from the planner's guide, but those numbers are generic. What is really needed are the loads that will be *actually* be experienced by the payload that will be mounted on the LV. The actual process is shown in Fig. 11.51.

In Fig. 11.51, we see that the process begins with constraints and requirements in the upper row of boxes. The second row includes loads and natural frequency constraints from the planner's guide, which lead to a preliminary spacecraft design. This design is mathematically modeled and examined for a number of predicted (by the spacecraft designers) loads and environments (thermal, dynamic, stress, etc.). Problems are fixed, and the math

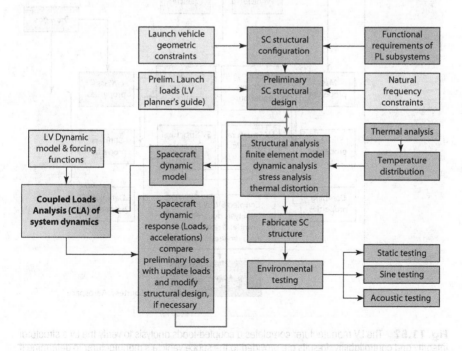

Fig. 11.51 The spacecraft's structure design verification process. Source: [25].

model of the SC is updated. At this time, the model is ready to be exposed to the actual set of loads, and that occurs in the box that says "Coupled Loads Analysis (CLA)." The CLA procedure is described in the following section and mathematically at http://www.vibrationdata.com/tutorials2/Primer_on_the_Craig-Bampton_Method.pdf.

11.5.1 Coupled-Loads Analysis

In reality, each launch vehicle is a lightweight, flexible structure whose response changes depending on what is mounted onto its front end (i.e., spacecraft mass and dynamic properties). This means that every new payload that wishes to launch may have a new response that potentially could cause a problem for the launch vehicle.

In order to ensure that its LV will operate safety, a LV company mathematically simulates flight events separately with each payload; the simulation is called a coupled-loads analysis (CLA). The procedures that occur in the CLA are shown in Fig. 11.52.

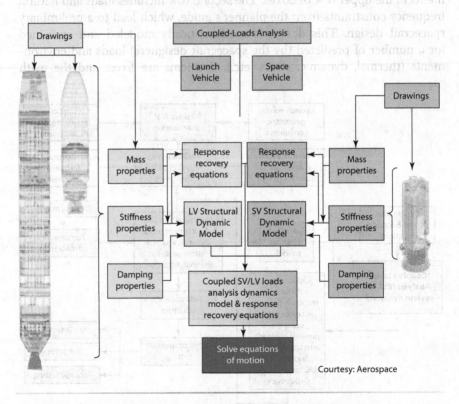

Fig. 11.52 The LV manufacturer completes a coupled-loads analysis to verify the LV's structural integrity and controllability. Results are provided to the space vehicle's manufacturer to determine if the SV can survive the imposed environments. Source: [26].

One ingredient of the CLA is a launch vehicle FEM and load environments (liftoff, maneuvering, max-q, staging, etc.) that are mature and well understood. The other ingredient comes from the spacecraft manufacturer, who provides a mathematical model of its payload when it's ready. The CLA is a production analysis that is done with the addition of each new payload to an existing, well-understood, and mature launch vehicle model.

The figure shows how the CLA mathematically couples or joins the launch vehicle's FEM with the payload's FEM, using a method called the Craig-Bampton method, or something similar. The LV and the payload likely use different coordinate systems, so there must be a complete understanding about both coordinate systems to ensure the coupling is correct. The coordinate system agreement may be specified by an *interface control document (ICD)* that is signed by both the LV and spacecraft companies.

When the joined FEM has been assembled, a number of flight events or environments are imposed in order to simulate various different events that occur during liftoff and flight. The responses of both the LV and the payload are recorded, and stresses in the LV's parts and all payload components are obtained.

In addition, a similar procedure is done where the *stability* of the joined-vehicle (launch vehicle and payload) control system is assessed. For this assessment, closed-loop attitude control with the actual payload's flexibility and natural frequencies is simulated. The usual requirement is that there be sufficient spacing between the bandwidth of the attitude control system and the structural natural frequencies.

The results of the mechanical CLA, in terms of the payload's stresses, are provided to the payload's company so that updated stresses and stability assessments go to the structures and controls design groups. The structure is redesigned as necessary to correct any stress and/or stiffness concerns, and once verified, can be manufactured and tested. The LV's controls group examines the results of the controls CLA, and if changes to the payload's stiffness are needed, the information is also fed back to the spacecraft company.

This leads to an observation regarding the launch vehicle company's payload planners' guide (or equivalent name). Guidelines are given for minimum fundamental natural frequency, as shown in Table 11.5. The frequencies are selected to ensure that there is no control-structure interaction (overlap of control bandwidth with structural frequencies). These are not hard rules; in fact they have been chosen so there is only a very small chance that there will be an issue with the launch vehicle's control system. Hence, they are conservative and may be "violated," provided they satisfy the actual controls CLA. Of course, each LV has its own set of frequency criteria.

A comment on coupled-loads analysis: imagine having to run an analysis program *every time* you wanted to transport something in the bed of a pickup

Table 11.5 Minimum Fundamental Frequency for Several Launch Vehicles

Launch System	Fundamental Frequency, Hz	
	Axial	Lateral
Atlas II, IIA	15	10
Ariane 4	31, dual PLs; 18, single PL	10
Delta 6925/7925	35	15
Long March 2E	26	10
Pegasus	20	20
Proton	30	15
Space Shuttle	13	13

Source: Payload planners' guides.

truck. Does this seem practical? Hardly. Efforts have been made to *decouple* the launch vehicle and the spacecraft through an isolation system. Information on this idea is presented later in this chapter.

11.5.2 After Coupled Loads Analysis: What Happens?

Looking at the bottom of the design verification flowchart, we see that the CLA results are fed back to the spacecraft company and updated loads go to the spacecraft design group. The CLA results are compared to the preliminary loads used for design, and the spacecraft's structure is updated as necessary to alleviate stress or stiffness concerns. The in-house model testing is repeated and, when adequate, the spacecraft structure is fabricated. Once hardware exists, it is tested, and the spacecraft's math model is adjusted to agree with the test results.

11.5.3 Payload Natural Frequencies ≠ LV Natural Frequencies

The fundamental stiffness requirement to be dealt with in spacecraft design is that there is no resonant coupling of SC modes with LV modes. To do this, we normally observe a very simple principle: if the SC structure has natural frequencies that are greater than the LV frequencies, there will usually not be a problem. "Greater than" in this case means at least a factor of three, but larger values are even better.

A simple way to look at this situation is to examine a single-degree-of-freedom (SDoF) system, namely, a mass m connected to a shaking system via a spring and dashpot. The mass, spring, and dashpot represent the mass, stiffness, and damping of a spacecraft, respectively, and the launch vehicle it's mounted on has the motion $x(t)$. This simple model is shown in Fig. 11.53a.

The spring stiffness k and the dashpot damping constant c may be used to calculate the undamped angular natural frequency $\omega_n^2 = k/m$ (rad/s), the undamped natural frequency $f_n = \omega_n/2\pi$ (Hz or cycles/s), and the damping ratio $\zeta = \dfrac{c}{2\sqrt{km}}$. Using these definitions, we can plot the amplification or attenuation of the system [ratio of mass motion $z(t)$ ÷ base motion $x(t)$] as a function of frequency ratio. In this simple model, the spacecraft possesses the mass, stiffness, and damping; its motion is $z(t)$. This results in a plot like the one shown in Fig. 11.53b.

If the shaking frequency Ω is increasing steadily, we wish to get through frequency ratio $\Omega/\omega_n = 1$ as fast as possible, or the response (proportional to TR) will become very large. The transmissibility of an SDoF system near resonance ($\Omega \approx \omega_n$) is approximately $1/(2\zeta)$. Because most spacecraft have damping ratios ζ that are *very* small, between 0.001 and 0.05, there may be magnification of factors of 10 to 500 of base motion at the spacecraft's natural frequency. Obviously this is not a good situation, and we need to think about ways to reduce it. There is no attenuation until the shake frequency $\Omega > \sqrt{2}\,\omega_n$.

11.5.4 Payload Isolation: Helps with Much of the Shock and Vibration

Can you imagine an automobile where the wheel axles were bolted directly to the frame, and there were no shock absorbers? It would be a rough ride, for sure. Yet, that is the situation on many launch vehicles: the payload is attached directly to the PAF or LV, and the PAF or LVA is bolted directly onto the top of the launch vehicle's upper stage.

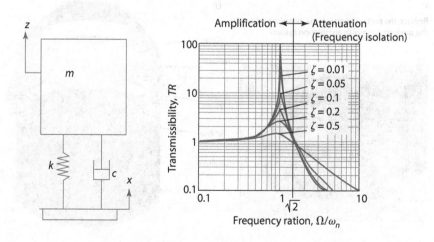

Fig. 11.53 a) SDoF system; b) a plot of its transmissibility (ratio of response amplitude to input amplitude, z/x) as a function of frequency ratio and damping ratio ζ. Source: [3].

Suppose that we instead placed the equivalent of a shock absorber for a car atop the PAF. The lower stiffness of the shocks would allow relative motion between the payload and its adapter, which means that the shocks could absorb shock. This idea is shown in Fig. 11.54, where an isolation system is placed to attenuate vibration from the launch vehicle. Figure 11.54a shows a drawing of such an isolation system; Fig. 11.54b shows an actual isolation system designed by Moog-CSA installed atop the PAF.

So, does an isolation system really help? Refer to Fig. 11.55, which shows the effect of payload isolation on vibrations experienced during an actual launch of a Taurus LV. The dark gray trace is the ambient acceleration acceleration measured below the payload attach fitting without isolation. The light gray trace shows measured acceleration *above* the isolator, and is about an order of magnitude—one tenth—of the acceleration shown in dark gray. That's a very large reduction.

There are several issues with such an isolation system: its added mass penalty, the increased flexibility between the payload and the LV, the system's damping, and its cost (which would be determined by the hardware and installation costs).

The mass of any isolation system must directly subtract from the allowable payload mass. This could be a problem, but the isolator shown in the figure has a mass of only about 41 kg, a cost in mass that would likely be saved in structural mass due to reduced dynamic loads.

Why is isolation system flexibility an issue? Remember that most payloads are long and skinny, so even a small rocking motion translates to much more movement at the other end of the payload, possibly interfering with the wall of the payload fairing (which of course is a no-no). So, an

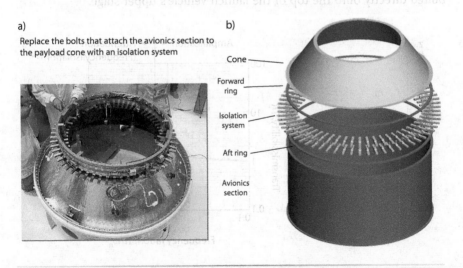

a)
Replace the bolts that attach the avionics section to the payload cone with an isolation system

b)
Cone
Forward ring
Isolation system
Aft ring
Avionics section

Fig. 11.54 By placing an isolation system between the payload and the launch vehicle, large reductions in payload vibration may result. Source: [4].

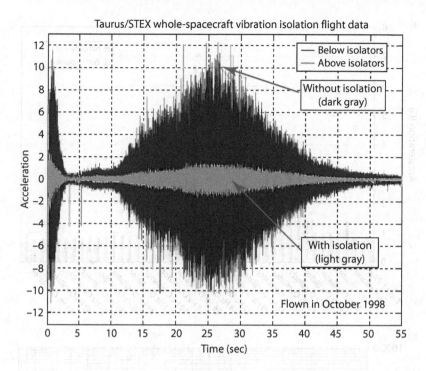

Fig. 11.55 Large reductions in vibration are possible with a well-designed vibration isolation system. This is a plot of accelerations measured on STEX flying on a Taurus launch vehicle. Source: [4].

ideal isolator would be soft in the axial direction and stiff in the lateral directions. This is not easy to achieve.

The isolator's damping is certainly important. We saw earlier that a system with small damping would have very high amplification, or gain, of the motion of its base when near the natural frequency of the spring-mass system. So, it would be ideal to add damping to the isolator in addition to its other traits.

The cost? Again, the savings in structural loads (and perhaps the cost of coupled-loads analyses) would probably make up for the cost, and then some.

The author, as an employee of McDonnell Douglas, carried out a study of what combination of stiffness and damping (or equivalently natural frequency and damping) would result in the "best" performance; see Edberg [27]. On a Delta II 7925 launch vehicle, it was found that an isolation system with a 6-Hz lateral natural frequency and 6% damping performed well, and *was able to pass the control system CLA procedure* without any modifications. Because the payload planner's guide suggests a Delta II 792X minimum frequency of *15* Hz (see Table 11.5), the level of conservatism of the values given in the PPG is easily seen. A plot showing the reduction in responses due to the isolation system is shown in Fig. 11.56.

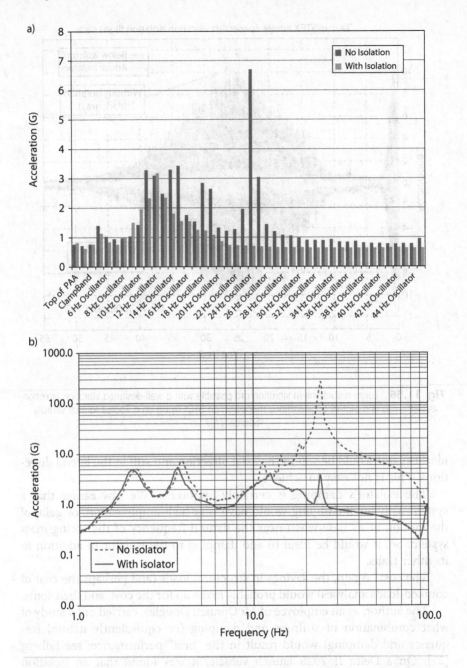

Fig. 11.56 a) Peak lateral acceleration levels on the spacecraft frame and simulated oscillators, isolated and unisolated. b) Shock response spectrum lateral responses for 24-Hz secondary structure show a large reduction in levels above 12 Hz comparing the isolated case (bottom) to the unisolated case (top). Source: [27].

11.5.5 Free-Free Natural Frequency Calculations

A launch vehicle in flight could be idealized as a free-free (F-F) beam because there are no boundary supports: the ends are free. The first and second natural vibrational frequencies of a *uniform* F-F beam may be calculated from

$$\omega_{n1} = 22.4\sqrt{\frac{EI}{ml^3}}$$

$$\omega_{n2} = 61.7\sqrt{\frac{EI}{ml^3}}$$

where ω_{n1} and ω_{n2} are the beam's first and second angular natural frequencies (in rad/s), EI = bending stiffness, m = mass of beam, and l = beam length. These can be converted into cycles per second by dividing by 2π.

These formulae are not appropriate for actual LVs whose local stiffness and mass values change along their length, but they could be used to find *rough order of magnitude (ROM)* values for the frequencies. For instance, a uniform F-F beam could approximate a uniform-diameter LV with propellants along most of its length, so these formulae would provide rough values for the first and second vibration modes. When using these formulae, be sure to use consistent units throughout—a dimensional analysis is recommended.

11.5.6 LV Frequency Considerations

Usually, we want to have vibration frequencies that are about an order of magnitude lower than the LV's control system frequencies, so that there is no chance of coupling between the two, which could cause instability. Also, the flexibility of an LV can be a problem: a phenomenon called *tail wags dog* results when commands from the thrust vector control (TVC) system used for attitude control cause the LV body itself to move in the opposite direction. This will be discussed in the next chapter.

11.5.6.1 Example Stiffness Design Factors: Atlas V

The Atlas V payload planner's guide stipulates that the axial direction natural frequency is ≥ 15 Hz, and the lateral direction natural frequency is ≥ 8 Hz. Generally speaking, any kind of secondary structure or payload mounts have to be sized by the *noncoupling* or *octave* rule: this equipment must have natural frequencies that are at least double the frequency of existing frequencies in order to preclude coupling and load amplification.

Let's do an example of analyzing stiffness requirements (ignoring what we learned earlier, that the PPG values are very conservative). Suppose you are launching your equipment on board an Atlas V, and the equipment's primary structure just meets the Atlas's 15-Hz axial requirement.

Considering the situation illustrated in Fig. 11.57, what are the axial stiffness requirements for unit A and unit B brackets, if the support deck shown has a 50-Hz natural frequency?

From the PPG, we know that the minimum axial stiffness results in frequency $f_n = 15$ Hz. Consider *unit A*, which is mounted directly above a support column that is primary structure. Applying the octave rule, unit A's brackets should be designed for $f_n = 2 \times 15$ Hz = 30-Hz stiffness. This allows you to solve for the bracket stiffness if unit A's mass is known.

The 50-Hz support deck on which it is mounted dictates *unit B's* behavior. Applying the octave rule, unit B's supports should be designed for $f_n = 2 \times 50$ Hz = 100 Hz. Again, the bracket system's stiffness may be calculated if unit B's mass is known.

Finally, note that the support deck itself also just satisfies the octave rule, because its natural frequency of 30 Hz is twice the minimum axial frequency of 15 Hz.

11.5.7 Payload Acoustic Environments

The launch vehicle's payload planner's guide or equivalent provides predicted acoustic environment curves inside the payload fairing, which indicate the maximum expected sound pressure levels (SPLs). The SPLs tend to vary, depending on two factors: the configuration of the launch vehicle and the payload geometry. The latter is very noticeable when a very full PLF is compared to a not-so-full PLF, as may be seen in Fig. 11.58 [28].

Note that the two-stage curves, plotted in dashed lines, are considerably higher than the three-stage (solid-line) curves. This may be attributed to the fact that much more of the volume inside the PLF is filled with spacecraft or

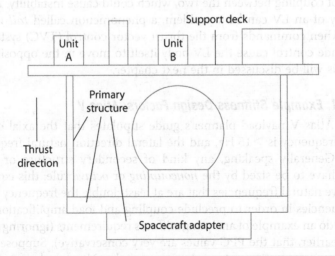

Fig. 11.57 Example of primary and secondary structure to determined stiffness requirements.

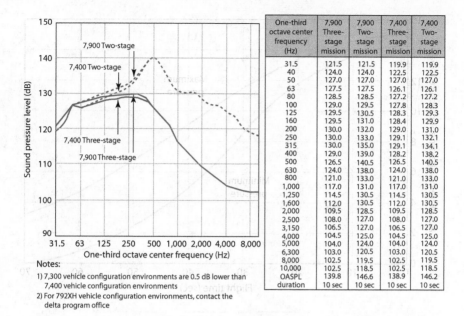

One-third octave center frequency (Hz)	7,900 Three-stage mission	7,900 Two-stage mission	7,400 Three-stage mission	7,400 Two-stage mission
31.5	121.5	121.5	119.9	119.9
40	124.0	124.0	122.5	122.5
50	127.0	127.0	127.0	127.0
63	127.5	127.5	126.1	126.1
80	128.5	128.5	127.2	127.2
100	129.0	129.5	127.8	128.3
125	129.5	130.5	128.3	129.3
160	129.5	131.0	128.4	129.9
200	130.0	132.0	129.0	131.0
250	130.0	133.0	129.1	132.1
315	130.0	135.0	129.1	134.1
400	129.0	139.0	128.2	138.2
500	126.5	140.5	126.5	140.5
630	124.0	138.0	124.0	138.0
800	121.0	133.0	121.0	133.0
1,000	117.0	131.0	117.0	131.0
1,250	114.5	130.5	114.5	130.5
1,600	112.0	130.5	112.0	130.5
2,000	109.5	128.5	109.5	128.5
2,500	108.0	127.0	108.0	127.0
3,150	106.5	127.0	106.5	127.0
4,000	104.5	125.0	104.5	125.0
5,000	104.0	124.0	104.0	124.0
6,300	103.0	120.5	103.0	120.5
8,000	102.5	119.5	102.5	119.5
10,000	102.5	118.5	102.5	118.5
OASPL	139.8	146.6	138.9	146.2
duration	10 sec	10 sec	10 sec	10 sec

Notes:
1) 7,300 vehicle configuration environments are 0.5 dB lower than 7,400 vehicle configuration environments
2) For 792XH vehicle configuration environments, contact the delta program office

Fig. 11.58 Predicted Delta II acoustic environments for 9.5-ft fairing missions. Source: [28].

third-stage hardware compared to the two-stage curves (which obviously don't have a third stage).

The bottom line of the tabulation on the right of Fig. 11.58 is labeled "OASPL," for overall sound pressure level. The OASPL is calculated as the area under the curve of interest

$$OASPL = \int_{31.5\ Hz}^{8\ kHz} SPL\ d(frequency)$$

It is an indication of the total sound energy a payload might be exposed to. Note that the OASPLs of the two-stage missions are consistently higher ($\sim$7 Hz) than for the three-stage missions.

A table or chart similar to those in Fig. 11.58 would be used to acoustically test a spacecraft and the PLF on the ground, to ensure that the high sound levels do not damage the spacecraft. This is discussed in Chapter 15.

11.5.8 Payload Pressure Environment

The atmospheric pressure decreases during a launch due to the vehicle's climb out of the atmosphere. This suggests that any parts of the spacecraft that have trapped volumes of air will need to have some sort of a vent system incorporated in order for them to vent and not burst due to internal pressure. A typical pressure vs time plot for the Delta II is shown in Fig. 11.59.

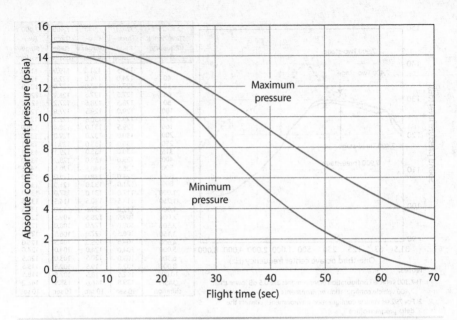

Fig. 11.59 Static pressure within Delta II payload fairing during launch. Source: [28].

A simple way to ensure that there is no trapped volume is to drill vent holes, or locate vent hatches, so that the trapped air can expand without building up pressure. The internal volume and the desired leak rate can determine the number or area of the holes.

The internal pressure within a payload fairing will vary due to changing patterns of shock waves as the vehicle passes through the transonic speeds. Figure 11.60 shows the internal and external pressures within a Brazilian launch vehicle (VLS). The plot also shows the difference between the two, which is especially interesting in the transonic area.

The plot in Fig. 11.60 shows how the pressure and pressure difference can vary depending on the Mach number M, which may be thought of as time scale. (It increases similar to time accumulating.) We see in Fig. 11.61 that the maximum negative pressure coefficient $C_p = (p - p_\infty)/q$ occurs at transonic speeds at line A's (solid curve's) corner. (We call the time period around $M = 0.7$ to 0.8 "transonic," because some of the flow over the vehicle is *supersonic*, and around other parts it's *subsonic*.) Remembering that we have to have a vent to allow air to exit the PLF as the rocket climbs out of the atmosphere, we have to consider how the pressure changes with Mach number.

Placing a vent at location A, the shoulder of the conical nose, reduces the pressure load there to low values, but increases the net pressure at location B. These relative differences may be very important when pressure vents need to be located.

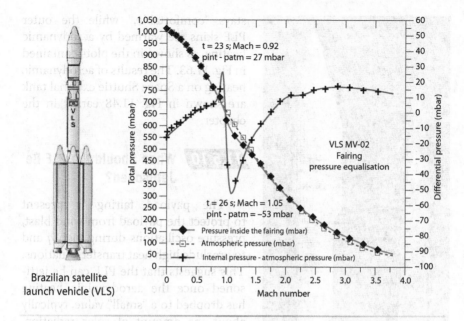

Brazilian satellite
launch vehicle (VLS)

Fig. 11.60   Internal/external fairing pressure vs Mach. Source: scielo.br.

11.5.9 Payload Thermal Loads

Once on the launch pad before launch, the payload inside the fairing is internally air-conditioned with dry air or nitrogen in a certain temperature range. One or more umbilical hoses supply the conditioned air, as shown in Fig. 11.62. Once the launch begins, the umbilicals disconnect, and the payload fairing and its isolation typically manage to keep the payload at approximately the same temperature during launch and ascent. The payload

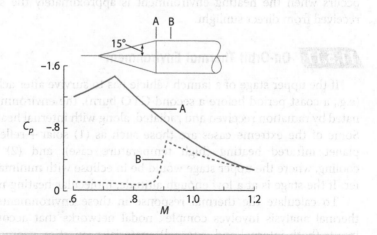

Fig. 11.61 C_p vs Mach number M at two locations near a conical nose. Source: [19].

Fig. 11.62 The payload is supplied with air-conditioning from hoses until liftoff. Source: Lockheed Martin.

stays "comfortable," while the outer PLF skins are warmed by aerodynamic heating, as shown in the plots contained in Fig. 11.63. The results of aerodynamic heating on a Space Shuttle external tank are shown in Fig. 11.48 earlier in the chapter.

11.5.10 When Should the PLF Be Jettisoned?

The payload fairing is present to protect the payload from wind blast, pressure oscillations during high q and to preclude high heat transfer situations. This suggests that the PLF can be jettisoned once the aerodynamic heating has dropped to a "small" value, typically about the amount of solar radiation, approximately 1,135 W/m^2 (0.1 BTU/ft^2/s), from the specifications for the Proton [29]. Of course, the mass reduction due to jettisoning the PLF will certainly improve launch performance. Figure 11.64 shows the flight trajectory values when a Delta II jettisons its PLF. Figure 11.65 shows the Ariane 5 thermal flux and total energy absorption after PLF jettison. The "Aerothermal flux" line shows that jettison occurs when the heating environment is approximately the same as that received from direct sunlight.

11.5.11 On-Orbit Thermal Environment

If the upper stage of a launch vehicle has to survive after achieving orbit (e.g., a coast period before a second GTO burn), the environment is dominated by radiation received and radiated, along with internal heat generation. Some of the extreme cases are those such as (1) solar + reflected solar + planet infrared heating (high temperature case), and (2) deep space cooling, where the upper stage would be in eclipse with minimal heat transfer. If the stage is at a low enough altitude, molecular heating may occur.

To calculate the thermal response in these environments, a detailed thermal analysis involves complex nodal networks that account for heat inputs (both internal and external), material properties, geometries, orbital considerations, and so forth.

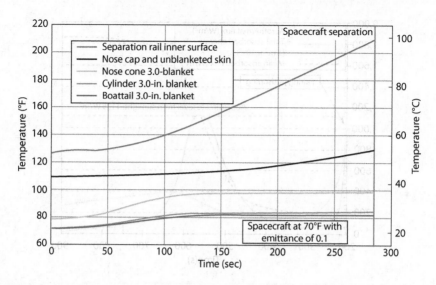

Fig. 11.63 Delta II predicted maximum internal wall temperature for a 10-ft fairing. Note the small temperature rise shown by the bottom curve: only about 6°C over the entire launch period. Source: [28].

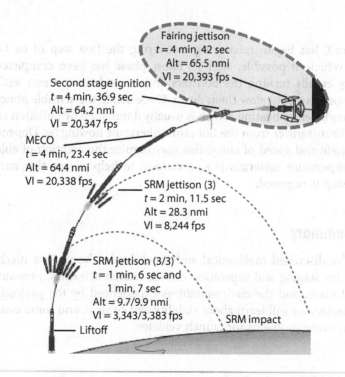

Fig. 11.64 Delta II payload fairing jettison occurs as soon as the external heating rate is low enough to not cause any damage to the payload, typically around 1,100 W/m². VI is an abbreviation for "velocity, inertial". Source: [28].

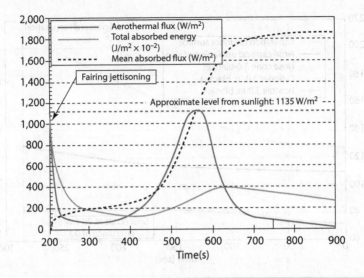

Fig. 11.65 Ariane 5 thermal fluxes after PLF jettison. Jettison occurs at about 200 s, when the aerothermal flux heating environment is approximately 1,135 W/m², about the same as that received from direct sunlight. Source: [13].

SpaceX has begun recovering recovering the first step of its Falcon 9 launch vehicle if possible, after its boost phase has been completed. That recovery entails turning the boosters around and firing them against the orbital speed so as to slow them down. Once inside the sensible atmosphere, there may be entry heating, which is usually dominated by radiation and convective heat transfer from the hot atmospheric air flowing by. Depending on the attitude and speed of entry, this may require the addition of ablative or high-temperature materials (e.g., ceramics) to help ensure that survival of rocket step is required.

11.6 Summary

We've discussed mechanical and shock loads, the various mechanisms needed for staging and separation events, launch acoustics, pressure loads, thermal loads, and the environment as experienced by the payload. In the next chapter, we will learn about stability and control, and some instabilities that can cause problems for launch vehicles.

References

[1] Perl, E., Do, T., Peterson, A., and Welch, J., "Environmental Testing for Launch and Space Vehicles," Crosslink Magazine, Aerospace Corp., Fall 2005.

[2] Gordon, S., "Delta II MECO Transient: A Payload Perspective," SC and LV Dynamic Environments Workshop, 25–27 June 2002, Los Angeles, CA.

[3] Sarafin, T. P., and Larson, W. J., *Spacecraft Structures and Mechanisms: From Concept to Launch*, Hawthorne, New York, 2007.

[4] Johal, R. S., Wilke, P. S., and Johnson, C. D., "Satellite Component Load Reduction Using Softride," AIAA 6th Responsive Space Conference, Los Angeles, CA, 30 April 2008.

[5] Fritz, J. E., "Separation Joint Technology," 39th AIAA/ASME/SAE/ASEE Joint Propulsion Conference, paper no. AIAA 2003-4436, 20–23 July 2003.

[6] Goldstein, S., "Exploding into Space: Explosive Ordnance for Space Systems," Crosslink Magazine, Aerospace Corp., Fall 2006.

[7] Stockinger, J., Shen, F., Bowers, G., Pan, M., and Whalley, I., "Delta IV Evolved Expendable Launch Vehicle Ordnance Overview," AIAA 2005-3655, 2005.

[8] Whalley, I., "Development of the STARS II Shroud Separation System," AIAA 2001-3769, 2001.

[9] Boeing, "MDAC Payload Fairings," presentation, Oct. 1987.

[10] Space Exploration Technologies, *Falcon 9 Payload User's Guide*, rev. 2, SpaceX, 2015.

[11] Wiesel, W. E., *Spaceflight Dynamics*, Aphelion Press, Beavercreek, OH, 2012.

[12] Hobbs, M., *Fundamentals of Rockets, Missiles and Spacecraft*, John F. Rider, New York, 1964.

[13] Fortescue, P., and Swinerd, G., *Spacecraft Systems Engineering*, Wiley, New York, 2011.

[14] Pisacane, V., *Fundamentals of Space Systems*, Oxford University Press, Oxford, England, 2005.

[15] Conley, P., Packard, D., and Purdy, W., *Space Vehicle Mechanisms: Elements of Successful Design*, Wiley, New York, 1998.

[16] Wijker, J., *Random Vibrations in Spacecraft Structures Design: Theory and Applications*, Springer Netherlands, Dordrecht, 2009.

[17] NASA, "Acoustic Loads Generated by the Propulsion System," NASA SP-8072, 1971.

[18] NASA, "Structural Response to Inflight Acoustic and Aerodynamic Environments," NASA CR-88211, 1967.

[19] Boswinkle, R. W., "Aerodynamic Problems of Launch Vehicles," NASA SP-23, Dec. 1962.

[20] NASA Engineering and Safety Center (NESC), "New Approaches to Refining the Launch Vehicle Buffet Environment," NESC Technical Update 2015, p. 32.

[21] NASA, "Aerodynamics of Space Vehicles," NASA SP-23, Dec. 1962.

[22] Barret, C., "Review of Our National Heritage of Launch Vehicles Using Aerodynamic Surfaces and Current Use of These by Other Nations," NASA TP-3615, April 1996.

[23] Huzel, D. K., Huang, D. H., and Arbit, H., *Modern Engineering for Design of Liquid-Propellant Rocket Engines*, American Institute of Aeronautics and Astronautics, Washington, D.C., 1992.

[24] Gomez, R. J., "Viton Debris Transport to SSME Hatband," part of STS-129 Agency Flight Readiness Review, Oct. 2009.

[25] Maini, A.K., and Agrawal, V., *Satellite Technology: Principles and Applications*, 3rd ed., John Wiley & Sons, New York, 2014.

[26] Kabe, A. M., Kim, M. C., and Spiekermann, C. E., "Loads Analysis for National Security Space Missions," Crosslink Magazine, Aerospace Corp., Winter 2003–2004.

[27] Edberg, D., and Fan, H., "Numerical Studies of the Performance of a Whole-Spacecraft Vibration Isolation System," 42nd Structures, Structural Dynamics and Materials Conference and Exhibit, 15–19 April 2001, Seattle, WA.

[28] Boeing, "Delta II Payload Planners Guide," 06H0214, July 2006.

[29] International Launch Services, "Proton Launch System Mission Planner's Guide," July 2009, https://www.ilslaunch.com/launch-services/proton-mission-planners-guide/.

Further Reading

Bejmuk, B., "Are We Learning from Past Programs? Are We Applying Lessons Learned?," 2012, https://www.nasa.gov/sites/default/files/atoms/files/bobejmuk.pdf [retrieved 8 March 2018].

Lange, O. H., and Stein, R. J., *Space Carrier Vehicles: Design, Development, and Testing of Launching Rockets*, Academic Press, New York, 1963.

Ryan, R. S., "A History of Aerospace Problems, Their Solutions, Their Lessons," NASA TP-3653, 1996.

11.7 Exercises

1. **Launch Vehicle Vibration.** As a *very rough* approximation, assume that your LV is a *uniform* free-free beam; in other words, its mass and stiffness properties don't change over its entire length, and there are no supports. *Clearly the uniform properties assumption is an approximation!*

   The first and lowest natural vibration frequency for a *uniform* free-free beam may be calculated by $f_1 = \dfrac{\beta_1^2}{2\pi}\sqrt{\dfrac{EI}{\rho A}}$, where $\beta_1 L = 4.73004$. In this relation, f_1 is the vibration frequency in Hz, L is the beam's length, EI is the beam's uniform bending stiffness, A is the beam's cross-sectional area, and ρA is the beam's (average) linear density (mass per unit length). Hence the mass of the beam $m = \rho A L$. Use this relation to make an estimate of your LV's first natural frequency f_1 in Hz for the following flight conditions (calculations should be neat and *include all units*):

   a) Liftoff, fully loaded first and second steps ($\rho A L_{total} = m_{total}$)
   b) After first-step jettison, fully loaded second step and payload only ($\rho A L_{step\text{-}2} = m_{PL} + m_{step\text{-}2}$)

   Note that if either of your natural frequencies is less than 1 Hz, your vehicle will probably be difficult to control because it is too flexible, and your vehicle resembles a javelin or piece of spaghetti. This means your slenderness ratio or fineness ratio is too small.

   Notes:
   (i) If your vehicle has two different materials, assume (for part 1a) that the value of E for the material for the entire two-step vehicle is the one used on the lower (first) step. For part 1b, just use the second-step material.
   (ii) You will need to assume a radius R_{mean} and a wall thickness t_{ave} in order to calculate the *area* moment of inertia $I = \pi R_{mean}^3 t_{ave}$ for your vehicle. Let R_{mean} be the geometric mean radius, $R_{mean} = \sqrt{R_1 R_2}$. For part 1a's t_{ave}, use the average of the maximum and minimum wall thickness for the entire vehicle. For part 1b, use t_{ave} = average of maximum and minimum wall thicknesses for the second step, and use the second-step's radius R_2.

(iii) Assume any internal propellants provide *zero* stiffness contribution (but do contribute to the uniform mass distribution as given after 1a and 1b).

2. **Payload Fairing Separation.** Refer to Fig. 11.22 in the text. A launch vehicle is accelerating in a vacuum with acceleration a_v. Figure 11.22 shows one half of a PLF separating from the LV. The PLF first pivots about the hinge at H, and then at a given angle ϕ_0, it separates. It is quite obvious that the PLF should *not* collide with the LV after separation.

   a) During the hinged phase, note that the apparent weight of the PLF is ma_v. Show that $ma_v(r \sin \phi) = I_h \ddot{\phi}$, where I_H is the PLF's moment of inertia about its hinge.

   b) After separation, show that (1) the PLF tumbles about its center of mass at a constant rate $\dot{\phi}$, and (2) the center of mass follows a parabola with respect to the (accelerating) LV reference.

Chapter 12

Space Launch Vehicle Stability and Control: Higher-Order Dynamic Effects

Moving forward from vibration, shock, acoustics, and thermal, we show the design roadmap again in Fig. 12.1 with the highlighted signpost indicating the information to be covered in this chapter: *assess vehicle stability and control*. We'll also discuss some common instabilities that occur, the problems they cause, and how they are dealt with.

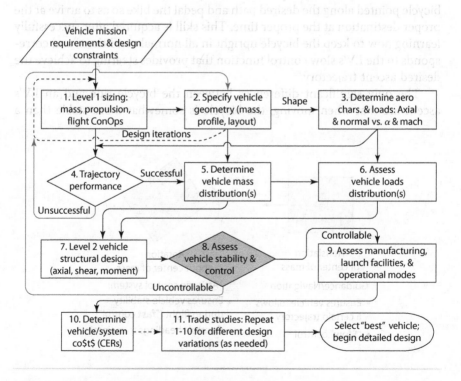

Vehicle mission requirements & design constraints

| 1. Level 1 sizing: mass, propulsion, flight ConOps | Shape | 2. Specify vehicle geometry (mass, profile, layout) | | 3. Determine aero chars. & loads: Axial & normal vs. α & mach |

Design iterations

4. Trajectory performance — Successful → 5. Determine vehicle mass distribution(s) → 6. Assess vehicle loads distribution(s)

Unsuccessful

7. Level 2 vehicle structural design (axial, shear, moment) → 8. Assess vehicle stability & control — Controllable → 9. Assess manufacturing, launch facilities, & operational modes

Uncontrollable

10. Determine vehicle/system co$t$ (CERs) ---→ 11. Trade studies: Repeat 1-10 for different design variations (as needed) → Select "best" vehicle; begin detailed design

Fig. 12.1 This chapter will look at stability and control, using data from previous chapters.

12.1 Guidance and Navigation vs Attitude Control

In order for a space launch vehicle to be useful, it has to be able to fly its payload to the proper orbit (velocity and altitude). To do this, the vehicle must have two traits. It must (1) be either stable or able to be stabilized, and (2) have a control system that can control headings in order to maintain a desired trajectory. As shown in Fig. 12.2, the functions of the vehicle's attitude control system (ACS) may be split into two portions—a "slow" control that steers the vehicle and ensures it follows the desired trajectory, and a "fast" control that keeps the vehicle from becoming unstable during flight.

To make this clear, the situation of successfully operating a launch vehicle (LV) has an analogy in an experience that nearly every person on the planet has experienced—the process of riding a bicycle. To successfully operate a bicycle, the rider must be able to carry out two separate functions. First, the rider must be able to maintain his or her balance while riding the bicycle and not tip over. The bicycle balancing skill is usually acquired during childhood and becomes an *automatic* function once acquired. This balancing skill is analogous to the LV's fast control, which is executed by the vehicle's attitude control system and keeps the vehicle from going unstable or *swapping ends* during flight.

The second function required for a bicycle rider is to be able to keep the bicycle pointed along the desired path and pedal the bike so as to arrive at the proper destination at the proper time. This skill is acquired after successfully learning how to keep the bicycle upright in all normal conditions, and corresponds to the LV's slow control function that provides steering to achieve the desired ascent trajectory.

The only significant differences between the bicycle ride and an LV's ascent are that when moving, the bicycle is somewhat more stable than a

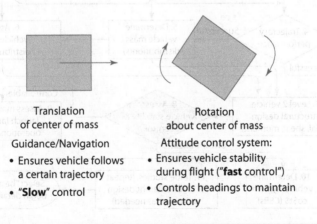

Translation of center of mass

Guidance/Navigation
- Ensures vehicle follows a certain trajectory
- "**Slow**" control

Rotation about center of mass

Attitude control system:
- Ensures vehicle stability during flight ("**fast** control")
- Controls headings to maintain trajectory

Fig. 12.2 The vehicle's ACS has a fast loop that can stabilize an unstable vehicle and a slow loop that ensures the vehicle follows the desired trajectory.

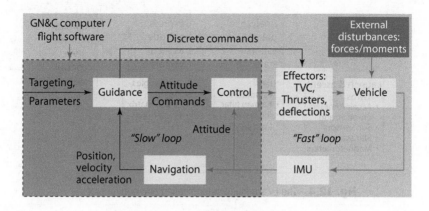

Fig. 12.3   The operation of a vehicle's ACS.

launch vehicle and can only lose control by tipping sideways, and the bicycle's motion is constrained to a (mostly) horizontal plane, whereas the launch vehicle is unconstrained in two lateral directions.

A typical launch vehicle guidance, navigation, and control system (GN&C) or attitude control system (ACS) has a block diagram, as shown in Fig. 12.3.

This ACS or GN&C system must have three components:

1. A sensor system, shown in Fig. 12.3 as an inertial measurement unit (IMU), whose outputs may be used to ascertain the position and speed of the vehicle.
2. A flight computer (dashed box in Fig. 12.3) that can determine whether the trajectory is ok and can tell the controls (effectors) on the vehicle what to do to maintain the desired trajectory.
3. A set of effectors—typically a thrust vector control (TVC) system and/or an independent thruster system and/or aerodynamic surfaces (within an atmosphere). The flight computer commands the effectors to move to control the vehicle.

The guidance system uses the vehicle's current position to determine what corrections are needed to follow the desired trajectory, and then the control system determines the appropriate engine steering angles. This is the slow loop in Fig. 12.3.

Maintaining the vehicle's stability is separate from the GN&C system. The vehicle may be stable on its own, but more likely than not, it will be unstable. For this reason, we "close the loop" and feed back the vehicle's attitude to be used by the fast loop (lower-right loop of arrows in Fig. 12.3) to maintain the vehicle's stability. Clearly, the attitude control system (fast loop) has to operate much faster than the guidance (outer) loop.

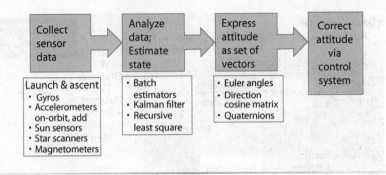

Fig. 12.4 The LV's attitude determination and control process.

We can add a bit more detail to the operation of the ACS. Figure 12.4 shows a three-step method to determine what corrections need to be made to a vehicle's attitude. From this diagram, it's easy to see that four processes must take place before the system can work properly.

The four steps are: collect measurements, convert the measurements into the estimated vehicle state, convert the vehicle's state into position and heading vectors, and then command the effectors to achieve the desired actual state. The boxes underneath the arrowed boxes contain methods used to carry out the instructions and will be described later in the chapter.

It can be seen that for successful control, we need an accurate attitude sensing/measurement system, a way to accurately translate measurements into the vehicle's attitude, a system that will interpret the current position and issue the commands to try to achieve the desired position, and finally, a set of effectors or actuators that can be positioned precisely enough to move the vehicle itself.

12.1.1 Vehicle Coordinate System

We will utilize a coordinate system that is similar to a common airplane coordinate system, shown in Table 12.1. These coordinates are illustrated in Fig. 12.5. Pitching motion corresponds to the action of pointing the vehicle's nose upwards and downwards. Yawing motion is motion to the left or right,

Table 12.1 Commonly Used Aircraft-Style Coordinates

Coordinate	Rotation About Coordinate
Forward = velocity v direction = $+x$-axis	θ = pitch (theta, about wings, + = nose-up)
Down towards attracting body = $+z$-axis	ψ = yaw (psi, about vertical, + = nose left)
Out left wing = $+y$-axis	ϕ = roll (phi, about velocity direction, + = right roll)

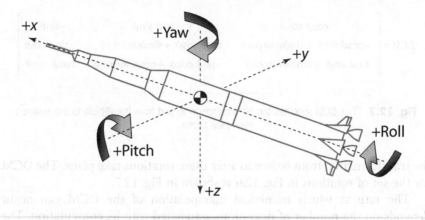

Fig. 12.5 A commonly used coordinate system, similar to that used for aircraft.

similar to that of a boat whose rudder is offset or an automobile whose steering wheel is rotated. Rolling motion corresponds to rotating or banking towards the right or left.

12.1.2 Rotations

In order to specify our vehicle's orientation with respect to some reference point, we need to specify numerically the relation between the selected reference frame and the launch vehicle's frame. This is usually defined by three angles specifying the changes made by one frame relative to the other frame. These angles are *not* unique, so the order of the rotations has to be agreed upon by all parties to keep things consistent, or to keep everyone on the same playing field.

One common set of frame rotation specifications is called *Euler* angles. Figure 12.6 shows a commonly used sequence of Euler rotations. Once the order of rotations is specified, a *direction cosine matrix (DCM)* specifies

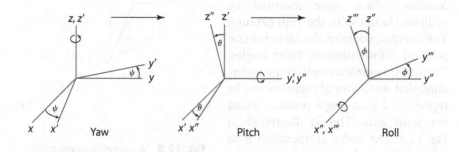

Fig. 12.6 One commonly used set of Euler angle rotation orders. Make sure all parties involved agree to the same order.

$$DCM = \begin{bmatrix} \cos\theta\,\cos\psi & \cos\theta\,\sin\psi & -\sin\theta \\ -\cos\phi\,\sin\psi + \sin\phi\sin\theta\,\cos\psi & \cos\theta\,\cos\psi + \sin\phi\sin\theta\,\sin\psi & \sin\phi\,\cos\theta \\ \sin\phi\,\sin\psi + \cos\phi\sin\theta\,\cos\psi & -\sin\phi\,\cos\psi + \cos\phi\sin\theta\,\sin\psi & \cos\phi\,\cos\theta \end{bmatrix}$$

Fig. 12.7 This DCM specifies the transformation to and from the vehicle to the selected reference frame.

the transformation from before to after these rotations take place. The DCM for the set of rotations in Fig. 12.6 is shown in Fig. 12.7.

The rate at which numerical manipulation of the DCM can occur depends on the number of operations associated with its computation. The DCM requires 16 multiplications, 4 adds, and 6 trigonometric operations (sine and cosine). Because of all the operations tabulated, a large amount of memory is required, and some of the operations (typically trig functions) are computationally intensive.

It very conveniently turns out that if we want to reverse the transformation (go from the other coordinates to the first), all we have to do is multiply by the *transpose* of the DCM, abbreviated as $[\text{DCM}]^{\text{T}}$. The transpose simply exchanges the numbers above its diagonal with those below the diagonal (e.g., the number in the second column, first row is exchanged with the number in the second row, first column). However, a negative associated with DCM matrices is that for certain attitudes and rotations, the DCM becomes singular, and special measures have to be taken to prevent the flight computer from getting confused by this situation. Are there any alternatives? Yes, there are.

12.1.3 Quaternions

Another way of tracking rotations relative to another frame is called *quaternions*, which were invented by William Hamilton in the 19th century. This method contains the same relative rotation information as Euler angles. The way quaternions work is the realization that *any* series of rotations can be represented as a *single* rotation about *one fixed axis*. This is illustrated in Fig. 12.8. The order of operation is as follows: *first* is the rotation amount e_1; then, *three* numbers (e_2, e_3, e_4) describe the *direction* or orientation of the

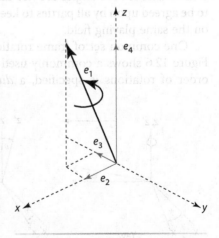

Fig. 12.8 A quaternion representation of a rotation specifies the amount of rotation e_1 and the orientation of the rotation vector (e_2, e_3, e_4).

Table 12.2   Comparison of Euler Angles and Quaternion Attitude Representations

Method	Advantages	Disadvantages
Euler angles (yaw, pitch, roll)	• No redundant parameters • Clear physical interpretation • Convenient product rule using DCM (direction cosine matrix)	• Singularities • 6 trig functions, 16 multiplies, and 5 adds slow computation
Quaternions	• Rapid computation: 15 multiplies, 12 adds, and no trig functions • No singularities	• Taught infrequently in schools

rotation vector with respect to the reference frame. Because four numbers are needed to specify this, the method is referred to as *quaternions*. An explanation of quaternions may be found in the text by Curtis [10], as well as a pair of presentations, "Introduction to Quaternions," available from NASA [29]. Diebel [11] has also provided a very detailed exposition.

It turns out that quaternion representations are always nonsingular, which is a big advantage over Euler angles. Their evaluation requires 16 multiplies, 12 adds, and *zero* trig functions, which means that they have a significant computational advantage over the DCM formulation. Because of these advantages, most flight-oriented systems use quaternions for their flight control systems. The two methods of tracking rotations are compared in Table 12.2.

12.1.4 Measuring Rotation Angles and Calculating Attitude and Position

We've discussed the use of rotation angles to track the motion of a vehicle relative to a reference coordinate system. However, we have not discussed *how* to measure these angles. There are several methods that have been used for this purpose.

Before starting on measuring instrumentation, we need to briefly discuss their history. In the early days of intercontinental ballistic missiles (ICBMs), radio guidance was used for vehicle steering; however, it was soon realized that radio signals could be jammed or interfered with by the enemy, so it was desired to have the ICBMs carry an autonomous way of measuring and navigation. This led to the development of inertial platforms, which could manage all navigation without any contact from outside, making them impervious to enemy tampering.

12.1.4.1 Method 1: Inertial/Stabilized Platform

The first method that was used in ballistic missiles as well as aircraft and launch vehicles is called an *inertial platform*. An inertial platform is a mechanism that uses gyroscopes (motors spinning flywheels that tend

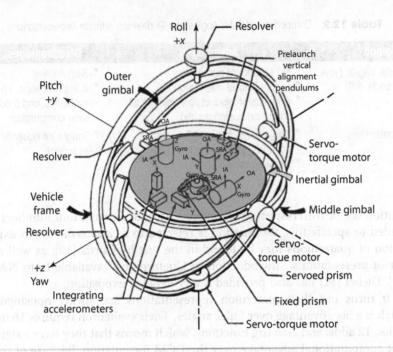

Fig. 12.9 A Saturn V ST-124-M3 inertial platform system's components. The inertial platform itself is shaded. Each of the servo-torque motors rotates its gimbal to maintain the platform's orientation with respect to inertial space. IA = input axis, OA = output axis, SRA = spin axis. Source: [23].

to keep the spin axis constant), torque motors, and resolvers (precision angle measuring devices) to infer its rotational position relative to the vehicle it's mounted on. This mechanism includes three *gimbals*, which are rings that allow the items within them to rotate in one axis. When in flight, each gimbaled direction rotates on its axis to preserve the gyros' initial spin direction. With three gimbals, it's possible to monitor motion in three rotational axes (pitch, yaw, and roll or equivalent). The Saturn V's inertial platform is shown in Fig. 12.9. The gimbal whose orientation is maintained is called the *inertial gimbal*, and it is pointed out in the figure.

Before launch, the platform is aligned with local vertical with the *vertical alignment pendulums* shown. This can level the platform to an accuracy of ± 2.5 arc-seconds (0.0007 deg). Azimuth alignment is also important, and is provided by on-board and ground-based theodolites, providing an accuracy of ± 5 arc-seconds (0.00139 deg). The platform is released to maintain its inertial reference from the launch point.

The inertial platform system works because the platform maintains its *angular* orientation with respect to inertial space, while the vehicle rotates around it. By keeping track of the angular changes of the vehicle as measured in time by the resolvers, and separately monitoring the vehicle's accelerations using the three integrating accelerometers shown, an on-board computer can infer the current position and angular attitude of the vehicle.

Notice that the signals from the inertial platform (sometimes known as a stabilized platform), where all the gyros, accelerometers, and sensing hardware are mounted, have to pass through the three rotational joints of the three platforms to get to the "outside" where the flight computer is located. This means that all the data from these parts have to pass through the pitch gimbal or resolver, then the yaw gimbal or resolver, and finally the roll gimbal or resolver. This means a number of wires carrying sensitive signals and power passing through moving contacts, and potentially issues with electrical noise, rotational stiffness, and others. This makes the stabilized platform scheme expensive, complex, and no longer commonly used.

By the way, if you're interested in the specifications of the launch vehicle digital computer (LVDC), here are some numbers: processing speed = 512 kbit/s, 32,768 words of memory, each 28 bits. Computers have sure come a long way since the 1960s!

12.1.4.2 Method 2: Inertial Measurement Unit

A second way to measure the motion of a vehicle is to use an inertial measurement unit, (IMU). The IMU measures acceleration and rotation rates using accelerometers and gyros as the inertial platform did; however, the IMU operates in a different way than the inertial platform: it has no moving parts, so the accelerometers and gyros are fixed in the case, which is in turn attached or strapped down to the vehicle frame. For this reason, they are sometimes called *strapdown* sensors or platforms.

A typical IMU is shown in Fig. 12.10. This LN-200S unit from Northrop Grumman has three (one per axis) silicon micro electromechanical systems (MEMS) accelerometers to sense accelerations experienced during launch and orbit insertion. In addition, it has three (one per axis) fiber-optic gyros to sense angular rates. The IMU shown has a mass of 0.75 kg, power consumption of 12 watts, and size of 8.89 cm diameter × 8.51 cm high. Each of these specifications is at least an order of magnitude better than the inertial platform shown earlier.

The IMU directly integrates the measured accelerations and rotational rates to generate an estimate of its current position and attitude relative to the starting point. A major disadvantage of using IMUs for navigation is that they suffer from accumulated error. Because the guidance system is continually integrating acceleration with respect to time to calculate velocity and position, measurement errors are accumulated over time. This creates *drift*, an error that grows larger and larger between the calculated and actual positions.

Fig. 12.10   IMU model LN-200S from Northrop Grumman 88.9 mm diameter by 85.1 mm high.

12.1.4.3 Method 3: Global Positioning System (GPS)

The GPS satellite network (or its equivalent) can be used to correct drift errors. Currently, radar and the vehicle's internal IMU navigation data are used to verify that the vehicle is safely on course. If there is on-board GPS, it provides a separate independent navigation measurement that does not accumulate error or drift. Launch vehicles do not use GPS by itself, but do use it to get rid of drift errors. In addition, the GPS information helps to eliminate scrubbed launches due to tracking station failures or unavailability, which improves the probability of launching on a particular day. Figure 12.11 shows a diagram of how GPS can augment tracking station data to carry this out.

We know from everyday experience that GPS receivers can provide accurate position and velocity measurements, and can be used to complement an on-board IMU to provide more accurate and reliable position measurement, thereby reducing error propagation in a dead reckoning system. Both IMU and GPS systems have instrument and environmental error sources that contribute to navigation errors, but the two systems use completely different methods to navigate. Because it's based on time integration, IMU navigation error grows with time, but GPS errors do not. Thus, the IMU is more accurate than GPS over short time periods, whereas the GPS system is more accurate over longer time periods. Table 12.3 summarizes how the two navigation systems, GPS and IMU, complement each other. Of course, this assumes that there is no jamming or other interference with GPS signals, an assumption that may not be valid during times of conflict.

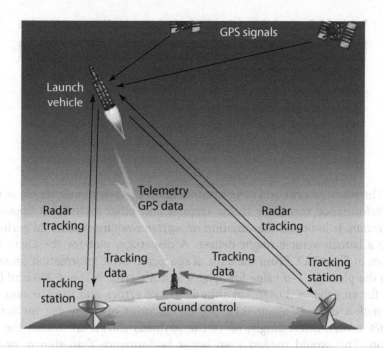

Fig. 12.11 The GPS may be utilized for launch vehicle navigation and to improve launch reliability. Source: [5].

However, be aware that IMUs can have faults of their own. The following article, taken from *Aviation Week*, explains how the crash of a Martian lander spacecraft was caused by a fault with its IMU.

2016 Mars Lander Crash Sparked By Bad Inertial Data

LYON, France: European Space Agency (ESA) engineers believe they've found the root cause of the Oct. 19 crash of *Schiaparelli* Mars lander—erroneous information from IMU resulting in incorrect altitude reading.

Shortly after the lander's parachute deployed, the IMU "saturated"—it could not measure higher rotation rates—and stayed that way for about 1 second. When fed into the navigation system, the data generated an estimated altitude that was below ground level, according to ESA.

This triggered a premature release of the parachute and backshell. Braking thrusters then fired for the minimum preselected time of about 3 seconds, instead of the nominal 60 seconds. Finally, on-ground systems were activated as if *Schiaparelli* had already landed, whereas it was still 3.7 km (12,000 ft) from the surface.

The scenario has been "clearly reproduced in computer simulations of the control system's response to the erroneous information," ESA says.

Source: [13].

Table 12.3 The Drift from the IMU May Be Reduced or Eliminated Using GPS

Item	IMU	GPS
Navigation method used	Dead reckoning	Line-of-sight
Error sources?	Yes	Yes
Error growth with time?	Yes	No
Best system accuracy?	Over short periods	Over long periods
Subject to interference	No	Yes

The result of drift and error accumulation of navigation systems, as well as performance variations in the engines and other systems, is known as *dispersion*. It indicates the variation or *scatter* away from nominal performance a launch vehicle might deliver. A dispersion plot for the Delta II is shown in Fig. 12.12. From the table of spacecraft orbit information provided with the plot, the 1-σ value for apogee altitude error is about 203 nmi (376 km) for an orbit of 39,486 km altitude (0.95% error), and the 1-σ value for inclination error is about 0.12 deg for an orbit of 20.74 deg inclination (0.58% error). Interestingly, all of the payloads shown are within the 1-σ region. This would suggest even better performance than shown, or that some "outliers" are not being shown.

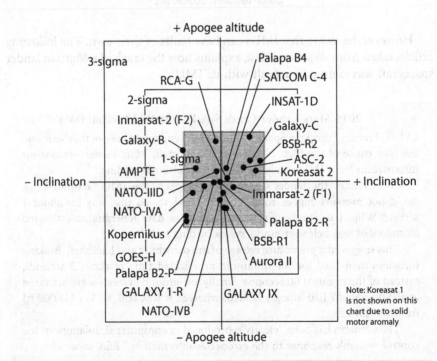

Fig. 12.12 Dispersion or scatter in geostationary transfer orbits for Delta II. Source: [8].

How sensitive are baseline design results to small changes in design inputs or uncertainties in the models? Find out by varying parameters in one of two ways:

* One-at-a-time (OAT) analysis
  * One parameter (e.g., thrust level, I_{sp}, burnout altitude, speed, pitch angle, azimuth, or other variable) is varied, while keeping all others constant.
  * Parameter is varied up and down by a percentage based on confidence of knowledge of its accuracy.
  * Neglects any potential interactions between the parameters.
* Monte Carlo (MC) analysis
  * All parameters are randomly varied at the same time, using a Gaussian distribution with mean error μ and standard deviation of the error σ.
  * Results show average values and their standard deviations. Magnitude of uncertainty in inputs may significantly affect the parameters of interest.

12.2 Stability and Control

Let's discuss stability and control. Both are important to the flight of a launch vehicle. We may obtain information on the stability of an LV through hand calculations, wind tunnel testing, or computational fluid dynamics (CFD) calculations. All of these provide some sort of distributed pressures on an LV that may be integrated along the length of the LV to get the location of the *center of pressure (CP)* (not to be confused with C_P, pressure coefficient).

Figure 12.13 shows the CP location on a Saturn V, both predicted and measured. As with many engineering calculations, the predicted CP differs from the measured CP. In this case, the CP is more aft (towards the tail) than predicted, so the predicted location will yield a conservative control system design. Also in the figure, one can see that the vehicle's center of gravity (CG) varies with burn time, as expected.

The location of the center of pressure may be thought of as the effective location of axial and normal aerodynamic forces; in other words, the distributed aerodynamic forces appear to act at this one location. The distance between the locations of the CP and center of mass (CM, also called center of gravity, CG) is called the *static margin* and denoted by l_{sm}. The static margin is sometimes given in units of *calibers*. A caliber is simply the reference diameter D of the LV, so the static margin may be stated as a number of diameters $= l_{sm}/D$. If the CP is *aft* of the CM, the vehicle is stable; if the CP is *forward* of the CG, the vehicle is unstable. This situation is shown in Fig. 12.14, which shows the locations of the CP and CM, and the definition of the static margin (SM).

It turns out that most launch vehicles are statically unstable. Although this in itself might be considered a problem, it's easy to add a closed-loop

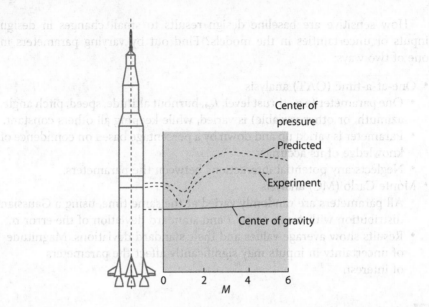

Fig. 12.13 Saturn V static aerodynamic stability and center of pressure movement vs Mach number. Source: NASA.

control system to stabilize the vehicle. We call it closed loop because the control system senses error and *feeds it back* to the vehicle's steering system, as depicted in Fig. 12.3.

The more unstable a vehicle is, the more authority the control system has to have to control it. This may lead to large thrust vector control actuators or excessive movement of the actuators. In this case we might want to decrease the instability. This can be done in one of two ways:

• Move the CP *aft* by adding tail fins (e.g., the Saturn V in Fig. 12.13, the Saturn IB, or the arrow shown in Fig. 12.15)
• Move the CG *forward*, that is, add ballast to nose (usually not desirable for vehicles whose performance depends on minimal mass)

Unfortunately, the situation becomes more complicated. The location of the CP changes with differing angles of attack and Mach numbers. The good

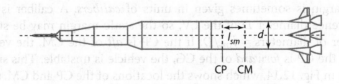

Fig. 12.14 The location of the center of pressure (CP), center of mass (CM), and static margin. The vehicle is unstable: the CP is ahead of the CM.

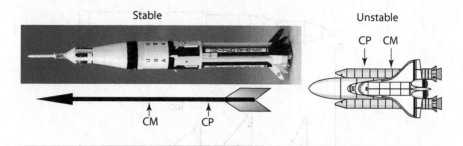

Fig. 12.15   The relative location of the CM and CP determine stability.

news is that the increase in distance aft of the nose is stabilizing, at least for the Vanguard missile shown. This behavior is shown in Fig. 12.16.

12.2.1 Locating Center of Pressure

Similar to other aerodynamic quantities, the location of the center of pressure may be found in one of three ways:

1. Use an analytical method: run CFD.
2. Measure in the wind tunnel.
3. Use an empirical method (e.g., James Barrowman method).

For initial studies, it probably doesn't make sense to make a detailed model of the airframe for CFD calculations, nor does it make sense to build a wind tunnel model. Therefore, the easiest way to locate the CP is to use the Barrowman method [4] or a similar method.

Figure 12.17 shows the pressure distribution C_p obtained on a vehicle at an angle of attack of 8 deg for the top and bottom surfaces. We see that at each increasing-diameter region along the vehicle's length, there is an upward force on the vehicle. The running force can be integrated longitudinally to get the pitching moment. In most cases, this is a nose-up

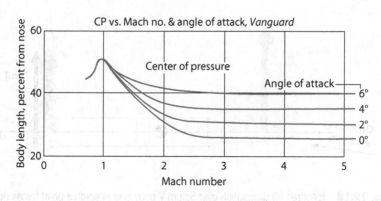

Fig. 12.16   CP location varies with Mach number and angle of attack. Source: [20].

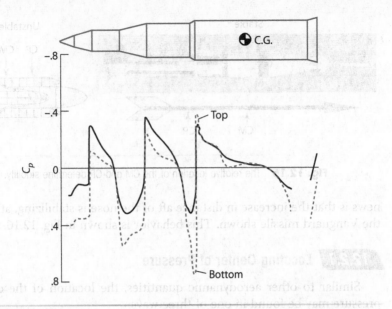

Fig. 12.17 Pressure distribution on a vehicle at +8-deg angle of attack. Source: [9].

(unstable) pitching moment with positive angle of attack $(+\alpha)$ for vehicles without fins. This instability may be dealt with by the vehicle's attitude control system.

Figure 12.18 shows an idealized plot of the running lift (longitudinally distributed lift) on a Saturn V. The two dashed curves, when integrated, produce point loads like those shown with the large arrows. Static stability may be enhanced by adding tail fins or increasing the cross-sectional area behind the CM, which produces the larger arrow near the tail, a stabilizing moment. If the fin moment is smaller than the forward moment, the resulting aerodynamic pitching moment will rotate the vehicle unless counteracted.

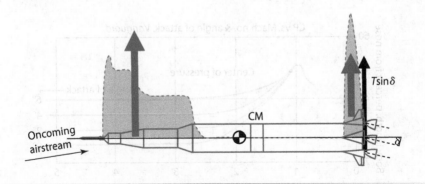

Fig. 12.18 Idealized lift distribution over Saturn V body and integrated point forces (large arrows). Note the vehicle is trimmed (net pitching moment is zero) but *not* in lateral equilibrium. The forces shown all accelerate the vehicle in the direction towards the top of the page.

This unstable situation is dealt with by gimbaling the aft engines so as to provide a stabilizing moment.

This situation can be summarized mathematically as follows. We produce an aerodynamic moment $M_{y_{aero}}$ from the body aerodynamics (as discussed previously) and any stabilizing surfaces, such as fins. We also have a moment $M_{y_{control}}$ from the control system (thrust vector control or other type). Using these, we can write the equation of pitch for the vehicle

$$I_y\ddot{\alpha} = M_{y_{total}} = M_{y_{aero}} + M_{y_{control}} \qquad (12.1)$$

Linearizing and using the definitions for stability derivatives, we get

$$\Delta\ddot{\alpha} \approx \frac{1}{I_y}\left[\left(\frac{\partial M_{y_{total}}}{\partial\dot{\alpha}}\right)\Delta\dot{\alpha} + \left(\frac{\partial M_{y_{total}}}{\partial\alpha}\right)\Delta\alpha\right] \qquad (12.2)$$

Small perturbations of the vehicle are stable if both $\frac{\partial M_{y_{total}}}{\partial\dot{\alpha}} < 0$ and $\frac{\partial M_{y_{total}}}{\partial\alpha} < 0$. However, if $\frac{\partial M_{y_{total}}}{\partial\dot{\alpha}} > 0$, we say that the vehicle is *dynamically unstable*, meaning that if it is disturbed from its neutral position, it will oscillate back and forth with amplitude growing each cycle—*oscillatory divergence*. If $\frac{\partial M_{y_{total}}}{\partial\alpha} > 0$, we say that the vehicle is *statically unstable*, meaning that if it is disturbed from its neutral position, it will steadily move away from that neutral position, called *static divergence*. To deal with the latter two cases, we need to add either stabilizing fins, thrust vector control, or both.

12.2.2 LV Flight Control System Elements

For a satisfactory flight control system, four components are needed. These are illustrated in Fig. 12.19. These four elements are:

1. *Control generation:* We need a means to produce a moment or torque on the vehicle, to overcome aerodynamic torques. This is usually a system that deflects the thrust to produce side forces and therefore torque. The most common version is called thrust vector control (TVC).
2. *Control actuation:* This is a mechanism that can take an electrical signal and use it to drive a mechanical component. In many cases, this is a servo that converts a signal into a motion that changes the rocket nozzle's angle.
3. *Measurement:* We need accurate knowledge of the vehicle's current state—attitude, angular rate, and control displacement—so it can be compared with the desired state of the vehicle.
4. *Computation:* We must have a flight computer or something similar that determines the proper actuation based on the on-board vehicle measurements and the desired vehicle state.

These four elements will be described in detail in the following sections.

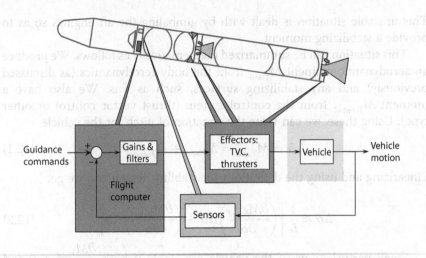

Fig. 12.19 The necessary vehicle control system elements. Source: [30].

12.2.3 **Thrust Vector Control**

As we mentioned earlier, we can control even an unstable vehicle by using a mechanism to offset the thrust direction by an angle δ from the centered position. This is called *gimbaling* the rocket nozzle. The control moment $M_{control}$ due to sideways thrust from angle δ is equal to the sideways thrust $\times$ l_c, the length from the engine gimbal to the vehicle's CM, which may be written $(T \sin \delta) \times l_c$. This control scheme is also known as *thrust vector control (TVC)*. This gimbaling process is shown in Fig. 12.20. The equation of pitch motion is:

$$I_y \ddot{\theta} = M_{y_{total}} = M_{y_{aero}} + M_{y_{control}} = qdSC_M + Tl_c \sin \delta \qquad (12.3)$$

When trimmed, $M_{y_{total}} = 0$ and $\delta_{trim} = qdSC_M / Tl_c$ (assuming small values of δ_{trim}. It is recommended that the value of δ_{trim} not exceed 2 deg.

One of the challenges of gimbaling an engine (or several of them) is that the propellants have to be delivered to the engine regardless of the gimbal angle. This means that some sort of flexible plumbing is required, or else the engine will not be movable. Figure 12.21 shows a way to connect the gimbaled portion of the engine (the combustion chamber and nozzle) to the rest of the vehicle while still transferring propellants.

12.2.3.1 Types of Control Effectors (Actuators)

Three ways to effect a change in the thrust vector are hydraulic, geared electric motors, and gas injection via electric valves.

The Saturn V F-1 engine actuator shown previously in Fig. 12.20 is a hydraulic servo, meaning it is driven by hydraulic fluid under pressure. Some details are given in the diagram shown in Fig. 12.22. There were two

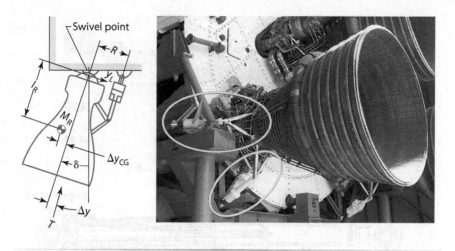

Fig. 12.20   a) How an engine nozzle can be swiveled or gimbaled to provide a side force. Source: [19]. b) One of the Saturn V's F-1 engines, including two hydraulic actuators (circled) to provide gimbaling in two axes for pitch and yaw control. Source: Edberg.

of these actuators on the S-IC stage for each of the four outer engines. (The center engine was fixed.) Interestingly, the designers, wishing to reduce launch mass, used the RP-1 rocket fuel as hydraulic fluid, rather than having a separate hydraulic system.

The details of a hydraulic actuator are shown in Fig. 12.23. An electric solenoid is used to drive a valve system that causes the actuator rod to move back and forth.

A complete TVC system for the small ICBM is shown in Fig. 12.24. This is a self-contained *blowdown* system, meaning that the low-pressure

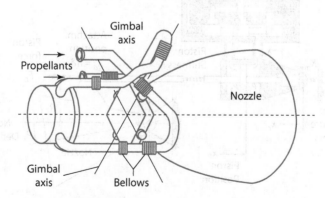

Fig. 12.21   Gimbal axes and propellant line bellows for the lunar module descent engine allow the nozzle to rotate while delivering propellants. Flexible joints are located so that only rotation occurs. The centers are located on the throat-mounted gimbal axes. Source: [17].

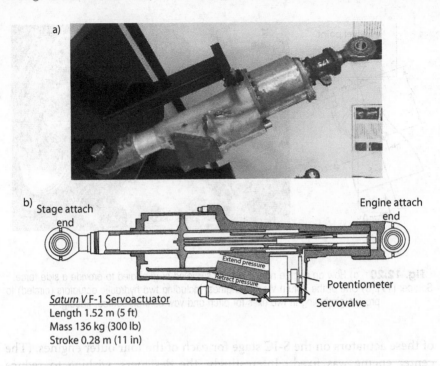

Saturn V F-1 Servoactuator
Length 1.52 m (5 ft)
Mass 136 kg (300 lb)
Stroke 0.28 m (11 in)

Fig. 12.22 a) Saturn V F-1 engine TVC actuator. This massive hydraulic actuator was able to move a single engine producing 6,670,000 N (1,500,000 lb) of thrust! b) Cross-section of F-1 actuator. Source: [23].

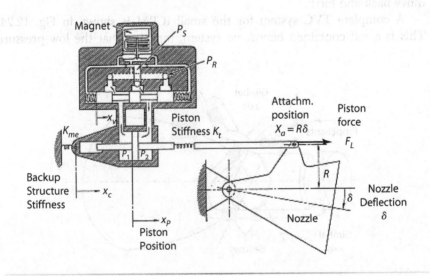

Fig. 12.23 Hydraulic actuator (servo) details. The parameters shown are used to estimate the dynamic performance of the servo, which is assumed to have flexibility and therefore its own set of dynamics. Source: [14].

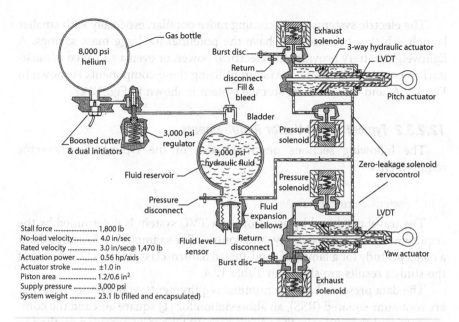

Stall force .... 1,800 lb
No-load velocity .... 4.0 in/sec
Rated velocity .... 3.0 in/sec@ 1,470 lb
Actuation power .... 0.56 hp/axis
Actuator stroke .... ±1.0 in
Piston area .... 1.2/0.6 in²
Supply pressure .... 3,000 psi
System weight .... 23.1 lb (filled and encapsulated)

Fig. 12.24 S-ICBM cold gas hydraulic blowdown TVC system. Actuation force is 6,670 N (1,800 lb), but the mass of the entire filled and encapsulated system is only $m = 10.5$ kg (23.1 lb). Source: Moog (technical brochure).

hydraulic fluid is dumped overboard after use. It also means that the needed amount of fluid must be in the reservoir, or it will run out early and lose control.

A second type of actuator uses electric power instead of hydraulic fluid to drive its actuation system. This is known as an *electromechanical actuator* or servo. An example of this type of actuation is shown in Fig. 12.25.

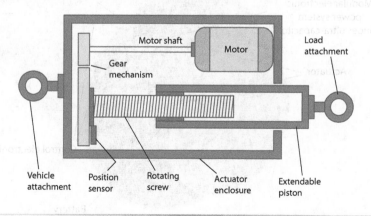

Fig. 12.25 Electromechanical actuator (servo). A geared electric motor is used to drive a threaded rod to provide fore and aft movement of the shaft. Source: [14].

The electric systems are becoming more popular, especially with smaller launch vehicles, because they have the potential for large mass savings. A lightweight battery provides the electrical power, or even a so-called *ultracapacitor* could do the same. A system utilizing these components is shown in Fig. 12.26, and the electrical storage system is shown in Fig. 12.27.

12.2.3.2 Typical Thrust Vector Requirements

The following sections describe some of the typical thrust vector requirements.

Thrust Vector Control System Angular Motion

The travel produced by actuators of a TVC system is determined by the requirements for trim, stability, and control. For example, NASA carried out a concept study for a family of solid-fueled Saturn-class vehicles [25]. Some of the study's results are shown in Table 12.4.

The data presented are the nominal requirements; you'll notice that they are root-sum-squared (RSS), an abbreviation for (1) square and add the components, and then (2) take the square root. RSSing is an accepted method to estimate the "effective" value of a group of measurements or components, under the assumption that they do not all occur simultaneously. If they all happened to be simultaneous, we would simply total the columns, and the deflection angles would be larger: 2.40, 3.46, and 2.20 deg, respectively. Although most TVC systems have travel of at least ±5 deg, Falangas [14] suggests a rule of thumb: steady-state gimbal deflections should not exceed 2 deg. This provides margin for steering and wind corrections.

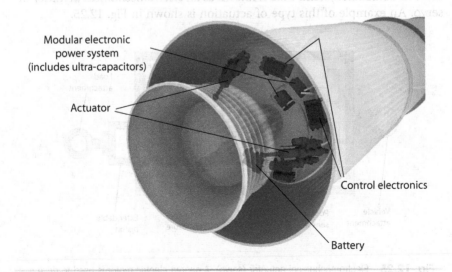

Fig. 12.26 Reduced mass electrical TVC system. Source: [18].

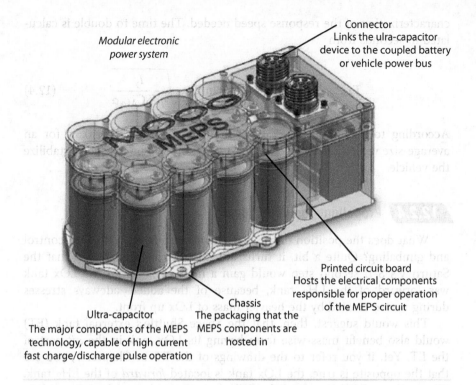

Fig. 12.27 Moog ultracapacitor system offers reduced mass over batteries. Source: [18].

Thrust Vector Control System Response

Another consideration of the effector is how *fast* it must operate. If one considers the attitude differential equation, the *time to double* is an important

Table 12.4 Thrust-Vector Deflection Angle Requirements

Parameter	Variation	260-in. Solid/S-IVB Apollo	260-in. Solid/ S-IVB Voyager	SSO PM 7 × 260-in. Solids/1 solid
		Deflection angle (degrees)		
A. Steady state winds	99%	1.35	2.30	1.17
B. Wind gusts	3σ	0.15	0.26	0.13
C. Thrust misalignment	3σ	0.25	0.25	0.25
D. Thrust & weights	3σ	0.15	0.15	0.15
E. Pitch program	Maximum	0.50	0.50	0.50
Total = A + $(B^2 + C^2 + D^2 + E^2)^{1/2}$		1.68	2.69	1.49

characterization of the response speed needed. The time to double is calculated in seconds from

$$T_{\text{double}} = \ln(2) \cdot \sqrt{\frac{I_y}{M_A}} = \ln(2) \cdot \sqrt{\frac{I_y}{qdSC_{M\alpha}\alpha}} \tag{12.4}$$

According to Falangas [14], T_{double} should be greater than 0.5 s for an average-size vehicle; otherwise, very fast actuators will be needed to stabilize the vehicle.

12.2.4 Propellant Tank Positioning

What does the position of the propellant tanks have to do with control and gimbaling? Quite a bit, it turns out. Recall from Chapter 10 that the Saturn V's S-II second step would gain a mass of 796 kg if the LOx tank was forward of the LH$_2$ tank, because of the added sideways stresses during max-q created by the heavier mass of LOx up front.

This would suggest, then, that the Space Shuttle's external tank (ET) would also benefit mass-wise from having its LOx tank in the *aft* part of the ET. Yet, if you refer to the drawings of the ET in Chapter 7, you see that the opposite is true: the LOx tank is located *forward* of the LH$_2$ tank. Why is this so? What caused the ET's designers to locate the LOx tank in the forward position? It turns out that vehicle control via TVC had at least a partial role in the decision.

Recall that for a hydrolox engine, the mass of LOx is about six times that of LH$_2$. This means that as propellant is consumed, the location of the center of mass will shift. The motion of the CM is important, because the thrust line of the engines must pass through the instantaneous CM location at all times. We have to consider three phases of a Space Shuttle launch. The first phase is liftoff, where both the main engines and the solid rocket boosters are firing. The second phase is after the SRBs are jettisoned, and the Shuttle/ET continue under main engines alone. The third phase is when the ET is empty and the main engines are about to shut down. The main engines' thrust lines must pass through the CM during and between these three phases. These phases are shown in Fig. 12.28, showing the situation with forward LOx in Fig. 12.28a and aft LOx in Fig. 12.28b (the as-flown configuration).

Recalling that the thrust line of the main engines must pass through instantaneous CM at all times, we have to accommodate the entire travel of the CM. The CM travel associated with the *aft* LOx tank (Fig. 12.28b) requires prohibitively large (22-deg) engine gimbal angles. This makes the TVC system (not to mention the bellows and plumbing required) more complicated, expensive, and possibly heavier, all reasons *not* to adopt the aft LOx configuration.

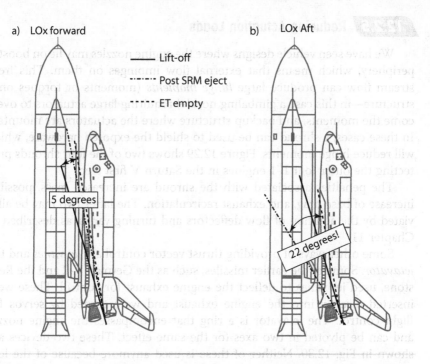

Fig. 12.28   Main engine thrust line actuation angles needed for a) aft vs b) forward (as flown) LOx tank on the Space Shuttle. Source: [15].

Another fact to consider is that there is a crossbeam between the two front attachment points of the SRBs onto the ET. The as-flown forward LOx tank conveniently provides an empty intertank section that can provide the volume needed for the crossbeam. Moving the LOx tank aft, assuming the same crossbeam is required, puts the crossbeam across the internal volume of the LH_2 tank instead of an open intertank volume. This could mean additional mass is needed to accommodate the crossbeam in the LH_2 tank.

So this brings up a final remark: referring to Figure Ariane 5 from Chapter 8, we see that the Ariane 5's first step is also hydrolox; however, that particular vehicle is an in-line vehicle, so there is no asymmetry as there is in the Space Shuttle (where the orbiter is offset from the ET and SRBs). Why, then, is the Ariane 5 designed with the LOx tank forwards?

We know the following: it is *not* designed this way to reduce stress levels in the lower portion of the step, as the Saturn V S-II step is. We know it's *not* because of a thrust-line gimbaling angle issue as with the Space Shuttle. If the Ariane 5 has a crossbeam for its two SRMs, it could pass over the front of the forward LOx tank. This leaves the issue of stability: we are aware that a forward CM is more stable than an aft CM. Could this be why the Europeans chose the forward LOx tank for the Ariane 5?

12.2.5 Reducing Actuation Loads

We have seen vehicle designs where the engine nozzles may lie on booster periphery, which means that external flow impinges on them. This free-stream flow can produce large *hinge moments* (moments or torques on a structure—in this case a gimbaling nozzle) requiring large actuators to overcome the moments, and backup structure where the actuators are mounted. In these cases, a shroud can be used to shield the exposed hardware, which will reduce hinge moments. Figure 12.29 shows two of the four shrouds protecting the outer four F-1 engines in the Saturn V first step.

The penalties associated with the shroud are increased mass, possible increase of base drag, and exhaust recirculation. The latter two may be alleviated by the addition of flow deflectors and turning vanes, as described in Chapter 11.

Some other ways of providing thrust vector control are *jet vanes* and the *jetavator*. Some of the earlier missiles, such as the German V-2 and the Redstone, used jet vanes to deflect the engine exhaust for control. These were inserted directly into the engine exhaust and were turned by servos for flight control. The jetavator is a ring that encompasses the engine nozzle and can be pivoted in two axes for the same effect. These two devices are shown in Fig. 12.30. Neither of these is used anymore because of the loss of performance involved with slowing down the engine exhaust.

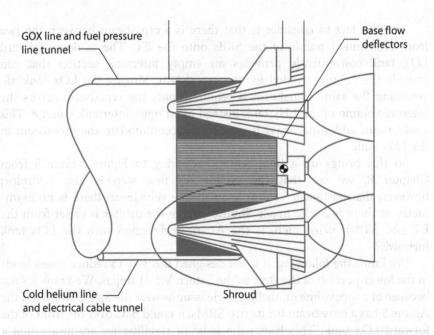

Fig. 12.29 The shroud upstream of the Saturn V's F-1 engines reduced the power necessary to pivot the engines for stability and control. Source: [3].

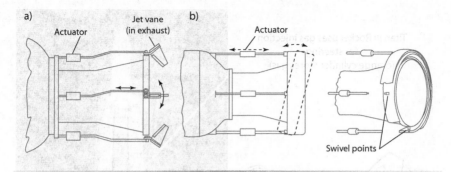

Fig. 12.30   a) The Redstone missile used thrust vanes inserted into the engine's exhaust for control. The missile also used air rudders for control within the atmosphere. b) The jetavator, encompassing the engine exhaust nozzle, could pivot for the same effect. Source: [20].

Another way of controlling a rocket is the *grid fin*. This is a flat panel that allows air to pass through it, as shown in Fig. 12.31. One advantage of grid fins is that when they are exposed to crosswinds, they don't develop much side force compared to conventional fins. Another advantage is that they can be folded back along the body so they are out of the way when not needed. Several Russian LVs use grid fins, and SpaceX Falcon 9s use the grid fins for control when they are recovering their boosters. Of course, it should be noted that grid fins are aerodynamic controls and will not work outside the atmosphere.

Another way to provide thrust vector control is to inject gases directly into the exhaust nozzle. This was used in the Titan III solid rocket motors, as shown in Fig. 12.32. The small tank near the bottom of the SRM is a high-pressure gas bottle containing the injection gases. Figure 12.33 shows three different methods for lateral gas injection TVC. When the gas or liquid is

Fig. 12.31   Grid fins for aerodynamic control. They are commonly used on munitions and on Russian rockets. Source: SpaceX.

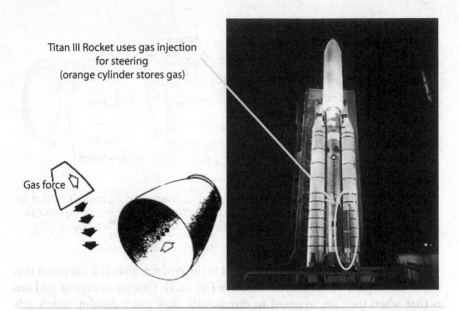

Titan III Rocket uses gas injection
for steering
(orange cylinder stores gas)

Gas force

Fig. 12.32 The Titan III LV uses gas injection for steering. The circled cylinder stores compressed gas. Source: U.S. Air Force.

injected, it creates a shock on one side of the nozzle, which causes asymmetric thrust.

We've seen how the engine(s) can be gimbaled (or equivalent). How do we actually carry out the process of steering? Because launch vehicles are often symmetrical, we have to be very careful defining the coordinate system. Referring to Fig. 12.34, it's pretty straightforward to get moments in pitch and yaw. If the vehicle is assumed to be in a level position as shown, then for pitch up and down, the engine gimbals up and down.

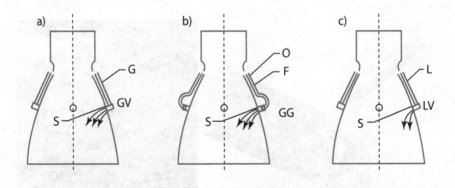

Fig. 12.33 Three gas injection TVC systems: a) a gas chamber tap-off system; b) a bipropellant gas-generator system; c) a liquid system. Source: [22].
S = shock front, G = hot gas duct, O = oxidizer, F = fuel duct, L = liquid duct, GV = gas valve, GG = gas generator, LV = liquid injection valve.

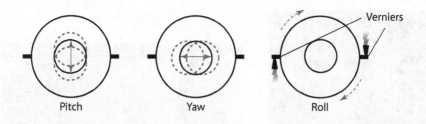

Pitch   Yaw   Roll

Fig. 12.34 Pitch and yaw moments are available by gimballing the single engine; however, roll is not available. Two small rocket engines, called *verniers*, provide roll control by pointing them in opposite directions.

For yaw left and right, the engine gimbals left and right. But, how do we control roll?

It's not possible to control roll with a single engine. What's commonly done in that situation is to add small engines, called *verniers*, which are offset from the LV's centerline. Then, to get roll control, the two verniers would gimbal in opposite directions.

Figure 12.35 shows a Thor intermediate-range ballistic missile (IRBM) on display at Edwards Air Force Base. The main engine, in the center, has a plastic cover. The two verniers, on either side of the main engine, are circled. The turbopump exhaust line is above the main engine.

Fig. 12.35 Thor missile with two verniers circled. Source: Edberg.

Fig. 12.36 a): Atlas with vernier engines; b) closeup of 1,000-lb$_f$ thrust vernier engine.
Source: U.S. Air Force.

The Atlas missile, shown in Fig. 12.36, also had two verniers. Both had to be mounted above the half-stage that was jettisoned partway through its launch.

With multiple engines, things can get more interesting. A twin-engine LV, like the Titan II, could achieve pitch and yaw by moving the two engine nozzles the same way, and accomplish roll control by moving them in opposite directions relative to the line connecting the two. This is shown in Fig. 12.37.

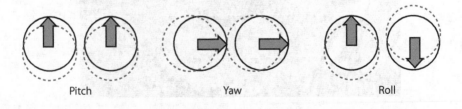

Fig. 12.37 Engine movements needed for a twin engine to pitch, yaw, and roll.

Naturally, things will get complicated the more engines there are. There can be situations where two engines "fight" each other and accomplish nothing. The controls engineer has to look at all possible situations and make sure that none will experience any problems.

The arrangement of engines, actuators, and nozzles for the Saturn V rocket is shown in Fig. 12.38. This only shows the main engines on steps 1,

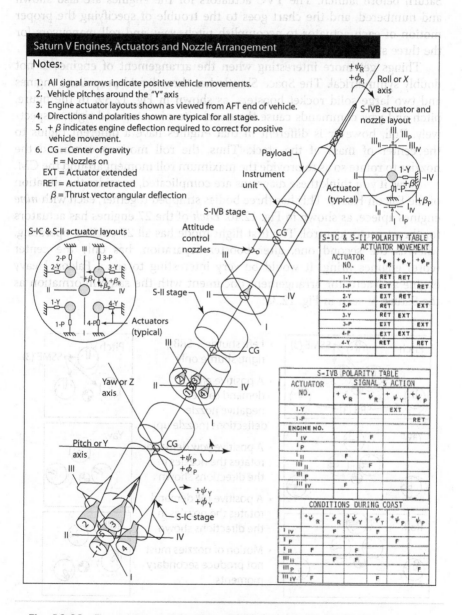

Fig. 12.38   The arrangement of engines, actuators, and nozzles for the Saturn V rocket.
Source: NASA.

2, and 3. The first two steps, the S-IC and S-II, each has five engines. The third step has a single engine and is controlled similarly to the Thor, mentioned earlier in the chapter. The controls engineer has to look.

To make sure there is no confusion, you will notice that the engines on the first two steps are numbered, and the vehicle itself is broken into four numbered quadrants. You may have seen these numbers on photos of the Saturn before launch. The TVC actuators for the engines are also shown and numbered, and the chart goes to the trouble of specifying the proper motion of each actuator to accomplish pitch, yaw, and roll maneuvers for the three steps.

Things get more interesting when the arrangement of engines is not doubly symmetrical. The Space Shuttle has three liquid-propellant engines and two large solid rocket boosters, as shown in Fig. 12.39. In the figure, pitch and yaw commands cause the nozzles to move up and right, respectively. Roll, however is different in that engines have a different radius to the center of mass of the stack. Thus, the roll motions command the nozzles to rotate so as to provide the maximum roll moment around the CM.

Now, if you think these diagrams are complicated, consider the situation of the Falcon Heavy. This has three bodies strapped together, each with *nine* engines apiece, as shown in Fig. 12.40. *Each* of the 27 engines has actuators for thrust vector control! The first flight mode has all 27 engines operating, whereas the second one, after booster separation, has only the center (core) engines firing. It would be very interesting to see a Falcon Heavy engine and actuator arrangement document with the same information as the Saturn V seen in Fig. 12.38!

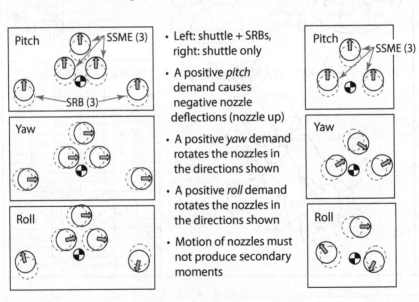

- Left: shuttle + SRBs, right: shuttle only

- A positive *pitch* demand causes negative nozzle deflections (nozzle up)

- A positive *yaw* demand rotates the nozzles in the directions shown

- A positive *roll* demand rotates the nozzles in the directions shown

- Motion of nozzles must not produce secondary moments

Fig. 12.39 The different nozzle positions for the Space Shuttle to pitch, yaw, and roll.

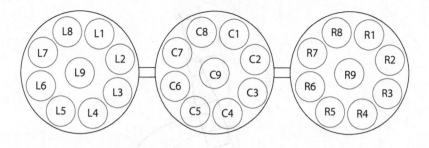

Fig. 12.40   The Falcon Heavy engine arrangement has 27 engines, each with TVC actuators.

12.3 Controlled Vehicle Equations of Motion

Consider the vehicle shown in Fig. 12.41. It is flying at an angle of attack, which generates aerodynamic forces and moments as we've seen earlier. We will consider motion of the vehicle in the pitch plane (the $x-z$ plane). The origin is located at the vehicle's center of mass. Body-mounted sensors measure quantities in the moving vehicle's system $(x-z)$, so the vehicle's IMU or equivalent can be used to calculate vehicle position and attitude. The variables of this motion are defined in the next section. Deviations of all angles from nominal values are assumed to be small.

12.3.1 Vehicle Coordinate Systems

We will be dealing with three different coordinate systems, as we did when calculating flight loads in Chapter 9. Those systems are:

1. Vehicle body system: x, y, z origin at G, vehicle CM
   x-axis = vehicle roll axis, positive forward
   z-axis = vehicle yaw axis, perpendicular to x, in pitch plane
   y-axis = vehicle pitch axis, positive out of paper
2. Wind system
   velocity direction = wind axis
3. Inertial coordinates
   vertical, horizontal = inertial coordinates

12.3.2 Variable Definitions

Some common variables that are typically used to describe a launch vehicle's trajectory are given below and illustrated in Fig. 12.41.

$\mathbf{v}$ = velocity vector (magnitude v)
α = angle of attack = $\gamma - \theta$
γ = flight path angle
θ = pitch angle

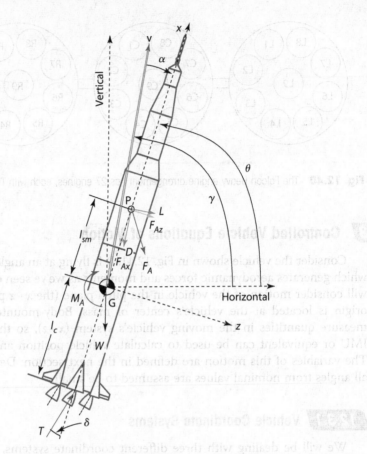

Fig. 12.41 Coordinates, forces, and torques to be used for stability analyses. Coordinate system and variable definitions are given in the text.

G = Center of Mass location

P = Center of Pressure location

W = mg = weight

δ = thrust offset angle (TVC gimbal angle)

T = total engine thrust

F_A = total aerodynamic force

F_{Ax} = aerodynamic force, axial (along vehicle x-axis)

F_{Az} = aerodynamic force, lateral (along vehicle z-axis, perpendicular to x-axis)

M_A = aerodynamic pitching moment

D = drag force (anti-parallel to wind axis)

L = lift or normal force (perpendicular/normal to wind axis)

l_c = length, engine gimbal to CM (G)

l_{sm} = length, static margin: CP to CM (P to G)

q = dynamic pressure = $^1/_2\,\rho v^2$

d = vehicle reference diameter
S = vehicle reference area = $\pi d^2 / 4$

12.3.3 Vehicle Force and Torque Definitions

Because we are dealing with several coordinate systems, it's convenient to define them independently of Fig. 12.41. The coordinates by themselves may be seen in Fig. 12.42.

We can write the aerodynamic forces F_{Aero} as two perpendicular forces,

$$F_{Az} = C_z \, qS \text{ (normal to } x\text{-axis)} \tag{12.5}$$

$$F_{Ax} = C_x \, qS \text{ (parallel)} \tag{12.6}$$

where C_z and C_x are nondimensional normal and parallel force coefficients. Often, we neglect the parallel force (drag) because the product of its magnitude and moment arm is small compared to the normal/lift force multiplied by the separation of the CM and the location of the center of pressure (CP), which is the point of application of the aerodynamic forces, P.

The aerodynamic moment M_A is created by the offset between G and C and is defined as

$$M_A = F_{Az} \, l_{sm} = C_z \, qS l_{sm} \tag{12.7}$$

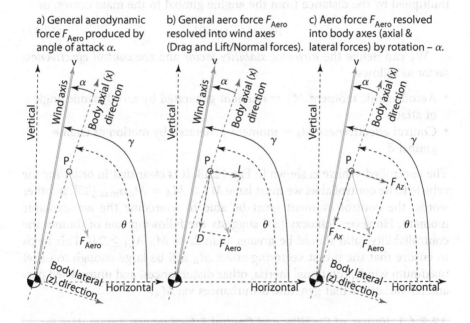

a) General aerodynamic force F_{Aero} produced by angle of attack α.

b) General aero force F_{Aero} resolved into wind axes (Drag and Lift/Normal forces).

c) Aero force F_{Aero} resolved into body axes (axial & lateral forces) by rotation $-\alpha$.

Fig. 12.42 The coordinate systems used for stability analysis. a) The general aerodynamic force F_{Aero} is shown. b) It is resolved into lift or normal L and drag forces D. c) The two components are transferred into the vehicle's body axis by a coordinate rotation of $-\alpha$. This is the same procedure followed in Chapter 9 for flight loads analysis.

If the vehicle is not symmetrical in the pitch plane, we add a second inherent aerodynamic moment defined by

$$M_A = F_{Az}\, l_{sm}(+C_M\, qSd) = C_z\, qSl_{sm}(+C_M\, qSd) \qquad (12.8)$$

The added C_M terms in parentheses are due to asymmetry of the vehicle. An example of such asymmetry is the Space Shuttle, whose side or pitch outer shape is not symmetrical.

In general, the aerodynamic force and moment coefficients are nonlinear functions of angle of attack and Mach number. To simplify things, we'll assume that there is symmetry and that we can use a "local slope" representation to linearize with respect to angle of attack a

$$C_Z = \left(\frac{\partial C_Z}{\partial \alpha}\right)\alpha \overset{\text{def}}{=} C_{Z_\alpha}\alpha \qquad (12.9)$$

$$C_M = \left(\frac{\partial C_M}{\partial \alpha}\right)\alpha \overset{\text{def}}{=} C_{M_\alpha}\alpha \qquad (12.10)$$

12.3.4 Vehicle Control

We can control the vehicle by offsetting the thrust line by an angle δ by *gimbaling the engine*, also known as *thrust vector control*. The control moment or torque M_δ is the product of the sideways component of thrust multiplied by the distance from the engine gimbal to the mass center, or

$$M_\delta = (T \sin \delta)l_c \qquad (12.11)$$

We can define the *airframe stability factor* and the *control effectiveness factor* as follows:

- Aerodynamic moment M_A = moment generated by aerodynamic angle of attack
- Control effectiveness M_δ = moment generated by motion of engine gimbal δ

The controlled vehicle is shown in Fig. 12.43. It is clear that in order for the vehicle to be controllable, we must have $M_\delta > M_A = M_\alpha\alpha_{max}$ [33]. In other words, the control moment must be able to overcome the aerodynamic moment. However, Rakoczy [30] suggests the following rule of thumb: the controllability ratio should be greater than 2, or $M_\delta/M_A \geq 2.0$. This helps to ensure that the thrust vectoring effect M_δ will be large enough to offset maximum winds, rotational inertia, other disturbances, and unmodeled parameter variations that generate disturbances via M_A.

12.3.4.1 Values of Stability and Control Effectiveness Parameters for Saturn V

Let's compare the stability and control effectiveness factors M_α and M_δ for the Saturn V at max-q that we analyzed for loads in Chapter 9 using

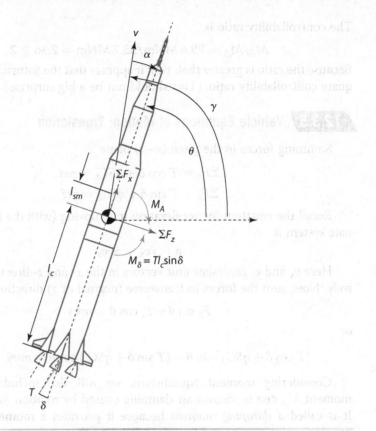

Fig. 12.43 The aerodynamic torque disturbing the vehicle motion (M_A) and the controlling torque due to engine gimballing ($M_\delta = Tl_c \sin\delta$).

aero loads data. From NASA [23], the maximum gimbal angle δ for the Saturn V F-1 engine nozzle was ± 5.0 deg in a square pattern. From Chapter 9, the max-q condition had a thrust level $T = 4/5 \times 38.55$ MN. (The $4/5$ factor comes from the fact that only the four outside engines of the five on the S-IC step gimbal.) The distance from the gimbal block to the CM at the time of max-q is $l = 34.5 - 5.47$ m $= 29.03$ m. Now we may calculate the stability and control effectiveness factors, and the controllability ratio.

The control effectiveness factor is calculated as

$$M_\delta = lT \sin\delta = (29.03\,\text{m})\frac{4}{5}(38.55\,\text{MN})\sin(5.0\,\text{deg})$$
$$= 79.6\,\text{MNm} \tag{12.12}$$

The Aerodynamic moment M_A is easily calculated using the required gimbal force (which is present to trim the vehicle under the moment of the aerodynamic forces taken from Chapter 9, section 9.2.5)

$$M_A = F_{\text{gimbal}}\, l = 1,161\,\text{kN} \times 29.03\,\text{m} = 33.7\,\text{MNm} \tag{12.13}$$

The controllability ratio is

$$M_\delta / M_A = 79.6\,\text{MNm}/33.7\,\text{MNm} = 2.36 \geq 2 \qquad (12.14)$$

Because the ratio is greater than two, it appears that the Saturn V has an adequate controllability ratio. (This should not be a big surprise.)

12.3.5 Vehicle Equations of Motion: Translation

Summing forces in the pitch $(x-z)$ plane

$$\Sigma F_x = T\cos\delta + qSC_x = m\ddot{x}$$
$$\Sigma F_z = T\sin\delta + qSC_z = m\ddot{z} \qquad (12.15)$$

Recall the equation for acceleration in a moving (with the body) coordinate system is

$$\mathbf{a} = \dot{v}\mathbf{e_x} + \dot{\gamma}v\mathbf{e_z} \qquad (12.16)$$

Here $\mathbf{e_x}$ and $\mathbf{e_z}$ represent unit vectors in the x- and z-directions, respectively. Now, sum the forces in transverse (normal or z) direction

$$F_x \sin\theta + F_z \cos\theta = mv\dot{\gamma} \qquad (12.17)$$

or

$$(T\cos\delta + qSC_X)\sin\theta + (T\sin\delta + qSC_z)\cos\theta = mv\dot{\gamma} \qquad (12.18)$$

Considering moment equilibrium, we will also include a damping moment $M_{\dot\theta}$ due to viscous air damping caused by rotation (usually small). It is called a *damping moment* because it provides a moment or torque opposing rotation rate and therefore damping the rate, and is defined as

$$M_{\dot\theta} = C_{M_{\dot\theta}} \frac{qd^2 S}{2v} \dot{\theta} \qquad (12.19)$$

Another type of rotational damping is called *jet damping*, and it occurs as a result of conservation of momentum for exhaust gases exiting a vehicle. For more information on jet damping, please refer to Ball [2] or Thomson [32].

Including all moments, the sum of the moments about the CM in the pitch plane or parallel to the y-axis = MoI × angular acceleration, or

$$\Sigma M_Y = Tl_c \sin\delta + qdSC_{M_\alpha}\alpha + C_{M_{\dot\theta}}\frac{qd^2 S}{2v}\dot{\theta} = I_y\ddot{\theta} \qquad (12.20)$$

From the figure, the angle of attack $\alpha = \theta - \gamma$. Now we can linearize this relation using small angle approximations ($\sin\theta \approx \theta$, $\cos\theta \approx 1$ for $\theta << 1$) and look at axial and rotational motions

$$(T + qSC_x)\theta + (T\delta + qSC_z) = mv\dot{\gamma}$$

$$I_y\ddot{\theta} = Tl_c\delta + C_{M_\alpha}qdS\alpha + C_{M_{\dot\theta}}\frac{qd^2 S}{2v}\dot{\theta}. \qquad (12.21)$$

We will use the following definitions to simplify these equations:

$$\text{Drag} = D = qSC_x \tag{12.22}$$

$$\text{aero moment: } M_\alpha = C_{M_\alpha} \frac{qdS}{I_y} \tag{12.23}$$

$$\text{Sideways acceleration: } Z_\alpha = C_{Z_\alpha} \frac{qS}{m}$$

$$\tag{12.24}$$

$$M_{\dot\theta} = C_{M_{\dot\theta}} \frac{qd^2S}{2vI_Y}$$

With these abbreviations, the two equations become the following: a set of two coupled differential equations that describe the vehicle's behavior.

$$\left(\frac{T+D}{m}\right)\theta + \frac{T}{m}\delta + Z_\alpha\alpha = v\dot\gamma \tag{12.25}$$

$$\ddot\theta = \frac{Tl_c}{I_Y}\delta + M_\alpha\alpha + M_{\dot\theta}\dot\theta \tag{12.26}$$

12.3.5.1 Dealing with Equations of Motion and Block Diagrams

To simplify working with equations such as these, controls analysis is done in the Laplace domain. A mathematical operation called the *Laplace transform* changes differential equations into algebraic ones involving complex numbers, making solution and simulation much simpler. The Laplace transformation, indicated by the script $\mathcal{L}$, of a function of time $f(t)$ is defined by

$$\mathcal{L}[f(t)] = \int_0^\infty f(t)e^{-st}dt = f(s) \tag{12.27}$$

The variable s is a complex number with real and imaginary parts of the form $s = \sigma + j\omega$, where j is the *imaginary unit*, $\sqrt{-1}$, also abbreviated as i.

Time derivatives are Laplace transformed into *multiplications* by s, and time integrals are transformed into *division* by s: $1/s$. Thus, a time derivative of position $x(t)$ becomes $sx(s)$, and the integration of acceleration $a(t)$ is $a(s)/s$. Further Laplace transform theory is beyond the scope of this book, but most texts on engineering mathematics have useful descriptions and practical examples of the Laplace techniques. We will have more to say about complex numbers later in this chapter. The control system block diagram and closed-loop controls analysis presented next follows the formulation presented in Heitchue [16].

A block diagram of these force and moment equations is shown in Fig. 12.44. The rectangular symbols with letters inside represent *multiplication*. For example, the top-left block in the figure contains the fraction T/m. This means that with the input of the box being δ, the output is the input quantity δ multiplied by the contents of the box T/m to get $\delta T/m$.

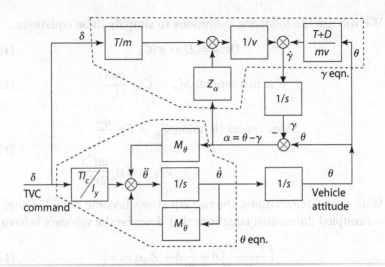

Fig. 12.44 The block diagram for the vehicle's equations of motion. The blocks corresponding to the two pitch-plane equations of motion, eqs. 12.25 and 12.26, are shown inside dotted lines.

A *summing junction*, a junction of two input arrows where variables are summed, is indicated by the $\otimes$ symbol. The incoming summation components are understood to be positive, unless a negative sign is associated with an incoming variable at the summing junction such as that on the lower-right of the figure.

The vehicle's two equations of motion in block format are enclosed in dotted lines; the $\alpha = \theta - \gamma$ equation occurs at the summing (actually subtracting) junction in the lower right.

As we have seen, this vehicle is probably unstable by itself, but even if it is stable, it needs to be steered on a certain path or trajectory. The important variables are the pitch attitude θ and its derivatives, thrust vector angle δ, angle of attack α, and flight path angle γ. In a typical control system, attitude error, attitude rate, and control position *feedback* are used. This means we measure the quantities and *feed them back* into the control system to get accurate control. The feedback is shown in Fig. 12.45.

The value for θ, the vehicle's attitude with respect to inertial space, is available from a stable platform or IMU. Its attitude rates (derivatives) are obtained from body-mounted gyros. The control position or thrust vector angle δ comes from a position sensor attached to the actuator.

It should be mentioned at this point that this simplified analysis ignores many real-life issues that must be dealt with, such as nonlinearities, servo or actuator dynamics (an actual actuator does not respond instantaneously), saturation and roll-off (reduced response at higher frequencies) of controls, flexible body and slosh effects, and several others. So the procedures shown here should be considered those that would be used to make a simple, first-order controls analysis.

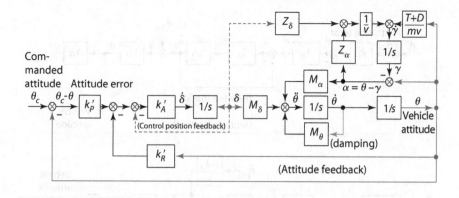

Fig. 12.45 A typical attitude control system feeds back vehicle attitude and control position (dotted lines).

Now let's simplify things. If we assume linearity and only consider basic short-period motion, Z_α, Z_δ, $M_{\dot\theta}$, and $T - D$ may be neglected. Then, the parameters of the vehicle (M_α, M_δ) and controller (k_A', k_R', k_P') may be absorbed into three new parameters k_A, k_R, k_P, as shown here:

$$k_A = k_A' \tag{12.28}$$

$$k_R = k_R' M_\delta - M_\alpha / k_A' \tag{12.29}$$

$$k_P = k_A'(k_P' M_\delta - M_\alpha)/(k_A' k_R' M_\delta - M_\alpha) \tag{12.30}$$

These assumptions and changed parameters are shown in the two equivalent block diagrams in Fig. 12.46.

12.3.5.2 Pitch-Plane Motion Simplifications

Looking at the upper block diagram in Fig. 12.46, it may be seen that $\ddot\theta$ results from the flow of the four variables θ_c, θ, $\dot\theta$, and $\ddot\theta_{(\delta)}$ weighted by the three gain parameters k_A, k_R, and k_P. From the block diagram, we can write

$$\ddot\theta_{(\delta)} = k_A k_R k_P(\theta_c - \theta) - k_A k_R \dot\theta - k_A \ddot\theta_{(\delta)} \tag{12.31}$$

Using Laplace transforms and assuming zero initial conditions, this reduces to

$$s^3 \theta = k_A k_R k_P \theta_c - k_A k_R k_P \theta - k_A k_R s\theta - k_A s^2 \theta \tag{12.32}$$

Treating the Laplace variable s as an ordinary algebraic variable, and separating terms in θ_c and θ,

$$k_A k_R k_P \theta_c = (k_A k_R k_P + k_A k_R s + k_A s^2 + s^3)\theta \tag{12.33}$$

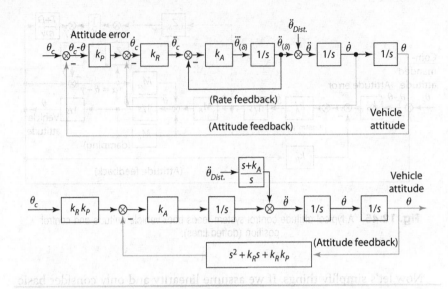

Fig. 12.46 Two simplified block diagrams of a linearized short-period control system.

Solving for the ratio θ/θ_c yields the *transfer function*

$$\frac{\theta}{\theta_c} = \frac{k_A k_R k_P}{s^3 + k_A s^2 + k_A k_R s + k_A k_R k_P} \tag{12.34}$$

The transfer function may be used to find the impulse or step response of the system. Because the equation resulting from setting the denominator equal to zero is *third order* (order s^3), number theory tells us that it must have three roots: either three real roots, or one real and one pair of complex conjugate roots. To understand the ramifications of the transfer function on system behavior, it's necessary to understand the concept of *poles and zeros*.

12.3.5.3 Understanding Poles and Zeros

Controls people use the *complex plane* or *s*-plane to study the behavior of a system. Here the variable *s* has *real* and *imaginary* parts σ and ω, respectively: we write $s = \sigma + j\omega$. For example, consider a simple function $F(s)$, where

$$F(s) = a/(s + a) = a/[(\sigma + j\omega) + a] \tag{12.35}$$

The magnitude of F can be plotted on the *s*-plane as a function of σ (real, or *x*-axis) and ω (imaginary, or *y*-axis). It's easy to see that there is a real *pole* of F at $s = -a$: when $s = -a$, F goes to *infinity* or becomes *singular*. The pole located at $-a$ is shown by the $\times$ in Fig. 12.47a.

We may also see that when $s = 0$, $F = 1$, and when ω increases or decreases from 0 (above or below the real axis), the magnitude decreases (as shown by the line in Fig. 12.47b).

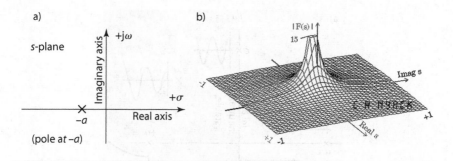

Fig. 12.47 a) Real pole at $s = -a$; b) a plot of the magnitude of $F(s)$, where the peak is truncated at 15 (it actually goes to infinity at $-a$). Source: http://cnyack.homestead.com/files/alaplace/laptr1.htm.

The *phase* is the angle between the $+s$-axis (the positive real axis) and the line to a point $j\omega_1$ on the $j\omega$-axis.

The locations of the roots of the transfer function define the dynamics of the system's response. If there is a *negative* real root at $(-a)$, the system has a *stable* response, because $e^{-at} \to 0$ as $t \to \infty$. If the system has roots that are conjugate pairs (conjugate means the roots have the same real part, but their imaginary parts are the same magnitude but opposite sign) with *negative* real parts $s_{1,2} = (-a \pm jb)$, then we have a *stable damped oscillatory* response: $e^{-at} \sin bt \to 0$ as $t \to \infty$.

On the other hand, if any of the roots have *positive* real parts $s_{1,2} = (+a \pm jb)$, the system will be *unstable* because amplitude grows like $\sim e^{+at}$ or $e^{+at} \sin bt$; the positive exponential term means that both approach infinity as time increases. It can be seen that if all of the roots are in the left-half plane, the system will be stable. *If any of the roots are in the right-half plane, the system will be unstable.*

The root locations and concepts just described are illustrated in Fig. 12.48a.

Figure 12.48b shows some useful information for the complex plane. Circles centered on the origin show a constant natural frequency, and as the radius increases, so does the circle radius. Straight lines passing through the origin produce damping, which gets smaller as the lines approach the vertical (imaginary) axis.

The system in Fig. 12.46 may be reduced to a single loop, as shown in Fig. 12.49. This block diagram is called *actuation loop stability* because the actuation loop gain k_A has the largest effect.

The closed-loop transfer function for θ/θ_c can be derived from the block diagram in Fig. 12.49 using

$$\frac{\theta}{\theta_c} = \frac{k_1 k_2 G(s)}{1 + k_2 G(s) H(s)} \qquad (12.36)$$

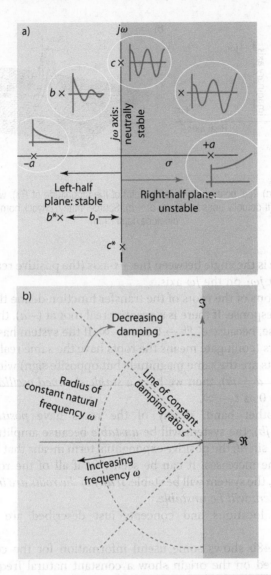

Fig. 12.48 a) A system is stable if all of its roots are in the left-half plane and is unstable if any of the roots are in the right-half plane. b) The plot shows how frequency increases away from the origin, and damping decreases when the imaginary axis is approached.

where $G(s)$ is the forward loop transfer function

$$G(s) = 1/s^3 \qquad (12.37)$$

and $H(s)$ is the feedback transfer function

$$H(s) = s^2 + k_R \cdot s + k_R k_P \qquad (12.38)$$

The *characteristic equation* is defined as $1 + k_2\, G(s)\, H(s)$, and its roots determine the system's stability. A solution occurs when

$$k_2 G(s)\, H(s) = -1 \qquad (12.39)$$

Considering the open-loop transfer function as

$$k_2 G(s)\, H(s) = k_A \cdot \left(\frac{1}{s} \cdot \frac{1}{s} \cdot \frac{1}{s}\right) \cdot \left(s^2 + k_R s + k_R k_P\right) \qquad (12.40)$$

there are three poles ($\times$) at the origin, from the $1/s^3$ term in the left parentheses, and a pair of complex zeros ($\bigcirc$) from the numerator term $s^2 + k_R s + k_R k_P$ (right parentheses) This situation is shown in Fig. 12.50. (The poles are staggered slightly so they are visible as three rather than one visible pole symbol.)

It appears that if the gain k_A is small, the three roots (light gray color) stay near the origin and the open-loop poles. As the gain k_A becomes very large, the real root moves to the left with increasing frequency, making it a stable root whose influence is rapid and short-lived. The other two roots, the complex ones, get arbitrarily close to the off-real axis open-loop zeros and control the system's transient response. So, it would appear that we should try to use as high gain (value of k_A) as possible for best system performance.

This low-order analysis is somewhat misleading, because there is always a limit to how much gain can be used. This limitation primarily comes from the physical behavior of actuator dynamics, which are known to make the actuator servo have realistic (rather than instantaneous) response and performance.

To include realistic actuator behavior, we add the second-order dynamics of a TVC actuator to the previous block diagram. [The actuator behaves like a second-order harmonic oscillator, characterized by a transfer function of $\omega^2/(s^2 + 2\zeta\omega s + \omega^2)$.] This makes the behavior of the system more realistic, because the TVC servo cannot instantaneously respond to commands from

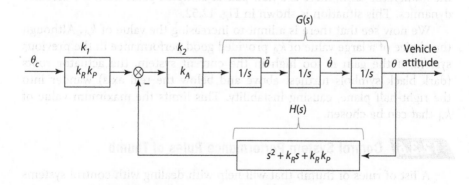

Fig. 12.49 Actuation loop stability block diagram.

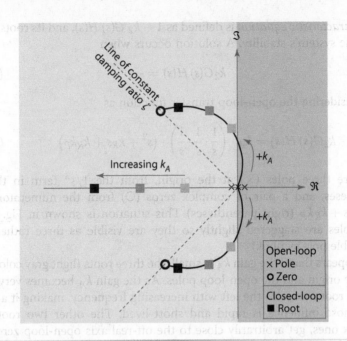

Fig. 12.50 Root locus for determining allowable range for actuation loop gain k_A. The closed-loop poles are shown as squares that become darker (from gray to black) as gain k_A increases.

the guidance system. The revised block diagram is shown in Fig. 12.51, with the added actuator dynamics block shaded.

The open-loop transfer function, including the actuator block, is

$$k_2 G(s)H(s) = k_A \cdot \left(\frac{1}{s} \cdot \frac{1}{s} \cdot \frac{1}{s}\right) \cdot \frac{\omega^2}{s^2 + 2\zeta\omega s + \omega^2} \cdot (s^2 + k_R s + k_R k_P) \quad (12.41)$$

In addition to the three poles at the origin and the pair of complex zeros from the numerator term, there is a new pair of complex poles from the actuator dynamics. This situation is shown in Fig. 12.52.

We now see that there is a limit to increasing the value of k_A. Although the choice of a large value of k_A provided good performance in the previous system, if the gain is too high in the current system, the actuator roots (dark black symbols furthest above and below the real axis) wander into the right-half plane, causing instability. This limits the maximum value of k_A that can be chosen.

12.3.6 Control System Performance Rules of Thumb

A list of rules of thumb that will help with dealing with control systems for rigid space launch vehicles follows. The reader should also read the recommendations provided at the end of the section on slosh later in the

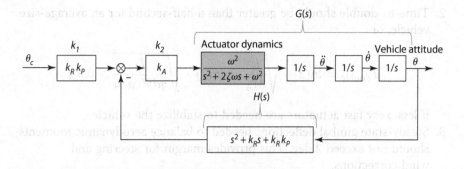

Fig. 12.51 The shaded block represents the dynamics of the actuator, to make the system behavior more realistic.

chapter. Note that item 8 is not discussed in this book, but is mentioned for the interested reader.

1. The controllability ratio $M_\delta/M_A \geq 2.0$. It is typically >1, although values less than $1/2$ have been observed in Delta III vehicles. The upper limit is dictated by vehicle flexibility and control-bandwidth separation.

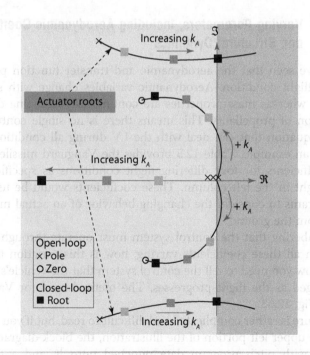

Fig. 12.52 Root locus for determining allowable range for actuation loop gain k_A, including the second-order actuator dynamics labeled "actuator roots." Too high a value of k_A will cause the system to go unstable, thereby limiting the maximum value of feedback.

2. Time-to-double should be greater than a half-second for an average-size vehicle, or

$$T_{\text{double}} = \ln(2) \cdot \sqrt{\frac{I_y}{M_A}} = \ln(2) \cdot \sqrt{\frac{I_y}{qdSC_{M_\alpha}\alpha}} > 0.5 \text{ s}$$

if less, very fast actuators are needed to stabilize the vehicle.
3. Steady-state gimbal deflections needed to balance aerodynamic moments should not exceed 2 deg. This provides margin for steering and wind corrections.
4. The amount of gain in a system is limited by the system's stability.
5. The simplified, short-period motion of the launch vehicle depends on three parameters: the gain of actuation loop, rate loop, and attitude loop.
6. Low feedback values result in roots close to origin, providing a system that is poorly damped and has sluggish performance.
7. For more realistic analysis, the effects of sloshing, flexible body behavior, and nonlinearities need to be incorporated. These effects will be discussed in the following sections.
8. The standard LV rigid-body stability margin requirements are ±6-dB gain margin and 30-deg phase margin [33].

12.3.7 Varying Parameters, Including Aerodynamic Coefficients and Structural Dynamics

We have seen that the aerodynamic and transfer function parameters vary with flight condition. Aerodynamic variables change with speed and air density whereas mass properties are continuously changing due to the consumption of propellants. This means there is no single control law or feedback equation that can deal with the LV during all conditions during ascent. As an example, Table 12.5 provides the Vanguard missile's aerodynamic coefficients for four differing flight conditions as specified by the time of flight in the left column. These coefficients would be used in the block diagrams to consider the changing behavior of an actual missile as it ascends from the ground.

Remembering that the control system must operate through complete ascent with all these coefficients varying, how is this variation taken care of? Somehow you need to tell the control system that the vehicle's dynamics have changed as the flight progresses. The method used for Vanguard is shown in Fig. 12.53.

The figure is rather complicated and difficult to read, but if you look carefully in the upper left portion of the illustration, the block diagram shows a steel tape with pitch program data punched onto it, and a timer that moves the metal tape as the mission progresses. The tape contains control system gain information for different times of flight, and had to move to a new position with new gains at the appropriate time of flight.

Table 12.5 Aerodynamic Coefficients for the Vanguard Missile, Assumed Rigid

t (s)	h (ft)	V (ft/s)	M	q (lb/ft^2)	m (slugs)	Θ (deg)	C_{m_α}	C_{z_α}	$\frac{d}{2U}C_{m_q}(s)$	C_{m_δ}	C_{z_δ}	I_y (slug ft^2)
48	12,200	604	0.57	300	547	81.7	10.55	−3.08	−0.683	−60.83	−8.69	121,400
75	36,000	1,285	1.26	585	445	68.5	11.27	−3.13	−0.321	−34.25	−4.63	115,000
100.4	76,000	2,400*	2.41*	300	354	57.2	16.59	−2.88	−0.172	−70.15	−9.17	104,000
139*	185,000	5,600	5.02	12	218	46.5	8.32	−2.3	−0.0736	−2140	−229	75,400

*First-stage burnout, $l = 27$ ft, $S = 11.04$ ft^2, $d = 3.75$ ft.
Source: [7].

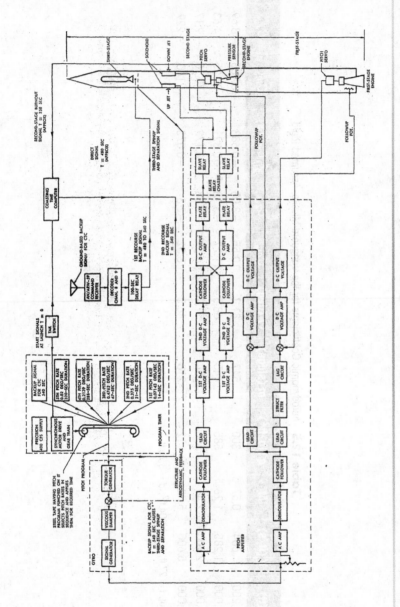

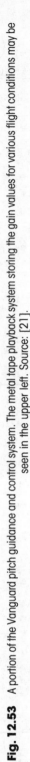

Fig. 12.53 A portion of the Vanguard pitch guidance and control system. The metal tape playback system storing the gain values for various flight conditions may be seen in the upper left. Source: [21].

This "open-loop" timer could not sense the vehicle's actual flight condition, so lots of work was needed to figure out the times of flight in which it would move to a new position to change the mode of the control system.

Progress in avionics has made the metal tape system obsolete. Modern electronics allows fast-processing computers that can sense the vehicle's state and automatically switch to different sets of stored control system gains as needed, so the clumsy and inflexible metal tape system is no longer needed.

12.3.8 Engine Angle Response to Wind Shear

To get a feel for the response of a TVC system to a wind gust, Fig. 12.54 provides the time histories for large launch vehicle studied in the early 2,000s. The LV consisted of a solid first step with 12.5 MN of thrust, attached to a liquid-propellant second step with a capsule on top. The overall length was about 90 m, and total mass was about 600 T. Like most LVs, the vehicle

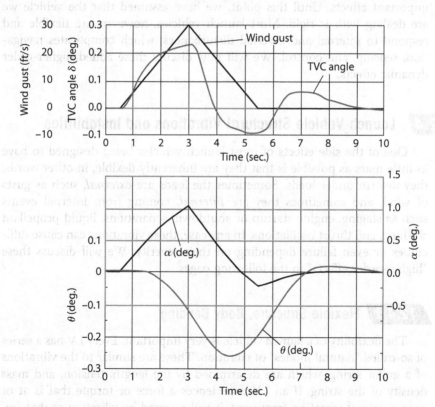

Fig. 12.54 The time histories of the wind gust input, the pitch angle θ response, the angle of attack α, and the thrust vector control gimbal angle δ of a large launch vehicle subjected to a triangular wind gust of 30 ft/s maximum speed at max-q conditions.

was statically unstable and had a load-relief control system using midbody acceleration feedback.

The LV encountered a horizontal triangular-shaped wind gust with a 30 ft/s (9.1 m/s) peak speed (see the straight-line time history in the top part of Fig. 12.54) while passing through max-q (840 psf or 40.2 kPa, 1,500 ft/s or 460 m/s speed).

In the plots, the angles α and θ are changes relative to the trimmed angles before the gust hit. The gust was perpendicular to the vehicle's x-axis (longitudinal direction) and generated a positive angle of attack α (lower figure). The attitude θ (lower figure) barely started to go positive, but the load-relief system responded quickly, and the engine gimbal rotated an angle δ (upper figure) in the positive direction to generate a negative pitch rate that turned the attitude θ negative, into the wind. Once the wind gust began tapering off, the TVC system moved the other way to cancel out the rate and stop near its initial, trimmed position.

We have seen that the performance of a launch vehicle's attitude control system is limited by system stability as well as some other important effects. Until this point, we have assumed that the vehicle we are dealing with is rigid. Most launch vehicles, however, are flexible and respond to internal and external disturbances, which complicates navigation, sensing, and control. We will now discuss these added higher-order dynamic effects.

12.4 Launch Vehicle Structural Vibrations and Instabilities

One of the side effects of most launch vehicles being designed to have as little mass as possible is that they are inherently flexible, in other words, they deform under loads. Sometimes the loads are *external*, such as gusts of wind, and sometimes they are *internal*, coming from internal events such as staging, engine startup or shutdown, separations, liquid propellant sloshing, and thrust oscillations. In any case, these vibrations can cause difficulties or even failure depending on their severity. We will discuss these "higher-order effects" in the following pages.

12.4.1 Flexible Structure, Body Bending

The flexibility of a launch vehicle is very important. Every LV has a series of so-called "natural modes" of vibration. These are similar to the vibrations of a guitar string, which are determined by the length, tension, and mass density of the string. If an LV experiences a force or torque that is at or near a natural vibration frequency, it will respond by vibrating at that frequency. The ensuing vibration can interfere with IMUs, gyros, and sensors, and can cause other problems such as encouraging propellant sloshing and other negative effects.

The very first McDonnell Douglas Delta III and the third SpaceX Falcon 1 were both lost because of vibrations interacting with their control systems. The Delta was lost when a torsional (twisting) vibration was not filtered out of the control system; instead, the control system caused it to oscillate, leading to a premature loss of control due to running out of hydraulic fluid. The Falcon was lost because a recontact during stage separation caused its propellants to slosh and also run out of hydraulic fluid.

The U.S. first satellite launcher, the Jupiter-C/Explorer 1 had a motorized spin table that was designed to average out any thrust inequalities that may have occurred during the firing of solid rockets clustered for upper stages. The Jupiter-C and its spin table are shown in Fig. 12.55. In order to keep the table's spin frequency from coinciding with the bending frequencies of the launch vehicle, the system spun at 550 RPM at launch, went to 650 RPM after 70 s of flight, and then to 750 RPM after 155 s of flight.

The structural frequencies increase with flight time because the vehicle's stiffness is nearly unchanged, whereas its mass decreases as propellants are

Fig. 12.55   The Jupiter-C launched the first U.S. satellite. It had a spin table whose spin rate had to change so as to avoid coinciding with a bending frequency of the LV. Sources: Left: NASA. Right: Author.

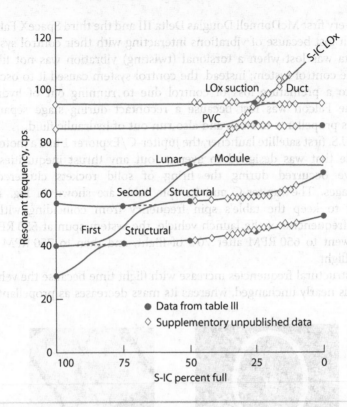

Fig. 12.56 Saturn V natural frequencies as a function of the S-IC (first step) percent full. Source: NASA.

consumed. This can be seen in the Saturn V data in Fig. 12.56. Note that all of the frequencies rise as propellants are consumed.

Every vehicle possesses what are characterized as rigid and flexible modes of vibration. The rigid-body modes can be used to describe unrestricted motion, which we call six degrees of freedom. Hence there are six rigid body modes, three in translation (x, y, z) and three in rotation (yaw, pitch, roll). Two rigid body modes (vertical translation and clockwise rotation) are shown in Fig. 12.57.

We are interested in flexible body bending vibration. Figure 12.57c shows the first of an infinite series of body bending shapes (dashed line) along with an equivalent spring-mass-damper system, which is mounted where the bending mode has amplitude = 1.0. By "equivalent," we mean that the spring-mass-damper system has the same natural frequency and damping as the body-bending mode. Force F applied to the first modal mass M_1 causes it to oscillate with amplitude A_1, defined as the generalized coordinate of the first bending mode.

In Fig. 12.57c, one can see that the first bending mode has two *nodes*, n_1 and n_2. These are locations where there is no displacement in the bending

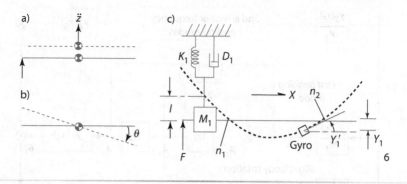

Fig. 12.57 a) Rigid translation, b) rigid rotation, and c) body bending mode. Source: [16].

beam (but there *is* rotation). Also, note that a gyro is shown on the bending beam. A gyro located here would measure *both* angular motion due to the beam oscillating (shown on the figure as y_1') *and* the rigid body rotation θ shown in Fig. 12.57b. This means that the beam oscillating signals *must be filtered out of the gyro's measurement* so as to leave only the rigid-body rotation.

The deflections of the entire system are calculated by summing rigid and flexible modes, as shown in the block diagram in Fig. 12.58. (Note that this block diagram is a *structural* block diagram, which is not related to the *controls* block diagrams shown previously. The idea is to show the calculation of the deflections symbolically by adding up the contributions of each of the rigid- and flexible-body modes.)

Earlier, in Fig. 12.54, we showed the response of an Ares I–like vehicle to a triangular wind gust. In addition to the rigid-body responses shown in that figure, the vehicle's flexibility was also excited by the gust. Figure 12.59 shows the flexible-body responses of the vehicle's three gyros and three accelerometers, which originated from sensors placed in three different locations: at the vehicle's interstage, near its top, and near its bottom near the SRB gimbal. The z-accelerometers initially measured negative acceleration because the vehicle was pushed along the z-direction by the wind gust. Note that the higher frequency body-bending vibrational modes can be seen superposed on the lower-frequency rigid-body responses in the earlier figure. In this case, it can be seen that structural filtering needs to be employed so that higher-frequency vibrations do not interact with the TVC system.

As an example of the importance of including flexible body effects into controls design, two Atlas ballistic missiles proved unsatisfactory because of an instability involving coupling between the automatic control system and a body-bending vibration, as described by Bisplinghoff [6]. Quoting a letter from the manufacturer regarding the telemetry traces seen in Fig. 12.60:

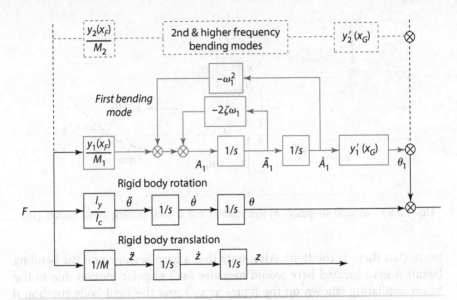

Fig. 12.58 Block diagram representation of force F driving rigid-body (black color) and flexible-body (blue) modes. Source: [16].

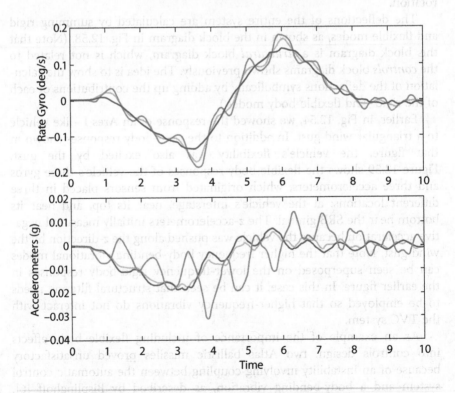

Fig. 12.59 Lateral acceleration and rate gyro responses of the large vehicle exposed to the max-q wind gust, including several of its bending vibration modes. The accelerometers and rate gyros were located at the vehicle's interstage, near its top, and at near its bottom near the nozzle gimbal.

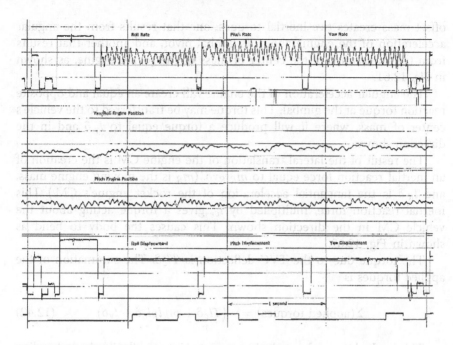

Fig. 12.60   Data recorded during an unstable flight of an Atlas missile. Source: [6].

The top trace is the rate gyro information. Roll, pitch and yaw rate outputs are shown in that order ... for one second for each channel. The lower two traces are the engine position traces in the pitch and yaw channels. The last trace is the displacement gyro data that are presented in the same manner as the rate gyro information. The high frequency (17 Hz) oscillation is a limit cycle resulting from the autopilot coupling with the third lateral bending mode. The lower frequency (1 Hz) is rigid body motion. Pre-flight simulations including the first three lateral bending modes did not uncover this problem because of the use of a linear third-order hydraulic actuator simulation, which did not incorporate the effect of missile vibration on the hydraulic servo system.

As a result of this flight, a highly non-linear hydraulic servo simulation was incorporated which permitted duplication of the flight test results. Changes were then made to the autopilot to attenuate the third lateral bending mode. All succeeding flights showed that the coupling of this mode with the autopilot was completely eliminated.

The details of the inclusion of flexible vibration modes into vehicle control systems are beyond the scope of this book. The reader is encouraged to consult with references such as Falangas [14] or Blakelock [7] for more detailed treatments of flexible body vibrations and their interactions with flight control systems.

12.4.2 Tail Wags Dog Motion

Let's look at what happens when a vehicle's thrust vector control system commands the engine nozzle to slew to a new angle. Gimbaling the engine's offset mass creates *two* inertial torques: one that results from the angular acceleration of the engine about its gimbal pivot, and another that results from the translation of the center of mass of the rocket engine, as shown in Fig. 12.61.

The angular acceleration of the engine will create an equal and opposite reaction torque at the gimbal. This torque may be transferred to the vehicle's center of mass, where it will produce a torque equal to $I_y \ddot{\delta}$ and in the direction shown.

The result of the lateral translation of the engine CM is the creation of an inertial reaction force equal to $m_R l_R y \ddot{\delta}$. (m_R is the moving engine mass, and $l_R \ddot{\delta}$ is the tangential acceleration of the rocket engine's CM.) This inertial reaction force, multiplied by l_c, gives a torque acting about the vehicle CM in the direction shown. This causes the body to bend as shown in Fig. 12.62.

The torque due to the TVC deflection is $T l_c \delta$. The summation of the applied torques is

$$\Sigma(\text{applied torques}) = -\left(T l_c \delta + m_R l_R l_c \ddot{\delta} + I_y \ddot{\delta}\right) \qquad (12.42)$$

This applied torque has to be incorporated into the flexible body-bending equations, which we have only presented in schematic form. The result is that the flexibility adds *complex zeros* to the overall vehicle transfer function, with values of

$$\omega_n = \pm \sqrt{\frac{T}{m_R l_R + I_y / l_c}} \qquad (12.43)$$

This is the frequency where the inertial forces resulting from the gimbaling of the rocket engine cancel the component of thrust normal to the missile axis because of deflection of the engine due to TVC commands (which is the meaning of a *zero*: where there is no response at all). For more details

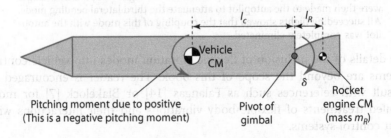

Pitching moment due to positive
(This is a negative pitching moment)

Vehicle CM

Pivot of gimbal

Rocket engine CM (mass m_R)

l_c l_R δ

Fig. 12.61 *Tail wags dog* occurs when the torque needed to gimbal the engine creates an opposite torque on the rear of the body, causing bending that reduces the desired pitching moment from the TVC system.

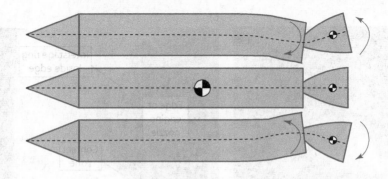

Fig. 12.62 The gimbal torque causes body bending that reduces the control moment from the TVC system and creates a second-order spring (body stiffness) mass (engine mass) system.

on how this and flexible body modes are calculated, the reader is referred to Blakelock [7] or Siefert & Brown [31].

12.4.3 Propellant Slosh

Sloshing describes the lateral wavelike motion of liquid propellant in partially empty tanks, as shown in Fig. 12.63. This motion induces internal forces that can excite structural modes, which in turn cause problems when resonance with flight motions occurs.

The results of slosh can be catastrophic, as still frames from the two videos in Fig. 12.54 demonstrate. Fig. 12.64a shows structural failure, and Fig. 12.64b shows slosh induced by recontact of the upper stage's engine bell with the interstage structure.

The Juno failure was caused by either air loads or maneuvering loads that exceeded the structural capabilities of the vehicle. It may be seen to very slowly fishtail in its ascent until such time as the nose section folds into the lower portion of the rocket, just before range safety presses the "red button."

The Falcon 1 failure was also due to sloshing, but in this instance, the sloshing was initiated when the upper stage's engine nozzle contacted the inside edge of the interstage ring. The ensuing slosh can be seen to grow with time, as evidenced by the larger and larger movement of the TVC system until it either runs out of hydraulic fluid or loses authority to correct for the sloshing.

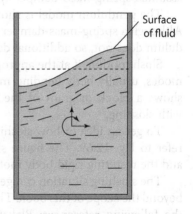

Fig. 12.63 Sloshing occurs when a fluid moves laterally back and forth in a partially filled tank.

a)   b)

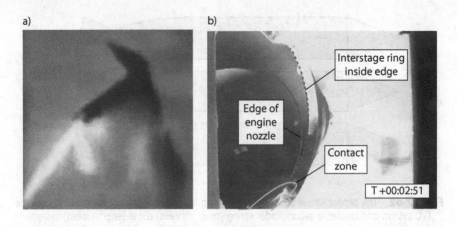

Fig. 12.64   a) Frame from video of early Juno flight; b) frame from Falcon 1 flight video. Note contact between interstage ring and engine bell.

SpaceX learned that preventing recontact was important, and increased the time before separation to ensure that the thrust of the lower step had tailed off adequately. Several other methods of preventing recontact are provided in Chapter 11.

12.4.3.1 Modeling Slosh Effects

There are two common ways to model sloshing liquids in a launch vehicle. They can be modeled as a spring-mass-damper system or as a pendulum with varying frequency (based on remaining propellant). A spring-mass-damper system has inherent damping. Figure 12.65 shows the assumed spring-mass-damper system.

The pendulum model is shown being carried by a vehicle in Fig. 12.66. Although a spring-mass-damper system has inherent damping, an ideal pendulum does not, so additional damping will likely be needed.

Slosh motion, if at the appropriate frequency, can couple with structural modes, usually body-bending modes, as shown in Fig. 12.67. Figure 12.68 shows a block diagram of the mechanical behavior of a flexible vehicle with sloshing.

To get an idea of how sloshing can affect the maneuvering of a vehicle, refer to Fig. 12.69. The figure shows a vehicle doing a change in attitude, and the resulting rigid body motion and total motion.

The sloshing situation can get very complicated, and analytical analysis is beyond the scope of this book. The interested reader is encouraged to consult the following references: Blakelock [7], Abramson [1], and its successor Dodge [12].

The following guidelines are provided for analysis purposes by Falangas [14]:

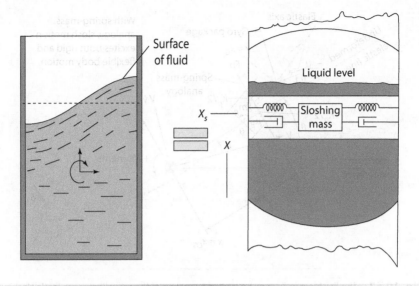

Fig. 12.65   Spring-mass-damper system used to model sloshing fluids.

- *For structural bending, or tail wags dog:* The engine/nozzle masses should be removed from the vehicle total mass and moments of inertia, because they are being treated as separate bodies interacting with the vehicle.
- *For slosh:* The vehicle mass and moments of inertia should not include the slosh masses. They are removed because in the equations, they are treated as separate bodies interacting with the rest of the vehicle.

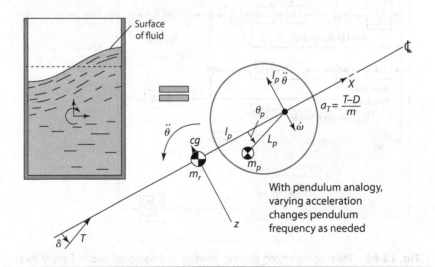

Fig. 12.66   The pendulum analogy for sloshing propellants.

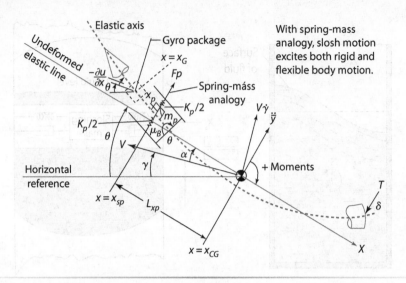

Fig. 12.67   The flexible missile may be excited by motion of the propellant mass that's sloshing. Source: [19].

12.4.3.2 *Alleviating Slosh*

Sloshing is reduced with the placement of *slosh baffles* inside the offending tank(s). Some slosh baffles may be seen in Fig. 12.70a. The annular baffles are placed at various heights above the bottom of the propellant tank so as to discourage sloshing at multiple tank fill levels.

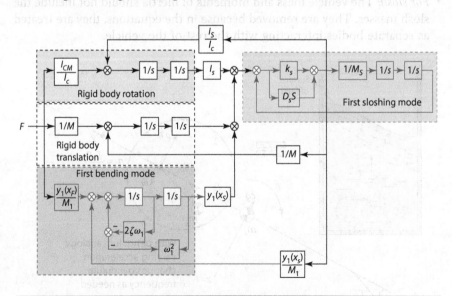

Fig. 12.68   The coupling among sloshing, bending, and rigid-body modes. Each of these additional modes, and the tail-wags-dog zeros, add *poles* and *zeros* to the vehicle's root locus diagram. Source: [16].

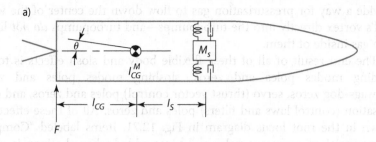

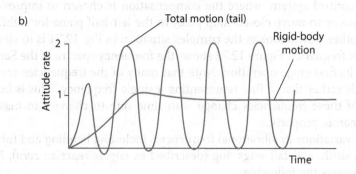

Fig. 12.69 a) A vehicle with a propellant tank having slosh dynamics is simulated with a spring-mass-damper system. b) When the vehicle is commanded to change attitude suddenly, the apparent rigid-body motion is smooth as desired, but the sloshing fuel causes the total motion of the vehicle (rigid + propellant mass) to oscillate. Motion would be even larger if the slosh frequency coupled with any of the flexible body modes. Source: [16].

Designers also may place an *antivortex baffle* at the bottom of the propellant tank, as shown in Fig. 12.70b. This baffle serves to prevent a whirlpool from forming at the bottom of the tank. If allowed, a whirlpool would

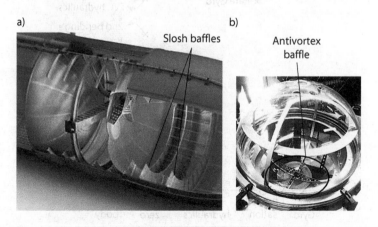

Fig. 12.70 a) Ring-shaped *slosh baffles* placed around the circumference of the sides of a tank; b) a cross-shaped *antivortex baffle*. Source: NASA.

provide a way for pressurization gas to flow down the center of the whirl-pool's vortex directly into the turbopumps—and turbopumps *do not* like to have gas inside of them.

The end result of all of these flexible body and slosh effects is to add bending modes poles and zeros, sloshing modes poles and zeros, tail-wags-dog zeros, servo (thrust vector control) poles and zeros, and compensation (control laws and filters) poles and zeros. All of these effects are shown in the root locus diagram in Fig. 12.71. Items labeled "Comp" or "Compensation" represent poles and zeros added when the loop is closed with a control system, where the compensation is chosen to improve performance or to move closed-loop poles to the left-half plane for stability.

Another way to look at the complex situation in Fig. 12.71 is to view it in terms of frequency. Figure 12.72 shows the frequency spectra of the Saturn V during its first-stage operation. Note that many of the frequencies are given as bands rather than a line representing a single frequency. This is because many of these frequencies change with time due to changes in mass and aerodynamic properties.

For variations in vibrational frequencies including bending and torsional modes, slosh, and tail wags dog (described as *engine reaction zero*), NASA [26] suggests the following:

* The rigid-body control frequency should be sufficiently below the frequency of the lowest bending mode to allow the design of filters that

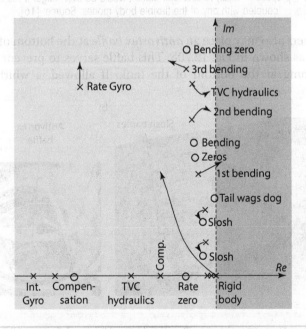

Fig. 12.71 Root locus of a launcher in a single axis. Source: [31].

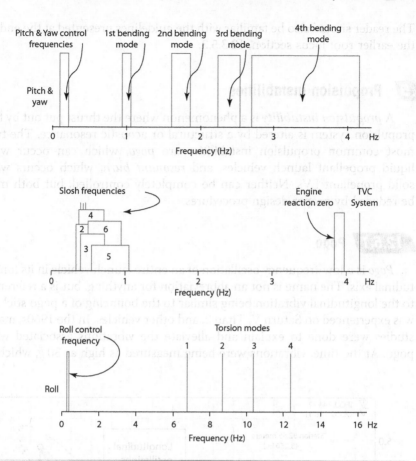

Fig. 12.72 The frequency ranges for different types of vibration occurring during first-stage Saturn V operations. Source: [26].

stabilize the bending modes without adversely affecting the control mode. Falangas [14] suggests, "Try to keep a factor of 10 separation between bandwidth and structural flexibility."

- The control frequency must be high enough to provide adequate vehicle response and to minimize trajectory dispersions caused by thrust vector misalignment.

- The control gains, which determine the control frequency, must be high enough to guarantee that the operating point does not go below the static stability margin when tolerances in coefficients are included. For the control system bandwidth f_b, Falangas [14] suggests using the following:

$$f_b = 2\pi \sqrt{\frac{M_A}{I_{yaw}}}.$$

The reader should also be familiar with the guidelines presented at the end of the earlier root locus section, 12.3.5.3.

12.5 Propulsion Instabilities

A *propulsion instability* is a phenomenon where the thrust put out by the propulsion system is altered by a structural or acoustic resonance. The two most common propulsion instabilities are *pogo*, which can occur with liquid propellant launch vehicles, and *resonant burn*, which occurs with solid propellant LVs. Neither can be completely controlled, but both may be reduced by proper design procedures.

12.5.1 Pogo

Pogo is a low-frequency oscillation of an entire launch vehicle in its longitudinal axis. The name is not an abbreviation for anything, but is a reference to the longitudinal vibration being similar to the bouncing of a pogo stick. It was experienced on Saturn V, Titan 2, and other vehicles. In the 1960s, many studies were done to explain and alleviate the vibrations associated with pogo. At the time, vibrations were being measured as high as 30 g, which is

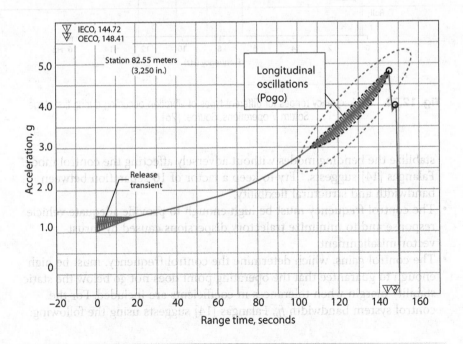

Fig. 12.73 Longitudinal acceleration time history of a pogo vibration instability occurring between 110 and 150 s range time on the Saturn V–Apollo 6 launch. Source: [24].

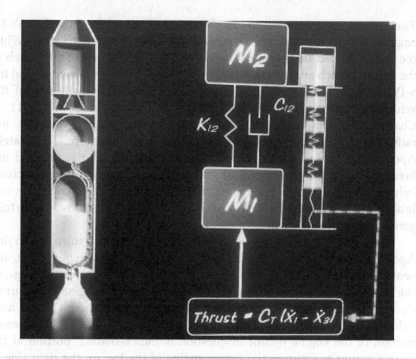

Fig. 12.74   Pogo caused by thrust oscillations that in turn are caused by axial structural vibrations that are caused by thrust oscillations that are caused by axial structural vibrations. ...
Source: Still frame from NASA video.

a *very* high level; however, the phenomenon only occurs at a certain time of flight, as shown by the acceleration plot shown in Fig. 12.73.

12.5.1.1 Pogo Instability Explained

Pogo is caused by the interaction of the vehicle's axial structural vibration mode and the propulsion system. Vibrations cause flexing of the propellant-feed pipes ... which induces pressure fluctuations ... which cause thrust variations ... which drive vibrations, causing more pipe flexing ... causing more thrust variations ... and so on. This is a classic unstable feedback situation. If the thrust oscillations couple with the axial natural vibration frequency, the amplitude can be very large. The situation is illustrated in Fig. 12.74.

Pogo is more of a problem with human-carrying vehicles. An uninhabited vehicle may provide a high-vibration environment to its payload, but the payloads carried can usually survive higher levels than humans. With human payloads, vibration levels can interfere with crew vision and therefore crew function, or even can cause injury. There are a number of references to pogo oscillation in NASA reports [27, 28].

Pogo nearly caused a failure of two Saturn V vehicles. During the unmanned flight of Apollo 6, pogo oscillations began after 1 min of flight.

The shaking caused the automatic shutdown of one of the second step's J-2 engines, which also shut down a second engine due to a wiring error. While the Saturn limped along on its three remaining engines, part of the panels of the Saturn launch adapter, the structure between the service module and the S-IVB upper skirt, began to fail, and some pieces eventually fell off the vehicle. (A falling piece may be seen just below the S-IVB in Fig. 11.45, Chapter 11.) This failure was not caused by the pogo oscillations, but rather by trapped air in the adapter's sandwich structure. Fortunately, Apollo 6 was far enough out of the atmosphere that the vehicle did not break up, but there was one more chilling event: because of the structural failure, one of the three emergency detection system (EDS) trip wires was broken. Had a second wire broken, the launch escape system would have triggered, ending the mission.

The second noteworthy pogo event occurred on the Saturn V carrying Apollo 13 to the moon. During the firing of the S-II second stage, the center engine vibrated so strongly that the pressure fluctuations triggered the engine's shutdown command. It's said that if the engine had burned during one more cycle of oscillation, the center engine mount would have failed, and a catastrophic failure would have ensued. The highly stressed portion of the engine mount in question is the "x-shaped" portion of the engine mount supporting the center engine of the five on the right side of Fig. 12.75.

If you have ever experienced a *water hammer* with household plumbing, the pogo occurs for a similar reason. In water hammer, the household

Fig. 12.75 Pogo nearly caused the failure of the center J-2 engine mount on the Apollo 13 Saturn V. Source: Still frame from NASA video.

plumbing oscillates due to the flow of water from the faucet causing a pressure variation that is at the same frequency as the piping system.

12.5.1.2 Pogo Suppression

Pogo occurs because of pressure pulses in the liquid propellant system causing thrust oscillations that excite structural vibrations in the LV structure, which cause pressure pulses that cause thrust oscillations, and so on. So the way to eliminate it is to add a "compliance" to the system so that its natural frequency is changed. Figure 12.76, taken underneath a sink, shows two cylinders that hold air under pressure that can expand or contract to absorb pressure variations in the plumbing system. The same principle is used to alleviate pogo in liquid propellant rockets.

Fig. 12.76 Under-sink *accumulators* (vertical cylinders in the photo) provide extra compliance to change the natural frequency of the plumbing and stop the water hammer vibrations, which are similar to those that occur during pogo.

The cure to pogo oscillations is to add pressurized gas–filled cavities known as *accumulators* to the propulsion system's plumbing. In this application, they are called *pogo suppressors*. This addition is similar to the act of adding capacitance to an electric circuit to smooth out voltage variations. In a hydraulic circuit, the accumulators absorb or filter the pressure oscillations so as to minimize them.

Schematically, a pogo suppressor acts as shown in Fig. 12.77. Pogo-induced pressure waves traveling through the plumbing in tube a can move sideways at the accumulator and, instead of propagating forwards, they compress and expand the trapped gas at b, usually helium. This compression and expansion tend to damp or dissipate the unwanted hydraulic pressure fluctuations.

The pogo problem was dealt with by a large number of researchers, leading to many alternate solutions. Figure 12.78 shows a number of pogo correction devices used in Titan II (launcher for Gemini manned spacecraft), Titan III (uninhabited), and Saturn V.

As an example of a pogo suppressor installed into a propulsion system, please see Fig. 12.79. This appears to be a trapped-gas standpipe system installed near the oxidizer tanks, and is similar to that in Fig. 12.78b. Interestingly, a similar suppressor does not appear in the fuel system.

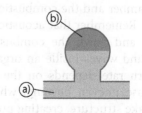

Fig. 12.77 Schematic of a pogo suppressor, also known as an accumulator. Source: http://www.xmission.com/ ~ jwindley/techsvpogo.html

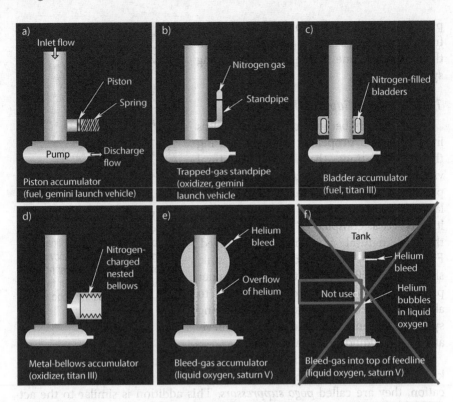

Fig. 12.78 A number of ways to address the pogo oscillations. The last method was not used in favor of method e. Source: [5].

12.5.2 Resonant Burn Oscillations

Next, we discuss the phenomenon of *resonant burn*. This event is similar to pogo in its vibration production, but resonant burning occurs in solid motors, not liquids. You can think of resonant burn as an interaction between the acoustics of the internal volume of an SRM combustion chamber and the combustion rate of the solid propellant.

Remember that acoustic waves are pressure waves, and the waves move up and down the combustion chamber in a manner similar to that of sound waves inside an organ pipe (or any wind musical instrument). The burn rate depends on the internal pressure, so transient pressure waves cause uneven burning, which in turn produces thrust transients, which shake structure, creating pressure waves, and so on.

Figure 12.80 shows a time history of acceleration for a resonant burn condition, where the peak accelerations of about 4 *g* are added to the steady or quasi-static acceleration. We saw in Chapter 11 that the vibrations on board a Taurus LV imposed *significant* axial loads that were attenuated with the use of a payload isolation system. Without a practical way to

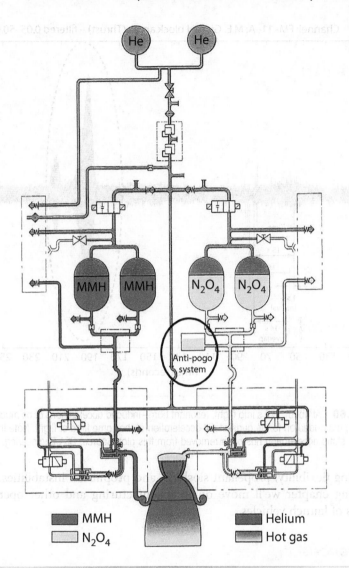

He   He

MMH   MMH N_2O_4 N_2O_4

Anti-pogo system

■ MMH   ■ Helium
■ N_2O_4 ■ Hot gas

Fig. 12.79 Ariane 5 Aestus engine diagram showing a pogo suppressor near the N_2O_2 (dinitrogen tetroxide) oxidizer tanks. Why there is no similar item in the fuel system is not known. Source: SNECMA.

prevent resonant burn, one must build the LV and payload to survive it, or provide payload isolation as discussed in that chapter.

12.6 Summary

We've seen how to guide a launch vehicle and how to control and stabilize it. We have learned about other effects that can cause problems with control,

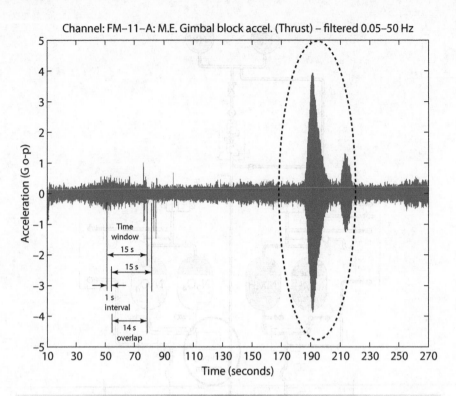

Channel: FM–11–A: M.E. Gimbal block accel. (Thrust) – filtered 0.05–50 Hz

Fig. 12.80   At about 190 s into flight, resonant burn–induced accelerations of as much as 4 g_0 (or more) are added on to the quasi-static acceleration at the same time of flight. Note that quasi-static acceleration has been removed from this plot. Source: CSA Engineering.

including flexibility, propellant sloshing, and propulsion instabilities. In the following chapter we'll move on to manufacturing and other operational aspects of launch vehicles.

References

[1] Abramson, H. N., "The Dynamic Behavior of Liquids in Moving Containers with Application to Space Vehicles," NASA SP-106, 1966.

[2] Ball, K. J., and Osborne, G. F., *Space Vehicle Dynamics*, Oxford University Press, London, 1967.

[3] Barret, C., "Review of Our National Heritage of Launch Vehicles Using Aerodynamic Surfaces and Current Use of These by Other Nations," NASA TP-3615, April 1996.

[4] Barrowman, J. S., "The Practical Calculation of the Aerodynamic Characteristics of Slender Finned Vehicles," M.S. Dissertation, Catholic University of America, Washington, DC, Mar. 1967.

[5] Blestos, N.A., "Launch Vehicle Guidance, Navigation, and Control," *Crosslink*, Winter 2003/2004, The Aerospace Corporation.

[6] Bisplinghoff, R., and Ashley, H., *Principles of Aeroelasticity*, 2nd ed., Dover Publications, Mineola, NY, 1962.

[7] Blakelock, J. H., *Automatic Control of Aircraft and Missiles*, 2nd ed., Wiley-Interscience, 1991.

[8] Boeing, "Delta II Launch Services," 05619VEU6, McDonnell Douglas Aerospace, Space Transportation Division, revised 12 July 1996.

[9] Boswinkle, R. W., "Aerodynamic Problems of Launch Vehicles," NASA SP-23, 1962.

[10] Curtis, H. D., *Orbital Mechanics for Engineering Students*, Elsevier, Oxford, UK, 2014.

[11] Diebel, J., "Representing Attitude: Euler Angles, Unit Quaternions, and Rotation Vectors," 20 Oct. 2006, https://www.astro.rug.nl/software/kapteyn-beta/_down loads/attitude.pdf [retrieved 4 May 2019].

[12] Dodge, F.T., *The New Dynamic Behavior of Liquids in Moving Containers*, Southwest Research Institute, San Antonio, 2000.

[13] Dubois, T., *Aviation Week*, Nov. 23, 2016. http://aviationweek.com/space/esa-s-mars-lander-crash-sparked-bad-inertial-data.

[14] Falangas, E. T., *Performance Evaluation and Design of Flight Vehicle Control Systems*, Wiley-IEEE Press, New York, NY, 2015.

[15] Greenberg, S., "Launch Vehicle Structural Analysis," UCLA short course, Feb. 2011.

[16] Heitchue, R., *Space Systems Technology*, Reinhold, Grigny, France, 1968.

[17] Huzel, D. K., and Huang, D. H., *Modern Engineering for Design of Liquid-Propellant Rocket Engines*, AIAA, New York, NY, 1992.

[18] Kasper, J., and Semrau, G., "A Power System – Reimagined", *Aerospace America*, Oct 2014, pp. 18–19.

[19] Koelle, H. H., *Handbook of Astronautical Engineering*, McGraw-Hill, New York, NY, 1961.

[20] Lange, O. H., and Stein, R. J., *Space Carrier Vehicles*, Academic Press, New York, NY, 1963

[21] Martin Company, "Vanguard Satellite Launching Vehicle—An Engineering Summary," Report 11022, 1960.

[22] Meyer, R. X., *Elements of Space Technology for Aerospace Engineers*, Academic Press, New York, NY, 1999.

[23] NASA, "Saturn V SA-503 Flight Manual," NASA TM-X-72151, 1968.

[24] NASA, "Saturn V Launch Vehicle Flight Evaluation Report AS-502 Apollo 6 Mission," NASA TM-X-61038, 1968.

[25] NASA, "Thrust-Vector Control Requirements for Large Launch Vehicles with Solid Propellant First Stages" NASA TN-D-4662, 1968.

[26] NASA, "Description and Performance of the Saturn Launch Vehicle's Navigation, Guidance, and Control System," NASA TN D-5869, July 1970.

[27] NASA, "On the Shoulders of Titans—A History of Project Gemini," NASA SP-4203B, 1977.

[28] NASA, "Chariots for Apollo-A History of Manned Lunar Spacecraft," NASA SP-4205, 1979.

[29] NASA, "Introduction to the Theory and Application of Quaternions," Parts 1 and 2, NESC Academy, https://nescacademy.nasa.gov [retrieved 4 May 2019].

[30] Rakoczy, J., "Fundamentals of Launch Vehicle Flight Control System Design," NESC Academy, https://nescacademy.nasa.gov [retrieved 4 May 2019].

[31] Seifert, H., and Brown, K., *Ballistic Missile and Space Systems*, John Wiley & Sons, New York, 1961.

[32] Thomson, W. T., *Introduction to Space Dynamics*, John Wiley & Sons, New York, 1961.

[33] Wie, B., Du, W., and Whorton, M., "Analysis and Design of Launch Vehicle Flight Control," AIAA paper no. 2008-6291, presented at the AIAA GN&C Conference, 18–21 Aug. 2008, Honolulu, HI.

Further Reading

Greensite, A. L., *Analysis and Design of Space Vehicle Flight Control Systems*, ISBN 0876715544, Spartan, New York, 1970.

Kim, C. H., *Commercial Satellite Launch Vehicle Attitude Control Systems Design and Analysis*, CHK, Fountain Valley, CA, 2007.

Pinson, L. D., and Leonard, H. W., "Longitudinal Vibration Characteristics of 1/10-Scale Apollo/Saturn V Replica Model," NASA TND-5159, 1969.

Rubin, S., "Prevention of Coupled Structure-Propulsion Instability," NASA SP-8055, 1970.

Seifert, H. S., *Space Technology*, John Wiley & Sons, New York, 1959.

12.7 Exercises: Vibration and TVC Analysis

Please use SI units for all calculations. *Provide no more than four significant digits for your calculations and answers; round numbers as necessary.*

1. **Launch Vehicle Stability.** We've seen that a TVC system can be used to steer a vehicle and keep it pointed in the "right" direction. As shown in the figure that depicts a side and top view of the same vehicle, the "small" engine gimbal angles are δ_1 and δ_2. The engine's thrust is T, and the distance between the engine gimbal and the vehicle's CM is the length l. The pitch angle θ_1 is rotation about $\mathbf{b_1}$, the yaw angle θ_2 is rotation about $\mathbf{b_2}$, and the roll angle is θ_3 about $\mathbf{b_3}$. The mass moments of inertia about the CM are J_{pitch}, J_{yaw}, and J_{roll} about the $\mathbf{b_1}$, $\mathbf{b_2}$, and $\mathbf{b_3}$ axes, respectively. (Symmetry makes the first two equal, so set $J_{\text{pitch}} = J_{\text{yaw}} = J$.) The vehicle is in a vacuum, so there are no aerodynamic forces.

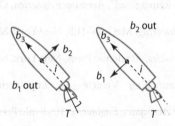

a) Assuming small angles, so that $\sin \delta_k \approx \delta_k$, write down the vector expression for the three components of thrust $\mathbf{T}$ of this vehicle/coordinate system, in other words

$$\mathbf{T} = (\ldots)\mathbf{b_1} + (\ldots)\mathbf{b_2} + (\ldots)\mathbf{b_3}$$

b) Determine expressions for the three torques or moments M_1, M_2, and M_3 that the engine produces in all three axes:

$$\mathbf{M} = M_1\,\mathbf{b_1} + M_2\,\mathbf{b_2} + M_3\,\mathbf{b_3}$$

c) Assume that we have a feedback control system that commands the engine's control angles to be proportional to the angular rates $\omega_1 = d\theta_1/dt$ and $\omega_2 = d\theta_2/dt$ of the vehicle: $\delta_1 = K\,\omega_1$ and $\delta_2 = K\,\omega_2$, where K is a gain constant. Write down the Euler equations (see the following box) for the assumed inertias and a *nonspinning* vehicle ($\omega_3 = 0$), simplified as much as you can. Hint: One of the three equations can be used to make the other two very simple.

d) Determine whether the magnitude of K should be *positive* or *negative* ($K > 0$ or $K < 0$) to ensure that the vehicle is stable in both axes. Do this by separating variables in one equation and integrating. The variable K is known as the "rate feedback gain."

The Euler equations are derived using the conservation of angular momentum equation: $\mathbf{M} = d\mathbf{H}/dt$, where the derivative is in an inertial frame, but we are evaluating it in the set of body-fixed vehicle axes. For a body with a moment of

inertia matrix $I = \begin{bmatrix} A & 0 & 0 \\ 0 & B & 0 \\ 0 & 0 & C \end{bmatrix}$, this analysis leads to the following set of

three coupled, nonlinear first-order differential equations:

$$M_1 = A\dot{\omega}_1 + (C - B)\omega_2\omega_3$$
$$M_2 = B\dot{\omega}_2 + (A - C)\omega_1\omega_3$$
$$M_3 = C\dot{\omega}_3 + (B - A)\omega_1\omega_2.$$

For bodies with an axis of symmetry, such as we have, two of the three inertia values are equal.

2. **Time to Double Calculation.** What is the worst-case condition for time to double? Determine your design's time to double at the worst-case condition before main engine cutoff (MECO).

3. **Launch Vehicle in Atmosphere.** Now consider a TVC system to steer a vehicle's pitch angle θ in an atmosphere. (Assume $\theta =$ angle of attack for this problem.) As shown in the figure, the small engine gimbal angle is δ. The engine's thrust is T, and the distance between the engine gimbal and the vehicle's CM (point C) is the length l. The mass moment of inertia of pitching about the CM is J. Note that the *mass* moment of inertia J is different from the *area* moment of inertia I.

We have seen that when traveling through an atmosphere, a vehicle encounters aerodynamic forces that tend to cause instability in pitch. From the control standpoint, the basic concern is the lateral force of the air, which tends to rotate the

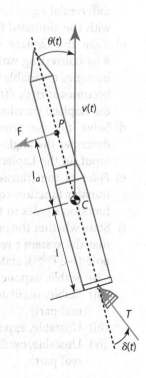

vehicle around its mass center. If the vehicle's centerline is not aligned with the direction in which the mass center is traveling, as shown in the figure, the angle of attack θ produces a side force $F = F_\theta \theta$, where $F_\theta = \partial F / \partial \theta$ is the force *gradient*. The side force F may be assumed to act at the center of pressure, point P. The side force has a tendency to cause the vehicle to pitch whenever the center of pressure P is in front of the mass center C.

The angular acceleration of the vehicle about C is $\alpha = \dot{\omega} = \ddot{\theta}$. (Don't confuse this α with the one we normally use for angle of attack.) Angular momentum balance says that $M = J\alpha$, where the moment (torque) $M = F\, l_a$. (We used M for moment so as not to confuse with thrust T.) Here l_a is the distance between C and P.

The main objective of this control system is to provide a stabilizing action to counteract the destabilizing effect of the side force. One way to do this is to gimbal the engine an angle δ to provide side thrust.

a) We've seen (previous problem) that if there is no steady rotation, the Euler equations simplify to be $J\ddot{\theta} = \sum$(external torques) in each of the three axes. Write the moment (torque) differential equation to relate rotational motion caused by thrust T, gimbal angle $\delta(t)$, angle of attack $\theta(t)$, and the other given parameters in the figure.

b) Assume small angles, so that $\sin \theta \approx \theta$. Write down the (now linear) differential equation for $\theta(t)$, with the forcing function associated with the gimbaled thrust on the right-hand side.

c) Take the Laplace transform of the linear differential equation for θ by converting variables to capital letters, and each time derivative becomes a variable multiplied by the Laplace variable s. Then $\ddot{\theta}$ becomes $s^2 \Theta(s)$, $\delta(t)$ becomes $\Delta(s)$. Assume zero initial conditions for the Laplace transforms. Write down the resulting equation.

d) Solve for the *transfer function* $\Theta(s)/\Delta(s)$. This transfer function describes the angle-of-attack response to the gimbal angle as an input, in the Laplace domain.

e) *Poles* of the transfer function occur where the denominator of the transfer function equals zero. Give the value(s) of the transfer function's poles in the form of an equation something like $s = (\ldots)$.

f) State whether the pole(s) is/are real, imaginary, or complex. Assuming that the system's response is proportional to the exponential of the pole ($\sim e^{(\text{pole})}$), state whether this system is one of these:

   (i) Stable, exponential decrease (all poles are negative real)

   (ii) Stable, oscillatory decrease (all poles are complex with negative real part)

   (iii) Unstable, exponential increase (any poles are positive real)

   (iv) Unstable, oscillatory increase (any poles are complex with positive real part)

Note that you must consider *all* of the poles, if more than one. If any single pole is unstable, the entire system is unstable, even if other roots are stable.

g) To add stability to the system, feed the pitch angle $\theta(t)$ back to the engine gimbal controller angle $\delta(t)$, so that $\delta(t) = k\,\theta(t)$, and k represents a feedback *gain*. Using this relation, the differential equation in part 3b becomes a differential equation in only θ. Write the equation. The variable k is known as the *proportional feedback gain* (related to position, not rate as in the previous problem).

h) Assume a solution of the equation in the form $\theta(t) = \theta_0\,e^{ist}$, where θ_0 is an assumed amplitude, $i = \sqrt{-1}$ is the imaginary unit, and s is a *real* parameter. Plug this solution into the equation you got for part 3g, and divide out the ($\theta_0\,e^{ist}$) that occurs in each term. The remaining equation is known as the *characteristic equation*. Write the characteristic equation, and find the value(s) of k that make the roots $= 0$. This value of k is the stability boundary. Provide the dimensions (units) of k.

Congratulations! If you've completed the previous tasks, you've stabilized an unstable vehicle without doing any controls analysis or root locus analyses! The simplified analysis in this problem does not include the dynamics of the TVC servo controlling the gimbal angle, nor does it incorporate any of the lags associated with measurement of states or computing needed gimbal angles. Nevertheless, it provides a reasonable first look at stabilization.

Chapter 13 / Launch Vehicle Manufacturing

Moving forward from stability, control, and instabilities, we show the design roadmap in Fig. 13.1 with the highlighted signpost indicating the information to be covered in this chapter.

13.1 Launch Vehicle Fabrication

Launch vehicles are often pushing the state of the art of manufacturing techniques; they are also costly. In addition, they must reliably carry and inject to orbit an even more expensive satellite or payload. For these reasons, there are lots of recommended and proprietary practices for fabrication to prevent problems later; as designers, you must observe them. Most manufacturers have a "tribal knowledge" of how things must be done; in most cases there is an engineering or operational basis for them. The data

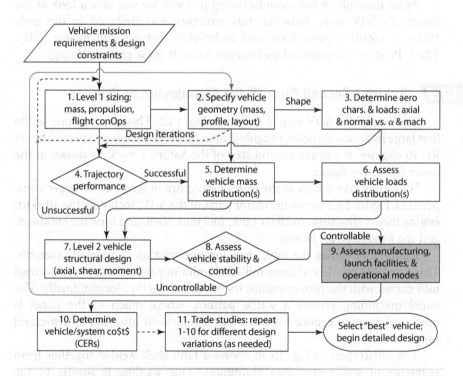

Fig. 13.1 This chapter will consider the manufacturing and logistics associated with producing space launch vehicles.

are hopefully recorded in the manufacturer's archives in the form of something like a handbook, or the equivalent electronic or online file.

Many, but not all, of the space launch vehicle (SLV) manufacturers began as aircraft suppliers, and for them, a set of procedures for metals and metallic structures such as bending, welding, machining, drilling, tapping, and so forth was already available. Of course it grew with the incorporation of new metallic materials (e.g., aluminum-lithium alloys augmenting regular aluminum) and new fabrication techniques (e.g., friction-stir welding coming online alongside conventional welding techniques).

After World War II and into the Cold War period, there was a large push for higher performance structures and also reduced weight. This led to the incorporation of *composite* materials for fabrication purposes. Composites are materials that are formed using a structural fiber (such as fiberglass, carbon fiber, or aramid fiber) embedded into an adhesive matrix (like epoxy or thermoplastic materials). For composites, practices needed to be followed regarding storage, layup schedule (how thick, and in what direction the fibers are laid), handling, curing, and so on. Along with these practices, new methods of inspection had to be developed for very large parts and assemblies in order to ensure that they were fabricated properly and would survive the environments for which they were designed.

As an example of the manufacturing process, we will take a look at the Saturn I's S-IV step. Although this structure was designed in the early 1960s, it required procedures and techniques that are still in use today. The S-IV step was provided by Douglas Aircraft (now part of Boeing).

13.2 Saturn I Second Step (S-IV) Manufacturing Process

The assembled S-IV step is shown in Fig. 13.2. This step was one of the first large hydrolox vehicles, roughly 7 m in diameter, and was powered by six RL-10 engines. It was the second step of the Saturn I stack, as shown in the lower left of the figure.

We'll now take a look at the overall structure in terms of its larger components. Figure 13.3 shows the major parts of the S-IV, including the aft skirt, engine thrust structure, oxidizer tank, fuel tank shell, and forward bulkhead, and the interstage on its top.

Figure 13.4 shows the tank shell location and its completed assembly. This assembly consists of three milled aluminum panels that are roll-formed into curves with the proper radius, then welded together longitudinally. The panel machining creates a waffle pattern, where much of the panel is machined away to reduce mass without significantly affecting its structural properties.

The illustration in Fig. 13.4b shows a tank shell welded together from segments of waffle-machined aluminum. This waffling is similar to the triangular webbing seen in Fig. 7.55 in Chapter 7, and the machining process is shown in Fig. 13.5a. Before joining, the machined panels must

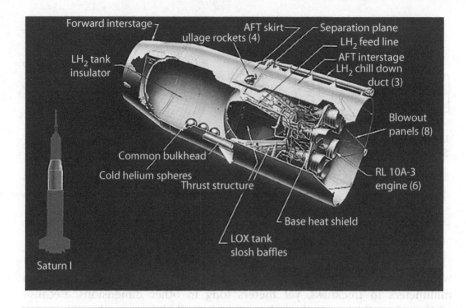

Fig. 13.2   The LH_2/LOx upper stage of the Saturn I. Source: NASA.

be roll-formed to the proper curvature. This roll-forming process is shown in Fig. 13.6.

Once roll-formed, the pieces must be welded together. The machining of the waffle pattern and the roll-forming operation may be seen in the top row of the tank shell manufacturing sequence shown in Fig. 13.7.

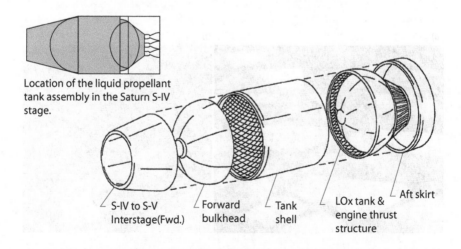

Location of the liquid propellant tank assembly in the Saturn S-IV stage.

S-IV to S-V Interstage(Fwd.)   Forward bulkhead   Tank shell   LOx tank & engine thrust structure   Aft skirt

Fig. 13.3   The main components of the S-IV propellant tank assembly. Source: [2].

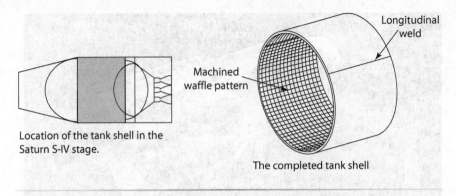

Location of the tank shell in the Saturn S-IV stage.

The completed tank shell

Fig. 13.4 S-IV tank shell layout and position in the step. Source: [2].

Large metal structures, such as domes and tank shells, are joined by a welding process; however, we've seen that many structures are just a few millimeters in thickness, yet meters long in other dimensions. Because of their low thickness, these large pieces will deform under their own weight. Yet, proper leak-free operations with structural integrity require extremely long, tight-tolerance welds. For example, the Saturn V S-IC step had 10 km (6.2 *miles*) of welds, and each centimeter was inspected.

The best practices to obtain weld quality are to:

• Clean components before welding.
• Use custom equipment to maintain part alignment and pressure during welding.
• Tightly control humidity and temperature during the weld process.
• Operate in a clean-room environment.

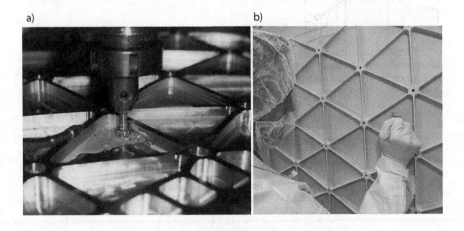

a) b)

Fig. 13.5 a) Isogrid panel being machined on a mill; b) isogrid panel inspection. Source: NASA.

Fig. 13.6   Orthogrid panel being roll-formed to obtain the proper curvature needed for welding.
Source: Revolution Machine Tools.

Figure 13.8 shows Saturn V bulkheads being prepared for welding. Note the size of the technicians in Fig. 13.8b. Moving pieces of metal like this around is not trivial!

The welding process has changed significantly since the original launch vehicles were fabricated in the 1950s and 1960s. During this period, a

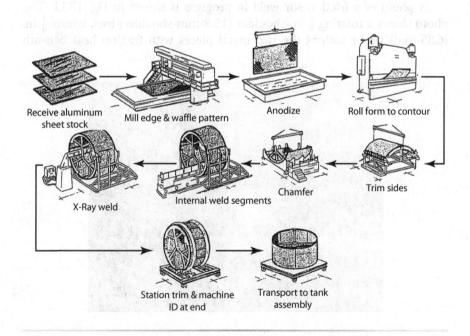

Receive aluminum sheet stock

Mill edge & waffle pattern

Anodize

Roll form to contour

X-Ray weld

Internal weld segments

Chamfer

Trim sides

Station trim & machine ID at end

Transport to tank assembly

Fig. 13.7   S-IV steps in the tank shell manufacturing sequence. Source: [2].

a) b)

Fig. 13.8 Saturn V tank domes. (a) assembled dome. (b) tank domes on welding fixtures in varying states of completion. Source: NASA.

machine designed to carefully apply weld metal at the joint between two metal pieces was used to weld pieces together. This process is shown in Fig. 13.9.

The previous style of welding has been pretty much replaced wherever possible with *friction stir* welding, as shown in Fig. 13.10. A *rotating-shoulder* tool is inserted while spinning between two sheets of metal to be joined. The heat created by friction softens the metals to the consistency of butter, allowing them to be "stirred" together as shown.

A photo of a friction-stir weld in progress is shown in Fig. 13.11. The photo shows a rotating $\frac{5}{8}$-in.-shoulder (15.8-mm-shoulder) tool, whose $\frac{1}{4}$-in. (6.35-mm) probe softens the two metal pieces with friction heat beneath

Fig. 13.9 Saturn tank welding, the old way. Source: NASA.

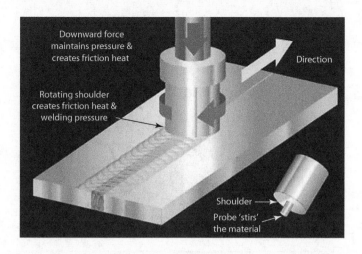

Fig. 13.10 The new way of welding, *friction-stir welding*. Source: [6].

the surface and "stirs" the two sheets' metals together with a solid joint. (The terms *shoulder* and *probe* are explained in Fig. 13.10).

Figure 13.12 shows a completed friction-stir weld, the result of stirring with a rotating probe forced in with 22.5–45 kN (5,000–10,000 lb$_f$) of force. The probe is turning at 180–300 RPM, and travels at a rate of 8.9–12.7 cm/min (3.5–5 in./min). The friction from rotation heats the materials to be joined and "plasticizes" them, so no actual melting occurs, and the crystalline state doesn't change. This means that after the stir weld is complete, heat treatment is *not needed*. In addition, the result is a more defect-free weld. This process saves a lot of time and expense.

NASA's Space Launch System (SLS) uses tanks that are friction-stir welded together. A time-lapse video of the process of stacking and joining the tank cylinder pieces is available on the Internet.

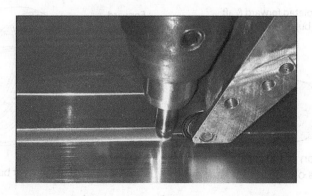

Fig. 13.11 Friction-stir welding in progress. Source: Boeing.

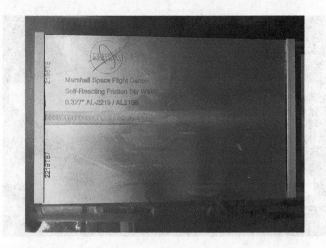

Fig. 13.12 The result of friction-stir welding is a joint of great strength that requires no heat treatment afterwards, saving considerable expense. Source: Edberg.

Figure 13.13 shows the location and makeup of the S-IV forward and aft bulkheads. Figure 13.14 shows the manufacturing process for these two bulkheads.

Next we look at the *common bulkhead* between the two tanks, which is shown in Fig. 13.15. As mentioned in Chapter 7, the S-IV, S-IVB, and S-II steps on the Saturn IB and Saturn V used a common bulkhead to separate their liquid oxygen propellant from the liquid hydrogen fuel, which must provide insulation between $-252°C$ ($-422°F$) LH_2 and $-182°C$ ($-295°F$) LOx, a very large $70°C$ ($127°F$) difference.

Figure 13.16 shows the common bulkhead manufacturing sequence. Note that the common bulkhead is a sandwich structure with insulating honeycomb material bonded between two aluminum domes. A photo of the

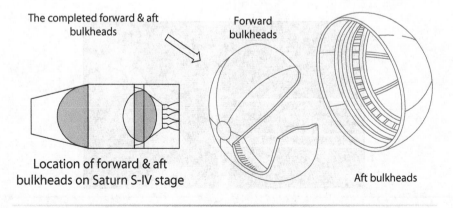

Fig. 13.13 S-IV forward and aft bulkhead cutaway drawings. Source: [2].

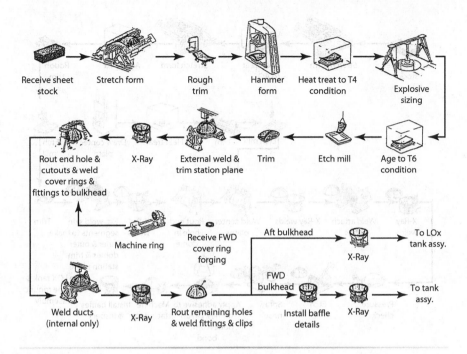

Fig. 13.14   The S-IVB forward and aft bulkhead manufacturing sequence. Source: [2].

cross-section of the S-II common bulkhead may be found in Fig. 7.27, Chapter 7.

Figure 13.17 shows the location of the engine thrust structure, a skin-stringer assembly, and its completed appearance. Figure 13.18 shows the location of the engine thrust structure, a skin-stringer assembly, and its

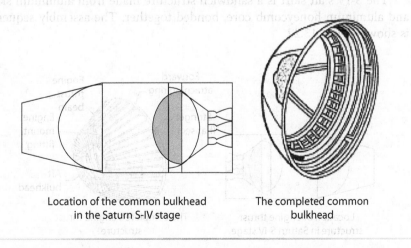

Location of the common bulkhead
in the Saturn S-IV stage

The completed common
bulkhead

Fig. 13.15   The S-IV common-bulkhead location on the stage and the completed assembly.
Source: [2].

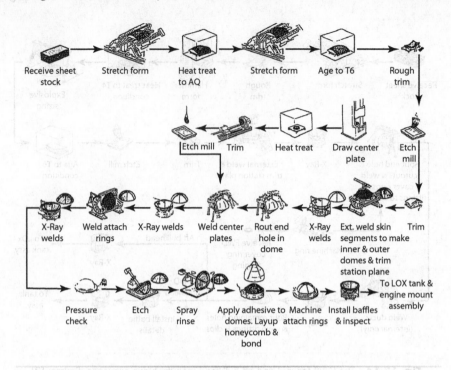

Fig. 13.16 The S-IV common-bulkhead manufacturing process. Source: [2].

completed appearance. Figure 13.19 shows the location of the forward skirt (or interstage) and its completed appearance. Figure 13.20 shows the manufacturing process used for the forward skirt (or interstage). The S-IV's aft skirt location and completed configuration are shown in Fig. 13.21.

The S-IV's aft skirt is a sandwich structure made from aluminum skins and aluminum honeycomb core, bonded together. The assembly sequence is shown in Fig. 13.22.

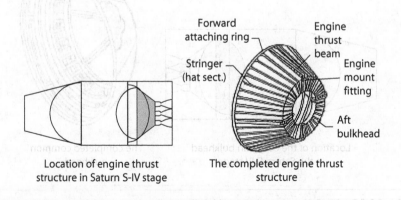

Fig. 13.17 S-IV thrust structure location and assembled configuration. Source: [2].

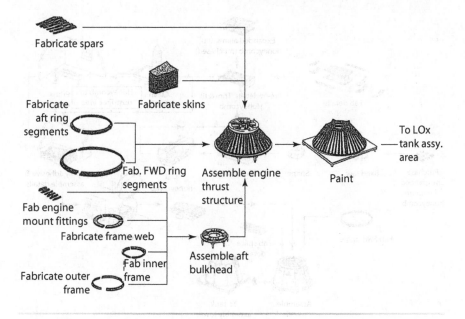

Fig. 13.18 S-IV thrust structure's manufacturing process. Source: [2].

The final piece of external structure is the S-IV's aft interstage structure, which is a sandwich structure made from aluminum skins and aluminum honeycomb core, bonded together. The completed configuration is shown in Fig. 13.23.

Figure 13.24 shows the manufacturing sequence for the S-IV's aft interstage structure. Figure 13.25 shows the S-IV LOx tank and thrust structure locations, and their principal parts.

The baffle in the LOx tank is an important measure to prevent sloshing. Figure 13.26 shows the S-IV LOx baffle manufacturing sequence.

Finally, the LOx tank and thrust structure are joined. Figure 13.27 depicts the assembly sequence joining the S-IV LOx tank and its thrust structure.

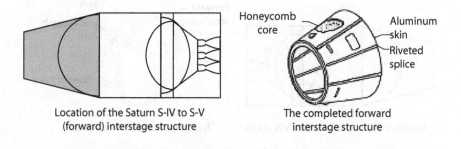

Fig. 13.19 S-IV forward skirt/interstage. Source: [2].

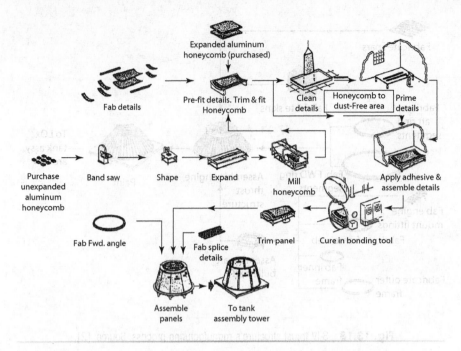

Fig. 13.20 S-IV forward skirt manufacturing sequence. Source: [2].

Domes and bulkheads are attached to tanks by *Y-rings*, which allow the pieces to be joined without large, heavy lap joints. Figure 7.20 in Chapter 7 shows a Y-ring sample and how it was used in the Saturn V to connect domes to the tank walls.

Figure 13.28 shows the tank assembly sequence for the S-IV. Remember that these steps must be executed with a part that has a 7-m diameter!

The S-IV has a heat shield to protect its rear end from the radiation from its exhaust plume. Figure 13.29 shows the heat shield's assembly sequence.

The last step is the S-IV stage final assembly sequence, shown in Fig. 13.30.

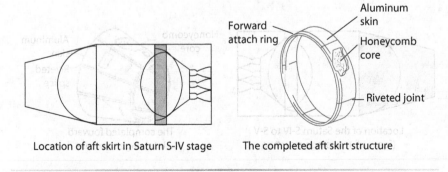

Fig. 13.21 S-IV aft skirt location and completed configuration. Source: [2].

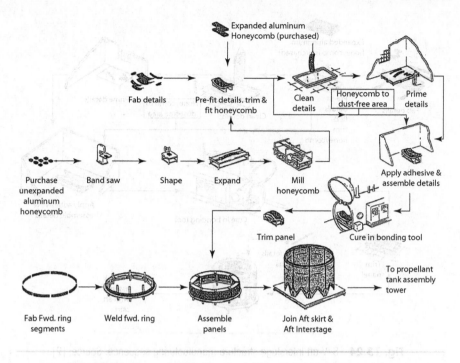

Fig. 13.22 The assembly sequence for the S-IV's aft skirt. Source: [2].

The result of this assembly process, a completed S-IV step, is shown in Fig. 13.31.

The manufacturing sequence for the Saturn V's S-IVB stage is shown in Fig. 13.32. This was the successor to the S-IV and had a single J-2 engine to replace the six H-1 engines on the S-IV, as well as other improvements.

Figure 13.33 shows part of the Atlas V assembly area at the ULA plant in Decatur, Alabama. The vehicle in the center of the photo has been *scarred* (structure added) for the addition of solid rocket motors to its girth. Just behind is another Atlas V with the Russian RD-180 engines attached.

Figure 13.34 shows two views of SpaceX's Falcon 9 production line.

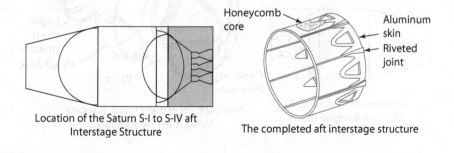

Fig. 13.23 S-IV aft interstage structure location and completed configuration. Source: [2].

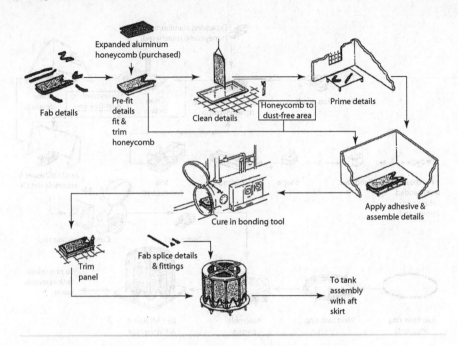

Fig. 13.24 S-IV aft interstage structure manufacturing sequence. Source: [2].

13.3 Composite Structure Fabrication

A *composite* is a part or structure that is a combination of structural fibers, such as carbon/graphite, fiberglass, or aramids, and an adhesive. Layers or *plies* of fibers are combined with a so-called *matrix* or adhesive glue, or it may already be combined with adhesive as a *prepreg*, or preimpregnated material. The layers are stacked on top of each other inside of a *tool*

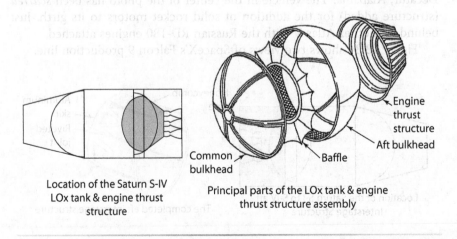

Location of the Saturn S-IV LOx tank & engine thrust structure

Principal parts of the LOx tank & engine thrust structure assembly

Fig. 13.25 S-IV LOx tank and thrust structure. Source: [2].

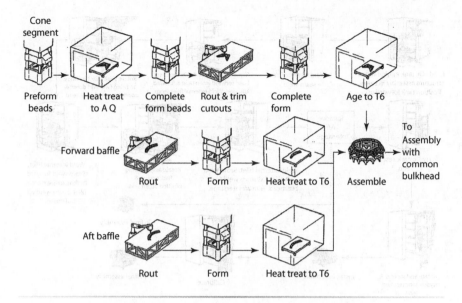

Fig. 13.26 S-IV LOx baffle manufacturing sequence. Source: [2].

(a mold or part) having the desired shape of the *outer mold line*. The stack of materials is bagged and then placed into an *autoclave* (a fancy name for an oven) for curing. The assembly is then cooled down, and the part is removed from the tool.

Figure 13.35 shows the process of laying down the layers of fibers and then using a hand roller to distribute the matrix (epoxy resin in this case) into the fibers. A prepreg material avoids the hand process and has much

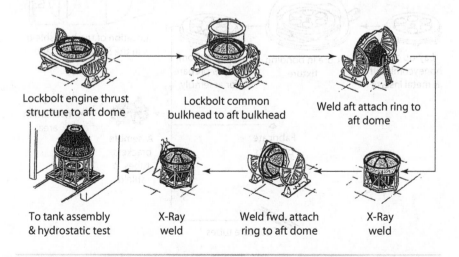

Fig. 13.27 Assembly sequence for S-IV LOx tank and thrust structure. Source: [2].

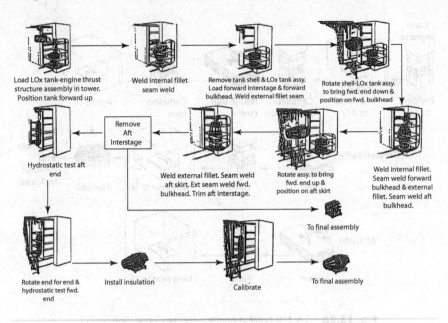

Fig. 13.28 S-IV tank assembly process. Source: [2].

more uniform matrix distribution, usually resulting in a reduced-mass part compared to a hand layup.

A composite's *stacking sequence* provides the fabricator with the required layer-by-layer sequence of the composite material layers. Note that not only

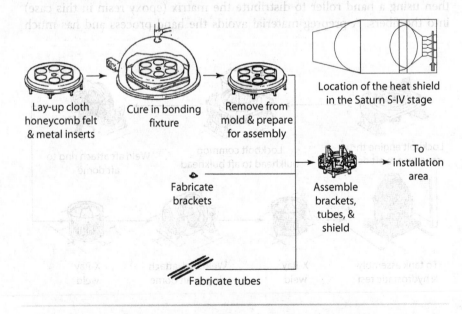

Fig. 13.29 S-IV heat shield (top right), and its assembly sequence. Source: [2].

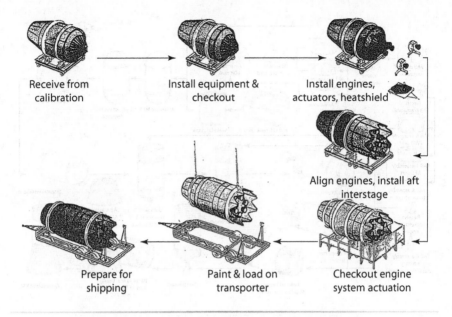

| Receive from calibration | Install equipment & checkout | Install engines, actuators, heatshield |

Align engines, install aft interstage

| Prepare for shipping | Paint & load on transporter | Checkout engine system actuation |

Fig. 13.30 S-IV stage final assembly sequence. Source: [2].

are the fibers specified, but also the direction or orientation of the fibers. This is done in order to get the desired structural properties with a composite, something that cannot be done with a metallic structure.

Fig. 13.31 The completed S-IV step, the upper stage of the Saturn I launch vehicle. Note the size of the technicians in the photo for scale. Source: NASA.

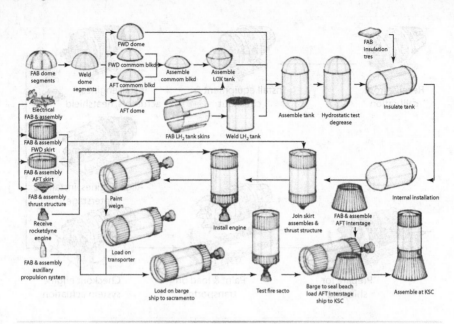

Fig. 13.32   The manufacturing sequence for the S-IVB step on the Saturn V. Source: [7].

As an example, the stacking sequence might be given as $[0_3/90_2/45/-45_3/Z_c]_s$. In this sequence, the *numbers* represent the orientation angle of the fibers with respect to a defined reference direction. The *subscripts*

Fig. 13.33   Atlas V production line in Decatur, Alabama. Source: [4].

Fig. 13.34   Two views of the Falcon 9 V1.1 production line in Hawthorne, California.
Source: SpaceX.

represent number of plies or layers to be put down. The term Z_c indicates the layup will have a honeycomb core, and finally the s subscript indicates the layup is symmetric; in other words, the bottom half is a mirror image of the top half. A diagram showing the cross-section of this example is provided in Fig. 13.36.

Usually designers employ composites to make parts with compound curves that would be difficult to fabricate from metals, such as payload fairings, aerodynamic shapes, and the like. Another reason to use composites is to reduce mass, because most composites have a higher specific modulus, meaning a structure made with composites would have less mass than a metallic structure with the same stiffness.

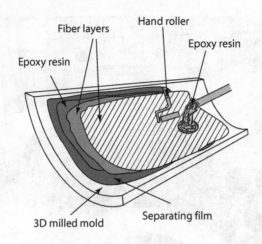

Fig. 13.35 The process used to lay up a composite part, before it's cured in an autoclave. Source: [3].

Figure 13.37 shows the major steps in forming a composite part, in this case, a payload fairing. Figure 13.37a shows the composite materials being carefully placed upon a fabrication tool; Fig. 13.37b shows the tool completely covered. Figure 13.37c shows how a bag is wrapped around the tool and composite materials. Next, the assembly is placed into an autoclave (Fig. 13.37d).

Once inside the autoclave with its door sealed, air is withdrawn from the *inside* of the bag, and pressure is simultaneously applied inside the autoclave but *outside* of the bag, which forces the composite to have complete contact with the tool. The heat is then turned on for several hours, and the stack is allowed to cure until the resin has solidified. Once the part has cooled down, the molded part is removed from the tool, ready to be trimmed and fitted with mounting hardware. Finally, after more attachments, the finished part is complete. These steps are shown in Fig. 13.38.

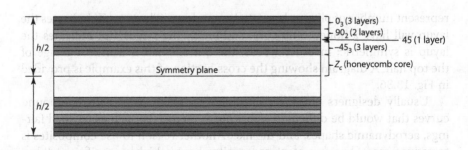

Fig. 13.36 A composite cross-section corresponding to $[0_3/90_2/45/-45_3/Z_c]_s$ layup schedule.

Fig. 13.37   A composite layup process. a) Composite materials applied to fabrication tool. b) The tool completely covered. c) The assembly is enclosed within a bag. d) The autoclave will contain the bagged part, where curing will begin. Source: (a–c) McDonnell Douglas; (d) [5].

The result of the molding process is a part that is ready to be prepared for flight. Instead of hundreds of metal parts fastened together with rivets or panels machined, bent, and then welded together, this is a monolithic part with minimal part count. The only drawback is that a molded composite part must be built with hard points in order to attach to other hardware or mounting.

This process is known as *hand layup* because it is done manually by technicians. Two other ways to build a composite structure are *fiber placement* and *filament winding*. Fiber placement uses a machine to apply composite fibers and matrix in varying directions and the placed composite fibers are immediately cured with a laser. Filament winding is commonly used to make cylindrical pressure vessels by wrapping unidirectional fibers around a *mandrel* before the curing process. The winding process allows the fibers to be applied at any desired angle and results in a very lightweight part because all of the composite fibers are "working" in the direction where maximum strength is needed. A filament winding operation is shown in Fig. 13.39.

At the present time, many vehicles utilize composite parts for intertanks, interstages, payload fairings, and attach fittings. Metallic liquid propellant

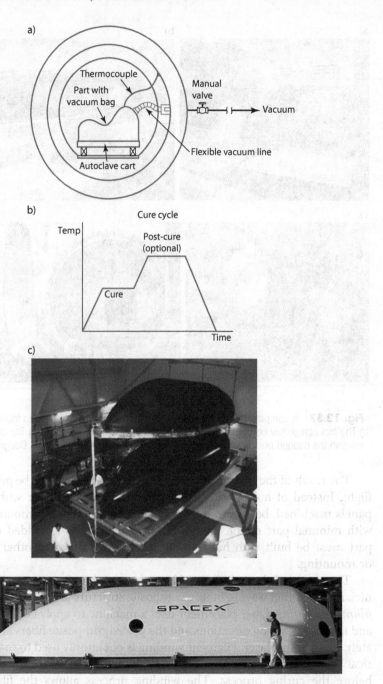

Fig. 13.38 Composite layup process, continued. a) The part, enclosed in a vacuum bag, is placed in the autoclave. Vacuum is used to evacuate the bag, and elevated temperature is applied. b) The cure cycle includes elevated temperatures for several hours to cure the composite. c) The cooled-off molded part is removed from the tool, edges cleaned, and mounting hardware attached. d) Finished composite payload fairing, ready for integration. Source: a) & b) [5] c) McDonnell Douglas d) SpaceX.

Fig. 13.39 The filament winding of a part for the OmegA launch vehicle. Source: Northrop Grumman.

tanks are standard, although a few designs, such as Virgin Orbit's Launcher-One, have adopted composites for liquid propellant tanks as well. This is an attractive option because of potential savings of mass and fabrication costs, but designers must be careful when using composites along with cryogenic propellants. Many solid rocket casings are also made from composites.

13.4 Manufacturing: The Future

We cannot continue looking at vehicle manufacturing without mentioning the *additive printing* process, often referred to as *3D printing*. The process is called additive printing because it fabricates three-dimensional shapes by *adding* material, rather than *subtracting* it like typical shop operations (mill, lathe, drill), which remove materials to achieve the final shape. Additive printing has been used to produce prototypes rapidly, but can also be used for rocket parts.

Additive printing machines are able to handle metals as well as plastics and composite materials. In the launch vehicle business, additive manufacturing has been used primarily to build rocket engines, specifically the combustion chambers, out of metal. These have intricate geometries and ducting to properly introduce the propellants into the combustion chamber and burn to produce pressure and thrust. In the past, it was very expensive and time-consuming to produce these parts, but now it can be done in days instead of months. Additive manufacturing may also be used to produce tanks. Relativity Space claims the process speeds development through operations and hopes to be able to "print a new rocket design from raw material to flight

Fig. 13.40 A Lockheed Martin 3D-printed, 1.16-m (46-in.) diameter titanium dome intended for use in satellite propellant tanks. According to LM, the printing process reduced the total delivery timeline from 2 years to 3 months. Source: Lockheed Martin.

in less than 60 days." Printed tanks and a printed rocket engine may be seen in Figs. 13.40 and 13.41, respectively.

For more information on additive manufacturing, the reader is referred to the website http://www.3dprinting.com.

13.5 Vehicle Stacking and Assembly

After the major assemblies are complete, there are still many details to be worked. Launch vehicles have electrical lines and conduit for power and data, pressurization lines to each of the tanks, and a special set of explosive cutters for range safety. These lines are usually routed through a *systems tunnel* or ducting where they are protected from the launch environment. Figure 13.42a shows the systems tunnel (circled) for a Saturn V S-IVB upper step, and Fig. 13.42b shows one for Atlas V (tunnel cover installed).

Fig. 13.41 The Relativity Space 3D-printed tank (upper left) and rocket engine (lower left, underneath bottom mouth of tank), along with the 3D printing room on the right, said to be the largest 3D printer in the world. Source: Relativity.

a)   b)

Fig. 13.42 The systems tunnel on an LV is used to route wires, pressure lines, and range safety on the LV body. a) Saturn S-IVB; b) Atlas V (tunnel covered). Sources: NASA, Lockheed Martin.

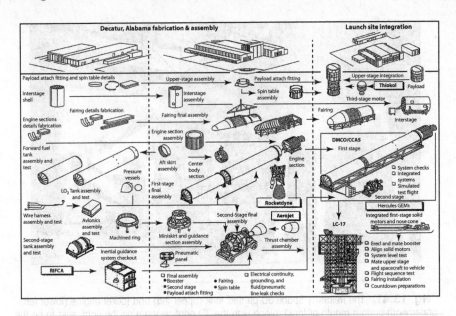

Fig. 13.43 Delta II production and launch site ops. Source: [1] (Fig. 05626VEU6)

Figure 13.43 gives an abbreviated version of the entire process of assembling a Delta II and its launch site integration with the launch pad and subsequent operations. Figure 13.44 shows the flow of hardware at the launch site.

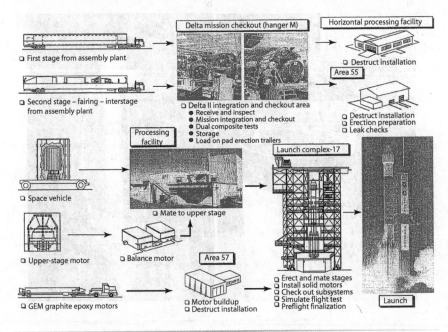

Fig. 13.44 Delta II hardware flow at launch site. Source: [1] (Fig. 05629VEU6)

a)   b)

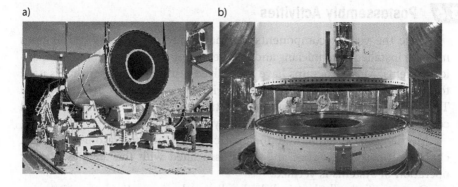

Fig. 13.45 a) Removal from horizontal fixture; b) vertical stacking. Joining the two segments involves inserting a steel pin into each of the holes along the circumference, 180 pins total. The cross-section of the joint may be seen in Fig. 16.14, Chapter 16. Source: NASA.

Many launch vehicles use solid rocket motors for thrust augmentation, and vehicles such as the Space Shuttle and Ariane 5 use very large solid rocket boosters. The latter are typically too large to ship in one piece, and instead are shipped as components that need to be stacked. Figure 13.45 shows two stages in the stacking of the Shuttle's SRB.

Next, we show the stacking sequence of the S-IC step of the Saturn V in Fig. 13.46. Starting with the aft fairing and engine support structure, we see the RP-1 tank, the intertank ring, the LOx tank, and the upper skirt. The photo on the right shows the stack being moved. Note the size of the people on the bottom.

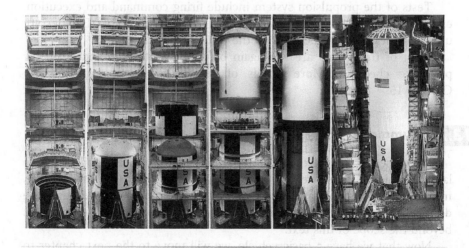

Fig. 13.46 The stacking sequence of the first step of the Saturn V. Note the people in the right photo on the bottom for scale. Source: NASA.

13.6 Postassembly Activities

Once the vehicle components are made, internal systems and engines must be installed. Technicians and engineers test the electrical and mechanical systems. Most programs have specialized testing equipment that is used to demonstrate the systems integrated with the vehicle part are acceptable. This includes calibration of the telemetry sensors and systems, electrical continuity checks, and radio frequency systems evaluation.

Mechanical, hydraulic, and pneumatic systems are tested to perform leak checks, propellant systems, and parts associated with the engines. Proper operation of systems is verified.

Once verified, all electrical, hydraulic, and pneumatic connections are made to the vehicle, and a physical examination is made. The environmental control system is checked for proper operation, electrical power is applied to the vehicle in sequential steps, and its distribution is monitored. The electrical checks include power distribution circuitry as well as the systems associated with heater power, destruct, sequencing, separation, and emergency detection.

Range safety systems have a complete end-to-end checkout including transmitting commands to the range safety command receiver and verifying the arm, cutoff, and destruct signals it generates.

Instrumentation is also inspected: data channels are identified, signal conditioners are adjusted for proper range of inputs, and telemetry and operational radiofrequency (RF) systems are verified.

Pressure and leak tests are carried out on propellant tanks and engines, propellant delivery systems and lines, and the control pressure system. Steering and attitude control are verified. Checks are made on the fill and drain operation and prevalve operation on all propellant systems.

Tests of the propulsion system include firing command and execution, engine shutdown, normal sequences, and malfunction cutoff. Engines are usually static-tested, sometimes on the vehicle, sometimes not.

Most of these tests are run again before static testing, and again for poststatic checkout. More details of vehicle testing are provided in Chapter 15.

13.7 Summary

We've seen some of the operations needed to fabricate and assemble launch vehicles. These vehicles are large, as much as 10 m in diameter, meaning their components are also large. When doing detailed design, the designer should take into account the handling of the parts, a topic that is beyond the scope of this text.

Now that we have a basic vehicle, we will move to the next chapter to discuss some internal details of launch vehicles and how they are integrated with the launch pad and its services.

References

[1] McDonnell Douglas, "Delta II Launch Services," Doc. 05619VEU6, 1996.

[2] Lange, O. H., and Stein, R. J., *Space Carrier Vehicles: Design, Development and Testing of Launching Rockets*, Academic Press, New York, NY, 1963.

[3] Ley, W., Wittmann, K., and Hallmann, W., *Handbook of Space Technology*. Wiley, Chichester, England, 2009.

[4] ULA, https://www.ulalaunch.com/docs/default-source/evolution/vulcan-centaur-overview-17may2018.pdf.

[5] Campbell, F. C., *Manufacturing Processes for Advanced Composites*, Elsevier Science, 2004.

[6] https://www.esabna.com/us/en/education/blog/what-is-friction-stir-welding-of-aluminum.cfm.

[7] http://heroicrelics.org/info/s-ivb/s-ivb-voverview/s-ivb-v-manu-flow-seq.jpg. Adapted from page 4 of the Saturn Third Stage, S-IVB Manufacturing. Located in the Saturn V Collection, Dept. of Archives/Special Collections, M. Louis Salmon Library, University of Alabama in Huntsville. Scan and adaptation by heroicrelics.org..

Further Reading

McDonnell Douglas Aerospace, "Delta II Launch Services," document 05619VEU6, 1966.

Launch Vehicle Systems and Launch Pad Facilities

oving forward from manufacturing, we show the design roadmap in Fig. 14.1 with the highlighted signposts indicating the information to be covered in this chapter: internal launch vehicle subsystems and interfaces with the launch pad.

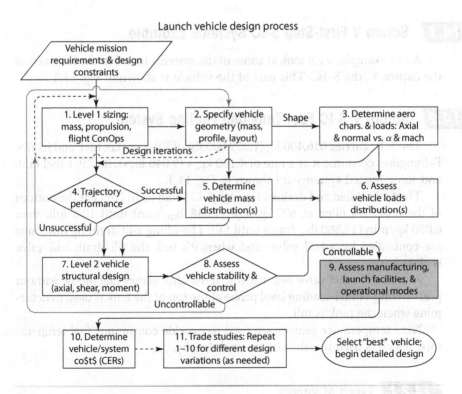

Launch vehicle design process

Fig. 14.1 This chapter will consider the internal systems in launch vehicles and their interface with the launch pad.

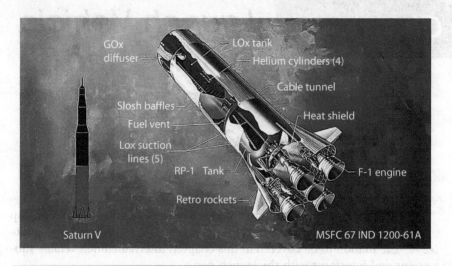

Fig. 14.2   The first step of the Saturn V. Source: NASA MSFC.

14.1   Saturn V First-Step S-IC Systems Example

As an example, we'll look at some of the systems found in the first step of the Saturn V, the S-1C. This part of the vehicle is shown in Fig. 14.2.

14.2   Saturn V S-IC Fuel Tank and Fueling System

The S-IC carries 616,400 kg (1,359,000 lb) of RP-1 rocket fuel, and its five F-1 engines consume it at a rate of 4,100 kg/s (9,040 lb/s). Its RP-1 fuel tank and some related systems are shown in Fig. 14.3.

The fuel is filled and drained through a 15.2-cm (6-in.) duct at the bottom of the tank. It's filled at 607 kg/min (1,338 lb$_m$/min) until 10% full, then 6,070 kg/min (13,380 lb$_m$/min) until full. The filling and draining functions are controlled by a ball valve, and when it's full, the fill/drain ball valve is closed.

A vent and relief valve near the top of the tank allows fuel inflow without pressurizing. A fuel-loading level probe at the top of the tank is used to determine when the tank is full.

Nine temperature sensors are used to provide continuous fuel temperature, which is used to calculate the fuel density.

14.2.1   Types of Valves

A number of types of valves are used for propellants; examples are shown in Fig. 14.4. A ball valve, as used in the fill and drain system, is shown in the bottom right of the figure. Other types of valves are also shown in the figure.

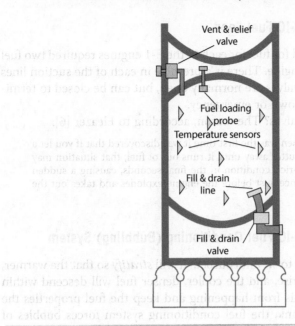

Fig. 14.3 S-IC fuel tank and services. Source: [10].

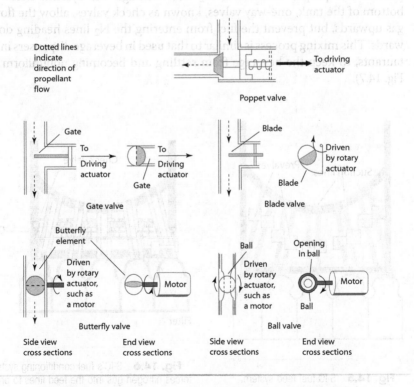

Fig. 14.4 Selection of valve types used for propellants. Source: [8].

14.2.2 Saturn V S-IC Fuel Feed

The prodigious need for fuel for each of the F-1 engines required two fuel suction lines for *each* engine. There is a prevalve in each of the suction lines (see Fig. 14.5). The prevalves are normally open, but can be closed to terminate fuel flow for shutdown or emergency.

Why *are* there prevalves? The reason, according to Eleazer [6]:

> It's ... not known when was the first time it was discovered that if you let a rocket engine just sputter away until it runs out of fuel, that situation may result in an oxidizer-rich condition in the final seconds, causing a sudden increase in performance, just before the engine explodes and takes out the whole vehicle.

14.2.3 Saturn V S-IC Fuel Conditioning (Bubbling) System

If the fuel is allowed to sit undisturbed, it will *stratify* so that the warmer, lower density fuel will rise, and the cooler, denser fuel will descend within the tank. To prevent this from happening and keep the fuel properties the same throughout the tank, the fuel conditioning system forces bubbles of gaseous N_2 (nitrogen) through feed lines. A schematic of the fuel conditioning system is shown in Fig. 14.6. Orifices located near the *check valves* provide the proper amount of N_2 flow through the system. Below the bottom of the tank, one-way valves, known as check valves, allow the flow of gas upwards, but prevent the fuel from entering the N_2 lines heading downwards. This mixing process is similar to that used in beverage dispensers in restaurants, to keep the beverage from settling and becoming nonuniform (see Fig. 14.7).

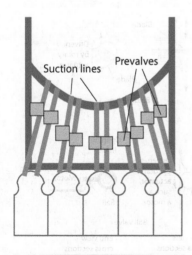

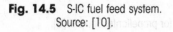

Fig. 14.5 S-IC fuel feed system. Source: [10].

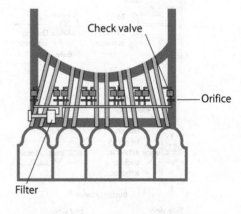

Fig. 14.6 S-IC's fuel conditioning system forces nitrogen gas into the feed lines to prevent the propellant from stratifying. Source: [10].

Fig. 14.7 This restaurant beverage container uses fluid conditioning to prevent the beverage from settling and becoming stratified (nonuniform).

14.2.4 S-IC Fuel-Level Sensing

The fuel-level sensing system is shown in Fig. 14.8. A cutoff sensor was mounted on the bottom of the fuel tank to signal engine shutoff when the fuel level met a predetermined level of depletion. This cutoff sensor was a backup to the step's LOx cutoff system, because LOx depletion was expected to occur first.

The fuel level is measured electronically by four slosh probes as well as a liquid-level probe, and then telemetered to the ground.

14.2.5 S-IC Fuel Pressurization

The fuel pressurization system maintains internal tank pressure so the fuel is properly sucked into the F-1 engines' turbopumps without cavitation, and also to reduce stresses within the tank walls. Four high-pressure helium bottles that are stored within the S-IC's LOx tank maintain the pressure. The bottles are stored in the low-temperature LOx because they can withstand higher pressures (and thus carry more helium) when stored at the low temperature.

Cold helium gas comes from the bottles and is routed via solenoids to the F-1 engines' cold helium manifolds to be heated. The hot

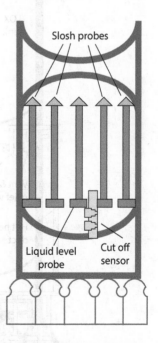

Fig. 14.8 The S-IC's fuel-level sensing and engine cutoff system. Source: [10].

helium is then distributed by the hot helium manifold up the side of the tank to provide pressurization in the fuel ullage.

A diagram of the fuel pressurization system is provided in Fig. 14.9.

S-IC LOx Fill and Drain

The S-IC LOx tank holds 1,442,000 kg (3,178,000 lb) of liquid oxygen at −183°C (−297°F). The LOx fill system is designed to deliver LOx to five F-1 engines at the enormous rate of 3,950 kg/s (8,710 lb/s); it is shown in Fig. 14.10. The LOx tank is loaded, and its oxidizer level is maintained via a 15.2-cm (6-in.) duct at the bottom of the tank. Note that the super-cold LOx boils off and must be replenished periodically.

The LOx is filled and drained through two 15.2-cm (6-in.) fill and drain lines at the bottom of the tank. It's filled at a slow rate of 4,550 kg/min

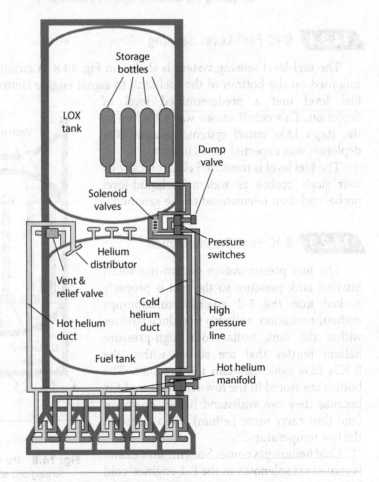

Fig. 14.9   S-IC fuel pressurization system. Source: [10].

(10,000 lb$_m$/min) until 6.5% full to avoid splash damage to tank components, and then 30,350 kg/min (66,900 lb$_m$/min) until 95% full, and finally at 4,550 kg/min (10,000 lbm/min) until a full tank is sensed. A third line can be used to fill the tank via the inboard suction line.

The LOx tank contains internal baffles for wall support and slosh suppression. If drained, the pressure in the helium storage bottles must be reduced (because the low temperature supplied by the LOx keeps pressure low).

14.2.7 S-IC LOx System

The LOx required by the S-IC's F-1 engines passes through the fuel tank inside of five 43.2-cm (17-in.) suction lines connected to prevalves, as shown in Fig. 14.11. Because of the vibration and thermal expansion and contraction, the five ducts are fitted with gimbals and sliding joints. To shut engines down safely, there are cutoff sensors at the top of the suction lines to indicate when LOx flow has ended. Prevalves near the bottom of the suction lines close to stop flow in case of shutdown or emergencies.

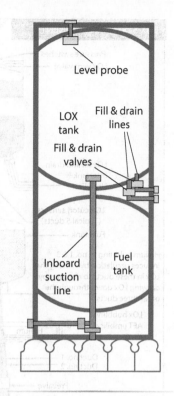

Fig. 14.10 LOx fill and drain system. Source: [10].

14.2.8 S-IC LOx Conditioning

The temperature of LOx cannot exceed −187°C (−297°F), or it will boil into gaseous oxygen. Boiling cannot be allowed because it can lead to *geysering*, which can begin when a portion of propellant in the propellant line boils. The bubble created by the boiling causes a slug of fluid to rise up into the tank. When the propellant line refills due to gravity and tank ullage pressure, a hydraulic hammer can develop with enough strength to damage the propellant delivery system. Also, high temperatures near the LOx turbopumps can cause cavitation and interfere with engine starting. More information on geysering may be found in [20].

The largest heating in the LOx system is in the outside surface of the suction lines passing through the (not cryogenic) RP-1 tank. The LOx conditioning system uses helium bubbles to cool the LOx rapidly and prevent boiling, and is shown in Fig. 14.12.

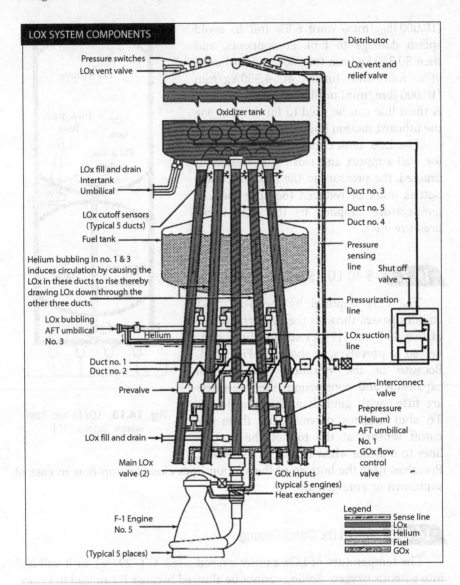

LOX SYSTEM COMPONENTS

Pressure switches
LOx vent valve

Distributor

LOx vent and
relief valve

Oxidizer tank

LOx fill and drain
Intertank
Umbilical

Duct no. 3

LOx cutoff sensors
(Typical 5 ducts)

Duct no. 5
Duct no. 4

Fuel tank

Helium bubbling In no. 1 & 3
induces circulation by causing the
LOx in these ducts to rise thereby
drawing LOx down through the
other three ducts.

Pressure
sensing
line   Shut off
valve

Pressurization
line

LOx bubbling
AFT umbilical
No. 3

Helium

LOx suction
line

Duct no. 1
Duct no. 2

Prevalve

Interconnect
valve

Prepressure
(Helium)
AFT umbilical
No. 1

LOx fill and drain

GOx flow
control
valve

Main LOx
valve (2)

GOx inputs
(typical 5 engines)
Heat exchanger

F-1 Engine
No. 5

(Typical 5 places)

Legend
─ ─ ─ ─ Sense line
▐▐▐▐ LOx
═══ Helium
▨▨▨▨ Fuel
▧▧▧▧ GOx

Fig. 14.11   S-IC LOx system components and function. Source: [13].

Other means to prevent geysering are to use a LOx recirculation system or an antigeyser line, as discussed in [7]. Of course, the simplest method to avoid geysering is to locate the LOx tank at the bottom of the step, if this is permitted by the vehicle dynamics as well as stability and control considerations. Of course, this information applies to any cryogenic propellant, including liquid hydrogen and liquid methane, in addition to LOx.

14.2.9 S-IC LOx Pressurization

Three gases are used to pressurize the S-IC's LOx tank: helium for prepressurization, gaseous oxygen for flight pressurization, and nitrogen for pressure during storage. The helium pressure is maintained between 167 and 179 kPa (24 and 26 psia) before flight, and the LOx tank must be pressurized 45 s before ignition so there is sufficient tank ullage pressure for engine starting. During flight, gaseous oxygen is added at a rate of 18 kg/s (40 lb/s) to maintain tank ullage pressure of 124–159 kPa (18–23 psia). The LOx pressurization system is shown in Fig. 14.13.

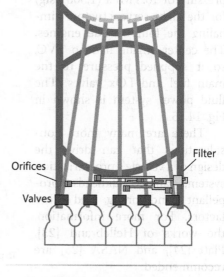

Fig. 14.12 LOx conditioning system. Source: [10].

14.2.10 S-IC Pressure Control System

This supplies pressurized gaseous nitrogen to operate valves via manifolds and tubing, and to purge engine systems to expel propellant leakage. Pressure is supplied by a titanium bottle at 22.4 MPa (3,250 psig). The pressure control system is shown in Fig. 14.14.

14.2.11 S-IC Fluid Power System

In Chapter 12 on stability and control, we noted that the Saturn's RP-1 fuel was also used to power hydraulics. This use of fuel eliminated the need for (and mass of) a separate pressure system.

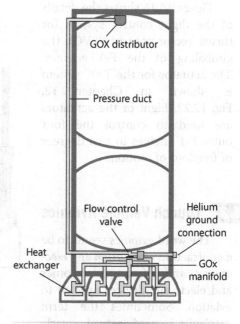

Fig. 14.13 S-IC LOx tank pressurized using gaseous oxygen (GOx). Source: [10].

The system provides a pressure of 10.3 MPa (1,500 psig) for the eight servoactuators gimbaling the four outside engines. The center engine had no TVC, so it supplied pressure to the main fuel and LOx valves. The fluid power system is shown in Fig. 14.15.

There are many more considerations that can drive the design of a vehicle's pressurization system, tanks, bulkheads, propellant conditioning, and other factors. For more information, the works of Hellebrand [23], Platt [24], and NASA [25] are recommended.

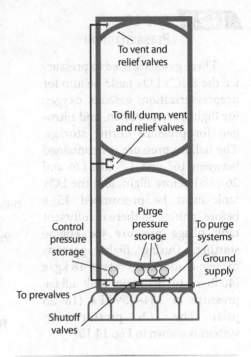

Fig. 14.14 Pressure control system. Source: [10].

14.2.12 S-IC Flight Control System Details

Figure 14.16 shows the details of the flight control system for thrust vector control (TVC), the gimbaling of the F-1 engines. The actuator for the TVC system is shown in Chapter 12, Fig. 12.22. Eight of the actuators are used to control the four outer F-1 engines in two degrees of freedom of motion.

14.3 Launch Vehicle Avionics

The word *avionics* seems to be a contraction of *aviation* and *electronics*, and refers to electronics and electronic devices applied to aviation. Sometimes the term *astrionics* is used instead, clearly referring to astronautics and electronics, such as launch vehicles and spacecraft.

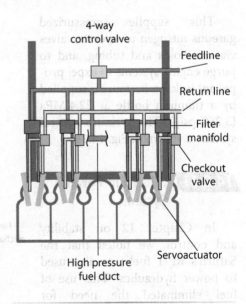

Fig. 14.15 The fluid power system provided power for TVC actuators and propellant valves. Source: [10].

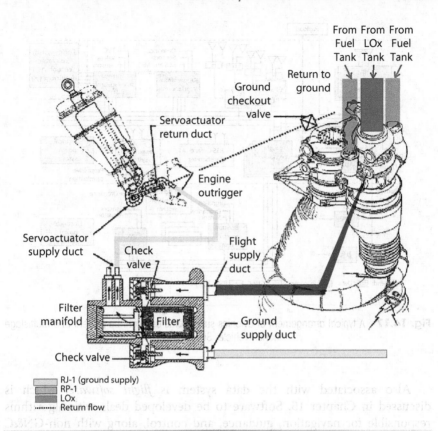

Fig. 14.16 The TVC actuator was powered by high-pressure RP-1 fuel. High-pressure fluid routing from the turbopump to the flight supply duct is shown. Source: [13].

A launch vehicle needs a number of electronic systems in order to function properly. A drawing illustrating these different systems is shown in Fig. 14.17. For avionics (astrionics), we can see there are five major categories of equipment: a data system, radio-frequency communications system (RF Comm), guidance navigation and control subsystem (GN&C), range safety system (RSS), and electrical power system (EPS).

14.3.1 Data System

The data system portion of the LV's avionics is the "brain and nervous system" of the LV. It includes the flight computer, which takes in data, receives and transmits commands, controls the steering system (both stability and navigation), receives sensor data, and interfaces with actuators, ordnance, and other devices. Not shown in Fig. 14.17 are other data system components, such as mass storage devices, redundant multiple processors, and the like.

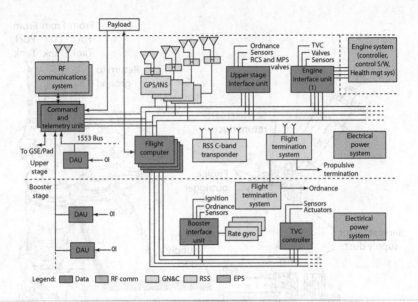

Fig. 14.17 A typical arrangement of avionics subsystems as might be present in a multistage launch vehicle. Source: [17].

Also associated with the data system is *flight software*, which is discussed in Chapter 15. Software to be developed deals with algorithms responsible for navigation, guidance, and control, along with non-GN&C software dedicated to prelaunch functions, telemetry, operating system, and so on.

14.3.2 RF and Communications System

The communications system is used to transmit housekeeping data about the LV to the ground. These data can include telemetry from any sensors on the LV, monitoring temperatures, pressures, voltages, rotation rates, acceleration, and the like. Telemetry (TM) of vehicle status is very important because it provides information on the state of the LV, as well as for troubleshooting failures that (hopefully don't) occur.

Instrumentation must provide monitoring of all key features in the LV's flight termination system to the ground station. There must be a dedicated hard line through the LV umbilical as well as an RF link through TM. We discuss flight termination systems in Chapter 15.

RF power must be sufficient to overcome range and field-of-view restrictions of on-board antennas. Antenna locations depend on LV configuration, ground station locations, and the trajectory profile. Line length between transmitters and antennae should be minimized, as should the number of connectors.

14.3.3   Guidance, Navigation, and Control

These items deal with the trajectory and stability of the launch vehicle. Guidance and navigation are "slow" variables as compared to control, which deals with both stabilizing an unstable LV and maneuvering it when necessary, both requiring a fast thrust vector control system (or equivalent). Guidance and navigation are covered in more detail in the trajectory chapter (Chapter 6); control (and stability) are discussed in Chapter 12.

14.3.4   Range Safety System/Flight Termination System

This system, intended to "render the vehicle nonpropulsive", has several names: in addition to range safety and flight termination, we see the term *propellant dispersion system*, because explosive charges are used to open up propellant tanks so they cannot feed the LV's engines. Another euphemism is the name *inadvertent separation destruct system (ISDS)*, which triggers the destruct mechanism if it detects a break in a trigger wire, indicating two vehicle parts have moved relative to each other. These systems are covered in detail within Chapter 16.

14.3.5   Electrical Power System

The electrical power system (EPS) must provide electrical power to be used by all parts of the LV as needed. It must include enough energy storage to cover the LV's power requirements with adequate margins. Therefore, one has to carefully estimate the required power budget for the launch process.

Most EPS use conventional batteries as a power source, but unconventional *thermal batteries* have also been used. A thermal battery is stored in a dry condition until it's needed. Then, the electrolyte is squirted into the cells of the battery, which now gets hot and produces electrical power

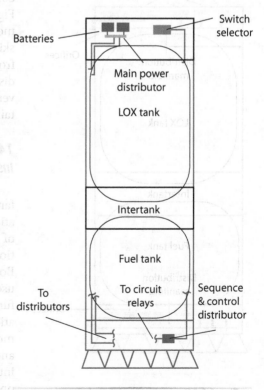

Fig. 14.18   Electrical system. Source: [10].

for a few minutes, hopefully long enough to complete the launch process. Regardless of the type of batteries or cells that are chosen, it is necessary to design and purchase batteries, *including margin for holds and scrubs*. Otherwise, it will be necessary to replace the batteries after a scrub or long delay.

The EPS will include a power distribution system that must provide short-circuit protection and other fault protection. If the system doesn't have enough current capacity to handle the required peak currents, the power drain can cause an undervoltage to occur, thus increasing the probability of a failure of an electronic component.

14.3.5.1 S-IC Electrical System

The S-IC electrical system has two independent 28-V batteries and power distribution, shown in Fig. 14.18. For other information, refer to the previous Section 14.3.1 on avionics.

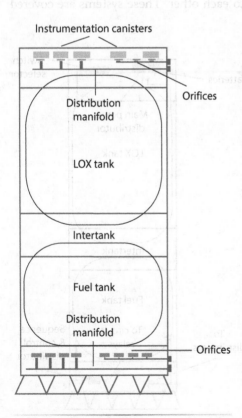

Fig. 14.19 Environmental control system. Source: [10].

14.3.5.2 S-IC Environmental Control System

The S-IC environmental control system, depicted in Fig. 14.19, was to protect equipment installed in the forward skirt and thrust structure areas from temperature extremes. A distribution manifold was used to vent air and gaseous N_2 to maintain proper temperatures.

14.3.5.3 S-IC Visual Instrumentation

Most everyone has seen the famous Saturn interstage separation movie, and it comes courtesy of the S-IC's visual instrumentation system, shown in Fig. 14.20. Both film and TV cameras were used to monitor critical stage functions. During stage separation, there were two cameras monitoring the oxidizer sloshing and motion in the LOx tank interior via 2.7-m (9-ft) fiber optics. The camera capsule was ejected 25 s after staging and was

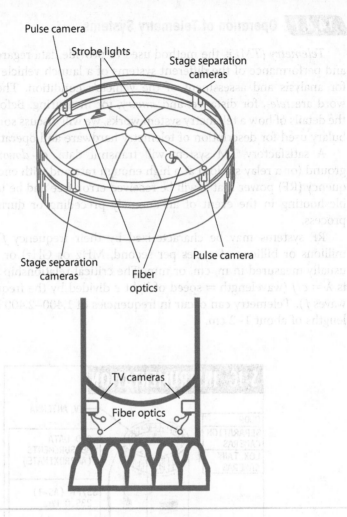

Fig. 14.20   S-IC visual instrumentation. Source: [10].

recovered in the ocean. The TV cameras provided four views of engine operations filmed at 15 FPS, and video was downlinked to ground by a 2.5-W transmitter.

14.4 Instrumentation and Telemetry

Some of the instrumentation on the S-IC is shown in Fig. 14.21, and Table 14.1 provides a detailed list of the sensors that were included. Almost 900 measurements were taken—no wonder there are so many data available! All of these data were either telemetered to the ground or recovered from devices that ejected from the step after stage separation.

14.4.1 Operation of Telemetry Systems

Telemetry (TM) is the method used to provide data regarding the status and performance of the different systems of a launch vehicle to the ground for analysis and assessment of the vehicle's condition. The roots of the word are *tele-*, for distance, and *-metry*, for measuring. Before we get into the details of how a telemetry system works, we will discuss some of the vocabulary used for description of telemetry hardware and operations.

A satisfactory TM system will transmit data, or *downlink* it to the ground (or a relay station) at a high enough rate and with enough radio frequency (RF) power that it will be received error-free and be useful for troubleshooting in the event of an anomaly preceding or during the launch process.

RF systems may be characterized by their frequency f (measured in millions or billions of cycles per second, MHz or GHz) or wavelength λ, usually measured in m, cm, or mm. The critical relationship between them is $\lambda = c/f$ (wavelength = speed of light c divided by the frequency of radio waves f). Telemetry can occur in frequencies of 1,400–2,400 MHz or wavelengths of about 1–2 cm.

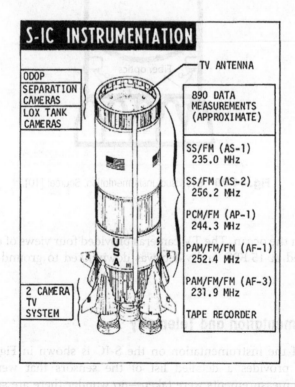

Fig. 14.21 Instrumentation, including cameras, sensors, and frequencies utilized. Source [13].

Table 14.1   Electrical Transducers Used on the First Step

STAGE MEASUREMENTS	
TYPE	QTY
ACCELERATION	3
ACOUSTIC	4
TEMPERATURE	252
PRESSURE	164
VIBRATION	150
FLOWRATE	35
POSITION	1
DISCRETE SIGNALS	147
LIQUID LEVEL	22
VOLTAGE, CURRENT, FREQUENCY	11
TELEMETRY POWER	19
ANGULAR VELOCITY	6
STRAIN	71
RPM	5
TOTAL	890

The telemetry system is mostly used to convey engineering data from the vehicle to the ground. These data can be hundreds to thousands of channels of information that may be used to assess the entire state of the vehicle. Data can include component temperatures, pressures, voltages, currents, rotational positions, attitude control system sensor outputs, propellant levels, and much more. It can also provide information on the status of items such as valves, switches, relays, and safe/arm mechanisms. All of this information is used by the operators on the ground to assess vehicle subsystem status and health.

14.4.2 **The Telemetry Process: Getting Information to the Ground**

Now we will describe how the telemetry system can provide the needed data to the ground. The process begins with that group of physical properties mentioned previously that needs to be known on the ground. Many of these measurements are analog physical quantities, and sensors convert them to electrical signals. Some quantities, such as voltages (e.g., battery voltages), need no conversion; however, the majority of sensors need to convert non-electrical physical quantities such as acceleration, temperature, pressure, and the like into electrical signals. Such a sensor is called a *transducer*. The end result or output from a transducer is also a time-varying voltage signal.

Now, we need to convert the signal into something that can be transmitted. The usual method is to *digitize* the signal. This means that we convert the analog signals into digital numbers on a regular periodic basis, for example every 0.001 s, or 1,000 times each second. Later we will discuss how often the conversion has to occur to accurately represent a time-varying signal.

The individual voltage signals are sampled in a sequence that is specified in advance, so that it can be "decoded" upon receipt. For timekeeping and bookkeeping reasons, we add additional digital data to include such things as clock time, date, and so forth. The data are transmitted to the receiving station, usually by an RF system.

The RF system consists of a transmitter (sometimes written as "xmitter"), which amplifies the signal, and a transmitting ("xmitting") antenna. The transmitting antenna is usually a low-gain antenna (LGA), because we don't always know the relative orientation between the vehicle and the ground station, and LGAs work in all orientations, also known as *omnidirectional*. *Gain* refers to the amplification of the system as described by its output divided by its input. The receiving (Rx) station often utilizes a parabolic-shaped high-gain antenna (HGA) that collects and amplifies the incoming signal. The HGA is very directional, which means it must track the vehicle's motion in order to produce a good signal. Additional amplification is provided by the receiver, after which the signal is decoded. Another system separates the telemetered data into separate streams that represent each analog signal that was measured. From this point, the data are usually stored and may also be processed and displayed either graphically or numerically. This process is shown symbolically in Fig. 14.22.

The airborne portion of a generic TM system is shown in in the top portion of Fig. 14.22. The top row of the figure depicts the analog or continuous signals that are being sampled. The sensor outputs are conditioned so as to be compatible with a device called an *analog multiplexer*, which processes the signals and feeds them into a device called a *formatter*. The formatter combines the digital signals that have been output from the digital multiplexer with the analog signals into a single stream of digital data that is then encoded and sent to the vehicle's transmitter, where it is converted into radio waves that can be received on the ground.

14.4.3 Multiplexing of Data: Commutation

Now we will discuss how a large number of separate signals gets combined into a single signal that contains all the same information but can also be transmitted down as a single channel. This process, called *commutation*, is performed by the multiplexers and is illustrated in Fig. 14.23. A sampling system, represented by the shorter thick arrow,

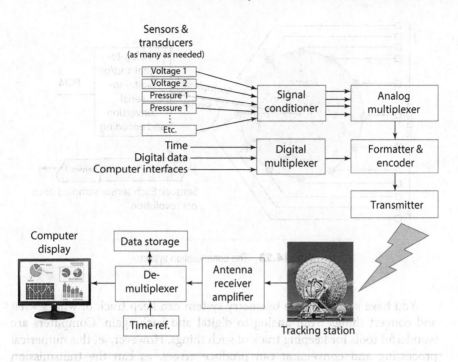

Fig. 14.22   The process of telemetry transfers a vehicle's sensor data to the ground, where it can be used to assess the system's condition.

rotates around the face of the circle similar to the hands of a clock (but *much* faster). Each time the "hand" touches one of the black dots, it samples the signal and sends it to the analog-to-digital converter (ADC). Then it moves on to the next signal, and so on. The multiplexed or muxed signal consists of multiple signals placed end-to-end.

Commutation produces a train of signals as depicted in Fig. 14.24. A *minor frame* contains blocks of converted data shown as "Word 1," "Word 2," and so on. Each word represents one sample of one channel or sensor, and the words are preceded by a sync pattern, which is the computer's way of saying "Ok, we're starting a new set of data now."

14.4.4   Decommutation

The data are received in frames on the ground, and we've seen the frames contain numerous signals placed end-to-end, and then a new frame arrives. The frames have to be split or decommutated in order to split the data stream up into separate channels. The sync patterns are used to start the conversion of a new string of separate, unrelated data into separate channels of data. This process is called decommutation, and is shown in Fig. 14.25.

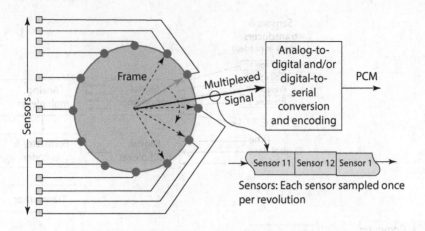

Fig. 14.23 The commutation system.

You have seen how the telemetry system can keep track of what's what and convert things from analog to digital and back again. Computers are wonderful tools for keeping track of such things. However, all this numerical processing and conversion can produce errors, as can the transmission system itself. Now we will look into some error sources in a data acquisition system.

14.4.5 Error Due to Aliasing

Before we talk about how the multiplexers and formatter process the data, we need to discuss data sampling. It should be intuitive that the signals have to be sampled at some minimum rate in order for rapid changes to register in the measured data. A rule of thumb is that the time-varying data must be sampled at least 2.2 × the maximum frequency component. You can think of approximating a sine wave. It would take at least a point near the maximum and minimum points to capture its oscillating frequency. If the

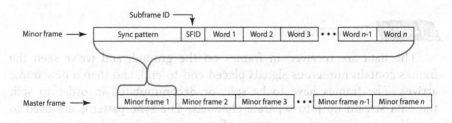

Fig. 14.24 A minor frame. Minor frames are assembled into master frames that contain descriptions of the time-varying data.

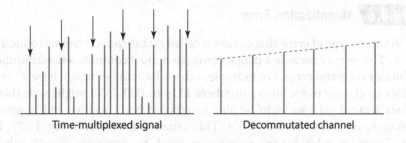

Fig. 14.25 The muxed signals split up based on their distance from the sync pattern. One such signal is shown by the sequence of bars under red arrows above the time-muxed signal. That particular signal's values are set aside and placed end-to-end. This results in the decommutated signal that represents what was happening on the vehicle before transmission to the ground.

data are not sampled rapidly enough, a phenomenon called *aliasing* occurs. An example of a signal that is sampled well above its fundamental frequency along with the same signal sampled at too low a frequency (called *undersampling*) is shown in Fig. 14.26. The top trace shows a sine wave being sampled at a rate high enough to clearly indicate its actual shape, a sine wave that is 10 cycles long. The bottom trace shows the same sine wave being sampled at too low a rate, where connecting the sampling dots produces an apparent sine wave that is just 2 cycles long, not 10 as it should be. This "low frequency" sine wave is very different from the actual signal's shape.

A good example of aliasing may be seen in some automobile commercials on television. When a vehicle is shown driving by, it appears to the observer that the vehicle's wheels are rotating backwards. This is an artifact of the frame rate of the television image, which is too low to properly measure the motion of the wheels.

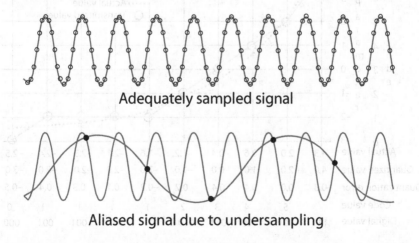

Fig. 14.26 Top: An adequately sampled sine wave. Bottom: An undersampled signal.

14.4.6 Quantization Error

Another type of error that occurs is called *quantization* error or rounding error. This occurs because a digital conversion must consume a fixed number of bits, or ones and zeros. For example, a three-bit number would have $2^3 = 8$ values to choose from, binary numbers 111_2 to 000_2. We might split these values around zero so as to be able to convert both positive and negative numbers, say from −3 to 0 to +4. This situation is shown in Fig. 14.27. In this figure, a 3-bit binary number is used to represent actual values between +4 and −3. Because of the coarseness of the scale, every measurement has to be rounded up or down to the nearest integer, producing the more jagged line. Clearly this line is not an accurate representation of the actual value. The difference is due to rounding or quantization error.

So, how does this system represent the value of the number 0.6? Because 0.6 is not an integer, it cannot be represented exactly, so it must be rounded off. The system in Fig. 14.27 shows this situation in its third column; here the value of 0.6 has been rounded to a value of 1.

There is a way to minimize rounding error, and that is to use more bits to represent the numbers being measured. Clearly, the more bits being used, the smaller the spaces between the bits and the better accuracy that can be achieved. For example, a 10-bit system with $2^{10} = 1,024$ levels is going to be four times as accurate as an 8-bit system with $2^8 = 256$ levels over the same positive–negative range of input numbers. Although increasing the number of bits seems like an easy fix, it also requires more bits to be processed, meaning analog-to-digital conversions take longer, and more data

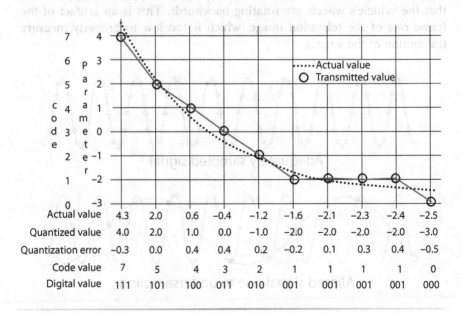

Actual value	4.3	2.0	0.6	−0.4	−1.2	−1.6	−2.1	−2.3	−2.4	−2.5
Quantized value	4.0	2.0	1.0	0.0	−1.0	−2.0	−2.0	−2.0	−2.0	−3.0
Quantization error	−0.3	0.0	0.4	0.4	0.2	−0.2	0.1	0.3	0.4	−0.5
Code value	7	5	4	3	2	1	1	1	1	0
Digital value	111	101	100	011	010	001	001	001	001	000

Fig. 14.27 A smooth curve cannot be sampled exactly with a system having too few digits.

need to be sent through the TM system for each measurement. This makes the system require a faster conversion system along with a faster transmission speed.

14.4.7 Effective TM System Operations

In order for a TM system to work properly, it must be able to (1) convert data to and from analog and digital formats, and (2) transmit to the ground with minimal errors. We have discussed analog-to-digital and digital-to-analog conversions already. The successful transmission requirement specifies both the system's transmit and receive power and the rate at which the data are transmitted. That data transmission rate depends on the data encoding scheme used and the characteristics of the transmitting and receiving equipment.

The discussion of both the data encoding and decoding scheme and the RF transmission is beyond the scope of this book; however, we can briefly say that the encoding scheme may be used to provide a *data check* that is embedded into the data stream that helps to detect and recover any data that are lost. A simple coding scheme would be to repeat everything a second time, but this would obviously double the transmission time or halve the effective data rate. Better encoding schemes, such as those used in cellular telephones, provide the same function with much less overhead or lost transmission speed.

RF transmission analysis involves developing what's called a *link budget* to determine the system performance. The link budget is nothing more than a list of all the gains (strengthening) and losses (attenuation) that the signals must pass through during their trip from the vehicle to the ground. This includes signal strength loss due to cabling, gains from the transmit and receive amplifiers and antennae, and attenuation occurring because of the distance between the vehicle and ground station. Of course, the transmitting and receiving antennae need to be properly oriented so that no loss of signal (LOS) occurs regardless of the relative orientation and distance between the vehicle and the ground station. The numerical performance of the link is often specified in decibels, which are a way to quantify output over input, or whether the input is strong enough to be properly sensed at the output.

Decibels and their use for acoustic signals are discussed in the acoustics portion of Chapter 11. They are also used extensively in RF systems to describe how the systems operate in terms of *ratios*, typically the ratio of (output P_2) $\div$ (input P_1). In the case of RF transmission systems, the units are often electrical power in terms of watts or milliwatts. Our systems need to receive a minimum number of milliwatts to function properly, and the link margin expresses how much more—or less—power is received than needed, typically in dB. As a review, decibels (dB) are computed from the expression

$$dB = 10 \log_{10}(P_2/P_1)$$

For example, if $P_2 = 100$ mW and $P_1 = 10$ mW,

$$dB = 10\log_{10}(100/10) = 10\log_{10}(10) = 10$$

Another way this can be stated is to use a power reference of milliwatts, just as we used a sound pressure level for a reference in the acoustics world. If we (for example) used a reference level of 1 mW, then the dB_m ratio would be $dB_m = 10\log_{10}[P_2/(1\text{ mW})]$. There are also dB_W (1 W reference) and other references for calculations.

Also useful are the "shortcuts" associated with doubling and halving a signal. Because $\log_{10}(2) \approx 0.3$, doubling a signal is equivalent to a change of approximately $+3$ dB, and halving is approximately -3 dB. These dB

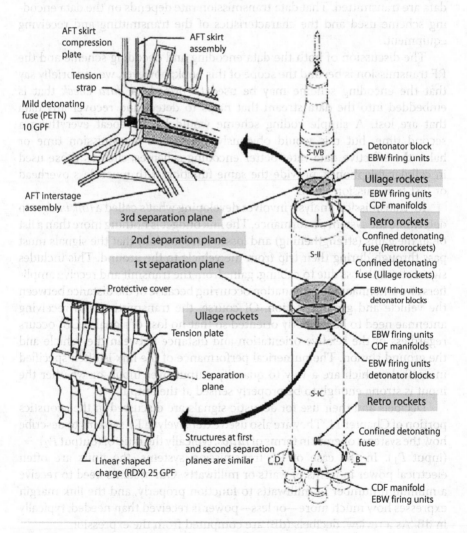

Fig. 14.28 Separation systems for the Saturn V. Three separation planes needed to operate for the vehicle to successfully carry out its mission. Three sets of retrorockets and two sets of ullage motors may also be seen. Source: [13].

Fig. 14.29   When the Saturn V's first step (S-IC) was jettisoned, the interstage ring remained attached to the S-II while the ullage motors fired to move the propellant to the tank bottom. After approximately 30 s, it was jettisoned as shown. Source: Still image from staging video (NASA).

values can be useful to compare different values of data. A common figure of a successful link budget is a margin of 3 dB, meaning the signal is about twice the strength needed for successful operation.

The reader interested in learning more about encoding and link budgets is referred to the text by Brown [2]. Other spacecraft design texts should also contain information on these topics.

14.4.8   Saturn V Staging: Separation Systems

Figure 14.28 shows the separation systems on the Saturn V. Below the spacecraft, there are three separation planes. Two of the separation planes are on either side of the interstage ring between the first and second steps. The first one released the S-IC from the interstage ring, and the second released the interstage ring from the S-II step. Those two separations are commonly seen in the famous video showing Saturn stage separations (see Fig. 14.29).

The third separation plane was between the S-II and the S-IVB, and occurred on the top of the tapered skirt connecting the two.

The "circular" interstage separations used pyrotechnic cords to sever connecting straps or cut through tension plates, as described in Chapter 11. *PETN* and *RDX* are two types of explosives used. The cords were triggered by other explosives known as exploding bridge wire (EBW) or detonating fuse.

There are a few more parts of the separation system. At the bottom of the S-IC, you can see a loop of detonating fuse that was used to ignite eight

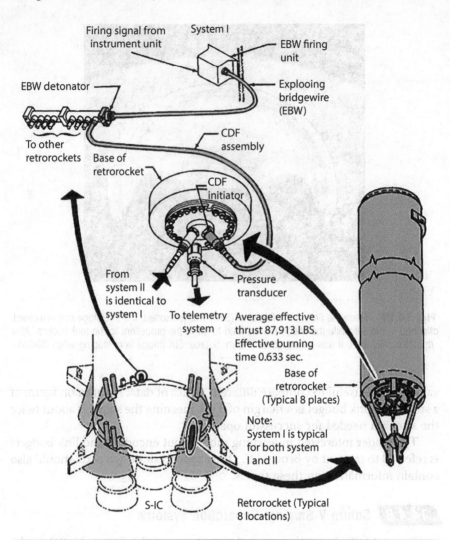

Fig. 14.30 Retrorockets used on the S-IC. They fired when the first separation system was activated, to push the S-IC downward with eight × 391 kN (87,900 lb$_f$) of thrust. Source: [10].

retrorockets that fired to pull the S-IC downwards, away from the climbing S-II. On the S-IC/S-II interstage ring, there are four *ullage rockets* that fired to accelerate the S-II step upwards so that its propellants would reach the bottom of their tanks and be used to start the five J-2 engines. Finally, there are some additional ullage rocket motors at the base of the S-IVB and some retrorockets on the skirt between the S-II and S-IVB steps.

Figure 14.30 shows the retrorockets used on the S-IC. Each of the eight retrorockets put out 391 kN (87,900 lb$_f$) of thrust for 0.633 s. The designers did not bother with a hatch or opening above the rockets. They simply let them burn through the engine fairings like a blowtorch!

14.4.9 Saturn V S-II Step Ullage and Retrorockets

The S-II's ullage motors are designed to momentarily accelerate the second step after first step burnout, to make certain that the propellants are in the proper position to be drawn into the pumps prior to starting the S-II's engines. The ullage motors provided 101 kN (22,700 lb$_f$) of thrust for 3.7 s. Some Saturn launches used eight of these ullage motors; others used four.

The four retrorockets burned as the S-IVB separated from the S-II. They provided 155 kN (34,810 lb$_f$) of thrust for 1.52 s.

Both the ullage motors and the retrorocket motors are shown in Fig. 14.31.

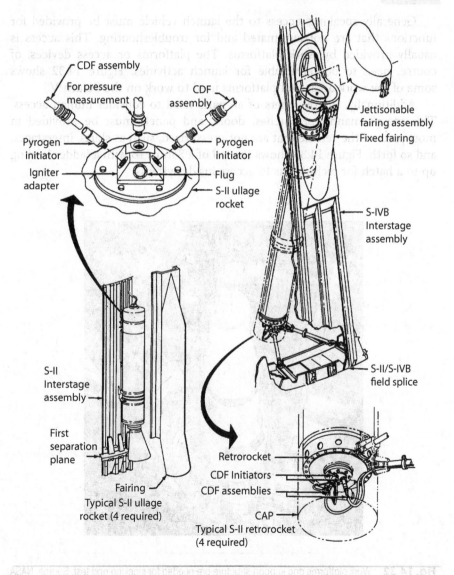

Fig. 14.31 Ullage and retrorockets on Saturn V's S-II step. Source: [13].

14.5 Launch Pad Facilities and Ground Accommodations

Next, we will discuss services that the launch pad provides. This includes access to the launch vehicle for personnel, access to the vehicle for services such as propellant and gas loading, electrical connections, propellant storage, loading and on-board management, protection from lightning strikes, and preliftoff considerations such as vehicle hold-downs. These topics will be discussed in the following sections.

14.5.1 Vehicle Access

Generally speaking, access to the launch vehicle must be provided for functions that are not automated and for troubleshooting. This access is usually provided by work platforms. The platforms or access devices, of course, need to be retractable for launch activities. Figure 14.32 shows some of the work areas and platforms used to work on the Saturn V.

Additionally, the designers of all parts need to consider human access. This means manholes, hatches, doors, and panels must be provided in most parts of the vehicle that are not pressurized, such as skirts, interstages, and so forth. Figure 14.33 shows the tail of a Saturn IB with a ladder leading up to a hatch for technicians to access inside systems.

Fig. 14.32 Work platforms and support structure are needed for stacking and test. Source: NASA

Fig. 14.33   An access door for technician access (circled) is at the top of the ladder.
Source: NASA.

14.5.2   Logistics at Base of Saturn V

While on the launch pad, a number of subsystems provided lock-down, noise suppression, and many other services. These systems are described in detail in the following sections.

14.5.2.1 Tail Service Connections

The Saturn V had three tail service masts, as shown in Fig. 14.34. A photo of the tail service masts is provided in Fig. 14.35. These three service masts were similar in function and together carried RP-1 fuel fill and drain lines, LOx fill and drain lines, air conditioning and deluge purge ducts, as well as numerous electrical, pneumatic, hydraulic, and cryogenic connections. The tail service masts connected umbilicals to the Saturn V's base, as shown in Fig. 14.35.

Disconnection and retraction were done with a pneumatically charged hydraulic system. If there were a pneumatic failure, a mechanical linkage would release the umbilical carrier, and the counterweighted masts would complete the operation. Normal disconnect took 3.0 s; emergency disconnect took 3.2 s.

14.5.2.2 Swing Arm Umbilicals

In addition to the tail service masts, an additional set of umbilical connections is provided by a number of swing-arm umbilicals, which provide access

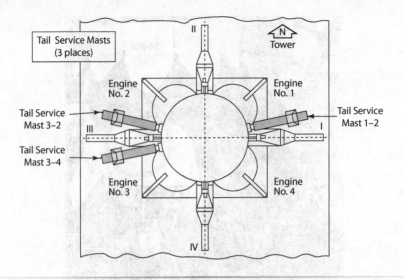

Fig. 14.34 Tail service masts for the Saturn V pad. Source: [14], shading added by author.

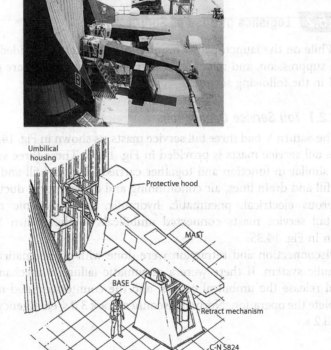

Fig. 14.35 Saturn V tail service masts. Source: NASA.

to the first and upper steps of the vehicle for the same services as the tail service masts. The Saturn V pad had nine swing arms attached to its mobile launch platform (see Fig. 14.36). According to the Saturn V flight manual, four of the arms were for preflight activities, and the remaining "in-flight" arms were separated at liftoff.

The four preflight arms (1, 2, 3, and 9) were retracted before liftoff. Of these, 1 and 2 were for preflight loading; 3 and 9 (top) arms were for access to the vehicle, the latter being used for the crew to access the spacecraft.

The remaining five in-flight swing arms (4, 5, 6, 7, 8) loaded LOx and LH_2 and carried air conditioning as well as numerous electrical, pneumatic, and hydraulic connections. For protection, these arms retracted away from the launch vehicle upon liftoff detection, taking between 6.4 and 9.0 s to retract.

Fig. 14.36 Saturn V's mobile launch platform provided umbilical connections via swing arms. Arms 1, 2, 3, and 9 retracted before launch; the remainder retracted upon liftoff detection. Source: NASA.

14.5.2.3 More on Umbilicals

To ensure positive retraction, methods such as counterweights, springs, or other mechanisms are used. Many masts and arms have covers that close to protect the item's services from the launch vehicle's high-temperature exhaust. They may also be coated with ablative coverings, which burn to absorb energy rather than transfer it to the object's structure. Additional umbilical references are at the end of the chapter.

14.5.2.4 Pad Support Subsystems

For a launch vehicle that is on the pad, ready to be loaded, there are a number of requirements. Some of these are:

* All tanks need to be purged before loading.
* Cryogenic propellants (LOx and LH$_2$) require propellant tank cooldown before loading.
* Hazardous propellants impose additional environmental and safety requirements.
* The duration of the planned (and unplanned) hold time prior to launch is important. It affects propellant boil-off and internal energy storage (battery) requirements.

All of this information helps set requirements on the LV architecture design, including the amount of propellants at the launch site, other fluid requirements, safety issues, and so forth. Additional information may be found in [3].

14.5.2.5 Propellant Loading Operations

Significant volumes of propellants must be loaded onto the launch vehicle in a reasonable amount of time (and may need to be unloaded as well). This may be done using large electric pumps or high-pressure gases. Either can provide high loading rates. Small pumps may be used to filter and unload, if necessary. A cooler may be used to cool the LOx and pressurant gas in order to increase density and reduce needed tank volume. Figure 14.37 shows a generic propellant loading system.

Fast propellant loading operations would proceed as follows (letters in parentheses refer to locations on Fig. 14.37):

1. Connect propellant (two fuel, one LOx) and gas feed umbilicals to vehicle.
2. Close vent valve EE; gaseous Ox (GOx, shown as light blue) fills storage ullage.
3. Open valve V and low-flow valve X; feed system begins cooling.
4. High-pressure regulator DD feeds GOx at 10–20 atm to storage tank. LOx flow builds but still passes through low-flow valve X.
5. Open full-flow valve W; LOx fills at high rate (~7,500 LPM or ~2,000 GPM).

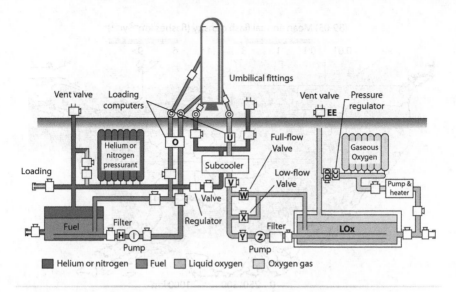

Fig. 14.37 The launch pad propellant loading system uses pressurized gases and pumps to transfer propellants into (and out of, if necessary) the launch vehicle. Source: [4], colored by author.

6. When nearly full, close W; low-rate top-off using X.
7. Pump I forces fuel through filter H into LV.
8. Pump Z removes LOx through filter to storage if needed.

14.5.2.6 Weather

In Chapter 9, we discussed ground winds in the ground wind loads subsection, as well as high altitude winds in the max-q loads analysis. But there is other weather that needs to be dealt with—lightning, which clearly is a possible issue with launch vehicles.

Figure 14.38 shows the mean annual lightning flash density for the continental United States. Similar maps can be found for other regions. Interestingly, it appears that the United States has located its Eastern Test Range (ETR)/Cape Canaveral Air Force Station (CCAFS)/Kennedy Space Center (KSC) at the worst possible location as far as lightning flash density is concerned.

Because of the energetic materials that are used for rocketry, it's important that provisions be made for lightning protection. The current state of the art appears to be to erect one or more grounded *lightning towers* that are somewhat higher in elevation than the rocket and its gantry, and connect multiple electrically conductive cables between the tower(s) and ground to encourage any lightning to strike the grounded towers. The idea is to make the lightning's very large current, sometimes in the millions of amperes, bypass the rocket itself and flow down the tower or the connected cables to ground.

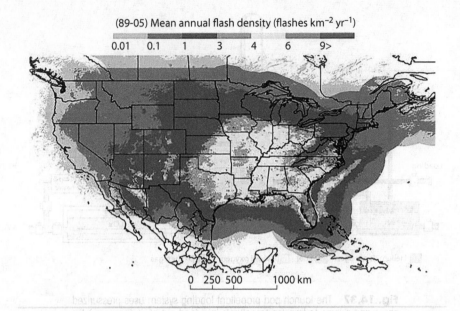

(89-05) Mean annual flash density (flashes km^{-2} yr^{-1})

| 0.01 | 0.1 | 1 | 3 | 4 | 6 | 9> |

0   250 500   1000 km

Fig. 14.38   The U.S. lightning flash density in units of flashes/km^2/year. A launch site in Florida appears to be a bad choice, as it lies in the maximum category of Mean Annual Flash Density. Source: [19].

One example of lightning protection is shown in Fig. 14.39, the lightning "protection" for the Space Shuttle at KSC. This system consisted of a tall lightning mast on top of the support structure adjacent to the Shuttle, and cables connecting the mast to ground. The reason the word *protection* above is in quotation marks is the photo on the right of the figure, showing a lightning bolt striking the launch pad next to the Shuttle, apparently missing the protection system!

Figure 14.40 shows the lightning protection used by SpaceX for its Falcon 9 launch pad in CCAFS. This system has four tall masts and grounding wires. Note the square "window" that the launch vehicle must pass through just after liftoff.

14.5.3   Launch and Liftoff Considerations

This section discusses the various events and procedures that occur during ignition, launch, and liftoff. Examples and explanations will be provided for the following topics:

• Staggered engine ignition
• Hold-downs, release mechanisms, and liftoff loads
• Hydrogen suppression
• Acoustic suppression
• Exhaust considerations
• Tower clearance

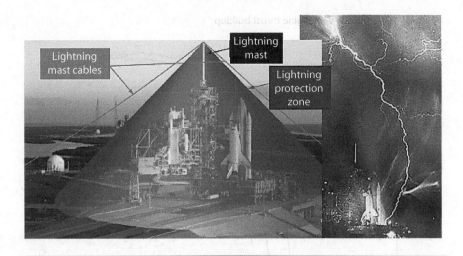

Fig. 14.39   The Space Shuttle lightning protection system. Its effectiveness is questionable considering the lightning strike on the right, when the Shuttle's launch pad was struck before mission STS-8. Sources: [19], http://www.nasa.gov/images/content/57894main_Shuttle_and_Lightning.jpg.

14.5.3.1 Staggered Engine Ignition Reduces Loads

As shown in Fig. 14.41, The Saturn V's five F-1 engines start in a staggered 1-2-2 sequence so that there is no single large load increase applied to rocket's structure. The center engine (#5) starts first.

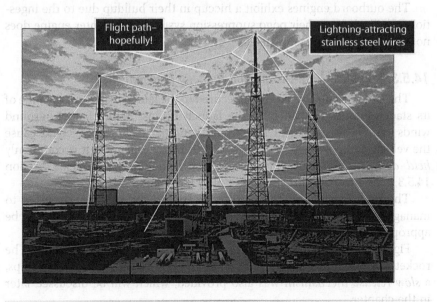

Fig. 14.40   Falcon 9 and its lightning protection wires. Source: https://www.spacex.com/gallery/2009-0#slide-5; edit, wires, and labels added by author.

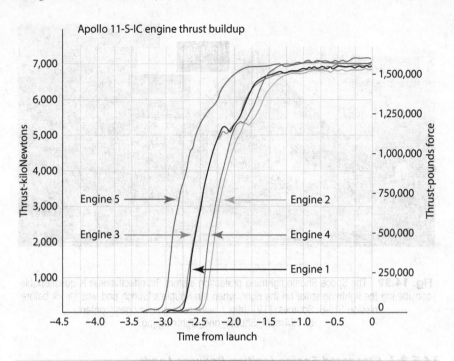

Fig. 14.41 Saturn V first step ignition sequence. The 1-2-2 sequence started with the center engine (5), then lit 1 and 3, and then 2 and 4. The hiccup in 1–4 is due to the pogo suppression systems in these engines. Source: [18].

The outboard engines exhibit a hiccup in their buildup due to the inges-tion of helium from their pogo suppression systems. The center engine does not have pogo hardware installed.

14.5.3.2 Vehicle Hold-Downs and Release Mechanisms

The launch vehicle needs to be secured to the launch pad from the time of its stacking onto the pad until the launch begins. Before launch, ground winds can cause significant lateral and bending loads, which could cause the vehicle to tip over. When the vehicle is ready to launch, it is commonly *held down* to the pad until satisfactory engine operation (see Section 14.5.3.1 for adequate thrust buildup plot) is attained.

The next portions of this chapter deal with the mechanisms used to manage the launch of a Saturn V. Many of the details shown here will be appropriate for other launch vehicles, albeit on a smaller scale.

Figure 14.42 is a drawing of the Saturn V pad hold-down clamps. The rocket was held down in four places. In addition to the hold-down clamps, a *slow release* mechanism was also provided, which will be discussed later in the chapter.

The layout of the hold-down clamps is shown in Fig. 14.43. In the clamped position, the *upper link* clamps the rocket down at point

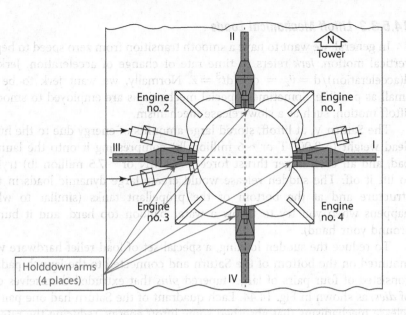

Fig. 14.42 Saturn V pad hold-down clamps. Source: NASA, shading added by author.

C. The center and lower links are nearly aligned in the vertical direction, positions A and B. When the mechanism is commanded to release, the explosive release fires the pneumatic separator, which drives the links to positions A′, B′, and C′, where they remain latched. At the same time, the blast hood rotates forward to protect the mechanism from the hot rocket exhaust.

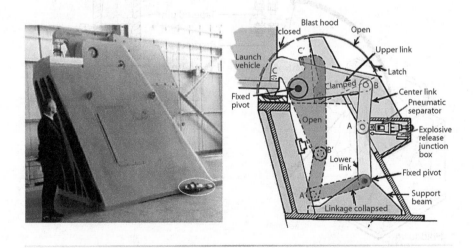

Fig. 14.43 Four hold-downs clamp and then release the 3,000-T Saturn V. Note controlled-release mechanism at the base of the support structure circled on the left (see Fig. 14.44). Source: NASA.

14.5.3.3 Liftoff Mechanical Loads

In general, we want to have a smooth transition from zero speed to begin vertical motion. *Jerk* refers to time rate of change of acceleration. Jerk = $d(\text{acceleration})/d = \dddot{v}_z = d^3z/dt^3 = \dddot{z}$. Normally, we want jerk to be as small as possible. Sometimes, special mechanisms are employed to smooth liftoff motion, such as a slow-release mechanism.

The Saturn V, at liftoff, stored large amounts of energy due to the huge dead weight ($\sim$3,000 T or $\sim$6 million lb) compressing it onto the launch pad, and an even larger thrust force ($\sim$33 MN or $\sim$7.5 million lb) trying to lift it off. The sudden release would create large dynamic loads in the structure and at the bottom of the propellant tanks (similar to what happens when you try to throw a water balloon too hard, and it bursts around your hand).

To reduce the sudden loading, a special set of load relief hardware was mounted on the bottom of the Saturn and connected to the launch pad. It consisted of four pairs of large, tapered *pins* that extruded themselves out of *dies*, as shown in Fig. 14.44. Each quadrant of the Saturn had one pair of release mechanisms that absorbed some liftoff energy, reducing the rate of acceleration over the time it took to extrude the pin from the die (see Fig. 14.45). Extensive design details for this system may be found in [15].

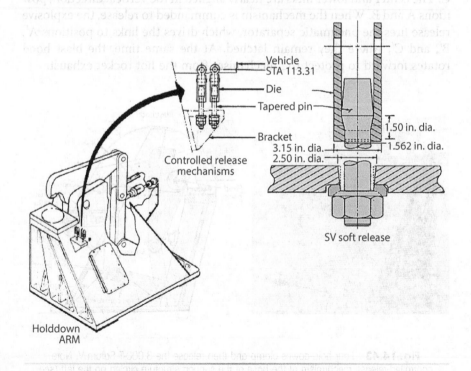

Fig. 14.44 Saturn V slow-release hardware. Source: [21], shading added by author.

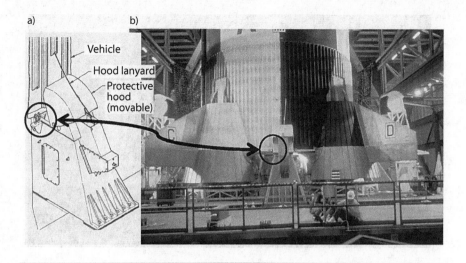

Fig. 14.45   a) Slow release mechanism location on hold-down mechanism (circled); b) location of two pin/die assemblies (circled). Source: NASA.

In case you were wondering: we said earlier that jerk = d(acceleration)/dt = $\dot{v}_z$ = d^3z/dt^3 = $\dddot{z}$.

d(jerk)/dt = *jounce* = d^4z/dt^4 = fourth derivative

And, the next few terms are referred to as

snap = fifth derivative

crackle = sixth derivative

pop = seventh derivative

Source: http://en.wikipedia.org/wiki/Jounce

Figure 14.46 shows a still frame taken from a slow-motion Saturn V liftoff. In the figure, we can see that the hold-down mechanism we looked at earlier has now released the rocket, so the hold-down is retracted and the cover is closed (center left), and the tail-servicing swing arms have retracted (right).

14.5.3.4 How Was the Space Shuttle Held Down?

Unlike the Saturn V, the Shuttle stack was held down by eight *frangible nuts* (also known as explosive nuts), four holding down each SRB. The frangible nuts have two lengthwise slots containing small explosive charges. When the charges are fired, the nut is split into two pieces, so that a release can be commanded electronically. The arrangement used to hold down the SRB is shown in Fig. 14.47.

The nuts were fired at a special time during the ignition process: the Shuttle's main engines were ignited first, causing the whole stack to bend over

Fig. 14.46 Saturn V liftoff. Note the closed hold-down (center left) and retracted service arms (right). Source: Frame taken from NASA Apollo 11 launch video.

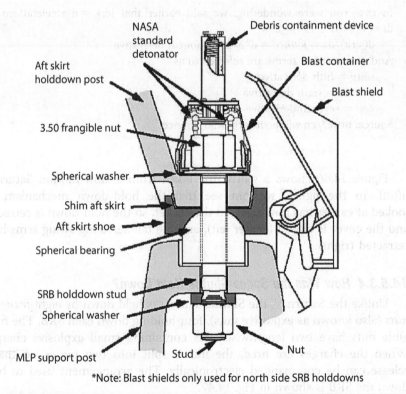

NASA standard detonator

Debris containment device

Aft skirt holddown post

Blast container

Blast shield

3.50 frangible nut

Spherical washer

Shim aft skirt

Aft skirt shoe

Spherical bearing

SRB holddown stud

Spherical washer

MLP support post

Stud

Nut

*Note: Blast shields only used for north side SRB holddowns

Fig. 14.47 Shuttle solid rocket booster frangible nut and SRB hold-down details. Source: [9].

Fig. 14.48   One-half of a (used) frangible nut. For scale, the hole diameter is 89 mm (3.5 in.). Source: Edberg.

(because of the engines' offset thrust line), and then *twang* backwards. At the instant when the stack had come back to the vertical position, the two SRBs were ignited and the nuts were fired simultaneously.

A piece of a fired frangible nut is shown in Fig. 14.48. To prevent damage to the Shuttle, the nut fragments were captured by *nut catchers* installed around the nuts. A photo of the bottom of the SRB, including the nut catchers, is shown in Fig. 14.49.

14.5.3.5 Taking Care of Loose Propellants

In most cases, when cryogenic propellants are involved, it's necessary to *precool* the engine and its equipment so there is reduced thermal shock when the engine begins firing. But the precooling process naturally releases the propellants to the outside, where they can collect and possibly explode when ignited.

Figure 14.50 shows a Delta IV Heavy near the moment of ignition, and it's pretty clear that there are lots of flames around, which seem to scorch

Fig. 14.49 Nut catchers were installed over the frangible nuts in order to capture the explosive nut debris at liftoff, preventing damage to the Shuttle. Source: NASA.

Fig. 14.50 Why hydrogen suppression is needed. The white areas near the booster's bottom are actually hydrogen burning flames. Scorched insulation caused by burning rocket propellants.
Source: ULA.

the spray-on insulation (the hydrogen flames are the white "smoky" trails near the bottom of the booster). Because the Delta IV uses hydrogen and oxygen as propellants, one would expect that mixture to produce the flames shown. However, hydrolox burns very cleanly, and the flame is almost invisible. Perhaps the orange flames are due to a very fuel-rich mixture used to pre-cool the engine hardware before ignition (in other words, more than two hydrogen molecules for each oxygen molecule). A *stoichiometric* mixture (exactly two-to-one ratio, as in H_2O_1) would burn invisibly, and would likely make an explosion rather than the flames you see in the photo.

Figure 14.51 shows the aft end of the Space Shuttle and its three main engines. The SSMEs also use hydrolox, and the *sparklers* under the engines provide two functions: first, they cause the excess propellant to burn below the rocket rather than next to it. Second, they ignite the SSMEs for flight.

14.5.3.6 Water at Launch to Reduce Overpressure

We introduced the liftoff and flight acoustic environments in Chapter 11. In this section, we will focus mainly on measures used on launch pads to reduce ignition overpressure and liftoff acoustic loads.

The first launch of the Space Shuttle had significant ignition overpressure (IOP) that caused damage to its thermal tiles. Later flights used water spray to reduce IOP levels, as shown in Fig. 14.52. Figure 14.53 shows the water system that was used underneath the Space Shuttle during launch. More information on the testing of water suppression systems can be found in Chapter 15.

Fig. 14.51   The Shuttle used sparklers called radially outward firing initiators (ROFI) to burn off excess H_2 and are also the ignition source for the SSMEs. Source: https://vimeo.com/21326605.

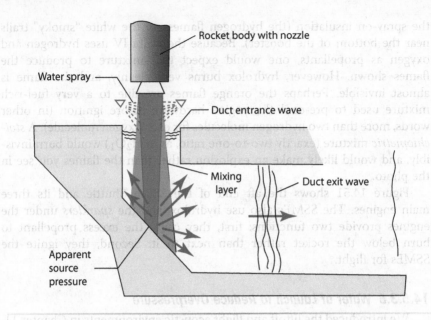

Fig. 14.52 Water spray can help absorb acoustic pressure or IOP. Source: [22].

The water system had a significant effect, reducing the Shuttle IOP by a factor of five between STS-1 and STS-2, as shown in Fig. 14.54.

Many launch vehicles use this method to reduce the acoustic environment, and it has the added benefit of also protecting the equipment on the pad from the hot exhaust of the rocket.

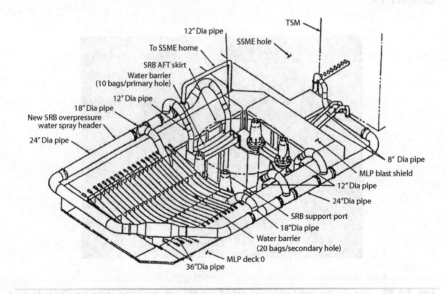

Fig. 14.53 Space Shuttle water troughs. Source: NASA.

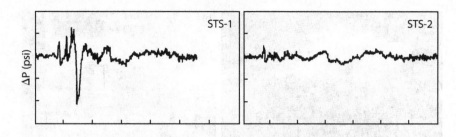

Fig. 14.54   Water suppression reduced Shuttle IOP by a factor of five between first (left) and second (right) Shuttle launches. Source: [26].

A water deluge system is shown schematically in Fig. 14.55. The water spray into the exhaust provides the acoustic suppression we desire, and is pretty much standard for all launches these days.

When Launch Complex 39 was used for Shuttle launches, an extensive sound suppression system was used. Photos of the system in operation are shown in Fig. 14.56. Some of the water sprayers are called *rainbirds*, after the popular lawn watering hardware.

Figure 14.57 shows a still frame from an Atlas V launch video. Note the jets of water issuing from the sides of the launch pad opening, similar to the concept shown in Fig. 14.55.

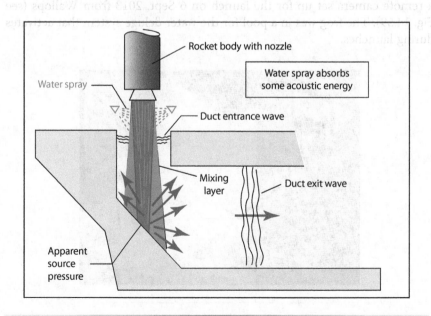

Fig. 14.55   Sound suppression: water deluge. Source: [22], shading added by author.

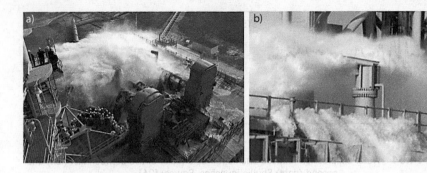

Fig. 14.56 a) Thousands of gallons of water are used to protect the pad and suppress sound. b) A rainbird jets the water out horizontally. Source: [16].

14.5.3.7 Exhaust and Flame Bucket

The launch pad area must be fitted to withstand the high speed, high temperature, and noisy engine exhaust, which must be channeled away from the rocket. Most launch pads have one or more ducts that direct the exhaust off to the sides. A *flame deflector*, as shown in Fig. 14.58, is used to change the exhaust direction.

14.5.3.8 Unintended Suborbital Flight

One frog of unidentified species was in "uncertain condition" after the LADEE launch, according to NASA. It was last seen in this photo taken by a remote camera set up for the launch on 6 Sept. 2013 from Wallops (see Fig. 14.59). The frog was in a pool for the water deluge system that activates during launches.

Fig. 14.57 The Atlas V pad has water jets that begin just after liftoff. Source: Still frame from video of Atlas V first launch, Lockheed Martin.

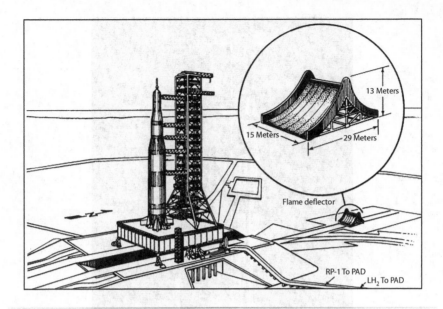

Fig. 14.58   The vehicle's launch pad must be configured to direct exhaust away from the vehicle. Source: [1].

14.5.3.9 Liftoff Service Tower Clearance

LVs with lower T/W_0 ratios tend to take a long time to get going, one example being the Saturn V with $T/W_0 = 1.25$. In these situations, there are concerns that ground winds coming from the wrong direction might cause the LV to drift sideways and contact a nearby structure. The Saturn V dealt with this concern by yawing 1.25 deg away from the pad starting at

Fig. 14.59   This frog had the bad luck of being present in a pool for the water deluge system that activates during launches. Source: [5].

Fig. 14.60   To clear the tower in case of unfavorable winds, the Saturn V yawed 1.25 deg during the first part of its launch. Source: [12], lines and label added by author.

T + 3.6 s and ending at T + 10.0 s to ensure adequate tower clearance during the first few seconds of launch. This yawing (tilting towards the left) motion may be seen in Fig. 14.60.

This yaw maneuver was also quite noticeable for the astronauts on board; one can hear the astronauts' exclamations in soundtracks of some Saturn V launches, particularly the launches of Apollo 12 (Pete Conrad) and Apollo 17 (Gene Cernan).

14.5.3.10 Tracking and Radar

We have not discussed tracking and radar. This subject is very specific to the particular launching location. It will be mentioned in Chapter 16 on flight termination systems.

14.6 Summary

In this chapter we've shown some of the internal subsystems and launch pad facilities needed for integration and to operate launch vehicles. When doing detailed design, the designer should consider the detailed design and layout of the launch pad, a topic that is beyond the scope of this text.

Now that we have a basic vehicle, we will move to the next chapter to discuss some details of *testing* launch vehicles, as well as considering reliability and redundancy to improve mission success probability.

References

[1] Benson, C., and Faherty, W., "Moonport—A History of Apollo Launch Facilities and Operations," NASA SP-4204, The NASA History Series, 1978.

[2] Brown, C., *Elements of Spacecraft Design*, AIAA, Reston, VA, 2002.

[3] Brown, K., and Weiser, P., *Ground Support Systems for Missiles and Space Vehicles*, McGraw-Hill, New York, 1961.

[4] Burgess, E., *Long-Range Ballistic Missiles*, Macmillan, New York, NY, 1962.

[5] Barber, E., "Frog Photobomb: NASA Launches Rocket, Frog," *Christian Science Monitor*, 12 Sept. 2013, https://www.csmonitor.com/Science/2013/0912/Frog-photobomb-NASA-launches-rocket-frog [retrieved 16 Aug. 2018].

[6] Eleazer, W., "Launch Failures: New Discoveries," *The Space Review*, 2018, http://www.thespacereview.com/article/3170/1 [retrieved 12 Jan. 2019].

[7] Fisher, M.F., "Propellant Management in Booster and Upper Stage Propulsion Systems," NASA TM-112924, 1997.

[8] Hobbs, M., *Fundamentals of Rockets, Missiles and Spacecraft*, John F. Rider, New York, 1964.

[9] Johnson Space Center, "Shuttle Booster SRB Overview," JSC-19041, 7 July 2003.

[10] Johnson Space Center, "Saturn V Stage 1 (S-IC) Overview," JSC-17237-24, 2009.

[11] Marshall Space Flight Center, "Umbilical Systems: V-2 to Saturn V," MSFC Report no. M-P & VE-M-7-63, 1963.

[12] Marshall Space Flight Center, "Saturn V Launch Vehicle Flight Evaluation Report-AS-502 Apollo 6 Mission," MSFC report no. MPR-SAT-FE-68-3, NASA TM-X-61038, 1968.

[13] Marshall Space Flight Center, "Saturn V SA-503 Flight Manual," Report no. MSFC-MAN-503, NASA TM-X-72151, 1968.

[14] NASA, "Interaction with Umbilicals and Launch Stand," NASA SP-8061, 1970.

[15] Phillips, J. D., and Tolson, B. A., "Holddown Arm Release Mechanism Used on Saturn Vehicles," NASA TM-X-3274, Paper 23, *Proceedings of the 9th Aerospace Mechanisms Symposium*, 17–18 Oct. 1974.

[16] NASA, "Sound Suppression Test Unleashes a Flood," 2004, http://www.nasa.gov/missions/shuttle/f_watertest.html [retrieved 16 Aug. 2018].

[17] NASA, "NASA's Exploration Systems Architecture Study Final Report," NASA TM-2005-214062, 2005.

[18] NASA, "Apollo 11 Flight Journal," 2006, https://history.nasa.gov/afj/ap11fj/pics/buildup.gif [retrieved 16 Aug. 2018].

[19] NASA, "Wings in Orbit," NASA SP2010-3049, 2010.

[20] Ring, E., *Rocket Propellant and Pressurization Systems*, Prentice Hall, Englewood Cliffs, NJ, 1964.

[21] Ryan, R., "Dynamic Challenges of Saturn Apollo," AIAA-00-1672, 41st Structures, Structural Dynamics, and Materials Conference and Exhibit, Structures, Structural Dynamics, and Materials and Co-located Conferences, 2000, https://doi.org/10.2514/6.2000-1672.

[22] Sarafin, T. P., and Larson, W. J., *Spacecraft Structures and Mechanisms: From Concept to Launch*, Springer, New York, 2007.

[23] Hellebrand, E. A., "Structural Problems of Large Space Boosters," Conference Preprint 118, presented at the ASCE Structural Engineering Conference, Oct. 1964.

[24] Platt, G. K., Nein, M. E., Vaniman, J. L., and Wood, C. C., "Feed System Problems Associated with Cryogenic Propellant Engines," Paper 687A, National Aero-Nautical Meeting, Washington DC, April 1963.

[25] McCool, A. A., and McKay, G. H., "Propulsion Development Problems Associated With Large Liquid Rockets," NASA TM X-53075, Aug. 1963.

[26] Bejmuk, B., "Space Shuttle Integration Lessons Learned," presentation in PDF format, 2006. https://www.nasa.gov/sites/default/files/atoms/files/bobejmuk.pdf, accessed 3 Aug. 2018.

Further Reading

For more information on umbilicals, refer to:

MSFC M-P&VE-M 7-63, "Umbilical Systems: V-2 to Saturn V"

NASA, "Apollo/Saturn V Space Vehicle Selected Structural Element Review Report, AS-503," NASA CR-105779, 1968.

NASA SP-8061, "Interaction with Umbilicals and Launch Stand"

Valkema, D., "Umbilical Connect Techniques Improvement Technology Study," Q19 Convair Aerospace Division, General Dynamics, Report No. NAS 10-7702, 1972.

For information on pyrotechnics, see:

Walker, N., *Rocket Science*, 2009, http://www.lulu.com/shop/norm-walker/rocket-science/ebook/product-17476181.html [retrieved 23 June 2019].

Chapter 15 Testing, Reliability, and Redundancy

e're moving ahead in our studies. We assume that a *verification matrix* containing analysis, inspection, demonstration, and test has been created, and all but *test* has been executed. Our vehicle is designed, production is complete, and the vehicle and its components are available for test. This chapter will discuss some of the testing that is done before launching. It will also cover what happens when testing or flight goes wrong and how safety is ensured during the process.

15.1 Testing

Different forms of testing are in use for different disciplines; however, all forms of testing are used to locate errors in analytical models and problems in already-built hardware. Types of testing can include:

* Mechanical (structural, usually static or quasi-static, pressure)
* Vibration, shock, and acoustics (VS&A)
* Thermal
* Radio frequency (RF), electrical, and the like
* Software

Of course, if any item fails a test, the failure must be investigated, and any defects must be corrected.

15.1.1 Levels of Testing

In spaceflight disciplines, mechanical, fluid, and electrical designs are exposed to testing at different levels. Those levels are as follows:

* *Qualification level*: Before a new component can ever fly, it must be tested to a qualification level that *qualifies the design to fly*. This level is the highest expected load multiplied by a factor of 1.4 or 1.5, or expected load with an added 3 dB, depending on the type of testing.
* *Protoflight level*: Protoflight comes from combining *prototype* and *flight hardware*. A protoflight component has been tested to qualification levels and then flown. In case something goes wrong, we prefer to use engineering prototype models [i.e., structural test article (STA), discussed

in the next section], but sometimes we only build a single protoflight component to save money.

- *Acceptance level*: Once a design has been qualified, other components built using that design need only be tested to lower acceptance levels to *ensure the workmanship is good*. This is also called *workmanship level*.

15.1.2 Classical Approach to Testing

In the past, when budgets were not as strict and testing was understood to be critical to success, testing was conducted with a number of test models. The most common three were:

1. *Structural test article (STA)*: This is *representative* of a flight structure, including mass-matched dummy electronic boxes. The STA may be used for *qualification* testing.
2. *Flight hardware*: This article has to withstand testing (be qualified) at higher than expected *flight* levels. If it passes, the design has been proven and can fly. This type of testing was retired.
3. *Engineering test article*: This is still used for testing things such as electrical or software systems, simulations, or troubleshooting. The engineering test article is commonly used for troubleshooting a vehicle that's in-flight, because it's as close to flight hardware as one can get. This is often made from the qualification-level tested (qual-tested) STA equipped with the flight hardware, after its testing is completed.

15.1.3 Protoflight Testing Approach

In today's world, budgets are tight, so there is emphasis on cost cutting, which means fewer testing articles. *Protoflight* refers to a strategy where no *test-dedicated* qualification article exists, and all production hardware (flight hardware) is intended for flight. Protoflight is the NASA standard for payloads (per NASA-STD-7002). Here, we have basically one test article, the prototype test article, similar to STA. It is used for qual-testing, and then the article is refurbished as needed to make it a flight model. The idea is that having one fewer model equals lower program cost.

It's a good idea to make a second article that could be used for troubleshooting when the flight article is in flight.

15.1.4 Testing vs Failure Mechanisms

All testing is designed to reveal flaws in an item. Launch vehicles, because they carry very valuable and costly payloads, must have high reliability in order to even be considered as a carrier vehicle. Hence, there are many tests that are done to try to uncover potential flaws. Table 15.1 shows a list of possible failures in the left column, followed by 10 columns of testing

Table 15.1 Testing Types vs Failure Mechanisms

Potential Failure Mechanism	Primary Tests to Induce Failure									
	Functional	Wear-In	Vibe/Acoustic	Shock	Thermal Cycle	Thermal-Vac	Leakage	Proof Pressure	Proof Load	EMC
Parameter drift	X		X		X	X				
Electrical intermittants			X	X	X	X				X
Solder joints			X							X
Loose wires			X							X
Connectors			X							X
Latent defective parts	X		X	X	X	X				
Parts shorting			X							
Chafed/pinched wires			X							X
Adjacent PWB contact			X	X						
Parameter change from deflection			X		X	X				X
Loose hardware			X	X						X
Binding of moving parts			X			X				
Leaky gaskets/seals					X	X	X			X
Lubricants characteristic change		X			X	X				
Material embrittlement				X	X	X				
Outgassing/contamination			X	X	X	X				
Degradation of insulation						X				X
Corona discharge/arcing						X				X
Defective pressure vessels								X		
Structural defects			X						X	
Defective wiring	X									X
Defective tubing							X			

EMI = electromagnetic interference, EMC = electromagnetic compatibility, PWB = printed-wire board (also known as PCB = printed-circuit board).

types that are intended to induce failure. An X in any column is meant to indicate that the test type on the top of the column is intended to reveal flaws in the article on the same line.

The tests in the table are all *physical* tests, so they are not going to be useful for testing nonhardware items like software (software testing is described later in this chapter). In addition, if the testing is not administered correctly, the item can either pass the test when it shouldn't have or be damaged in the testing process. *Testing is an art*, and if done improperly it can cause unnecessary damage or failures.

15.1.5 Types of Testing

Table 15.1 indicates 10 different tests that can be applied. Some of these tests are intended for the entire assembled vehicle or subcomponent-level testing; others are for components individually before they are integrated into the vehicle.

The subcomponent testing is for individual parts, subassemblies, and so forth. Similar principles may also be applied to blocks or subroutines of software that will eventually become the flight software.

Whole-vehicle testing involves an assembled or integrated vehicle, or the outer mold line in the case of scale model testing. This is typically done with wind tunnel or acoustic scaled models; other testing must be done on an integrated vehicle, for instance on-pad tests. Still others are done as simulations, where the testing may be partly flight hardware along with computer-simulated input signals and flight software.

15.1.6 Typical Test Sequence

Figure 15.1 shows a generic sequence of qualification testing that might be done on a flight vehicle. A quick look reveals that some of the tests are repeated, such as leak testing (done *three* times). Some items (e.g., centrifuge testing) would be done only on smaller vehicles.

Leak testing is done multiple times, because the testing after the first leak test could possibly damage the item and make it leaky. Low-level modal testing *after* the thermal-vacuum (thermal-vac) testing could also indicate damage if any changes in modal characteristics (frequencies or modal shapes) are detected.

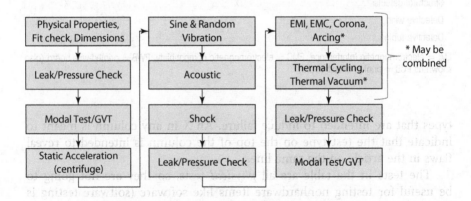

Fig. 15.1 A typical vehicle qualification testing sequence. Note: GVT = Ground vibration test.

Fig. 15.2 Testing of four very different launch vehicle components or operations. a) Atlas V fuel depletion probe; b) Atlas V master data unit; c) Atlas V launch hold-down testing; d) Saturn V swing-arm and umbilical test. Sources: [2], NASA.

15.1.7 Component Testing

It is standard practice to test a component for proper operation before installing it onto a vehicle. Component tests can follow the test sequence in the previous section, or only a few tests may be singled out depending on the nature and usage of the component. Avionics and propulsion components must be qualified to MIL-STD-1540C (military standard test requirements for launch, upper-stage, and space vehicles) equivalent levels for shock and vibration. The requirements may be found in [1].

Figure 15.2 shows three very different components under test. The first (a) is a fuel depletion probe, designed to send a signal to shut down the engine(s) when the fuel level is nearly empty. The second (b) is a master data unit, where considerable electrical and software testing would occur. The third (c) shows a swing arm connecting an umbilical to a mockup of the upper stage of a Saturn vehicle. Finally, (d) shows a test verifying operation of a pad hold-down device for an Atlas V.

The component testing varies greatly, depending on the component's tasks and its perceived weaknesses (which can be affected by location, such as shock). A fuel depletion probe should operate under cryogenic temperatures, pressure, and vibration. A master data unit would be shaken to predicted vibration levels and electrically tested at the same time.

The Saturn swing-arm testing was very unique. The swing-arms had to be in place to the very end of the countdown, but in the last moments had to be turned away so as to clear the rising rocket. They also had to be able to be reconnected in case of an abort and the vehicle would be "safed" by removing all its propellants. The testing was very strenuous: aircraft engines swinging propellers were used to create driving rain to drench the swing-arms and interconnecting plugs. At the same time, the rocket was made to sway back and forth to simulate wind-induced motion that might occur during a storm.

Hold-downs are very important. A launch vehicle should be clamped down to the launch pad until full thrust is detected, at which time the clamps should be released—quickly and simultaneously. If a vehicle's engines do not achieve the proper thrust in a few seconds, the vehicle *should not* be able to launch, and the engines should be commanded to shut down.

15.1.8 Wind Tunnel Testing for Ground Wind Response

Figure 15.3 shows the ground configurations of an Atlas and SLS being wind tunnel tested. Simulated ground testing helps to characterize how the vehicle responds to ground winds, using a scale model of the rocket and its service structure and any other buildings in a wind tunnel. Notice that the Atlas ground simulation in Fig. 15.3a is built on a turntable, so that winds from any direction can be simulated. Such testing determined that the

a)   b)

Fig. 15.3 Wind tunnel testing of Saturn IB for influence of ground winds. Note turntable allows 360° testing directions and b) SLS. Source: [15], NASA.

Saturn V had to have a damper installed to be sure it did not resonate in the wind (see Fig. 9.25 in Chapter 9, for a photo). The technician in Fig. 15.3b is releasing a small smoke stream in order to watch how the flow behaves near certain objects. If the flow becomes turbulent, it can cause problems with forces on the LV.

15.1.9 Wind Tunnel Testing for Flight

Wind tunnel testing is a way to gather aerodynamic data for use with stability and control, performance, and also to verify mathematical models used for computational fluid dynamics (CFD). (Some CFD folks believe it's the other way around.)

Wind tunnels are classified by the speeds they can attain (usually given as Mach number) and the size of their test section. Wind tunnels can go from subsonic (M < 0.8) to transonic (0.8 < M < 1.2) to supersonic (1 < M < ~4) to hypersonic (M > 4). It's also usually true that the higher the Mach number, the more power the wind tunnel will require, so the smaller the test section.

Figure 15.4 shows several wind tunnel tests of LVs. Figure 15.4a shows a large Saturn V model being tested in a supersonic wind tunnel. Figure 15.4b shows a much smaller model, about 10 cm (4 in.) long used in a hypersonic wind tunnel. Figure 15.4c shows a model of NASA's space launch system (SLS) with its shock waves visible via Schlieren photography. Figure 15.4d shows the aerodynamics occurring when the solid rocket boosters (SRBs) separate from the SLS core stage.

15.1.10 Mass Properties

It is very necessary to know the mass properties of the vehicle. Possibly most important is the center of mass (CM). Of course, the CM varies as propellants and consumables are used. Dry mass can be measured directly; wet mass is more difficult, so it may be inferred by measuring propellants and other liquids as they are added. Mass moments and products of inertia are particularly critical for successful control system operation. Careful calculation and/or measurement of these parameters is critical to mission success.

15.1.11 Structural Testing

One of the most visible and most important sets of testing is structural testing. It is imperative that the analysis used to design a launch vehicle or its parts be verified by test. A proper testing regimen will verify that the structure can withstand all predicted ground and flight loads without failure.

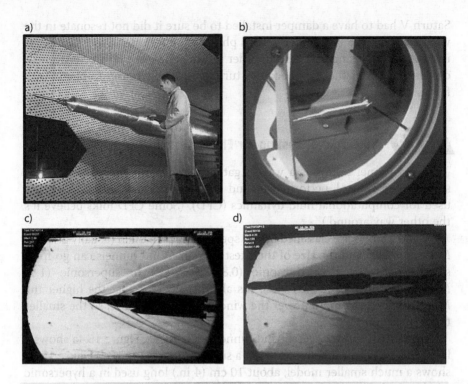

Fig. 15.4 Wind tunnel testing: a) Saturn V model in transonic wind tunnel; b) 10-cm (4-in.) hypersonic Saturn model; c, d) SLS model in wind tunnel with shock waves visible, aerodynamics and forces during separation of SLS and its strap-on SRM are measured. Sources: a) NASA, b, c, d) D. Edberg.

Stiffness testing must be performed on all major structural elements in the vehicle, and *pressure testing* must be done on tanks and bottles.

Figure 15.5 shows Atlas V components being tested with applied mechanical loading. The results of the testing are compared to numerical predictions from finite-element models (FEMs). Good agreement indicates the math models are reasonably accurate; poor agreement indicates more work needs to be done.

Pressure testing is particularly important for pressurized components. LV tanks are subjected to proof pressure testing using water, because pressurized air stores far too much energy that is released in the event of a tank failure. Typically, the component is placed into a steel-reinforced concrete test chamber and internal pressure is applied, as shown in Fig. 15.6.

15.1.12 Modal/Vibration Testing

Modal testing is a way to learn a structure's natural frequencies. This knowledge is necessary because there can be serious problems created

Fig. 15.5   a) Atlas V LOx tank under static structure test; b) Atlas V fuel (RP-1) tank under structure test. Source: [2].

by interactions between aerodynamic buffeting, structural response, and the vehicle's attitude control system. For modal testing, the structure (either a model or a full-scale vehicle) is excited through a range of frequencies by an electromagnetic shaker, and its response is usually measured with accelerometers. The data are typically converted to frequency response functions (FRFs), and these results are used as inputs for vibration analyses such as aeroelasticity, buffeting, and flutter analysis. This can also be used to verify a mathematical or FEM of the vehicle or structure.

Figure 15.7 is a still frame from a NASA video describing a test that became famous as the "tennis shoe test." A full-size mockup of a Saturn V was assembled in a tall building at Marshall Space Flight Center (MSFC), but the engineers did not know how to excite it. With a vehicle that size, a sledgehammer would barely move it and could do some serious damage. So it was decided that the technicians would excite the stage by gently pushing on it with their tennis shoes. When they stopped pushing, pulled back their shoes, and allowed it to vibrate freely, they could count cycles and estimate the natural frequency, and watch the decay in order to estimate the LV's damping.

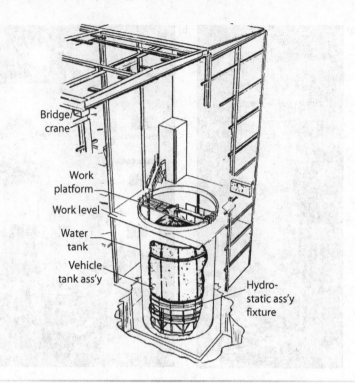

Fig. 15.6 Hydrostatic pressure is applied to the Saturn V's S-IVB stage to verify its pressure integrity. Source: [3].

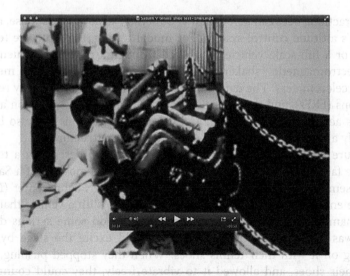

Fig. 15.7 The infamous "tennis shoe test" where engineers and technicians pushed directly on a Saturn V test article to excite its cantilevered bending vibration. Source: Still frame from NASA video.

15.1.13 Shock Testing

Shock testing is used to characterize the mechanical shock levels that occur from such events as solid rocket motor release, payload fairing separation, payload separation, and so on. We have seen that many of these become the design drivers for the design. In most cases, it's necessary to actually fire the pyrotechnic device in order to observe the system's response.

Figure 15.8 shows some details of the Centaur forward load reactor (CFLR), a structural member that serves to keep the somewhat flexible Centaur from rocking too hard and making itself or the payload come into contact with the payload fairing (PLF). The left side shows the assembled Centaur, CFLR, and part of the PLF. The upper right photo shows the CFLR, which has to separate from the upper stage when the PLF is jettisoned. The lower right photo shows a Falcon PLF separation test.

15.1.14 Slosh Testing

As discussed in Chapter 12, slosh has the potential to destabilize a launch vehicle. Each design is unique, so it's necessary to test and verify predictions made by computer code. The testing serves to characterize the natural slosh frequencies of liquid tanks and the damping associated with each. Figure 15.9a shows a photo of a slosh-testing tank at MSFC; Fig. 15.9b

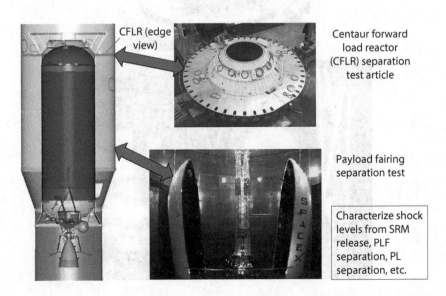

CFLR (edge view)

Centaur forward load reactor (CFLR) separation test article

Payload fairing separation test

Characterize shock levels from SRM release, PLF separation, PL separation, etc.

Fig. 15.8   CFLR separating the upper stage (top right) from the payload fairing (top left). Tests were done to measure the separation shock level. Sources: Left and top right, [2]; bottom right, still frame from https://www.youtube.com/watch?v=Ltl1V624vWM.

shows a still frame from a movie on the development of the Centaur upper stage.

15.1.15 Radio Frequency Testing

Radio frequency (RF) testing is done to ensure that electrical systems do not interfere with other systems and that they are compatible with other systems. The former is called *electromagnetic interference (EMI) testing*; the latter is called *electromagnetic compatibility (EMC) testing*. Testing of these parameters is usually done in an antenna test range or *anechoic* chamber, as shown in Fig. 15.10. The wedges of foam on the floor, walls, and ceiling serve to absorb RF energy rather than echoing it (thus *anechoic*).

a)

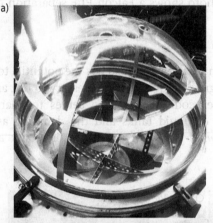

b)

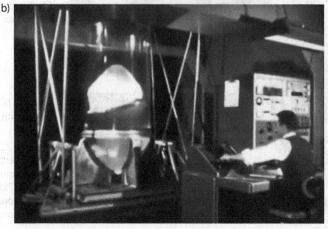

Fig. 15.9   a) Slosh-testing tank at MSFC; b) still frame from a video on the Centaur upper stage development. Sources: a) D. Edberg; b) https://www.youtube.com/watch?v=me3GOvLWaXE.

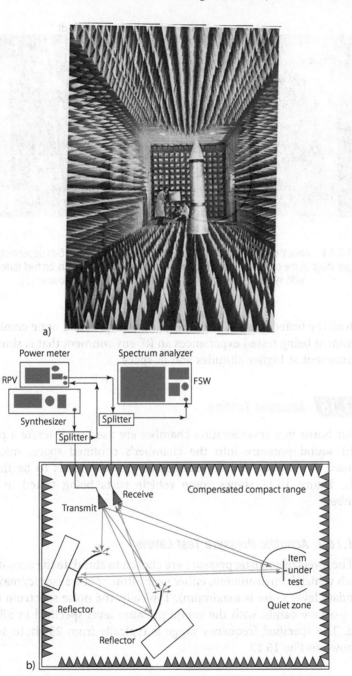

Fig. 15.10 a) Vehicle tests for EMI and EMC in anechoic chamber; b) typical setup for RF testing for equipment on LV and range safety receivers, antennas, and so on. Source: a) https://www.gallerym.com/collections/ralph-morse?page=2; b) Rohde & Schwarz.

a)   b)   c)   d)

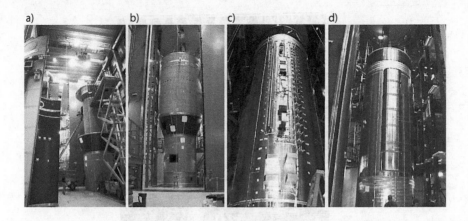

Fig. 15.11   Atlas V components being tested in a reverb chamber. a) 5-m upper stage assembly; b) upper stage in the chamber with PLF; c) avionics boxes and simulators on fuel tank; d) fuel tank with systems ducting covering boxes and simulators. Source: [2].

With all the tested items' RF energy going outwards and none coming in, the equipment being tested experiences an RF environment that is similar to the environment at higher altitudes and in space.

15.1.16 Acoustic Testing

Air horns in a reverberation chamber are used to generate a prescribed (high) sound pressure into the chamber's confined space; microphones are used to monitor and control sound pressure levels to be the desired levels. Figure 15.11 shows some vehicle parts being tested in a reverb chamber.

15.1.16.1 Acoustic Pressure Test Levels

The levels of acoustic pressure are chosen to simulate the acoustics of the launch vehicle's environment, either near liftoff or at transonic/max-q (when boundary layer noise is maximum). Typically, the noise spectrum is divided into $\frac{1}{3}$-octave bands, with the sound pressure level specified in dB for each band. The specified frequency range is typically from 25 Hz to 10,000 Hz, as shown in Fig. 15.12.

15.1.16.2 Acoustic Test Criteria

The sound pressure levels for an acoustic test inside of a PLF are usually listed in the LV's payload planner's guide (or user's guide, or similar information from the LV manufacturer). Both the levels and test durations depend on the type of test being carried out. The three levels of testing are typically acceptance/workmanship, protoflight, and qualification. The levels are related to each other as specified here:

- *For acceptance testing*: Add 0 dB to the maximum predicted environment.
- *Protoflight testing*: Add 3 dB to the maximum predicted environment.
- *Qualification testing (qual testing)*: Add 6 dB to the maximum predicted environment.

What is the duration of an acoustic test? Just enough to verify the item can survive—too long a test can cause fatigue and premature failure.

15.1.16.3 Example Acoustic Test

A payload acoustic testing example is taken from the 2006 edition of the *Delta II Payload Planner's Guide*. The maximum predicted environment is provided in Section 4.2.3.3 and shown graphically in Fig. 15.13.

> Now, according to Section 4.2.4.2:
> The maximum flight level acoustic environments defined in §4.2.3.3 are **increased by 3.0 dB** for acoustic **qualification** and **protoflight** testing. The acoustic test duration is **120 s** for qualification testing and **60 s** for protoflight testing. For spacecraft acoustic **acceptance** testing, acoustic test levels are equal to the maximum flight level acoustic environments defined in §4.2.3.3. The acoustic acceptance test duration is **60 s**. The acoustic qualification, acceptance, and protoflight test levels for the *Delta II* launch vehicle configurations are defined in Tables 4-9, 4-10, and 4-11. verbatim: [Emphasis added].

This indicates that, at least for the Delta II, qual- and protoflight-testing levels are the flight levels plus 3 dB. The only difference is the duration of the testing: 120 s for qual and 60 s for protoflight. For acceptance testing, it's flight levels for 60 s.

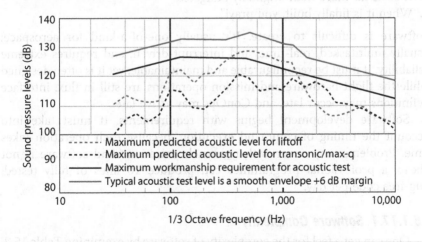

Fig. 15.12 Four different sound pressure level spectra corresponding to different flight events (liftoff and transonic/max-q), workmanship, and qualification testing. Source: Adapted from [4].

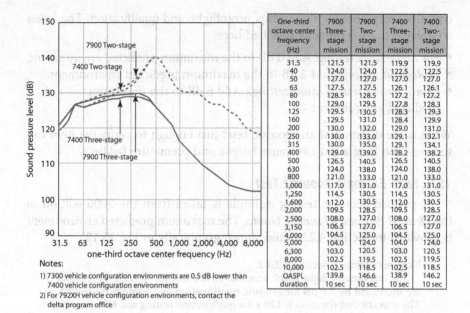

One-third octave center frequency (Hz)	7900 Three-stage mission	7900 Two-stage mission	7400 Three-stage mission	7400 Two-stage mission
31.5	121.5	121.5	119.9	119.9
40	124.0	124.0	122.5	122.5
50	127.0	127.0	127.0	127.0
63	127.5	127.5	126.1	126.1
80	128.5	128.5	127.2	127.2
100	129.0	129.5	127.8	128.3
125	129.5	130.5	128.3	129.3
160	129.5	131.0	128.4	129.9
200	130.0	132.0	129.0	131.0
250	130.0	133.0	129.1	132.1
315	130.0	135.0	129.1	134.1
400	129.0	139.0	128.2	138.2
500	126.5	140.5	126.5	140.5
630	124.0	138.0	124.0	138.0
800	121.0	133.0	121.0	133.0
1,000	117.0	131.0	117.0	131.0
1,250	114.5	130.5	114.5	130.5
1,600	112.0	130.5	112.0	130.5
2,000	109.5	128.5	109.5	128.5
2,500	108.0	127.0	108.0	127.0
3,150	106.5	127.0	106.5	127.0
4,000	104.5	125.0	104.5	125.0
5,000	104.0	124.0	104.0	124.0
6,300	103.0	120.5	103.0	120.5
8,000	102.5	119.5	102.5	119.5
10,000	102.5	118.5	102.5	118.5
OASPL	139.8	146.6	138.9	146.2
duration	10 sec	10 sec	10 sec	10 sec

Notes:

1) 7300 vehicle configuration environments are 0.5 dB lower than 7400 vehicle configuration environments

2) For 792XH vehicle configuration environments, contact the delta program office

Fig. 15.13 The predicted acoustic environments from the *Delta II Payload Planner's Guide*. Source: [5].

15.1.17 Software Is Becoming More and More Important

An old joke about software development goes as follows: "Building software is like building a cathedral:

1. It takes longer than originally expected.
2. It costs far more than originally budgeted.
3. When it is finally built, you pray!"

Software is difficult to create. It's usually one-of-a-kind for aerospace; usually multitasked, real-time, and interrupt-driven; and requires extreme reliability. It must be reconfigurable and maintainable, yet it is often designed while the flight hardware and mission operations are still in flux: interface definitions may occur late, and ConOps may arrive late.

Software development begins with requirements. It must take into account the timing of command execution, because each operation takes time. Problems may occur with nonsynchronization. And software is not cheap: a professional programmer can produce 10 lines of fully tested, bug-free code per day.

15.1.17.1 Software Complexity

One can get a feel for the complexity of software by examining Table 15.2, which lists software complexity for different vehicles. The trend is more and more lines of code.

Table 15.2 Software Complexity, As Measured By Lines of Code, For Some Vehicles

System	Lines of Code (LOC)*
Vega (ESA Launch Vehicle, 2013)	100,000[†]
A310 control system	400,000
Curiosity Mars Rover (2015)	500,000[§]
Automobile	2 million
Space Shuttle	3 million
B-2 Stealth Bomber	4 million (1.8 million[‡])
A340 control system	20 million
SDI and NMD (estimate)	25–100 million

*Most values courtesy [3].
[†]ESA Bulletin.
[‡]Private conversation, Thomas Colangelo, Boeing.
[§]NASA Entry, Descent, and Landing video.

15.1.17.2 Software Testing

After software coding has been completed and successfully compiled, the software must be tested. Table 15.3 provides a series of common software testing methods. Besides the items listed in the table, the following should also be carried out:

* Verification and validation (V&V), preferably by an independent organization (IV&V)
* Test software with actual spacecraft hardware or simulator with flight-like harnesses.

15.1.17.3 Software Reviews Pay Off

Figure 15.14 shows the errors found in 6,877,000 source lines of debugged code (including comments) on 28 projects. *Eighty-three percent of the errors* made in the categories marked with an asterisk (*) were detectable by review, showing how important and effective the software review process can be.

15.1.18 External Acoustic Testing

The acoustic signature of a launch vehicle near the ground is difficult to predict. For this reason, organizations carry out testing of scale models and extrapolate the results to the full-size vehicle. Testing is generally done with either *cold-flow* or *hot-flow* model testing. Cold-flow testing produces thrust and noise by forcing cold compressed gas through a rocket nozzle. Hot-flow testing uses actual rocket engines powered by real rocket propellants to create thrust and noise. Suggestions on model testing may be found in NASA SP-8072, Acoustic Loads Generated by the Propulsion System.

Table 15.3   Methods Used to Verify Proper Software Operation

White box	Based on detailed knowledge of design Example: programmer testing his or her own module
Black box	Based on functional requirements (specification) only Example: a *Red Team* conducting a test
Software defect testing	Design tests to cause the system to perform incorrectly and expose a defect Interface tests: use knowledge of functional specification, structure, and implementation to design tests that will exercise each object and message type in the system. Never permit defect testing to replace static verification (code walkthroughs, formal methods).
Code walkthrough/ Fagan inspection	Walkthrough/inspection requires a detailed study of the requirements, design, and code *prior* to the actual review, after the first good compilation. The walkthrough process includes some or all of the following participants: Presenter (lead reader, usually the designer/programmer) Moderator (coordinator, chairman) Recorder (scribe, secretary) One or two other technical reviewers Maintenance oracle (optional) Standards bearer (optional) User representative (optional) System liaison/system engineer (optional) Walkthrough should be performed software module by software module, and can be highly effective.
Software error seeding	The process of *intentionally* adding known faults to a program, It may be used to monitor the rate of bug detection and removal, and yields an estimate of the number of faults remaining in the program. Don't forget to *remove* the test faults (red tag items)!

Source: [6].

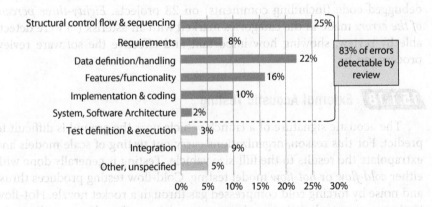

Fig. 15.14   Eighty-three percent of the errors found in approximately 7 million source lines of debugged code were detectable by review. Source: [7].

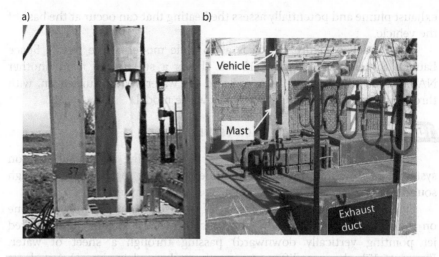

Fig. 15.15 a) 3.3% Atlas V model using cold high-pressure GN_2 to simulate noise due to burning; b) Atlas V model hot-fire testing facility includes scaled model of launch pad and exhaust ducting. Source: [2].

Figure 15.15 shows testing of cold-flow and hot-flow models. Figure 15.15a is an Atlas V 3.3% scale model in a cold gas test using gaseous nitrogen. Figure 15.15b is an Atlas V suspended above a simulated launch pad, including the pad's service mast and an exhaust duct underneath.

15.1.19 Hot Gas and *Exhaust* Plume Testing

As seen previously, the hot gas testing is used to estimate acoustic properties of the launch vehicle, but it can also be used to characterize the LV's

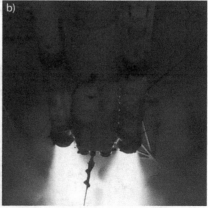

Fig. 15.16 a) A 5-s hot fire of 5% SLS model on test stand at MSFC; b) still frame of a video showing the hot firing of a 5% model of the NASA SLS. Source: [8].

exhaust plume and potentially assess the heating that can occur at the base of the vehicle.

The model in Figure 15.16a is a 5% scale model of the NASA Space Launch System (SLS). Figure 15.16b shows a still frame from another NASA video showing the burn and also the water system turned on, with three Rainbird water-ejecting heads below the model.

15.1.20 Acoustic Suppression Testing

Scale models may also be used for investigating sound suppression systems. These spray out large quantities of water to attenuate the high sound levels associated with a launch, as seen in Chapter 14.

Figure 15.17 shows some photos of acoustic suppression work done on the Atlas V. Figure 15.17a shows a cold exhaust stream (light-colored jet pointing vertically downward) passing through a sheet of water. Figure 15.17b shows a different water spray located in the exhaust ducts leading away from the launch pad.

15.1.21 Full-Scale Engine Testing

Before flight, it's necessary to have a good understanding of the engine vibration environment. The engines may be attached to a test stand, or better yet, the actual LV could be placed on its pad and static-fired. This allows one to obtain vibration data for the startup transient, steady operation, and the shutdown transient. Two photos of the Atlas V's Russian RD-180 engines firing statically are shown in Fig. 15.18. Figure 15.18a shows a close-up, and Fig. 15.18b shows the engine test stand in operation.

15.1.22 Result of Successful Testing Sequence

The result of a successful test sequence, the delivered vehicle, is shown in Fig. 15.19. This photo shows an Atlas V arriving at Cape Canaveral.

Fig. 15.17 a) Acoustic suppression water system where water system dispenses sheetlike spray; b) view of in-duct acoustic suppression water spray. Source: [2].

Fig. 15.18 a) Russian RD-180 engines installed on an Atlas V; b) the test occurring on a static test stand at MSFC. Source: [2].

15.1.23 After Hardware Delivery

Once the hardware is delivered, it must be assembled (*stacked*) and then tested again before flight. The test sequence for the Saturn V is shown in Fig. 15.20. After assembly, the vehicle undergoes a series of tests including power, systems testing, umbilical and swing arm testing, spacecraft testing, simulated flight testing, radio-frequency testing, and propellant loading. Once successfully completed, the vehicle is ready for flight.

Fig. 15.19 The result of a successful test program is the delivery of hardware. Here an Atlas V arrives at Cape Canaveral. Source: [2].

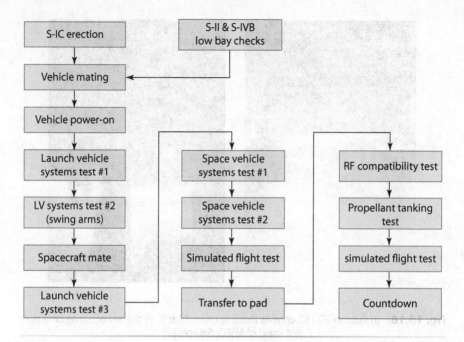

Fig. 15.20   These are the milestones in the checkout process for the Saturn V. Source: [9].

15.1.24   Flight Test

The first flight of an LV allows one to obtain actual data on liftoff transient loads, acoustics, engine vibration, ignition and shutdown transients, and separation transients. Figures 15.21 and 15.22 show four different flight phases for the first Atlas V, as recorded by a camera on board.

Once the vehicle is flying regularly, flight data can be collected, statistically analyzed, and used to generate required test levels for different

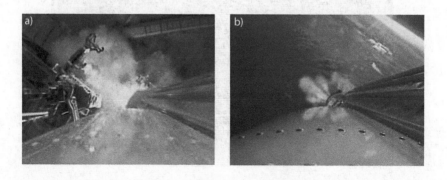

Fig. 15.21   a) View from an Atlas V looking downward shortly after ignition and liftoff; b) just before main engine cutoff. Source: [2].

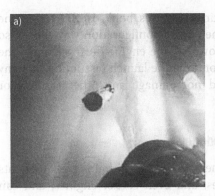 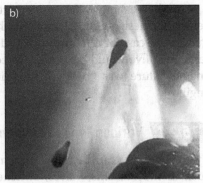

Fig. 15.22   a) Stage separation of the first Atlas V. The lower step can be seen in the center of the photo. b) Just after the Atlas V's payload fairing separation. The two halves can be clearly seen, and the lower step is still visible near the center. Source: [2].

on-board equipment. Also, the time histories can be used for the coupled-loads analysis discussed in Chapter 11.

Flight data are obtained by telemetry systems, which have hundreds of channels of data or more. The topic of telemetry is rich and complex, and is briefly introduced in Chapter 14.

15.1.25   Summary of Testing Practices for LVs

We've looked at a few of the tests that are done on prototype and in-production launch vehicles. For additional information, general data on testing launch vehicles (and upper stages) may be found in the references and within the most current edition of [10]. Next, we will consider several ways to increase a vehicle's operational reliability.

15.2   Redundancy

Redundancy is a means of improving reliability by adding extra components so that if one component fails, another can continue doing what the failed component did, and the mission can still proceed. One example of redundancy is your automobile. If you have a flat tire, you simply unmount the wheel with the flat and replace it with the spare wheel/tire in your trunk (or wherever it happens to be stored).

Note: some automobiles are being sold without a spare! According to *Consumer Reports* magazine [11], the quest for higher gas mileage makes shedding weight an interest, and the spare tire/wheel and a car jack can weigh 23–27 kg (50–60 lb). The elimination of that item can slightly improve fuel economy—and save the manufacturers money and time, too! So, the tire redundancy example is going to eventually not be true anymore.

Another example, far more pertinent to this book, is the six-engine vehicle we designed in Chapter 8. The engine configuration was chosen so that in the case of a failure or shutdown of one engine during launch, the remaining five would be enough to continue the launch (assuming that any engine failure was "benign" and did not damage any other systems or engines).

15.2.1 Various Types of Redundancy

Redundancy can be applied to mechanical and plumbing components such as mechanisms, valves, switches, and even rocket engines, and can also be applied to electronic components, such as batteries, switches, black boxes, and so on. In all these cases, additional parts are provided to carry on in the case of a failure.

It is very important to consider redundancy in great detail when designing a vehicle's subsystems. Additional cost, mass, and volume are required, and the very act of adding more parts can *increase* the likelihood of a part failure. So, an analysis has to be made trading increased mass and cost vs increased reliability.

15.2.2 Redundancy: Fluids and Hydraulics

Several types of fluid routing components are available, but it usually takes a combination to provide redundancy and properly safeguard a system, as shown in Fig. 15.23. *Parallel* redundancy (Fig. 15.23a) eliminates a failure due to a valve or part that is unable to open, but because there are now two parts, there is the possibility of more leakage if the parts leak. *Series redundancy* (Fig. 15.23b) reduces leakage, but now the failure-to-open probability is increased. Finally, *series-parallel redundancy* allows protection from failures to both open and close, and also reduces leakage. However, its negatives are higher mass, more testing required, and additional cost.

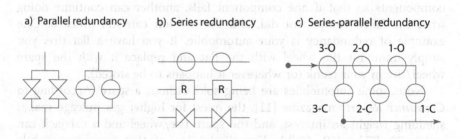

a) Parallel redundancy   b) Series redundancy   c) Series-parallel redundancy

Fig. 15.23 a) Parallel redundancy protects against failure to open, but can increase leakage. b) Series reduces leakage, but doesn't protect against failure to open. c) Combined series-parallel redundancy protects from both failures.

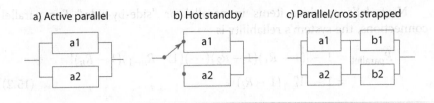

a) Active parallel   b) Hot standby   c) Parallel/cross strapped

Fig. 15.24 Three types of redundancy applied to electrical or electronic circuitry.

In Fig. 15.23c, the "-O" labels mean normally open, and the "-C" labels mean normally closed. Note that if valve 3-O fails to open, the system is useless.

15.2.3 Redundancy: Electronics

Redundancy may also be used in electrical/electronic systems. Some schemes are illustrated in Fig. 15.24. Figure 15.24a shows *active parallel*, where two components are active at the same time; it is useful if failure is either improbable or not common to both. Figure 15.24b shows *hot standby*, where the system can switch from unit a1 to unit a2 without any delays; however, the switch must be *very* reliable. Figure 15.24c shows *parallel/cross-strapped*, where there are four different paths from left to right: a1–b1, a1–b2, a2–b1, and a2–b2. This scheme tolerates many more failures, but adds complexity and requires additional fault analysis and increased testing time.

15.2.4 Reliability

When multiple items are selected to increase redundancy, it's clear that one must consider the fact that because more items are present, the probability of a failure of any one of them is larger, and one has to weigh whether the additional redundancy and its accompanying change in reliability are worth the added expense in parts, testing, and space required.

This section is concerned with the calculation of reliability when multiple items are being introduced, such as multiple rocket engines or additional valves in a propulsion system.

Let's first consider items in series, or "end-to-end." For series connections, the system's reliability R_{Series} is just the product of the individual items' reliabilities R_i

$$R_{\text{Series}} = R_1 \times R_2 \times R_3 \times R_4 \times \cdots \times R_{N-1} \times R_N = \prod_{i=1}^{n} R_i \quad (15.1)$$

Next, let's consider items in parallel or "side-by-side." For parallel connections, the system's reliability is

$$R_{\text{parallel}} = 1 - [(1 - R_1)(1 - R_2)(\cdots)(1 - R_{n-1})(1 - R_n)]$$
$$= 1 - \Pi_{i=1}^{n}(1 - R_i) \tag{15.2}$$

From these two basic configurations, series and parallel, it's possible to construct many different types of systems with series and parallel components. We will now look at some examples.

15.2.5 Redundancy Example: Heater Strip

Here's an example of how redundancy can help to protect against a component failure. Suppose you have an item that needs to be maintained within a given temperature range between a low and high value. That heater circuit might appear as shown in Fig. 15.25. If it's too cold, a heater strip is turned on using a switch module. If it's too hot, the switch module turns off the heater, and the component is allowed to cool. The question is: what happens if the switch module fails?

The switch module could fail in one of two ways: it could fail *closed* (a closed switch is a "short circuit" that always passes current and never stops, causing the item to overheat), or it could fail *open* (the switch will never close to provide power to the heater strip, and the item is cold). Let's consider a fail-closed situation first.

If the switch fails closed, then the item will overheat, unless we provide a second switch that we can open to allow it to cool off. The second switch would have to be in *series* with the first one, so that it could interrupt

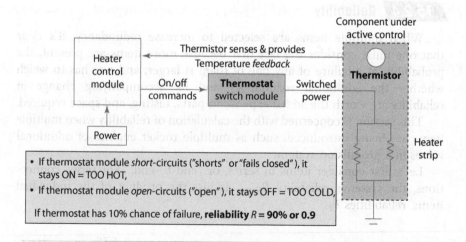

Fig. 15.25 Heater circuit.

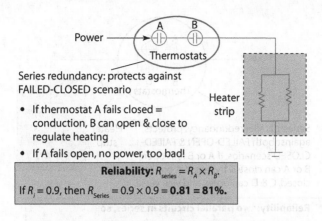

Power ——— A   B
Thermostats

Series redundancy: protects against
FAILED-CLOSED scenario

- If thermostat A fails closed =
  conduction, B can open & close to
  regulate heating
- If A fails open, no power, too bad!

Heater
strip

Reliability: $R_{series} = R_A \times R_B$.
If $R_i = 0.9$, then $R_{Series} = 0.9 \times 0.9 = \textbf{0.81} = \textbf{81\%}$.

Fig. 15.26 Series redundant heater circuit. A second switch in series protects against the other switch failing closed.

the flow of electricity that would be produced by the failed-closed switch. This *series-redundant* circuit is shown in Fig. 15.26. If switch 1 fails closed, switch 2 can be opened to control the circuit, and vice versa. But this scenario won't help if either switch 1 or 2 fails *open*.

Now suppose that a switch fails open. The item will be too cool because the power is interrupted. The only way to get power to the heater strip is to provide a second switch *in parallel* that we can close to heat up the item and also allow it to cool off. This *parallel-redundant* circuit is shown in Fig. 15.27. If switch 1 fails open, switch 2 can be closed to provide power,

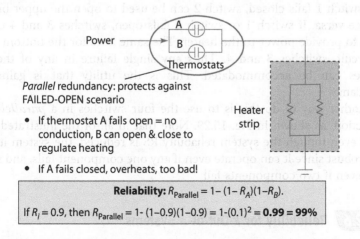

Power ——— A
B
Thermostats

Parallel redundancy: protects against
FAILED-OPEN scenario

- If thermostat A fails open = no
  conduction, B can open & close to
  regulate heating
- If A fails closed, overheats, too bad!

Heater
strip

Reliability: $R_{Parallel} = 1 - (1 - R_A)(1 - R_B)$.
If $R_i = 0.9$, then $R_{Parallel} = 1 - (1 - 0.9)(1 - 0.9) = 1 - (0.1)^2 = \textbf{0.99} = \textbf{99\%}$

Fig. 15.27 Parallel redundant heater circuit will not help if switch fails closed.

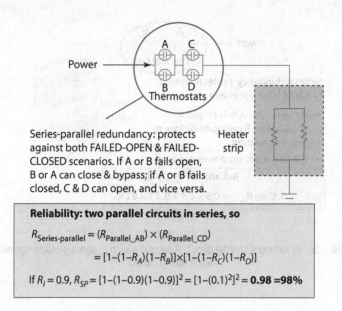

Series-parallel redundancy: protects against both FAILED-OPEN & FAILED-CLOSED scenarios. If A or B fails open, B or A can close & bypass; if A or B fails closed, C & D can open, and vice versa.

Reliability: two parallel circuits in series, so

$$R_{\text{Series-parallel}} = (R_{\text{Parallel\_AB}}) \times (R_{\text{Parallel\_CD}})$$

$$= [1-(1-R_A)(1-R_B)] \times [1-(1-R_C)(1-R_D)]$$

If $R_i = 0.9$, $R_{SP} = [1-(1-0.9)(1-0.9)]^2 = [1-(0.1)^2]^2 = \mathbf{0.98} = \mathbf{98\%}$

Fig. 15.28 A *series-parallel* redundant heater circuit can accommodate a single failure in any of its four switches.

and vice versa. But if either switch 1 or 2 fails closed, we cannot control the circuit.

Because we don't know whether the switch will fail open or closed, we have to take both into consideration. The simplest way to handle both fail-closed and fail-open failures is to employ an additional *three* switches, as shown in Fig. 15.28, to produce a *series-parallel* redundant heater circuit.

If switch 1 fails closed, switch 2 can be used to open the upper branch, and vice versa. If switch 1 or switch 2 fails open, switches 3 and 4 can be closed to provide power to the heater. The same goes for the bottom leg of the circuit, switches 3 and 4. Thus any single failure in any of the four switches can be accommodated. This is the utility that is gained by redundancy.

Another way to do this is to use the four switches in a *parallel-series* connection, as shown in Fig. 15.29. Note that in all of the illustrated cases above, even though the system reliability RS is reduced, the system itself is more robust since it can operate even if any one component fails, and sometimes even if two components fail.

15.2.6 Reliability for *k*-Out-of-*n* Systems

Let's consider the reliability of a *k*-out-of-*n* system. The definition of such a system is that the system *works* if *k* or more of the *n* components

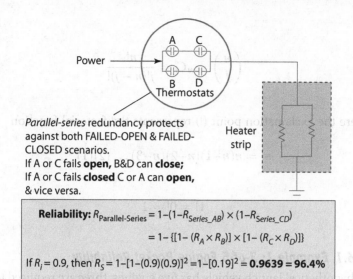

Parallel-series redundancy protects against both FAILED-OPEN & FAILED-CLOSED scenarios.
If A or C fails **open**, B&D can **close**;
If A or C fails **closed** C or A can **open**, & vice versa.

Reliability: $R_{\text{Parallel-Series}} = 1-(1-R_{\text{Series\_AB}}) \times (1-R_{\text{Series\_CD}})$

$$= 1-\{[1- (R_A \times R_B)] \times [1- (R_C \times R_D)]\}$$

If $R_i = 0.9$, then $R_S = 1-[1-(0.9)(0.9)]^2 = 1-[0.19]^2 = \textbf{0.9639} = \textbf{96.4\%}$

Fig. 15.29   A *parallel-series* redundant heater circuit can also accommodate a single failure in any of its four switches, although with slightly less reliability.

work. This is exactly a description of a multiengined vehicle, as we saw with the Saturn V and Falcon 9 earlier. Here are some examples of *k*-out-of-*n* systems in multiengine rockets with three engines:

- *3-of-3:* A three-engine rocket requires *all three* engines to succeed. No failures allowed.
- *2-of-3:* A three-engine rocket requires *two* engines to succeed. A single failure is allowed of any of the three engines.
- *1-of-3:* A three-engine rocket requires *one* engine to succeed. Two failures are allowed of any of the three engines.

It can be shown that for *n* identical components in parallel (i.e., clustered or parallel rocket engines as in the previous examples), the reliability R_S of *m* or more successes from among the *n* components is calculated from

$$R_S = \sum_{j=k}^{n} \binom{n}{j} R^j (1-R)^{n-j} \triangleq \sum_{j=k}^{n} P(j)$$

where

$$P(j) = \binom{n}{j} R^j (1-R)^{n-j}$$

and

$$\binom{n}{j} = {}_nC_j = \frac{n!}{j!(n-j)!}$$

Here the exclamation point (!) represents the factorial function

$$n! = n(n-1)(n-2)(n-3)\dots(2)(1)$$

and

$$1! = 0! = 1$$

15.2.6.1 Example 1: 3-of-5 Engine Rocket Reliability

A hypothetical launch vehicle has five engines; three are required to reach orbit. Assuming each engine's reliability is 95% = 0.95 or 1-in-20 chance of failure, what is the probability of reaching orbit?

Solution: $n = 5$, $k = 3$, $R = 0.95$, and system reliability is calculated as follows:

$$R_S = \sum_{j=k}^{n} P(j) = P(j=3) + P(j=4) + P(j=5) = \binom{5}{3}(0.95)^3(1-0.95)^{5-3}$$

$$+ \binom{5}{4}(0.95)^4(1-0.95)^{5-4} + \binom{5}{5}(0.95)^5(1-0.95)^{5-5}$$

$$= \frac{5!}{3!(5-3)!}(0.95)^3(0.05)^2 + \frac{5!}{4!(5-4)!}(0.95)^4(0.05)^1 + \frac{5!}{5!(5-5)!}(0.95)^5(0.05)^0$$

$$= \frac{120}{6(2)}(0.95)^3(0.05)^2 + \frac{120}{24(1)}(0.95)^4(0.05)^1 + \frac{120}{120(1)}(0.95)^5(0.05)^0$$

$$= (10)(0.95)^3(0.05)^2 + (5)(0.95)^4(0.05) + (0.95)^5 = 0.99884 = 99.884\%$$

Thus, we see that with 95% individual engine reliability, hypothetical rocket #1 would have 99.88% reliability, or a failure rate of 1 in 863. Clearly, the multi-engine design has improved reliability over a single engine.

15.2.6.2 Example 2: 7-of-9 Engine Rocket Reliability

A second hypothetical launch vehicle has nine engines; seven are required to reach orbit. Assuming each engine's reliability is 95% = 0.95, what is the probability of orbit for LV #2?

Solution: $n = 9$, $k = 7$, $R = 0.95$, and system reliability is as follows:

$$R_S = \sum_{j=k}^{n} P(j) = P(j=7) + P(j=8) + P(j=9) = \binom{9}{7}(0.95)^7(1-0.95)^{9-7}$$

$$+ \binom{9}{8}(0.95)^8(1-0.95)^{9-8} + \binom{9}{9}(0.95)^9(1-0.95)^{9-9}$$

$$= \frac{9!}{7!(9-7)!}(0.95)^7(0.05)^2 + \frac{9!}{8!(9-8)!}(0.95)^8(0.05)^1 + \frac{9!}{9!(9-9)!}(0.95)^9(0.05)^0$$

$$= (36)(0.95)^7(0.05)^2 + (9)(0.95)^8(0.05) + (1)(0.95)^9 = 0.99164 = 99.164\%$$

With 95% individual engine reliability, hypothetical rocket #2 would have 99.16% reliability or a failure rate of 1 in 120, a bit worse than the five-engine hypothetical design 1. It would seem that the reliability of nine engines is poorer than that of five engines, because nine engines provides more likelihood of failure than five engines does.

15.2.6.3 Actual Engine Reliability

The two examples above selected 95% as their engine probability, but this number was chosen arbitrarily and is actually quite a bit lower than achieved reliability values. Some actual engine reliability statistics are provided in Table 15.4 below [12].

If we applied the 99.88% reliability of the Merlin 1-D to our two above examples, we would obtain:

* 3 of 5 engine design has system failure rate of 1 in *58 million*
* 7 of 9 engine design has system failure rate of 1 in *6.9 million*

These are pretty respectable numbers, and speak well for the potential of success for multi-engined launch vehicles.

15.2.6.4 Example 3: Destruct System Reliability

Whalley [13] provides an analysis of the reliability of a vehicle's payload fairing separation system; the system's main parts are shown in Fig. 15.30.

Table 15.4 Selected Liquid-Propellant Engine Reliability

Engine	Qty	Fails/Uses	Reliability	Vehicle
Merlin 1D	9	1/801	0.9988 = 99.88%	*Falcon 9*
RS-25	3	1/405	0.9975 = 99.75%	*Shuttle*
RD-180	1	1/86	0.9883 = 98.83%	*Atlas V*
RD-107/108	5	1/1,335	0.9992 = 99.92%	*Soyuz*, post-2000
F-1	5	0/65	1.0 = 100%	*Saturn V*

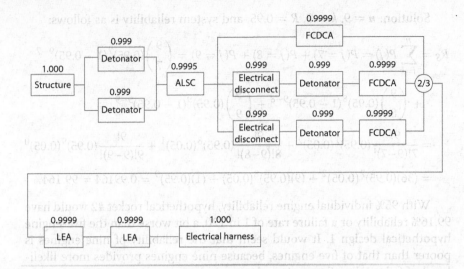

Fig. 15.30 Reliability analysis diagram for an LV flight termination system. This system had a calculated reliability of 0.9993, or 99.93%. The specification was 0.9985, or 99.85%. Requirement satisfied. Source: [13].

The analysis of the model indicated that its 99.93% reliability would meet the required reliability of 99.85% with a 95% confidence level. We will not discuss the analysis shown in Fig. 15.30 any further except for noting that a proper analysis requires knowledge of the reliability of each of its components, as well as the topography of the items' connections.

The two hypothetical examples in the previous sections (five- and nine-engine designs) used *fictitious* reliability data for their analyses, whereas the third used actual numbers. For a more informed and realistic analysis, the reader is referred to [14]. This reference contains component and propulsion system reliabilities, as well as results of failure analysis.

15.3 Summary

So, now we have seen the process of testing for launch vehicles, and discussed reliability and redundancy. In the next chapter, we will consider several different launch vehicle failures, as well as how to disable them in case something goes wrong.

References

[1] Space and Missile Systems Center, "Standard Test Requirements for Launch, Upper-Stage, and Space Vehicles," MIL-STD-1540C, 1994.

[2] Sowers, G., "Atlas V Development—Overview of Dynamics Testing," Presented at the 2001 Spacecraft and Launch Vehicle Dynamic Environment Workshop, El Segundo, CA, June, 2001.

[3] Ley, W., Wittmann, K., and Hallmann, W., *Handbook of Space Technology*, AIAA, 2009.

[4] "Environmental Testing for Launch and Space Vehicles," *Aerospace Crosslink*, Vol. 6, No. 3, Fall 2005, p. 13.

[5] Boeing, "Delta II Payload Planner's Guide," 2006, Fig. HB00956REU0.4.

[6] Hoffman, E., "Spacecraft QA Integration & Test," ATI short course preview, 2009.

[7] Blum, B., *Software Engineering: A Holistic View*, Oxford Press, Oxford, England, 1992.

[8] NASA, "NASA Ramps Up Space Launch System Sound Suppression Testing," 2014, https://www.nasa.gov/exploration/systems/sls/sls-acoustic-testing-2014.html [retrieved 19 Aug. 2018].

[9] NASA, "Moonport—A History of Apollo Launch Facilities and Operations," NASA SP-4204, 1978.

[10] Department of Defense, "Test Requirements for Launch, Upper Stage, and Space Vehicles, Vol I: Baselines; Vol II: Applications Guidelines," MIL-HDBK-340A, 1 April 1999.

[11] *Consumer Reports*, "Your Next Car May Not Have a Spare Tire," 2014, https://www.consumerreports.org/cro/news/2014/08/your-next-car-may-not-have-a-spare-tire/index.htm [retrieved 9 April 2018].

[12] Everyday Astronaut, "Engine Reliability as of Dec. 15, 2019," https://everydayastronaut.com/starship-abort/ accessed 2020 MAR 06.

[13] Whalley, I., "Development of the STARS II Shroud Separation System," AIAA 2001-3769, 2001, 37th Joint Propulsion Conference and Exhibit, Joint Propulsion Conferences.

[14] NASA, "State-of-the-Art Reliability Estimate of Saturn V Propulsion Systems," NASA CR-55236, General Electric Report RM 63TMP-22, 1963.

[15] Coupe, G., "Space Launch System – Next Giant Leap," Aerospace Testing International, Nov./Dec. 2013, http://www.ukimediaevents.com/publication/6fa6daa8/1 [retrieved 18 Aug. 2012].

Recommended Reading

Cheng, P. G., "Strategic Testing Lessons from Satellite Failures," Aerospace Corp., 4 Feb. 2003.

Fagan, M., "Design and Code Inspection," IEEE Trans. Software Engr., July 1986.

NASA, "Acoustic Loads Generated by the Propulsion System," NASA SP-8072, June 1971.

NASA, "Payload Test Requirements—NASA Technical Standard," NASA-STD-7002, July 1996.

15.4 Exercise

1. Consider the failure of tires on a four-wheel automobile.

   a) How many tires are required to operate the automobile? Which of the two situations shown in Figures 16.9 (series reliability) and 16.10 (parallel reliability) does the auto correspond to? How many items "n" are in the auto's system?

   b) Suppose you have an automobile tire that goes flat one day per year, on average. What is its reliability R in terms of days per year (round to the nearest integers)?

   c) Using the "k-out-of-n" terminology, what is the system description if a fifth tire, the "spare", is included in the analysis? what is the system reliability using the value of R in part B?

d) We often assume that vehicles are symmetrical, but it's interesting to consider whether there might there be variations in the flat-tire probability from side-to-side or front-to-back. Provide as many reasons as possible that could explain how the tires' flat probability could vary geometrically and not just randomly.

Chapter 16

Failures, Lessons Learned, Flight Termination Systems, and Aborts

There is a saying that says something to the effect of "those who don't learn from the past are condemned to repeat it." This is certainly true for launch vehicles and space flight. The reason for this chapter is to provide a few examples of mistakes that have been made, to instill the idea that a well-read, experienced engineer or designer should be able to recognize when something is not going to end well. Some folks say that the definition of an expert is someone who realizes she or he is going to repeat a mistake *just before* she or he repeats it!

16.1 Causes of Expendable Launch Vehicle (ELV) Failures

A presentation to the Commercial Space Transportation Advisory Committee (COMSTAC) in the U.S. Federal Aviation Administration by Demidovich [1] provided a breakdown of the known causes of worldwide ELV launch failures between 1957 and May 2007. The breakdown is shown as a pie chart in Fig. 16.1.

Barnaby Wainfan is a Northrop Grumman Technical Fellow and a colleague, and kindly volunteers to judge my students' vehicle design presentations. One of his more famous quotes: "It's always propulsion's fault." True or false? Well, perhaps not always, but definitely a majority of the time.

The illustrations following demonstrate that propulsion issues do cause the majority of launch failures, at least up to May 2007. Figure 16.1 shows a distribution of launch failures as determined by COMSTAC. It was commissioned by the U.S. Federal Aviation Administration and presented in 2007. Looking at the figure, it's very apparent that

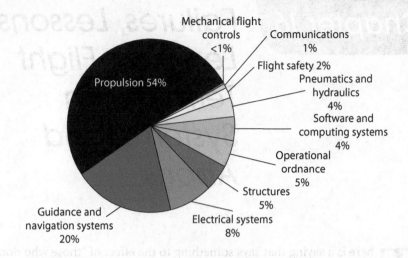

Fig. 16.1 Known causes of launch failures worldwide between 1957 and May 2007. Source: [1].

propulsion is responsible for the majority of failures between 1957 and May 2007.

Technology is supposed to improve with time, so we might expect that an analysis of a later period of time would show fewer failures of propulsion. Figure 16.2, however, shows a similar analysis for a later period. It shows that at 54% of known failures during the period of 1980 to May 2007, propulsion remains the largest failure cause. The presentation suggests: "Propulsion anomalies remain salient failure mode to date."

A similar study by Futron, commissioned by SpaceX and containing only U.S.-built vehicles, provides another assessment [4]. Launch failures by subsystem root cause for vehicles between 1984 and 2004 are shown in Fig. 16.3. Although the categories are different, the sum of the solid and liquid propulsion rates is 52%, virtually identical to the COMSTAC results given in the previous figures. A list of the 25 failures used for Futron's analysis is provided as an appendix to the report.

16.1.1 Summary of Launch Failure Findings: Trends

The 2004 Futron report [4] provides an interesting observation regarding U.S. launch failures: Separation events and engine/motor failures are the cause of approximately 80% of launch vehicle (LV) failures.

The 2007 COMSTAC presentation provides additional interesting information about worldwide launches:

- Thirty-nine percent of failures occur during operation of final stage.
- Propulsion anomalies—liquid, solid, and combination—have caused and continue to cause most known failures in expendable launch vehicle (ELV) launches.

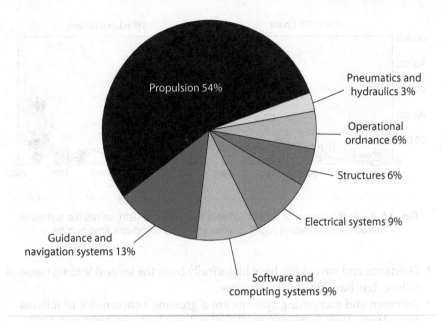

Fig. 16.2   Known causes of launch failures worldwide, 1980–May 2007. Source: [1].

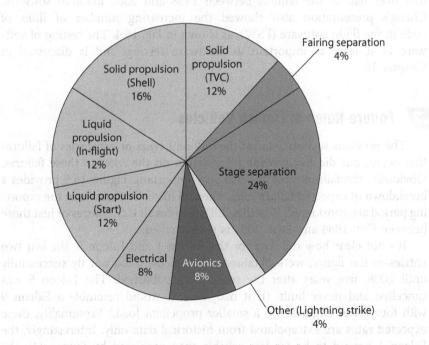

Fig. 16.3   Root causes of launch failures of U.S.-built vehicles from Oct. 1980 to Sept. 2004.
Source: Data from [4].

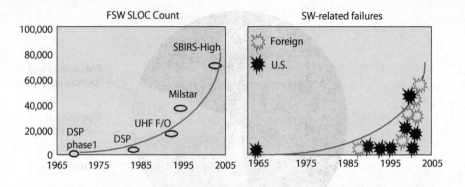

Fig. 16.4 As the number of lines of software has increased (left), so has the number of software-related failures (right). Source: [5]. SLOC = software lines of code

- Guidance and navigation have historically been the second leading cause of failure, but have dropped over time.
- Software and computing systems are a growing concern: 8% of failures from 1990–1999 increased to 21% of failures between 1999 and 2007.

The concerns about software failures that are mentioned in the COMSTAC presentation are echoed by Cheng [5], whose work mentioned that over half of the failures between 1998 and 2000 involved software. Cheng's presentation also showed the increasing number of lines of code in the flight software (FSW), as shown in Fig. 16.4. The testing of software is at least as important as hardware testing, and is discussed in Chapter 15.

16.2 Failure Rates of Launch Vehicles

The previous section detailed the different *types* or categories of failures that occur, but did not provide information on the *rates* of these failures. Obviously, the failure rates are also very important. Figure 16.5 provides a breakdown of expected failure rates when all historical failures in the reporting period are considered. Whether "all" includes all U.S. failures or just those between Oct. 1984 and Sept. 2004 is not specified.

It's not clear how the data for the Falcon 1 and Falcon 5, the last two entries in the figure, were obtained. The Falcon 1 did not fly successfully until 2008, five years after the report was published. The Falcon 5 was cancelled and never built. (If it had been, it would resemble a Falcon 9 with four fewer engines and a smaller propellant load.) Presumably, these expected rates are extrapolated from historical data only. Interestingly, the Falcon 1 proved to be far less reliable than predicted by Futron [4]: the first three flights ended in failure, leading to a reliability of 40% over five flights total. However, due to initial "teething problems," one would expect that with continued use and more flight experience, the reliability figure would grow much better. In general, it seems that one can expect a success rate of around 95–99% for many developed launch vehicles.

16.2.1 Design Implications of Multiple Engines

Space Exploration Technologies Corp. (SpaceX) has taken the failure cause information in Figs. 16.1, 16.2, and 16.3 to heart. A SpaceX 2013 brochure [6] states:

> ... The Falcon launch vehicles have been designed to eliminate the main causes of launch vehicle failures—separation events and engines. Our vehicles have only two stages for minimum staging events and make use of either one engine per stage for simplicity or multiple engines for propulsion redundancy. To ensure ... system performance, we have ... a hold-before-launch system to prevent a liftoff with an underperforming first stage.

Launch vehicle designs with multiple engines have both advantages and disadvantages. Advantages include simplification of the control system—instead of having to provide vernier engines for attitude control, multiple engines can be gimbaled differentially for roll control, and in case of a failure, the remaining engines can provide enough thrust to continue flight, albeit at a reduced acceleration level. Of course, the earlier an engine failure occurs, the more serious the situation is, because the thrust-to-weight ratio is lowest at liftoff and increases with time of flight. Disadvantages include the cost of multiple engines over the cost of a single large engine, and the simple fact that the chance of a failure in a group of k engines is k times the probability of a single engine's failure.

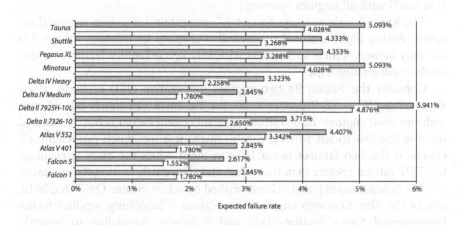

Fig. 16.5   Expected failure rates due to (1) propulsion and separation events (gray bars, based on the historical average) and (2) all causes (white bars, based on historical average subsystem failures). Source: Data from [4].

As we mentioned earlier in this chapter, one of the explanations SpaceX provides for the Falcon 9 design having multiple engines is that the most common failures occur in the propulsion system. The use of multiple engines allows for the loss of the thrust of an engine without mission failure, provided there are enough remaining operating engines to maintain a thrust-to-weight (T/W) ratio greater than 1.0 for positive acceleration. In the aircraft business, the loss of one engine is referred to as *one engine inoperative (OEI)*, and we shall use the term here as well.

The OEI T/W ratio may be easily calculated for multiengine vehicles with k engines as follows. If *one* engine produces a thrust of T_k, the liftoff thrust-to-weight ratio ψ of a vehicle with k engines can be written

$$\frac{T}{W_0} = \frac{kT_k}{W_0} = \psi \tag{16.1}$$

Then for $k-1$ engines, or an OEI situation, the thrust-to-weight ratio at liftoff T_{OEI}/W_0 is calculated as

$$\frac{T_{OEI}}{W_0} = \frac{(k-1)T_k}{W_0} = \frac{kT_k}{W_0} - \left(\frac{1}{k}\right)\frac{kT_k}{W_0} = \psi - \frac{\psi}{k} = \psi\left(\frac{k-1}{k}\right) = \psi\left(1-\frac{1}{k}\right) \tag{16.2}$$

Consider the design of the Falcon 9. Data from SpaceX's website (1 Dec. 2018) indicate at liftoff, $m = 549{,}054$ kg, nine Merlin engines thrust $9 \times T_9 = 7{,}607$ kN, and the thrust-to-weight ratio is

$$\psi = \frac{kT_k}{W_0} = \frac{(7{,}607 \text{ kN})(546\text{E3 kg})}{9.81 \text{ m/s}^2} = 1.412 \tag{16.3}$$

For OEI, $\psi_{OEI} = T_{OEI}/W_0 = (8/9)\,\psi = 1.256$, about the same as Saturn V at liftoff with all engines running.

The SpaceX webpage says Falcon 9 "can sustain up to two engine shutdowns during flight and still successfully complete its mission. Falcon 9 is the only launch vehicle in its class with this key reliability feature." Is this *truth or advertising hype*?

Consider the Falcon 9's two engines inoperative (2EI) scenario. Here, $\psi_{2EI} = T_{2EI}/W_0 = (7/9)\psi = 1.09$. So, the Falcon 9 can *slowly* lift off even with *two* dead engines! In this situation, a successful ascent is not guaranteed because the low thrust-to-weight ratio might use too much propellant. Of course, if the two failures occur later in flight when the mass is reduced, the T/W ratio is greater than this value, and the vehicle can accelerate faster.

The SpaceX claim has also been verified by actual events. On 7 Oct. 2012, one of the nine first-step engines on a Falcon 9 launching supplies to the International Space Station (ISS) had a failure. According to SpaceX, although the engine did not explode, pressure was lost, and the engine was shut down. In response, the flight computer computed a new ascent profile to ensure orbit entry. Because the Falcon 9 first step is designed to handle an engine-out situation and complete its mission, its payload was delivered to ISS as planned.

An additional factor to consider is the selected thrust line of the multiple engines. Instead of being vertical, it is said that the Saturn V's four outboard S-IC engine nozzles were tilted toward the outside of the vehicle. In the event of a premature outboard engine shutdown, the remaining engines would still be thrusting through the vehicle's center of mass, reducing any destabilizing torque caused by unbalance of the thrust of the remaining engines. Otherwise, there will be lateral acceleration or the vehicle has to fly with its long axis not aligned with the flight path.

Two Saturn V missions were also saved by a multiengine design. One of Apollo 6's J-2 engines on the S-II step was shut down due to pressure fluctuations caused by pogo oscillation; interestingly, incorrect engine control wiring accidentally shut down a second J-2 engine. Even with two out of five engines disabled, the rocket limped into orbit.

Another example occurred during Apollo 13's ascent. This time, the Saturn's S-II stage's central J-2 engine (#5), shut down 2 min and 12 s early, again due to severe pogo vibrations. The remaining four engines burned for an extra 4 min to compensate for the lost engine; however, the spacecraft achieved desired orbit.

Here is a final comment: note that the Falcon 9 and all other vehicles with single-engine *upper stages* have *no* capability to achieve orbit in an engine-out situation.

16.3 Some Examples of Launch Vehicle Failures

16.3.1 The *Challenger* Incident: A Propulsion System Failure

The U.S. Space Shuttle *Challenger* was destroyed, and its seven crew-members killed after an explosion 73 s into flight on 28 Jan. 1986. The essence of the event, which should not be characterized as an accident, is that the Shuttle was *deliberately* launched in conditions that were known to be outside its envelope of performance. It had been exposed to subfreezing conditions overnight, and this was known to reduce the performance of the O-rings that maintained the pressure seal inside of the Shuttle's solid rocket boosters.

When the Shuttle's solid rocket boosters (SRBs) were ignited, the cold temperature of the O-rings prevented them from sealing properly, which allowed high-temperature combustion gases to leak out. The leaking high-temperature gases, in turn, burned through the Shuttle's external tank, leading to its collapse and explosion, which also destroyed the Shuttle.

The events leading up to the decision to launch are quite extraordinary; without going into details, an engineering manager was encouraged by his NASA customer to "take off his engineering hat and put on his management hat," so he would overrule the objections of his direct reports to the launch that day. NASA management abandoned its standard "prove to us that the vehicle is safe to fly" and pressed for launching anyway, taking on a "prove to us why we *shouldn't* launch" attitude.

There is a great deal of literature on this event; the author recommends McDonald and Hansen [12] as well as the Rogers Commission's official report of its investigation of the event [13].

Let's look at the engineering that caused the problem. Figure 16.6 shows the original joint cross-section that utilized *two* O-rings on the left, and the redesigned joint on the right with an additional O-ring.

The geometry of the original SRB joint, as shown in Fig. 16.6, was such that it caused the O-ring seal to be *less* effective when the motor pressurized, thus leading to leaks and burning of O-rings. This joint rotation flaw, shown in Figure 16.7, was well known, but not corrected. Instead, management continued to provide "waivers" of safety issues so the Shuttle could continue to fly.

This flaw, along with the poor O-ring performance at low temperatures, led to the *Challenger*'s destruction and loss of its crew. This event cannot be characterized as an *accident*, because it was the result of deliberate actions and pressure on a contractor coming from NASA Marshall Space Flight Center's management.

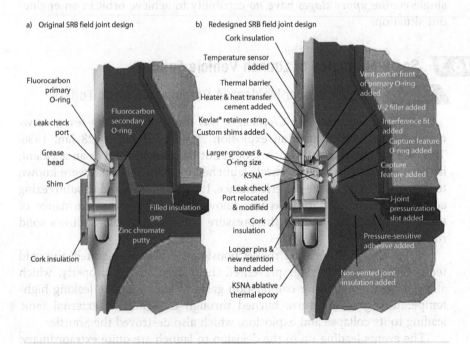

Fig. 16.6   a) Before *Challenger* and b) after *Challenger* configurations of the Space Shuttle's SRB field joints. Source: http://science.ksc.nasa.gov/shuttle/technology/images/ srb_mod_compare_6.jpg.

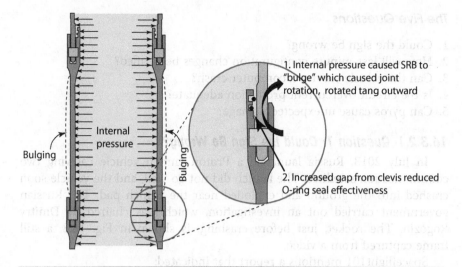

Fig. 16.7 How the internal pressure rotated the joints of the SRB segments, causing O-rings to leak. Source: [14].

The design solution to the O-ring problem was to add a third O-ring and its associated "capture tang," as depicted on the right of Fig. 16.6. In addition, an electric heater was added to make sure the O-rings' temperature would never get low enough for them to not seal properly. These corrections cured the blow-by problem, and the remainder of the Shuttle flights until retirement were not troubled with SRB sealing problems, as far as the authors know.

This incident regarding the Space Shuttles' SRBs is one that involved the performance of a system that was being exposed to an environment that was out of its certification. That alone raised red flags, but NASA's management ignored the concerns. Instead of demanding proof that the launch was safe, NASA demanded reasons why the launch should be delayed. This is opposite of the common flight readiness procedures.

Now, we will discuss common mistakes that are worth looking for.

16.3.2 Common Mistakes to Look For

Cheng [15] provided many common mistakes in a report titled "Five Common Mistakes Reviewers Should Look Out For." The report includes a reviewer's checklist titled "100 Questions for Technical Review" and all of the lesson bulletins published through June 2007. Unfortunately, this report is available *only* to U.S. government agencies and contractors.

However, the five common mistakes are available with unlimited distributions, so here they are. We will follow them with some examples to demonstrate how they caused failures.

The Five Questions

1. Could the sign be wrong?
2. How will last-minute configuration changes be verified?
3. Can the vehicle survive a computer crash?
4. Is the circuit overcurrent protection adequate?
5. Can pyros cause unexpected damage?

16.3.2.1 Question 1: Could the Sign Be Wrong?

In July 2013, Russia launched a Proton launch vehicle carrying two communication satellites. The launch did not go well, and the vehicle soon crashed into the ground and exploded near the launch pad. The Russian government carried out an investigation, which was chaired by Dmitry Rogozin. The rocket, just before crashing, is shown in Fig. 16.8, a still frame captured from a video.

Spaceflight101 mentions a report that indicated:

> ...all pre-launch activities had been proceeding normally until around 0.4 seconds before liftoff, when the emergency flight algorithm was activated. Around 6.8 seconds after the liftoff signal, the telemetry showed a sharp increase in the movement of steering mechanisms in engines numbers 1, 3, 4 and 6, with their actuators reaching maximum angles. At T + 7.7 seconds in flight, gimbal angles along the yaw axis reached their maximum possible angle of 7.5 degrees. Practically from the beginning of the flight, an unstable process of deviation from the correct yaw axis was observed. At T + 12.7 seconds in flight, a signal indicating the exceeding of the maximum allowable angle was issued, as the stabilization system (of the rocket) was no longer able to control the yaw. As a result, at T + 12.733, the "launch vehicle failure" command had been generated."

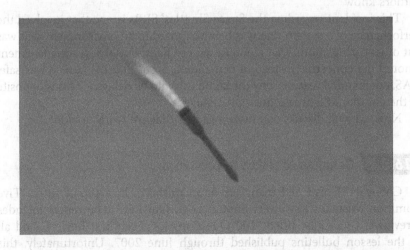

Fig. 16.8 A Russian Proton launch vehicle spins out of control just before crashing. The cause: control system rate gyros were installed *backwards*. Source: Still frame from video from Spaceflight101 (http://www.spaceflight101.net/proton-m-block-dm-03-glonass-launch-2013.html).

"... the telemetry analysis had confirmed that the anomaly in the movement of the vehicle along the yaw axis had been caused by an abnormal operation of angular velocity sensors in the PV-301 instrument unit. A total of six PV-301 units are mounted in two groups on a platform in the aft section of the second stage of the Proton rocket. Three units are responsible for the pitch and three for yaw axis of the flight trajectory."

It was also confirmed that " ... two out of three incorrectly installed yaw DUS sensors had been identified after their recovery from the crash site, thanks to remnants of red and yellow paint. The third such instrument was not positively identified because its paint cover had been completely burned."

NPR [16] provided information from Russian Space Web [17] that indicates that investigators traced the problem to a series of sensors that were apparently installed *upside down*. The angular velocity sensors were a critical part of the circuitry that kept the rocket upright during launch. They were so important that they even had little arrows on them that were *supposed to point toward the top of the rocket*.

That seemingly was not enough to prevent a technician from installing them upside down: "The upside-down sensors misinformed the rocket's flight control system." More discussion may be found in NASASpaceflight [18].

In the crash video, you can actually see the massive machine wobble back and forth as the system appears to try to correct the problem. Note that this was not actually a case of "getting the sign wrong," which would occur as a wiring error or software error. This was actually an installation error that had the same effect as an incorrect sign. The next example is also not a wrong sign error, but it is a misplaced decimal point error causing another failure.

Titan IV B32/Centaur (MILSTAR II-1) Failure

A Titan-Centaur launched a spy satellite into orbit on 30 April 1999. Unbeknownst to all, a roll rate filter constant that was manually entered into Centaur's avionics database had a decimal point placed one numeral to the left of where it was supposed to be, cutting the roll rate constant to one tenth of its correct value [19]! A test one week before launch produced data showing something was amiss, but the test had been done but *had not been monitored!* In addition, there were some anomalous indications in the final hours of the countdown, but they were misinterpreted.

During the first burn of the Centaur, an anomalous roll caused the rocket to use up 85% of its attitude control propellant, which ran dry during a subsequent maneuver. The result of this apparently tiny error was a spacecraft placed into an incorrect orbit, as shown in Fig. 16.9. The incorrect orbit made the spacecraft unusable and resulted in a US$1.32 billion loss (total cost of spacecraft and launch vehicle), the costliest for DoD up to that time!

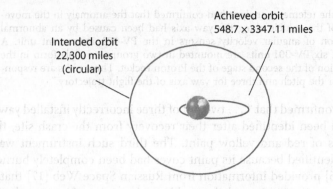

Intended orbit
22,300 miles
(circular)

Achieved orbit
548.7 × 3347.11 miles

Fig. 16.9 A slipped decimal point caused the spacecraft to miss its geosynchronous orbit by a very long way, making the spacecraft useless.

Technically speaking, this was not a failure caused by a sign error. But, it occurred as a result of carelessness that could have just as easily omitted a minus sign that should have been present.

16.3.2.2 Question 2: How Will Last-Minute Configuration Changes Be Verified?

Launch vehicles (and spacecraft) are frequently modified after factory testing, particularly if they've been disassembled and reassembled after being shipped to the launch site. During the process of integration with the launch vehicle, a number of things have to occur:

* Flight connectors must be mated, stored databases are updated, and brackets may be installed to secure the hardware.
* Nonflight items, such as test plugs and dust covers, are removed.
* Last-minute changes may be made in the heat of a countdown. In the past, some of these changes have caused several failures; this is because late installations and removals are difficult to track and verify.

Mistakes are often repeated due to incomplete requirement implementation, improper changes to the hardware being made, items being reused or recycled, or just plain *inadequate configuration management processes.*

Incident at Lockheed

A NOAA-N spacecraft being worked on at Lockheed Martin's Sunnyvale, California, plant in 2003 was *dropped* while being turned from vertical to horizontal! It happened because the 24 fasteners that were supposed to be holding the spacecraft onto the turn-over cart were *missing*. The fallen spacecraft is shown in Figs. 16.10 and 16.11.

The drop was caused by a sequence of events. First, technicians from another satellite program removed the hold-down fasteners from the

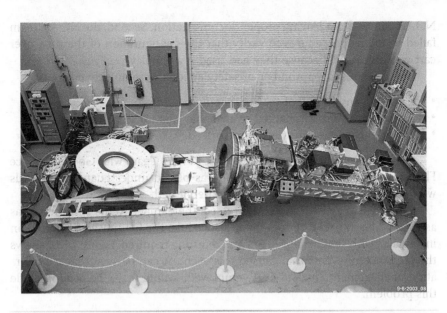

Fig. 16.10 Because technicians from another program had removed the spacecraft's hold-down fasteners while it was vertical on the fixture and didn't properly document their removal, and because the NOAA techs didn't check to see that the hold-down fasteners were holding down before rotating the spacecraft, it fell off and tipped over. Source: NASA.

Fig. 16.11 The resulting fall and damage provided a US$135 million repair bill for the US$233 million spacecraft, which came out of Lockheed's profits. Ouch! Source: NASA.

NOAA cart without leaving proper documentation. Second, the NOAA team failed to follow procedures to verify that the configuration of the turn-over cart was corrected, because they had used it a few days earlier.

After the mishap, a number of safety actions were quickly taken:

* Preventing any rolling of the spacecraft.
* The batteries were discharged so that they had no stored energy.
* The propulsion system was depressurized.

Lockheed Martin formed an Accident Review Team with Goddard Space Flight Center (GSFC). The spacecraft was placed under guard, all records were impounded, and personnel were interviewed.

After the safety issues were addressed, the damage to the spacecraft was assessed. The impact on program and schedule: the fall caused $135 million in damage, with significant rework and retesting required. Figure 16.12 shows the sequence of events that caused this event. It can be seen that configuration management, had it been employed, would have helped preclude this problem.

What Is Configuration Management?

Configuration management has to do with keeping track of what has actually been done and when it was done. It includes modifications to drawings, procedures, processes, plans, and generally lots of paperwork. The common statement is, "A vehicle is not ready to fly until the weight of its

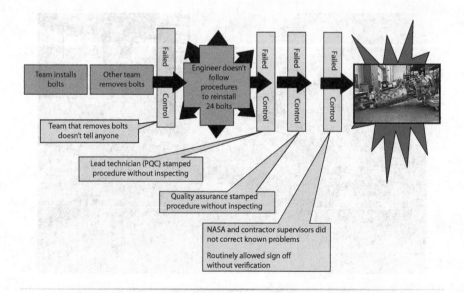

Fig. 16.12 The unfortunate sequence of events leading to severe damage to an NOAA-N spacecraft. Source: https://www.slideshare.net/NASAPMC/chandler-faith.

documentation is equal to the vehicle weight!" Although this might be an exaggeration, it certainly refers to the masses of paperwork necessary for configuration management, some of which are listed here:

• Design specifications	• Change control board	• Nonconformances
• Purchase specifications	• Software problem	• Deviations and waivers
• Interface control	reports	• Material review board
documents	• Software unit	• Configuration accounting
• Design reviews	development folders	• Test plans, procedures, and
• Drafting standards	• Drawing numbers and	data sheets
• Content and format	serial numbers	• Configuration audits
• Checking	• Fabrication controls	• Functional
• Release	• Processes	• Physical
• Changes	• Fabrication control	• As-built documentation
• Change control and	cards	
incorporation	• Workmanship	
	standards	
	• Parts and material	
	traceability	

16.3.2.3 Question 3: Can the Vehicle Survive a Computer Crash?

Computer errors are often caused by subtle timing mismatches, memory glitches, or power interruptions. Every vehicle should be able to gracefully handle a computer fault in some manner, like:

* Reverting to the "last known good state"
* Rebooting without being stuck in endless reset cycles
* Remaining in a safe mode
* Recovering from low bus voltage

Computer Crash Example 1: Ariane 5 First Flight Failure

The incident: in June 1996, about 40 s after liftoff, a software bug in the Ariane 5's flight controller made the rocket veer off course, leading to a destruct command. How did this happen?

Ariane 5 reused software from Ariane 4 without proper testing. The new software used a 64-bit floating-point number relating to the vehicle's horizontal velocity, which was converted to a 16-bit signed integer. The number was larger than 32,767, the largest integer possible in a 16-bit signed integer, so the conversion failed. This caused the controller to go "hard over."

Run-time range checking (which would have revealed the overflow) had been turned off due to processor limitations. In addition, the backup channel had failed just milliseconds earlier because of the same coding defect, making its backup capability useless. Some photos of the vehicle are shown in Fig. 16.13.

Fig. 16.13   The first Ariane 5 launch failed due to a software error resulting from reusing software from Ariane 4 to save costs. The flight software commanded an abrupt pitch change due to a numerical overflow. Source: ESA.

Delta III: Missing? What Happened?

If you follow rocketry, you have no doubt heard of Delta II (retired in 2018) and Delta IV (the Boeing EELV launcher). The family is shown in Fig. 16.14. What happened to Delta III? Well, it was not the victim of a computer crash. Instead, it was the victim of a control system design error.

According to FlightGlobal [20], after the first Delta III launched on 26 Aug. 1998, it began developing a slow oscillating roll in its first minute of flight. Several of the SRM strap-ons had thrust vector control, so the TVC attempted to control or damp-out the torsional oscillations. Unfortunately, the system had chosen an improper filter value, which caused destabilization instead of stabilization. The continuing motion of the TVC system used up the hydraulic steering fluid. (The system was a blowdown system; it did not recirculate its hydraulic fluid, but instead tossed it overboard.) The lack of hydraulic fluid caused loss of control, and the Delta III was destroyed by range safety.

Eventually, the Delta III had two more flights that reached orbit but were not "perfect," and satellite companies lost confidence in the design, so production ceased. Author Edberg was told that parts for 10 vehicles had been built, so there must be 7 in storage somewhere. There is a Delta III upper step on display just off the Santa Ana Freeway, Interstate 5, in Santa Ana, near the California Science Center.

But, not all was lost. The cryogenic upper step developed for Delta III went on to be the upper step for some Delta IV models as well, so some of the design work did not go to waste.

16.3.2.4 Question 4: Is the Circuit Overcurrent Protection Adequate?

Short circuits may cause many failures and may be due to breaks in wire insulation, a workmanship issue, or a quality issue. Most electric circuits have either fuses or circuit breakers that are intended to protect components from failures due to high transient current values.

Short Circuit Example: USAF Titan IVA Loss

A U.S. Air Force Titan IVA was launched 12 Aug. 1988. At $T + 39.4$ s into the flight, a wiring short-circuit caused the guidance system to lose electrical power for a fraction of a second. The connection to power was restored almost immediately, but ... the power loss erased the stored inertial references in the flight computer, and power restoration caused it to reset to T-0. As a result, the guidance computer sent a preprogrammed pitch maneuver command. Normally this command is done just after launch to begin the vehicle's eventual tilt to horizontal, but at supersonic speed, the pitch command resulted in the booster pitching downwards and beginning to break up at $T + 41.3$ s, triggering the vehicle's auto-destruct sequence [21]. Figure 16.15 shows a still photo of the failure taken from a video of the Titan's explosion.

16.3.2.5 Question 5: Can Pyros Cause Unexpected Damage?

Brazil's technicians were carrying out some launch readiness work on the Brazilian VLS-1 V03 launch vehicle in 2003. Unfortunately, an electrostatic

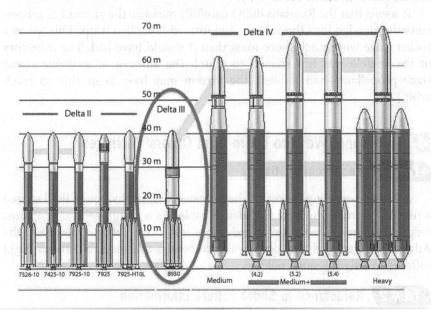

Fig. 16.14 Delta IIs had, and Delta IVs are having, a long service life. What happened to Delta III? See text for answer.

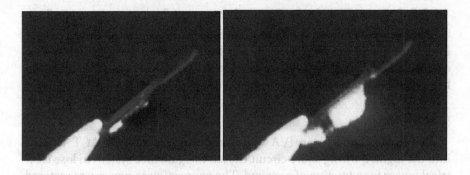

Fig. 16.15 Two still frames from a video of the range-safety destruction of a Titan IV due to a short circuit in 1988. The destruct sequence is just beginning the detonation along the SRBs in the left frame; on the right the destruct system's blast is expanding. Source: US Air Force.

discharge introduced an electrical arc into the unshielded pyro circuits, setting off the initiator for solid rocket motor. Because this motor had no *safe-arm* devices to contain an accidental flash, the vehicle caught fire and exploded on its pad, killing 21 workers.

16.3.3 The Newest Failure Category: Incorrect Mass Properties

The Russians win the "honor" of a new failure category. In 2010, a Proton booster incorporating additional upper-stage propellant tankage for longer burns lifted off. The booster's systems apparently worked properly, but the vehicle failed to attain orbit. Why?

It seems that the Russians didn't carefully measure the amount of propellants they loaded into the vehicle, and instead just filled it up. This led to a rocket stage with much more mass than it should have had. The trajectory of the vehicle was not shaped to match the amount of available upper stage propellant—had it been, the Proton may have been able to reach orbit [22].

16.4 Additional Ways to Learn from Others' Mistakes

16.4.1 Researching History

A Maxus sounding rocket crashed because drained hydraulic fluid started a fire that burned through a guidance cable. As a result, several programs redesigned fluid drains or added more cable insulation. However, the Athena program did not do so; unfortunately, the maiden Athena flight suffered exactly the same failure.

16.4.2 Reluctance to Share Failure Information

The test meter that was being used to check the solid motor's firing circuitry on the 1993 Orion sounding rocket was missing a critical voltage-clamp component that had not been installed. So, when the meter was turned on, its test voltage ignited the rocket, killing a technician.

After the Orion accident was made public, two other facilities revealed that the same test meter had inadvertently set off their rockets!

16.5   Range Safety and Flight Termination Systems

No text on launch vehicles would be complete without a mention of safety of flight. As we've seen, designers and engineers go out of their way to make their designs as reliable as possible by doing as much analysis and testing as possible. However, it's pretty well known that rockets do have problems from time to time.

16.5.1   Why We Have Flight Termination Systems

China attempted to launch a Long March launch vehicle on Valentine's Day, 1996. The vehicle went out of control almost immediately after liftoff and crashed into a town near Sichuan, the launch site. The resulting explosion and fire essentially flattened the town and is said to have killed some 80 local residents. Reports of the incident stated that foreigners were not allowed to depart the launch site for several hours, and the speculation was that the hours were spent cleaning up at least some of the damage that occurred. Despite the supposed cleanup, clandestine videos taken of the area showed extensive damage. A video on YouTube shows this unfortunate incident: https://www.youtube.com/watch?v=FBJ9ue6GKek.

To prevent such a horrendous event, or at least to mitigate the damage that occurs, almost all space launch vehicles carry a system called a *flight termination system (FTS)*. The FTS must be designed to meet the following list of requirements. The FTS must:

* Be capable of stopping the acceleration of any part of the vehicle. It must render the vehicle nonpropulsive and stop thrust, lift, and yaw.
* Reduce the vehicle to an unpowered, nonhazardous mass of materials in ballistic flight.
* Be commanded by RF transmission.
* Include status in telemetry stream.
* Physically separate FTS components on the vehicle to prevent detonations from disabling other parts of the FTS before they work.
* Be independent of all other vehicle systems.
* Have either a commanded or an automatic form of FTS on each powered stage of the vehicle. The automatic form is triggered if there is an *unscheduled stage separation* before burnout. The automatic system is called *inadvertent separation destruct system (ISDS)*.
* Have an antenna system that covers 95% of the vehicle's field of view with a 12-dB margin over the entire trajectory, and must pass mandatory pattern testing.

- Have internal redundancy in its firing circuitry.
- Use shielded, filtered wiring with current-limiting resistors.
- Must propagate the explosive stimulus from the explosive transfer system throughout the LV simultaneously. This is done by flexible confined detonating cord or similar.
- Have destruct charges, which can be block, pancake, or shaped charges.
- Have instrumentation to provide monitoring of all key features in FTS to the ground station. There must be a dedicated hard line through the LV umbilical and an RF link through TM.
- Have batteries dedicated to the FTS system that are redundant and continuously monitored, with 150% margin in worst-case scenario.

The following requirements are concerned with the FTS system's reliability:

- System must be single fault tolerant.
- System must have 99.9% reliability with 95% confidence.
- Acceptance and qualification tests must be used to augment reliability predictions.

The system's reliability must be documented in a *range safety data package (RSDP)*, which includes:

- Failure mode, effects, and criticality analysis (FMECA)
- Reliability test plan
- Reliability requirements analysis

16.5.2 Typical Flight Termination System Make-Up

A generic flight termination system is shown in Fig. 16.16. Note that there are duplicates of all elements for redundancy. The box marked "LOGIC" contains components so that either of the two systems can trigger the FTS.

16.5.3 Method of Termination

As we saw in Chapter 11, explosive charges, also known as *shaped charges* or *linear shaped charges (LSCs)*, can be used to separate (stage) vehicle parts, or they can *open up* structures. This opening up process goes along with one official designation for an FTS, a *propellant dispersal system*. Flexible LSCs of various sizes are used to cut or puncture structures for propellant release. Explosive initiation causes the chevron to expand until the concave side (bottom side) turns inside out to form a hot metal plume or *jet*, which is focused by the chevron shape. The jet extrudes down towards the structure below, penetrating and severing it. This process is shown in Fig. 11.9 of Chapter 11. Shaped charges turn out to be a more effective cutting tool than a plain circular cross-section charge. Figure 16.17 shows the result of firing shaped charges to destruct a rocket. Note that the initiation of the shaped charges on the top right opens the

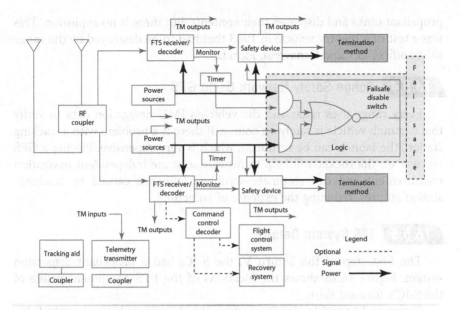

Fig. 16.16   Components of a typical FTS. Note all parts are duplicated except for those marked "optional," and either system can trigger, even if the other one does not trigger. Source: [7].
Note: TM = telemetry; RF = radio frequency.

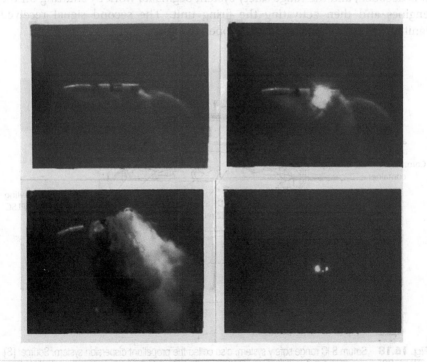

Fig. 16.17   Disassembly of Titan I missile as a result of firing shaped charges: propellants dispersed. Source: US Air Force.

propellant tanks and disperses their contents, but there is no explosion. This was a test of a reentry vehicle in 1963 that had to be destroyed by the range safety officer at Cape Canaveral, Florida.

16.5.4 Range Safety Network Using GPS

Every range uses radar and the vehicle's IMU navigation data to verify that a launch vehicle is safely on course. If there is a problem with a tracking station, the launch can be scrubbed, which is very expensive. Placing a GPS receiver on the rocket will help to provide separate independent navigation measurement and can eliminate scrubbed launches caused by tracking-station failures, reducing the expense of launching.

16.5.5 FTS System Details

The first step of the Saturn V, the S-IC, had a propellant dispersion system. Figure 16.18 shows the locations of the FTS equipment inside of the S-IC's forward skirt.

Figure 16.19 provides details of where the charges are located on the S-IC, and how the charges are initiated. There are two identical independent systems, as is required. If a signal from range safety on the ground is received, it is decoded, and the range safety system begins its work by shutting off the engines and then activating the firing unit. The second signal received ignites explosive trains, which then open the tanks.

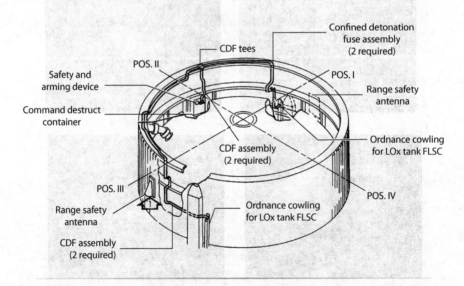

Fig. 16.18 Saturn S-IC range safety system, also called the propellant dispersion system. Source: [8]. Note: CDF = confined detonating fuse; FLSC = flexible linear shaped charge. The designation "POS" represents the location of four quadrant position markers, I, II, III, and IV, used to specify wiring and equipment locations in the structure.

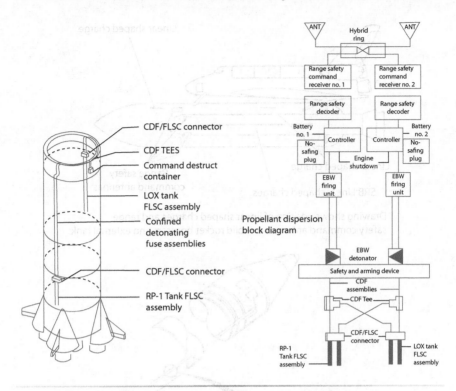

Fig. 16.19   Front end of the Saturn S-IC range safety system. Source: [8].
Note: EBW = exploding bridgewire, ANT = antenna.

In Chapter 13, we saw that a *systems tunnel* was placed on the outside of the launch vehicle to carry wiring, pressurization lines, and the like. In some cases, the same tunnel is used to carry the LSCs used by the flight termination system. Figure 16.20 shows that the LSCs for the Space Shuttle solid rocket boosters are contained in the motors' systems tunnel.

The Shuttle's external tank (ET) has a separate destruct system and set of antennas, as shown in Fig. 16.21. The reason for this is, of course, that the ET tanks carried liquid propellants, hydrogen and oxygen. The external tank's range safety package was deleted after mission STS-78.

16.5.6 Abort Modes

Crewed vehicles have *abort modes*, where the vehicle is commanded to follow certain procedures in case of emergency, which is usually caused by a propulsion problem. If there is still control of the vehicle, it may be commanded to follow a different procedure than originally planned. The Space Shuttle's abort modes are provided in Fig. 16.22. Note that abort procedures range from *return to launch site* to *abort once around* to *abort to orbit*, depending on the speed and location of the vehicle when the emergency occurs.

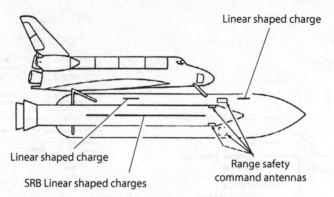

Drawing shows position of linear shaped charges and range
safety command antennas on solid rocket boosters and external tank.

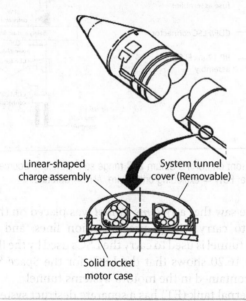

Fig. 16.20 Propellant dispersion or destruct system on Shuttle SRBs is placed in the systems tunnel on the outside of the motor casing. Sources: [9, 10].

If the vehicle appears to be leaving its safety zone (tracking), the range safety officer (RSO) is responsible for pressing the "red button" to trigger the propellant dispersement system, which of course fires explosives to release propellants and pressures to stop powered flight. Figure 16.23 illustrates Cape Canaveral's launch range, which is bounded by a solid line. If the vehicle were to cross any lines, the RSO would command the breakup of the vehicle.

In case you were wondering what the RSO's console might look like, we have provided Fig. 16.24, a photo of the abort switch at NASA's Wallops Flight Facility (WFF). Notice the switch has a plastic flip cover: it

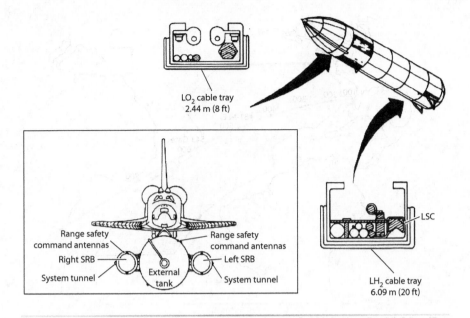

Fig. 16.21   Top right: RSS installed in the cable trays on the Space Shuttle's external tank. Lower left: Location of RSS antennae on the Shuttle stack. Source: [9].

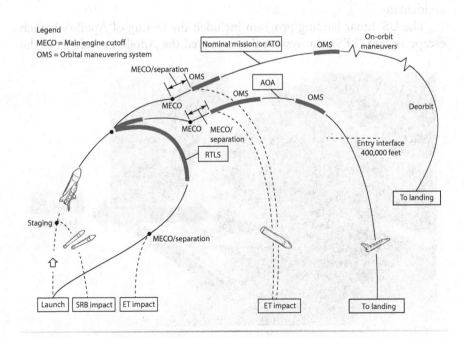

Fig. 16.22   Shuttle launch abort modes. Actions taken depend on the vehicle's energy, based on its speed and altitude. Source: [9].

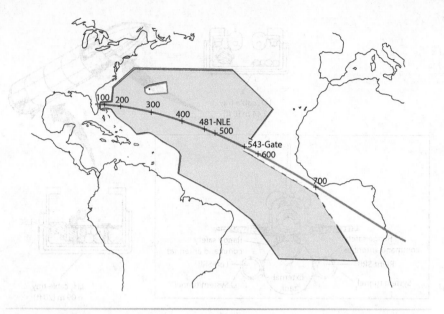

Fig. 16.23 The USAF 45th Space Wing launch range at Cape Canaveral for Titan IV-B25. Numbers indicate time after launch in seconds. If the tracking system shows the rocket crossing the border of the shaded area, the RSO commands the destruct system to operate. Source: [11].

takes two actions to execute a command, to make sure it doesn't happen accidentally.

The US' lunar landing program included the testing of Apollo's launch escape system (LES), the system that removed the Apollo capsule from the

Fig. 16.24 The abort switch at NASA's WFF. Source: Edberg.

top of the Saturn launcher, in case of emergency. A test launch vehicle known as Little Joe II was used to accelerate the capsule and escape system up to max-q speed, after which the system would be activated. Unbeknownst to the test operators, a roll gyro on the vehicle was miswired, and the vehicle's roll rate accelerated until it broke up, at which time the LES operated as designed and removed the Apollo capsule. Figure 16.25 shows still frames from the video, one at liftoff and four showing the Little Joe breaking up and the capsule's removal.

The LES is the thin white pole extending from the top of the Apollo capsule, shown in Fig. 16.26. It is a solid rocket motor that has enough thrust to rapidly pull the entire Apollo capsule away from an exploding booster. The U.S. Mercury and Apollo capsules used them (although the LES was never needed in service for U.S. flights), as well as the Russian Soyuz capsule, which has had two occasions to use them: one mission where the booster rocket caught fire on the launch pad (1983), and a second occasion when one of the Soyuz strap-ons did not separate properly (2018). The U.S. Gemini capsule had ejection seats, as did the Space Shuttle for its first few missions.

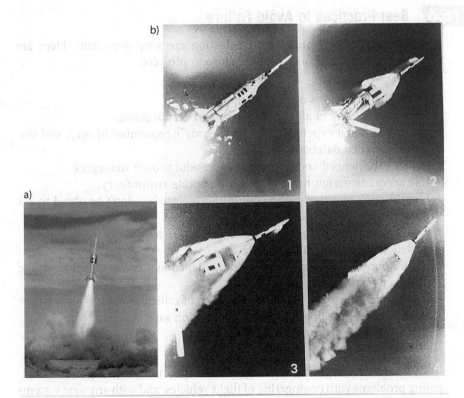

Fig. 16.25   a) Liftoff of a Little Joe II; b) breakup of another Little Joe II carrying an Apollo boilerplate capsule and its launch escape system. Sources: http://images-assets.nasa.gov/image/S63-15701/S63-15701~orig.jpg; http://www.apolloarchive.com/apollo_gallery.html.

Fig. 16.26 The Apollo capsule being removed during an abort test. Source: https://www.nasa.gov/images/content/618283main_LAS_226x170.jpg.

16.6 Best Practices to Avoid Failure

Obviously, testing and failure detection are very important. Here are some suggestions for how to avoid failure in projects:

* Know the limitations of your equipment.
* Know the validity of all specifications and certifications.
* Use institutional expertise (i.e., "graybeards"); remember history, and use lessons-learned databases.
* Utilize experienced and previously successful project managers.
* Use adequate design margins and incorporate *redundancy*.
* Identify risks effectively, and communicate frankly—don't be afraid to ask questions and offer opinions.
* Contractors should notify the customer of project risk, as well as any deviations from acceptable practice.
* Have competent analysis, verification, independent reviews, oversight, and thorough testing.
* During flight, provide telemetry coverage of critical events.
* Don't let schedule pressure outweigh mission safety.
* Demand proof: *do not launch* unless ready.

It is clear that a good knowledge of history can be very helpful in anticipating problems with engineering of flight vehicles, and with any new systems in general. The quotes from Bismarck and Ballhaus at the beginning of this chapter are a confirmation of this fact. The reader is encouraged to do as much reading and research as possible in order to gather knowledge of others' work experiences. One excellent source of historical information on space failures is the book by Harland and Lorenz [23].

16.7 Summary

We have considered a number of launch vehicle failures and their causes, as well as ways to "disassemble" a launch vehicle in-flight if it begins to malfunction or stray towards an off-limits area on the map. The next chapter will consider the processes of estimating the cost of a LV program, and its administration and management.

References

[1] Demidovich, N., "Launch Vehicle Failure Mode Database," presented to the COMSTAC RLV Working Group, 17 May 2007.

[2] Fossage, E., "The Effect of Job Performance Aids Quality Assurance," Sandia Report SAND2014-4762, Sandia National Laboratories, June 2014, https://prod-ng.sandia.gov/techlib-noauth/access-control.cgi/2014/144762.pdf, accessed 2020 March 06.

[3] Federal Aviation Administration, *Guide to Probability of Failure Analysis for New Expendable Launch Vehicles, Version 1.0*, 2005, https://www.faa.gov/about/offi ce_org/headquarters_offices/ast/licenses_permits/media/Guide_Probability_Fai lure_110205.pdf [retrieved 26 May 2018].

[4] Futron Corporation, *Design Reliability Comparison for SpaceX Falcon Vehicles*, Futron, Bethesda, MD, Nov. 2004.

[5] Cheng, P. G., "Ground Software Errors Can Cause Satellites to Fail Too—Lessons Learned," presentation 4 March 2003, The Aerospace Corp.

[6] http://forum.nasaspaceflight.com/index.php?action=dlattach;topic=25597.0;attach= 324249)

[7] U.S. Army, "Flight Termination Systems Commonality Standard," Range Safety Group, U.S. Army White Sands Missile Range Document 319-10, Oct. 2010.

[8] NASA, "Saturn V SA-503 Flight Manual," MSFC-MAN-503, NASA TM-X-72151, Nov. 1968.

[9] Boeing, "Press Information: Space Shuttle Transportation System," Rockwell International, March 1982.

[10] NASA, "Report of the Presidential Commission on the Space Shuttle Challenger Accident," 1986, https://history.nasa.gov/rogersrep/v1p185.htm [retrieved 19 Aug. 2018].

[11] 45th Space Wing, "Eastern Range and Western Range Collective Risk and Associated Data," Patrick Air Force Base, Florida, 13 Aug. 1999.

[12] McDonald, A. J., and Hansen, J. R., *Truth, Lies, and O-Rings: Inside the Space Shuttle 'Challenger' Disaster*, University Press of Florida, Gainesville, 2012.

[13] Rogers Commission, "Report of the Presidential Commission on the Space Shuttle Challenger Accident," 1986, https://apps.dtic.mil/dtic/tr/fulltext/u2/a171402.pdf [retrieved 22 May 2019].

[14] Walker, N., "Rocket Science," 2009, https://www.lulu.com/shop/norm-walker/ rocket-science/ebook/product-17476181.html, visited 23 June 2019.

[15] Cheng, P. G., "Five Common Mistakes Reviewers Should Look Out For," Aerospace Report No. TOR-2007(8617)-1, 29 June 2007. [Available to U.S. government agencies and contractors only.]

[16] https://www.npr.org/sections/thetwo-way/2013/07/10/200775748/report-upside-down- sensors-toppled-russian-rocket

[17] http://www.russianspaceweb.com/proton_glonass49.html#culprit

[18] https://forum.nasaspaceflight.com/index.php?topic=32282.380

[19] Eleazer, W., "Launch Failures: The 'Oops!' Factor," The Space Review, 31 Jan. 2011, http://www.thespacereview.com/article/1768/2 [retrieved 26 May 2018].

[20] https://www.flightglobal.com/news/articles/boeing-confirms-control-system-contribu ted-to-delta-iii-failure-42864/

[21] Eleazer, W., "Launch Failures: The Predictables," The Space Review, 14 Dec., 2015, http://www.thespacereview.com/article/2884/1 [retrieved 26 May 2015].

[22] Eleazer, W., "Launch Failures: New Discoveries," The Space Review, 13 Feb. 2017, http://www.thespacereview.com/article/3170/1 [retrieved 7 April 2018]

[23] Harland, D. M., and Lorenz, R. D., *Space Systems Failures*, Springer, Berlin, 2005.

Recommended Reading

National Academy of Sciences, *Streamlining Space Launch Range Safety*, National Academy of Sciences, Washington DC, 2000.

Pappalardo, J., "As Shuttle Lifts Off, NASA Will Man Destruct Switch—Just in Case," *Popular Mechanics*, 2008, http://www.popularmechanics.com/science/space/nasa/4262479 [retrieved 18 Aug. 2018].

U.S. Air Force, "Eastern and Western Range 127-1 Range Safety Requirements," U.S. Air Force EWR 127-1, 31 Oct. 1997.

Chapter 17 Launch Vehicle Financial Analysis and Project Management

I n the minds of many, the aerospace industry is characterized by seemingly unlimited budgets, minimal accountancy, and endless spending, all in an attempt to maintain technical superiority over one's political or commercial adversaries. For many years, the overarching figure of merit (and in some cases, the only figure of merit) for a launch vehicle was maximum performance for minimum weight, costs be damned. These days, however, notions such as "maximum performance at any cost" are woefully inadequate; even in the "Nifty Fifties," when levels of technical and operational risk that would be career-ending in modern times were allegedly accepted as a matter of course, the economics of a proposed design were always under consideration.

The financial concerns of a flight vehicle's development are often treated derisively by engineers who worry about focusing on "solving the real problems," but this aversion to fiscal and actuarial matters is illogical and professionally damaging. Maximizing performance and minimizing weight have always been expensive propositions, and the practicing engineer will likely discover that the most successful launch vehicles employ a systems-level approach to design, such that no one subsystem is unduly favored (i.e., costs more to develop and implement) relative to another unless there is a direct and measurable improvement to an originating or derived requirement. Put simply, technology must "buy its way" onto a launch vehicle, often by improving performance while also reducing cost or risk.

In the commercial world, this approach is often utilized in order to maximize the profitability of each launch and thus maximize the stakeholder return on investment; for research and military missions, financial considerations are crucial to ensure that the best value is obtained for each taxpayer dollar spent. It may therefore be argued that *all* launch vehicles are "commercial" vehicles regardless of their payload or mission; it is the wise engineer who remembers that design calculations are not performed in a vacuum, and the most successful launch vehicle is the one that is affordable enough to actually be used.

The design roadmap first presented in Chapter 1 is presented again as Fig. 17.1, with the highlighted "signpost" (determining vehicle cost) indicating the information to be covered in this chapter.

All launch vehicle designs must begin with the mission the vehicle is intended to perform, often called the design reference mission (DRM), from which a "womb-to-tomb" project lifecycle is developed. A typical project lifecycle is shown in Fig. 17.2. The entire lifecycle is broken down into discrete phases, each of which has specific exit criteria that must be met prior to the project moving on to the next phase. Although the terms can vary somewhat between organizations and projects, the general format is universal. Multiple types of reviews are conducted throughout the lifecycle in order to ensure that necessary developments are achieved in a timely and correct fashion. These reviews, colloquially known as "scrub-downs" or "murder boards," are usually conducted by committees of project managers and subject matter experts to aggressively review a proposed design, often without pleasantries.

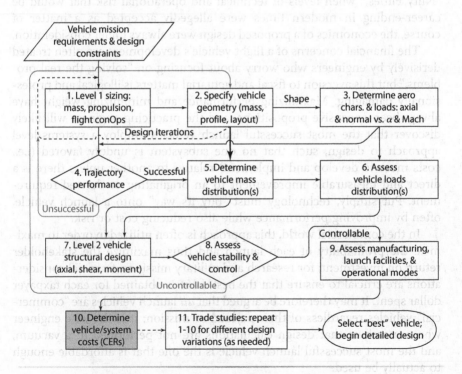

Fig. 17.1 The launch vehicle design process.

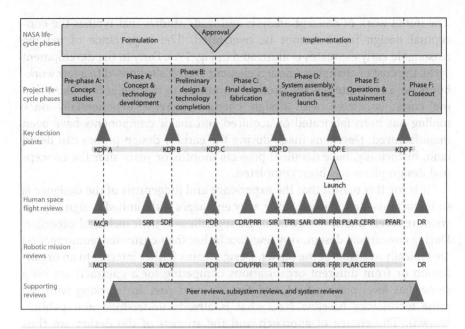

Fig. 17.2   Stages of mission development. Source: [1].

17.2 The Design Cycle

Laying out the project lifecycle in this fashion ensures that the technical products are matured (as needed) from the conceptual phase to fielding and, finally, retirement. Broadly speaking, the design process is divided into three phases:

1. Conceptual design
   * Requirements definition
   * Initial configuration concepts defined and evaluated
   * Basic design trade studies
   * Initial weight, cost, and performance estimates
2. Preliminary design
   * Configuration downselect and freeze
   * Testing and analysis
   * Major component layout and design
   * "Real" cost
3. Detail design
   * Design parts to be built; create shop drawings and assembly instructions
   * Design and fabricate tooling
   * Major component testing
   * Refinement of financial and risk analyses

The initial work of concept analysis and trade studies that typifies the conceptual design phase cannot be overstated. The importance of making reasonable early estimates is illustrated in Fig. 17.3. Early in the development cycle, the cost of changing a design element is typically limited to a few workhours on a computer; as the product continues to mature, it is evident that the costs associated with design changes increase exponentially once tooling has been fabricated or acquired and major components have been manufactured. Decisions made during the earliest design phases can derail (and, historically, have derailed) projects months or years after the conceptual design phase has been completed.

It is for this reason that the experience and judgement of the designer is so important—and, consequently, why engineers with limited design experience are rarely put in charge of much, regardless of their technical expertise. During conceptual design, very few data (other than customer requirements) are typically available at the outset. Design teams (either internal to an organization or from different organizations competing for a contract) are on a somewhat level playing field, and for the most part are working with the same technology baseline (i.e., what is already understood and publicly known). The choice of approach and the success of the design are thus rooted in the designers' knowledge and experience, and the methodologies available to them.

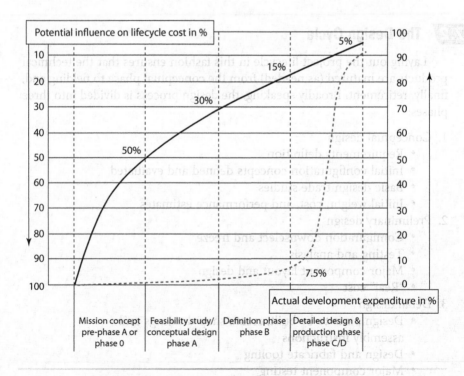

Fig. 17.3 The importance of early estimates. Source: [2].

17.3 **Design Decision Making**

All design decision making is rooted in a document that is often known as the basic assumptions memorandum (BAM). This document is usually created by the project management team, and it defines the ground rules to be used throughout the design process. There is no universal template for a BAM, nor are there universal rules of what a BAM must contain, but almost all BAMs will have the following elements at a minimum:

* *Mission statement*: A short and formal summary of the aims and goals of the project. Note that the mission statement also underscores the tone of the project and often provides philosophical and practical guidance to the engineering team. Some mission statements can be unofficial; for instance, the F-15 Special Project Office heavily opposed using the aircraft for ground attack roles, giving rise to the unofficial (but quite catchy) mission statement "Not a pound for air-to-ground."
* *Mission profile*: Also called the concept of operations (CONOPS), this is a brief (and usually graphical) description of the vehicle's overall operating mission, broken down into mission segments. The mission profile describes what the vehicle must do in terms of both time and space: begin gravity turn at 35,000 ft MSL and T + 350 s, fire stage separation charges at 2.5 s after previous stage burnout, and so on. These mission phases will usually trace back to one or more of a customer's originating requirements.
* *Requirements*: A list of customer demands that serve as the final word for judging the performance of the vehicle. Unless requirements are officially changed by the customer, they should be considered gospel. Customers usually provide *originating requirements*, which define *what* the vehicle must do at different points in the mission; from these come *derived requirements*, which define *how* the vehicle is to meet the originating requirements. *Functional requirements* define how a system must behave when certain inputs are given; *nonfunctional requirements* represent criteria used to judge how a system operates (rather than specific behaviors) and provide *constraints* to the functional requirements.
* *Constraints*: Conditions that serve to bound a design in one or more ways, thus defining an acceptable subset of all possible designs. For instance, a rocket motor may not be able to exceed a certain form factor, a bearing must have the same bore diameter as the shaft it supports, or a payload may only be subjected to a certain g level and vibratory environment during launch.
* *Goals*: The desired outcomes that the project team plans and commits to attaining. Note that goals are included in the mission statement, but are not in and of themselves a mission statement. The key feature to remember about goals can be thought of in this fashion: a company can fail to meet some (or most) of its goals for a project, but so long as the resulting vehicle meets the customer's demands, the company has a chance of remaining economically viable. For instance, developing new analysis and fabrication

methods to leverage as a competitive advantage on future projects is a goal, but is not necessarily something the customer will care about (and hence won't pay to have) if it doesn't allow the vehicle to measurably meet a performance requirement.

- *Assumptions*: A collection of factors that do not fall into one of the other categories but can still affect the design process. How should designers compare different configurations in a trade study? What software should be used for things such as computational fluid dynamics (CFD), and how should the analysis cases be set up? Which turbulence model should be used? Which analyses should be verified with hand calculations or other computational methods, and which mathematical models should be employed for those verifications? What system of units should be used? These questions seem trivial and unnecessarily constraining, but it is of critical importance that all members of the design team are "playing in a shared sandbox"—the same methodologies, the same assumptions, and the same principles—in order for the design phase to progress smoothly and correctly.

Note that a BAM is a living document and should reflect changes in assumptions as the design process proceeds. The value of a BAM is in its ability to quickly communicate information to all team members; as such, a BAM is usually accessible to all design team members.

Design decisions must be based solely on how they allow the vehicle to better fulfill the customer's requirements, and features should be evaluated (whenever possible) in mission-specific environments. This is commonly known as the *mission and merit* design philosophy: the mission comes first, and then the configuration. This forces features and components to "buy their way onto the vehicle," in that it forces engineers to justify their design choices by demonstrating how a design element improves the performance of the vehicle in ways that trace back to originating requirements. This means that there is no single "best" configuration for all mission types, but launch vehicles are instead a collection of compromises that were required in order to perform a specific task.

This should also serve as a warning for engineers to not fall in love with one particular concept and thus turn a blind eye to the mission: concepts can end up driving the entire design of a launch vehicle (and at times, historically, have), with poor performance and/or unnecessary risk—from the standpoint of both cost and safety—usually being the end result. Engineers should also be wary of incorrectly assuming the ease with which certain lifecycle processes—such as maintenance—can be performed. As shown in Fig. 17.4, the proposed maintenance and payload integration process (Fig. 17.4a) bears little resemblance to the actual process (Fig. 17.4b)!

Designers must also be wary of requirements creep. Also called *scope creep* or *feature creep*, this refers to uncontrolled changes or continuous growth in a project's scope. Scope creep is usually the result of one or more of the following:

Fig. 17.4   a) Proposed maintenance operation concept of the Space Shuttle. The Space Shuttle Orbiter is designed for a two-week ground turnaround, from landing to relaunch. About **160 hours** of actual work will be required; b) actual maintenance operation of the Space Shuttle From 1990 to 1997, the average time spent in the Orbiter Processing Facility (OPF) was **88 days** (2112 hours). The nose of the Orbiter can just be seen in the center right of the photo. Source: NASA.

- Poor configuration management and/or change control
- Not properly identifying customer needs and correctly addressing originating requirements (i.e., "What does the customer *actually* want?") before work begins
- Poor communication among sponsors, customers, and design teams
- Weak or ineffective project managers and executives
- Lack of product viability

More often than not, this occurs when the scope of the project is not well-defined, properly documented, or adequately controlled; however, requirements creep is often an unintended side effect of the real economic situations present in the industry: a company cannot be held accountable for changes in requirements unless those changes are documented in an engineering change order (ECO), a documentation packet that details proposed changes and areas of the design to be affected, and is agreed upon by both customer and company; however, managers must also balance changing customer needs with the practicalities of ensuring that development continues in a timely fashion. This is a delicate balance, because managers must accept the reality that ECOs and a rescoping of work usually involve change fees, which in many cases can be substantial and are of obvious financial benefit to the company.

In short, fear and/or unwillingness to put a contract at risk for the sake of seemingly minor changes can often lead to project managers approving more ECOs than they should, until the vehicle itself becomes so bloated with changes that it can't adequately perform *any* mission, let alone the one for which it was originally intended. This is colloquially known as "death by a thousand cuts": small changes that are difficult to justifiably deny continually aggregate into large-scale consequences that affect the entire vehicle. Although requirements creep in small amounts can sometimes be seen as an advantage (every ECO comes with a change fee, after all), in most cases it is a risk that can kill an entire project. Scope creep almost always ends in cost overruns and delays to delivery schedules, and can also create insurmountable technical risks for the design team.

So what, then, is considered a "good" design? Certainly, the ability to safely perform a mission while maintaining structural integrity is always necessary, but there are other factors that good designs always have and poor designs always lack (c.f. [3]):

* Simplicity
* Timeliness
* Flexibility

The successful design is *simple* because it is within the contemporary state of the art and does not require any untested, research-grade developments (in either its configuration or its subsystems) to work properly. This is in direct opposition to the pursuit of "moreness" that plagues poor design efforts: more of something that is not needed, whether it be capability or performance, is useless. The successful design is *timely* because it represents a step-change in the operational capability of a customer to perform a mission; in other words, the vehicle offers capabilities that are in excess of its forebears and competitors. All designs are, by definition, a compromise, and tradeoffs are inevitable; however, it is vital to maintain an appropriate scope during the design process. The successful design is *flexible* because as requirements change, proper management will permit the vehicle to change alongside them without compromising its abilities to meet requirements. That means accounting for some percentage of growth (of either required capabilities or future needs) and meeting as many technical figures of merit as possible. The meeting of requirements is critical, because no launch vehicle will ever receive approval if it cannot meet the customer's requirements, but the ability of a design to be easily adapted as requirements change is often what differentiates a signed contract from a forgotten development effort.

17.4   Cost Engineering

The common view of engineering is limited and disregards a large portion of what engineering actually entails. The most obvious (and true) interpretation is that engineering is concerned with addressing the technical issues pertaining to a product, such as the design of the outer mold line of an aircraft or the fabrication of a rocket motor; however, beyond the obvious physical manifestation of engineering effort is an entire business ecosystem that must be considered, one that concerns itself with time, money, physical assets, available capital, and other resources that must be consumed to build the physical product. These are collectively referred to as *costs*.

Cost engineering is the practice of managing project costs, which includes cost estimation, control, and forecasting, as well as investment appraisal and risk analysis. This assists midlevel and top-level management in the control and allocation of resources, and supports project-level decision making. Stated another way, it is focused on the relationships between the development of a physical product and the cost dimensions of that product. A cost engineer will typically create a project budget and monitor costs over the lifetime of a project, seeking an ideal balance of performance and quality for a fixed amount of budget and time. One of the most important tasks of cost engineering, therefore, is the correct and accurate estimation of project costs in order to avoid overruns and schedule slips; this means that cost engineering encompasses a broad array of engineering and business disciplines, including project management, business management, engineering, accounting, and financial analysis, among others.

In the interest of brevity and focus, only those aspects of cost engineering that are immediately applicable to the design and construction of a launch vehicle will be covered in this text. For other aspects of cost engineering, such as earned value management, financial analysis, and total cost of ownership, the reader is directed to [4, 5, 6].

17.4.1   Cost-Estimating Relationships

All financial models used for the development of an aerospace vehicle are rooted in cost-estimating relationships (CERs), colloquially known as a *cost model.* As the name implies, a CER is used to estimate the cost of an item by using established relationships against an independent variable; for instance, the cost of a launch vehicle will be directly dependent upon its payload capacity or its intended orbital insertion capabilities. Strictly speaking, a CER is not a quantitative technique, but a framework for specifying the relationships between vehicle characteristics and development costs.

There is no one, single CER that is inherently superior to all others, although companies and research institutions certainly have several CERs upon which they regularly rely. So long as one can identify an independent variable that demonstrates a measurable relationship with a development cost, a CER can be produced. CERs can be mathematically simple or may involve very complex equations; this is especially true of CERs based on statistical regression, meaning that the projected cost of a future vehicle is estimated by examining several characteristics of previous vehicles within the same range of gross weight and performance.

Note that CERs are only useful when bounded, meaning that the applicable range of vehicle characteristics is expressly stated. For instance, a CER developed for estimating the cost of a 100,000- to 500,000-lb commercial passenger jet (such as the RAND DAPCA IV model) might not necessarily produce accurate estimations for a lightweight, propeller-driven primary trainer aircraft. CERs are typically bounded by all-up vehicle weight, maximum speed, maximum payload, fuel type, primary structural material, average labor costs, or some combination thereof. Typically, the more parameters that are included in a CER, the more accurate and reliable the final estimation will be.

CERs are typically developed through a six-step process. The process outlined herein is adapted from the current Defense Procurement and Acquisition Policy of the U.S. Department of Defense. Most organizations will have their own CERs developed internally, either modified from publicly available documentation or developed based on previous project experience developed within the organization; as such, the information presented in this chapter is generic unless explicitly denoted otherwise. The practicing design engineer is duly advised to become intimately familiar with this process, because the acquisition manager deciding whether to fund development of that engineer's work most certainly will be!

1. *Define the dependent variable.* Define what must be estimated by the CER. Normally, a CER is used to estimate the all-up vehicle cost or break-even sell price, but a CER can also be created to estimate material cost, labor hours, and so forth.
2. *Select independent variables.* The factors that affect cost, and their impact upon cost, are usually taken from personal experience, the collective experience of a design team, published sources, company policy, cost-estimating software, or some combination thereof. Note that most of these assume a design is using current state-of-the-art technology; the

incorporation of untested or radically new technology into a vehicle may reduce the accuracy of a CER's initial result. When selecting dependent variables, recall that parameters must be measurable to be useful. This means that data must also be available: if historical data for a variable are unavailable, it will be impossible to use that variable as part of the prediction process. Most importantly, when faced with the choice of using either a performance variable or a physical characteristic in a CER, choose the performance variable, because performance requirements are typically known long before physical ones.

3. *Quantify relationships.* This is normally the most time-consuming element of developing a CER; however, it is also the most important, because it is what allows the establishment of relationships between vehicle characteristics and cost. Normally, this is done via statistical regression, which can be both a boon and a hinderance: it can assist the design engineer in estimating the cost of a (semi-)conventional design (i.e., "If you build what's been built before, it will cost what it always has"), but also constrains the engineer to assumptions based only on available data (i.e., "If you design something never seen before, you're on your own"). As with most statistical data sets, there should be as many *relevant* samples incorporated into the analysis as possible (i.e., "Quantity *is* quality"), and said data set should be free of outliers that will unduly affect the regression outcome. A more thorough discussion of regression is given in [7].

4. *Explore relationships between variables.* Common statistical tools and metrics will assist the engineer in determining the strength of the relationships between dependent and independent variables. These tools include moving averages, linear regression, ratio analyses, and more complex graphical and mathematical relationships. Not all characteristics will have the same effect on cost, which can assist the engineer in deciding how to best optimize the vehicle configuration in subsequent phases of the design process.

5. *Select the best prediction.* A high correlation between independent and dependent variables will typically indicate the usefulness of the independent variable as a predictive tool. Of course, it is necessary that a value for the independent variable is actually available during a given stage of the design process; if a value for said independent variable cannot be determined until later, an alternative predictive variable will be needed.

6. *Document findings.* Not only is it good engineering practice to document everything, but proper CER documentation will permit others involved in the cost-estimation process to expand and improve upon the initial CER. This documentation should, at a minimum, include the independent and dependent variables used, all gathered data, all applicable references where said data were found, time periods of the data (if applicable), and any adjustments or corrections used.

17.4.2 Cost-Estimating Software

More often than not, practicing engineers will find themselves in an organization where CERs, and the processes for using them, have already been established by the company. This is usually accomplished through the use of external consulting firms (such as RAND Corporation or Booz Allen Hamilton) that have already worked with the engineering organization to collect, normalize, and analyze the appropriate data. The end result of this arrangement is typically a piece of software (or spreadsheet plug-in), which is usually updated as more missions are incorporated into the cost database. For the practicing engineer, this results in a turn-key arrangement, in which much of the inner workings of the CER are obfuscated (to some extent) in exchange for usability of the model. Some common cost models the practicing engineer is likely to encounter include:

- *Project Cost Engineering Capability (PCEC)*: A development of the NASA/Air Force Cost Model (NAFCOM), which is no longer widely used, analyzes previous space missions to estimate vehicle cost based on mission type, vehicle type, and mission duration. The end result is a large collection of historical mission data (updated regularly) incorporated into a Microsoft Excel plug-in for maximum usability.
- *TRANSCOST*: A collection of graphs and tables used to determine approximate lifecycle costs on an average basis, developed by Dr. Dietrich Koelle.
- *USAF Cost Model*: Contained in the U.S. Air Force Space Planner's Guide, published in 1965. Originally a handbook of manual calculations, it has since been incorporated into a variety of software packages. Despite its year of publication, this model still produces acceptable results (within 10–15% of actual costs), even with modern launch vehicles such as the Ariane 5.
- *TRASIM*: A program that processes approximately 380 input values to estimate lifecycle costs on an annual basis.
- *FINANCE 1.0*: A standalone program used to determine the economic performance of a selection of previous launch vehicles; not typically used on its own, but can be used to develop a business case for a launch vehicle.

Note that although the purpose of a CER is to serve as a method of estimating cost, the output of a CER is not necessarily a dollar amount. Often, a CER will output the cost of a development effort as *work-years (WY)* or *man-years (MY)*, based on the number of hours that one full-time employee can work in one year. (The common standard is 2,000 hours, 50 × 40 hour weeks (includes 2 weeks of vacation). Historically, *man-years* has been used, although in the modern era it is equally common to find the term *work-years* used instead. From a mathematical point of view, they are equivalent. Work-years or man-years is often a more useful metric than a lump-sum dollar amount, because it allows managers to adjust head count and make hiring decisions based on the customer's desired completion date. The monetary cost is then estimated based on the number of work-years required for a specific phase of development:

* Development: 1 WY = $205,000
* Production: 1 WY = $200,000
* Operation: 1 WY = $220,000
* Other tasks: 1 WY = $208,000

Note that these figures are given for FY2000 (fiscal year 2000) constant dollars and should be inflated to the appropriate year for a given application.

Another common output of a CER is the hourly rates by trade (e.g., engineering, tooling, quality control, and manufacturing). The cost is usually calculated graphically by finding the hourly rate (y-axis) for a given trade for a given fiscal year (x-axis). This is more commonly used in the design of aircraft, but the same principle can be applied to the design of launch vehicles to potentially improve the performance of a CER. An example of this is shown in Fig. 17.5, from [8]. Note that these trade-specific work-year estimates are not what the *practitioner* of the trade receives as their pay (few engineers in 2010 were making $120 per hour—almost $250,000 per year—to work on launch vehicles), but instead are estimates of the costs the company can reasonably expect to incur as part of the hiring, training, administration, and benefits activities that accompany each employee. The hourly rates shown in Fig. 17.5 can therefore be considered estimates of

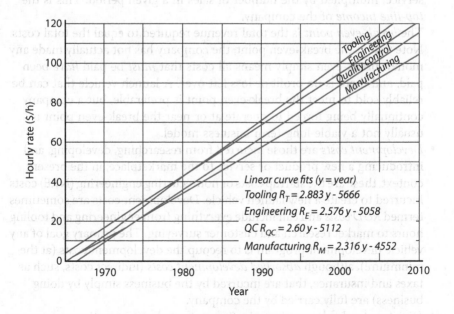

Fig. 17.5   Hourly rates by trade for fiscal years 1965–2010. Source: [8].

"all-up" costs at the company level. This is usually known as the *overhead* or *burden* rate, although the latter term is more appropriate to avoid confusion, because *overhead* has very specific meanings in the accounting field.

17.5 Cost Considerations

CERs are useful in estimating the development cost of a launch vehicle, but by themselves are rarely useful for analyzing the business metrics of an operation. There are several types of costs to consider when planning a new launch vehicle; these cost types may be broadly defined as *development*, *fixed*, and *recurring* costs. Although this is not a financial text, the practicing engineer must still be aware of certain basic terminology, because incorporating an understanding of business practices into the foundation of launch vehicle design gives the company an advantage against competitors. Note that cost and price are not the same thing: the former is the money paid to create the launch vehicle, whereas the latter is the money charged to the customer for the purchase of the launch vehicle, and can be more than or less than or equal to the cost, depending on the need to generate business or get one's "foot in the door" with an artificially low introductory price.

- *Revenue* should be understood first and foremost, because it drives a great number of business decisions. Revenue is the amount of money a business receives during a specific period, including discounts and reductions for returned items. Stated simply, revenue is the "sticker price" of a good or service, multiplied by the number of sales in a given period. This is the *top-line income* of the company.
- The *break-even point* is the total revenue required to equal the total costs. Note that at the break-even point, the company has not actually made any money: break-even simply means all costs that *must* be paid *have* been paid, and there is no profit or loss left over. A launch vehicle that can be reliably sold to meet the break-even point is preferable, but a company continually being forced to operate at or near the break-even point is usually not a viable long-term business model.
- *Development costs* are those incurred from researching, developing, and introducing a new product or service to the marketplace; in the present context, these are the startup costs or nonrecurring engineering (NRE) costs incurred to create a new launch vehicle. Development costs are sometimes termed *R&D costs* and can include everything from engineering and tooling hours to market research and customer surveying. The primary goal of any vehicle development program is to recoup the development costs (at the minimum!), although *absorbed development costs* (indirect costs, such as taxes and insurance, that are incurred by the business simply by doing business) are fully carried by the company.
- *Direct costs*, also known as *operating costs* or *direct operating costs*, are costs necessarily incurred by the business merely to maintain its own

existence. Direct costs may be fixed or variable, but the key is that a direct cost is *directly attributable* to a *cost object* (literally, an object that costs money). It should therefore stand to reason that an *indirect cost* is one that is not directly attributable to a cost object. For instance, the fuel used for a mission is a direct cost. Overhead; sales, general, and administrative (SG&A) expenses; and the like are examples of indirect costs. Note that the lines between direct and indirect costs are sometimes debatable and may vary by organization; an in-depth discussion of the nuances of financial and managerial accounting is beyond the scope of this text, but the interested reader is directed to [9] for more detailed information.

* *Fixed costs* are those costs that depend on neither the number of goods or services produced nor the rate at which they are produced. Examples of fixed costs include insurance, property taxes, and interest expenses. Fixed costs, together with variable costs, form the total *cost structure* of the company. The higher the percentage of the total cost structure that is dominated by fixed costs, the higher the total revenue will have to be to break even; of course, this also means that the *marginal* cost is typically low, implying that the company can scale to meet demand quickly and cheaply. Note that the actions of a business (e.g., previous launch attempts ending in success or failure) can sometimes influence these fixed costs (like insurance), and thus fixed costs must usually be reanalyzed and reoptimized periodically.

* *Variable costs* are those that change in proportion to the number of units produced. Intuitively, this is easy to understand: fuel for 10 launch vehicles will clearly be cheaper than fuel for 100 launch vehicles. The key factor with variable cost is that it is also the sum of the *marginal cost* (i.e., the additional cost to produce one additional unit) over all units produced. This means that variable cost depends on *both* the total number of units produced and the *rate* at which units are produced. For instance, it may be more profitable to utilize overtime production (at a higher per-hour rate) to produce a given number of launch vehicles or services faster.

* *Amortization* is the paying off of a debt with a fixed payment schedule. The most easily relatable example is an auto loan, where a buyer will pay off a vehicle over time, rather than absorbing the entire cost of the vehicle up-front. The financial mechanics of buying a launch vehicle are no different, although amortizing the cost of an expendable launch vehicle (which, by definition, is a single-use asset) is not always as directly apparent as amortizing a reusable one. Amortization is similar to *depreciation* (allocating the cost of a tangible asset over its useful life), but the details and intricacies differentiating the two are best left to an accounting text such as [9].

* *Return on investment (ROI)* refers to the efficiency of an investment, usually the estimated future value of an investment (such as a new launch vehicle) less the cost of the investment, divided by the cost of the investment. For instance, if a launch vehicle costs $80 million to build and

field and is expected to produce $100 million in revenue (laughably small numbers for a launch vehicle, to be sure!), the return on investment would be 25% [(100–80)/80]. Typically, a higher ROI is preferable, but it is not the only measure of interest: a 95% ROI won't be of interest to an educated manager if it only nets the company an additional $5 in revenue. When proposing engineering or developmental changes, or when working on a multidisciplinary team tasked with doing so, presenting a case that maximizes ROI will usually be viewed favorably by managers who ultimately make the financial decisions.

The aforementioned points are illustrated in Fig. 17.6. The development cost of a launch vehicle is fixed once development has concluded, and is independent of the number of units built (hence the "non-recurring" part of "non-recurring engineering" or NRE, described earlier). Development cost is therefore a fixed cost. Once units begin to be produced, each unit will cost the same amount of money to produce (in theory), but building more units results in more costs being incurred; this is why production costs are considered variable costs: they increase in direct proportion to the number of units built. Total costs are simply the sum of development and production costs; this is an example of a *mixed* cost, because it is composed of both fixed and variable costs. Note that no matter how large total cost becomes, the development cost stays constant: development cost may decrease as a *percentage* of total cost, but the development cost *as a dollar amount* does not change. Manufacturers will sometimes amortize the development cost over the production run as an "average fixed cost" – this is technically incorrect (because a fixed cost is precisely that: **fixed**), but it is still common practice.

Note that once production begins, units are sold. In a perfect world, every unit built would be sold; this is not entirely accurate, but this simplifying

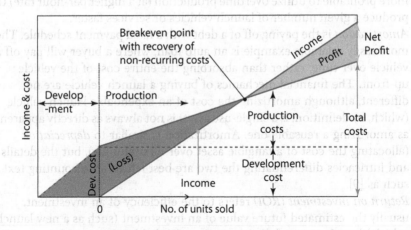

Fig. 17.6 Costs vs. Income vs. Breakeven for a production line.

assumption is sufficient for now. It must be assumed that the company will sell each unit at a price that's high enough to at least cover the cost of producing it, but this is not the whole story: there should be some "extra" profit left over after the production costs are covered.

Notice in Fig. 17.6 that although the company begins selling units, it is still operating at a loss. This is because the money to pay all of the engineers, designers, etc. necessary to develop the launch vehicle has already been paid by the company: the development cost has already been incurred, and the only way for the company to recover that cost is by selling additional units. Eventually, the company sells enough units so that the "extra" profit (above and beyond what's necessary to cover production costs) is equal to the development cost. This is called the **breakeven point**, and it is only after this point that the development program can be said to be profitable. Every unit after the breakeven amount increases the net profitability of the program.

Notice that the breakeven point is not dependent on the "quality" or capability of the launch vehicle: it is purely based upon the development and production costs relative to the sell price of the launch vehicle. This means that, yes, it is entirely possible to sell hundreds or thousands of the same launch vehicle for decades and never actually make a profit. This isn't realistically *probable*, but it is theoretically *possible*.

This is one of the key reasons why companies typically avoid taking risks when developing a new launch vehicle: older, proven technology will oftentimes be preferred, because the cost to develop those new technologies has already been paid. The frank reality is that a company must make money to remain in existence and to keep its employees paid, and it is as much the engineer's duty as the manager's to ensure that a design is financially realistic. Anything that can lower development cost brings the breakeven point closer to the start of the production run, and companies tend to prefer breaking even sooner rather than later.

17.5.1   Examples of Launch Vehicle Costs

With the aforementioned cost considerations in mind, what is the actual driver for using an expendable launch vehicle instead of a reusable one? The decision is often rooted in matters that are financial, rather than technical. As an example, consider a comparison between two three-step launch vehicles, as shown in Table 17.1. For this example, the third step is considered expendable for both vehicles.

From Table 17.1, it might seem that an expendable vehicle might be the obvious choice; however, the total mission *value* of the vehicle (of which cost is only one part) makes this decision more difficult. This is especially true when a customer request for proposals (RFP) does not mandate that a launch vehicle be expendable or reusable. When considering whether to design a reusable or expendable launch vehicle, the design team is essentially trading development cost for direct operating cost; this requires

Table 17.1   Sample Cost Breakdown of Launch Vehicles

Expendable	Reusable
Vehicle Costs	
1st step recurring cost	1st step cost amortization
1st step engine(s) recurring cost	1st step engine(s) cost amortisation
2nd step recurring cost	2nd step cost amortization
2nd step engine(s) recurring cost	2nd step engine(s) cost amortization
3rd step recurring cost	3rd step recurring cost
3rd step engine(s) recurring cost	3rd step engine(s) recurring cost
	1st step refurbishment
	1st step engine(s) refurbishment
	2nd step refurbishment
	2nd step engine(s) refurbishment
Other items	Other items
Add: Direct Operating Costs (DOG)	
Prelaunch ground operations	Prelaunch ground operations
Mission & flight ops/range cost	Mission & flight ops/range cost
Propellants & consumables	Propellants & consumables
Ground transportation & storage	Ground transportation & storage
Facility per-launch user fees	Facility per-launch user fees
Mission abort & vehicle loss charges	Mission abort & vehicle loss charges
Insurance and misc. costs	Insurance and misc. costs
Add: Indirect Operating Costs (IOC)	
Proj. admin & mgmt	Proj. admin & mgmt
Marketing	Marketing
Customer relations	Customer relations
Contracts	Contracts
Technical support & improvements	Technical support & improvements
Taxes & fees	Taxes & fees
Fixed annual launch site costs	Fixed annual launch site costs
Total: Cost per flight	
Add: Profit & Miscellaneous	
Development cost amortization	Development cost amortization
Profit	Profit
Depreciation	Depreciation
Miscellaneous costs	Miscellaneous costs
Total: Price per flight	

an in-depth knowledge of the operating requirements and preferences when the end user is a government agency, and a sound understanding of market forces and opportunities when developing a commercial vehicle. The savvy engineer is therefore able to make themselves uniquely valuable to top management in ways that may be ignored by the conventionally minded.

As a general rule, expendable vehicles offer the lowest technical risk and lowest development cost, and are preferable for missions with a low launch rate and predictable time horizon; conversely, the high development cost and technical risk of a reusable system is offset by the low operating cost for launch rates above 20 per year [10]. The maintenance and logistics footprint of the system must also be considered: a surge in required launch rates would require expendable systems to be stockpiled in their entirety, whereas a reusable system will likely require only a stockpile of payloads to support an operational surge.

Note that when designing a launch vehicle for a government contract, profitability is tightly controlled via statutory limitations; currently, profit is limited by U.S.C. 2306(d) and U.S.C. 254(b) to 15% for research, developmental, or experimental work and 10% otherwise. The fixed fee does not vary with the actual development costs incurred by the contractor (under 48CFR16.306), but may be adjusted based on performance. This type of cost-plus contract is commonly used for research or preliminary exploration, when the actual development effort is unknown. Financially, this type of contract protects contractors from development efforts that might otherwise present a high programmatic risk, but provides the contractor with few incentives to control costs.

For a cost-plus-fixed-fee contract, the *fixed fee* is negotiated at the inception of the contract and is based on six factors addressed in Part 15.404-4 of the Federal Acquisitions Regulations:

1. Contractor effort
2. Contractor cost risk
3. Federal socioeconomic programs
4. Capital investments on the part of the contractor
5. Cost control and prior performance
6. Independent development

In addition to these factors, individual agencies may include additional factors as part of their analysis, which may in turn provide additional opportunities for profit. The U.S. Department of Defense uses a method called Weighted Guidelines to calculate profit, as detailed in DD Form 1547 [11]. The Weighted Guidelines method is based on analysis of the following contractor risks:

• Technical
• Performance
• Management cost control
• Contract type

Additionally, the contractor's working capital, facilities, and cost efficiency are considered. The astute reader will notice that both the FAR15.404-4 method and the Weighted Guidelines method take into account a contractor's capital assets and past performance, thus biasing these methods toward larger, established contractors with a history of successful vehicle development and launches.

The development of a purely commercial vehicle may avoid this type of government oversight and have its profitability dictated purely by market dynamics, but that does not guarantee its legality and operability. Commercial launch operations (i.e., those that are not by and for the government) within the United States are regulated by the Office of Commercial Space Transportation (OCST), a branch of the Federal Aviation Administration. Any rocket that exceeds 200,000 lb_f-s of total impulse and that can reach an altitude of at least 150 km above sea level is considered *licensable* and falls under the jurisdiction of the OCST.

Licensable rockets must obtain a commercial launch license before being launched from a U.S. territory or by a U.S. citizen. At a minimum, a commercial launch vehicle will need to pass an expected casualty analysis (probability of casualty to any and all groups of people and infrastructure along the flight path, based on launch vehicle failure modes and potential causes of injury/death), a system safety process [both "top-down" analyses, such as fault trees, and "bottom-up" analyses, such as failure modes and effects analysis (FMEA), are required], and meet other operating restrictions (such as collision avoidance for operations above 150 km). Each of these requirements, and subrequirements, results in a go/no-go decision. It may be safely assumed that any "unknown" automatically results in a no-go.

To give the reader an idea of the costs involved, the per-launch cost of a SpaceX Falcon 9 rocket in 2013 was $56.5 million to low Earth orbit (LEO), which was already considered the least expensive in the industry [12]. Although the internal accounting of commercial launch vehicle manufacturers is not the focus of this text, it should become readily apparent that developing a new launch vehicle is an extraordinarily expensive endeavor that requires not only access to large amounts of capital assets, but also access to an extensive (and expensive) logistical network. This may present a significant barrier to entry for launch vehicle startups, which will usually have little (or zero) of either.

17.5.2 Inflation Factors

Any prolonged development effort that requires multiple years to complete will require an estimation of inflation costs. *Inflation* is the rate at which the price of goods and services increases (or, conversely, the rate at which purchasing power decreases) over a fixed period of time. For instance, if the inflation rate is 3% per annum, an item that costs $1.00 one year will cost $1.03 the subsequent year. Inflation is the economic embodiment of someone saying, "A dollar isn't what it used to be."

Inflation factors are necessary in order to make an adequate comparison among historical, existing (or near-horizon), and proposed launch vehicles. The most commonly used inflation factor is based on the consumer price index (CPI), which is a weighted average of prices paid by consumers for retail goods and services. Note that the CPI is hardly the only index to be used—there is a producer price index, an export price index, a gross domestic product (GDP) deflator, Paasche-Laspeyres indices, Lowe indices, Fisher and Marshall-Edgewood indices, and more—but the CPI is one of the most commonly used indices. The formula given by the U.S. Bureau of Labor Statistics is

$$I_f = 1.031^n \tag{17.1}$$

where n is the number of years since the reference year. This formula represents an average inflation of 3.1% from 1926 to 1992. For example, if the cost of an item in 2010, C2010, is $1.00, its cost in 2020 ($n = 10$), C2020, would be

$$C_{2020} = C_{2010}(1.031)^n$$

$$C_{2020} = \$1.00(1.031)^{10} \tag{17.2}$$

$$C_{2020} = \$1.36$$

The use of inflation factors is usually noted with the phrase *constant dollars* (e.g., 2015 constant dollars). Because the purchasing power of a dollar changes over time, the use of constant dollars is what permits a fair comparison between alternatives. This is critical when a company must decide whether to, for example, develop a new launch vehicle from scratch or refurbish and modify an existing launch vehicle to meet perceived (or actual) changes in customer needs. A handy on-line inflation calculator is the CPI Inflation Calculator at www.bls.gov/data/inflation_calculator.htm.

17.5.3 Recommendations for Initial Cost Estimation

For the purposes of initial planning, historical data suggest that a new launch vehicle will cost between $6 million and $600 million each, or about $2,000–$120,000 per pound to orbit. Needless to say, a customer must foot the bill for an *integer number* of launch vehicles, unless payload can be piggy-backed onto other launches. The ability to launch larger, more complex payloads (such as weather and telecommunications satellites) will obviously skew estimates toward the more expensive end of this range.

The general method for initial cost estimation is to assume an average cost of $5,000 per pound to orbit, although a strict dollar-per-pound assumption is deceptive, because the per-pound launch cost is entirely dependent upon the capabilities and specifics of the vehicle in question. For instance, in 2008 NASA signed a $1.9 billion contract with Orbital Sciences to launch 20 metric tons of cargo over eight launches; this equates to a cost of approximately $43,000 per pound. That same year, SpaceX signed a contract with NASA for $1.6 billion for 12 resupply flights, which equates to a cost of approximately $27,000 per pound, or $9,100 per pound if the SpaceX capsule is full of supplies both at launch and at recovery [13]. By comparison, despite the massive cost of the Space Shuttle (at least $500 million per launch), the cost of payload delivery was only on the order of $10,000 per pound—for 50,000 pounds of cargo and seven astronauts.

17.6 Cost Modeling Examples

The efficacy of statistically based CERs is best illustrated via example. The cost breakdown of the Atlas V 401 will be examined first, and then a more detailed investigation will be presented of the TRANSCOST CER (see [14, 15]). It should be noted that there are several other CERs in use today, including [16] and [17], which can be used to estimate vehicle costs as necessary. For TRANSCOST examples, all costs are presented in 2010 constant dollars unless otherwise indicated; cost values from other sources are given where available. All TRANSCOST images referenced are from [14, 15] unless otherwise indicated.

17.6.1 Atlas V 401 Cost Breakdown

A breakdown of the Atlas V costs is shown in Figs. 17.7–17.9. Although these costs are by no means identical for all rockets, they are representative of mature technology and make for useful estimation guidance when developing a new launch vehicle. It should be immediately apparent that engine costs make up the majority of the first-step costs; this shouldn't be surprising given the dependence on Δv to achieve orbit. This is further reinforced in Fig. 17.8, which shows that the first step, by mass, is very much the "flying gas can" that one might imagine: an overwhelming majority of the first step mass is propellant. Given the previous discussions on orbital energy requirements and the rocket equation, this is logical. Far more surprising, perhaps, is the fact that the relationship between mass and cost is not anywhere near one-to-one: in other words, a pound (or kilogram) of mass does not equate monotonically to a dollar spent. Some of this is due to the fine-tuning required to maximize performance (optimization is expensive), and some is due to simple market forces: propellants, as discussed later, constitute such a small percentage of a launch vehicle's cost that the money spent on propellants and gases is hardly worth even considering.

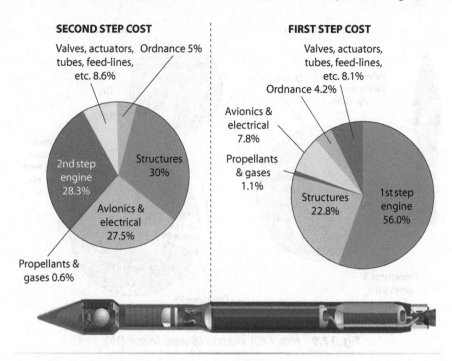

SECOND STEP COST
- Valves, actuators, tubes, feed-lines, etc. 8.6%
- Ordnance 5%
- 2nd step engine 28.3%
- Structures 30%
- Avionics & electrical 27.5%
- Propellants & gases 0.6%

FIRST STEP COST
- Valves, actuators, tubes, feed-lines, etc. 8.1%
- Ordnance 4.2%
- Avionics & electrical 7.8%
- Propellants & gases 1.1%
- Structures 22.8%
- 1st step engine 56.0%

Fig. 17.7   Atlas V 401 overall cost. Source: [19].

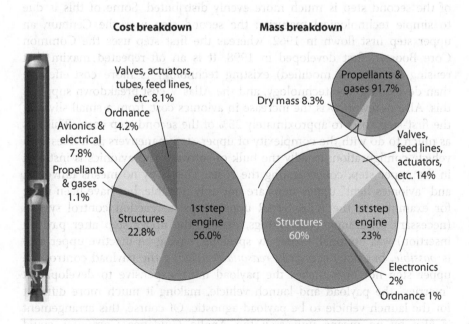

Cost breakdown
- Valves, actuators, tubes, feed lines, etc. 8.1%
- Ordnance 4.2%
- Avionics & electrical 7.8%
- Propellants & gases 1.1%
- Structures 22.8%
- 1st step engine 56.0%

Mass breakdown
- Propellants & gases 91.7%
- Dry mass 8.3%
- Valves, feed lines, actuators, etc. 14%
- Structures 60%
- 1st step engine 23%
- Electronics 2%
- Ordnance 1%

Fig. 17.8   Atlas V 401 first-step costs. Source: [19].

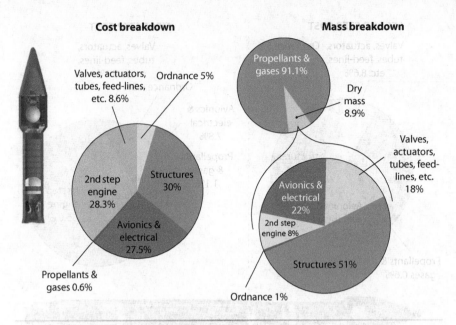

Fig. 17.9 Atlas V 401 second-step costs. Source: [19].

The mass breakdown of the second step shown in Fig. 17.9 mirrors that of the first step, in that propellants and gases once again make up the majority of the second step weight. More interestingly, the cost structure of the second step is much more evenly distributed. Some of this is due to simple technological maturity: the second step uses the Centaur, an upper step first flown in 1962, whereas the first step uses the Common Core Booster, first developed in 1998. It is an oft-repeated maxim that reusing (sometimes modified) existing technology is more cost-effective than developing new technology, and the Atlas V cost breakdown supports this. Also noteworthy is the increase in avionics cost, from a small sliver of the first-step costs to approximately 25% of the second-step costs. This has as much to do with the complexity of upper-step maneuvers as it does with vehicle configuration: usually the bulk of software and avionics is installed in the upper step, concentrating the costs. This is by no means universal, and "avionics-light" upper steps are not only possible, but have been used; for example, on the ATK STAR upper steps, a reaction control system (necessary to, among other things, deorbit the upper step after payload insertion) was optional. Generally speaking, having an inactive upper step is *possible*, but not necessarily *reasonable*: having the payload control the upper step not only makes the payload more expensive to develop, but "marries" the payload and launch vehicle, making it much more difficult for the launch vehicle to be payload-agnostic. Of course, this arrangement is also by no means universal: the Apollo Guidance Computer could control the entire Saturn V stack, not just the upper steps.

17.6.2 TRANSCOST Cost-Estimating Relationships

TRANSCOST is one of the more commonly used CERs in the launch vehicle development community. It provides a top-down cost analysis (costs are determined at the system level), which can be used by the designer to create first-order cost estimates. Based on historical data, the accuracy of TRANSCOST is usually within ± 20%. Note that TRANSCOST estimates are based on *actual* vehicles, and therefore include the costs associated with unforeseen development problems and their subsequent resolution; thus, TRANSCOST estimates will almost always be higher than estimates derived from a bottom-up approach. TRANSCOST also provides plots that may be of use to the design engineer, such as the relation between work-years and total launch mass by vehicle type, as shown in Fig. 17.10. The basic structure of TRANSCOST is shown in Fig. 17.11. Each of the TRANSCOST submodels is interconnected to estimate costs for the entire lifecycle of a launch

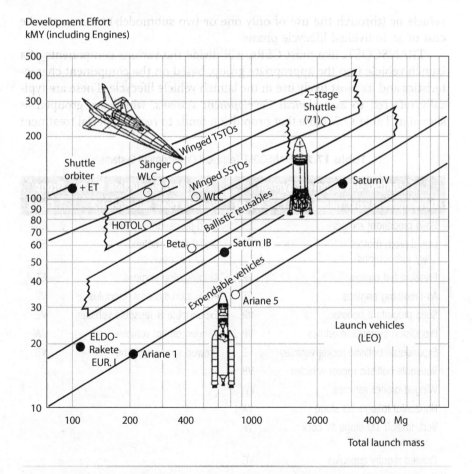

Fig. 17.10 Example TRANSCOST regression plot.

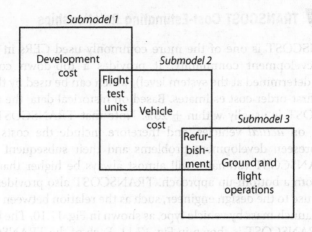

Fig. 17.11 TRANSCOST basic structure.

vehicle or (through the use of only one or two submodels) to estimate the cost of an individual lifecycle phase.

TRANSCOST, like most CERs, will divide the various components of a launch vehicle into the appropriate group, based on the component characteristics and its most likely use in the launch vehicle lifecycle. These are typically referred to as *technical development systems*, which are grouped as shown in Table 17.2. Note that propulsion tends to receive special treatment

Table 17.2 TRANSCOST Technical Development Systems

Technical Development System		Technical Production System Group	
Development cost, submodel 1	Code	Vehicle cost, submodel 2	Code
Solid propellant motors	ES	Solid-propellant motors	ES
Liquid-prop. motors w/ turbopumps	EL	Liquid prop. motors w/ turbopumps	EL
Pressure-fed motors	EP	Air-breathing engines	ET
Air-breathing engines	ET	Propulsion systems/modules	VP
Solid propellant booster	VR	Expendable & reusable vehicles	VE
Propulsion systems/modules	VP	Winged orbital vehicles	VW
Expendable ballistic rocket vehicles	VE	Crewed space systems	VS
Reusable ballistic rocket vehicles	VB		
Winged orbital vehicles	VW		
Horizontal takeoff 1st stage	VA		
Vert, takeoff 1st stage fly back vehicles	VF		
Crewed reentry capsules	VC		
Crewed space systems	VS		

because launch vehicles (or steps) can have a mix of motors, some of which are currently in production and others of which require new development. For each of the technical areas in TRANSCOST, the most direct CER is mass-driven and has the basic form of

$$C = aM^x \qquad (17.3)$$

where C is the cost, a is some system-specific value, M is the system mass (in kg), and x is a system-specific sensitivity factor that correlates cost to mass. Note that this is a highly empirical approach to developing a CER, and thus the aforementioned constraints and advisories with regard to regression-based models apply.

As previously stated, realism is the most critical component of any CER. The first step in the derivation of any CER is the establishment of a large and accurate database from which relationships may be calculated. This database need not necessarily be an expensive software package: often, some hours with a search engine and Microsoft Excel (which provides searchability and statistical analysis tools) will suffce. By establishing the relationships between cost and mass, the factors a and x from Eq. (17.3) can be calculated using a best-fit curve.

However, the use of historical data to derive a cost relationship is, in and of itself, insufficient to estimate *development* costs of an all-new launch vehicle. Three factors heavily influence the cost of a development process; using TRANSCOST terminology, these factors are:

- *Technical development status factor (f_1)*: The similarity of the current development effort to previous projects. For first-of-its-kind projects, f_1 will increase the development cost. Truly first-of-its-kind development efforts can increase the estimated cost by up to 20%.
- *Technical quality factor (f_2)*: A factor derived for each system independently, based on figures of merit. As expected, higher precision, higher reliability, and higher performance will lead to higher cost.
- *Team experience factor (f_3)*: A factor encompassing the team's level of experience with similar projects.

For production projects, a fourth factor—f_4, the cost reduction factor for series production, or learning curve—is given. These factors are incorporated multiplicatively into the TRANSCOST CER derivation process, shown in Fig. 17.12. The total cost is then all lower costs, with factors applied to each, summed together. A detailed explanation of each development factor, along with guidance with regard to the numerical value that should be assigned to each, is presented in [15].

Recall that the ultimate goal of a CER is to trace a relationship back to cost; more often than not, this is defined as a work-year effort, but

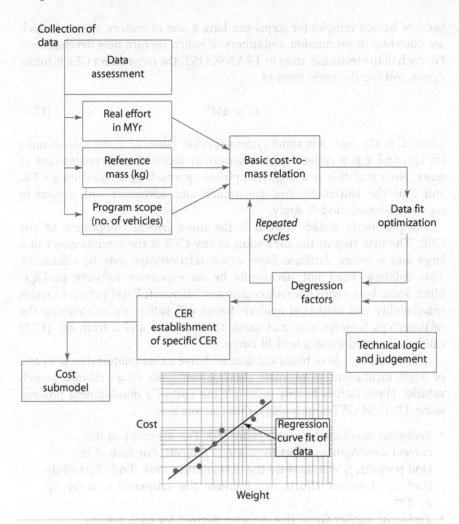

Fig. 17.12 TRANSCOST CER derivation process.

how much is a work-year actually worth in terms of a dollar amount? For the aerospace industry, the work-year dollar amounts are assumed to include secondary costs such as overhead, SG&A, travel, taxes, and profits. The cost of management and support staff is dependent on the accounting scheme of the company and how much management and support time is accounted to a given project. The historical dollar value of a work-year in the United States, Europe, and Japan is shown in Table 17.3, with U.S. dollar figures based on NASA's cost escalation factors and European values based on the ESA's annual growth cost values. The BOLD value in Table 17.3 represents the NASA reference value for one work-year from 533 contracts.

Table 17.3   Work-Year Dollar Amounts by Year

Year	US$	Europe, €	Japan, M¥
1970	38,000	31,000	–
1980	92,200	79,600	–
1990	156,200	139,650	18.1
1997	**191,600**	177,650	21.5
2000	208,700	190,750	23.2
2010	296,000	261,000	28.9

Source: [15]

One of the most useful features of TRANSCOST is the ability to use plotted data to estimate costs based on launch mass. These plots are based on the power-law relationship shown in Eq. (17.3) based on historical data. A detailed derivation of these relationships is given in [15]. Development effort vs mass for various vehicle and component types is given in Figs. 17.13–17.17.

17.6.3 TRANSCOST Cost Shares and CER Verification

As previously mentioned, an expendable launch vehicle will tend to have much higher production costs than a reusable launch vehicle, traded against

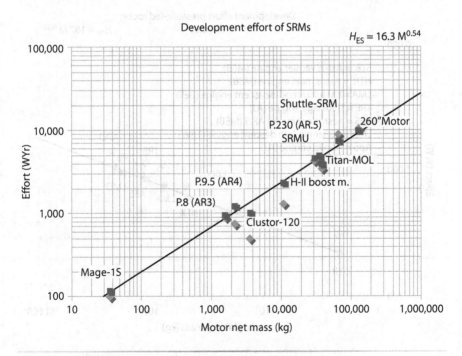

Fig. 17.13   Solid-propellant motor development effort.

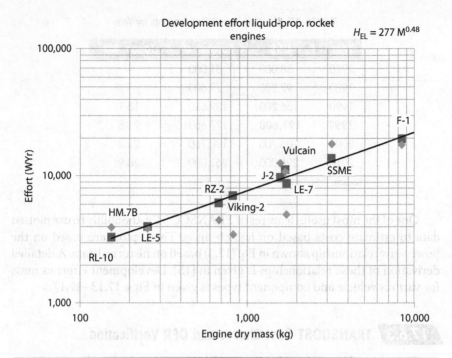

Fig. 17.14 Liquid-propellant motor development effort.

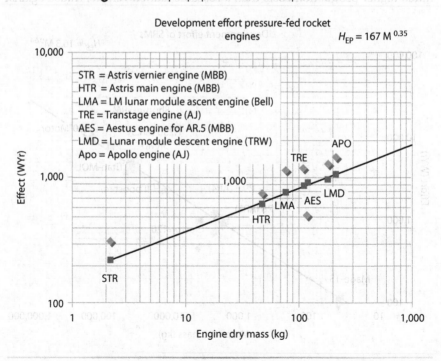

Fig. 17.15 Pressure-fed motor development effort.

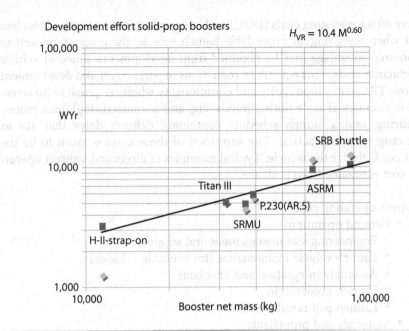

Fig. 17.16   Solid-propellant booster development effort.

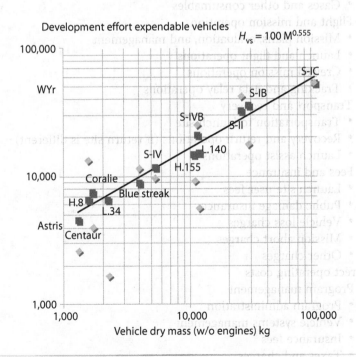

Fig. 17.17   Expendable vehicle development effort.

lower direct operating costs (DOC) and lower refurbishment costs. It is clear that when designing an expendable launch vehicle, the greatest amount of economic advantage may be obtained from developments aimed at vehicle production costs. This equates to reductions in parts count and development of low-TRL items, modularity and commonality wherever possible (to leverage economies of scale during purchasing and manufacturing), lean manufacturing, and a launch schedule containing delivery dates that are as unchanging as is practical. The sum total of these costs is taken to be the cost per flight for the vehicle. Typical examples of direct and indirect operating cost elements, from [15], are:

- Direct operating costs
  - Ground operations
    - Engineering site management and support
    - Launch vehicle maintenance (for reusable vehicles)
    - Assembly integration and checkout
    - Launch preparation
    - Launch pad refurbishment
  - Materials and propellants
    - Single-use elements
    - Fuel, including boil-off losses
    - Oxydizer, including boil-off losses
    - Gases and other consumables
  - Flight and mission operations
    - Mission plans, evaluation, and management
    - Launch and flight operations
    - Crewed mission operations
    - Tracking and data relay operations
  - Transport and recovery
    - Transportation to launch site
    - Recovery and return operations (if return site is different)
    - Launch assist operations
  - Fees and insurance
    - Launch site user fees
    - Public damage insurance
    - Vehicle loss charges
    - Mission abort charges
    - Other charges
- Indirect operating costs
  - Program management
    - Program administration
    - Vehicle systems management
    - Insurance fees
    - Taxes and charges
    - Profit

- Marketing and contracts
  - Marketing and customer relations
  - User manual distribution
  - Exhibitions
  - Publications
  - Contracts handling
  - Financing support
- Technical system support
  - Vehicle procurement
  - Production control
  - Technical standard and performance
  - Failure analyses
  - Parts management and storage
  - Crew support
- Launch site infrastructure
  - Administration
  - Facilities maintenance
  - Safety and security
  - Supplies
  - Range facilities

Note that payload integration and checkout are traditionally *not* considered part of the ground operations phase. These costs are typically accounted for as part of the payload cost. Maintenance is also not equivalent to refurbishment, the latter being considered an offline activity that is performed after a certain number of flights. For *expendable* launch vehicles, insurance costs are normally paid by the customer to an insurance company (however, government entities usually self-insure); for *reusable* launch vehicles, the launch provider must ensure (and insure) the capability, reliability, and performance of the vehicle.

Given the amount of dependency and importance that TRANSCOST (and other cost models) place upon regression and historical trends to estimate the cost of future launch vehicles, the question must be asked: just how accurate are some of these predictions? The utility of a CER is most effectively demonstrated via comparing the actual cost of a real-world program to a CER's estimated cost for that same program, with gross launch masses and vehicle performance metrics fixed. By way of example, the estimate of TRANSCOST for the Ariane 5 is compared to the actual costs of the program, as reported by the European Space Agency, and is presented in Table 17.4. It may be seen that the difference between the TRANSCOST estimate for Ariane 5 and the program's actual cost is less than 4%, thereby demonstrating the applicability of a robust cost model when properly implemented.

Another commonly used cost model is found in the U.S. Air Force (USAF) Space Planner Guide. From the parameters found in [17], the development cost of the Ariane 5 is estimated at $1.74 billion in 1970 dollars, or $38,000 per work-year from Table 17.3. The USAF model predicts 45,790

Table 17.4 TRANSCOST Model Performance Verification, Ariane 5

Element	Ref. Mass, kg	WYr$_{CER}$	f_1	f_2	f_3	f_8	WYr
SRB	39,300	5,398	1.1	–	0.9	0.86	5,056
Vehicle + shroud	11,210	19,692	1.6	1.16	0.85	0.86	18,368
Vulcain motor	1,685	9,800	1.1	0.79	0.9	0.86	6,592
3rd stage + equip. bay	1,200	7,689	0.9	1.09	0.9	0.86	5,838
Aestus motor	119	890	0.8	–	0.8	0.77	438
Total development effort of elements, WYr							36,292
Total program cost, with system engineering effort $f_0 = 1.04^3$							40,827
Actual program cost per ESA, WYr							39,300
Δ, CER - Actual							3.89%

work-years to develop, which is approximately 11% higher than the actual cost of 39,300 work-years presented in Table 17.4; this is certainly less accurate than the TRANSCOST prediction (no doubt due to the age and lack of maturity of the USAF model, which was developed in 1965), but sufficiently adequate for an initial prediction.

17.6.4 Specific Transportation Cost

The specific transportation cost is a different (and, it may be argued, better) method of examining launch cost. Specific transportation cost is a measure of a launch vehicle's cost *efficiency*, and equates launch vehicle size and performance with payload capability. Simply put, larger launch vehicles will lower specific transportation costs, according to economies of scale. This concept has been found to hold true across all manner of transportation vehicle types, from trucks to aircraft. This is commonly known as the *cost of transport*, a dimensionless quantity that quantifies the energy E or power P required to transport a fixed payload of mass m over a given distance d or at a specific speed v

$$\text{COT} = \frac{E}{mgd} = \frac{P}{mgv} \qquad (17.4)$$

For the cost of transportation metric, a higher value is desirable. The velocity of a space transportation system v is predetermined by the orbital requirements; the cost of transportation therefore decreases logarithmically with increasing mass. This is due to the fact that payload ratio increases as launch mass increases, as well as the fact that launch operations costs are not one-to-one with launch mass. Historical data, as shown in Fig. 17.18, generally hold to these tenets. The specific transportation costs in Fig. 17.18

differ from the trendline by +100%/−50% due to variations in launch rate, launch vehicle complexity, and the efficiency of ground operations.

Note that most cost models assume that the payload capability of a launch vehicle is fully utilized. In reality, this is rarely the case, and a utilization factor of 60–80% is more likely; this increases the specific transportation cost 20–60% above the theoretical value in order to account for this unused payload capability [14, 15]. This utilization factor can be increased by launching multiple satellites; however, it is sometimes difficult to find multiple satellites that are the correct combination of size and mass with similar-enough orbital requirements to fully utilize the payload capability. This makes the concept of allowing small satellites (under 100 kg) to piggy-back onto the launch of larger satellites, often at a steep discount to the small satellite customer, an attractive proposition: the launch provider enjoys additional revenue, and the primary satellite customer enjoys a lower specific transportation cost. Cost and payload capability information for several global launch vehicles are given in Table 17.5.

Of course, the cost models presented in [15] and [17] are far from the only CERs available. A great deal of other regression-based cost models exist, all of which attempt to perform the basic function of predicting the relationship between launch costs and payload mass. In essence, this is the only economic metric that matters in space transportation: maximum pounds lofted for minimum dollars spent. One alternate CER is shown in Fig. 17.19, relating cost per pound of payload to maximum payload capacity; note that this

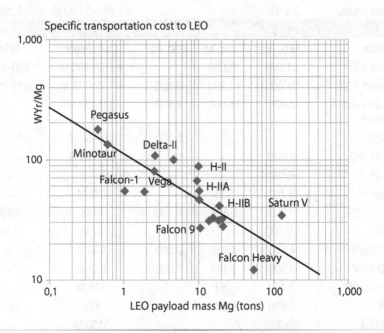

Fig. 17.18   Specific transportation cost. Source: [15].

Table 17.5 Launch Vehicle Cost and Payload Data, ca. 2013

U.S. Launch Vehicles					
Vehicle	LEO, kg	Polar, kg	GTO, kg	Cost/Flight	WYr
Atlas V 401	9,600	7,300	4,750	$130–140 M	420–460
Atlas V 431	15,400	11,300	7,500	$140–160 M	460–530
Atlas V 551	18,700	15,000	8,900	$165–180 M	550–600
Delta II 7320	2,690	1,580	–	$75–90 M	250–300
Delta II 7925	–	3,200	1,840	$110–130 M	570–400
Delta IVM	8,100	–	4,200	$150 M	500
Delta IVM+4,2	10,400	–	5,800	$170 M	570
Delta IVH	23,000	–	13,180	$360 M	1,200
Falcon 1	420	420	–	$8.9 M (2009)	57
Falcon 1e	1,010	–	–	$10.9 M (2010)	62
Falcon 9	10,450	7,000	4,500	$54–59.5 M (2011)	290
Minotaur	600	335	–	$21–24 M	70–80
Pegasus XL	450	200	–	$20–25 M	80–100
Taurus XL	1,590	860	557	$45–50 M	150–160
Taurus II	5,000	–	–	$45 M	150
Saturn IB	15,000	–	–	$43 M (1963)	1,500
Saturn V	127,000	–	45,000	$220 M (1974)	4,280
Space Shuttle	24,000	–	–	$1,250–3,000 M	2,200
International Launch Vehicles					
Vehicle	LEO, kg	Polar, kg	GTO, kg	Cost/Flight	WYr
Ariane 5 ES	21,000	10,000	–	€152–160 M	620–650
Ariane 5 ECA	25,000	14,200	9,500	€165–175 M	675–720
CZ–2C Long March	2,800	15,000	8,900	$25–30 M	–
CZ–3B	–	–	5,100	$70–80 M	–
CZ–2E/HO	8,800	–	–	$60–76 M	–
CZ–2F	8,400	–	–	–	–
Dnepr	3,700	300	–	$12 M	–
H–IIA–202	9,900	3,600	4,100	¥16B	540
H–IIB	19,000	–	8,000	¥23B	730
ISRO GSLV	4,500	1,800	2,300	$40 M	–
ISRO PSLV	3,500	1,350	1,050	$32 M	–
Kosmos–3M	1,400	760	–	$15 M	–
M–5	1,850	600	–	¥7B	280
Proton–K	20,700	–	–	$100 M	

(Continued)

Table 17.5   Launch Vehicle Cost and Payload Data, ca. 2013 *(Continued)*

International Launch Vehicles					
Proton–M Briz-M	–	–	6,150	$120 M	–
Rockot	1,850	1,300	–	€30 M	–
Soyuz/Molniya	6,200	2,700	1,700	$65 M	–
Soyuz–2 Kourou	6,200	3,900	2,800	€75–80 M	–
Tsyklon–3	3,600	2,100	–	$30–35 M	–
Vega	2,400	1,500	–	€50 M	–
Zenit 3SL	–	–	6,000	$150 M	–
Zenit 2	13,700	5,000	–	$120 M	–

Source: [15].

CER examines payload lofted to a 100 NM (185 km) circular orbit (by far the most common reference orbit used when designing a launch vehicle) at a fixed inclination. It should be noted that launch costs vary by orbital inclination, and therefore by launch site; it is perfectly legitimate for the engineering team to keep this fact in mind during the design phase, because a launch vehicle might quickly find itself incapable of *profitably* lofting a desired payload mass to some orbits from available launch sites.

Another common CER is a quick table lookup, shown in Table 17.6, that allows a rapid (if not always terribly accurate) method of comparing launch vehicle performance by class, target orbit, and country of origin. Again, the trend of increasing cost efficiency (decreasing specific transportation cost) with increasing vehicle gross launch mass is apparent; however, it should be

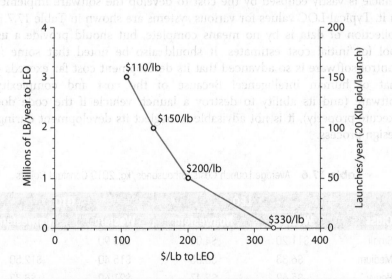

Fig. 17.19   Launch cost per pound of payload. Source: NASA.

noted that specific transportation cost does not tell the entire story: specific transportation cost is a measure of cost *efficiency*, not necessarily cost *magnitude*. Often, as mentioned earlier it is difficult to cost-share with additional payloads because of operational requirements, requiring a payload customer to buy the entire launch. At that point, they are paying the full launch cost, specific transportation cost be damned. Cost is therefore *an* important metric, but far from the only one.

17.6.5 Software Cost

The most obvious application of software for any space mission is that found on the launch vehicle itself, which governs flight control, guidance, and navigation. Although these are necessary for any modern launch vehicle, they are not the only application of software, nor are they the only source of software cost. Software is commonly used for manufacturing, ground operations, mission planning, trajectory optimization, performance simulation, station-keeping, and a host of other functions. The basic cost element of software is a line of code, which accrues cost from its being written, tested (at the subroutine, function, subsystem, and system level), and maintained. The software cost estimation is then the simple multiplication of the number of lines of code (LOC) for a vehicle (or mission function) multiplied by the lifecycle cost to develop, test, and maintain each line of code.

Because this text is concerned with the design of launch vehicles, the number of lines of code at the vehicle level is of principal interest. The lifecycle cost of software is between $10 and $25 per fully tested line of code [18]; it is readily apparent that for some systems, the cost to develop the physical vehicle is vastly eclipsed by the cost to develop the software implemented in it. Typical LOC values for various systems are shown in Table 17.7. This collection of data is by no means complete, but should provide a useful tool for initial cost estimates. It should also be noted that some flight control software is so advanced that its development cost far exceeds even that of human intelligence! Because of the cost and complexity of software (and its ability to destroy a launch vehicle if the code doesn't execute properly), it is not advisable to neglect its development during the design process.

Table 17.6 Average Launch Costs, $thousands/kg, 2010 Constant Dollars

	LEO		GTO	
Class	Western	Nonwestern	Western	Nonwestern
Small	$11.20	$4.06	$23.90	-
Medium	$6.33	$3.05	$15.40	$12.50
Heavy	$5.62	$2.47	$21.60	$8.83

Source: Futron Corp.

Table 17.7   Examples of Software Complexity

System	Lines of Code (LOC)	Cost at $25/LOC
Human comprehension	~5 K	$125 K
Apollo 11	145 K	$3.6 M
Space Shuttle	400 K	$10 M
F-22 Raptor	1.7 M	$42 M
Hubble Space Telescope	2 M	$50 M
U.S. Military Drone (control SW only)	3.5 M	$87 M
Curiosity Mars Rover	5 M	$125 M
B-787 (avionics, on-line support only)	6.5 M	$162 M
B-787 (total flight software)	14 M	$350 M
F-35 Lightning II (2013)	24 M	$600 M

Source: [20], [21]

17.6.6   Propulsion Cost

General William Shelton, Commander of the U.S. Air Force Space Command, gave a keynote speech at the 15th Annual FAA Commercial Space Transportation Conference in 2012. During his speech, General Shelton had this to say regarding rocket propulsion: "Each of the Atlas and Delta [RL-10] upper stage engines requires almost 8,000 man-touch-hours—more than goes into putting together a hand-built Lamborghini." Note that at $108 per hour (in 2012 dollars), that amounts to a cost of $862,000 *per engine.* Clearly, the cost of developing a propulsion system is not to be overlooked. The cost of a propulsion system is likely to always remain high due to the complexity and precision of the subsystems involved, but the astute engineer will quickly realize that (per Fig. 17.3) prudent design efforts early in the project lifecycle will help reduce the production cost of the propulsion system as much as possible. It would not be an exaggeration to claim that the propulsion system drives much of the cost of a launch vehicle, as per Fig. 17.7, and that a great deal of attention should be paid to ensuring that intelligent decisions are made with regard to propulsion, lest a company find itself with a rocket that is insufficiently economical to operate.

The cost of the actual propellants, however, is significantly less important, and in many cases, is not even worth considering. The cost of typical launch vehicle propellants is listed in Table 17.8, from [14]. Note that for a typical SpaceX Falcon 9 launch cost of $56.5 million to LEO, and using only Table 17.8 and publicly available data for the propellant mass fraction of the Falcon 9, it is seen that the total cost of propellants for a typical launch is only on the order of $200,000. That's approximately 0.3% of the total cost of the launch, which is small enough to be a rounding error. Barring the use of hydrazine (and its considerable safety and logistical footprint)

Table 17.8 Propellant Costs, 2019 Costs Inflated from 2009 Constant Dollars

Propellant	Cost, $/kg
Liquid hydrogen (LH_2)	8.38
Liquid oxygen (LOX, LO_2)	0.25
Kerosene (RP-1)	3.75
Dinitrogen tetroxide	103.00
Hydrazine (N_2H_4, UDMH, MMH, Aerozine 50)	398.00
Solid propellant (typical)	10.53
Hydrogen peroxide (85%)	5.50
Helium gas	18.00
Nitrogen gas	0.11
Hydrogen gas	10.60

Source: [15].

or the development of some truly exotic propellant, the cost of propellants and other consumables may be safely ignored during the initial design process.

17.7 Concluding Remarks

A certain amount of the seemingly large sums of money associated with space missions must simply be accepted as a sunk cost: physics demands fixed minimums from a launch vehicle in terms of performance and structural integrity, and the practicing engineer tempts those margins at great risk to the survivability of the mission and his or her own career. As previously explained, an expendable launch vehicle has the advantage of a potentially lower development cost relative to a reusable one. For a predictable and stable launch schedule, an expendable vehicle will have a cost advantage over a reusable one for up to 20 launches per year. It will therefore deliver a better value. As is the case with conventional airliners, economic efficiency improves with additional payload capacity, at the expense of a potentially reduced customer base and the risk of needing to spread the per-pound cost over multiple smaller payloads.

As with most engineering practice, it is critical to understand the customer's actual needs and requirements in as great a depth as possible during the early stages of design. It is often quoted that 70% of a system's total lifecycle cost is locked in during the initial phases of design, and although planning for weight growth is essential throughout the aerospace industry, weight growth in rocketry is particularly damning due to the exponential nature of Δv: small increases in mass can quickly render a design incapable of performing a customer's mission. As always, it is critical to understand the data referenced and the assumptions used, especially when contemplating costs and attempting to maximize customer value.

References

[1] National Aeronautics and Space Administration, "Systems Engineering Handbook," NASA SP-2016-6105 Rev2, 2016.

[2] Ley, W., Wittman, K., and Hallmann, W., "Handbook of Space Technology, York: Wiley, 2009.

[3] Locke, F. W. S, Jr., "Some Ideas on What Makes a Successful Operational Navy Aircraft," U.S. Navy, Weapon Systems Analysis Division, Report # R-5-61-16, 1961.

[4] U.S. Department of Defense, "Cost Analysis Improvement Group Operating and Support Cost-Estimating Guide," May 1992, https://www.cape.osd.mil/files/OS_Guide_v9_March_2014.pdf [retrieved 22 May 2017].

[5] U.S. Navy, "NAVAIR Acquisition Guide," 1 Oct. 2013, http://www.acqnotes.com/Attachments/NAVAIR%20Acquisition%20Guide%20Oct%202013.pdf [retrieved 22 May 2017].

[6] U.S. Navy, "Total Ownership Cost Guidebook," June 2014, http://www.acqnotes.com [retrieved 22 May 2017].

[7] Ott, R. L., and Longnecker, M. T., *An Introduction to Statistical Methods and Data Analysis*, Brooks Cole, Salt Lake City, UT, 2010.

[8] Nicolai, L., and Carichner, G., *Fundamentals of Aircraft and Airship Design*, AIAA, Reston, VA, 2010.

[9] Williams, J., Haka, S., Bettner, M., and Carcello, J., *Financial and Managerial Accounting*, McGraw-Hill, New York, 2014.

[10] Gstattenbauer, G. J., "Cost Comparison of Expendable, Hybrid, and Fully Reusable Launch Vehicles, Thesis, Air Force Institute of Technology, AFIT/GSS/ENY/06-M06, 2006.

[11] Department of Defense, *Defense Federal Acquisition Regulation Supplement; Weighted Guidelines Form*, DFARS Case 2002-D012, Document Number 02-19083, 67 FR 49254, 30 July 2002.

[12] Belfiore, M., "The Rocketeer," *Foreign Policy*, 9 Dec. 2013, https://foreignpolicy.com/2013/12/09//the-rocketeer/ [retrieved 3 June 2017].

[13] Mosher, D., and Kramer, S., "Here's How Much Money It Actually Costs to Launch Stuff Into Space," *Business Insider*, http://www.businessinsider.com/spacex-rocket-cargo-price-by-weight-2016-6 [retrieved 5 June 2017].

[14] Koelle, D. E., "Economics of Fully Reusable Launch Systems (SSTO vs. TSTO Vehicles)," 47th IAF Congress, Beijing, 7–11 Oct. 1996.

[15] Koelle, D. E., *Handbook of Cost Engineering*, TCS, Ottobrunn, Germany, 2011.

[16] Northrop Grumman, *Space Handbook*, Northrop Grumman, El Segundo, CA, 2003.

[17] United States Air Force Air Force Systems Command, "Space Planners Guide," 1 July 1965. https://www.worldcat.org/title/space-planners-guide/oclc/1404743.

[18] Parker, J. S., and Anderson, R. L., "*Low-Energy Lunar Trajectory Design*," NASA Deep Space Communications and Navigation Series, July 2013.

[19] Bruno, S. T., "*Tory*," *Interesting infographic on some of the Systems Engineering characteristics of LVs*," Twitter, https://twitter.com/torybruno/status/561933155951595520 [tweeted 1 February 2015].

[20] https://informationisbeautiful.net/visualizations/million-lines-of-code/.

[21] Apollo: https://www.itworld.com/article/2725085/curiosity-about-lines-of-code.html.

17.8   Exercises: LV Cost Estimation

1. Estimate the minimum development cost for an expendable launch vehicle with a total launch mass of 8,000,000 kg in 2030 constant dollars.

2. A two-stage expendable launch vehicle is needed to deliver a 8,000 kg payload to a circular orbit at an orbital altitude of 1,000 km. For reasons that are not actually relevant to this problem, no existing launch vehicle

anywhere in the world is capable of performing this mission. The launch will occur at sea level from Cape Canaveral, and will not require an inclination change. You will be using off-the-shelf engines/motors as you deem appropriate, so Figs. 17.12–17.15 will be useful, although you may need to look up thrust and I_{sp} values. You may use whatever method you deem appropriate to estimate gravity and aerodynamic losses. No learning curve is applicable. Calculate the **total** cost of this launch vehicle (development, production, operation, and other costs) in 2025 constant dollars. How does this vehicle compare against other U.S. launch vehicles, such as the launch vehicles given in Table 17.5?

3. In reality, program costs involve much more than simply estimating the cost of the launch vehicle: engineers need space in which to work, technicians need space in which to manufacture and assemble parts, etc. Oftentimes, there is a tradeoff between taking longer to complete a project (at decreased potential revenue) and hiring more people (at increased overall cost) to try and complete the project sooner.

You are the founder of an aerospace startup company, and from preliminary calculations you have estimates of development costs (both as a dollar amount and in terms of work-years) for the launch vehicle described in Problem #2. Because you are not independently wealthy, you must ask your investors for money to "stand up" (hire and staff) the project, and **you must build facilities** at a cost of $130 per square foot of floor space – the incumbent aerospace companies see you as a threat and have established relationships with all nearby real estate companies so that no existing buildings are available. The actual planform of this facility space is irrelevant: you're concerned with total floor space required. The following constraints apply:

* Each engineer requires 110 ft.^{2} of floor space
* Each manager requires 150 ft.^{2} of floor space, and there must be one manager for every 30 engineers (rounded up)
* Each executive requires 200 ft.^{2} of floor space, and there is one executive for every 10 managers (rounded up)
* The company president (you!) has an office of 400 ft.^{2} of floor space
* The company must have at least two conference rooms of 350 ft.^{2} each
* The company must have one reception area of 300 ft.^{2}
* As defined by 29 CFR 1910.141(c)(1)(i), you must have the following number of restrooms (55 ft.^{2} each) based on employee headcount:
    * 1–15 employees: 1
    * 16–35 employees: 2
    * 36–55 employees: 3
    * 56–80 employees: 4
    * 81–110 employees: 5
    * 111–150 employees: 6

- Over 150 employees: 1 additional bathroom for each additional 40 employees (rounded up)

Your investors want the launch vehicle developed in ten years, and they estimate that it will take three years to actually build it and have it ready for launch, but production cannot begin until the engineering is completed. It is estimated that each launch should yield $250 M in revenue, and that there is sufficient market demand for 26 launches at a rate of 6 launches per year. It is also estimated that, if the launch vehicle is not ready for production in ten years' time, the total acquirable market (i.e. how much money can be made) will decrease by $10 M per year.

You are not responsible for launch costs, production costs, or operating costs – only development costs. Assume the following flat rates are in 2020 constant dollars:

- Engineer: $150,000 per work-year
- Manager: $175,000 per work-year
- Executive: $225,000 per work-year
- Company President: $300,000 per work-year

Assume your design will have a steering loss of 0.40 km/s.
Given these criteria:

- Calculate the development cost necessary to meet the ten-year development deadline
- How does the cost of your launch vehicle compare with the TRANSCOST estimations presented in Fig. 17.9?
- Is this project economically feasible? That is to say, will the project be able to recover its development cost?
- If the project is economically viable, calculate the total headcount necessary to maximize the revenue of this project
- If the project is economically viable, what is the total expected revenue in 2020 constant dollars, ignoring inflationary effects?
- If the project is economically viable, what is the expected ROI, given the expected revenue and the development costs you've calculated?

Hint: this problem requires you to use information from multiple chapters, starting with Ch.03. You may ignore the cost of developing engines (assume any currently-existing engine is available and will magically fit); you may also ignore the cost of propellants. You're more than welcome to design your own launch vehicle from scratch, but finding "good-enough" surrogates based on performance requirements is acceptable (e.g. if your design is "close to" a Saturn V, assume it's a Saturn V to calculate aerodynamic losses, K_g and K_{gg} for gravity losses, etc.). You will also need to specify the engines and/or motors you're going to use, in order to estimate performance as well as thrust and aerodynamic loss factors such as K_a and K_D, respectively.

Problem 4: LV Cost Estimation

Determine the **cost** of your launch vehicle using Chapter VII of the U.S. Air Force **Space Planner's Guide** cost model. A PDF version of this chapter may be downloaded from the book's webpage at arc.aiaa.org.

For your estimate, complete Tables VII.C-1 (Input Parameters to Cost Curves) and VII.C-3 (LV Cost Summary Worksheet). In Table VII.C-1, items a(1) to a(11) depend on your particular design, but provide them on your worksheet. Be sure to include values for liquids or solids as appropriate. For the remainder, assume:

* items a(12) and a(13) = 2 units,
* a(14) = 6/year,
* a(15) = sea/train,
* a(16) = 1/month,
* a(17) = 1 launch site,
* a(18) = 5 years,
* a(19) = 24 vehicles.

Make (and state) any other necessary assumptions or estimates. Note that the USAF SPG's convention of "stage" refers to the item we've characterized as "step." Remember this model uses 1965 US$; inflate your estimate to *current-year* US$. Provide the costs for both 1965 and current year.

Compare your costs to at least one existing (or retired) similar (in mission or size) launch vehicle. Discuss your results: is your predicted cost within the range of similar vehicles?

GLOSSARY AND ABBREVIATIONS

A

A-4	Developmental designation for V-2 rocket
A5	*Ariane 5* launch vehicle
	Atlas V launch vehicle
A6	*Ariane 6* launch vehicle
AA	Associate Administrator (of NASA)
ABMA	Army Ballistic Missile Agency (US Army)
Ablative	a material that absorbs heat by turning directly to gas (Ch. 11)
Accelerometer	Transducer (sensor) that can measure acceleration (Ch. 12)
Access Door	Opening in vehicle to provide access to internal areas
Accumulator	A gas-filled fluid component added to a fluid system to attenuate pogo oscillations (Ch. 12)
Acoustic panels	Panels made to reduce the transmission of sound to the payload
ACS	Attitude Control System
Adapter	Structure joining two parts of vehicle
ADC	Analog-to-Digital Converter
A_e	Engine exhaust area
AEB	Brazilian Space Agency
Aero	Short for *Aerodynamics*
Aerozine 50	a 50/50 mix by weight of hydrazine and UDMH used as hypergolic fuel for Titan II missiles and others; FORMULA: 50% N_2H_4, 50% $(CH_3)_2NNH_2$
AF	Acceleration Factor (thermal testing)
	Air Force
AFB	Air Force Base
AFT	Autonomous Flight Termination
AIRS	Advanced Inertial Reference Sphere, MX guidance system
AJ	Aerojet (company)
AK	Alaska, U.S.
AKM	Apogee kick motor, used on a GTO to circularize the orbit at the top of the transfer.
Al	Aluminum
Al-Li	Aluminum-Lithium alloy
Alcolox	Refers to Alcohol-LOx propellant system (usually methanol or ethanol)
ALTO	Air-Launch-To-Orbit
Amatol	High explosive used on German WWII V-2 warhead
Ames	NASA Ames Research Center, Mountain View, CA, U.S.

AN	Ascending node of orbit, location where orbit passes *upwards* through reference plane from S to N
Anechoic	A special chamber with energy-absorbing walls used to do acoustic and RF testing (Ch. 15)
AoA	Angle of Attack, abbreviated Greek alpha (α)
AOS	Acquisition of signal
APCP	Ammonium Perchlorate Composite Propellant (propellant that was used in the Space Shuttle SRBs)
APS	Auxiliary propulsion system used on *Saturn V* S-IV & S-IVB steps (Ch. 7)
AR	Aspect ratio of a tank dome. $AR = 1$ is a hemispherical dome, others are ellipsoidal
AFB	Air Force Base (USAF)
AFRC	NASA Armstrong Flight Research Facility located at Edwards Air Force Base, U.S.
ANT	Antenna
Antares	Northrop Grumman LV used for ISS re-supply
	Red giant star in constellation of Scorpius
ARC	NASA Ames Research Center, Mountain View, CA, U.S.
Ariane	Series of European LVs, launch from French Guiana
ASLV	Augmented Satellite Launch Vehicle (India)
Astrionics	Contraction of 'Astronautic electronics'
A_t	Engine throat area
ATK	short for Alliant Techsystems, originally Thiokol, then Orbital ATK, now Northrop Grumman
Atlas	U.S. ICBM evolved into LV used for U.S. manned Mercury spacecraft and many NASA and DOD launches
ATM	Pressure of one standard atmosphere. 1 ATM = 101325 Pa = 101.325 kPa.
ATO	Abort to orbit
AU	Astronomical Unit (mean distance from Earth to Sun, 149, 597, 870.7 km)
AV	*Atlas V* launch vehicle
Autogenous	Pressurization using heating liquid propellant rather than externally-supplied high pressure gas (i.e. Helium). Used in *New Glenn* vehicle by Blue Origin
Avionics	Contraction of 'Aviation electronics'
Azimuth	Angle measured clockwise from north that a LV either lifts off or achieves orbit

B

B	Billion currency units
Baffles	Ring structures placed inside of propellant tanks to attenuate effects of sloshing liquids
	confuses, bewilders, or perplexes
Bake	Refers to thermal laboratory testing
BAM	Basic Assumptions Memorandum: ground rules and assumptions for a design (Ch. 1)

BE-3U	Hydrolox vacuum engine from Blue Origin, 2.4 MN thrust (550 klbf)
BE-4	Methalox engine from Blue Origin, 1.06 MN thrust (240 klbf)
Beanie Cap	fixture used to top off tanks in Space Shuttle External tank until just before launch (Ch. 7)
BECO	Booster Engine Cutoff
Black Zone	Any period of flight when an abort would be unsafe for occupants, usually excess acceleration (Ch.
_BO	Burnout condition
Boattail	Transition from larger to smaller diameter in the direction of flow (tapers down)
Booster	Name for the initial, high-thrust step of a multi-step launch vehicle
Bottle	A storage vessel usually containing pressurants such as gaseous Helium or Nitrogen at very high pressure
BTU	Unit of thermal energy (British Thermal Unit)
Bulkhead	A structural shell that forms the front or rear of a propellant tank (AKA dome)
Buran	Soviet space shuttle launched by Energia LV. Flew one time. ("Snowstorm" in Russian)

C

c	exhaust speed
	speed of light, approximately 3×10^5 km/s
C&DS	Command and Data Systems (Spacecraft design)
C3	Orbital specific energy (per unit mass), v_∞^2
CA	California, U.S.
CAD	Computer Aided Design
Caliber	Diameter, usually the rocket's reference diameter (Ch. 12)
CBC	Common Booster Core^{TM} (*Delta IV*). See also CCB, URM.
CCAFS	Cape Canaveral Air Force Station (Florida, U.S.)
CCB	Common Core Booster^{TM} (*Atlas V*). See also CBC, URM.
CCW	Counter-Clockwise
C_D	Drag coefficient; $C_D = D/(\frac{1}{2}\rho V^2 S_{ref})$ (nondimensional)
CDF	Confined detonating fuse used to trigger separation systems, retro rockets, and range safety systems
CDR	Critical Design Review
Centaur	Upper step to Atlas, Titan III/IV, and Vulcan
	Greek mythological creature with human upper body and lower body and legs of a horse
Centerbody	Structure between two tanks on a rocket step (intertank structure)
CECO	Center Engine Cutoff
CEP	Circular Error Probability: the radius of a circle encompassing 50% of all rounds fired. Describes accuracy of weapon delivery
CER	Cost Estimating Relationship
CF	Carbon Fiber (graphite fibers)
CFD	Computational Fluid Dynamics

CFLR..........................	Centaur Forward Load Reactor (*Atlas V* only). Because of the structural flexibility of the *Centaur* upper step, this reduces lateral motion of the payload during launch
CFPAR......................	Constant Flight Path Angle Rate (simple form of PEG)
CFRP.........................	Carbon fiber reinforced plastic
CG............................	Center of Gravity, identical to CM unless there is a gravity gradient
CH_4........................	Methane
C_L...........................	Lift coefficient; $C_L = L/(\frac{1}{2}\rho V^2 S_{ref})$ (nondimensional)
CLA..........................	Coupled Loads Analysis
....	Abbreviation for Alcantara, Brazil launch site
Clamp Band...............	Device used to secure payload onto PAF and then pyrotechnically release it, basically the mother of all hose clamps. See Marmon clamp.
CM..........................	Center of mass, identical to CG unless there is a gravity gradient
....	Configuration Management
CoE..........................	Conservation of Energy
COEs........................	Classical Orbital Elements
Common Bulkhead ..	Structure separating volumes of two adjacent propellants in a rocket step, insulated for minimum heat transfer if propellants differ in temperature
CONOPS...................	Concept of Operations, used for mission planning and design purposes
Controllability Ratio..........................	Ratio of available control torque to destabilizing aerodynamic torque (M_δ/M_A)
CONUS......................	Continental United States
COPV........................	Composite Overwrapped Pressure Vessel
cos...........................	cosine of an angle (trigonometric function)
Cosmonaut................	Russian/Soviet astronaut
COT..........................	Cost of Transport, quantifies cost to transport a given mass (Ch. 17)
COTS........................	Commercial Off-The-Shelf
CP............................	Center of Pressure, location where aerodynamic forces appear to be concentrated
CPI...........................	Consumer Price Index, used for estimating inflation
CPP...........................	California State Polytechnic University, Pomona, where the authors taught or attended
Cranking...................	Orbit inclination change using "cross" Δv.
CRS..........................	Commercial Resupply Services. NASA contract to supply materials and crew to ISS.
CS............................	Cold Start (refers to thermal testing)
CSC..........................	Conical shaped charge, a pyrotechnic used for flight termination system to punch a hole in a tank (Ch. 11)
CSG..........................	Centre Spatial Guyanais (ESA launch site in French Guiana)
CSM..........................	Command/Service Module of Apollo program
CTE..........................	Coefficient of Thermal Expansion

CW............................. Clockwise
CX*nn*........................... Launch Complex *nn*

D

D..................................... vehicle diameter
D2, D3, D4................. *Delta II, III, IV* LVs
D6AC........................... Grade of high-strength steel used for SRM casings
DAC............................. Design Analysis Cycle
.... Digital-to-Analog Converter
DAU............................. Data Acquisition Unit
dB .... decibel, measuring unit of 1) acoustic sound pressure or 2) radio transmission strength
DC............................... Direct Current (batteries, electric motors)
DCM.......................... Direction Cosine Matrix (Ch. 12)
DC-X......................... Delta Clipper Experimental, McDonnell Douglas' reusable VTVL rocket demonstrator
Dead Reckoning........ the process of calculating current position using previous position and advancing based on estimated speeds over elapsed time. Ch. 11.
DFRC........................... NASA Dryden Flight Research Center located at Edwards AFB, now known as Armstrong Flight Research Facility
Dispersion................... The result of random drift and errors in navigation applied to nominal performance (Ch. 12)
DLF .... Design Load Factor
DMLS.......................... Direct Metal Laser Sintering (additive manufacturing)
DMSP.......................... Defense Meteorological Satellite Program (USAF)
DN............................... Descending node of orbit, location where orbit passes *downwards* through reference plane from N to S
DNTO.......................... DiNitrogen Tetroxide or N_2O_4 is a chemical oxidizer used with hydrazine for hypergolic engines (often called Nitrogen Tetroxide, see NTO). AKA red fuming nitric acid.
DOC .... Direct Operating Cost
Dogleg......................... an inefficient curved ascent trajectory used to reach a lower orbit inclination or to avoid overflying certain surface locations
Dome........................... A structural shell that forms the front or rear of a propellant tank (AKA bulkhead)
DPAF.......................... Dual payload attach fitting can carry two payloads to orbit at the same time (Ch. 7)
DRM........................... Design Reference Mission: vehicle's intended mission
DSP .... Defense Support Program (Ch. 17)
DUF............................. Dynamic Uncertainty Factor

E

E..................................... Modulus of Elasticity (Young's modulus)
.... Eccentricity of elliptical hemispheroid tank dome (Ch. 8)

e	Eccentricity of orbit (Ch. 3)
E	East direction (Ch. 3)
EBW	Exploding bridgewire (used to trigger separation systems and range safety)
ECLSS	Environmental Control & Life Support Systems (Ch. 8)
ECO	Engineering Change Order (documentation of proposed changes to a design)
EDS	Emergency Detection System (*Saturn V*)
EDW	Edwards Air Force Base (California, U.S.)
EELV	Evolved Expendable Launch Vehicle, a U.S. government program leading to vehicles with identical launch capabilities and interfaces, *Delta IV* and *Atlas V*. Name change to NSSL (National Security Space Launch in 2019.
EGSE	Electrical Ground Support Equipment
EIRP	Effective Isotropic Radiated Power is a measure of power density a receiver intercepts
ELV	Expendable Launch Vehicle
EMC	Electromagnetic Compatibility
EMI	Electromagnetic Interference
EMR	Engine Mixture Ratio
Energiya	Soviet heavy-lift vehicle used for Buran shuttle
EOM	End of Mission
EOR	Earth Orbit Rendezvous mission mode (not used by Apollo program)
EP dome	elliptical hemispheroid dome
EPDM	Synthetic rubber used to insulate solid rocket motor casings from the high temperatures created by burning propellant (ethylene propylene diene methylene)
EPO	Earth Parking Orbit
EPS	Electrical Power System
ERB	Engineering Review Board
ESA	European Space Agency
ESD	Electrostatic Discharge
ESPA	EELV Secondary Payload Adapter
ET	External Tank (space shuttle)
Ethanol	Ethyl alcohol, an edible rocket fuel
ETR	Eastern Test Range (CCAFS, Cape Canaveral Air Force Station, Florida, U.S.)

F

f	Fuel-to-Oxidizer mixing ratio, usually by mass. $f = 1/r$.
	frequency or clock rate measured in Hz or cycles/s (or MHz or GHz)
F-1	Largest single combustion chamber Rocketdyne kerolox engines used on S-IC, 1st step of *Saturn V* (1.5 million pounds or 6700 kN of SL thrust)
F9, F9H	*Falcon 9*, *Falcon 9 Heavy* LVs

FAA .... Federal Aviation Administration (U.S.)

FAR .... Federal Acquisition Regulations (U.S.)

Fairing .... structure to protect payload from aerodynamic forces and heating, AKA shroud

FBD .... Free-Body Diagram

FCDCA .... Flexible Confined Detonating Cord Assembly

FCS .... Flight Control System

Feedline .... a pipe that transmits a fluid between two locations

.... a pipe that transmits radio frequency energy between two locations (waveguide)

FECO .... First Engine Cutoff

FEM .... Finite Element Model (computational math model)

F-F, FF .... Free-Free usually refers to vibration of a beam with unsupported ends (no boundary conditions, or "free", at ends)

.... full-functional tests (testing)

FG .... Fiberglass

FH .... *Falcon Heavy* LV

Fineness .... ratio of vehicle length to diameter (L/D)

Fire in the hole .... staging scheme where upper step engines ignite before separating from the lower step.

FL .... Florida, U.S.

FLOX .... Mixture of LOx and liquid fluorine to improve performance of LOx engines

FLR .... Forward Load Reactor, a structural support used within *Atlas* PLFs to reduce bending of *Centaur* upper step (Ch. 7)

FLSC .... Flexible linear shaped charge

FMEA .... Failure Mode & Effects Analysis (Ch. 17)

FMECA .... Failure Mode, Effects, & Criticality Analysis (required for Range Safety System)

FMH .... Free Molecular Heating, rate of heating in upper atmosphere determines when to jettison the PLF.

FPR .... Flight Performance Reserve

fps, FPS .... feet per second. 1 FPS = 0.3048 m/s = 1.09728 km/h

.... frames per second (video)

FRF .... Frequency Response Function: mathematical description of structural response used for analysis

FRSI .... Felt Reusable Surface Insulation used on Space Shuttle (Ch. 11)

FS .... Factor of Safety

FSW .... Flight software

ft .... foot; feet: unit of length. 1 ft = 0.3048 m

FTS .... Flight Termination System

Fuel .... Propellant that combusts to provide thrust

G

G .... Shear modulus

.... Gravitational constant, 6.674×10^{-11} $m^3 \cdot kg^{-1} \cdot s^{-2}$

g .... Local value of gravity

g_0	Standard Earth gravity, acceleration of $9.80665 \text{ m/s}^2 = 32.17405 \text{ ft/s}^2$
G	Giga or 10^9 engineering units
Gain	amplification of system = output/input. Antenna gain tends to increase with diameter.
gal	unit of acceleration, 1 gal = 1 cm/s². Named after Galileo Galilei.
....	Gallon, unit of volume. 1 gal = 0.0037854 m³.
GE	Glass (fiberglass) epoxy
GEM	Graphite-Epoxy Motor, a solid-rocket motor with a casing made from graphite-epoxy composite materials
GEO	Geosynchronous Earth Orbit
GF	Glass Fibre (U.K.)
GH_2	Gaseous Hydrogen
GHe	Gaseous Helium
Gimbal	special mechanical frame allowing rotational motion in one axis.
GLL	Galileo: spacecraft mission to Jupiter
GLOM	Gross Liftoff Mass
GLOW	Gross Liftoff Weight
GMT	Greenwich Mean Time (replaced with Universal Time UT)
GN&C	Guidance Navigation and Control
GN_2	Gaseous Nitrogen
GO_2, GOx	Gaseous Oxygen
GPF	Grams per foot, specifies amount of explosive used for pyrotechnic separations
GPOPS-II	next-generation software intended to solve general nonlinear optimal control problems, runs in MATLAB®
GPS	Global Positioning System
GRC	NASA Glenn Research Center, Cleveland, Ohio, US
Gr/E	Graphite epoxy, AKA CFRP
Grain	The solid propellant cast inside a solid rocket motor casing
Grid Fin	Mechanical "waffle" structure used to steer rockets and missiles aerodynamically. Used for control when recovering F9 first steps.
GSE	Government Supplied Equipment
....	Ground Support Equipment
GSLV	Geosynchronous Satellite Launch Vehicle (India)
GSO	Geostationary Orbit
GTO	Geostationary Transfer Orbit; Geosynchronous Transfer Orbit
GVT	Ground vibration test (modal test)

H

H_0, h_0	Atmospheric exponential model scale height (change in height where atmospheric property drops by $1/e = 1/2.71828...$)
H_2	Hydrogen

Hammerhead.............	A LV with nose or PLF larger diameter than body below it (Ch. 11)
HAPS.........................	Hydrazine Auxiliary Propulsion System (Minotaur I)
He.............................	Helium
HEO..........................	High Earth Orbit (higher than GEO)
HFFJ.........................	Hollow Form Frangible Joint separation system (Ch. 11)
HGA	High gain antenna
HITL.........................	Hardware In The Loop
HLV	Heavy Lift Vehicle
HM-CFRP	High modulus Carbon fiber reinforced plastic (graphite-epoxy)
Hover slam................	Powered landing with $T/W \gg 1.0$ used to decelerate Falcon 9 1st steps rapidly just before touchdown (AKA "suicide burn")
HS.............................	Hot Start (refers to thermal testing)
HT-CFRP	High Tensile (strength) Carbon Fiber Reinforced Plastic, AKA HT graphite-epoxy
HTP	high-test peroxide; highly-concentrated hydrogen peroxide
HTPB........................	Hydroxyl Terminated Polybutadiene rubber, part of a solid propellant
Hybrid......................	rocket using a combination of solid & liquid propellants
Hybrid, conventional..............	Hybrid using liquid oxidizer & solid fuel
Hybrid, reverse	Hybrid using liquid fuel & solid oxidizer
Hybrid, quasi............	Hybrid using liquid fuel with solid oxidizer & solid fuel
Hydrazine.................	N_2H_4, dangerous chemical used as fuel for hypergolic engines
Hydrolox...................	Refers to Hydrogen-LOx propellant system
Hydyne.....................	mixture of 60% unsymmetrical dimethylhydrazine (UDMH) and 40% diethylenetriamine (DETA)
Hypergolic................	Describes combustion that occurs spontaneously, no ignition system needed
hypersonic................	Mach no. M greater than approx. 4

I

i...............................	Orbital inclination (tilt of orbit relative to equatorial plane)
....	the imaginary unit $(-1)^{1/2}$
I..............................	total impulse (area under the thrust-time curve)
IA.............................	Input axis for gyro (Ch. 12)
ICBM........................	Intercontinental Ballistic Missile
ICD...........................	Interface control document (Ch. 11)
I_D	Integral used to characterize drag loss (Chapter 3, section 3.3.3)
IGM	Iterative Guidance Mode, a closed-loop guidance mode for *Saturn V* S-II step
ILS............................	International Launch System
IMU	Inertial Measurement Unit
in..............................	inch, unit of length. 1 in = 25.4 mm.

Injection errors	Errors in speed, altitude, or inclination at instant of orbital injection
INS	Inertial Navigation System
INU	Inertial Navigation Unit
Interstage	Structure connecting two rocket steps
Intertank	Structure between two tanks on a rocket step (centerbody)
IOP	Ignition Overpressure (Ch. 11)
IR	Infrared
IRBM	Intermediate Range Ballistic Missile
IRFNA	inhibited red-fuming nitric acid, alternate oxidizer for DNTO
IRIG	Inter range instrumentation group. Used for telemetry standards.
Iron waffle	Compact "waffle-iron" aerodynamic control surface; also, "grid fin"
ISA	Interstage Adapter
ISAS	Institute of Space & Astronautical Sciences (Japan)
ISDS	Inadvertent Separation Destruct System
Isogrid	Refers to a structure with stiffening ribs in an equilateral triangular orientation possessing *isotropic* mechanical qualities
Isotropic	mechanical properties (stress, strain, etc.) identical in all directions
....	antenna transmits or receives equal radiation, all directions
I_{sp}	Specific impulse: thrust force ÷ (mass of propellants/unit time). Exhaust speed in SI system; exhaust speed ÷ g_0 in English units.
ISRO	Indian Space Research Organization
ISS	International Space Station
ITAR	International Traffic in Arms Regulations: U.S. regulations limiting info access to U.S. government and its contractors and controlling export of sensitive items.
ITE	Integral throat entrance (throat of SRM)
I_{tot}	Total impulse: $\int T(t)\, dt$ (area below the thrust-time curve, in N-s)
IUS	Inertial Upper Stage, solid propellant step used by Titan and Shuttle.
IV&V	Independent Verification and Validation

J

J-2	Rocketdyne hydrolox engines used on 2nd and 3rd steps of Saturn V.
JATO	"Jet" Assisted Takeoff: a rocket motor used to help aircraft take off faster or with heavier loads (correct term would be RATO)
JAXA	Japan Aerospace Exploration Agency: ISAS, NASDA, and National Aerospac Lab combined (similar to NASA)
Jimsphere	Weather/sounding balloon used to measure wind speeds at altitude before launch

JPL	NASA Jet Propulsion Laboratory (Caltech-operated, Pasadena, California, U.S.)
JSC	NASA Johnson Space Center (Houston, Texas, U.S.)
Jupiter	Series of IRBMs developed by US Army in 1950s, culminating in Jupiter-C/Juno 2

K

K	kelvins, absolute temperature in SI system. 0 K = −273.15 C = −459.67 F.
Kapton®	Trade name for plastic tape used for thermally-stressful environments
kerolox	Kerosene-LOx (or LOx-RP) propellant system
Kevlar®	Trade name for high-strength, energy-absorbing aramid fibers
kg	kilogram, unit of mass. 1 kg = 2.204623 lb_m
kip	kilopound(s), thousands of pounds-force
klb_f	kilopound(s), thousands of pounds-force
KLC	Kodiak Launch Complex, Alaska, U.S. (see PSCA)
km	kilometer(s), unit of length, 10^{+3} m. 1 km = 0.6213712 mile.
km/s	speed in km per second. 1 km/s = 3280.84 ft/s = 2236.936 mi/h (MPH).
kN	kilonewton(s). 1 kN = 224.809 lb_f.
kPa	kilopascal(s). Pressure unit: kPa = 1,000 N/m^2
KSC	NASA Kennedy Space Center (adjacent to CCAFS)

L

L, l	Length
L_0	Latitude above or below Equator.
LAIOP	Liftoff Acoustic Ignition Overpressure
LASRM	Liquid Augmented Solid Rocket Motor (type of hybrid rocket)
lb_f	pound(s)-force. 1 lb_f = 4.448 N.
lb_m	pound(s)-mass, a mass that weighs 1 lb_f at 1 standard gravity. 1 lb_m = 0.4535924 kg.
LC	Launch Complex
LCH_4	liquid methane
LEO	Low-Earth Orbit, <1,000 km altitude
LeRC	NASA Lewis Research Center (now Glenn Research Center), Ohio, U.S.
LES	Launch escape system
LGA	Low Gain antenna
LH_2	Liquid Hydrogen
LHe	Liquid Helium
LHP	Left-half plane (stable side of complex plane)
Link budget	Calculation of effectiveness of a data transmission system (Ch. 14)

Lithobraking .... See SEO

LM .... Lunar Module (Apollo program lunar landing spacecraft, formerly called LEM)

LM- .... Prefix to Chinese Long March rocket series

LN_2 .... Liquid Nitrogen

ln .... natural log, $\log_e(\ldots)$

LNG .... Liquified Natural Gas (methane, CH_4)

LO_2 .... Liquid Oxygen

LOC .... Lines of Code

log .... log base 10, $\log_{10}(\ldots)$

LOR .... Lunar Orbit Rendezvous mission mode used by Apollo program

LOS .... Loss of signal

LOx .... Liquid Oxygen

LPRE .... Liquid Propellant Rocket Engine

LRB .... Liquid Rocket Booster

LSA .... Launch Services Agreement, USAF-issued contract for EELV

LSC .... Linear Shaped Charge, a pyrotechnic system that can slice through metal to separate structures (Ch. 11) or release propellants for FTS (Ch. 16)

l_{sm} .... Static margin (distance between CM and CP) (Ch. 12)

LSMA .... Launch Services Mission Assurance (SMC contract)

LST .... Local Sidereal Time, used for launch windows. LST is time since vernal equinox passed over local longitude line.

LUT .... Launch Umbilical Tower

LV .... Launch Vehicle

LVA .... Launch Vehicle Adapter, same as Payload Attach Fitting, PAF (Ch. 7)

LVLH .... Local Vertical, Local Horizontal coordinate system fixed in moving vehicle.

LWST .... Launch Window Sidereal Time, time for orbital plane to pass over after vernal equinox crossed local longitude line.

M

m .... length unit. 1 m = 39.37 in = 3.28084 ft.

M .... bending moment

.... Mach number (vehicle speed ÷ speed of sound)

M .... Million (engineering units or currency units)

MC .... Monte Carlo: a statistical analysis where many parameters are allowed to vary randomly at the same time (Ch. 12)

m_0 .... Initial mass (with payload)

M-1 .... Large hydrolox engine developed by Aerojet during Apollo program, >1 million pounds of thrust. Never flown.

M1D .... Merlin 1D engine used on SpaceX Falcon 1 & 9

Mandrel .... Tool used during solid propellant casting process to deliver desired grain cross-section

Marmon Clamp .... Trade name of clamp band-like device used to secure payload onto PAF

MARS	Mid-Atlantic Regional Spaceport, located at the southern tip of the NASA Wallops Flight Facility
Max-air	Refers to both max-q and max $q\alpha$
Max-αq	Maximum (q = dynamic pressure) × (α = alpha, AoA)
Max-q	Maximum Dynamic Pressure
MBB	Messerschmitt-Bölkow-Blohm (company)
MECO	Main Engine Cutoff
MEMS	Micro electromechanical systems: generally, a miniaturized sensor or mechanical item made like an integrated circuit (Ch. 12)
MEO	Medium Earth Orbit (between LEO and GEO)
MER	Mass Estimating Relationship
	Mars Exploration Rover
Merlin	Kerolox engines used on lower & upper steps of SpaceX *Falcon 1* & *Falcon 9*
MES	Main Engine Start
MET	Mission Elapsed Time
m_f	Final mass (burnout mass, includes payload)
Methalox	Refers to LCH_4 or liquid Methane-LOx propellant system
Methanol	Methyl alcohol, a poisonous rocket fuel
MFCO	Missile Flight Control Officer (in charge of activating destruct system)
MGA	Medium Gain Antenna
Miniskirt	Cylindrical structure used to support Delta rocket 2nd step
MIRV	Multiple Independently-Targetable Reentry Vehicle (warhead)
mm	millimeter, unit of length, 10^{-3} m. 1 mm = 0.03937 in.
MM	Minuteman missile
MMC	Metal matrix composite
MMH	Monomethyl Hydrazine (hypergolic fuel), Formula: CH_3NHNH_2
m_0	Initial or liftoff mass
MoI	Moment of Inertia (usually one of the diagonal terms of the item's MoI tensor). See also PoI.
Molniya	Special inclined orbit useful for communication to/from high Earth latitudes ("lightning" in Russian)
MON-xx	mixed oxides of nitrogen, a mixture of DNTO and nitric oxide to lower freezing point. 'xx' indicates % nitric oxide by weight
Monte Carlo analysis	a group of analyses where multiple parameters are simultaneously randomly varied to determine effects of off-nominal parameters. (Ch. 12)
m_p	Propellant mass of a step
MPa	megapascal(s). Pressure unit: MPa = 10^6 N/m^2
m_{PL}	Payload mass (can also be the sum of the loaded masses of the steps above)
MPE	Maximum predicted environment
MPG	Mission Planner's Guide: a "user's manual" supplied by the LV manufacturer containing technical data on the LV's performance, environment, and payload accommodations. A.K.A. MPG, PUG, etc.

MPS	Main propulsion system
MR	Mass Ratio for vehicle performance calculation, AKA μ
MR	Mixture Ratio (for propellants)
MRC	Moments Reference Center (a reference point fixed in the vehicle about which aerodynamic coefficients are calculated, then translated to the moving CM location)
m_S	Step "structure" (inert) mass (no payload)
MS	Margin of safety
m/s	meter per second. 1 m/s = 3.28084 ft/s.
MSFC	NASA Marshall Space Flight Center (Huntsville, Alabama, U.S.)
MUG	Mission User's Guide: a "user's manual" supplied by the LV manufacturer containing technical data on the LV's performance, environment, and payload accommodations. A.K.A. MPG, PUG, etc.
MUX	multiplexer, a device which combines and processes digital signals
MVAC	Merlin Vacuum engine (M1D engine with much larger nozzle for higher expansion ratio)
MW	Molecular weight of a gas
MX	Missile Experimental, a missile program to replace Minuteman
MY	Man-Year = person-year. One full-time employee works 40 h/week × (52 −2) weeks/year = 2,000 hours in one person-year (subtract 2 weeks for vacation).
Mylar®	Trade name for plastic film used for thermal protection blankets

N

n	Load factor, T/W, often stated in g_0. For axial loads, n_x is used; for lateral loads, n_z is used.
N	newton, SI unit of force. 1 N = 0.2248 N
	North direction
N-1	Soviet Union moon launch rocket
N_2	Nitrogen
N_2H_4	Hydrazine (storable liquid fuel)
N_2O_4	Dinitrogen Tetroxide, a storable toxic chemical oxidizer used with hydrazine for hypergolic engines (often called 'Nitrogen Tetroxide'). Creates nitric acid when mixed with water.
NA	not applicable or not available
NAA	North American Aviation, contractor for Saturn V S-II step
NASA	National Aeronautics and Space Administration (U.S.)
NASDA	National Space Development Agency (Japan)
NASTRAN	A finite-element modeling (FEM) analysis program used for static, thermal, and dynamic analysis
NCS	Nutation Control System: a thruster used on Delta II third step to control nutation (Ch7)
NM	nautical mile = 6,076 ft = 1.1508 statute mile = 1,852 m
NMD	Nuclear Missile Defense

Nova	Heavy lift moon rocket intended for direct lunar landing. Never built, Saturn V was used instead.
Nozzle	The "rear end" of a rocket engine, where hot gases exhaust to produce thrust
NPSH	Net Positive Suction Head (minimum turbopump pressure needed at engine start & for running)
NRE	Non-Recurring Engineering: startup costs incurred in creating a new vehicle (Ch. 17)
NSD	NASA Standard Detonator, used for frangible nuts and other pyro devices
NSSL	National Security Space Launch, formerly EELV. USAF changed name to NSSL in 2019.
NTO	Nitrogen Tetroxide, properly called Dinitrogen Tetroxide N_2O_4, a toxic chemical oxidizer used with hydrazine for hypergolic engines.
n_x	Axial load factor (lower-case n, see next entry)
N_x	running load or edge load, axial (x) dir. (*CAPITAL N*, see previous entry)

O

O_2	Oxygen
OA	Output axis for gyro (Ch. 12)
OAT	One-at-a-time: a statistical analysis where one parameter only is varied (Ch. 12)
OASPL	Overall Sound Pressure Level: integrated sound pressure level across a sound spectrum (Ch. 11)
OCONUS	Outside Continental United States
Octave rule	Guideline suggesting support equipment should be mounted on supports with at least twice the natural frequency of applied excitation (Ch. 11)
OECO	Outboard Engine Cutoff
OEI	One Engine Inoperative
OI	Operational instrumentation
OMS	Orbital Maneuvering System (Space Shuttle)
Orthogrid	A thin metal structure with stiffening ribs in a rectangular pattern ("waffle") possessing orthotropic material properties.
OST	Office of Commercial Space Transportation (branch of U.S. FAA)
OTIS	*Optimal Trajectories by Implicit Simulation*: general-purpose optimization program
Oxidant	Propellant providing oxygen for combustion (UK)
Oxidizer	Propellant providing oxygen for combustion (U.S.)
OV-	Orbiter Vehicle (designation for Space Shuttle orbiter)

P

p	pressure (*lower-case p*)
p_∞	Ambient pressure

P	axial load (*CAPITAL P*)
Pa	pascal, unit of pressure. 1 Pa $= 1$ N/m^2. 1 MPa $= 145.04$ psi. 1 ATM $= 101, 325$ Pa.
PAF	Payload Attach Fitting, same as Launch Vehicle Adapter, LVA (Ch. 7)
Payload	The spacecraft or satellite the rocket carries that "pays" to fly
PBAN	Polybutadiene acrylonitrile rubber, part of a solid propellant mix used on Minuteman missiles and Space Shuttle SRBs
Pc	Combustion chamber pressure (Ch. 8)
PCB	Printed-circuit board
PCEC	Project Cost Engineering Capability, a cost estimation model
PDR	Preliminary Design Review
PDS	Propellant Dispersion System, euphemism for vehicle destruct mechanism
p_e	Exhaust pressure
PE	Propellant Excess
PEG	Powered Explicit Guidance, automatic steering usually after leaving atmosphere
Pegasus	Air-dropped LV operated by Northrop Grumman
PEO	Polar Earth Orbit
Peroxide	Short for hydrogen peroxide, H_2O_2. Relatively safe oxidizer (if rocket fuels can use the adjective 'safe')
PERT	Program Evaluation and Review Technique management system for Polaris program
PETN	PentaErythritol TetraNitrate: high explosive used for stage separation
PFJ	Payload Fairing Jettison
PIFS	Plume Induced Flow Separation. LV exhaust plume causes flow near aft end to separate, drawing hot gases forward along the skin (Saturn V, Ch. 11)
PIGA	Pendulous Integrating Gyroscopic Accelerometer used on V-2 measured acceleration and simultaneously integrated to produce speed
Pitch	a motion where a vehicle rotates upwards or downwards in a vertical plane
PK	Peacekeeper ICBM
PL	Payload
PLF	Payload Fairing, structure to protect payload from aerodynamic forces and heating (Ch. 7)
PMD	Propellant management device or diaphragm: used to manage propellants within tanks experiencing microgravity
Pogo	Sustained unwanted axial vibration of a launch vehicle, can be damaging. Not an acronym, name came from child's pogo stick.
Pogo suppressor	An accumulator used to attenuate dangerous pressure waves in propulsion system feed lines (Ch. 12)
PoI	Product of Inertia (non-diagonal terms of inertia tensor). See also MoI.

Polaris	2-step solid propellant SLBM (submarine-launched ballistic missile)
POS	represents the location of four quadrant position markers, I, II, III, and IV, used to specify wiring and equipment locations (Ch. 15)
Poseidon	2-step solid propellant SLBM successor to Polaris
POST	*Program to Optimize Simulated Trajectories*: general-purpose optimization program
PPG	Payload Planner's Guide: a "user's manual" supplied by the LV manufacturer containing technical data on the LV's performance, environment, and payload accommodations. A.K.A. MPG, PUG, etc. (Ch. 7)
P_{RMS}	Effective value of fluctuating pressure
pre-preg	Pre-impregnated composite fibers (unhardened resin or 'matrix' has been applied to fibers)
Proton	Russian heavy-lift LV with extensive history and high reliability, from the 1960s still used 2020
PSCA	Pacific Spaceport Complex Alaska, U.S. (see KLC)
PSD	Power Spectral Density, a measure of magnitude as a function of frequency
psf	pounds-force per square foot. 1 psf = 47.88 Pa.
psi	pounds-force per square inch. 1 psi = 6894.76 Pa.
PSLV	Polar Satellite Launch Vehicle (India)
PSW	Payload Systems Weight
PSWC	Payload Systems Weight Capability
PUG	Payload User's Guide: a "user's manual" supplied by the LV manufacturer containing technical data on the LV's performance, environment, and payload accommodations. A.K.A. MPG, PUG, etc.
PWB	Printed-wire board
PV	Pressure vessel
PY	Person-year. One full-time employee works 40 h/week × (52−2) weeks/year = 2,000 person-hours in one person-year (subtract 2 weeks for vacation).
Pyro	Pyrotechnic (separation or valve system driven by small explosive charge)
Pyrophoric	A substance which ignites spontaneously on contact with air, such as TEA-TEB, used to ground- and air-start rocket engines (and SR-71 Blackbirds)

Q

Q, q	Dynamic Pressure, pressure due to speed of vehicle in atmosphere.$q = 1/2\ \rho\ v^2 = 1/2 \times$ air density × speed^2
Q	"Quality" (in mechanical vibrations)
QA	Quality assurance
quincunx	a pattern of five points arranged with four in a square pattern, and the fifth at its center (like Saturn V's first & second steps)

QS............................	Quasi-static (rapid time-varying signals removed)
QSA	Quasi-static acceleration
QSL...........................	Quasi-static Loads (Ch. 11)

R

R, r............................	Radius (of a cylinder, or planetary radius)
R...............................	Reliability
....	Thrust-to-weight ratio T/W (AKA ψ)
R (script)......................	Universal gas constant, 8.314 J/(mol K)
r.................................	Oxidizer-to-fuel mixture ratio, usually by mass. $r = 1/f$.
R-7.............................	Soviet Union's first ICBM that evolved into Soyuz space launch vehicle
R&D............................	Research and Development
RAAN..........................	Right Ascension of the Ascending Node: the longitude or angle from the vernal equinox where the orbit passes through the equatorial plane towards *North*.
Raptor..........................	Methalox rocket engine developed by SpaceX
RATO..........................	Rocket Assisted Takeoff: a rocket motor used to help aircraft take off faster or with heavier loads
RCC	Reinforced Carbon-carbon high-density ablator used to protect structures and engine nozzles from heat (Ch. 11)
RCO............................	Range Control Officer
RCS.............................	Reaction Control System (small rocket engines to control vehicle attitude)
RD-180........................	Twin-nozzle Russian engine used on ULA *Atlas V*
RDX............................	'Research Department Explosive' used for separation systems & C-4 plastic explosive
RE...............................	Reynolds number, parameter describing ratio of inertial to viscous forces of a fluid
Redstone......................	U.S. IRBM evolved into LV used for U.S. first satellite and the manned Mercury spacecraft
....	US Army base hosting MSFC in Alabama, US
Reverberant.................	A special chamber with sound-reflecting walls used to do high strength acoustic testing (Ch. 15)
RF	Radio Frequency
RFNA..........................	Red Fuming Nitric Acid, a dangerous liquid propellant, AKA dinitrogen tetroxide
RHP............................	Right-half plane (unstable side of complex plane)
RIFCA	Redundant Inertial Flight Control Assembly (used on *Delta II* second step)
RK4	Fourth-order numerical integration method (Runge-Kutta) in MATLAB®
RL-10..........................	Pratt & Whitney hydrolox upper step rocket engine
RLV............................	Reusable Launch Vehicle
RMS............................	Root-Mean Square (effective value of a time history or spectral distribution). (Ch. 11)
RN	Reynolds number, a dimensionless parameter describing ratio of inertial to viscous forces of a fluid: $RN = vL/v$

ROFI	Radially Outward Firing Initiators ("sparklers" under SSMEs and hydrolox engines)
Rohacel®	Solid foam used between face sheets for sandwich structures
ROI	Return on Investment (Ch. 17)
Roll	a motion where a vehicle banks left or right
ROM	Rough order of magnitude, a crude but useful estimate
RP-1	Refined kerosene storable liquid fuel (Rocket Propellant-1, or Refined Petroleum-1)
R_{system}	System reliability (percent or fraction)
RS-25	Aerojet Rocketdyne *Space Shuttle* hydrolox engines used for SLS
RS-27A	Aerojet Rocketdyne *Delta II* 1st step engine
RS-68	Aerojet Rocketdyne *Delta IV* hydrolox engines with ablative nozzles
RSDP	Range Safety Data Package
RSO	Range Safety Officer
RSRM	Reusable Solid Rocket Motor (redesigned SRM for *Space Shuttle*)
RSS	Range Safety System (flight termination system)
	Root-Sum Square, the square-root of $\Sigma[(\text{component})^2]$ used for numerical analysis
RTLS	Return To Launch Site abort mode (Space Shuttle)
RUD	Rapid Unscheduled Disassembly, Rapid Unplanned Disassembly, Rapid Unintended Disassembly
RX	Receiver
R_x	Reliability of part x (percent or fraction)

S

S	South direction
S-IB	*Saturn IB* first step, built by Chrysler
S-IC	*Saturn V* first step, built by Boeing
S-II	*Saturn V* second step, built by North American (now Boeing)
S-IV	*Saturn 1* hydrolox upper step, built by Douglas (now Boeing)
S-IVB	*Saturn 1B* and *Saturn V* hydrolox upper step, built by McDonnell Douglas (now Boeing)
S/C, SC	Spacecraft
Saturn	Series of U.S. heavy-lift LVs used to support the Apollo moon-landing program
SBIRS	Space Based Infrared System
SBL	Shock-Boundary-Layer (buffeting)
	Space-Based Laser (weapons system)
Scar, scarring	Added structure designed and fabricated into a part to accommodate an optional structure (typically a mount for a "strap-on" SRM)
SDI	Strategic Defense Initiative ("Star Wars")
SE	Systems Engineering

SECO	Sustainer Engine Cutoff
	Second engine cutoff (2nd stage)
SEO	Subterranean Earth Orbit (intersects Earth surface, terminated by lithobraking)
Sep	Separation
Separation plane	Plane where two parts separate (between steps or payload and mounting structure)
Separation spring	Compression springs used to separate the payload from its mounting structure when released
SES	Second Engine Start
SEZ	South-East-Zenith Earth-based coordinate system
SFIR	Specific Force Integrating Receiver, an accelerometer similar to PIGA used on MX
SG	Specific gravity (density relative to water; $SG_{water} = 1.0$)
SG&A	Sales, General, and administrative costs
Shake & Bake	Refers to vibration and thermal testing
Shroud	Payload fairing used to protect payload during launch, AKA "fairing."
Shutdown	When rocket engines are commanded to end thrusting
SIGINT	Signals Intelligence
sin	sine of an angle (trigonometric function)
Skirt	Structure transitioning from one diameter to another, either larger or smaller (can also be constant dia.)
SL	Sea level
SLA	Super Lightweight Ablator: heat shield material designed to burn away to extract heat
SLA	Saturn Launch Adapter (jettisonable structure enclosing the Lunar Module between the Apollo Service Module and the S-IVB upper skirt)
SLBM	Submarine launched ballistic missile
SLC	Space Launch Complex (U.S.). Pronounced "slick".
Slenderness	ratio of vehicle length to diameter (L/D). AKA "fineness"
SLOC	software lines of code
SLS	Space Launch System, a bland name NASA chose for the agency's successor to the bland name Space Transportation System (Space Shuttle)
	Selective Laser Sintering (additive manufacturing)
Slug	unit of mass in English system. 1 slug = 32.174 lb_m = 14.5939 kg
SLV	Space Launch Vehicle
	Satellite Launch Vehicle (India)
SM	Static Margin (Ch. 12)
SMC	USAF Space and Missile Command
SNOPT	Optimization program by Phil Gill, U.C. San Diego., used for trajectory optimization
SOFI	Spray-on foam insulation (Ch. 11)
Soyuz	Russian LV with extensive history and high reliability, from the 1950s still used 2020

Spin Table...................	Device used to provide spin to portion of vehicle for spin stabilization
SPL.............................	Sound Pressure Level (Ch. 11)
Squatcheloid...............	Plot of $q\alpha$ and $q\beta$ limits for a space shuttle vehicle
SRA............................	Spin axis for gyro (Ch. 12)
SRB............................	Solid Rocket Booster as used on Space Shuttle
S_{ref}	Reference cross-sectional area
SRM...........................	Solid Rocket Motor
SRR............................	System Requirements Review
SRS............................	Shock Response Spectra (Ch. 11)
SSBN..........................	Ship, Submersible, Ballistic, Nuclear: US Navy submarine class
SSME..........................	Space Shuttle Main Engine (also designated as RS-25)
SSO............................	Sun-Synchronous Orbit
SSTO..........................	Single Stage to Orbit: a hypothetical one-stage rocket that does not stage during flight
stage...........................	A flying rocket consisting of one or more steps and a payload
step............................	Refers to one propulsive "piece" of a rocket. A *Saturn IB* has two steps, the S-IB and the S-IVB, in addition to its payload. Incorrectly referred to as "stages."
Strap-on.....................	A SRM, SRB, LRB, or combination of propellant tanks and rocket engine that is "strapped on" to the side of a vehicle and is jettisoned during flight
Static margin.............	Distance between the CP and CM, often measured in "calibers" (reference diameters) (Ch. 12)
STS.............................	Space Transportation System: a bland name NASA chose for the *Space Shuttle* system. Also used as a prefix to a particular shuttle flight, i.e. STS-73.
Subsonic.....................	Mach no. $M < 0.8$
Suicide burn.............	Powered landing with $T/W > 1.0$ (AKA "hover slam")
Supersonic..................	Mach no. M between 1 and approx. 4
Sustainer....................	Name for an engine that burns all the way through a first-step burn (i.e. *Atlas* missile)
SV..............................	Space Vehicle
SW, S/W....................	Software

T

t.................................	thickness
T................................	Thrust (from engine or motor)
....	Temperature (absolute for most calculations)
T	1000 kg (metric tonne, 1.1023 English tons = 2204.6 lb$_m$, about 110% of an English ton)
Taikonaut...................	Chinese astronaut
TAL............................	Trans-Atlantic Landing (after shuttle abort decision)
tan	tangent of an angle (trigonometric function)
TBL............................	Turbulent Boundary Layer (Ch. 11)

TEA-TEB	TriEthylAluminum $Al_2(C_2H_5)_6$ – TriEthylBorane $B(C_2H_5)_3$. Pyrophoric chemicals used to ignite rocket engines, spontaneously burns with green flames
TECO	Third [step] Engine Cutoff
TEI	Two Engines Inoperative
TER	Torque Estimating Relationship
Thermal-Vac	Thermal-Vacuum testing, done in a vacuum chamber with internal heaters
Throat	narrowest part of a rocket engine, where flow becomes supersonic
Thrust Structure	Structure connecting main propulsion system to the rest of the vehicle
Ti	Titanium (metal)
Time-to-Double	Time needed for amplitude of unstable system to double
Titan	Series of U.S. ICBMs and LVs used for the manned Gemini spacecraft and many NASA interplanetary missions
TLV	Target Launch Vehicle
TM	Telemetry
TNT	Trinitrotoluene $C_7H_5N_3O_6$, a high explosive used on German V-2 warhead (WWII)
ToF	Time of Flight
Tonne	1,000 kg (1.1023 English tons = 2204.6 lb_m, about 110% of an English ton)
TPS	Thermal Protection System
TR	Transmissibility Ratio (base shake transferred to suspended mass)
Track & Crack	Refers to range safety operations
TRANSCOST	Cost estimation model, author Koelle
transonic	Mach no. M between 0.8 and 1.2
tribrid	tribrid rocket uses liquid oxidizer & liquid fuel with solid fuel (see other "hybrid" entries)
Trident	Solid propellant SLBM successor to Poseidon
TRL	Technology Readiness Level, used to assess readiness of a technology for flight
TRR	Test Readiness Review
TRW	Company now part of Northrop Grumman
T_{SL}	Sea-level thrust
TSTO	Two-Step to Orbit
TT&C	Telemetry, Tracking & Control
Turbopump	High-power pump used to force propellants into the combustion chamber, often powered by the rocket's propellants
T_{vac}	Vacuum thrust
TVC	Thrust Vector Control. A system used to *gimbal* (pivot) a rocket engine from side-to-side in order to produce side forces that allow vehicle steering
Twang	A springback that occurs after the shuttle's engine thrust leans the stack away from the SRBs (Ch. 14)

TWR............................ Thrust-to-Weight Ratio $T/W = \psi$
TX............................... Texas, U.S.
.... Transmitter

U

UDMH....................... Unsymmetrical Dimethyl Hydrazine, a type of hydrazine with lower freezing temperature than pure hydrazine; Formula: $(CH_3)_2NNH_2$
UHF F/O.................... Ultra High Freq. Follow-On: DOD program to provide satellite communications (Ch. 17)
UHM-CFRP............... Ultra-high modulus carbon fiber
ULA........................... United Launch Alliance, a partnership between Boeing and Lockheed Martin
Uplink........................ data sent UP to vehicle
URM.......................... Universal Rocket Module (Angara). See also CBC & CCB for related.
USSF.......................... U.S. Space Force
USAF......................... U.S. Air Force
UT.............................. Universal Time

V

v .... Velocity magnitude (speed)
$\mathbf{v}$ .... Vector velocity: speed AND direction (bold face)
v_{circ} .... Circular orbit speed
V-2 .... Ballistic missile developed by Nazis in WWII, first practical launch vehicle
V-block .... Block with "V" shaped cutout used with clamp band to secure payload to payload attach fitting
V&V........................... Verification and Validation
VA.............................. Virginia, U.S.
VAB .... Vehicle Assembly Building (KSC)
VAFB......................... Vandenberg Air Force Base, California, U.S.
VDC........................... volts direct current
Verification Matrix... A table containing vehicle items and how their proper operation is verified, i.e. by analysis, inspection, demonstration, or test (Ch. 15)
VI............................... an abbreviation for "velocity, inertial" (Ch. 11)
VLC............................ Verification Loads Cycle
VLS .... Brazilian Satellite Launch Vehicle (Veiculo Lancador de Satelites)
Vol............................. Volume
VS&A .... Vibration, Shock, and Acoustics
v_{SF} .... Space-fixed velocity
VTVL......................... Vertical Takeoff, Vertical Landing

W

W	West direction
W_0	Initial or liftoff weight, or gross liftoff weight (GLOW)
Waffle	A metal structure with stiffening ribs in a rectangular pattern ("Orthogrid")
WBS	Work breakdown structure
WFF	NASA Wallops Flight Facility, Virginia, U.S.
WSMR	White Sands Missile Range, New Mexico, U.S.
WT	Wind Tunnel
WTR	Western Test Range (VAFB, Vandenberg Air Force Base, California, U.S.)
WY, WYr	Work-Year, amount of time one full-time employee works in one year (2,000 hours in US). See MY.

X

x	Axis of coordinate system, often in longitudinal direction
XBM	Excess Ballistic Missile (designates retired assets)
Xmit	transmit
Xmitter	transmitter

Y

y	Axis of coordinate system, often in lateral direction
Yaw	a motion where a vehicle rotates left or right in a horizontal plane
Yaw-torquing	flying an inefficient curved ascent trajectory used to 1) reach a lower orbit inclination or 2) to avoid overflying certain surface locations (Ch. 3)

Z

z	Axis of coordinate system, often in vertical direction
Z	Zenith (vertical) direction

Subscripts/Numbers

$_0$	Quantity referring to initial or starting condition
	value in combustion chamber
$_1$, $_2$	referring to locations "1", "2"
$_\infty$	ambient or far-away conditions
$_a$	apoapsis
	axial direction
$_{BO}$	burnout condition
$_c$	combustion chamber
$_{del}$	delivered

$_e$	"effective" condition
	Exit or exhaust condition
$_f$	final or end condition
	force
	fuel
$_{HE}$	hyperbolic escape
$_m$	mass
$_{nozzle}$	variable for nozzle
$_{ox}$	oxidizer
$_p$	periapsis
	propellant
$_{PL}$	*payload*
$_{REF}$	Reference quantity
$_{SL}$	Sea level condition
$_s$	Quantity referring to structure (inert)
$_t$	tensile or tension
$_u$	ultimate stress (failure)
$_{vac}$	vacuum
$_x$	axial (longitudinal) direction
$_y$	referring to y-coordinate direction
	material yielding
$_z$	lateral (transverse) direction
2EI	two engines inoperative
2STO	Two Stage to Orbit

Greek Symbols

α	Angle of attack, angle between reference line and incoming flow direction (Ch 3)
	Coefficient of Thermal Expansion (CTE)
	Angular acceleration, rad/s^2 ($\alpha = d\omega/dt = d^2\theta/dt^2$)
	Value between 0 & 1 to divide burnout speeds
$\alpha_{conv\ section}$	convergent section angle
β	Burnout angle (Ch. 3)
	Azimuth angle (measured clockwise from North)
	Value between 0 & 1 to divide ascent losses
	Burn rate pressure coefficient (Ch. 4)
γ	flight path angle (Ch. 3)
	Gas specific heats ratio c_p/c_v (Ch. 4)
$\Delta(\ldots)$	Change in $(\ldots)$
Δv	change in speed v (Ch. 3)
δ	Thrust vector control engine angle (Ch. 3)
	Launch site auxiliary angle (Ch. 3)
ϵ	Nozzle expansion ratio, A_{exit}/A_{throat}
ζ_p	Propellant mass fraction, m_p/m_0
θ	Angle between two orbits (Ch. 3)
	Angle with respect to horizontal (Ch. 3)

λ, Λ .... Latitude (Ch. 3)

λ .... wavelength of radio waves used to transmit data, usually cm or mm. $\lambda = c/f$ (wavelength = speed of light ÷ frequency of radio waves)

μ .... Vehicle mass ratio = m_0/m_f

.... Gravitational constant = $g_0\, r^2$ (Ch. 3)

ν .... kinematic viscosity of atmosphere

π .... payload ratio (payload mass/total mass)

ρ .... Material density (mass/unit volume)

Φ .... Sound pressure spectral density function (Ch. 11)

φ .... equivalence ratio: actual fuel/ox ratio ÷ stoichiometric fuel/ox ratio (Ch. 4)

σ .... Structural fraction of a rocket step = $m_s/(m_s + m_p)$

.... Stress (subscripts used to indicate type: tensile, compressive, hoop, axial, etc.)

.... Standard deviation, where 69.3% of all measurements fall

τ .... Shear stress (Ch. 10)

τ_b .... Time to burnout (Ch. 3)

ψ .... Thrust-to-weight ratio T/W

Ω .... RAAN: right ascension (or longitude) of the ascending node of an orbit (Ch. 3)

.... Rotational speed, RPM or RPS

ω .... Argument of periapsis, the angle from the orbit's line of nodes to its periapsis (Ch. 3)

.... Rigid body angular velocity $d\theta/dt$, rad/s

ζ .... Damping coefficient for 2nd order spring-mass-damper or control system

Other Symbols

Ø .... Diameter

Υ .... Vernal equinox line (points to constellation Ares) (Ch. 3)

€ .... Euro (European currency)

¥ .... Yen (Japanese currency)

Index

Supporting Materials

To download supplemental material files, please go to AIAA's electronic library, Aerospace Research Central (ARC), at arc.aiaa.org. Use the menu bar at the top to navigate to Books > AIAA Education Series; then, sort alphanumerically by title to navigate to the desired book's landing page.

A complete listing of titles in the AIAA Education Series is available from AIAA's electronic library, Aerospace Research Central (ARC), at arc.aiaa.org. Visit ARC frequently to stay abreast of product changes, corrections, special offers, and new publications.

AIAA is committed to devoting resources to the education of both practicing and future aerospace professionals. In 1996, the AIAA Foundation was founded. Its programs enhance scientific literacy and advance the arts and sciences of aerospace. For more information, please visit www.aiaafoundation.org.